Air Pollution and Its Impacts on U.S. National Parks

Air Pollution and Its Impacts on U.S. National Parks

Timothy J. Sullivan

CRC Press
Taylor & Francis Group
Boca Raton London New York

CRC Press is an imprint of the
Taylor & Francis Group, an **informa** business

CRC Press
Taylor & Francis Group
6000 Broken Sound Parkway NW, Suite 300
Boca Raton, FL 33487-2742

First issued in paperback 2020

ISBN 13: 978-0-367-57394-2 (pbk)
ISBN 13: 978-1-4987-6517-6 (hbk)

Library of Congress Cataloging-in-Publication Data

Names: Sullivan, Timothy J. (Timothy Joseph), 1950-
Title: Air pollution and its impacts on U.S. national parks / Timothy J.
Sullivan.
Description: Boca Raton : CRC Press, 2017. | Includes bibliographical
references and index.
Identifiers: LCCN 2016055204 | ISBN 9781498765176 (print : alk. paper)
Subjects: LCSH: Air--Pollution--Environmental aspects--United States. |
Air--Pollution--United States. | Environmental monitoring--United States.
| National parks and reserves--United States. | Wilderness areas--United
States.
Classification: LCC TD883.2 .S85 2017 | DDC 363.6/80973--dc23
LC record available at https://lccn.loc.gov/2016055204

Visit the Taylor & Francis Web site at
http://www.taylorandfrancis.com

and the CRC Press Web site at
http://www.crcpress.com

To Deb, Laura, and Jenna, with whom I have shared countless national park adventures.

Contents

Section I   Background

Section II   Case Studies

List of Figures

List of Tables

Preface

Perhaps you like to fish for brook trout. Maybe you remember your grandparents tapping sugar maple trees and making maple syrup, and just maybe you enjoyed some of the maple candy that they made in their kitchen. Personally, I enjoy watching a pair of bald eagles raising one or two eaglet chicks each summer at a seaside cottage. It is important to me that a Native American tribe in Wisconsin can eat the fish they catch without fear of mercury poisoning; fish catching and consumption are important to their religion and culture. The next time I hike the Grand Canyon with my wife, I want to be able to clearly view the geological formations on the other side. Is wildflower diversity important to you, or is it acceptable to see fewer and fewer species in the coming years? These ecosystem services (services provided by nature to benefit humans) that we care about, and others, are threatened in multiple ways by various air pollutants. The effects are especially noteworthy in the national parks.

A variety of air pollutants are emitted into the atmosphere from human-caused and natural emissions sources throughout the United States and elsewhere. These materials are transported with prevailing winds and impact sensitive natural resources in wilderness and other natural areas, including national parks.

The system of national parks in the United States is among our greatest assets. Our parks attract tens of millions of visitors each year from this country and throughout the world. One important goal of the National Park Service, which manages these precious lands and resources, is to protect the highly valued ecological and cultural resources in the parks and ensure that air-quality-related values and ecosystem services are preserved for the enjoyment of current and future generations.

This book provides a compilation and synthesis of scientific understanding regarding the causes and effects of a suite of air pollutants that impact national parklands throughout the United States. One pollutant category of great concern that is evaluated here is ground-level ozone, which can damage sensitive vegetation. Ozone is not emitted directly from pollution sources but rather is formed in the atmosphere in response to photochemical reactions that involve nitrogen oxides, volatile organic compounds (both of which are at least partly human caused), and sunlight. Other human-caused pollutants considered here include atmospheric particles that cause haze and degrade visibility and also deposits from the atmosphere to the earth's surface as acids (mainly sulfur, but also nitrogen), nutrients (mainly nitrogen), and toxic compounds. Emitted nutrients include both oxidized (nitrogen oxides) and reduced (ammonia) forms of nitrogen. Toxics include mercury (a potent neurotoxin), a variety of historic-use and current-use pesticides, and other toxic materials. These contaminants are released into the air from such emissions source types as power plants, motor vehicles, agriculture, manufacturing, and others.

The degrees of impact on national park resources in response to air pollution and pollutant deposition are governed primarily by the amount and type of pollutant exposure and by the inherent sensitivity of park resources. Atmospheric deposition of acids, nutrients, and toxic materials varies considerably from park to park. Some parks receive substantial pollutant deposition loads; others are more remote from pollutant emissions sources and receive relatively low pollutant deposition. Vegetation in parks is exposed to varying levels of atmospheric ozone that can injure foliage and perhaps reduce plant growth.

Differing levels of haze impair the ability of visitors to experience the scenic vistas offered by many of the parks.

Nevertheless, harm to park resources is not governed solely, or in many cases even mostly, by pollutant exposure. Parks also differ greatly in their inherent sensitivity to air pollution damage. Park resources in some parks are highly sensitive to harm or change caused by inputs of acid precursors or nutrients. Some parks contain ecosystems that are highly sensitive to methylation of mercury by bacteria, a process that converts inorganic mercury to an organic form that can bioaccumulate to dangerously high levels in the food web. Some plant species are especially sensitive to injury from exposure to atmospheric ozone. Some parks have outstanding visual resources such that even relatively low levels of haze dramatically affect visitor enjoyment of the park experience.

This book describes these pollutant emissions, concentrations, deposition, and exposures and the reasons for differences in resource sensitivity. Recent trends in pollutant exposure and the extent of any adverse effects are summarized. Mechanisms of impact are elucidated, in some cases in the context of a changing climate. Where possible, this book identifies the critical (tipping point) loads or critical levels of pollutant deposition at which adverse impacts are manifested. Monitoring data are discussed along with mathematical model projections of future changes in resource conditions. It is important to note that pollutant emissions, atmospheric concentrations, exposures, and deposition amounts are always changing. Emissions and deposition of some pollutants are decreasing markedly across the eastern United States at the time of this writing. Exposure levels at the time that you read this book may be substantially higher or lower than they were when the available data were analyzed and when this book was written. Air pollution and its effects/impacts are moving targets.

This book is targeted at students and practitioners of environmental science and at those interested in air pollution, water pollution, soil science, water resources, biogeochemistry, and aquatic and terrestrial ecology. It will also be helpful to natural resource professionals and other scientists and natural resource managers. The information presented here is applicable not only to resources in the national parks but also to resources in the vicinity of the national parks, including in some U.S. Forest Service or Fish and Wildlife Service wilderness areas and other natural areas.

The book is divided into two sections. The first provides background on the topics covered and a general description of air pollution effects science. It is broken down into three chapters: an introduction, an explanation of pollutant exposures and deposition, and a description of air pollution impacts. The latter includes acidification, nutrient enrichment, ozone injury, visibility impairment, and effects of airborne toxics. The second section comprises 19 case studies of specific parks and/or park networks (small groups of parks that have similar resources and experience similar threats). They are organized by geographic region. The case studies focus on many of the larger parks that have been relatively well studied. Case studies summarize pollutant exposures and effects in national parks across the United States from Maine to California, Washington to the U.S. Virgin Islands, north to Alaska, and west to Hawaii. Affected ecosystems span the spectrum from rain forests to deserts, high mountains to coastal areas, and forests to grasslands and wetlands. The reader will likely need to master the material presented in the first three introductory chapters in order to fully understand the case study chapters.

It is my hope that the reader of this book will come away with an improved understanding of air pollution in the national parks and the extent to which sensitive resources have been impacted. Along the way, you will learn a great deal about eco-systems and biogeochemistry. Armed with knowledge of pollutant sensitivities, exposures, and effects, we can collectively enhance the ability of the National Park Service to protect park resources so that these national treasures will be available for the enjoyment of future generations. I hope that you enjoy this book and your future national park adventures.

Acronyms and Abbreviations

BRAVO	Big Bend Regional Aerosol and Visibility Observational study
cm	centimeters
cmol/kg	centimoles per kilogram
DDD	dichlorodiphenyldichloroethane (a DDT breakdown product)
DDE	dichlorodiphenyldichloroethylene (a DDT breakdown product)
DDT	dichlorodiphenyltrichloroethane
EPA	U.S. Environmental Protection Agency
ha	hectare
HCH	hexachlorocyclohexane
IMPROVE	Interagency Monitoring of Protected Visual Environments network
kg	kilogram
km	kilometer
km^2	square kilometers
L	liter
mg/kg	milligrams per kilogram
mg/L	milligram per liter
mi	mile
mi^2	square mile
mm	millimeters
Mm^{-1}	inverse mega meters
µeq/L	microequivalent per liter
µg/g	micrograms per gram
µm	micrometer
MOHAVE	Measurement of Haze and Visual Effects
N100	number of hours in a growing season during which the ozone concentrations are 0.100 ppm or greater
ng/g	nanograms per gram
ng/L	nanograms per liter
pH	relative acidity, or the molar concentration of H$^+$ in solution
PCB	polychlorinated biphenyl
ppb	parts per billion
ppm	parts per million
SUM06	seasonal sum of all hourly average ozone concentrations ≥0.06 ppm
W126	seasonal sum of hourly ozone concentrations, with higher concentrations weighted more heavily
tons/yr	tons per year
WACAP	Western Airborne Contaminants Assessment Project

Acknowledgments

The work that provided the basis for this book was funded by the National Park Service, Air Resources Division, through contracts to E&S Environmental Chemistry, Inc., located in Corvallis, Oregon. National Park Service ecologist Ellen Porter provided extensive guidance and support. She reviewed and offered very helpful suggestions on material that was drafted for the National Park Service and used throughout the entire book. Many other people assisted in various ways in the preparation of this book. Data were compiled and analyzed by Todd McDonnell, Tyler McPherson, Deian Moore, Jenna Sullivan, Sam Mackey, and Jayne Charles. The shaded relief base map was provided by ESRI © 2009. Maps and figures were prepared by Deian Moore and Todd McDonnell. Jayne Charles prepared and assisted with editing the document. This book would not have been possible without the sustained and substantial assistance of Ellen, Jayne, and Deian. Robert Kohut edited sections on the effects of ozone on park vegetation. Photographs were mostly provided by the National Park Service. Helpful suggestions that contributed to book organization and content were provided by Ellen Porter, Tamara Blett, and Tonnie Cummings. The book benefited greatly from review comments offered on an earlier draft by Gail Begley, David Gay, William Jackson, Robert Kohut, Dixon Landers, and Jason Lynch. My wife and daughters provided encouragement and accompanied me on visits to many of the parks discussed herein. The following additional National Park Service scientists provided suggestions that were very helpful in the preparation of this book: Bill Baccus, Jill Baron, Jeffery Bennett, Andrea Blakesley, Clare Bledsoe, Michael Bower, Paul Burger, Joffre Castro, Jim Comiskey, Jalyn Cummings, Robert Emmott, Annie Esperanza, Kirsten Gallo, Michael George, Johnathan Jernigan, Sarah Jovan, John Klaptosky, Greg Kudray, Michael Larrabee, Amy Larsen, Jim Lawler, Kristin Legg, Teresa Leibfreid, Rhonda Loh, Eathan McIntyre, Paul McLaughlin, Julie McNamee, Brian Mitchell, Peter Neitlich, Megan Nortrup, Matt Patterson, Dusty Perkins, Melanie Peters, Ashley Rawhouser, Regina Rochefort, Jed Redwine, Hildy Reiser, Jim Renfro, Ben Roberts, Judy Rocchio, Bill Route, Dave Schirokauer, Don Shepherd, Nita Tallent, Lee Tarnay, Kathy Tonnessen, and David VanderMeulen.

Views, statements, findings, conclusions, recommendations, and data in this book do not necessarily reflect views and policies of the National Park Service, U.S. Department of the Interior. Mention of trade names of commercial products does not constitute endorsement or recommendation for use by the U.S. government.

Author

Timothy Sullivan holds a BA in history from Stonehill College (1972); an MA in biology from Western State College, Colorado (1977); and a PhD in biological sciences from Oregon State University (1983) through an interdisciplinary program that included areas of focus in ecology, zoology, and environmental chemistry. He did his postdoctoral research at the Center for Industrial Research in Oslo, Norway, on the acidification of surface and ground water, episodic hydrologic processes, and Al biogeochemistry. His expertise includes the effects of air pollution on aquatic and terrestrial resources, watershed analysis, critical loads, ecosystem services, nutrient cycling, aquatic acid–base chemistry, episodic processes controlling surface water chemistry, and environmental assessment. He has been president of E&S Environmental Chemistry, Inc., since 1988. He began studies of the effects of air pollution on sensitive aquatic and terrestrial resources in the national parks in the 1980s and has worked more or less continuously on that research for nearly 30 years. He has served as project manager and/or lead author for a wide variety of projects that have synthesized the science of complex air and water pollution effects for diverse audiences. He was project manager of the work to draft a scientific summary and integrated science assessment of the effects of nitrogen and sulfur oxides on terrestrial, transitional, and aquatic ecosystem for the U.S. Environmental Protection Agency in support of its 2008 review of National Ambient Air Quality Standards for oxides of sulfur and nitrogen and has contributed to the 2017 review. He is the author of the National Acid Precipitation Assessment Program State of Science and Technology Report on past changes in surface water acid–base chemistry throughout the United States from acidic deposition. He served as project manager for the preparation of air quality reviews for national parks throughout California and coauthored similar reviews for the Pacific Northwest and the Rocky Mountain and Great Plains regions. He has summarized air pollution effects at all 272 inventory and monitoring national parks in the United States and has managed dozens of air and water pollution modeling and assessment studies throughout the United States for the National Park Service, the U.S. Forest Service, the U.S. Department of Energy, and the U.S. Environmental Protection Agency. He has authored a book on the aquatic effects of acidic deposition (Lewis Publishers, 2000), coauthored a book on sampling, analysis, and quality assurance protocols for studying air pollution effects on freshwater ecosystems (CRC Press, 2015), and authored a book on the impacts of air pollutant deposition on natural resources in New York State (Cornell University Press, 2015). He has also published more than 125 peer-reviewed journal articles, book chapters, and technical reports describing the results of his research.

Section I

Background

1

Introduction

Emissions into the atmosphere of a wide range of substances can impact sensitive resources in national parks at downwind locations. The air pollutants of concern for this book include gases and particles containing sulfur (S), nitrogen (N), mercury (Hg), and other metals; pesticides and other organic compounds; ozone; and particles that reduce visibility. Emissions sources are varied and in particular include power plants, industrial facilities, motor vehicles, incinerators, agriculture, fire, oil and gas development, and other human-caused and natural sources. Air pollutants can cause or contribute to numerous adverse environmental impacts. These include acidification of soil and water; nutrient over-enrichment (eutrophication) of terrestrial and aquatic ecosystems; biomagnification (increase in concentration within the food web) and toxicity of contaminants; inhibition of plant health, growth, or reproduction; changes in species composition (which species occur at a given location) and abundance; and visibility degradation caused by haze.

Some air pollutants, called primary pollutants, are directly emitted into the atmosphere from emissions sources. These primary pollutants, which include Hg and oxides of S and N, are released into the atmospheric environment from locations of human activities, including urban areas, power plants, agriculture, ground disturbance, oil and gas development, and industrial activities. Other air pollutants, called secondary pollutants, are formed in the atmosphere as a result of photochemical reactions among both natural and human-produced substances. Ground-level ozone is a good example. The various primary and secondary chemicals, particles, and gases can be carried in the atmosphere short or long distances from their source areas or atmospheric formation location and subsequently affect air quality in downwind areas, including within the national parks. There are also natural sources of some air pollutants, such as volcanos and forest vegetation.

An air quality value is a resource, generally identified by federal land managers for one or more parks, wildernesses, or other federal areas, that may be adversely affected by a change in air quality. The impacted resource may include visibility or a specific scenic, cultural, physical, biological, ecological, or recreational resource identified by the federal land managers as valuable for a particular area. These air quality values can include aspects of visibility, water quality, and the health and vitality of plants, animals, water bodies, and other life forms that reside in and around the federal lands. They also include elements of scenery (color, form, texture) and culture (rock art, artifacts). These air quality values can also be identified by Native American tribes for tribal lands.

National park visitors expect clean, clear air; healthy plants; and abundant wildlife. Visitors enjoy and remember beautiful scenes of mountains, valleys, water bodies, and the forms of landscapes and geological formations. Enjoyment and appreciation of air quality and ecosystems are integral parts of the visitor experience. The quality of this experience is closely linked to the visitor's ability to view the park environment unobstructed by haze and to the knowledge that park ecosystems are healthy, with the assumption that they will remain so into the future.

Monitoring and research in the national parks have shown that human-caused air pollution has adversely impacted air quality, visibility, lake and stream water chemistry, soil condition, and native terrestrial and aquatic biota in some parks. Despite improvements since the 1970s in air quality under the Clean Air Act and its amendments as well as other national and state legislation, the resources in some parks are still adversely impacted relative to estimated natural or preindustrial background conditions. Because of their high degree of sensitivity, some resources will remain affected even after full implementation of existing emissions control legislation, including the 1990 Clean Air Act Amendments and the Cross-State Air Pollution Rule.

Ecosystem impacts from air pollution in some of the national parks have been substantial. Efforts to characterize and quantify these impacts, and to examine more recent recovery as some air pollution levels decline in some areas, have focused primarily on visibility, surface waters, soils, and vegetation. Research on plants has included, in particular, work on the responses of lichens, forest health, alpine plant communities, and biodiversity. Although visibility is still considered degraded in many parks, interfering with visitor enjoyment of scenic vistas, conditions have generally improved in recent years. As S, N, and Hg emissions continue to be controlled, some more highly responsive surface waters and soils show signs of partial chemical recovery from previous acidification (Lawrence et al. 2011, 2012). Nevertheless, available data suggest that soil chemistry (which affects vegetation health) will continue to deteriorate in some eastern areas under expected future pollution levels (Warby et al. 2009), and these changes in soil condition may continue to affect vegetation and future lake and stream water chemistry for many decades (Sullivan et al. 2006). Eutrophication persists in many areas.

Biological dose-response functions* for some sensitive receptors, for example, brook trout (*Salvelinus fontinalis*) presence, fish species richness (number of species present), lichen distribution, stream macroinvertebrates, and others, have been developed (Cosby et al. 2006, Sullivan et al. 2006, 2013, Burns et al. 2011, Pardo et al. 2011). These data inform the calculation of chemical and ecosystem indicator variables for defining critical loads, or critical levels,† and recovery targets. Chemical and biological responses to decreases in pollution inputs are being quantified at some monitoring sites in parks as air pollution and atmospheric deposition of pollutants from the atmosphere to the earth's surface generally continue their recent decline. Resource managers, policy makers, and scientists focus on the status of terrestrial and aquatic sensitive elements or receptors, existing level of damage to sensitive receptors, extent to which current and projected future emissions reductions will lead to ecosystem recovery (both chemical and biological), deposition levels that will be required to promote recovery, and biogeochemical processes that govern elemental cycles and ecosystem responses to atmospheric deposition and other forms of air pollution. Synthesis and integration of existing data and knowledge and assembly of sensitive receptor inventories, for example, for lichens, diatoms, fish, macroinvertebrates, or plants, will provide a stronger basis for setting further emissions reduction goals, evaluating incremental improvements, conducting cost/benefit analyses, and prioritizing research and monitoring needs in the national parks.

---

* A dose-response function quantifies the biological response (such as mortality or the number of species of a given type present in the community) relative to the level of pollution input.
† A critical load, or critical level, of atmospheric deposition is the level of deposition, or pollutant exposure, below which adverse effects do not occur according to current knowledge.

There is a growing body of scientific research demonstrating that atmospheric deposition and other aspects of air pollution induce a range of effects on aquatic and terrestrial ecosystems and scenic views in the national parks. These effects range from subtle to severe. Such effects include those associated with acidification, nutrient enrichment, visibility impairment, physiological injury, and toxicity (Porter et al. 2005, 2011, Burns et al. 2011).

Congress assigned to the Federal Land Managers the responsibility to protect air quality–related values in federal Class I areas,* which are afforded the highest level of protection under the Clean Air Act. Carrying out these responsibilities generally involves identifying air quality–related values in each Class I area, establishing and prioritizing inventory and monitoring protocols, specifying a process for evaluation of air pollution effects on air quality–related values, and specifying adverse effects on each air quality–related value. An adverse effect is one that diminishes the Class I area's significance, impairs the structure or function of the protected ecosystem(s), and/or impairs the quality of the visitor experience. There are other mandates and tools for the protection of park resources that are not designated as Class I. These include the National Park Service Organic Act, the Wilderness Act, and the Clean Water Act.

As a result of syntheses provided by Pardo et al. (2011) and the Federal Land Managers' Air Quality–Related Values Work Group (FLAG): Phase I Report (U.S. Forest Service et al. 2000, 2010), the Federal Land Managers have made progress toward specifying the levels of air pollution that trigger concerns for air quality values and impacts in federally mandated Class I areas, as required by the Clean Air Act. Nevertheless, difficulties arise in assessing the impacts of multiple sources simultaneously, whereby a large number of minor pollution sources can collectively contribute to an unacceptable cumulative impact, or when a new potential emissions source applies for an emissions permit in an area that already has a high level of air pollution from existing emissions sources. A new source, regardless how small, can increase the cumulative impact of many sources that may be unacceptable even though any one individual source may be innocuous.

The National Park Service has a fundamental legislative mandate to protect park resources, including air quality values, from the adverse effects of air pollution so as to leave them unimpaired for the enjoyment of future generations (Shaver et al. 1994). National Park Service Management Policies (2006) state that

> The Service will actively promote and pursue measures to protect these values from the adverse impacts of air pollution. In cases of doubt as to the impacts of existing or potential air pollution on park resources, the Service will err on the side of protecting air quality and related values for future generations.

This book provides context for understanding some of the complexities of managing national parks for clean, clear air and healthy ecosystems, following a legacy of decades of air pollution that has impacted park resources.

In order to protect park resources for the enjoyment of future generations, the National Park Service must protect park resources from overuse and misuse of many kinds. Unfortunately, however, many resources in the national parks are adversely impacted by a variety of forms of air pollution, many of which originate outside the park boundaries, and the National Park Service has limited control over them. The first step in protecting the national treasures that comprise our national parks is to develop an understanding of the

---

* Class I areas include national park and wilderness areas of a certain minimum size and establishment date.

threats, the inherent sensitivities of the park resources to being damaged, and the existing levels of pollutant exposures and consequent damages.

Over the past several decades, various parks within the National Park Service system have been subjects of a wide variety of air pollution effects characterization and monitoring studies (http://www.nature.nps.gov/air/index.cfm), including

- Wet pollutant deposition (in rain and snow) monitoring
- Dry atmospheric deposition (gases and particles) monitoring
- Cloud atmospheric deposition monitoring
- Ground-level ozone monitoring
- Visibility optical monitoring (light extinction and light scattering that affect haze)
- Visibility particle monitoring
- Photographic visibility monitoring
- Ozone precursor* evaluation
- Surface water chemistry
- Soil chemistry
- Nutrient cycling
- Ecosystem and atmospheric modeling
- Ozone effects on vegetation
- Nitrogen and sulfur effects on lichens
- Acidification effects on fish and other aquatic biota
- Mercury effects on fish and other aquatic and piscivorous biota

Key information from these studies, along with results from various regional and national studies, was compiled at selected national parklands for this book. The National Park Service Inventory and Monitoring Program organizes the approximately 270 parks that contain significant natural resources into 32 networks. The current status of air pollution–sensitive resources is evaluated, along with available empirical data and dynamic and steady-state model simulations of critical load and future prognosis for ecosystem damage or recovery. Data were gathered for individual parks where available. Park-specific analyses focus primarily on those parks thought to contain the greatest concentration of air pollution–sensitive resources and those parks that have the most relevant data. These tend to be the larger parks. Since air pollutants can be widespread, reduction of pollution emissions affecting large parks will often result in the protection of smaller parks located nearby as well.

I synthesize here many of the available inventories and published findings regarding air quality–related values that are sensitive to impacts due to the atmospheric deposition of N, S, and toxic compounds, formation of ground-level ozone, and impacts on in-park visibility. Chapters 2 and 3 provide background information on emissions, deposition, and effects of air pollutants. In Chapters 4 through 22, site-specific research results provide indications of the status of the air quality–related values in many of the Inventory and Monitoring parks that are distributed across the nation. There are more than 400 national

---

* Ozone forms in the atmosphere in response to photochemical reactions between nitrogen oxides and volatile organic compounds in the presence of sunlight.

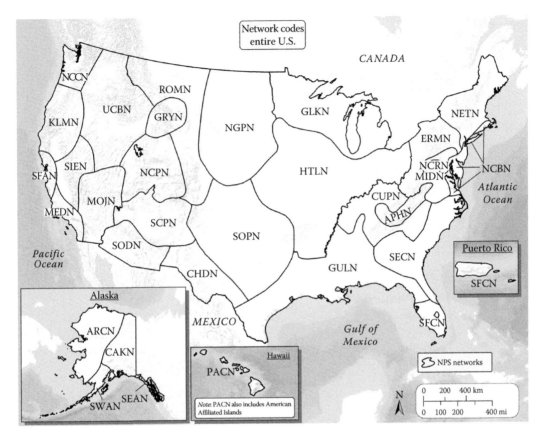

FIGURE 1.1
Locations of Inventory and Monitoring networks. The network names associated with the four-letter identification codes are given in Table 1.1.

parks, monuments, seashores, and other lands managed by the National Park Service. Forty-five of them, mainly the larger and more heavily visited ones, are discussed here in some detail.

The main objective of this book is to synthesize scientific understanding of the causes and impacts of air pollutants in the national parks. The focus is on pollution dynamics, biogeochemistry,* and ecology. The target audience includes students and practitioners of pollution effects science and those involved in natural resource management and policy. A good basic understanding of ecology and chemistry will be helpful to the reader. Detailed expertise is not needed.

Locations of the National Park Service Inventory and Monitoring networks that provide the organization of park lands for inclusion in this book are shown in Figure 1.1. The Inventory and Monitoring parks and networks are listed in Tables 1.1 and 1.2. These national-level data provide context for comparing exposures and sensitivities across selected parks and networks.

---

* Biogeochemistry is the study of the interactions among the biological, geological, soil, and chemical components of natural ecosystems.

TABLE 1.1

List of National Park Service Inventory and Monitoring Networks

Network Name	Network Code
Appalachian Highlands	APHN
Arctic	ARCN
Central Alaska	CAKN
Chihuahuan Desert	CHDN
Cumberland Piedmont	CUPN
Eastern Rivers and Mountains	ERMN
Great Lakes	GLKN
Greater Yellowstone	GRYN
Gulf Coast	GULN
Heartland	HTLN
Klamath	KLMN
Mediterranean Coast	MEDN
Mid-Atlantic	MIDN
Mojave Desert	MOJN
National Capital Region	NCRN
North Coast and Cascades	NCCN
Northeast Coastal and Barrier	NCBN
Northeast Temperate	NETN
Northern Colorado Plateau	NCPN
Northern Great Plains	NGPN
Pacific Island	PACN
Rocky Mountain	ROMN
San Francisco Bay Area	SFAN
Sierra Nevada	SIEN
Sonoran Desert	SODN
South Florida/Caribbean	SFCN
Southeast Alaska	SEAN
Southeast Coast	SECN
Southern Colorado Plateau	SCPN
Southern Plains	SOPN
Southwest Alaska	SWAN
Upper Columbia Basin	UCBN

TABLE 1.2

List of National Parklands and Associated Codes

Park Name	Park Code	Network Name	Network Code
Abraham Lincoln Birthplace	ABLI	Cumberland Piedmont	CUPN
Acadia	ACAD	Northeast Temperate	NETN
Agate Fossil Beds	AGFO	Northern Great Plains	NGPN
Alagnak	ALAG	Southwest Alaska	SWAN
Alibates Flint Quarries	ALFL	Southern Plains	SOPN
Allegheny Portage Railroad	ALPO	Eastern Rivers and Mountains	ERMN
American Memorial Park	AMME	Pacific Island	PACN
Amistad	AMIS	Chihuahuan Desert	CHDN
Aniakchak	ANIA	Southwest Alaska	SWAN
Antietam	ANTI	National Capital Region	NCRN
Apostle Islands	APIS	Great Lakes	GLKN
Appomattox Court House	APCO	Mid-Atlantic	MIDN
Arches	ARCH	Northern Colorado Plateau	NCPN
Arkansas Post	ARPO	Heartland	HTLN
Assateague Island	ASIS	Northeast Coastal and Barrier	NCBN
Aztec Ruins	AZRU	Southern Colorado Plateau	SCPN
Badlands	BADL	Northern Great Plains	NGPN
Bandelier	BAND	Southern Colorado Plateau	SCPN
Bent's Old Fort	BEOL	Southern Plains	SOPN
Bering Land Bridge	BELA	Arctic	ARCN
Big Bend	BIBE	Chihuahuan Desert	CHDN
Big Cypress	BICY	South Florida/Caribbean	SFCN
Big Hole	BIHO	Upper Columbia Basin	UCBN
Big South Fork	BISO	Appalachian Highlands	APHN
Big Thicket	BITH	Gulf Coast	GULN
Bighorn Canyon	BICA	Greater Yellowstone	GRYN
Biscayne	BISC	South Florida/Caribbean	SFCN
Black Canyon of the Gunnison	BLCA	Northern Colorado Plateau	NCPN
Blue Ridge	BLRI	Appalachian Highlands	APHN
Bluestone	BLUE	Eastern Rivers and Mountains	ERMN
Booker T. Washington	BOWA	Mid-Atlantic	MIDN
Boston Harbor Islands	BOHA	Northeast Temperate	NETN
Bryce Canyon	BRCA	Northern Colorado Plateau	NCPN
Buck Island Reef	BUIS	South Florida/Caribbean	SFCN
Buffalo	BUFF	Heartland	HTLN
Cabrillo	CABR	Mediterranean Coast	MEDN
Canaveral	CANA	Southeast Coast	SECN
Canyon de Chelly	CACH	Southern Colorado Plateau	SCPN
Canyonlands	CANY	Northern Colorado Plateau	NCPN
Cape Cod	CACO	Northeast Coastal and Barrier	NCBN
Cape Hatteras	CAHA	Southeast Coast	SECN
Cape Krusenstern	CAKR	Arctic	ARCN
Cape Lookout	CALO	Southeast Coast	SECN
Capitol Reef	CARE	Northern Colorado Plateau	NCPN

(Continued)

TABLE 1.2 (*Continued*)

List of National Park Lands and Associated Codes

Park Name	Park Code	Network Name	Network Code
Capulin Volcano	CAVO	Southern Plains	SOPN
Carl Sandburg Home	CARL	Cumberland Piedmont	CUPN
Carlsbad Caverns	CAVE	Chihuahuan Desert	CHDN
Casa Grande Ruins	CAGR	Sonoran Desert	SODN
Castillo de San Marcos	CASA	Southeast Coast	SECN
Catoctin Mountain	CATO	National Capital Region	NCRN
Cedar Breaks	CEBR	Northern Colorado Plateau	NCPN
Chaco Culture	CHCU	Southern Colorado Plateau	SCPN
Channel Islands	CHIS	Mediterranean Coast	MEDN
Chattahoochee River	CHAT	Southeast Coast	SECN
Chesapeake and Ohio Canal	CHOH	National Capital Region	NCRN
Chickamauga and Chattanooga	CHCH	Cumberland Piedmont	CUPN
Chickasaw	CHIC	Southern Plains	SOPN
Chiricahua	CHIR	Sonoran Desert	SODN
City of Rocks	CIRO	Upper Columbia Basin	UCBN
Colonial	COLO	Northeast Coastal and Barrier	NCBN
Colorado	COLM	Northern Colorado Plateau	NCPN
Congaree	COSW	Southeast Coast	SECN
Coronado	CORO	Sonoran Desert	SODN
Cowpens	COWP	Cumberland Piedmont	CUPN
Crater Lake	CRLA	Klamath	KLMN
Craters of the Moon	CRMO	Upper Columbia Basin	UCBN
Cumberland Gap	CUGA	Cumberland Piedmont	CUPN
Cumberland Island	CUIS	Southeast Coast	SECN
Curecanti	CURE	Northern Colorado Plateau	NCPN
Cuyahoga Valley	CUVA	Heartland	HTLN
Death Valley	DEVA	Mojave Desert	MOJN
Delaware Water Gap	DEWA	Eastern Rivers and Mountains	ERMN
Denali	DENA	Central Alaska	CAKN
Devils Postpile	DEPO	Sierra Nevada	SIEN
Devils Tower	DETO	Northern Great Plains	NGPN
Dinosaur	DINO	Northern Colorado Plateau	NCPN
Dry Tortugas	DRTO	South Florida/Caribbean	SFCN
Ebey's Landing	EBLA	North Coast and Cascades	NCCN
Effigy Mounds	EFMO	Heartland	HTLN
Eisenhower	EISE	Mid-Atlantic	MIDN
El Malpais	ELMA	Southern Colorado Plateau	SCPN
El Morro	ELMO	Southern Colorado Plateau	SCPN
Everglades	EVER	South Florida/Caribbean	SFCN
Fire Island	FIIS	Northeast Coastal and Barrier	NCBN
Florissant Fossil Beds	FLFO	Rocky Mountain	ROMN
Fort Bowie	FOBO	Sonoran Desert	SODN
Fort Caroline	FOCA	Southeast Coast	SECN
Fort Davis	FODA	Chihuahuan Desert	CHDN

(*Continued*)

TABLE 1.2 (*Continued*)

List of National Park Lands and Associated Codes

Park Name	Park Code	Network Name	Network Code
Fort Donelson	FODO	Cumberland Piedmont	CUPN
Fort Frederica	FOFR	Southeast Coast	SECN
Fort Laramie	FOLA	Northern Great Plains	NGPN
Fort Larned	FOLS	Southern Plains	SOPN
Fort Matanzas	FOMA	Southeast Coast	SECN
Fort Necessity	FONE	Eastern Rivers and Mountains	ERMN
Fort Point	FOPO	San Francisco Bay Area	SFAN
Fort Pulaski	FOPU	Southeast Coast	SECN
Fort Sumter	FOSU	Southeast Coast	SECN
Fort Union	FOUN	Southern Plains	SOPN
Fort Union Trading Post	FOUS	Northern Great Plains	NGPN
Fort Vancouver	FOVA	North Coast and Cascades	NCCN
Fossil Butte	FOBU	Northern Colorado Plateau	NCPN
Fredericksburg and Spotsylvania	FRSP	Mid-Atlantic	MIDN
Friendship Hill	FRHI	Eastern Rivers and Mountains	ERMN
Gates of the Arctic	GAAR	Arctic	ARCN
Gateway	GATE	Northeast Coastal and Barrier	NCBN
Gauley River	GARI	Eastern Rivers and Mountains	ERMN
George Washington	GWMP	National Capital Region	NCRN
George Washington Birthplace	GEWA	Northeast Coastal and Barrier	NCBN
George Washington Carver	GWCA	Heartland	HTLN
Gettysburg	GETT	Mid-Atlantic	MIDN
Gila Cliff Dwellings	GICL	Sonoran Desert	SODN
Glacier	GLAC	Rocky Mountain	ROMN
Glacier Bay	GLBA	Southeast Alaska	SEAN
Glen Canyon	GLCA	Southern Colorado Plateau	SCPN
Golden Gate	GOGA	San Francisco Bay Area	SFAN
Golden Spike	GOSP	Northern Colorado Plateau	NCPN
Grand Canyon	GRCA	Southern Colorado Plateau	SCPN
Grand Portage	GRPO	Great Lakes	GLKN
Grand Teton	GRTE	Greater Yellowstone	GRYN
Grant-Kohrs Ranch	GRKO	Rocky Mountain	ROMN
Great Basin	GRBA	Mojave Desert	MOJN
Great Sand Dunes	GRSA	Rocky Mountain	ROMN
Great Smoky Mountains	GRSM	Appalachian Highlands	APHN
Guadalupe Mountains	GUMO	Chihuahuan Desert	CHDN
Guilford Courthouse	GUCO	Cumberland Piedmont	CUPN
Gulf Islands	GUIS	Gulf Coast	GULN
Hagerman Fossil Beds	HAFO	Upper Columbia Basin	UCBN
Haleakala	HALE	Pacific Island	PACN
Harpers Ferry	HAFE	National Capital Region	NCRN
Hawaii Volcanoes	HAVO	Pacific Island	PACN
Herbert Hoover	HEHO	Heartland	HTLN
Home of Franklin D. Roosevelt	HOFR	Northeast Temperate	NETN

(Continued)

TABLE 1.2 (*Continued*)

List of National Park Lands and Associated Codes

Park Name	Park Code	Network Name	Network Code
Homestead	HOME	Heartland	HTLN
Hopewell Culture	HOCU	Heartland	HTLN
Hopewell Furnace	HOFU	Mid-Atlantic	MIDN
Horseshoe Bend	HOBE	Southeast Coast	SECN
Hot Springs	HOSP	Heartland	HTLN
Hovenweep	HOVE	Northern Colorado Plateau	NCPN
Hubbell Trading Post	HUTR	Southern Colorado Plateau	SCPN
Indiana Dunes	INDU	Great Lakes	GLKN
Isle Royale	ISRO	Great Lakes	GLKN
Jean Lafitte	JELA	Gulf Coast	GULN
Jewel Cave	JECA	Northern Great Plains	NGPN
John Day Fossil Beds	JODA	Upper Columbia Basin	UCBN
John Muir	JOMU	San Francisco Bay Area	SFAN
Johnstown Flood	JOFL	Eastern Rivers and Mountains	ERMN
Joshua Tree	JOTR	Mojave Desert	MOJN
Kalaupapa	KALA	Pacific Island	PACN
Kaloko-Honokohau	KAHO	Pacific Island	PACN
Katmai	KATM	Southwest Alaska	SWAN
Kenai Fjords	KEFJ	Southwest Alaska	SWAN
Kennesaw Mountain	KEMO	Southeast Coast	SECN
Kings Canyon	KICA	Sierra Nevada	SIEN
Kings Mountain	KIMO	Cumberland Piedmont	CUPN
Klondike Gold Rush	KLGO	Southeast Alaska	SEAN
Knife River Indian Villages	KNRI	Northern Great Plains	NGPN
Kobuk Valley	KOVA	Arctic	ARCN
Lake Clark	LACL	Southwest Alaska	SWAN
Lake Mead	LAME	Mojave Desert	MOJN
Lake Meredith	LAMR	Southern Plains	SOPN
Lake Roosevelt	LARO	Upper Columbia Basin	UCBN
Lassen Volcanic	LAVO	Klamath	KLMN
Lava Beds	LABE	Klamath	KLMN
Lewis and Clark	LEWI	North Coast and Cascades	NCCN
Lincoln Boyhood	LIBO	Heartland	HTLN
Little Bighorn Battlefield	LIBI	Rocky Mountain	ROMN
Little River Canyon	LIRI	Cumberland Piedmont	CUPN
Lyndon B. Johnson	LYJO	Southern Plains	SOPN
Mammoth Cave	MACA	Cumberland Piedmont	CUPN
Manassas	MANA	National Capital Region	NCRN
Manzanar	MANZ	Mojave Desert	MOJN
Marsh-Billings-Rockefeller	MABI	Northeast Temperate	NETN
Mesa Verde	MEVE	Southern Colorado Plateau	SCPN
Minute Man	MIMA	Northeast Temperate	NETN
Mississippi	MISS	Great Lakes	GLKN
Missouri	MNRR	Northern Great Plains	NGPN

(*Continued*)

TABLE 1.2 (*Continued*)

List of National Park Lands and Associated Codes

Park Name	Park Code	Network Name	Network Code
Mojave	MOJA	Mojave Desert	MOJN
Monocacy	MONO	National Capital Region	NCRN
Montezuma Castle	MOCA	Sonoran Desert	SODN
Moores Creek	MOCR	Southeast Coast	SECN
Morristown	MORR	Northeast Temperate	NETN
Mount Rainier	MORA	North Coast and Cascades	NCCN
Mount Rushmore	MORU	Northern Great Plains	NGPN
Muir Woods	MUWO	San Francisco Bay Area	SFAN
Natchez Trace Parkway and National Scenic Trail	NATR	Gulf Coast	GULN
National Capital Parks—East	NACE	National Capital Region	NCRN
National Park of American Samoa	NPSA	Pacific Island	PACN
Natural Bridges	NABR	Northern Colorado Plateau	NCPN
Navajo	NAVA	Southern Colorado Plateau	SCPN
New River Gorge	NERI	Eastern Rivers and Mountains	ERMN
Nez Perce	NEPE	Upper Columbia Basin	UCBN
Ninety Six	NISI	Cumberland Piedmont	CUPN
Niobrara	NIOB	Northern Great Plains	NGPN
Noatak	NOAT	Arctic	ARCN
North Cascades	NOCA	North Coast and Cascades	NCCN
Obed	OBRI	Appalachian Highlands	APHN
Ocmulgee	OCMU	Southeast Coast	SECN
Olympic	OLYM	North Coast and Cascades	NCCN
Oregon Caves	ORCA	Klamath	KLMN
Organ Pipe Cactus	ORPI	Sonoran Desert	SODN
Ozark	OZAR	Heartland	HTLN
Padre Island	PAIS	Gulf Coast	GULN
Palo Alto Battlefield	PAAL	Gulf Coast	GULN
Pea Ridge	PERI	Heartland	HTLN
Pecos	PECO	Southern Plains	SOPN
Petersburg	PETE	Mid-Atlantic	MIDN
Petrified Forest	PEFO	Southern Colorado Plateau	SCPN
Petroglyph	PETR	Southern Colorado Plateau	SCPN
Pictured Rocks	PIRO	Great Lakes	GLKN
Pinnacles	PINN	San Francisco Bay Area	SFAN
Pipe Spring	PISP	Northern Colorado Plateau	NCPN
Pipestone	PIPE	Heartland	HTLN
Point Reyes	PORE	San Francisco Bay Area	SFAN
Prince William Forest	PRWI	National Capital Region	NCRN
Pu'uhonua o Honaunau	PUHO	Pacific Island	PACN
Pu'ukohola Heiau	PUHE	Pacific Island	PACN
Rainbow Bridge	RABR	Southern Colorado Plateau	SCPN
Redwood	REDW	Klamath	KLMN
Richmond	RICH	Mid-Atlantic	MIDN
Rock Creek Park	ROCR	National Capital Region	NCRN

(Continued)

TABLE 1.2 (*Continued*)

List of National Park Lands and Associated Codes

Park Name	Park Code	Network Name	Network Code
Rocky Mountain	ROMO	Rocky Mountain	ROMN
Russell Cave	RUCA	Cumberland Piedmont	CUPN
Sagamore Hill	SAHI	Northeast Coastal and Barrier	NCBN
Saguaro	SAGU	Sonoran Desert	SODN
Saint Croix	SACN	Great Lakes	GLKN
Saint-Gaudens	SAGA	Northeast Temperate	NETN
Salinas Pueblo Missions	SAPU	Southern Colorado Plateau	SCPN
San Antonio Missions	SAAN	Gulf Coast	GULN
San Juan Island	SAJH	North Coast and Cascades	NCCN
Santa Monica Mountains	SAMO	Mediterranean Coast	MEDN
Saratoga	SARA	Northeast Temperate	NETN
Saugus Iron Works	SAIR	Northeast Temperate	NETN
Scotts Bluff	SCBL	Northern Great Plains	NGPN
Sequoia	SEQU	Sierra Nevada	SIEN
Shenandoah	SHEN	Mid-Atlantic	MIDN
Shiloh	SHIL	Cumberland Piedmont	CUPN
Sitka	SITK	Southeast Alaska	SEAN
Sleeping Bear Dunes	SLBE	Great Lakes	GLKN
Stones River	STRI	Cumberland Piedmont	CUPN
Sunset Crater Volcano	SUCR	Southern Colorado Plateau	SCPN
Tallgrass Prairie	TAPR	Heartland	HTLN
Theodore Roosevelt	THRO	Northern Great Plains	NGPN
Thomas Stone	THST	Northeast Coastal and Barrier	NCBN
Timpanogos Cave	TICA	Northern Colorado Plateau	NCPN
Timucaun Ecological and Historical Preserve	TIMU	Southeast Coast	SECN
Tonto	TONT	Sonoran Desert	SODN
Tumacacori	TUMA	Sonoran Desert	SODN
Tuzigoot	TUZI	Sonoran Desert	SODN
Upper Delaware	UPDE	Eastern Rivers and Mountains	ERMN
Valley Forge	VAFO	Mid-Atlantic	MIDN
Vanderbilt Mansion	VAMA	Northeast Temperate	NETN
Vicksburg	VICK	Gulf Coast	GULN
Virgin Islands	VIIS	South Florida/Caribbean	SFCN
Voyageurs	VOYA	Great Lakes	GLKN
Walnut Canyon	WACA	Southern Colorado Plateau	SCPN
War in the Pacific	WAPA	Pacific Island	PACN
Washita Battlefield	WABA	Southern Plains	SOPN
Weir Farm	WEFA	Northeast Temperate	NETN
Whiskeytown	WHIS	Klamath	KLMN
White Sands	WHSA	Chihuahuan Desert	CHDN
Whitman Mission	WHMI	Upper Columbia Basin	UCBN
Wilson's Creek	WICR	Heartland	HTLN

(*Continued*)

TABLE 1.2 (*Continued*)

List of National Park Lands and Associated Codes

Park Name	Park Code	Network Name	Network Code
Wind Cave	WICA	Northern Great Plains	NGPN
Wolf Trap National Park for the Performing Arts	WOTR	National Capital Region	NCRN
Wrangell-St. Elias	WRST	Central Alaska	CAKN
Wupatki	WUPA	Southern Colorado Plateau	SCPN
Yellowstone	YELL	Greater Yellowstone	GRYN
Yosemite	YOSE	Sierra Nevada	SIEN
Yucca House	YUHO	Southern Colorado Plateau	SCPN
Yukon-Charley Rivers	YUCH	Central Alaska	CAKN
Zion	ZION	Northern Colorado Plateau	NCPN

References

Burns, D.A., J.A. Lynch, B.J. Cosby, M.E. Fenn, J.S. Baron, and U.S. EPA Clean Air Markets Division. 2011. National acid precipitation assessment program report to congress 2011: An integrated assessment. National Science and Technology Council, Washington, DC.

Cosby, B.J., J.R. Webb, J.N. Galloway, and F.A. Deviney. 2006. Acidic deposition impacts on natural resources in Shenandoah National Park. NPS/NER/NRTR-2006/066. U.S. Department of the Interior, National Park Service, Northeast Region, Philadelphia, PA.

Lawrence, G.B., W.C. Shortle, M.B. David, K.T. Smith, R.A.F. Warby, and A.G. Lapenis. 2012. Early indications of soil recovery from acidic deposition in U.S. red spruce forests. *Soil Sci. Soc. Am. J.* 76:1407–1417.

Lawrence, G.B., H.A. Simonin, B.P. Baldigo, K.M. Roy, and S.B. Capone. 2011. Changes in the chemistry of acidified Adirondack streams from the early 1980s to 2008. *Environ. Pollut.* 159:2750–2758.

National Park Service. 2006. Management policies. National Park Service, Washington, DC. Available at: www.nps.gov/policy/mp2006.pdf.

Pardo, L.H., M.E. Fenn, C.L. Goodale, L.H. Geiser, C.T. Driscoll, E.B. Allen, J.S. Baron et al. 2011. Effects of nitrogen deposition and empirical nitrogen critical loads for ecoregions of the United States. *Ecol. Appl.* 21(8):3049–3082.

Porter, E., T. Blett, D.U. Potter, and C. Huber. 2005. Protecting resources on federal lands: Implications of critical loads for atmospheric deposition on nitrogen and sulfur. *BioScience* 55(7):603–612.

Shaver, C.L., K.A. Tonnessen, and T.G. Maniero. 1994. Clearing the air at Great Smoky Mountains National Park. *Ecol. Appl.* 4(4):690–701.

Sullivan, T.J., C.T. Driscoll, B.J. Cosby, I.J. Fernandez, A.T. Herlihy, J. Zhai, R. Stemberger et al. 2006. Assessment of the extent to which intensively-studied lakes are representative of the Adirondack Mountain Region. Final Report 06-17. New York State Energy Research and Development Authority, Albany, NY.

Sullivan, T.J., G.B. Lawrence, S.W. Bailey, T.C. McDonnell, C.M. Beier, K.C. Weathers, G.T. McPherson, and D.A. Bishop. 2013. Effects of acidic deposition and soil acidification on sugar maple in the Adirondack Mountains, New York. *Environ. Sci. Technol.* 47:12687–12694.

U.S. Forest Service, National Park Service, and U.S. Fish and Wildlife Service. 2000. Federal land managers' air quality related values workgroup (FLAG) phase I report. U.S. Forest Service-Air Quality Program, National Park Service-Air Resources Division, U.S. Fish and Wildlife Service-Air Quality Branch. National Park Service, Denver, CO.

U.S. Forest Service, National Park Service, and U.S. Fish and Wildlife Service. 2010. Federal land managers' air quality related values work group (FLAG): Phase I report—Revised (2010). Natural Resource Report NPS/NRPC/NRR—2010/232. National Park Service, Denver, CO.

Warby, R.A.F., C.E. Johnson, and C.T. Driscoll. 2009. Continuing acidification of organic soils across the northeastern USA: 1984–2001. *Soil Sci. Soc. Am. J.* 73(1):274–284.

2

Pollutant Exposure

2.1 Emissions

Human-caused emissions of S, N, and Hg have been relatively high in the past in some areas but have decreased over the past few decades throughout much of the United States in response to emissions controls that resulted from the Clean Air Act and subsequent federal and state regulations. Particularly important have been controls on large power plants and motor vehicles, changes in industrial processes, and changes in fuels (i.e., coal to oil and gas). In general, the states that have had the highest levels of S and N emissions have shown the largest reductions in response to national Acid Rain Program regulations and other U.S. Environmental Protection Agency (EPA) programs, including the Clean Air Interstate Rule (Burns et al. 2011).

Changes in climate can also have profound effects on air quality and air pollution impacts (Krabbenhoft and Sunderland 2013). Jacob and Winner (2009) provided a comprehensive review. For example, heat waves in the eastern United States (1988) and Europe (2003) were accompanied by air pollution episodes (Lin et al. 2001, Guerova and Jones 2007), which may have been partly due to increased cooling demands. Under a warming climate, heat waves are expected to become more common (Christensen et al. 2007). Ozone concentrations in atmospheric surface layers are positively correlated with air temperature (Jacob and Winner 2009). There may be multiple causes of this relationship, including associations among high temperature, high ozone, and air stagnation. Summertime, under high atmospheric pressure, is the primary ozone season. In addition, ozone formation is partly dependent on natural biogenic hydrocarbon emissions from vegetation, which are temperature dependent (Jacob et al. 1993, 2011). Isoprene is a common natural volatile organic compound released by vegetation.

Most of the biogeochemical processes that regulate elemental cycling and that are discussed in this book are influenced by temperature, precipitation, snowpack dynamics, and/or other aspects of climate. Thus, effects of pollutants are assessed in the context of a changing climate (Intergovernmental Panel on Climate Change [IPCC] 2014). Air pollution and climate change have been described as two sides of the same coin because both originate largely from the burning of fossil fuels. Increased atmospheric carbon dioxide, warming temperatures, and altered hydrologic regimes associated with climate change modify specific effects of air pollutants and will influence sensitive ecosystem acid–base chemistry, nutrient cycling, ozone exposure, and visibility degradation. In addition, many prominent air contaminants, including carbon dioxide, methane, S, and particulate matter among others, are important climate-forcing agents. Thus, it is important to consider air pollution from a global, holistic perspective and to effectively integrate goals focused collectively on land use, air quality, and climate in setting environmental policy (Rockstrom et al. 2009, Jacob et al. 2011, Val Martin et al. 2015).

Previous national-scale air pollution sensitivity and effects screening assessments conducted for the National Park Service-Air Resources Division (Sullivan et al. 2011a,b) relied upon emissions and deposition data for the period 2001–2003, as these were the most current data available at the time of preparation of those assessments. Since that time, emissions in proximity to national parklands and deposition within park borders have changed. At many locations, especially in the eastern United States, S and oxidized N emissions and deposition levels have decreased, in large part due to legislated emissions reductions and federal and state rules that controlled emissions from coal-fired power plants and other major pollutant sources (Haeuber 2013). Emissions and deposition of reduced N (ammonia, ammonium) have more commonly increased in recent years. Emissions were expressed by Sullivan et al. (2011a,b) as annual estimates; deposition was expressed as three-year annual averages to partially account for variation in intra-annual and inter-annual weather conditions. These estimates are updated in this book by calculating analogous estimates focused on the year 2011.

The National Emissions Inventory databases for 2002 and 2011 (https://www.epa. gov/air-emissions-inventories) were used as the basis for mapping county-level total emissions of oxidized N, ammonia, and sulfur dioxide for this book. The National Emissions Inventory is prepared every three years by the EPA, based primarily upon emissions estimates and emissions model inputs provided by state, local, and tribal air agencies for emissions sources in their jurisdictions, and supplemented by data developed by the EPA. Although there are uncertainties in the emissions estimates, they provide useful information on the amount and distribution of emissions quality and quantity and the kinds of emissions sources. Emissions source types that comprised more than 1% of total national emissions were used as the basis for the mapped values presented here. The individual source types were grouped into broad categories, including agricultural, electricity generation, industrial and residential, fire, mobile, and biogenic. Table 2.1 describes the types of sources in each category. Emissions from these aggregated source type categories were mapped for each county in units of tons per square mile per year (tons/mi^2/yr) for the years 2002 and 2011 and are shown in Figures 2.1 through 2.3.

The largest sources of sulfur dioxide emissions in the United States are coal-fired power plants. Coal contains varying amounts of S. Historically, the states having the largest sulfur dioxide emissions have included Illinois, Indiana, Kentucky, Missouri, Ohio, Pennsylvania, Tennessee, and West Virginia (see National Emissions Inventory). These states lie upwind of extensive national park and/or wilderness* areas that are sensitive to the acidifying effects of S. As human populations shift south and west, new electrical demands in these regions are leading to new sulfur dioxide emissions sources, and some of them are in proximity to national parks (http://www.nature.nps.gov/air/planning/index.cfm). The Acid Rain Program required substantial reductions in the atmospheric emissions of sulfur dioxide from selected electricity generating units. A cap was established on total electricity generating units' emissions of sulfur dioxide in the conterminous United States. By 2009, the 3572 electricity generating units covered by the Acid Rain Program had reduced their combined sulfur dioxide emissions by about two-thirds from 1980 levels (Burns et al. 2011). The Clean Air Interstate Rule also further reduced S emissions.

---

* Some national parks have portions that have been designated wilderness.

TABLE 2.1

Emissions Source Type Groups Used for Mapping 2011 National Emissions Inventory Data

Emissions Source Group	Emissions Source Type from National Emissions Inventory
Agriculture (fertilizer application)	Agriculture: fertilizer application
Agriculture (livestock waste)	Agriculture: livestock waste
Biogenics	Biogenics: vegetation and soil
Fire	Fires: prescribed fires
	Fires: wildfires
Electricity generation	Electric generation: coal
	Electric generation: natural gas
	Electric generation: oil
Industrial and residential	Industrial boilers, ICEs[a]: coal
	Industrial boilers, ICEs: natural gas
	Industrial boilers, ICEs: oil
	Residential: natural gas
	Residential: oil
	Residential: wood
	Industrial processes: chemical manufacturing
	Industrial processes—NEC[b]
	Industrial processes: nonferrous metals
	Industrial processes: oil and gas production
	Industrial processes: petroleum refineries
	Waste disposal
Mobile	Mobile: commercial marine vessels
	Mobile: locomotives
	Mobile: nonroad equipment—Diesel
	Mobile: nonroad equipment—Gasoline
	Mobile: on-road diesel heavy-duty vehicles
	Mobile: on-road gasoline heavy-duty vehicles
	Mobile: on-road gasoline light-duty vehicles

[a] ICE—internal combustion engine.
[b] NEC—not elsewhere classified.

Ambient spatial patterns in sulfur dioxide emissions are shown in Figure 2.1 for the three-year average period centered on 2011, the latest three-year period for which a national emissions inventory is available at the time of this writing. Highest sulfur dioxide emissions occurred in the eastern United States where emissions in many counties were higher than 50 tons of sulfur dioxide per square mile per year, but have been decreasing in recent years. The largest sulfur dioxide sources were associated with electricity generation. Other sulfur dioxide emissions sectors were generally relatively small contributors to total sulfur dioxide emissions.

Emissions of nitrogen oxides in the United States derive mainly from power plants, motor vehicles, industry, and more recently oil and gas development. Nitrogen oxide is a pollutant that is released during combustion and the unintended combination of N and oxygen (O) in the fuel and the air. Ammonia emissions derive mainly from agricultural sources, including fertilizer and livestock waste. The 960 electricity generating units subject to Acid Rain Program regulations governing oxidized N emissions decreased their oxidized N emissions by two-thirds between 1995 and 2009 (Burns et al. 2011). More recent decreases are reflected in EPA estimates at http://www.epa.gov/airmarkets. Nonpoint* sources of ammonia emissions remain largely uncontrolled, however, and deposition from the atmosphere to the ground surface of these emissions has increased at most park locations over the last decade (see National Emissions Inventory; https://www.epa.gov/air-emissions-inventories/national-emissions-inventory).

Spatial patterns in oxidized N emissions are shown in Figure 2.2 for the three-year period centered on 2002 and 2011. Emissions of nitrogen oxides were above 5 tons/mi²/yr in many counties scattered throughout the conterminous United States. Counties in the east generally had larger emissions than those in the west. There were hot spots of oxidized N emissions near large urban areas. The largest contributions to oxidized N

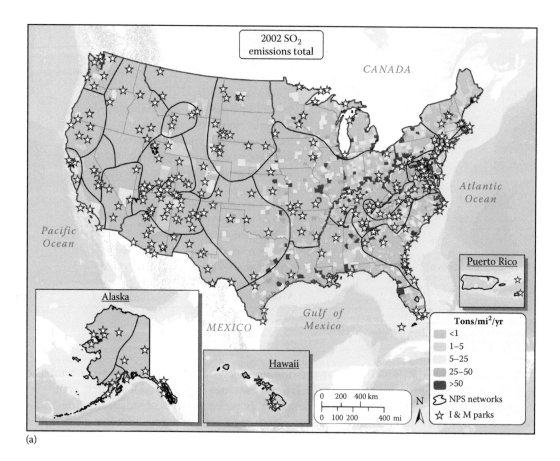

(a)

FIGURE 2.1

Total sulfur dioxide (SO$_2$) emissions, by county, for the years (a) 2002 and (b) 2011. (Based on data from the U.S. Environmental Protection Agency's National Emissions Inventory, https://www.epa.gov/air-emissions-inventories, accessed August, 2010.)   *(Continued)*

---

* Pollution sources can be classified as point sources or nonpoint sources. The former include generally large sources that occur at a particular location, such as a power plant, industrial facility, or animal feedlot. The latter include relatively small, and often numerous, sources that are spread across the landscape, such as motor vehicles.

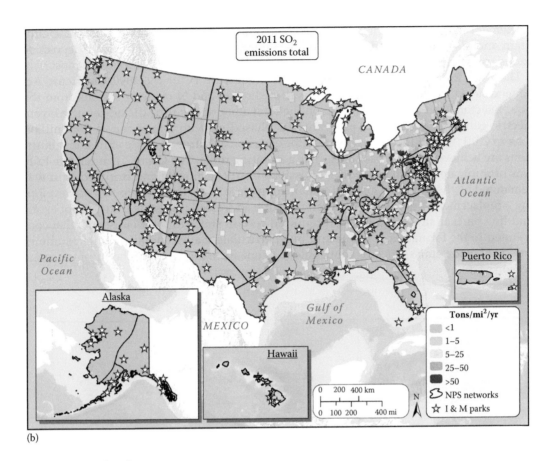

(b)

FIGURE 2.1 (*Continued*)
Total sulfur dioxide (SO₂) emissions, by county, for the years (a) 2002 and (b) 2011. (Based on data from the U.S. Environmental Protection Agency's National Emissions Inventory, https://www.epa.gov/air-emissions-inventories, accessed January, 2014.)

emissions were from the mobile source sector. The electricity generation and industrial and residential sectors were also important contributors.

In contrast to the generally decreasing trends in sulfur dioxide and oxidized N emissions throughout the United States between 2002 and 2011 (Figures 2.1 and 2.2), ammonia emissions have mostly stayed constant or increased (Figure 2.3). Ammonia emissions were concentrated mainly in the Upper Midwest, with additional pockets of relatively high ammonia emissions at locations along the East Coast and in California. The largest ammonia emissions sectors were agricultural (fertilizer application and livestock).

The Acid Rain Program and the Clean Air Interstate Rule were cap and trade programs, whereby emissions allowances were actively traded in financial markets, with a cap placed on total annual emissions. The former covered power plants across the contiguous United States. The latter covered power plants in 27 eastern states. In addition to the Acid Rain Program and the Clean Air Interstate Rule, a number of other air pollution control programs have contributed to recent reductions in oxidized N emissions in many areas, especially in the eastern United States. These have included a memorandum of understanding (1999–2002) among 11 states and the District of Columbia to implement oxidized N control technologies and an ozone-season cap and trade program and also the oxidized N State Implementation Plan call issued in 1998 to reduce ozone transport.

The Acid Rain Program was established under Title IV of the 1990 Clean Air Act Amendments to require reductions in sulfur dioxide and N oxide emissions from the power sector. It set a permanent cap on the total amount of sulfur dioxide emitted from electric generating units. The program was phased in, with a final cap of 8.95 million tons in 2010 (half of the 1980 power sector emissions). Nitrogen oxide emissions reductions are achieved for a subset of electric generating units. The Clean Air Interstate Rule was promulgated in 2005 to reduce emissions of sulfur dioxide and oxidized N. It was remanded by the courts in 2009 but remained in place while the EPA worked on developing a replacement. The Clean Air Interstate Rule addressed interstate transport of ozone and fine particulate air pollution. It required certain states to limit emissions that contributed to the formation of ozone and fine particulate matter and also limited summer season N oxide emissions that contribute to ozone formation during the summer ozone season.

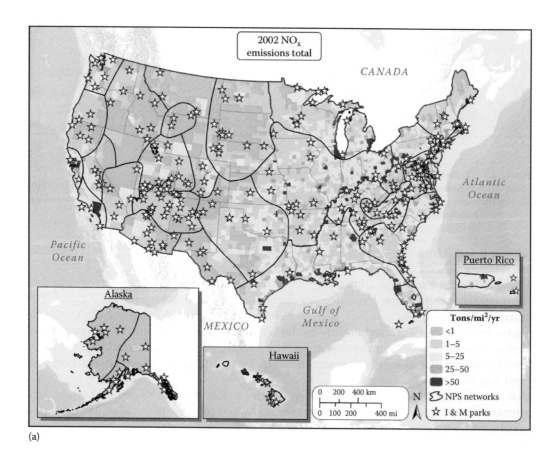

(a)

FIGURE 2.2

Total oxidized nitrogen (NO_x) emissions, by county, for the years (a) 2002 and (b) 2011. (Based on data from the U.S. Environmental Protection Agency's National Emissions Inventory, https://www.epa.gov/air-emissions-inventories, accessed August, 2010.)   (*Continued*)

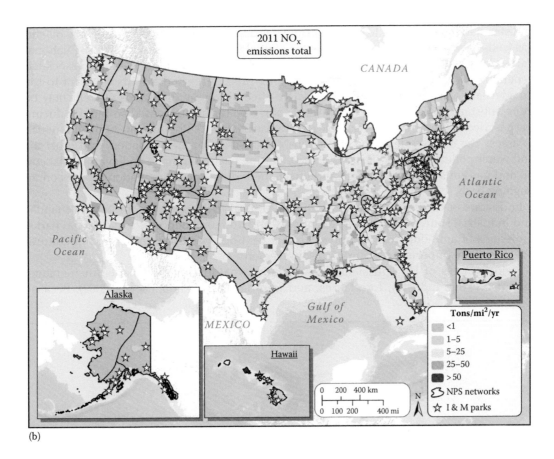

(b)

FIGURE 2.2 (*Continued*)
Total oxidized nitrogen (NO_x) emissions, by county, for the years (a) 2002 and (b) 2011. (Based on data from the U.S. Environmental Protection Agency's National Emissions Inventory, https://www.epa.gov/air-emissions-inventories, accessed January, 2014.)

In 2015, EPA's Cross-State Air Pollution Rule replaced the Clean Air Interstate Rule. The clean diesel rule further helped to control some mobile sources (http://www3.epa.gov/otaq/standards.htm). Emissions reductions under these, and other, air pollution control programs have contributed to improvement in air quality as documented in the measurements of ambient atmospheric concentrations of oxidized S, oxidized N, and ozone. Total annual national emissions of sulfur dioxide throughout the United States, as represented by the National Emissions Inventory, decreased from about 31 million tons in 1970 to about 5 million tons in 2012. Over that same time period, oxidized N emissions decreased from about 28 to 11 million tons. For background information on federal regulations, see http://www.epa.gov/airmarkets/historical-reports; http://www3.epa.gov/airmarkets/progress/reports/index.html. Analogous emissions reductions have not been achieved for ammonia.

Emissions of oxidized N from oil and gas development are rapidly emerging as an important threat to air quality in various regions of the United States (Sullivan and McDonnell 2014). Recent and ongoing oil and gas development in the United States is accomplished mainly by hydraulic fracturing (fracking), which uses pressure from water and steam to fracture shale to release oil and gas deposits. Many thousands of new oil and gas wells have been installed in parts of the United States, most since about 2008 (National Parks Conservation Association 2013). Emissions from an individual well and associated generators vary with equipment and operating procedures and are insignificant to the broader landscape. However, the cumulative effects of many thousands of wells may be more substantial. Because the oil and gas source sector has increased so rapidly during the past decade, few data are available (cf., Karion et al. 2013, Prenni et al. 2016) with which to assess the magnitude of the effects on N deposition, ozone formation, and other air and water pollution issues. There is an accelerating need to compile and evaluate data on oil and gas emissions, deposition, and effects in and around many national parks on an ongoing basis.

Emissions of oxidized N from traditional vertical drilling and from other vehicular and industrial sources also impact air quality and pollutant deposition in national parks. Emissions of ammonia from confined animal feeding operations and other agricultural operations are important N sources in some regions.

Major sources of Hg emissions have included coal-fired power plants, incinerators, manufacturing, industry, forest fires, and mining (Pacyna and Pacyna 2002). Because

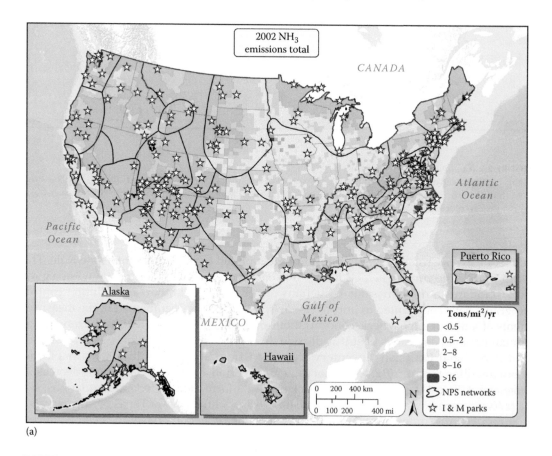

(a)

FIGURE 2.3
Total reduced nitrogen (NH_3) emissions, by county, for the years (a) 2002 and (b) 2011. (Based on data from the U.S. Environmental Protection Agency's National Emissions Inventory, https://www.epa.gov/air-emissions-inventories, accessed August, 2010.)   *(Continued)*

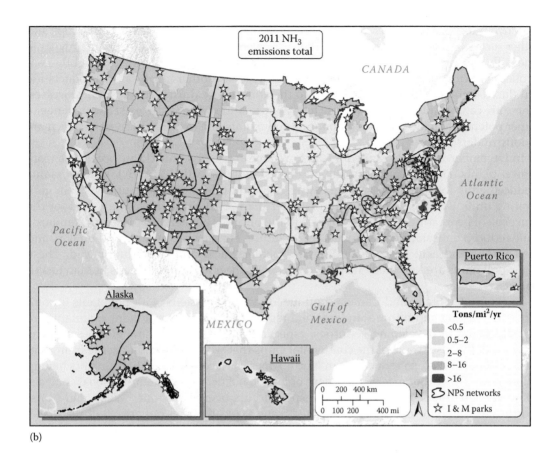

(b)

FIGURE 2.3 (*Continued*)
Total reduced nitrogen (NH₃) emissions, by county, for the years (a) 2002 and (b) 2011. (Based on data from the U.S. Environmental Protection Agency's National Emissions Inventory, https://www.epa.gov/air-emissions-inventories, accessed January, 2014.)

the atmospheric lifetime of Hg is between six months and one year, it is a global pollutant. Sources must be considered from a global perspective (i.e., everyone is in one Hg boat!), and Hg emissions are increasing rapidly in other parts of the world. The peak Hg deposition in much of the eastern United States occurred in the 1980s, with a substantial decrease since then (Drevnick et al. 2012). During the more recent decades, however, global human-caused Hg emissions increased 17% (Pirrone et al. 2010), to some degree negating the benefits of emissions reductions in the United States. Ambient Hg deposition in this country is about three to four times higher than preindustrial deposition, based on the analysis of the Hg content of lake sediment cores (Drevnick et al. 2012). However, Hg emissions decreased in the eastern United States from about 246 tons in 1990 to 101 tons in 2005. This decrease has largely been attributed to controls on municipal and medical waste incinerators and power plants (Butler et al. 2008). The largest single type of U.S. anthropogenic emissions source of Hg in 2005 was coal-fired power plants (52.3 tons emitted), followed by other boilers, cement manufacturing, and electric arc furnaces (Schmeltz et al. 2011).

Toxic substances other than Hg are also emitted into the atmosphere from a variety of industrial, agricultural, and other sources. These include many pesticides. Emissions of current-use pesticides derive mainly from agricultural sources such as row crops and other forms of intensive agriculture. There are also emissions, presumably mainly from the soil, of legacy pollutants, including pesticides that are now banned or tightly controlled. Dioxins, polycyclic aromatic hydrocarbons, and other chemical by-products are produced unintentionally and released to the atmosphere during combustion processes. Some are cancer causing and/or cause fetal abnormalities. There also exist a wide range of other potentially toxic organic compounds and metals, the toxicities of most of which are poorly understood.

In December 2001, EPA signed a rule to reduce emissions of toxic air pollutants from power plants. The new mercury and air toxic standards (called MATS) for power plants reduced emissions from new and existing coal- and oil-fired electric utility steam generating units. The rule applies to units larger than 25 MW that burn coal or oil for the purpose of generating electricity for sale to the public. The mercury air toxics standards rule reduced emissions of heavy metals, including Hg, arsenic (As), chromium (Cr), and nickel (Ni), and acid gases, including hydrochloric acid and hydrofluoric acid. These toxic pollutants are known or suspected to cause cancer and other serious human health effects.

As the United States and Canada continue to reduce emissions of air pollutants to comply with existing and future standards and regulations, emissions of many pollutants are rapidly rising in developing countries, especially in Asia. These new global emissions may partially offset some of the continental gains that would otherwise accrue from local and regional emissions reductions in North America (Jacob et al. 2011). Pollutants having short atmospheric lifetimes of a few days or less do not necessarily require global analysis. However, circumpolar atmospheric transport occurs over about four weeks, and transport from Asia to the conterminous United States typically takes only two to three weeks (Jacob et al. 2011). Thus, pollutants with atmospheric lifetimes longer than several weeks do warrant such a broad perspective.

Fine particulate matter (called $PM_{2.5}$) is often a secondary pollutant that forms within the atmosphere, rather than being directly emitted from a pollution point source. Secondary particulate matter largely derives from atmospheric gas-to-particle conversion reactions involving nucleation, condensation, and coagulation and from evaporation of water from contaminated fog and cloud droplets. Fine particulate matter may also contain condensates of volatile organic compounds, volatilized metals, and products of incomplete combustion, including polycyclic aromatic hydrocarbons and elemental carbon (C; soot; U.S. EPA 2004).

Coarse particulate matter (called PM_{10}) is mainly a primary pollutant, having been emitted from wildland fire and pollution sources as fully formed particles derived from abrasion and crushing processes, soil disturbance, desiccation of marine aerosols, hygroscopic fine particulate matter (fine particulate matter that absorbs water and expands with humidity), and/or gas condensation onto existing particles (U.S. EPA 2009b). Suspended primary coarse particulate matter may contain iron (Fe), silicon (Si), aluminum (Al), and base cations from soil, plant and insect fragments, pollen, fungal spores, bacteria, and viruses, as well as fly ash, brake lining particulates, debris, and automobile tire fragments. In many rural areas and some urban areas, the majority of the mass in the atmospheric coarse particle mode derives from the elements Si, Al, calcium (Ca), and Fe, suggesting a crustal origin as fugitive dust from disturbed land, roadways, agriculture tillage, or construction activities. Rapid sedimentation of coarse particles tends to restrict their direct effects on vegetation largely to roadsides and forest edges, which often receive the greatest coarse particle deposition (U.S. EPA 2004).

2.2 Deposition

Airborne particles, their gas-phase precursors, and their transformation products are removed from the atmosphere by wet, dry, and occult (cloud and/or fog) deposition processes. These deposition processes transfer pollutants to other environmental media such as vegetation, soil, and exposed rock or water surfaces where they can alter the structure, function, diversity, and sustainability of ecosystems.

There are important differences among the various media and how pollutant amounts are expressed. Emissions are commonly expressed to reflect the quantity of pollutant emitted from a source into the atmosphere in units such as tons per year (tons/yr). Once in the atmosphere, the amount of pollutant may be expressed as an atmospheric concentration. This may be, for example, in units that reflect the amount of pollutant per cubic meter of air.

Wet deposition is calculated as the concentration of the pollutant in precipitation (rain and snow) times the amount of precipitation. It is expressed in units that reflect the amount of wet deposition pollutant transfer as a flux from the atmosphere to the ground/vegetation/water surfaces per unit area per unit time (often as kilograms per hectare per year [kg/ha/yr]).

Dry deposition transfers pollutants (gases, aerosols, and particles) from the atmosphere to the ground/vegetation/water surfaces without precipitation as the transfer vehicle. It is estimated based on the pollutant concentration in the air and an estimated deposition velocity (or transfer rate), which is partly a function of the type of vegetation surface. The actual amount of dry deposition is expressed as the amount of pollutant that is dry deposited per unit ground or watershed area per unit time (often as kg/ha/yr).

There are differences in the deposition behavior of gaseous pollutants and fine and coarse particles. Coarse particles generally deposit due to gravity nearer their site of emission than do fine particles or gases. Fine particles and gases generally stay in the atmosphere longer and therefore move farther away from their sources. The chemical composition of individual particles is also related to particle size. For example, fine particulate matter can act as a carrier for materials that deposit to the particles such as herbicides that are phytotoxic (poisonous to plants). Fine particulate matter provides much of the surface area of particles suspended in the atmosphere and in many areas is primarily responsible for visibility impairment, or haze, whereas coarse particulate matter provides much of the mass of atmospheric particles. Surface area can influence ecological effects associated with the oxidizing capacity of fine particles, their interactions with other pollutants, and their adsorption of phytoactive organic compounds. Fine and coarse particles also respond to changes in atmospheric humidity, precipitation, and wind, and these can alter their deposition characteristics.

Atmospheric deposition can be the primary source of various metals to watersheds at remote locations. Metal inputs can include the primary crustal elements (Al, Ca, potassium [K], Fe, magnesium [Mg], Si, titanium [Ti]) and the primary anthropogenic elements (copper [Cu], zinc [Zn], cadmium [Cd], Cr, manganese [Mn], lead [Pb], and vanadium [V]). The crustal elements are derived largely from weathering and erosion, whereas the anthropogenic elements are derived from combustion, industry, and other human-caused sources (Goforth and Christoforou 2006). Many of the heavy metals (e.g., Cd, Cr, Pb, Hg) are toxic and can impair the functions of organisms and ecosystems.

Deposition of coarse and fine particulate matter is generally highest in urban areas. It is derived largely from industrial processes, vehicular traffic, and home heating

(Lu et al. 2003, Rocher et al. 2004). Urban settings, therefore, continue to be a major focus of atmospheric particulate research (U.S. EPA 2009a). Previous work tended to focus on the size of particles because epidemiologic studies found strong associations between particle size and human respiratory illness, with smaller particles able to penetrate deeper into the lungs and cause harm. Studies have also assessed the chemical composition of particles and the processes involved in their formation (Seinfeld and Pandis 2006).

Some of the most useful records of long-term atmospheric particulate matter deposition are recorded as accumulations of atmospherically deposited metals in ice, snow, peat, and lake sediments. Case studies presented by Norton (2007) focused on three elements (Cd, Pb, Hg), which are biologically active, have negative consequences for ecosystem and human health, and are dominated by atmospheric inputs. Sedimentary pollution records suggested that the atmospheric deposition of these elements peaked in about the period 1965–1975 but subsequently declined by 75% or more. High concentrations still reside in soil in some areas, but the flux through aquatic ecosystems has decreased in recent decades (Norton 2007).

The earlier national park atmospheric deposition estimates by Sullivan et al. (2011a,b) for the time period centered on 2001 were generated from interpolated wet deposition measured by the National Atmospheric Deposition Program–National Trends Network in 2002 and modeled dry deposition simulated by the Community Multiscale Air Quality model (Byun and Schere 2006). At that time, the most recent Community Multiscale Air Quality model data available were for 2002. At the time of this writing, the most recent wet plus dry deposition data available are for the year 2012. Thus, deposition is expressed here as the average of data for 2010, 2011, and 2012. To evaluate changes in deposition over time for the parks evaluated here, estimates of deposition are compared for two time periods, as three-year averages centered on 2001 and 2011.

2.2.1 Wet Deposition

Pollutants are deposited into ecosystems through rain and snow as wet deposition. Wet deposition of S, N, Hg, base cations (Ca, Mg, sodium [Na], and K), and other elements is measured at individual monitoring stations. Monitors are located throughout the United States, including in about 40 national parks. Rain and snow are collected on a weekly basis and analyzed in a central laboratory for a variety of chemical species, which are recorded as concentrations (e.g., milligrams per liter [mg/L]) or deposition (e.g., kg/ha/yr). Data are available from the National Atmospheric Deposition Program–National Trends Network, and the National Atmospheric Deposition Program–Mercury Deposition Network, with monitoring at some sites extending back to the early 1980s for S and N. Spatial patterns in sulfate and nitrate wet deposition across the United States are generally similar, with highest values in a band from the Midwest to the Northeast. This reflects the prevalence of power plants, human population centers, and motor vehicles in these areas. Ammonium wet deposition shows highest values throughout the Midwest and other areas of intensive agricultural development. Available data are generally assumed to well represent geographic patterns in wet deposition, although high-elevation areas (which often receive proportionally higher levels of wet deposition) are less well represented due to the logistical difficulties of operating monitoring stations at high elevation. Wet mercury deposition tends to be highest in summer (Mercury Deposition Network: http://nadp.sws.uiuc.edu/MDN/why.aspx). It is especially high in Indiana, Ohio, Illinois, and

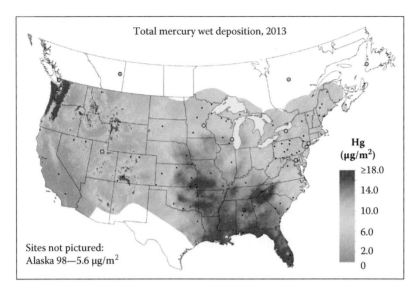

FIGURE 2.4
Interpolated wet mercury (Hg) deposition for 2013 determined by the Mercury Deposition Network. Black dots mark site locations that meet National Atmospheric Deposition Program completeness criteria. Open circles designate urban sites, defined as having at least 400 people/km^2 within a 15 km radius of the site. Urban sites do not contribute to the contour surface. (Modified from National Atmospheric Deposition Program/Mercury Deposition Network; http://nadp.isws.illinois.edu/maplib/pdf/mdn/hg_dep_2013.pdf.)

portions of Pennsylvania, Michigan, and Wisconsin (Figure 2.4; Risch et al. 2012, Wiener et al. 2012). Most of the Hg emitted to the atmosphere deposits as inorganic Hg. The reactive species of Hg are more readily removed from the atmosphere by precipitation than is elemental Hg. Particulate Hg is readily wet deposited due to cloud scavenging processes and precipitation. Elemental Hg can be converted in the atmosphere to a reactive form, which is more soluble. This makes it more easily deposited to land surfaces via wet deposition.

2.2.2 Dry Deposition

In general, wet deposition measurements collected in the various monitoring networks are considered to be relatively robust. Scientists generally place rather high confidence in these measurements. In contrast, measurements of dry deposition, described in the following text, take multiple forms and are much more uncertain. This can be problematic because most ecological effects reflect total (wet plus dry, and [where important] also occult [described in Section 2.2.3]) deposition. Furthermore, dry inputs can constitute half or more of the total amount deposited at many locations.

Mercury gases and particles, including elemental Hg, can be directly deposited to land upon contact as dry deposition (http://nadp.isws.illinois.edu/newissues/litterfall/). This is especially important in forested areas, where Hg is deposited to tree foliage and then transported to the soil through leaf fall and throughfall (water that flows down the stem or drips through the canopy). A portion of the Hg deposited to forest canopies may be photo-reduced and then reemitted to the atmosphere.

Gustin et al. (2006) showed that reemission of deposited Hg from soil can be affected by environmental conditions such as soil moisture, temperature, light, the presence of atmospheric oxidants, and Hg concentrations in the air (U.S. EPA 2009a). It can also be promoted by fire (Dicosty et al. 2006).

Pollutants are deposited as dry deposition into ecosystems through particles, aerosols, and gases. Dry deposition generally cannot be directly measured. Dry deposition rates and totals are often calculated as the product of the measured ambient atmospheric concentration and a modeled deposition velocity. This inferential method is widely used because atmospheric concentrations are easier to measure than is dry deposition directly, and models have been developed to estimate deposition velocities. The Clean Air Status and Trends Network (Clarke et al. 1997) was established in an effort to document the magnitude, spatial variability, and trends in dry deposition across the United States. Ambient pollutant concentrations and meteorological conditions required for the application of inferential models are routinely collected on a weekly basis at Clean Air Status and Trends Network dry deposition sites. Monitored chemical species include ozone, sulfate, nitrate, sulfur dioxide, and nitric acid, which are reported as concentrations in the atmosphere (e.g., micrograms per cubic meter [$\mu g/cm^3$]) or deposition loading (e.g., kg/ha/yr). Ammonium is monitored by the Ammonia Monitoring Network.

Dry particulate deposition, especially of heavy metals, base cations, and organic contaminants, is a complex, poorly characterized process. It appears to be controlled primarily by such variables as atmospheric stability, macro- and microsurface roughness, particle diameter, and surface characteristics (Hosker and Lindberg 1982). The range of particle sizes, diversity of vegetation canopy surfaces, and variety of chemical constituents in airborne particulate matter have made it difficult to estimate and to predict dry particulate deposition (U.S. EPA 2004).

Estimates of regional particulate dry deposition are based on measurements of variable and uncertain particulate concentrations in the atmosphere and even more variable and uncertain measured or modeled estimates of the deposition velocity. The latter is parameterized for a variety of specific surfaces (e.g., Brook et al. 1999). Even for specific sites and well-defined particles, uncertainties are large (U.S. EPA 2009a).

Collection and analysis of stem flow and throughfall can provide useful estimates of dry deposition when compared to directly sampled precipitation. This method is most precise when gaseous deposition is a small component of the total dry deposition and when the leaching or uptake of compounds of interest out of or into the foliage is not a significant fraction of the total depositional flux. This approach works relatively well for S but not for N because N is an important nutrient that is often strongly taken up by plant foliage.

The plant leaf surface has an important influence on the deposition velocity of particles and therefore on the rate of dry deposition to the terrestrial environment. Conifer needles are more efficient than broad leaves in collecting particles by impaction, reflecting the greater surface area, small cross section of the needles relative to the larger leaf laminae of broadleaves, and the greater penetration of wind into conifer, as compared with broadleaf, canopies (U.S. EPA 2004). Canopies of uneven age or with a diversity of species are typically aerodynamically rougher and receive larger inputs of dry-deposited pollutants than do smooth, low, or monoculture vegetation (Garner et al. 1989, U.S. EPA 2004). Canopies on slopes facing the prevailing winds receive larger inputs of pollutants than more sheltered, interior canopy regions.

2.2.3 Total Wet Plus Dry Deposition

Wet and dry deposition estimates were obtained for this assessment from the recent effort by federal agencies in conjunction with the Total Deposition Science Committee of the National Atmospheric Deposition Program, to develop improved estimates of total atmospheric deposition (Schwede and Lear 2014). The Total Deposition project estimates were calculated using a hybrid approach that combines measured and modeled estimates of wet and dry deposition to calculate total deposition. Occult deposition is not included. Monitoring data from Clean Air Status and Trends Network, the Ammonia Monitoring Network, the Southeastern Aerosol Research and Characterization Network, and the National Trends Network were used for generating Total Deposition project estimates. Table 2.2 provides information on the measurement data used from each monitoring network. The Community Multiscale Air Quality model was used as the basis for modeled estimates of deposition. The total deposition estimates provided by the Total Deposition project were produced at 4 km resolution using a method that combines monitoring data with output from the modeling system. This method gives priority to measurement data near the location of the monitors and priority to model simulations in areas where monitoring data are not available.

Model output is used for species such as peroxyacetyl nitrate, oxides of N, and organic nitrate that are not routinely measured in the monitoring networks but likely contribute a significant amount to the total N budget at some locations. However, recent Total Deposition project estimates of wet and dry organic N deposition derived from natural and/or human-caused sources are likely underestimated, and updated Total Deposition project releases are expected to have a more complete representation of organic N deposition. Total Deposition project estimates of total deposition are just that, estimates. However, they are available at a relatively fine scale (4 km) throughout the conterminous United States. In addition, it is hoped that they may be more correct than other available estimates (McDonnell and Sullivan 2014). There are insufficient monitoring data available for Alaska, the Pacific Islands, and the Caribbean to generate modeled deposition estimates for those areas. Also, organic N is not routinely measured by the national deposition monitoring networks but may contribute significantly to total N deposition in some areas (Cornell 2011). Efforts by the various monitoring networks and the modeling community continue to improve deposition estimates.

TABLE 2.2

Summary of Data from Monitoring Networks Used to Derive Total Deposition Project Estimates

Network	Chemical Species	Period of Record	Website
Clean Air Status and Trends Network	Concentration: nitric acid; sulfur dioxide; particulate sulfate, nitrate, and ammonium	2000–2012	http://epa.gov/castnet/javaweb/index.html
Ammonia Monitoring Network	Concentration: ammonium	2008–2012	http://nadp.sws.uiuc.edu/AMoN/
Southeastern Aerosol Research and Characterization	Concentration: nitric acid, sulfur dioxide, and ammonium	2005–2011	http://www.atmosphericresearch.com/studies/SEARCH/
National Trends Network	Wet deposition: sulfate, nitrate, and ammonium	2000–2012	http://nadp.sws.uiuc.edu/NTN/

TABLE 2.3

Distribution of Estimated Change from 2001 to 2011 in Total Sulfur Deposition across Each of the National Park Service Inventory and Monitoring Networks in the Conterminous United States, Derived by the Total Deposition Project

Network Code	Network Name	Number of Parks	Changes in Deposition (kg S/ha/yr) to Parks[a]		
			Minimum	Median	Maximum
APHN	Appalachian Highlands	4	−8.2	−6.7	−4.2
CHDN	Chihuahuan Desert	6	−0.1	0.3	0.9
CUPN	Cumberland Piedmont	14	−9.3	−4.8	−2.8
ERMN	Eastern Rivers and Mountains	9	−16.4	−6.5	1.7
GLKN	Great Lakes	9	−2.9	−0.3	5.7
GRYN	Greater Yellowstone	3	0.2	0.3	0.3
GULN	Gulf Coast	8	−2.6	−0.6	0.8
HTLN	Heartland	15	−9.1	−0.7	0.3
KLMN	Klamath	6	0.0	0.2	0.6
MEDN	Mediterranean Coast	3	−0.4	0.7	2.1
MIDN	Mid-Atlantic	10	−10.1	−6.3	−2.9
MOJN	Mojave Desert	6	0.0	0.0	0.2
NCBN	Northeast Coastal and Barrier	8	−9.5	−4.8	2.0
NCCN	North Coast and Cascades	7	−3.5	−0.6	0.1
NCPN	Northern Colorado Plateau	16	−0.3	−0.1	3.5
NCRN	National Capital Region	11	−9.1	−7.0	−5.6
NETN	Northeast Temperate	11	−7.3	−5.6	−3.3
NGPN	Northern Great Plains	13	−0.9	0.1	0.5
ROMN	Rocky Mountain	6	−0.3	−0.1	0.4
SCPN	Southern Colorado Plateau	19	−0.5	−0.1	0.5
SECN	Southeast Coast	17	−4.7	−2.8	0.6
SFAN	San Francisco Bay Area	6	−0.8	−0.1	3.0
SFCN	South Florida/Caribbean	4	−1.2	−0.9	−0.1
SIEN	Sierra Nevada	4	0.2	0.2	0.2
SODN	Sonoran Desert	11	−0.2	0.0	0.1
SOPN	Southern Plains	10	−3.4	−0.8	0.0
UCBN	Upper Columbia Basin	8	0.0	0.1	0.3

Source: Data from Schwede, D.B. and Lear, G.G., *Atmos. Environ.*, 92, 207, 2014.

[a] Deposition estimates are based on three-year averages centered on the years 2001 and 2011. The minimum represents the largest decrease or the smallest increase; the maximum represents to largest increase or smallest decrease.

Between 2001 and 2011, total S deposition at Inventory and Monitoring parks decreased by varying amounts in many park networks (Table 2.3; Figure 2.5). The median decrease was more than 5 kg S/ha/yr in the National Capital, Appalachian Highlands, Eastern Rivers and Mountains, Mid-Atlantic, and Northeast Temperate networks, all located in the eastern United States. The typical park in all networks either decreased in S deposition during that ten-year time period, stayed the same, or only increased by a small amount (e.g., +0.7 kg S/ha/yr in the Mediterranean Coast Network).

The median park among all parks in the network showed a decrease in oxidized N deposition in all networks (Table 2.4), with some decreases larger than 3 kg N/ha/yr (Mediterranean Coast, National Capital Region, Northeast Temperate, Cumberland Piedmont,

Eastern Rivers and Mountains, Mid-Atlantic networks). Conversely, all networks except the Northeast Temperate, Eastern Rivers and Mountains, and Great Lakes networks showed increases in ammonium deposition (and those three networks showed only very small decreases (−0.1 to −0.2 kg N/ha/yr; Table 2.5). Total oxidized N plus ammonium exhibited variable responses, with some networks showing decreases and some showing increases (Table 2.6). The largest decreases were in the Eastern Rivers and Mountains (−3.5 kg N/ha/yr), Northeast Temperate (−3.2 kg N/ha/yr), Mid-Atlantic (−2.9 kg N/ha/yr), and National Capital Region (−2.6 kg N/ha/yr) networks. The largest median increases were smaller and occurred in the Sierra Nevada (+1.4 kg N/ha/yr) and Heartland (+0.6 kg N/ha/yr) networks.

Estimates of total wet plus dry deposition of S and N throughout the conterminous United States are shown in Figures 2.5 and 2.6, respectively. Total S deposition is highest in the northeastern United States and along the Appalachian Mountains. Relatively high total N deposition occurs more sporadically throughout the eastern half of the country and in southern California. Current total N deposition in the Intermountain West may be somewhat higher than is represented in Figure 2.6 due to recent increased oil and

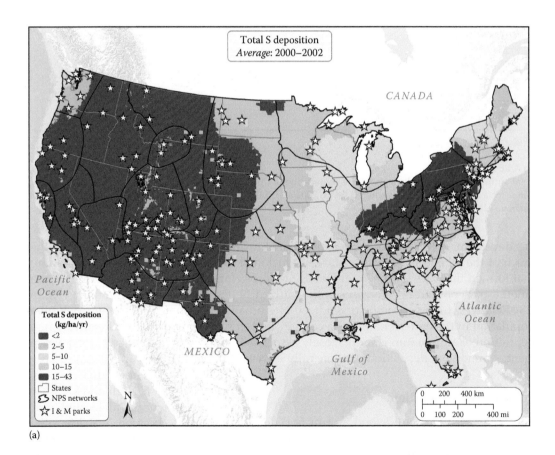

(a)

FIGURE 2.5
Total annual S deposition throughout the conterminous United States based on Total Deposition project estimates for a three-year average centered on (a) 2001 and (b) 2011. (From Schwede, D.B. and Lear, G.G., *Atmos. Environ.*, 92, 207, 2014.) *(Continued)*

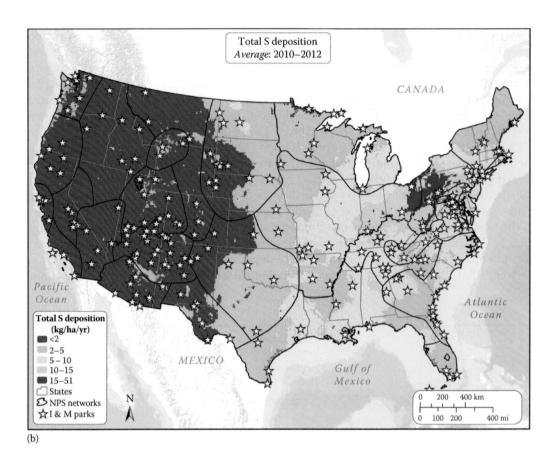

(b)

FIGURE 2.5 (*Continued*)
Total annual S deposition throughout the conterminous United States based on Total Deposition project esti-
mates for a three-year average centered on (a) 2001 and (b) 2011. (From Schwede, D.B. and Lear, G.G., *Atmos.
Environ.*, 92, 207, 2014.)

gas development that has occurred subsequent to 2012 in some regions and the fact that
much of the oil and gas emissions are not included in the National Emissions Inventory
that formed the basis for the Total Deposition project modeling presented here. The total
amount of deposited N comprises both oxidized N (Figure 2.7) and ammonium (Figure
2.8). Relatively high deposition of oxidized N is broadly associated with centers of human
population in the eastern half of the country and in southern California. Relatively high
deposition of ammonium is more closely associated with centers of agricultural development. Ammonium deposition has generally increased in recent years.

2.2.4 Occult Deposition

The occurrence of occult (cloud and/or fog) deposition tends to be more restricted geo-
graphically, mainly to coastal and high mountain areas. In some areas, typically along
foggy coastlines or at high elevations, occult deposition represents a substantial fraction of
total deposition of S and N to foliar surfaces (Fowler et al. 1991).

Several factors make occult deposition particularly effective, where it occurs, for the delivery of dissolved and suspended particulates to vegetation (U.S. EPA 2009a). Concentrations

TABLE 2.4

Distribution of Estimated Change from 2001 to 2011 in Total Oxidized Inorganic Nitrogen Deposition across Inventory and Monitoring Parks in Each of the Park Networks in the Conterminous United States, Derived by the Total Deposition Project

Network Code	Network Name	Number of Parks	Changes in Deposition (kg N/ha/yr) to Parks[a]		
			Minimum	Median	Maximum
APHN	Appalachian Highlands	4	−3.2	−2.6	−1.8
CHDN	Chihuahuan Desert	6	−0.7	−0.4	−0.1
CUPN	Cumberland Piedmont	14	−4.1	−3.0	−1.9
ERMN	Eastern Rivers and Mountains	9	−4.6	−3.4	−2.4
GLKN	Great Lakes	9	−3.0	−1.6	0.2
GRYN	Greater Yellowstone	3	−0.3	−0.3	−0.3
GULN	Gulf Coast	8	−5.4	−1.9	0.0
HTLN	Heartland	15	−4.3	−0.9	−0.3
KLMN	Klamath	6	−1.0	−0.4	−0.4
MEDN	Mediterranean Coast	3	−7.0	−4.3	0.1
MIDN	Mid-Atlantic	10	−4.6	−3.9	−2.3
MOJN	Mojave Desert	6	−1.4	−0.6	0.0
NCBN	Northeast Coastal and Barrier	8	−7.0	−2.5	2.8
NCCN	North Coast and Cascades	7	−1.2	−0.4	0.4
NCPN	Northern Colorado Plateau	16	−1.1	−0.5	0.1
NCRN	National Capital Region	11	−5.1	−3.7	−2.9
NETN	Northeast Temperate	11	−5.9	−3.5	−1.2
NGPN	Northern Great Plains	13	−0.8	−0.5	0.2
ROMN	Rocky Mountain	6	−0.6	−0.3	−0.1
SCPN	Southern Colorado Plateau	19	−1.4	−0.3	0.7
SECN	Southeast Coast	17	−6.4	−2.9	0.5
SFAN	San Francisco Bay Area	6	−8.1	−2.4	−0.6
SFCN	South Florida/Caribbean	4	−2.5	−2.1	−1.0
SIEN	Sierra Nevada	4	−0.3	−0.1	0.1
SODN	Sonoran Desert	11	−0.8	−0.4	0.1
SOPN	Southern Plains	10	−1.3	−1.0	0.1
UCBN	Upper Columbia Basin	8	−0.4	−0.3	0.0

Source: Data from Schwede, D.B. and Lear, G.G., *Atmos. Environ.*, 92, 207, 2014.

[a] Deposition estimates are based on three-year averages centered on the years 2001 and 2011. The minimum represents the largest decrease or the smallest increase; the maximum represents to largest increase or smallest decrease.

of particulate-derived materials are often manyfold higher in cloud or fog water than in precipitation or ambient air due to orographic effects* and gas–liquid partitioning. In addition, fog and cloud water deliver chemical particle constituents in a bioavailable hydrated form to foliar surfaces.

Cloud water can be 5–20 times more acidic than rain water (Mohnen 1990, Vong et al. 1991), and clouds can contribute as much, or more, acidic deposition as wet plus dry deposition processes at some high-elevation sites (Lovett and Kinsman 1990,

---

* As air flows over mountains, orographic lifting can create precipitation effects. The side of the mountain that causes the lifting typically receives additional precipitation. The opposite side of the mountain experiences a rain shadow and is dryer as the air descends and warms.

TABLE 2.5

Distribution of Estimated Change from 2001 to 2011 in Total Ammonium Deposition across Inventory and Monitoring Parks in Each of the Park Networks in the Conterminous United States, Derived by the Total Deposition Project

Network Code	Network Name	Number of Parks	Changes in Deposition (kg N/ha/yr) to Parks[a]		
			Minimum	Median	Maximum
APHN	Appalachian Highlands	4	0.7	0.9	1.0
CHDN	Chihuahuan Desert	6	−0.1	0.0	0.1
CUPN	Cumberland Piedmont	14	0.0	1.2	3.5
ERMN	Eastern Rivers and Mountains	9	−0.3	−0.1	0.1
GLKN	Great Lakes	9	−0.6	−0.1	1.2
GRYN	Greater Yellowstone	3	0.1	0.7	1.0
GULN	Gulf Coast	8	−0.5	0.2	0.9
HTLN	Heartland	15	−0.2	1.8	5.5
KLMN	Klamath	6	0.2	0.6	0.8
MEDN	Mediterranean Coast	3	0.1	2.6	3.1
MIDN	Mid-Atlantic	10	−1.3	0.8	2.3
MOJN	Mojave Desert	6	0.3	0.5	0.9
NCBN	Northeast Coastal and Barrier	8	−1.0	0.2	2.7
NCCN	North Coast and Cascades	7	0.2	0.3	1.0
NCPN	Northern Colorado Plateau	16	0.0	0.3	2.3
NCRN	National Capital Region	11	0.4	1.2	3.5
NETN	Northeast Temperate	11	−1.0	−0.2	0.9
NGPN	Northern Great Plains	13	0.0	0.5	2.4
ROMN	Rocky Mountain	6	0.2	0.3	2.3
SCPN	Southern Colorado Plateau	19	−0.1	0.3	1.5
SECN	Southeast Coast	17	−0.6	0.2	5.0
SFAN	San Francisco Bay Area	6	0.4	1.1	2.5
SFCN	South Florida/Caribbean	4	0.1	0.3	0.5
SIEN	Sierra Nevada	4	1.3	1.6	2.8
SODN	Sonoran Desert	11	0.1	0.4	1.7
SOPN	Southern Plains	10	−0.3	0.3	2.1
UCBN	Upper Columbia Basin	8	0.1	0.5	5.7

Source: Data from Schwede, D.B. and Lear, G.G., *Atmos. Environ.*, 92, 207, 2014.

[a] Deposition estimates are based on three-year averages centered on the years 2001 and 2011. The minimum represents the largest decrease or the smallest increase; the maximum represents to largest increase or smallest decrease.

Vong et al. 1991, Fuzzi and Wagenbach 1997). Anderson et al. (1999) presented the results of cloud chemistry measurements at three high-elevation sites in the eastern United States: Whiteface Mountain, New York; Whitetop Mountain, Virginia; and Clingmans Dome, Tennessee. The most prevalent ions in cloud water, in decreasing order of concentration, were sulfate, hydrogen, ammonium, and nitrate. Variations in ionic concentrations were largely due to differences in liquid water content of cloud deposition samples. See the review by Weiss-Penzias et al. (2012). Cloud water sampling was discontinued at most sites but continued as the Mountain Acid Deposition Program at Clingmans Dome in Great Smoky Mountains National Park through 2011, when the program was finally terminated (AMEC Environment & Infrastructure Inc. 2013).

TABLE 2.6

Distribution of Estimated Change from 2001 to 2011 in Total Inorganic Nitrogen Deposition across Inventory and Monitoring Parks in Each of the Park Networks in the Conterminous United States, Derived by the Total Deposition Project

Network Code	Network Name	Number of Parks	Changes in Deposition (kg N/ha/yr) to Parks[a]		
			Minimum	Median	Maximum
APHN	Appalachian Highlands	4	−2.2	−1.8	−0.8
CHDN	Chihuahuan Desert	6	−0.8	−0.5	0.0
CUPN	Cumberland Piedmont	14	−2.9	−1.8	1.7
ERMN	Eastern Rivers and Mountains	9	−4.5	−3.5	−2.7
GLKN	Great Lakes	9	−3.5	−1.4	0.5
GRYN	Greater Yellowstone	3	−0.2	0.4	0.6
GULN	Gulf Coast	8	−5.3	−1.2	−0.1
HTLN	Heartland	15	−3.5	0.6	5.0
KLMN	Klamath	6	−0.7	0.2	0.4
MEDN	Mediterranean Coast	3	−4.3	−1.2	0.2
MIDN	Mid-Atlantic	10	−5.1	−2.9	−0.8
MOJN	Mojave Desert	6	−0.8	0.0	0.6
NCBN	Northeast Coastal and Barrier	8	−8.0	−2.1	5.6
NCCN	North Coast and Cascades	7	−0.9	−0.2	1.4
NCPN	Northern Colorado Plateau	16	−0.8	0.0	2.2
NCRN	National Capital Region	11	−4.4	−2.6	0.5
NETN	Northeast Temperate	11	−6.3	−3.2	−1.3
NGPN	Northern Great Plains	13	−0.8	0.0	1.8
ROMN	Rocky Mountain	6	−0.3	0.1	1.7
SCPN	Southern Colorado Plateau	19	−1.4	0.1	1.7
SECN	Southeast Coast	17	−6.9	−2.2	2.1
SFAN	San Francisco Bay Area	6	−7.3	−1.4	1.9
SFCN	South Florida/Caribbean	4	−2.1	−1.9	−0.6
SIEN	Sierra Nevada	4	1.4	1.4	2.8
SODN	Sonoran Desert	11	−0.4	0.1	1.6
SOPN	Southern Plains	10	−1.5	−0.5	1.0
UCBN	Upper Columbia Basin	8	−0.1	0.4	5.4

Source: Data from Schwede, D.B. and Lear, G.G., *Atmos. Environ.*, 92, 207, 2014.

[a] Deposition estimates are based on three-year averages centered on the years 2001 and 2011. The minimum represents the largest decrease or the smallest increase; the maximum represents to largest increase or smallest decrease.

The Adirondack Lakes Survey Corporation continues to monitor cloud deposition at Whiteface Mountain in New York (Dukett et al. 2011).

2.3 Ozone Production and Dispersion

Ozone is a colorless gas component of smog, which can develop during the clear warm weather associated with high pressure systems and low wind speeds. It forms during the atmospheric transport of its precursors and can occur at high concentrations in areas remote from precursor sources. Ozone is an important atmospheric pollutant throughout substantial portions of the United States, with relatively high levels in urban areas and at many remote locations, including within some national parks. Ozone is not emitted directly into the atmosphere by pollutant sources. Rather, it is produced in the atmosphere by photochemical reactions that involve atmospheric oxidized N and volatile organic compound precursors. These reactions are driven by ultraviolet radiation from the sun. The ozone precursors are emitted by motor vehicles, power plants, and industry in the case of oxidized N, and by industries that use organic chemicals and by natural vegetation in the case of volatile organic compounds. Ozone builds up to relatively high levels under appropriate meteorological conditions primarily during summer, when abundant sunlight powers the photochemical reactions that produce ozone (Kohut 2007).

Wildland fires, including both wildfires and prescribed fires, emit N oxides and hydrocarbons and therefore can contribute to ozone formation. The concentration of ozone in the atmosphere in the nonurban western United States has increased since the late 1980s by about 5 parts per billion (ppb) by volume (Jaffe and Ray 2007). Jaffe et al. (2008) investigated the role of forest fire in this trend. The summer-burned area was significantly correlated with ozone concentration at Clean Air Status and Trends Network and National

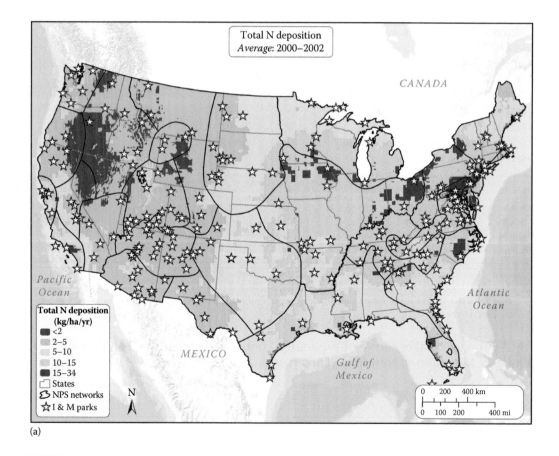

(a)

FIGURE 2.6
Total annual N deposition throughout the conterminous United States based on Total Deposition project estimates for a three-year average centered on (a) 2001 and (b) 2011. (From Schwede, D.B. and Lear, G.G., *Atmos. Environ.*, 92, 207, 2014.) *(Continued)*

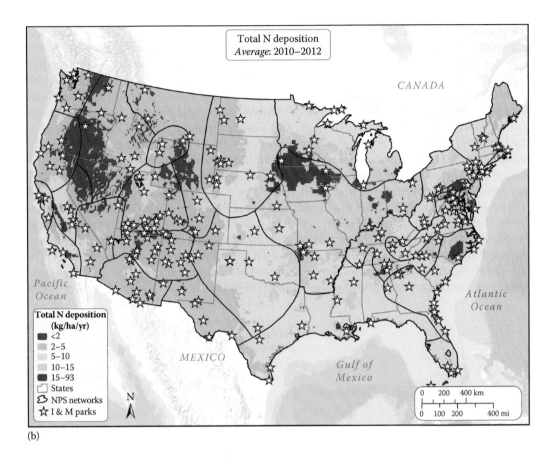

(b)

FIGURE 2.6 (*Continued*)
Total annual N deposition throughout the conterminous United States based on Total Deposition project estimates for a three-year average centered on (a) 2001 and (b) 2011. (From Schwede, D.B. and Lear, G.G., *Atmos. Environ.*, 92, 207, 2014.)

Park Service atmospheric chemistry monitoring sites. For mean and maximum fire years, the concentration of ozone in the atmosphere was increased by fire by an average of 3.5 and 8.8 ppb by volume, respectively. The estimated amount of biomass consumed by fire was a slightly better predictor of ozone concentration than burned area. This relationship between atmospheric ozone concentration and burned area or biomass consumed is especially important because the frequency of extreme fire years in the western United States appears to be increasing (Jaffe et al. 2004) due to past fire suppression, increased spring and summer temperature, earlier snowmelt, and dryer forest conditions (Cook et al. 2004, Westerling et al. 2006).

Jaffe et al. (2008) concluded that the increase in wildfires in the western United States has largely been responsible for the observed increase in summer ozone concentrations reported by Jaffe and Ray (2007). Increasing temperature in the future will likely further increase the prevalence of fire and therefore may increase the formation of ozone even as N oxide emissions are decreasing.

Understanding the contribution of N oxides to ozone formation during transport from pollutant source region to national parks and the effects of local photochemistry on ozone formation are also important in developing predictions of plant exposure to ozone. Future reductions in oxidized N emissions from upwind point and mobile oxidized N sources will contribute to decreased ozone exposure in many areas (Tong et al. 2005).

Ozone is a particular threat to relatively high-elevation plant species because concentrations and total exposure can be higher at high elevations under appropriate atmospheric conditions (Loibl et al. 1994, Sandroni et al. 1994, Brace and Peterson 1996, 1998). Exchange of ozone between the stratosphere and the troposphere contributes to the natural background ozone concentration at both high- and low-elevation locations (Lefohn et al. 2011). The background ozone concentration is not known with a high degree of certainty, partly because of different ways of defining background conditions and partly because the natural background is not a measurable condition. The level of ozone in a pristine area may have a weekly average as low as 10–25 ppb by volume (Altshuller and Lefohn 1996, Cooper and Peterson 2000), with maximum hourly ozone concentrations generally less than about 50–80 ppb by volume. Human activities influence ozone concentration throughout the Northern Hemisphere. Changes not normally associated with air pollution, such as in agriculture and landscape vegetation, have changed the background ozone level. The EPA (2006) defined a policy-relevant background ozone concentration as the amount of ozone that would be present in the atmosphere under a scenario of zero North American

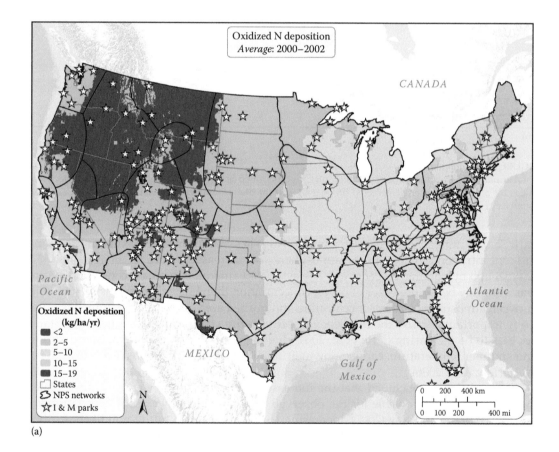

(a)

FIGURE 2.7

Total annual oxidized inorganic N deposition throughout the conterminous United States based on Total Deposition project estimates for a three-year average centered on (a) 2001 and (b) 2011. (From Schwede, D.B. and Lear, G.G., *Atmos. Environ.*, 92, 207, 2014.)   *(Continued)*

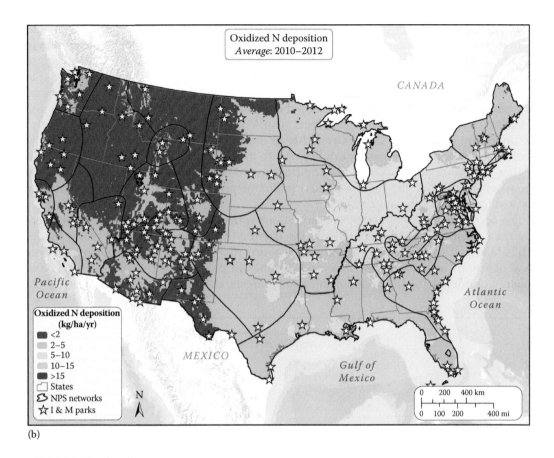

(b)

FIGURE 2.7 (*Continued*)
Total annual oxidized inorganic N deposition throughout the conterminous United States based on Total
Deposition project estimates for a three-year average centered on (a) 2001 and (b) 2011. (From Schwede, D.B. and
Lear, G.G., *Atmos. Environ.*, 92, 207, 2014.)

human-caused ozone precursor pollutant emissions. The EPA previously estimated the
policy-relevant background in the range of 20–40 ppb by volume of ozone (Fiore et al. 2003),
of which half or more was estimated to be of human-caused origin outside North America
(Mickley et al. 2001, Shindell and Faluvegi 2002, Lamarque et al. 2005). In the more recent
Policy Assessment for the ozone National Ambient Air Quality Standards (U.S. EPA 2014),
the EPA estimated that natural background levels range from approximately 15 to 35 ppb
(median value of 24 ppb), with the highest values at high-elevation sites in the western
United States. Natural background levels are higher at these high-elevation locations primarily because natural stratospheric (upper atmosphere) ozone and international transport
both increase with altitude (where ozone lifetimes are longer). Much of that human-caused
component of the policy-relevant background originates from Asia. Emissions of oxidized
N in Asia have increased dramatically in recent years and may account for about 3–7 ppb
by volume of ozone in portions of the United States (Zhang et al. 2008).

Nonattainment of ozone standards in the southeastern United States, where ozone concentrations tend to be relatively high, has been partially attributed to the oxidation of
atmospheric N in the presence of large amounts of volatile organic compounds, much of
which is derived as natural emissions of biogenic isoprene (Chameides et al. 1988). Within

southeastern rural areas, biogenic volatile organic compounds emitted by vegetation and nonbiogenic volatile organic compounds emitted by human activities are both common in the atmosphere (Hagerman et al. 1997, Fuentes et al. 2000, Kang et al. 2001). It is important to understand the roles of these volatile organic compounds in ozone formation in order to devise effective control strategies for ozone (Kang et al. 2003). Rural areas in the Southeast are generally considered to be limited by the amount of oxidized N for ozone formation (Hagerman et al. 1997); nevertheless, local volatile organic compound emissions can also be important (Kang et al. 2003). Ground-level ozone is a particular concern for national parklands because remote areas can experience greater cumulative ozone exposure and higher minimum and maximum concentrations than the upwind urban and industrial areas where most of the human-caused pollutant precursor sources originate and are located (Brace and Peterson 1998).

The occurrence of peak ozone concentrations (greater than or equal to 100 ppb) during the growing season is generally on the decline throughout the United States (Smith 2012). Moderate ozone concentrations are becoming more widespread in rural areas (Percy et al. 2003). In response to the observed declines in peak exposures, the occurrence of foliar symptoms has also declined in eastern forests, especially since 2002 (Smith 2012).

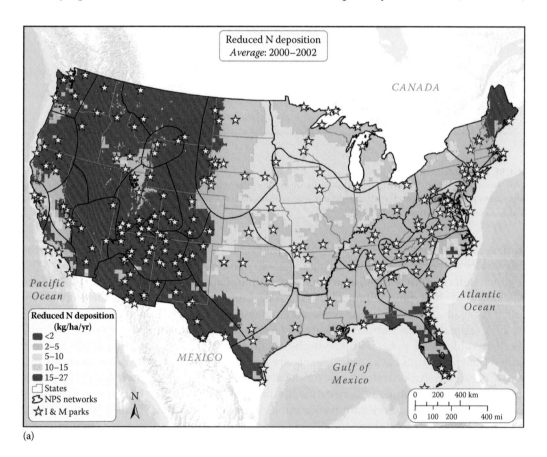

(a)

FIGURE 2.8
Total annual reduced inorganic N deposition throughout the conterminous United States based on Total Deposition project estimates for a three-year average centered on (a) 2001 and (b) 2011. (From Schwede, D.B. and Lear, G.G., *Atmos. Environ.*, 92, 207, 2014.) *(Continued)*

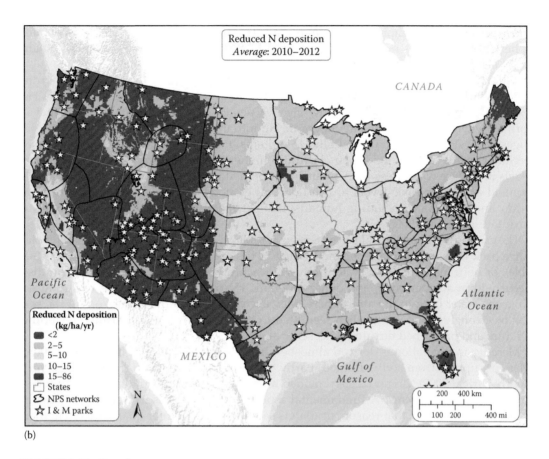

(b)

FIGURE 2.8 (*Continued*)
Total annual reduced inorganic N deposition throughout the conterminous United States based on Total Deposition project estimates for a three-year average centered on (a) 2001 and (b) 2011. (From Schwede, D.B. and Lear, G.G., *Atmos. Environ.*, 92, 207, 2014.)

However, quite high ozone concentrations have been observed at locations downwind of oil and gas development in Wyoming, Utah, and Colorado in recent years. For example, one-hour concentrations as high as 149 ppb and many days with 8-hour ozone exceedances of the National Ambient Air Quality Standards were recorded in the Uintah Basin of Utah in 2010 (Lyman and Shorthill 2013). The ozone National Ambient Air Quality Standard is currently set at 70 ppb, based on the annual fourth highest daily maximum 8-hour concentration, averaged over three years.

2.4 Mechanism of Exposure to Metals and Persistent Organic Pollutants

2.4.1 Mercury

Atmospheric Hg emissions and deposition increased with industrialization. Steinnes et al. (2005) found that Hg concentrations in dated peat samples were about 15 times higher in deposits from the last 100 years as compared with deposits laid down during preindustrial times. However, the complexities of atmospheric Hg chemistry and the typically low atmospheric concentrations make quantification of historical Hg deposition difficult (Lynam and Keeler 2005).

Although any form of Hg can be considered toxic, the greatest threat to the environment, wildlife, and people is low-level exposure to an organic form of Hg: methyl Hg (Schroeder and Munthe 1998). Methyl Hg is a persistent bioaccumulative toxin that is formed when bacteria methylate inorganic Hg into methyl Hg. It can enter through the digestive systems of animals and bind to proteins. It bioaccumulates as each successive predator in the aquatic food web consumes higher concentrations of methyl Hg. Low levels in surface water (less than about 1 part per trillion) can bioaccumulate over a millionfold and reach toxic levels in fish (Driscoll et al. 2007). Predatory fish, piscivorous birds, and mammals (including humans) that eat contaminated fish can be exposed to toxic levels of methyl Hg.

The cycling of Hg in natural ecosystems is complex. Deposited inorganic Hg can revolatilize back to the atmosphere, be methylated in the soil or sediment (making it bioavailable), or be transported to a water body via leaching and runoff. Methylation can also occur within a water body, and either inorganic Hg or methyl Hg can be reintroduced from the water back to the atmosphere. Mercury deposited to the O horizon* of the soil can be strongly retained and then reemitted to the atmosphere with burning, including wildfire and prescribed burning (Dicosty et al. 2006).

Mercury in vegetation is derived almost exclusively from the atmosphere (Grigal 2003). Mercury uptake from soil is limited, partly because plants transport Hg to foliage very poorly (Grigal 2002). Grigal (2003) provided a review of the sequestration of Hg in forest and peatland ecosystems. A fundamental aspect of Hg cycling is its strong relationship to organic matter. For that reason, peatlands sequester much larger quantities of Hg than would be expected on the basis of their land area. As a consequence, peatlands constitute important sources of Hg exposure in wildland ecosystems.

In anaerobic environments such as occur in wetlands and lake or stream sediments, inorganic Hg can be converted to methyl Hg, which biomagnifies and bioaccumulates in aquatic food webs (Wiener et al. 2003, Hall et al. 2005). Sulfate-reducing bacteria are largely responsible for the production of methyl Hg from inorganic Hg (Compeau and Bartha 1985, Gilmour et al. 1992). Iron reducers also play a role. Thus, atmospheric S deposition is an important driver of Hg methylation and subsequent biomagnification. Relatively low concentrations of inorganic Hg in surface water can contribute to substantial Hg bioaccumulation in aquatic ecosystems when and where environmental conditions favor methylation.

2.4.2 Persistent Organic Pollutants

Important toxic atmospheric contaminants deposited in national parks are generally those that are transported moderate to long distances in the atmosphere, are subsequently deposited into remote locations, and are bioaccumulated to sufficient levels that they can impact humans, fish, or wildlife (Swackhamer and Hornbuckle 2004). Certain physical and

---

* Upper temperate zone soils typically possess horizontal layering associated with characteristics of the climate, vegetation, and underlying soil and geologic materials. The layers are grouped under four categories: O, A, B, and C. Subdivisions of these layers are called horizons. The O group comprises organic horizons, which develop above the mineral soil in forested areas. The A group includes upper mineral horizons subject to maximum leaching. The B horizons comprise the zone of maximum accumulation of materials such as Fe, Al oxides, and silicate clays. The C horizon includes unconsolidated material under the A and B horizons.

chemical properties may facilitate the movement of these contaminants from land and water surfaces into the atmosphere, provide stability, and enhance accumulation in animal body fats (lipids) or muscle tissue.

Most persistent organic pollutants enter the environment as a consequence of human activities, including synthetic pesticide application, emissions of polychlorinated dibenzo-*p*-dioxins from incinerators, and accidental release of polychlorinated biphenyls (PCBs) from transformers (Lee et al. 2003). Once they are introduced into the environment, their accumulation and magnification in biological systems are determined by physiochemical properties and environmental conditions. Many are carcinogenic (cancer causing) or teratogenic (causing fetal abnormalities) and pose potential threats to aquatic and terrestrial biota.

Spatial patterns in exposure and effects of these air toxics in national park units throughout the United States are not well known. To better understand atmospheric transport and the potential risks to national park resources of atmospheric emissions of toxic substances, the National Park Service initiated the Western Airborne Contaminants Assessment Project (WACAP) in 2002 (Landers et al. 2010). Composed of a team of multidisciplinary researchers, WACAP was charged with determining the presence, identity, method of accumulation, contamination indicators, sources, and effects of airborne contaminants in 8 core national parks and 12 secondary national parks in the western United States, extending from Alaska south to the Texas border and the Sierra Nevada and southern Rocky mountains.

The WACAP study concluded that local sources, such as nearby agricultural and industrial activity, are more likely than global sources to control atmospheric deposition of current-use pesticides and historic-use pesticides in most western national parklands. Contaminant deposition in Alaska, which has few nearby industrial or agricultural sources, is attributed to atmospheric transport from outside the region (Landers et al. 2010).

Hageman et al. (2006) reported the results of analysis of snow samples collected in seven western national parks that were analyzed for current-use and historic-use pesticides. The most frequently detected current-use pesticides were dacthal, chlorpyrifos, endosulfan, and γ-hexachlorocyclohexane. The researchers concluded that agricultural practices were largely responsible for the distribution of pesticides in snow in many of the surveyed national parks. In the Alaskan parks, however, pesticide deposition was attributed to long-range atmospheric transport. In other parks, the percent of the total pesticide deposition attributable to regional sources was associated with regional cropland occurrence and the vapor pressure and atmospheric half-lives of the pesticides (Hageman et al. 2006). Among the parks studied, percent regional transport was highest for Sequoia/Kings Canyon, Glacier, and Rocky Mountain national parks and lowest for Mount Rainier National Park and the parks in Alaska. Percent regional, as opposed to long-range, transport was highest for dacthal and dieldrin due to their low vapor pressures and short half-lives in air.

The use of flame retardants has increased in recent decades due to fire product safety regulations. Some flame-retardant chemicals are toxic and are readily transported atmospherically. The use of some has been banned in Europe and parts of the United States because of their persistence and tendency to bioaccumulate (Hoh et al. 2006).

Polycyclic aromatic hydrocarbons include hundreds of different compounds that are characterized by possessing two or more fused benzene rings. They are widespread contaminants in the environment and are formed by incomplete combustion of fossil fuels and other organic materials, including during wildfire. Eight polycyclic aromatic hydrocarbons are considered carcinogenic and 16 are classified by the EPA as priority pollutants. The behavior of polycyclic aromatic hydrocarbons is strongly determined by their chemical characteristics, especially their nonpolarity and hydrophobicity. They readily adsorb to particulates in the air and to sediments in water. Srogi (2007) provided a review of polycyclic aromatic hydrocarbons concentrations in various environmental compartments and their environmental risks and possible effects on ecosystems and human health.

The polycyclic aromatic hydrocarbons contaminant group includes common air pollutants in metropolitan areas, derived in part from vehicular traffic. Other sources, in addition to gasoline and diesel combustion, include Al production industries, wood heating, forest fires, prescribed fires, and various forms of fossil fuel combustion (Sanderson et al. 2004). The analysis of WACAP data found that elevated levels of polycyclic aromatic hydrocarbons in Glacier National Park were traced to a nearby Al smelter, whereas sources to other national parks ranged from local point sources to diffuse regional and global sources (Usenko et al. 2010).

Experiments have shown a variety of adverse effects from exposure of animals to persistent organic pollutants, especially PCBs, polychlorinated dibenzo-*p*-dioxins, and polychlorinated dibenzofurans. Observed effects have included fetal deformities, DNA mutations, immune suppression, neurobehavioral effects, and cancer (Pohjanvirto and Tuomisto 1994, Clair et al. 2011). Such effects have not been found in the national parks.

References

Altshuller, A.P. and A.S. Lefohn. 1996. Background ozone in the planetary boundary layer over the United States. *J. Air Waste Manage. Assoc.* 46:134–141.

AMEC Environment & Infrastructure Inc. 2013. Clean Air Status and Trends Network (CASTNET) 2011 annual report. Contract No. EP-W-09-028. Prepared for the U.S. Environmental Protection Agency (EPA), Office of Air and Radiation, Washington, DC.

Anderson, J.B., R.E. Baumgardner, V.A. Mohnen, and J.J. Bowser. 1999. Cloud chemistry in the eastern United States, as sampled from three high-elevation sites along the Appalachian Mountains. *Atmos. Environ.* 33:5105–5114.

Brace, S. and D.L. Peterson. 1996. Spatial patterns of ozone exposure in Mt. Rainier National Park. *Proceedings, Meeting of the Air Waste Management Association*, Spokane, WA.

Brace, S. and D.L. Peterson. 1998. Spatial patterns of tropospheric ozone in the Mount Rainier region of the Cascade Mountains, U.S.A. *Atmos. Environ.* 32(21):3629–3637.

Brook, J.R., L. Zhang, F. Di-Giovanni, and J. Padro. 1999. Description and evaluation of a model of deposition velocities for routine estimates of air pollutant dry deposition over North America. Part I: Model development. *Atmos. Environ.* 33:5037–5051.

Burns, D.A., J.A. Lynch, B.J. Cosby, M.E. Fenn, J.S. Baron, and U.S. EPA Clean Air Markets Division. 2011. National acid precipitation assessment program report to congress 2011: An integrated assessment. National Science and Technology Council, Washington, DC.

Butler, T.J., M.D. Cohen, F.M. Vermeylen, G.E. Likens, D. Schmeltz, and R.S. Artz. 2008. Regional precipitation mercury trends in the eastern USA, 1998–2005: Declines in the Northeast and Midwest, no trend in the Southeast. *Atmos. Environ.* 42:1582–1592.

Byun, D. and K.L. Schere. 2006. Review of the governing equations, computational algorithms, and other components of the Models-3 Community Multiscale Air Quality (CMAQ) modeling system. *Appl. Mech. Rev.* 59:51–77.

Chameides, W.L., R.W. Lindsay, J. Richardson, and C.S. Kiang. 1988. The role of biogenic hydrocarbons in urban photochemical smog: Atlanta as a case study. *Science* 241:1473–1474.

Christensen, J.H., B. Hewitson, A. Busuioc, A. Chen, X. Gao, I. Held, R. Jones et al. 2007. Regional climate projections. *In* S. Solomon, D. Qin, M. Manning, Z. Chen, M. Marquis, K.B. Averyt, M. Tignor, and H.L. Miller (Eds.). *Climate Change 2007: The Physical Science Basis. Contribution of Working Group I to the Fourth Assessment Report of the Intergovernmental Panel on Climate Change.* Cambridge University Press, Cambridge, U.K.

Clair, T.A., D. Burns, I.R. Pérez, J. Blais, and K. Percy. 2011. Ecosystems. *In* G.M. Hidy, J.R. Brook, K.L. Demerjian, L.T. Molina, W.T. Pennell, and R.D. Scheffe (Eds.). *Technical Challenges of Multipollutant Air Quality Management.* Springer, Dordrecht, the Netherlands, pp. 139–229.

Clarke, J.F., E.S. Edgerton, and B.E. Martin. 1997. Dry deposition calculations for the Clean Air Status and Trends Network. *Atmos. Environ.* 31:3667–3678.

Compeau, G.C. and R. Bartha. 1985. Sulfate-reducing bacteria: Principal methylators of mercury in anoxic estuarine sediment. *Appl. Environ. Microbiol.* 50(2):498–502.

Cook, E.R., C.A. Woodhouse, C.M. Eakin, D.M. Meko, and D.W. Stahle. 2004. Long-term aridity changes in the western United States. *Science* 306(5698):1015–1018.

Cooper, S.M. and D.L. Peterson. 2000. Tropospheric ozone distribution in western Washington. *Environ. Pollut.* 107:339–347.

Cornell, S.E. 2011. Atmospheric nitrogen deposition: Revisiting the question of the importance of the organic component. *Environ. Pollut.* 159(10):2214–2222.

Dicosty, R.J., M.A. Callaham, Jr., and J.A. Stanturf. 2006. Atmospheric deposition and re-emission of mercury estimated in a prescribed forest-fire experiment in Florida, USA. *Water Air Soil Pollut.* 176(1–4):77–91.

Drevnick, P.E., D.R. Engstrom, C.T. Driscoll, E.B. Swain, S.J. Balogh, N.C. Kamman, D.T. Long et al. 2012. Spatial and temporal patterns of mercury accumulation in lacustrine sediments across the Laurentian Great Lakes region. *Environ. Pollut.* 161:252–260.

Driscoll, C.T., Y.-J. Han, C.Y. Chen, D.C. Evers, K.F. Lambert, T.M. Holsen, N.C. Kamman, and R.K. Munson. 2007. Mercury contamination in forest and freshwater ecosystems in the northeastern United States. *BioScience* 57(1):17–28.

Dukett, J.E., N. Aleksic, N. Houck, P. Snyder, P. Casson, and M. Cantwell. 2011. Progress toward clean cloud water at Whiteface Mountain New York. *Atmos. Environ.* 45(37):6669–6673.

Fiore, A.M., D.J. Jacob, H. Liu, R.M. Yantosca, T.D. Fairlie, and Q.B. Li. 2003. Variability in surface ozone background over the United States: Implications for air quality policy. *J. Geophys. Res.* 108:4787.

Fowler, D., J.H. Duyzer, and D.D. Baldocchi. 1991. Inputs of trace gases, particles and cloud droplets to terrestrial surfaces. *Proc. R. Soc. Edin. Sect. B Biol. Sci.* 97:35–59.

Fuentes, J.D., M. Lerdau, R. Atkinson, D. Baldocchi, J.W. Bottenheim, P. Ciccioli, B. Lamb et al. 2000. Biogenic hydrocarbons in the atmospheric boundary layer: A review. *Bull. Am. Meteorol. Soc.* 81:1537–2000.

Fuzzi, S. and D. Wagenbach. 1997. *Cloud Multiphase Processes and Pollutant Deposition in the High Alps.* Springer, Berlin, Germany.

Garner, J.H.B., T. Pagano, and E.B. Cowling. 1989. An evaluation of the role of ozone, acid deposition, and other airborne pollutants in the forests of eastern North America. General Technical Report SE-59. U.S. Department of Agriculture, Forest Service, Southeastern Forest Experiment Station, Asheville, NC.

Gilmour, C.C., E.A. Henry, and R. Mitchell. 1992. Sulfate stimulation of mercury methylation in freshwater sediments. *Environ. Sci. Technol.* 26(11):2281–2287.

Goforth, M.R. and C.S. Christoforou. 2006. Particle size distribution and atmospheric metals measurements in a rural area in the Southeastern USA. *Sci. Total Environ.* 356(1–3):217–227.

Grigal, D.F. 2002. Inputs and outputs of mercury from terrestrial watersheds: A review. *Environ. Rev.* 10:1–39.

Grigal, D.F. 2003. Mercury sequestration in forests and peatlands: A review. *J. Environ. Qual.* 32(2):393–405.

Guerova, G. and N. Jones. 2007. A global model study of ozone distributions during the August 2003 heat wave in Europe. *Environ. Chem.* 4:285–292.

Gustin, M.S., M. Engle, J. Ericksen, S. Lyman, J. Stamenkovic, and M. Xin. 2006. Mercury exchange between the atmosphere and low mercury containing substrates. *Appl. Geochem.* 21:1913–1923.

Haeuber, R. 2013. Energy-related environmental science and policy: Challenges and strategies. *In* Paper presented at *New York State Energy Research and Development Authority (NYSERDA): Environmental Monitoring, Evaluation, and Protection in New York: Linking Science and Policy Conference*, Albany, NY.

Hageman, K.J., S.L. Simonich, D.H. Campbell, G.R. Wilson, and D.H. Landers. 2006. Atmospheric deposition of current-use and historic-use pesticides in snow at national parks in the western United States. *Environ. Sci. Technol.* 40(10):3174–3180.

Hagerman, L.M., V.P. Aneja, and W.A. Lonneman. 1997. Characterization of non-methane hydrocarbons in the rural southeast United States. *Atmos. Environ.* 31:4017–4038.

Hall, B.D., H. Manolopoulos, J.P. Hurley, J.J. Schauer, V.L. St. Louis, D. Kenski, J. Graydon, C.L. Babiarz, L.B. Cleckner, and G.J. Keeler. 2005. Methyl and total mercury in precipitation in the Great Lakes region. *Atmos. Environ.* 39:7557–7569.

Hoh, E., L. Zhu, and R.A. Hites. 2006. Dechlorane plus, a chlorinated flame retardant, in the Great Lakes. *Environ. Sci. Technol.* 40:1184–1189.

Hosker, R.P., Jr. and S.E. Lindberg. 1982. Review: Atmospheric transport, deposition, and plant assimilation of airborne gases and particles. *Atmos. Environ.* 16:889–910.

Intergovernmental Panel on Climate Change (IPCC). 2014. Climate change 2014: Synthesis report. Contribution of Working Groups I, II and III to the fifth assessment report of the intergovernmental panel on climate change [Core Writing Team, R.K. Pachauri, and L.A. Meyer (Eds.)]. IPCC, Geneva, Switzerland.

Jacob, D.J., J.A. Logan, G.M. Gardner, R.M. Yevich, C.M. Spivakovsky, and S.C. Wofsy. 1993. Factors regulating ozone over the United States and its export to the global atmosphere. *J. Geophys. Res.* 98:14817–14826.

Jacob, D.J., D.L. Mauzerall, J.M. Fernández, and W.T. Pennell. 2011. Global change and air quality. *In* G.M. Hidy, J.R. Brook, K.L. Demerjian, L.T. Molina, W.T. Pennell, and R.D. Scheffe (Eds.). *Technical Challenges of Multipollutant Air Quality Management.* Springer, Dordrecht, the Netherlands, pp. 395–432.

Jacob, D.J. and D.A. Winner. 2009. Effect of climate change on air quality. *Atmos. Environ.* 43:51–63.

Jaffe, D., I. Bertschi, L. Jaegle, P. Novelli, J.S. Reid, H. Tanimoto, R. Vingarzan, and D.L. Westphal. 2004. Long-range transport of Siberian biomass burning emissions and impact on surface ozone in eastern North America. *Geophys. Res. Lett.* 31:L161606. doi:10.1029/2004GL020093.

Jaffe, D., D. Chand, W. Hafner, A. Westerling, and D. Spracklen. 2008. Influence of fires on O$_3$ concentrations in the western U.S. *Environ. Sci. Technol.* 42:5885–5891.

Jaffe, D.A. and J. Ray. 2007. Increase in ozone at rural sites in the western U.S. *Atmos. Environ.* 41(26):5452–5463.

Kang, D., V.P. Aneja, R. Mathur, and J.D. Ray. 2003. Nonmethane hydrocarbons and ozone in three rural southeast United States national parks: A model sensitivity analysis and comparison to measurements. *J. Geophys. Res. Atmos.* 108(D19):4604. doi:10.1029/2002JD003054:17.

Kang, D., V.P. Aneja, R.G. Zika, C. Farmer, and J.D. Ray. 2001. Nonmethane hydrocarbons in the rural southeast United States national parks. *J. Geophys. Res.* 106(D3):3133–3155.

Karion, A., C. Sweeney, G. Pétron, G. Frost, R. Michael Hardesty, J. Kofler, B.R. Miller et al. 2013. Methane emissions estimate from airborne measurements over a western United States natural gas field. *Geophys. Res. Lett.* 40(16):4393–4397.

Kohut, R. 2007. *Handbook for Assessment of Foliar Ozone Injury on Vegetation in the National Parks*, revised 2nd edn. Boyce Thompson Institute, Cornell University, Ithaca, NY. http://www.nature.nps.gov/air/Permits/ARIS/networks/docs/O3_InjuryAssessmentHandbookD1688.pdf.

Krabbenhoft, D.P. and E.M. Sunderland. 2013. Global change and mercury. *Science* 341(6153):1457–1458.

Lamarque, J.F., P. Hess, L. Emmons, L. Buja, W. Washington, and C. Granier. 2005. Tropospheric ozone evolution between 1890 and 1990. *J. Geophys. Res.* 110:D08304.

Landers, D.H., S.M. Simonich, D. Jaffe, L. Geiser, D.H. Campbell, A. Schwindt, C.B. Schreck et al. 2010. The western Airborne Contaminant Project (WACAP): An interdisciplinary evaluation of the impacts of airborne contaminants in western U.S. National Parks. *Environ. Sci. Technol.* 44(3):855–859.

Lee, W.Y., W.A. Iannucci-Berger, B.D. Eitzer, J.C. White, and M.I. Mattina. 2003. Plant uptake and translocation of air-borne chlordane and comparison with the soil-to-plant route. *Chemosphere* 53(2):111–121.

Lefohn, A.S., H. Wernli, D. Shadwick, S. Limbach, S.J. Oltmans, and M. Shapiro. 2011. The importance of stratospheric-tropospheric transport in affecting surface ozone concentrations in the western and northern tier of the United States. *Atmos. Environ.* 45:4845–4857.

Lin, C.-Y.C., D.J. Jacob, and A.M. Fiore. 2001. Trends in exceedances of the ozone air quality standard in the continental United States, 1980–1998. *Atmos. Environ.* 35:3217–3228.

Loibl, W., W. Winiwarter, A. Kopsca, J. Zueger, and R. Bauman. 1994. Estimating the spatial distribution of ozone concentrations in complex terrain. *Atmos. Environ.* 28:2557–2566.

Lovett, G.M. and J.D. Kinsman. 1990. Atmospheric pollutant deposition to high-elevation ecosystems. *Atmos. Environ.* 24A(11):2767–2786.

Lu, R., R.P. Turco, K. Stolzenbach, S.K. Friedlander, C. Xiong, K. Schiff, L. Tiefenthaler, and G. Wang. 2003. Dry deposition of airborne trace metals on the Los Angeles Basin and adjacent coastal waters. *J. Geophys. Res. D Atmos.* 108(2):11-1–11-24.

Lyman, S. and H. Shorthill. 2013. Final report. 2012 Uintah Basin Winter Ozone & Air Quality Study. Document No. CRD13-320.32. Utah State University, Logan, UT. http://rd.usu.edu/files/uploads/ubos_2011-12_final_report.pdf.

Lynam, M.M. and G.J. Keeler. 2005. Artifacts associated with the measurement of particulate mercury in an urban environment: The influence of elevated ozone concentrations. *Atmos. Environ.* 39:3081–3088.

McDonnell, T.C. and T.J. Sullivan. 2014. Total atmospheric nitrogen and sulfur deposition in forest service wildernesses and national forests throughout the conterminous United States. Report prepared for USDA Forest Service, Asheville, NC. E&S Environmental Chemistry, Inc., Corvallis, OR.

Mickley, L.J., D.J. Jacob, and D. Rind. 2001. Uncertainty in preindustrial abundance of tropospheric ozone: Implications for radiative forcing calculations. *J. Geophys. Res.* 106:3389–3399.

Mohnen, V.A. 1990. Project summary. An assessment of atmospheric exposure and deposition to high elevation forests in the eastern United States. EPA/600/S3-90/058. U.S. Environmental Protection Agency, Research Triangle Park, NC.

National Parks Conservation Association (NPCA). 2013. National parks and hydraulic fracturing. Balancing energy needs, nature, and America's national heritage. Washington, DC. Available at: http://www.npca.org/assets/pdf/Fracking_Report.pdf.

Norton, S.A. 2007. Atmospheric metal pollutants-archives, methods, and history. *Water Air Soil Pollut. Focus* 7(1–3):93–98.

Pacyna, E.G. and J.M. Pacyna. 2002. Global emission of mercury from anthropogenic sources in 1995. *Water Air Soil Pollut.* 137(1):149–165.

Percy, K.E., A.H. Legge, and S.V. Krupa. 2003. Tropospheric ozone: A continuing threat to global forests? *In* D.F. Karnosky (Ed.). *Air Pollution, Global Change and Forests in the New Millennium.* Elsevier, New York, pp. 85–118.

Pirrone, N., S. Cinnirella, X. Feng, R.B. Finkelman, H.R. Friedli, J. Leaner, R. Mason et al. 2010. Global mercury emissions to the atmosphere from anthropogenic and natural sources. *Atmos. Chem. Phys.* 10:5951–5964.

Pohjanvirto, R. and J. Tuomisto. 1994. Short-term toxicity of 2,3,7,8-tetrachlorodibenzo-p-dioxin in laboratory animals: Effects, mechanisms, and animal models. *Pharmacol. Rev.* 46:483–549.

Prenni, A.J., D.E. Day, A.R. Evanoski-Cole, B.C. Sive, A. Hecobian, Y. Zhou, K.A. Gebhart et al. 2016. Oil and gas impacts on air quality in federal lands in the Bakken region: An overview of the Bakken Air Quality Study and first results. *Atmos. Chem. Phys.* 16(3):1401–1416.

Risch, M.R., J.F. DeWild, D.P. Krabbenhoft, R.K. Kolka, and L. Zhang. 2012. Litterfall mercury dry deposition in the eastern USA. *Environ. Pollut.* 161:264–290.

Rocher, V., S. Azimi, J. Gasperi, L. Beuvin, M. Muller, R. Moilleron, and G. Chebbo. 2004. Hydrocarbons and metals in atmospheric deposition and roof runoff in central Paris. *Water Air Soil Pollut.* 159:67–86.

Rockstrom, J., W. Steffen, K. Noone, A. Persson, F.S. Chapin, E.F. Lambin, T.M. Lenton et al. 2009. A safe operating space for humanity. *Nature* 461(7263):472–475.

Sanderson, E.G., A. Raqbi, A. Vyskocil, and J.P. Farant. 2004. Comparison of particulate polycyclic aromatic hydrocarbon profiles in different regions of Canada. *Atmos. Environ.* 38(21):3417–3429.

Sandroni, S., P. Bacci, G. Boffa, U. Pellegrini, and A. Ventura. 1994. Tropospheric ozone in the pre-alpine and alpine regions. *Sci. Total Environ.* 156:169–182.

Schmeltz, D., D.C. Evers, C.T. Driscoll, R. Artz, M. Cohen, D. Gay, R. Haeuber et al. 2011. MercNet: A national monitoring network to assess responses to changing mercury emissions in the United States. *Ecotoxicology* 20:1713–1725.

Schroeder, W.H. and J. Munthe. 1998. Atmospheric mercury—An overview. *Atmos. Environ.* 32(5):809–822.

Schwede, D.B. and G.G. Lear. 2014. A novel hybrid approach for estimating total deposition in the United States. *Atmos. Environ.* 92:207–220.

Seinfeld, J.H. and S.N. Pandis. 2006. *Atmospheric Chemistry and Physics: From Air Pollution to Climate Change,* 2nd edn. John Wiley & Sons, Inc., Hoboken, NJ.

Shindell, D.T. and G. Faluvegi. 2002. An exploration of ozone changed and their radiative forcing prior to the chlorofluorocarbon era. *Atmos. Chem. Phys.* 2:363–374.

Smith, G. 2012. Ambient ozone injury to forest plants in northeast and northcentral USA: 16 years of biomonitoring. *Environ. Monitor. Assess.* 184:4049–4065.

Srogi, K. 2007. Monitoring of environmental exposure to polycyclic aromatic hydrocarbons: A review. *Environ. Chem. Lett.* 5(4):169–195.

Steinnes, E., O.Ø. Hvatum, B. Bølviken, and P. Varskog. 2005. Atmospheric supply of trace elements studied by peat samples from ombrotrophic bogs. *J. Environ. Qual.* 34:192–197.

Sullivan, T.J. and T.C. McDonnell. 2013. Mapping of nutrient-nitrogen critical loads for selected national parks in the intermountain west and great lakes regions. Report prepared for National Park Service. Natural Resource Technical Report NPS/ARD/NRTR—2014/895. National Park Service, Fort Collins, CO.

Sullivan, T.J., T.C. McDonnell, G.T. McPherson, S.D. Mackey, and D. Moore. 2011b. Evaluation of the sensitivity of inventory and monitoring national parks to nutrient enrichment effects from atmospheric nitrogen deposition. Natural Resource Report NPS/NRPC/ARD/NRR—2011/313. U.S. Department of the Interior, National Park Service, Denver, CO. Available at: http://www.nature.nps.gov/air/permits/aris/networks/n-sensitivity.cfm.

Sullivan, T.J., G.T. McPherson, T.C. McDonnell, S.D. Mackey, and D. Moore. 2011a. Evaluation of the sensitivity of inventory and monitoring national parks to acidification effects from atmospheric sulfur and nitrogen deposition. U.S. Department of the Interior, National Park Service, Denver, CO. Available at: http://nature.nps.gov/air/Permits/ARIS/networks/acidification-eval.cfm.

Swackhamer, D.L. and K.C. Hornbuckle. 2004. Assessment of air quality and air pollutant impacts in Isle Royale National Park and Voyageurs National Park. Report prepared for the U.S. National Park Service, Washington, DC.

Tong, D.Q., D. Kang, V.P. Aneja, and J.D. Ray. 2005. Reactive nitrogen oxides in the southeast United States national parks: Source identification, origin, and process budget. *Atmos. Environ.* 39:315–327.

U.S. Environmental Protection Agency. 2004. Air quality criteria for particulate matter. Volumes I and II. EPA/600/P-99/002aF. National Center for Environmental Assessment-RTP, Office of Research and Development, Research Triangle Park, NC.

U.S. Environmental Protection Agency. 2006. Air quality criteria for ozone and related photochemical oxidants. Volumes I–III. EPA 600/R-05/004a-cF. U.S. Environmental Protection Agency, Research Triangle Park, NC. Available at: http://cfpub.epa.gov/ncea/cfm/recordisplay.cfm?deid=149923.

U.S. Environmental Protection Agency. 2009a. Integrated science assessment for particulate matter (final report). EPA/600/R-08/139F. U.S. Environmental Protection Agency, Washington, DC. Available at: http://cfpub.epa.gov/ncea/cfm/recordisplay.cfm?deid=216546.

U.S. Environmental Protection Agency. 2009b. Risk and exposure assessment for review of the secondary national ambient air quality standards for oxides of nitrogen and oxides of sulfur: Final report. EPA-452/R-09-008a. Office of Air Quality Planning and Standards, Health and Environmental Impacts Division, Research Triangle Park, NC.

U.S. Environmental Protection Agency. 2014. Policy assessment for the review of the ozone national ambient air quality standards second external review draft. EPA-452/P-13-002. U.S. Environmental Protection Agency, Washington, DC.

Usenko, S., S.L. Simonich, K.J. Hageman, J.E. Schrlau, L. Geiser, D.H. Campbell, P.G. Appleby, and D.H. Landers. 2010. Sources and deposition of polycyclic aromatic hydrocarbons to western U.S. national parks. *Environ. Sci. Technol.* 44(12):4512–4518.

Val Martin, M., C.L. Heald, J.F. Lamarque, S. Tilmes, L.K. Emmons, and B.A. Schichtel. 2015. How emissions, climate, and land use change will impact mid-century air quality over the United States: A focus on effects at national parks. *Atmos. Chem. Phys.* 15(5):2805–2823.

Vong, R.J., J.T. Sigmon, and S.F. Mueller. 1991. Cloud water deposition to Appalachian forests. *Environ. Sci. Technol.* 25:1014–1021.

Weiss-Penzias, P.S., C. Ortiz, R.P. Acosta, W. Heim, J.P. Ryan, D. Fernandez, J.L. Collett, and A.R. Flegal. 2012. Total and monomethyl mercury in fog water from the central California coast. *Geophys. Res. Lett.* 39(3):L03804.

Westerling, A.L., H.G. Hidalgo, D.R. Cayan, and T.W. Swetnam. 2006. Warming and earlier spring increase western U.S. forest wildfire activity. *Science* 313(5789):940–943.

Wiener, J.G., D.C. Evers, D.A. Gay, H.A. Morrison, and K.A. Williams. 2012. Mercury contamination in the Laurentian Great Lakes region: Introduction and overview. *Environ. Pollut.* 161:243–251.

Wiener, J.G., D.P. Krabbenhoft, G.H. Heinz, and A.M. Scheuhammer. 2003. Ecotoxicology of mercury. *In* D.J. Hoffman, B.A. Rattner, G.A. Burton and J. Cairns (Eds.). *Handbook of Ecotoxicology*, 2nd edn. CRC Press, Boca Raton, FL, pp. 409–463.

Zhang, L., D.J. Jacob, K.F. Boersma, D.A. Jaffe, J.R. Olson, K.W. Bowman, J.R. Worden et al. 2008. Transpacific transport of ozone pollution and the effect of recent Asian emission increases on air quality in North America: An integrated analysis using satellite, aircraft, ozonesonde, and surface observations. *Atmos. Chem. Phys.* 8:6117–6136.

3

Impacts

Air pollutants, including gases, aerosols, and deposited materials, contribute to a range of environmental stressors in the national parks. Each pollutant has the potential to cause one or more effects on sensitive receptors (Table 3.1). Effects depend on both the type and amount of pollutant and the inherent sensitivity of the receptor or ecosystem. Background information on several types of air pollution–related ecosystem stress is presented here.

3.1 Acidification

Atmospheric deposition of S and/or N can cause acidification of soil, soil water, lakes, and streams. Soil and freshwaters can also be naturally acidified, largely by organic acids derived from wetlands and forest soils. Ocean acidification is different in that it is controlled primarily by increased carbon dioxide concentrations in the atmosphere and carbonic acid formation. It is not discussed in this book.

In most portions of the United States that have experienced soil and freshwater acidification attributable to air pollution, such effects have mainly been due to S inputs (Driscoll et al. 1998, 2007). There are, however, some regions, especially in the western United States, where resources are more threatened or have been more affected by the acidity of N inputs than by the acidity of S inputs. This is at least partially due to the low levels of S deposition received at most western locations. There are also regions where atmospheric S and N both contribute substantially to the observed acidification. These include portions of the Northeast, West Virginia, and high elevations in North Carolina, Tennessee, and Virginia. Historically, these areas received more S than N deposition but now receive about equal amounts.

The freshwater aquatic ecosystems within the national parks that are thought to be most sensitive to the effects of acidification from atmospheric S and N deposition include remote lakes that often occur at relatively high elevation and headwater (Strahler order 1–3) streams. Acid-sensitive waters most commonly occur in areas of steep terrain, on shallow soils, having shallow flow paths, and bedrock types that provide limited contribution of base cations that could, if present, buffer incoming acidity.

The soils that are thought to be most sensitive to acidification effects include shallow soils having low base saturation* (less than about 12%–20%). These low base saturation soils generally exhibit low weathering rates and tend to have low clay content. They are often found at high elevation and are exposed to relatively high levels of precipitation, which can leach nutrient base cations to drainage water, reducing the soil's ability to neutralize acidity from acidic deposition.

---

* Base saturation reflects the percent of exchangeable cations adsorbed to soil that are base cations (Ca, Mg, Na, K) rather than acid cations (H, Al).

TABLE 3.1

Matrix of Environmental Stressors and Selected Effects Caused by Air Pollutants

Pollutant	Stressor	Selected Effects
Acidic deposition	Soil acidification	Decreased soil base saturation, loss of plant nutrients, increased stress to sensitive plants, and loss of sensitive terrestrial species
	Surface water acidification	Decreased surface water acid neutralizing capacity and pH, increased stress to sensitive aquatic biota, and loss of sensitive aquatic species.
Nutrient N deposition	Nutrient enrichment	Eutrophication, changes in species composition, loss of biodiversity
Ground-level ozone	Plant foliage exposure to ozone	Increased stress to ozone-sensitive plants, reduced growth, foliar injury
Toxic substances	Exposure to toxic conditions	Bioaccumulation of toxic materials, neurotoxicity, reproductive effects, etc.
Atmospheric concentrations of sulfate and nitrate	Formation of haze	Visibility degradation
Hydrocarbons	Formation of ozone	Increased stress to ozone-sensitive plants

3.1.1 Terrestrial Effects

Acidification can affect soils and vegetation by depleting soil nutrients and releasing metals toxic to plant roots. Two tree species (red spruce [*Picea rubens*] and sugar maple [*Acer saccharum*]) are known to be highly susceptible to damage from acidic deposition (U.S. EPA 2008). This damage can take the form of reduced growth, canopy dieback, reduced regeneration, increased susceptibility of foliage to cold temperature, and other symptom manifestations. Some national parks have extensive coverage of vegetation types that include one or both of these sensitive tree species. Although soil acidification effects in the United States are expected to be especially pronounced in the plant communities that include these tree species, the same kinds of effects might also occur in other vegetation types but are poorly documented in this country.

Many lichen species are known to be especially sensitive to air pollution (Figure 3.1). To some extent, they represent the "canaries in the coal mine," or early warning signals, because lichens may be impacted by air pollution before other species. Effects seem to be more clearly associated with N inputs than with S inputs (Bobbink et al. 2003, Geiser and Neitlich 2007, Glavich and Geiser 2008). These effects may be driven by nutrient enrichment processes more than, or in addition to, acidification processes. It is also likely, however, that S air pollution has impacted the distribution of lichens, especially in the eastern United States (U.S. EPA 2008).

The release of base cations from soil to soil water through weathering, cation exchange, and mineralization contributes to neutralization of acidity derived both from acidic deposition and from natural processes (van Breemen et al. 1983). If the acidity is associated with anions that are mobile within the soil environment, such as sulfate or nitrate, cations can be leached to groundwaters and eventually to surface waters. Some loss of base cations from soil occurs naturally from the leaching of organic and carbonic acids. The limited mobility of anions associated with naturally derived organic acids and carbonic acid controls the rate of base cation leaching under conditions of low atmospheric deposition of S and N. Because inputs of S and N in acidic deposition supply anions that are often highly mobile in the soil, these mineral acid anions can accelerate base cation leaching

FIGURE 3.1
Lichens are among the terrestrial receptors that are most sensitive to N inputs. (National Park Service photo.)

(Cronan et al. 1978). In addition, depletion of nutrient base cations, especially Ca, can cause damage to acid-sensitive plants and those that require high Ca levels.

In regions of the eastern United States affected by acidic deposition, the total concentration of sulfate and nitrate in soil waters and surface waters has increased from historical conditions. In response to these changes in mineral acid anion concentrations, the concentrations of other ions in surface water must also have changed so that electroneutrality is maintained (the total cationic and anionic charges are balanced). The leaching of sulfate does not directly cause adverse environmental effects. Changes in other ions are responsible for environmental impacts caused by drainage water* acidification. As the sulfate concentration in drainage water has increased over time in the past, other anions (mainly bicarbonate or organic acid anions) have decreased and/or cations (e.g., base cations, hydrogen ion, or inorganic monomeric Al) must have increased in solution to maintain the charge balance (the sum of the cations equals the sum of the anions). Changes in the concentrations of these ions have affected the ability of soils to support acid-sensitive plant species and other biota.

Most acidification effects on plants are mediated through the soil and are governed by Al toxicity and nutrient base cation (Ca, Mg, K) deficiencies. Nitrogen saturation can also be involved. These three factors are often closely related.

3.1.1.1 Aluminum Mobilization

The key biogeochemical process impacting freshwaters that is altered by acidic deposition is the mobilization of Al from soils to drainage waters (Cronan and Schofield 1979,

---

* Drainage water is water derived from precipitation that drains through the soil and into groundwater or stream water.

Mason and Seip 1985), potentially causing toxicity to both terrestrial and aquatic organisms. Aluminum becomes more soluble at pH values below about 5.5. As a consequence, Al concentrations in drainage waters having pH below about 5.0 are often substantially higher than in waters having pH above 6.0. At high concentration in soil water, Al is toxic to plant roots and causes reduced root growth. This limits the ability of the plant to take up water and nutrients, especially Ca (Parker et al. 1989).

Red spruce trees died over large portions of the eastern United States in the 1980s. This mortality was linked to the exposure of foliage to acidic cloud water and to an increase in the amount of dissolved Al compared with dissolved Ca in soil water (U.S. EPA 2008). Some of the red spruce decline occurred at high-elevation sites that frequently experience cloud cover and often receive high levels of occult deposition.

The leaching of atmospherically deposited sulfate into soil waters, and eventually to surface waters, controls soil acidification and Al toxicity to plants at most acid-impacted areas in the northeastern United States. At such locations, much of the deposited S is leached to surface waters as sulfate. This contributes to soil and surface water acidification and Al mobilization. In contrast, much of the deposited N is commonly taken up from watershed soils by plants and microbes, reducing its acidification potential. Thus, the mobilization and toxicity of Al are largely controlled by sulfate mobility in most affected ecosystems. Nitrate mobility is also important at some locations, but the dominant mobile strong acid anion in U.S. national parks is usually sulfate.

3.1.1.2 Depletion of Base Cations from Soil

Base cations, some of which are necessary plant nutrients, are common in rocks and soils, but largely in forms that are unavailable to plants. There is also a pool of bioavailable exchangeable base cations that are adsorbed to negatively charged surfaces on soil particles. Base cations in this pool are gradually leached from the soil in drainage water but are constantly resupplied through weathering and atmospheric deposition of base cations. Weathering slowly breaks down rocks and minerals, releasing base cations to join the pool of adsorbed exchangeable base cations on the soil. The base saturation is a soil metric that reflects the percentage of the adsorbed cations that are base cations rather than acid cations. The balance between base cation supply and base cation loss determines whether the pool of available soil base cations is increasing or decreasing in size over time. Enhanced leaching of base cations by acidic deposition in some cases can deplete the soil of exchangeable bases faster than they are resupplied (Cowling and Dochinger 1980). Nutrient base cations, including Ca, Mg, and K, are taken up through plant roots from soil water to satisfy plant nutritional needs. In soils having low base saturation, exchangeable Ca, Mg, or K can be depleted so much that nutrient deficiencies develop in vegetation.

The hardwood tree species most commonly associated with acidification effects due to base cation depletion is sugar maple. It is distributed throughout the northeastern United States, Upper Midwest, and Appalachian Mountain region as a component of the northern hardwood forest. Several studies, mainly in Pennsylvania and New York, have shown that sugar maple decline is linked to the occurrence of relatively high levels of acidic deposition and base-poor soils. Decline results, in part, from Ca depletion (Horsley et al. 1999).

The health of sugar maple trees is strongly influenced by the availability of Ca and other base cations in soil. Sugar maple trees that grow on soils having low base cation supply are stressed and consequently often become more susceptible to damage from defoliating insects, drought, and extreme weather. The overall response can include canopy damage or death of mature trees and poor regeneration of seedlings (Horsley et al. 1999, Sullivan et al. 2013).

Soil acidification and depletion of soil base cations may be contributing to sugar maple mortality in some of the eastern national parks, especially on sites having marginal (low-nutrient) soils. Sugar maple dieback at 19 sites in northwestern and northcentral Pennsylvania and southwestern New York was correlated with combined stress from defoliation and soil deficiencies of Mg and Ca (Horsley et al. 1999). Dieback occurred pre-dominately on ridgetops and on upper slopes, where soil base cation availability was much lower than on middle and lower slopes (Bailey et al. 1999). Sugar maple in the Adirondack Mountains of New York exhibited essentially no regeneration on sites having an upper B soil horizon base saturation less than about 12% (Sullivan et al. 2013). This threshold may also be applicable to other areas.

3.1.2 Aquatic Effects

Surface water acidification produces chemical changes in lakes and streams, most notably a decrease in acid neutralizing capacity, usually a decrease in pH, and often an increase in the concentration of inorganic Al in lakes and streams (Driscoll et al. 2001b). Many species of aquatic biota are sensitive to water acidification.

Acid neutralizing capacity is the most widely used water chemistry indicator for both acidic deposition sensitivity and effects. It can be measured in the laboratory by Gran titra-tion or defined as the difference between the measured base cation and mineral acid anion concentrations in water:

$$\text{Acid neutralizing capacity} = (Ca^{2+} + Mg^{2+} + K^+ + Na^+ + NH_4^+) - (SO_4^{2-} + NO_3^- + Cl^-) \quad (3.1)$$

where
  NH_4^+ is ammonium
  SO_4^{2-} is sulfate
  NO_3^- is nitrate
  Cl^- is chloride

Surface water acid neutralizing capacity integrates the chemical, physical, and biological interactions that occur as atmospheric deposition and precipitation move from the atmo-sphere into or over the soil and eventually become surface water in a stream or lake. If the sum of the base cation concentrations (in equivalent units*) exceeds those of the strong acid anions, the water will have positive acid neutralizing capacity. Higher acid neutral-izing capacity is generally associated with higher pH and Ca concentration; lower acid neutralizing capacity is generally associated with lower pH and higher inorganic Al con-centrations and a greater likelihood of toxicity to aquatic biota.

Acid neutralizing capacity values were grouped by Cosby et al. (2006) into five major classes of biological concern: acute concern (less than 0 microequivalents per liter [μeq/L]), severe concern (0–20 μeq/L), elevated concern (20–50 μeq/L), moderate concern (50–100 μeq/L), and low concern (greater than 100 μeq/L), with each range representing a probabil-ity of ecological damage to the aquatic community. Biota are generally not harmed when acid neutralizing capacity values are above 100 μeq/L (U.S. EPA 2009c). In acid-sensitive regions, many surface waters commonly have acid neutralizing capacity below 100 μeq/L,

---

* Equivalent units are expressed in molar terms and then multiplied by the ionic charge.

and in some cases below 50 μeq/L, even in the absence of acidic deposition. Thus, although acidic deposition is responsible for the loss of acid neutralizing capacity in many sensitive water bodies, a relatively low acid neutralizing capacity value does not necessarily indicate that human-caused acidification has occurred if the acidic deposition levels are low.

A number of factors influence the sensitivity of aquatic ecosystems to acidification in response to S and N deposition. Geologic composition of the watershed of a water body largely determines base cation availability and plays a dominant role in influencing the sensitivity of surface waters to the effects of acidic deposition. Bedrock geology formed the basis for maps of surface water sensitivity (Norton et al. 1982, Dise 1984, Bricker and Rice 1989, Sullivan et al. 2007). Most of the major concentrations of surface waters having low acid neutralizing capacity are located in areas of the United States that are underlain by bedrock resistant to weathering, with consequent low supply of base cations (U.S. EPA 2008). Soil chemistry, land use, watershed slope, and hydrologic flow path also contribute to the sensitivity of surface waters to acidic deposition. Land disturbance and consequent exposure of S-bearing minerals to oxidation,* loss of base cations through erosion and timber harvesting, and change in N status of the forest through insect infestation, disease, or timber management can all influence the relative availability of mobile mineral acid anions (sulfate, nitrate) and base cations (Ca, Mg, K, Na) in drainage water. This affects the acid neutralizing capacity of lakes and streams.

Most watersheds in the central and southeastern United States are no longer acidifying but also are not exhibiting much recovery of surface water pH or acid neutralizing capacity in response to recent large decreases in atmospheric S deposition (Burns et al. 2011). Recovery has been somewhat more pronounced in the northeastern United States but still relatively modest (Driscoll et al. 2001a, 2007). This limited recovery is partly due to previous leaching losses of base cations from soils in the watersheds draining to surface waters.

3.1.3 Spatial Patterns in Acidification Risk

High-elevation lakes and streams are of particular interest with respect to potential aquatic impacts attributable to acidic deposition in the national parks. Many waters located at high elevation tend to be dilute (low ionic concentrations) and have low acid neutralizing capacity levels, and this contributes to the increased risk of acidification and biological harm from acid input. Because soils at high elevation are often shallow and poorly developed, with much exposed bedrock, the supply of base cations with which to neutralize acidity can be low (U.S. EPA 2008).

Streams and lakes vary in their sensitivity to acidification from acidic deposition. The surface waters that tend to be most sensitive to acidification are located on geological formations that contribute minimal quantities of base cations to drainage water. Figure 3.2 shows the locations of lakes and streams in the United States known to have low acid neutralizing capacity. These maps were compiled from available datasets and include a total of 31,662 sampling locations. Surface waters having acid neutralizing capacity ≤50 μeq/L are considered to be especially sensitive to acidification.

National and regional databases of measured water chemistry data were aggregated to generate a master database to represent surface water acid neutralizing capacity status throughout the National Park Service networks. Sources of data were mostly U.S. federal agencies (Table 3.2). Duplicate records among the data sources were identified and

---

* Oxidation of S-bearing minerals can release sulfuric acid to drainage waters.

removed. An attempt was made to also remove samples expected to be affected by acid mine drainage, sea salt spray, and road salt application. This was performed by applying chemistry screens that removed samples with observed chloride or sulfate greater than 300 μeq/L or acid neutralizing capacity <–100 μeq/L.

These samples were considered to represent acid neutralizing capacity results that were likely confounded by disturbances other than acidic deposition. The final database contained 196,997 acid neutralizing capacity values from 19,808 spatially unique sites sampled between January 8, 1980, and May 25, 2011 (Figure 3.2a). There were 6,065 sites with low (<100 μeq/L) median acid neutralizing capacity derived from 69,649 samples (Figure 3.2b). The distribution of surface water acid neutralizing capacity across the National Park Service networks is shown in Table 3.3.

Although these data cannot be considered to be statistically representative of spatial distributions of surface water acid neutralizing capacity across the nation, there are clear patterns that emerge. Acidic lakes and streams are widely distributed but are mostly confined to upstate New York, the Appalachian Mountains, New England, the Upper Midwest, and Florida. Low acid neutralizing capacity lakes and streams (0–100 μeq/L) occur in these

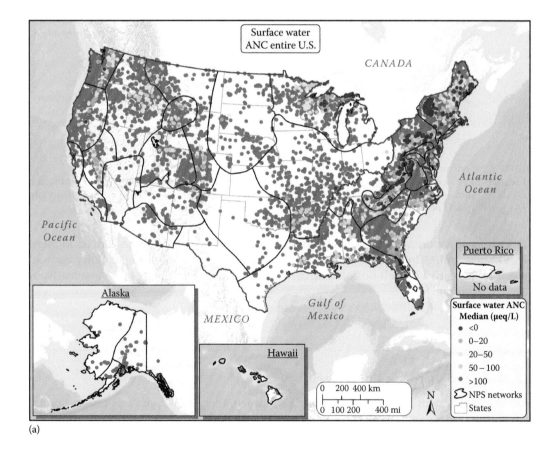

(a)

FIGURE 3.2
(a) Surface water acid neutralizing capacity (ANC), based on data compiled from the data sources listed in Table 3.2. (b) Surface water acid neutralizing capacity, shown as a subset of data represented in (a), but only sites having median acid neutralizing capacity ≤100 μeq/L.   *(Continued)*

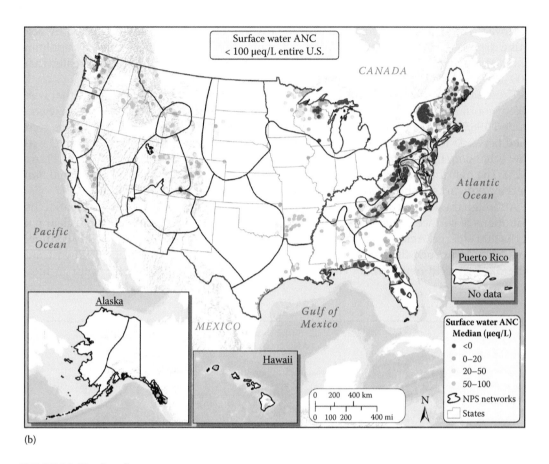

(b)

FIGURE 3.2 (Continued)
(a) Surface water acid neutralizing capacity (ANC), based on data compiled from the data sources listed in Table 3.2. (b) Surface water acid neutralizing capacity, shown as a subset of data represented in (a), but only sites having median acid neutralizing capacity ≤100 μeq/L.

same areas plus portions of Arkansas, the Gulf states, and the Rocky, Cascade, and Sierra Nevada mountains. In other regions of the United States, available water quality data are generally indicative of well-buffered streams and lakes that are likely to be insensitive to acidic deposition.

3.2 Nutrient Nitrogen Enrichment

Nutrient enrichment describes a host of environmental changes that can occur when the availability of a key nutrient is increased as a consequence of air and/or water pollution. Nutrient enrichment effects can occur in terrestrial, wetland, and aquatic environments. In all cases, the addition of a key nutrient (typically N) from atmospheric deposition can contribute to changes in the species mix of the plant, lichen, and algal communities. Nutrient enrichment can cause some species to thrive, at the expense of others. Thus, the mix of species present in an ecosystem can change as a consequence of nutrient addition.

TABLE 3.2

Data Sources for National Map of Surface Water Acid Neutralizing Capacity

Database	Source
Adirondack Lakes Survey	Adirondack Lakes Survey Corporation
Long-Term Monitoring Project	U.S. Environmental Protection Agency
Temporally Integrated Monitoring of Ecosystems	U.S. Environmental Protection Agency
Western Adirondack Stream Survey	U.S. Geological Survey
Eastern Lakes Survey	U.S. Environmental Protection Agency
Western Lakes Survey	U.S. Environmental Protection Agency
Environment Monitoring and Assessment Project	U.S. Environmental Protection Agency
Mid-Atlantic Streams Survey	U.S. Environmental Protection Agency
National Lakes Survey	U.S. Environmental Protection Agency
Wadeable Streams Assessment	U.S. Environmental Protection Agency
National Stream Survey	U.S. Environmental Protection Agency
STORET	U.S. Environmental Protection Agency
Regional Environmental Monitoring and Assessment Program	U.S. Environmental Protection Agency
High Elevation Lake Monitoring Program, Maine	Steve Kahl, Jason Lynch
Aquifer Lakes Project, Maine	Steve Kahl, Jason Lynch
Water Quality Monitoring in George Washington and Jefferson National Forests	U.S. Forest Service
Federal Land Manager Environmental Database	U.S. Forest Service
Monongahela National Forest Database	U.S. Forest Service
North Carolina National Forest Database	U.S. Forest Service
National Water Information System	U.S. Geological Survey
Alpine Hydrology Research Group	U.S. Geological Survey
Virginia Trout Stream Sensitivity Study	University of Virginia

3.2.1 Terrestrial Effects

In many terrestrial ecosystems, N is the most important nutrient limiting the growth of plants under a plant's natural water regime. If an appreciable amount of N is added from atmospheric deposition, plant growth rates may increase. Because some species are better able to take advantage of added N than others, some plant species grow well and others are crowded out. The species that benefit are often nonnative opportunistic species; those that are suppressed are sometimes rare native species (Sullivan 2015).

The extent to which ecosystems respond to nutrient addition depends in part on the extent to which the growth of plant communities is limited by N availability, as compared with the availability of light, water, phosphorus (P), or some other nutrient. Most north temperate terrestrial ecosystems are at least partially N-limited (U.S. EPA 2008). If the availability of N is increased, there can be a shift in competitive advantage to favor those species that grow faster. This can alter the species makeup of the plant community, decrease species diversity, and can eliminate some of the rare, or otherwise valued, species. In general, plant communities that are dominated by herbaceous plants appear to be more sensitive to N enrichment than plant communities dominated by woody plants. This may be partly because it takes a long time for scientists to document effects on trees that can grow for decades or centuries. Nutrient enrichment effects on plant communities in the United States have been most convincingly demonstrated for herbaceous arctic and alpine plant communities, grasslands, wetlands, and arid or semiarid lands

TABLE 3.3

Distribution of Measured Surface Water Acid Neutralizing Capacity (μeq/L) Values across the National Park Service Networks[a]

Network Name	Number of Sites	Minimum	5th Percentile	25th Percentile	Median	75th Percentile	95th Percentile	Maximum
Northeast Temperate	4300	−96.7	−23.1	12.8	80.0	199.0	853.6	6580.0
Northeast Coastal and Barrier	349	−71.4	−2.2	20.0	100.0	280.0	911.7	3600.0
Appalachian Highlands	1597	−22.0	2.0	43.0	106.0	228.7	1477.6	5602.2
Sierra Nevada	518	0.4	21.6	45.0	106.6	388.6	1380.5	4340.0
Eastern Rivers and Mountains	1636	−99.7	−23.7	64.5	194.9	500.0	1805.0	5792.2
Mid-Atlantic	834	−85.3	−10.4	77.7	220.0	539.0	1620.5	5020.0
Cumberland Piedmont	886	−20.4	16.8	132.8	340.0	1120.0	2754.9	4340.0
Gulf Coast	693	−58.7	0.0	120.0	368.0	910.0	2596.0	5200.0
Southeast Coast	1103	−87.0	20.0	200.0	370.0	550.0	1100.0	3810.0
National Capital Region	89	6.9	83.6	220.0	391.0	895.0	2455.4	3560.0
Great Lakes	1103	−48.6	7.0	81.5	391.5	1696.8	3738.0	6040.0
Southwest Alaska	64	140.0	147.5	260.0	410.0	560.0	797.0	940.0
Pacific Island	23	80.0	132.0	370.0	420.0	540.0	984.0	1040.0
North Coast and Cascades	815	−4.7	30.9	183.3	496.5	1005.0	2140.0	5844.9
Upper Columbia Basin	887	20.9	44.2	122.8	518.0	1320.0	3266.3	18720.0
Greater Yellowstone	495	12.0	35.4	88.8	518.2	1523.4	3860.0	6280.0
Klamath	599	0.0	26.7	219.1	654.6	1217.2	2862.0	8004.9

(Continued)

TABLE 3.3 (*Continued*)

Distribution of Measured Surface Water Acid Neutralizing Capacity ($\mu eq/L$) Values across the National Park Service Networks[a]

Network Name	Number of Sites	Minimum	5th Percentile	25th Percentile	Median	75th Percentile	95th Percentile	Maximum
Rocky Mountain	1186	5.0	42.2	200.9	744.6	1817.5	4305.0	10700.0
Southeast Alaska	52	60.0	173.0	512.5	920.0	1375.0	3209.0	15180.0
Central Alaska	85	260.0	360.0	670.0	960.0	2360.0	3924.0	5200.0
Northern Colorado Plateau	788	−5.7	59.3	138.3	1173.7	3400.0	6012.5	12020.0
Arctic	40	310.0	716.0	985.0	1300.0	1625.0	2143.5	3300.0
Southern Colorado Plateau	216	10.0	48.7	484.6	1480.0	2810.0	4645.0	9847.1
Southern Plains	140	28.9	185.8	648.4	1500.0	2713.8	4790.0	14200.0
Heartland	637	0.0	100.0	524.5	1652.2	2860.0	4920.0	6520.0
Mojave Desert	86	126.9	476.4	955.0	2000.0	3775.0	4548.9	5940.0
South Florida/Caribbean	108	−46.1	29.9	542.5	2050.0	2805.0	4388.0	5400.0
San Francisco Bay Area	48	354.5	476.0	1374.6	2115.0	2535.0	3432.0	6860.0
Mediterranean Coast	40	300.0	579.0	1492.5	2242.5	3490.0	6628.0	9540.0
Sonoran Desert	82	115.4	540.0	1047.5	2260.0	3552.5	4619.0	5922.0
Northern Great Plains	263	20.0	786.0	2149.0	3040.0	4686.7	9798.0	14400.0
Chihuahuan Desert	9	740.0	1396.0	2480.0	3320.0	3840.0	8712.0	11920.0

[a] Data are based on samples mapped in Figure 3.2a, arranged in order of increasing median acid neutralizing capacity.

(U.S. EPA 2009c). These are the primary plant community types that are discussed here. This does not mean that some of these same effects may not also be occurring in forest communities.

Alpine plant communities are known for their high diversity of vascular plant species (Bobbink et al. 2010) and have been shown to be especially sensitive to N enrichment. Some studies suggest that sedges benefit more from N addition to alpine herbaceous plant communities than do grasses or forbs (Bowman et al. 2006, Bobbink et al. 2010). Changes in the abundance of a species of alpine sedge (*Carex rupestris*) have been demonstrated at N deposition levels near 3 kg N/ha/yr (Bowman et al. 2012). More dramatic effects on alpine vascular plant community composition within a relatively short time period (a few years) probably require higher levels of N deposition, perhaps near 10 kg N/ha/yr (Bowman et al. 2006). Model simulations suggest that effects can develop over centuries in response to very low levels of N input (Sverdrup et al. 2012, McDonnell et al. 2014). There are a number of plant community characteristics in the alpine zone that appear to govern sensitivity to N input. These include low temperature, restricted growing season, low levels of primary production, and wide variation in moisture regimes from boggy meadows to the desiccating conditions of wind-swept ridges (Bowman et al. 1993, Bowman 1994, Bowman and Fisk 2001). Another contributing factor is the fact that soil formation is a slow process in the harsh alpine environment. Therefore, alpine plants have evolved under conditions of low nutrient supply (U.S. EPA 2008). In some alpine scrub habitats, ground-layer bryophytes and lichens are thought to be especially sensitive to N addition (Fremstad et al. 2005, Britton and Fisher 2007).

Arctic plants have not been well studied with regard to their sensitivity to modest levels of N enrichment that might correspond with potential future levels of atmospheric N deposition. Nevertheless, some arctic plant communities share many commonalities with alpine vegetation and are therefore thought to be highly sensitive. Arctic environmental conditions also contribute to slow soil development and low nutrient availability, thereby enhancing sensitivity to N deposition inputs. Low levels of N addition (1–5 kg/ha/yr) have been shown to change species composition and cover of grasses, forbs, and shrubs in some tundra ecosystems (Arens et al. 2008).

Increased N input has reduced plant biodiversity in grasslands in both Europe and North America (Bobbink et al. 2003, Stevens et al. 2004, Clark and Tilman 2008). Such effects have been documented across a range of soil conditions. Changes in species composition have been reported at N deposition levels as low as about 10 kg N/ha/yr (Bobbink et al. 2003). In the San Francisco Bay area, which receives total N deposition of about 10–15 kg N/ha/yr, nonnative nitrophilous (N loving) grasses have displaced native plant species. This effect has been attributed to greater N supply from the atmosphere combined with cessation of livestock grazing, which had previously removed some of the excess N from the ecosystem (Fenn et al. 2003, U.S. EPA 2008). A survey across acidic grasslands in the United Kingdom by Stevens et al. (2004) estimated a decrease of one plant species for every 2.5 kg N/ha/yr of deposition.

Experimental N fertilization studies have been conducted in arid and semiarid plant communities in southern California and the Colorado Plateau. Such vegetation communities are commonly dominated by shrubs but can also be dominated by grasses, forbs, and/or cacti. Results of the fertilization studies suggested increased biomass of some nonnative plant species and reduced biomass of some native species with sufficiently high levels of N input. There also appears to have been an increased fire risk where grasses replaced shrub cover. There are apparently multiple interactions involving N stimulation of nonnative grass species, competitive abilities of native forb and shrub species, and the incidence of fire (cf., Eliason and Allen 1997, Yoshida and Allen 2001). The coastal sage scrub plant community in California, in particular, has been declining for many decades. Current scientific understanding suggests that atmospheric N deposition plays a role in this decline and causes shrub replacement by Mediterranean annual grasses, which promote increased fire cycles (D'Antonio and Vitousek 1992, Minnich and Dezzani 1998, Padgett and Allen 1999, Padgett et al. 1999, Fenn et al. 2003).

Additional studies have been conducted in desert plant communities, suggesting that N additions in the range of 10 kg N/ha/yr or higher can cause invasion of nonnative plant species, including grasses (U.S. EPA 2008). Effects appear to be more pronounced at higher elevations, perhaps because the increased precipitation that occurs at higher elevation in arid lands can contribute to increased grass production. Increased grass biomass in desert plant communities is associated with increased fire frequency (Brooks 1999, Brooks and Esque 2002, Brooks et al. 2004). The effects of N deposition on arid and semiarid plant communities may vary with moisture levels. In Joshua Tree National Park, California, N loadings of 3–4 kg/ha/yr were estimated to significantly increase the cover of nonnative grasses, with subsequent increases in fire risk (Rao et al. 2010). In some cases, N may be limiting during wet seasons or years, whereas water may be limiting during dry seasons or years.

Lichens are typically among the most N-sensitive terrestrial receptors (Figure 3.1; Bobbink et al. 2003, Fenn et al. 2003, Vitt et al. 2003, Geiser et al. 2010). Most lichens can be classified into groups based on N requirements. Some additional lichen taxa exhibit broad tolerance to N supply. Lichens that inhabit trees are called epiphytes. They tend to be highly sensitive to N inputs and are widely recognized as good early warning indicators of terrestrial ecosystem effects from atmospheric N input (Fenn et al. 2011). Epiphytic lichens vary in their N requirements. Montane species tend to be adapted to low N availability and are classified as oligotrophic (low nutrient availability). They are commonly found on coniferous trees. Valley species tend to be adapted to more mesotrophic (moderate nutrient availability) conditions and often occur on hardwood trees. Eutrophic (high nutrient availability) species are adapted to high N supply, often in association with animal nesting or roosting sites. Atmospheric N deposition results in a shift from more oligotrophic toward more eutrophic conditions; this shift is typically accompanied by a change in the lichen community to favor the mesotrophic and eutrophic forms, with a decrease in the oligotrophic species (Geiser et al. 2010). Many lichen species that are prevalent in the western United States have relatively well-known N requirements. Critical loads of N have been well established, especially in the West (Pardo et al. 2011, Cleavitt et al. 2015). A change in lichen species distribution and abundance in response to a change in N supply can have broad ecosystem-level effects. This is due to the importance of various oligotrophic lichen species as forage, for nutrient cycling, and as nesting materials for wildlife.

3.2.2 Aquatic Effects

Estuaries and coastal marine waters are generally susceptible to nutrient enrichment effects from atmospheric N deposition because N tends to be the main growth-limiting nutrient in such environments. Growth of plants and algae in freshwaters, in contrast, is often limited by the availability of P, which is typically not an important component of air pollution and atmospheric deposition at most locations. There are some freshwaters, however, that are limited in their algal growth by N or by a combination of both N and P. Many of these occur at high elevation.

High-elevation lakes are of particular concern, among aquatic ecosystems, with respect to potential impacts attributable to atmospheric nutrient N enrichment. There are several reasons. Many high-elevation lakes tend to be dilute to ultradilute (low ionic concentrations) and nutrient-poor, contributing to increased risk of biological change from nutrient addition, even in relatively low quantities. Many high-elevation lakes have been shown to be N-limited. Diatom communities in several high-elevation lakes in Colorado and Wyoming have been shown to have been impacted by relatively low levels of atmospheric N deposition (Wolfe et al. 2001, Saros et al. 2003, Baron et al. 2011, Nanus et al. 2012). Because soils in the watersheds of high-elevation lakes are often poorly developed, with much exposed bedrock, the transport of N from atmospheric deposition to lake water can be quick and direct.

Research by Elser et al. (2009a,b) suggested that atmospheric inputs of N in some high-elevation areas have altered the relative proportional availability of N and P to phytoplankton. These changes in nutrient availability may have shifted lakes from N limitation toward P limitation, with consequent alterations to plankton community structure, species diversity, and trophic interactions (Elser et al. 2009b).

3.2.3 Nitrogen Saturation

An undisturbed forest commonly uses and incorporates, mostly through the soil, almost all of the small amounts of N that it receives from atmospheric deposition. This N is cycled between soil and vegetation, but the cycle can be disrupted by increased atmospheric N deposition (Aber et al. 1989). Forests have a maximum capacity to store N that they receive from outside the watershed. This capacity is determined by the plant species present on the site and the history of logging and other disturbances that removed some of the N previously stored on site. When N inputs exceed this storage capacity, the site becomes N-saturated, and an appreciable percentage of the incoming N leaches as nitrate to soil water and to streams and lakes. Leaching of nitrate can contribute to soil acidification. As N saturation progresses, tree health deteriorates and the forest may release to drainage water more N than is coming into the watershed from atmospheric deposition. Under conditions of advanced N saturation, tree growth declines and sensitive tree species die in response to acidification, Al toxicity, and/or base cation depletion (U.S. EPA 2008).

Atmospheric deposition of N has increased N availability in soils at some locations, which has led to increased nitrification and acidification of soil and soil water. Because the N retention capacity of soils is strongly dependent on land-use history, the relationships between N deposition and ecosystem N status are variable. In general, atmospheric deposition of about 8–10 kg N/ha/yr or higher is required in order for appreciable amounts of nitrate to leach from forest soils to surface waters in the eastern United States (U.S. EPA 2008). At high elevations in the West, the scarcity of soil and the dynamics of snowmelt contribute to substantial N leaching at lower levels of atmospheric N deposition (Williams et al. 1996, Nanus et al. 2012).

Many studies in the southern Appalachian Mountains (cf., Joslin et al. 1992, Van Miegroet et al. 1992a,b, Joslin and Wolfe 1994, Nodvin et al. 1995) have found high concentrations of nitrate in soil water and stream water at high-elevation, old-growth spruce-fir forest locations. This nitrate leaching is believed to have been caused by high N deposition, low N uptake by forest vegetation, and inherently high N release from soils. Forest age also affects N uptake by vegetation. Mature trees take up relatively small amounts of N for new growth and often show higher nitrate leaching than younger, faster growing stands (Goodale and Aber 2001).

3.3 Critical Load and Exceedance

The critical load is defined as the threshold of acid or nutrient pollutant deposition below which specified harmful ecological effects do not occur according to present knowledge (Nilsson and Grennfelt 1988, Porter et al. 2005). The critical load allows evaluation of relative resource sensitivity. Critical loads can be developed for any pollutant but are most often applied to N and S deposition and acidity or to N enrichment. Critical loads are commonly expressed as kg/ha of N or S or as equivalents per hectare (eq/ha) of combined acidity from N and S per unit time (usually per year). The critical load is generally determined for a specified indicator and endpoint for that indicator. For example, a critical load for S can be calculated that will prevent acid neutralizing capacity from decreasing to a level below 50 or 100 μeq/L in a lake, levels considered safe for most aquatic biota. A critical load for N can be calculated that will prevent loss of a certain amount of plant species diversity in a grassland or alpine meadow. Critical loads can be calculated using dose-response functions, ecological models, or empirical observations (Sullivan and Jenkins 2014). The critical load is assumed to be sustainable under long-term steady-state mass balance conditions (Henriksen and Posch 2001). In other words, the critical load specifies the pollutant load which, when applied to the ecosystem in question for a period of years to centuries, will eventually trigger a change in a chemical or ecological indicator of harm at the time that the ecosystem comes into steady state with respect to that particular pollutant input level.

A dynamic, as opposed to steady-state, pollutant load can also be defined, which is specific to a particular point in time. This dynamic pollutant load will specify, for example, the deposition load that can be tolerated in a certain lake or stream until the year 2100 without loss of brook trout populations. If a stream is already acidified, it can identify how much S or N loading should be reduced in order to restore the stream to a healthy acid neutralizing capacity by the year 2100, for example. This dynamic load, which includes consideration of the temporal component of ecosystem damage or recovery, is called a target load or a dynamic critical load. The target load concept can be used to describe not only the effects of time but also the inclusion of various management actions or scenarios (Sullivan and Jenkins 2014).

There is no single "definitive" critical load for a natural resource. The critical load depends on what is to be protected, to what level of protection, what indicator is used, and what is specified as the critical or threshold tipping point level of that indicator. Critical load estimates reflect the current state of knowledge and policy priorities, both of which can change. This process typically results in the calculation of multiple critical loads for a given pollutant at a given location, and those values can change over time. Multiple critical load values may also arise from an inability to agree on a single definition of significant harm (Sullivan and Jenkins 2014).

The critical load is affected by the heterogeneity of natural ecosystems. Because of high spatial and temporal variability of soils and surface waters, there may be a continuum of sensitivity and critical load values for a given sensitive indicator.

The critical load or target load provides an indication of the point at which the ecosystem may begin losing ecosystem services and transitioning from a sustainable functioning ecosystem for those services to one that is not functioning properly and is no longer considered sustainable. The critical load concept provides a tool that enables the implementation of decision-making based on ecosystem services (Sullivan 2012).

On its own, the critical load does not predict whether the ecosystem experiences, or will experience, biological harm. The ambient pollutant load must also be considered.

If the ambient deposition is higher than the critical load or target load that the ecosystem can tolerate, the ecosystem is in exceedance, which suggests an increased likelihood of biological harm. Transitioning between a condition of nonexceedance to a condition of exceedance does not mean a change in whether the ecosystem is currently experiencing damage. The transition from nonexceedance to exceedance, or vice versa, indicates a change in the probability that ecosystem services will be reduced, lost, or recovered, depending on whether the starting point is an undamaged or a damaged state (Sullivan 2012). For undamaged systems, exceedance signifies that if deposition is continued at the current exceedance level, damage will likely occur at the time point specified for the analysis (e.g., 2100, eventual steady-state condition). Transitioning from exceedance to nonexceedance suggests that recovery will occur at the future time specified in the analysis (Sullivan and Jenkins 2014).

In a watershed that receives a loading of acidic deposition that is lower than the critical load, a target load might be selected that is higher than or lower than the critical load. A higher target load might be justified if the damage projected to occur under the critical load will not be manifested for a very long time, allowing future opportunity to further mitigate pollution levels before the damage is realized. A lower target load might be justified to err on the side of resource protection or to protect against short-term damage prior to attaining a long-term steady-state protected condition. For a watershed that receives an acidic deposition loading that is higher than the critical load, an interim target load might be selected that is higher than the critical load in order to allow for partial resource recovery within a management timeline. For an already damaged watershed, a target load might be selected that is lower than the critical load if resource managers are unwilling to wait the decades or centuries that it might take to achieve full resource recovery under sustained loading at the level of the critical load (Porter et al. 2005, Sullivan 2012).

Pardo et al. (2011) compiled empirical data on nutrient N critical loads in ecoregions throughout the United States for a variety of N-sensitive resources, including lichens, mycorrhizal fungi, herbaceous plants and shrubs, and forests. These critical loads were generally reported by Pardo et al. (2011) as ranges for a given ecoregion, rather than absolute values. Because ecosystems vary, sometimes substantially, within an ecoregion, the critical load range for a given receptor (species or group of species) may not apply to all plant communities or all portions of the ecoregion. Some are more sensitive than others.

Sullivan and McDonnell (2014) evaluated empirical critical load estimates for ecoregions developed by Pardo et al. (2011) and analyzed available vegetation distribution data to estimate the appropriate critical load range for 12 selected national parks in the intermountain West. They also examined oxidized N emissions by county and total (oxidized and reduced) N deposition estimates in and near the selected parks. Deposition estimates for 2008 were compared to park nutrient N empirical critical loads to evaluate the likelihood that critical loads were exceeded in the national parks considered, given the presence of the sensitive element. The parks selected for study included many of those in the western United States that have been identified as having N-sensitive ecosystems. Some of the parks have been experiencing, in recent years, increasing levels of nearby energy, agricultural, mineral extraction, and/or transportation development, with associated increasing N emissions.

The critical load and exceedance estimates developed by Sullivan and McDonnell (2014) were based on the lower limits of the ranges of critical load, by ecoregion, reported by Pardo et al. (2011). This was done to conform with the Congressional mandate when considering critical loads for national parks and other protected areas, to err on the side of resource protection. Based on these lower limits of the critical load ranges reported by Pardo et al. (2011), terrestrial resources in most of the parks evaluated by Sullivan and McDonnell (2014) were either in exceedance of the nutrient N critical load or received ambient (year 2008) total wet plus dry N deposition that was below, but within 1 or 2 kg N/ha/yr of, the critical load. Thus, nutrient-sensitive terrestrial resources in some of these parks may be experiencing adverse impacts associated with critical load exceedance. In other cases, appreciable increases in oxidized N or ammonia emissions and deposition in the future may trigger exceedances.

Of the 12 parks that were evaluated in the study of Sullivan and McDonnell (2014), estimated nutrient N critical load exceedances were most pronounced in Voyageurs, Mesa Verde, Black Canyon of the Gunnison, and Saguaro national parks. Large portions of Grand Canyon, Arches, Badlands, Theodore Roosevelt, and Wind Cave national parks and Colorado and Dinosaur national monuments received N deposition in 2008 that was within 1 kg N/ha/yr or less below the nutrient N critical load. Much of Canyonlands National Park received N deposition in 2008 that was below, but within 2 kg N/ha/yr of, the nutrient N critical load. These results provide estimates of relative risk of nutrient enrichment to sensitive vegetative receptors among the parks selected for study.

Baron (2006) recommended a critical load of 1.5 kg N/ha/yr (wet deposition) to protect alpine lakes in Rocky Mountain National Park against eutrophication. Similarly, Saros et al. (2011) found that 1.4 kg N/ha/yr (wet deposition) caused nutrient enrichment effects (i.e., changes in diatom diversity) in high-elevation lakes in the eastern Sierra Nevada and the Greater Yellowstone ecosystems. Similar thresholds have been identified by Baron et al. (2011) and Nanus et al. (2012).

Nitrogen deposition to wetlands can alter competitive relationships among species, sometimes increasing the establishment of nonnative species at the expense of rare species. Such changes are thought to occur in Europe at N deposition levels in the range of 5–10 kg N/ha/yr for raised and blanket bogs, 10–20 kg N/ha/yr for poor fens, and at higher deposition levels for rich fens and salt marshes (Achermann and Bobbink 2003). Pardo et al. (2011) suggested a critical load range of 2.7–13 kg N/ha/yr to protect wetlands in the northeastern United States.

3.4 Ozone Exposure and Effects

The challenge in assessing ozone effects is to quantify the spatial distribution of ozone exposure relative to the location of sensitive plant receptors and the environmental conditions that promote ozone uptake by plants into the stomata. The fact that ozone concentrations are higher downwind from urban areas, and that concentrations tend to increase with elevation, suggests that some national parklands can be particularly vulnerable.

Despite known general relationships between regional sources of ozone (and its precursors) and wildland receptors, it has been difficult to estimate ozone exposure in those wildlands because of high variability. Ozone is monitored at about 40 national parks, although about twice that many urban national park units have nearby ozone monitors. The National Park Service-Air Resources Division also provides estimates of ozone concentrations and exposures at over 200 park units in the continental United States, using all available monitoring data to derive interpolated estimates (http://www.nature.nps.gov/air/Maps/AirAtlas/index.cfm). Trends are highlighted at http://www.nature.nps.gov/air/data/products/parks/index.cfm.

3.4.1 Effects of Ozone on Plants

Ozone injury to plants is determined primarily by three things: (1) the plant and its inherent sensitivity to ozone, (2) the level of ozone in the atmosphere at the location of the plant, and (3) site-specific environmental conditions (Kohut 2007b). In general terms, ozone injury will only occur if the plant is genetically predisposed to ozone injury, the level of atmospheric ozone exposure exceeds the threshold for injuring that plant, and the environmental conditions at the site facilitate the uptake of ozone through the stomata into plant leaves.

Injury occurs when all three key elements are satisfied. First, the plant must be predisposed to ozone sensitivity. Differences in sensitivity are often pronounced at the species level. For example, ponderosa pine (*Pinus ponderosa*) and quaking (also called trembling) aspen (*Populus tremuloides*) are two common tree species found in many national parks in the western United States that are known to be sensitive to ozone. However, differences in plant sensitivity can also be observed among clonal lines, subspecies, or individual plants. The reasons for such differences are not well understood (Kohut 2007b). Second, the threshold level of atmospheric ozone exposure that can produce injury on a plant must be exceeded at the location of the plant. Peak levels of exposure can produce acute effects; lower but sustained levels of exposure can produce chronic effects. Exposure indices that consider both atmospheric ozone peak levels and sustained exposure are used to evaluate the likelihood of the plant experiencing a high ozone exposure. Third, the plant must experience environmental conditions that facilitate ozone uptake into the leaf through the stomata. The environmental conditions that inhibit stomatal opening, and therefore also inhibit ozone uptake into the leaf, include mainly low soil moisture and high temperature. Conditions of optimum moisture availability, humidity, illumination, and temperature facilitate both photosynthesis and uptake of ozone into the leaf. Such conditions can be highly variable spatially. For example, low soil moisture and high temperature can limit stomatal opening broadly across a vegetative community, yet more favorable conditions for ozone uptake may occur at the microsite level, for example, in riparian zones or other areas that retain soil moisture. Thus, plants on favorable microsites may exhibit physiological responses, such as foliar ozone symptoms, whereas adjacent plants under water stress may not. Similarly, an ecosystem may express ozone symptoms during one year, but not the next, due entirely to differences in ozone exposures, soil moisture, and/or temperature. Foliar symptoms are most pronounced when all variables of the response triad are satisfied (Kohut 2007b).

Visible foliar symptoms provide an easily identified (with training) indicator of ozone injury to vegetation. Ozone injury of cells and tissues is essentially the same in woody and herbaceous plants (Bytnerowicz and Grulke 1992). Injury generally occurs first in the most photosynthetically active tissues, with disruption of chloroplasts in the palisade and mesophyll tissue. The loss of photosynthetic tissue results in visible chlorosis (bleaching) and necrosis (death of tissue). Visible symptoms do not constitute an early warning signal indicating damage to the plant; physiological impacts may have occurred prior to the production of foliar symptoms. Various effects may have occurred inside the plant before visible symptoms are evident. Furthermore, the presence of visible symptoms does not necessarily indicate that effects on growth, health, or reproduction have, or will, also occur.

3.4.2 Standards and Cumulative Exposure Indices

Evaluation of plant response to ozone exposure requires the use of an appropriate ozone exposure summary statistic. Exposure reflects the atmospheric concentration of ozone to which a plant is exposed; dose reflects the amount of ozone that is actually taken up by the plant. Factors that influence plant response are complex and vary with environmental conditions and the physiology of ozone-sensitive plants (Musselman et al. 2006). The degree of pollutant injury depends on the effective dose, which is a function of concentration, length of exposure, and stomatal aperture, as well as plant biochemical defense response to the ozone dose received (Kozlowski and Constantinidou 1986). Exposure indices are unable to consider the full range of processes that control ozone transport from the atmosphere into the leaf or the physiological response, including detoxification mechanisms, of the plant. As a consequence, there is no appropriate dose-based index for use in evaluating the effects of ozone on plants (Musselman et al. 2006). Rather, exposure-based indices are used.

In order to protect human health and welfare, the EPA has established primary (to protect human health) and secondary (to protect welfare and the environment) National Ambient Air Quality Standards for maximum allowable atmospheric ozone concentration levels. Prior to 1997, these standards were based upon 1-hour average ozone measurements. They were revised in 1997, when the EPA promulgated new standards, both primary and secondary, based upon an 8-hour average value. The standards were both set at 0.08 parts per million (ppm; or rounded to 85 ppb), a value that represented the annual fourth highest daily maximum 8-hour ozone concentration, averaged over three years for the protection of both human health (primary National Ambient Air Quality Standard) and vegetation (secondary National Ambient Air Quality Standard). The standards were strengthened in 2008 to 0.075 ppm and in 2015 to 0.070 ppm. This average is computed by first determining the highest 8-hour average ozone value for each day of the year at a site and then identifying the fourth highest of all daily maximum 8-hour ozone values that occurred during the year. These fourth highest values are then averaged over three successive years to determine the final concentration value that is compared to the standard.

The decision by the EPA to transition from an ozone standard based upon a 1-hour average to a standard based upon an 8-hour average was prompted by research indicating that prolonged exposure to ozone at concentrations lower than the 1-hour standard can have significant impacts. The agency has also recognized the potential importance of an exposure index in setting the secondary ozone standard to protect native vegetation and ecosystems. The EPA proposed a new secondary ozone standard in January 2010 (Federal Register Vol. 75, No. 11, 40 CFR Parts 50 and 58, National Ambient Air Quality Standards for Ozone, Proposed Rules, January 19, 2010, p. 2938). It was based on an index, called the W126, of the total plant ozone exposure during the daytime (8:00 AM–8:00 PM), whereby hourly values were weighted according to magnitude and then summed. Higher concentrations are weighted more heavily, as they are more likely to induce ozone injury. The W126-3-mo statistic is calculated as the three-month period having the highest cumulative exposure. The proposed standard was based on a three-year average of the annual W126-3-mo metric; nevertheless, the EPA administrator judged that the primary standard at that time was sufficient to protect sensitive vegetation.

In addition to the W126, another metric, the SUM06, is used to express ozone exposures (Heck and Cowling 1997). This index is calculated as a 90-day maximum sum of the hourly concentrations of ozone during daylight hours that are higher than or equal to 60 ppb (0.06 ppm). It is calculated over a running 90-day period. The maximum value can occur at any time of year during the ozone monitoring season but usually occurs during the summer months.

The W126 index (Lefohn et al. 1997) can be calculated as the weighted sum of the 24 one-hour ozone concentrations measured daily for seven months from April through October as originally proposed by Lefohn or calculated for daylight hours over a three-month period as modified by the EPA. When originally developed, the W126 was expressed in conjunction with the number of hours (called the N100) during which the ozone exposure is ≥0.100 ppm. This was because the N100 was associated with foliar effects noted in chamber studies.

It is important to note that calculated values of the W126 index differ depending on whether they are calculated over a seven-month period (W126-7-mo; original Lefohn version) or a three-month period (W126-3-mo; modified by the EPA in 2010). The ozone assessments of Kohut (2007a,b) used the Lefohn version; the assessment presented here uses the EPA version. The National Park Service has adopted the EPA version for assessing ozone exposure in parks with respect to potential injury to vegetation (NPS 2011).

3.4.3 Mode of Action

In the context of impacts of air pollutants on plants, damage and injury terminology is rather precise, as applied by Federal Land Managers (cf., Guderian 1977, Musselman et al. 2006, U.S. Forest Service et al. 2010). Injury refers to all physical and biological responses to air pollutants, including effects on processes and cycles. Damage is a reduction in the intended use or value of the resource, including reduction in aesthetic value, plant diversity, or other ecosystem services.

Ozone can cause injury to various plant tissues, including direct effects on leaves and indirect effects on stems and roots. Manifestation of foliar symptoms is the most visible indication of ozone injury, although without proper field training, one can sometimes confuse ozone symptoms with other biophysical injuries, including abrasion, desiccation, insect herbivory, and effects of fungal pathogens. It is difficult to equate visible ozone-induced foliar symptoms with other effects on vegetation, such as on growth or reproduction. The significance of foliar symptoms appears to depend on how much of the total leaf area is affected and also plant age, size, and developmental stage. Nevertheless, visible foliar symptoms have been shown in some studies to correlate with decreased plant growth (cf., Benoit et al. 1982, Peterson et al. 1987, Karnosky et al. 1996) and, in some cases, decreased reproduction (cf., Black et al. 2000, Chappelka 2002).

Ozone enters plant leaves as a gas and dissolves in the presence of water. The resulting free radicals oxidize proteins of cell membranes, including those of the thylakoid membranes where photosynthesis takes place. Injury includes leaf discoloration, reduced photosynthetic rates, and lowered sugar production.

Ozone exposure is believed to inhibit plant stomatal responses. This may explain the finding of McLaughlin et al. (2007a,b) that peak hourly ozone exposure increased water loss from several tree species and reduced late-season modeled streamflow in forested watersheds in eastern Tennessee. Ozone exposure was also found to decrease the net primary productivity of most forest types in the Mid-Atlantic region by 7%–8% but had a much smaller effect (~1%) on the net primary productivity of high-elevation spruce-fir forests (Pan et al. 2009).

The timing of ozone exposure relative to the life cycle of the affected plant is important in determining response. For instance, exposure to ozone may reduce the growth of the root system in an annual plant; as the plant repairs leaf injury and maintains photosynthesis, there is less carbon (C) available to grow roots (U.S. EPA 1996b). In fact, root systems may be reduced in growth long before deleterious effects are manifested on aboveground portions of the plant. In perennial plants, the process is more complex, as stored reserves are usually available for growth and the effect of an exposure may not be manifested for several growing seasons (Hogsett et al. 1989, Andersen et al. 1991, Laurence et al. 1994). However, in the most sensitive perennial species, a single season of exposure can be sufficient to reduce growth significantly (Wang et al. 1986, Woodbury et al. 1994). The EPA estimated that under ambient ozone exposures, biomass loss in quaking aspen could exceed 10% in some areas of the East; biomass loss in black cherry (*Prunus serotina*) seedlings could exceed 20% in many areas (U.S. EPA 2007). Impacts to seedlings may decrease long-term growth and survival, ultimately affecting both individuals and populations of sensitive species.

Because ozone enters the plant through stomata, the duration of stomatal opening greatly affects the pollutant dose assimilated. Once ozone enters a plant cell, there are many biochemical processes that can be affected (Wellburn 1988, Heath and Taylor 1997). Ozone increases the potential for the formation of free radicals, which are highly reactive and disrupt various metabolic processes through oxidation and toxicity.

The most common visible effects of ozone exposure on vegetation are stipple (pigmentation on leaves), fleck (tiny light colored markings on upper layers of leaves), mottle (blotchy appearance), and necrosis (Miller et al. 1983, Thompson et al. 1984a,b, Hogsett et al. 1989, Treshow and Anderson 1989, Stolte 1996, Brace et al. 1999). Oxidant stipple is highly specific to ozone injury. Other types of injury may also occur in response to ozone exposure, including chlorosis, premature senescence, and growth reduction. However, these types of injury are not specific to ozone and can be caused by a variety of stressors. Documentation of stipple between the veins on the upper leaf surface of older sun exposed leaves provides a strong indication of ozone injury; changes in growth, species composition, and diversity in response to ozone exposure are much more difficult to document and quantify in the field.

Visible symptoms on leaves usually occur after acute exposure to high concentrations of ozone. Exposure to moderate concentrations of ozone for periods of several days to several months can also cause chronic injury, accelerated aging, premature casting of foliage, or reduced growth (Pell et al. 1994a,b, U.S. EPA 1996a,b). Accelerated aging may result in premature color change and loss of foliage, an effect of considerable importance at some national parks where park visitation levels are relatively high during the fall color season. Growth reductions at ambient levels of ozone are often difficult to measure, although a cumulative stress over multiple growing seasons may significantly reduce the growth and productivity of trees and understory vegetation (Reich and Amundson 1985, U.S. EPA 1996b).

Ozone exposure typically does not kill plants outright. Rather, it weakens them by disrupting their C budgets. Exposure to ozone contributes to reduced C fixation. In addition, because C is needed to repair ozone injury to membranes and organelles, there is less C available in ozone-stressed plants for growth, reproduction, or responding to other stressors such as low temperature, insects, and disease. Plants must expend energy to counteract the effects of ozone. This energy might otherwise be used for growth or health maintenance. Injured plant cells may die if detoxification and cellular repair are outpaced by ozone uptake into plant tissue.

There is considerable genetic variability, both within and among plant species, in the amount of damage that occurs in response to ozone exposure (Miller et al. 1982). The most sensitive plants are injured by exposure to concentrations of 60 ppb by volume or less for several days. These species are often used as biological indicators of ozone exposure (U.S. EPA 1996b). Conversely, some plants are very tolerant of ozone and are unaffected, even with severe exposures. Stomatal responses to ozone are complex. There can be pronounced differences in stomatal response due to differences in species sensitivities, exposure levels, and duration of exposure (McAinish et al. 2002, McLaughlin et al. 2007a).

Oxidant stipple can occur on sensitive plant species at ozone concentrations considered to be near or at background levels (Singh et al. 1978), sometimes in response to naturally occurring stratospheric ozone intrusions. However, such intrusions typically occur during spring, and not during the major part of the growing season when the leaf uptake of ozone is more pronounced (Singh et al. 1980, Wooldridge et al. 1997).

3.4.4 Field Observations

Smith (2012) reported the results of ozone exposure and effects biomonitoring across the northeastern and northcentral United States, with data collected over a period of 16 years at some locations (biosites). Ozone-sensitive plants were evaluated for ozone-induced foliar symptoms. However, regression models were generally not successful in predicting the presence of foliar symptoms based on ozone exposure (SUM06 and N100), site moisture (Palmer Drought Severity Index), and a plant moisture availability index ($r^2 > 0.1$). She did, however, generate plausible clustering of the data that supported the following weak associations:

- Biosites showing no symptoms occurred across all SUM06 and N100 exposures.
- Biosites showing symptoms occurred across all SUM06 and N100 exposures.
- When the Palmer Drought Severity Index and plant moisture availability index indicated that moisture was limiting, the percent of sites not showing symptoms was much greater than the percentage showing symptoms.

Generally, similar results were reported by Campbell et al. (2007) for western forests. They found an association between foliar symptoms and exposure, but high symptom expression could occur with low ozone exposure, and vice versa.

Field studies and controlled exposure experiments continue to identify ozone-sensitive plant species (cf., Kline et al. 2008, 2009). The U.S. Forest Service's Forest Inventory Analysis program previously surveyed the incidence and severity of ozone-induced foliar symptoms on multiple sensitive species using standardized procedures at biosites throughout the United States (Coulston et al. 2003, Smith et al. 2003, USDA FS 2011). These monitoring data have confirmed the applicability of several species as effective biomonitors in the eastern United States, including sweetgum (*Liquidambar styraciflua*), black cherry, blackberry (*Rubus* spp.), and milkweed (*Asclepias* spp.). A more extensive list of bioindicator species is available in a National Park Service report (Porter 2003), and lists of both sensitive and bioindicator species, by park, are available through the National Park Service Data Store (https://irma.nps.gov/App/Portal).

Early studies of ozone foliar injury to conifers focused on ponderosa pine and eastern white pine (*Pinus strobus*; Miller and Millecan 1971, Pronos and Vogler 1981, Peterson et al. 1991, Peterson and Arbaugh 1992). Applicability to other coniferous species may vary considerably with respect to specific symptoms and pollutant exposure. Numerous studies have documented the sensitivity of quaking aspen to ozone under field and experimental conditions (Wang et al. 1986, Karnosky et al. 1992, Coleman et al. 1996), although there is considerable variability in sensitivity among different genotypes (Berrang et al. 1986).

Available plant symptom data do not suggest a broadly applicable ozone injury threshold. In general, there may be adverse effects with prolonged exposure to ozone concentrations greater than about 60 ppb, and there are likely effects with prolonged exposure to concentrations greater than about 80 ppb (Sullivan et al. 2003).

Efforts to take results of seedling ozone exposure studies and scale them up to whole trees have generally not been very successful (Samuelson and Kelly 2001). Trees and seedlings differ in energy budgets, canopy-to-root balance, and P allocation and cannot be assumed to exhibit similar levels of ozone sensitivity.

Kohut (2007c) evaluated the risk of ozone-induced foliar injury on sensitive plant species in 244 national park units as part of the Vital Signs Monitoring Network. The analysis examined plant response related to ozone exposure level and environmental conditions. He concluded that 27% of the parks had high risk of foliar injury, 19% moderate risk, and 54% low risk. Parks determined to have high risk included Gettysburg, Valley Forge, Delaware Gap, Cape Cod, Fire Island, Antietam, Harper's Ferry, Manassas, Wolf Trap Farm, Mammoth Cave, Shiloh, Sleeping Bear Dunes, Great Smoky Mountains, Joshua Tree, Sequoia and Kings Canyon, and Yosemite.

Because ozone exposure is both chronic and episodic and is difficult to predict, plant responses may be controlled in large part by the capacity of plants to maintain leaf antioxidant systems throughout the growing season. Burkey et al. (2006) assessed seasonal patterns in ascorbate pool size and redox status in leaves from natural populations of three wildflower species (cutleaf coneflower [*Rudbeckia laciniata* L.], tall milkweed [*Asclepias exaltata* L.], and crown-beard [*Verbesina occidentalis* Walt.]) in Great Smoky Mountains National Park to determine relationships between foliar injury and ozone exposure. Ascorbic acid is an important metabolite that protects plant leaves from ozone injury (Conklin and Barth 2004). This was perhaps the first study to characterize leaf ascorbate in natural wildflower populations under field conditions. The three species showed differences in the content of leaf ascorbic acid and oxidation state that corresponded with differences among the three species in foliar symptoms in response to ozone exposure. The aspects of ascorbic acid metabolism that contributed to plant defense against ozone injury were identified.

Evans et al. (1996) found that the percent of dead palisade parenchyma cells in plant leaves was positively correlated with the percent of leaf area showing visible ozone injury. This was found for sassafras (*Sassafras albidum*), cutleaf coneflower, and smooth blackberry (*Rubus canadenis*). Also, the percent of dead cells was positively correlated with cumulative ozone exposure for sassafras and smooth blackberry.

Research at the Aspen Free-Air Carbon Dioxide Enrichment site in Rhinelander, Wisconsin, demonstrated that exposure to ozone can change the composition of plant communities. Aspen clones with high ozone tolerance were dominant over more ozone-sensitive clones, which showed reduced growth and increased mortality as a result of ozone exposure. Additionally, ozone exposure resulted in a competitive advantage for birch and maple over aspen in mixed aspen-birch and aspen-maple communities (Kubiske et al. 2007) with aspen biomass decreasing by 13%–23% in response to elevated ozone over a seven-year study (King et al. 2005). Biomass (above- and belowground) and net primary productivity were measured at the Aspen Free-Air Carbon Dioxide Enrichment site after seven years of ozone exposure. Relative to biomass at the control plot, total biomass at the ozone-enriched sites decreased by 23%, 13%, and 14% in the aspen, aspen-birch, and aspen-maple plant communities, respectively (King et al. 2005). Ozone exposure reduced growth and increased the mortality of an ozone-sensitive aspen clone; as a consequence, the tolerant clone became dominant. In mixed aspen-birch and aspen-maple communities, ozone exposure decreased the competitive ability of aspen compared to birch and maple (Kubiske et al. 2007).

3.4.5 Environmental Interactions

Chappelka et al. (1996) judged that the influences of microsite factors are important in controlling responses to ozone-induced injury. Available soil moisture appeared to be the most important environmental factor regulating ozone uptake into foliage, and subsequently effects on plants, through the influence of ozone on stomatal function. Kohut et al. (2012) found significant ozone injury on coneflowers in Rocky Mountain National Park in surveys conducted from 2006 to 2010. Risk to vegetation at the park had been rated as low because higher ozone concentrations that occur during the summer often coincided with dry soil conditions (Kohut 2007c). However, the 2006–2010 surveys were conducted in riparian areas, where soil moisture was sufficient to allow the plants to remain physiologically active and to promote ozone uptake. Kohut recommended that ozone surveys in western parks that often experience dry conditions be conducted in riparian or moist areas (Kohut et al. 2012). These riparian areas are also some of the most important areas ecologically in arid regions, supporting high levels of biodiversity.

Ozone may also affect the growth of plants indirectly through interactions with pests and pathogens (Laurence 1981), or by altering the symbiotic relations between plants and associated organisms (McCool 1988, Stroo et al. 1988, Andersen and Rygiewicz 1991, Rygiewicz and Andersen 1994, Andersen and Rygiewicz 1995). The resulting changes in nutrient availability or uptake may also result in altered plant growth, mediated by ozone exposure (Weinstein et al. 1991, Weinstein and Yanai 1994, Andersen and Scagel 1997). Increased atmospheric concentration of carbon dioxide and N deposition may increase plant productivity; ozone exposure may reduce or counteract these effects (Ollinger et al. 2002, Felzer et al. 2004).

Interactions with insects, pathogens, and other pollutants (Bytnerowicz and Grulke 1992) can accentuate the stress complex for plants exposed to ozone. This has been well documented for the effects of ozone on ponderosa pine. The most common stress complex includes ozone exposure, drought stress, and bark beetle injury (especially mountain pine beetle [_Dendroctonus ponderosae_] and western pine beetle [_D. brevicomis_]; Stolte 1996, Pronos et al. 1999). This interaction is particularly prominent in the mixed conifer forests of southern California and the southern Sierra Nevada. For example, during the late 1980s and early 1990s, ozone-stressed trees were subjected to several years of low precipitation. This reduced xylem pressure, and many trees were sufficiently weakened that they were susceptible to bark beetles, which proliferated through local and regional outbreaks, resulting in high levels of tree mortality in many areas.

Because lack of adequate soil moisture contributes to decreased stomatal conductance and consequently limits ozone entry into the leaf tissues (Panek and Goldstein 2001, Grulke et al. 2003, Panek 2004, Matyssek et al. 2006), dry periods exhibit decreased incidence and severity of symptoms of foliar ozone injury. Such symptoms are therefore not always higher during years having higher atmospheric ozone concentrations (Smith et al. 2003). The relationship of pollutant exposure to seasonal variation in physiological activity can have a significant influence on ozone injury in plants (Grulke 1999). The potential for pollutant uptake is higher during periods when plants are most metabolically active. However, biological defense mechanisms may be less at night when some remote locations receive peak exposures. Therefore, at some locations, exposure may contribute disproportionately to injury during the spring when soil moisture and gaseous uptake are higher, and there may be stratospheric intrusions of ozone.

3.4.6 Spatial Patterns in Ozone Effects

Most (244 parks) of the Inventory and Monitoring national parks in the conterminous United States, plus one in Alaska, were ranked by Kohut (2007a) with respect to overall risk from exposure of vegetation to ozone. This risk assessment was based on the presence of bioindicator plant species, estimated levels of ozone exposure, and estimated soil moisture conditions during periods of exposure over five years in each park. Each park was assigned a risk ranking of low, moderate, or high. These estimated park risk levels are shown in Figure 3.3 and are used in this book as an important part of the basis for estimating spatial patterns in ecological risk in the Inventory and Monitoring parks from ozone exposure. Most of the parks that were classified by Kohut (2007a) as having high risk are located in the eastern United States or in California. The former group includes Mammoth

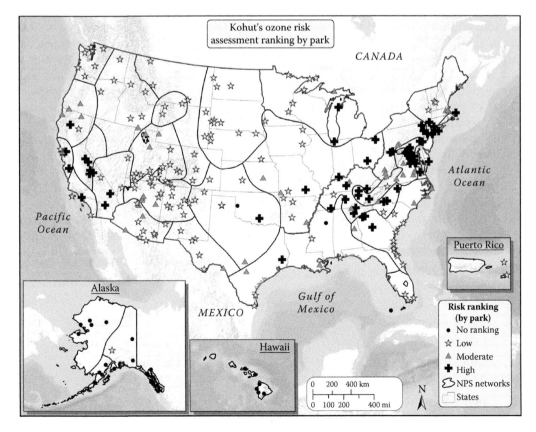

FIGURE 3.3
Risk ranking for exposure of vegetation in Inventory and Monitoring parks throughout the United States to ground-level ozone. (From Kohut, R., *Environ. Pollut.*, 149, 348, 2007.)

Cave, Delaware Water Gap, and Great Smoky Mountains. The latter includes Joshua Tree, Sequoia/Kings Canyon, and Yosemite. The process used by Kohut (2007a) for assigning relative risk was based on

1. The extent and consistency by which the SUM06, W126, and N100 ozone exposure thresholds were exceeded
2. The nature of the relationship between ozone exposure and soil moisture
3. The extent to which soil moisture levels constrained the vegetative uptake of ozone during high exposure years

Kohut (2007a) judged that parks ranked as high were likely to experience foliar symptoms due to ozone exposure in most years. A ranking of moderate suggested that injury was likely to occur at some point during a five-year period. Parks ranked as low were judged to be unlikely to experience injury during any year.

Kohut's analysis represented an examination of the major factors known to influence ozone uptake and injury and provided the best information to date on risk to plants in parks. However, it was based on data available at the time of the study, from the period 1995 to 1999. But to provide more updated information, the National Park Service-Air Resources Division developed a method for rating more recent ozone condition, based on exposure thresholds. This method does not take into account environmental factors like soil moisture. For this book, ozone data from 2005 to 2009, compiled by the National Park Service, were examined. These data were used to generate interpolated surfaces of SUM06 and W126-3-mo indices at 10 km resolution (D. Bingham, National Park Service, unpublished data, 2010). The SUM06 and W126-3-mo index values were determined for each Inventory and Monitoring park as a spatially weighted average of all the values that occurred within a park. Each was calculated as a three-month average of exposure during the 12 daytime hours. Thresholds for rating condition were based on information from the EPA's review of the ozone standards (2007). The review considered the recommendations of an expert working group (Heck and Cowling 1997) who suggested ozone thresholds for protecting natural vegetation against foliar injury. Based on these recommendations, the National Park Service determined that a W126-3-mo ≤ 7 ppm-h could be considered "good" condition, with impact unlikely, and a W126-3-mo > 13 ppm-h would warrant significant concern (NPS 2010). Exposure of W126-3-mo between 7 and 13 ppm-h would warrant moderate concern. Equivalent thresholds for the SUM06 index are 8 and 15 ppm-h, respectively (NPS 2010). Thus, the breakdown is as follows:

W126-3-mo:

    Good < 7 ppm-h

    Moderate Concern 7–13 ppm-h

    Significant Concern > 13 ppm-h

SUM06:

    Good < 8 ppm-h

    Moderate Concern 8–15 ppm-h

    Significant Concern > 15 ppm-h

Spatial patterns in estimated cumulative ozone exposure were similar between the W126-3-mo and SUM06 indices (Figures 3.4 and 3.5). However, the areas classified as having high ozone exposure were somewhat more extensive for the SUM06 index as compared with the W126-3-mo index. Areas classified as having low ozone exposure were nearly identical using the two indices and included most of the Pacific Northwest, southern Texas, southern Florida, and the northern tier of states except around the eastern Great Lakes region.

Note that there are differences between Kohut's (2007a) risk ranking (Figure 3.3) and the more recent National Park Service estimates (Figures 3.4 and 3.5). In particular, many parks in the arid west were estimated by Kohut to be at relatively low risk even though the exposure indices were relatively high. This difference reflects the low soil moisture conditions in these parks. Thus, there is not always a clear relationship between exposure and risk; low soil moisture limits ozone uptake, thereby reducing risk. Nevertheless, it is important to recognize that riparian vegetation communities in these arid western parks may, in fact, have sufficiently high soil moisture that risk is enhanced in locally moist areas.

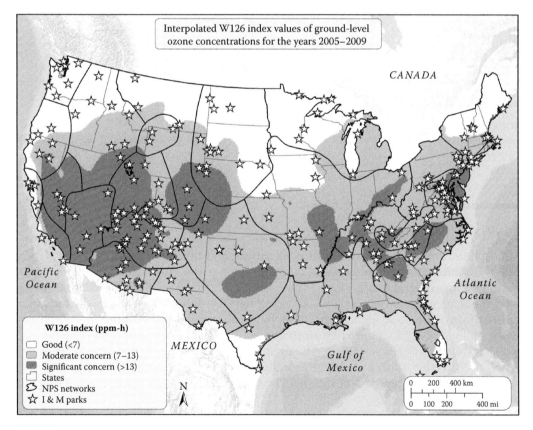

FIGURE 3.4
Interpolated values of cumulative ozone exposure during the five-year period 2005–2009 using the W126-3-mo exposure index.

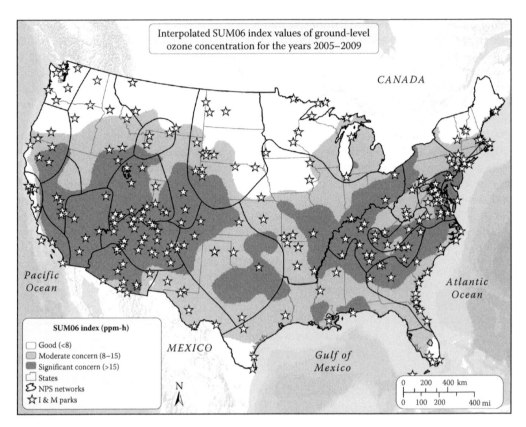

FIGURE 3.5
Interpolated values of cumulative ozone exposure during the five-year period 2005–2009 using the SUM06 exposure index.

3.5 Visibility Impairment

Atmospheric pollutants derived from both natural and human sources can degrade visibility and produce regional haze. In many national parks, this haze impairs the scenic vistas that are integral to a park or wilderness experience. Visibility degradation results from the scattering and absorption of visible light by gases and particles in the atmosphere. If there are no suspended particles in the air, the natural visibility is determined by the amount of light scattered by air molecules. Scattering of visible light from air molecules is called Rayleigh scattering. The light is scattered evenly in both the forward and backward direction, with a preferential scattering of shorter wavelengths; this scattering is responsible for the blueness of the sky. The value of the Raleigh scattering is a function of air density and elevation.

Visibility is a key air quality indicator. Haze can affect how far and how well we can see. National Park Service management policies prohibit the impairment of visibility in all national park units. The Clean Air Act set a specific goal for visibility protection in Class I areas:

the prevention of any future, and the remedying of any existing, impairment of visibility in mandatory Class I federal areas which impairment results from manmade air pollution (42 U.S.C. 7491).

Federal regulations require each state to develop a plan to improve visibility in Class I areas, with the goal of returning visibility to natural conditions in 2064. Natural background visibility assumes no human-caused pollution but varies with natural processes such as windblown dust, fire, volcanic activity, and biogenic emissions. Visibility is monitored by the Interagency Monitoring of Protected Visual Environments (IMPROVE) network and typically reported using the haze index measured in deciviews.*

Current visibility estimates reflect current pollution levels and also include natural background conditions, that is, conditions that would exist in the absence of human-caused pollution. Estimates were used to rank conditions at parks discussed in this book in order to provide park managers with information on spatial differences in visibility and air pollution. Rankings range from very low haze (very good visibility) to very high haze (very poor visibility). Only parks with on-site or representative IMPROVE monitors were used in generating the visibility ranking.

3.5.1 Sources of Visibility Degradation

Ambient visibility in the national parks is typically evaluated in the context of the natural background visibility at that location. The natural visibility condition is affected by wildland fire, volcanic emissions, wind, soil conditions, biogenic emissions from vegetation, and sea salt aerosols. Research is ongoing to further refine scientific understanding of these natural contributions to visibility impairment.

Particulate matter in the atmosphere, derived from both natural and human sources, scatters light. The amount of particle light scattering depends on the size and concentration of the particles, which are affected by their physical and chemical properties. Fine particles (particles less than 2.5 micrometers [µm] in diameter) have a greater scattering efficiency on a per mass basis than coarse particles (particles between 2.5 and 10.0 µm in diameter). Particles with sizes near the wavelength of visible light (0.4–0.7 µm) scatter light most efficiently. The chemical composition of particles in the air also plays a role in their relative light scattering efficiencies. The major categories of particulate chemical composition generally used to differentiate among fine particles that scatter light are sulfate, nitrate, organics, and soil. Sulfate and nitrate aerosols are hygroscopic. The growth of these aerosols at higher relative humidity conditions can dramatically enhance their effect on light scattering. Coarse particles are generally not speciated and are lumped together in terms of their scattering efficiency.

Sulfate and nitrate are secondary particulate matter species produced in the atmosphere, primarily from gaseous precursors. At sufficiently high humidity levels (relative humidity >85%), they are typically the most efficient particulate species contributing to haze (U.S. EPA 2009a). Particulate sulfate, as ammonium sulfate, is the dominant source

---

* The *deciview* visibility metric expresses uniform changes in haziness in terms of common increments across the entire range of visibility conditions, from pristine to extremely hazy conditions. Because each unit change in deciview represents a common change in perception, the deciview scale is like the decibel scale for sound. A one deciview change in haziness is a small but noticeable change in haziness under most circumstances when viewing scenes in Class I areas.

of regional haze in the eastern United States and is an important contributor to haze elsewhere in the country. Most atmospheric sulfate forms from gaseous emissions of sulfur dioxide from coal-fired power plants. In coastal areas, some sulfate is derived from sea spray.

Particulate nitrate is only a minor component of remote-area regional haze in the eastern United States and in the western United States outside California. It is, however, an important contributor to haze throughout most of California and in the upper midwestern United States, especially during winter. Both nitric acid (a reaction product of oxidized N emissions) and ammonia are needed to form ammonium nitrate. Urban particulate nitrate concentrations are significantly higher than remote-area background concentrations. Particulate ammonium nitrate concentrations in California and the Midwest are commonly an order of magnitude higher than estimated natural levels. Particulate nitrate concentrations are sensitive to changes in either oxidized N emissions (from a combination of human-caused mobile and point emissions sources) or ammonia emissions (principally from agricultural sources).

Elemental C and organic carbonaceous components of particulate matter are responsible for a large fraction of the haze that occurs in many western parks. This is especially the case in the northwestern United States (U.S. EPA 2009a). Both elemental C and organics are products of incomplete combustion of fuels, including gasoline and diesel emissions and smoke from wildland fire (Figure 3.6). Organic particulate matter is also produced by the atmospheric transformation of some precursor gaseous emissions. Smoke plume particulates from large wildfires dominate many of the worst haze periods in the western United States. Particulate matter from elemental and organic C is generally the largest component of urban excess fine particulate matter. Western urban areas have more than twice the average concentrations of carbonaceous particulate matter compared with remote areas in the same region. In eastern urban areas, fine particulate matter is dominated by sulfate and organic C components, though the usually high relative humidity in the East causes the hydrated sulfate particles to contribute about

FIGURE 3.6
Fire (prescribed, accidental, natural) is an important source of air pollutants, including coarse particulate matter, N oxides, and revolatilized Hg. This photo was taken of the 1988 fire in Yellowstone National Park. (National Park Service photo by Jim Peaco, 1988.)

twice as much of the urban haze as that caused by the carbonaceous particulate matter (U.S. EPA 2009a).

Both fine soil and coarse particles are significant contributors to haze in the arid southwestern United States where they typically contribute a quarter to a third of the haze. Coarse mass elements usually contribute twice as much haze as fine soil in this region. Coarse mass concentrations are also high in the central Great Plains. The relative contribution to haze by the high coarse mass in the Great Plains is smaller, however, because of the relatively high concentrations of sulfate and nitrate particulate matter in that region.

Dust is an important contributor to haze in the West. A comprehensive assessment (U.S. EPA 2009a) of the 610 worst sampled haze episodes over a three-year period in the western United States, where dust was the major contributor, categorized each site/ sample period into four causal groups: Asian dust, local windblown dust, transported regional windblown dust, and undetermined dust (i.e., not in one of the three other groups). Most dust days occurred at sites in Arizona, New Mexico, Colorado, western Texas, and southern California, and these were dominated by local and regionally transported windblown dust (U.S. EPA 2009a). Asian dust caused only a few of the worst dust days during the three-year assessment period, though it was an important source of dust for the more northern regions of the West and was responsible for 10%–40% of their worst dust periods.

Organic particulates in the atmosphere are diverse in their makeup and sources. Vegetation itself is an important source of hydrocarbon aerosols. Terpenes, released from tree foliage, may react in the atmosphere to form submicron particles. These naturally generated organic particles contribute significantly to the blue haze aerosols formed naturally over some forested areas (Geron et al. 2000, U.S. EPA 2004).

Some gases can also absorb light. Nitrogen dioxide is the only major visible light-absorbing gas in the lower atmosphere. It usually does not occur in sufficient concentrations in remote areas to make a major contribution to the total light absorption. Elemental C is the dominant light-absorbing particle in the lower atmosphere.

3.5.2 Visibility Metrics

To quantify visibility conditions and estimate the degree of degradation caused by various air pollutants, several visibility scales/metrics can be employed. Depending on the application, metrics may include a contrast and color difference index, light extinction coefficient, visual range, or degree of haziness.

The light extinction coefficient, often called extinction, represents the attenuation of light per unit distance of travel in some medium, such as air. It is commonly measured in inverse length (e.g., inverse kilometers [km^{-1}] or inverse mega meters [Mm^{-1}]). The extinction reflects attenuation of light due to scattering and absorption of light by aerosols/ particles or gases. The best possible visibility (extinction-free, or Rayleigh conditions) corresponds to an extinction of approximately 10 Mm^{-1}, although this value is spatially variable.

The extinction is not linear with respect to increases or decreases in perceived visual air quality. For example, a given change in extinction can result in a scene change either unnoticeably small or very apparent depending on the baseline visibility conditions. Therefore, another visibility metric, the haziness index (expressed as deciviews), was defined to index a constant fractional change in the extinction to visual perception (Pitchford and Malm 1994). The haze value in deciviews is a log-transformation of extinction that provides a perceptually uniform haze metric. The advantage of this characterization is that equal changes in deciview are equally perceptible to the human eye across different baseline conditions. Higher haziness index values signify poorer visibility. The EPA has adopted the deciview metric for determining progress in improving visibility under the Regional Haze Rule described in the following text. The deciview is often used in describing overall visibility condition or haze and in tracking changes in visibility. A 1-deciview change would be a small but likely perceptible change in uniform haze conditions, regardless of the baseline visibility level (Pitchford and Malm 1994).

Although the deciview value is widely used to describe haze conditions and monitor trends in visibility, extinction is the characterization most used by scientists concerned with the assessment of the causes of visibility degradation. Extinction can be directly calculated from light transmittance measurements (measured extinction) or derived from measured atmospheric particle concentrations using linear relationships between the concentrations of particles and gases and their contribution to the extinction coefficient (calculated extinction). Understanding these relationships provides a method of estimating how visibility would change with changes in the concentrations of each of these atmospheric constituents. This methodology, known as "extinction budget analysis," is important for assessing the visibility consequences of proposed pollutant emissions sources or for determining the extent of pollution control required to meet a desired visibility condition (Sullivan et al. 2003). Each haze-causing species in the atmosphere can be measured, and the amount of light extinction that it causes can be calculated at a given location and time.

Uptake of water by fine aerosol particles in the atmosphere has been investigated at remote natural settings, including within three national parks: Great Smoky Mountains, Grand Canyon, and Big Bend (Day and Malm 2001). Inorganic sulfate and ammonium salts can absorb considerable water in moist atmospheres. This uptake and release of atmospheric water in response to changes in relative humidity can influence the light scattering properties of aerosols (Horvath 1996, Tang 1996, Day et al. 2000). Light scattering can increase twofold or more at relative humidity above 80% (Rood et al. 1986, Day et al. 2000). Organic C is also an important vehicle for water uptake by atmospheric aerosols (Day and Malm 2001).

3.5.3 Visibility Monitoring

The IMPROVE Program is a cooperative monitoring effort that is governed by a steering committee comprising representatives from federal and regional-state organizations. It was established in 1985 to aid in the creation of federal and state implementation plans for the protection of visibility in mandatory Class I areas, as stipulated in the 1977 Clean Air Act Amendments. A complete IMPROVE monitoring station includes fine and coarse particle monitoring and optical monitoring and may have previously also included view monitoring with photography.

There are 187 sites included in the IMPROVE database (downloaded from http://vista.cira.colostate.edu/IMPROVE/Default.htm). Fifty-five of these sites are located in national park units, representing 53 individual Inventory and Monitoring parks. Average deciview values were calculated for some analyses presented in this book over a five-year period (2004–2008) for those IMPROVE sites that are located in or near the Inventory and Monitoring parks.

Effects on visibility vary spatially across the United States. Most impairment occurs as regional haze, which is generally most pronounced in the eastern United States and southern California. In the East, ammonium sulfate contributes at least half of the visibility impairment at most Class I areas, and ammonium sulfate is a major contributor to visibility impairment throughout the United States (Hand et al. 2011). The contribution from ammonium nitrate is highest in central and southern California and in parts of the Midwest. The contribution from organic C is highest in the Southeast. The contributions from coarse particles and fine soil are highest in the arid southwest (Debell et al. 2006, U.S. Forest Service et al. 2010, Hand et al. 2011).

Since the early 1990s, visibility throughout most remote areas in the conterminous United States has improved substantially. Hand et al. (2014) computed extinction on the haziest 20% of days following Regional Haze Rule guidelines for three major regions: East, Intermountain/Southwest, and West Coast. During the two-decade period, 1992–2011, the haze level on the regional mean 20% haziest days decreased by 52% in the East and 20% in the West Coast region and remained unchanged in the Intermountain/Southwest region. Improvements for the West Coast region accelerated to a –3.5%/yr decrease in extinction during the second decade of the monitoring period. Park-specific visibility condition and trends are available at http://www.nature.nps.gov/air/data/products/parks/index.cfm.

3.5.3.1 Particle Monitoring

Particle monitoring provides concentration measurements of atmospheric particles that contribute to visibility impairment. Four independent IMPROVE sampling modules are used to automatically collect 24-hour samples of suspended particles every three days by drawing in air and collecting suspended particles on filters. The filters are later analyzed to determine the chemical makeup of the suspended particles. Three of the four samplers (modules A, B, and C) collect fine particles with diameters <2.5 μm. The fourth sampler (module D) collects particles with diameters up to 10 μm. Module A filters are analyzed to determine the gravimetric mass and elemental composition of the collected particles. Module B filters are analyzed specifically for sulfate, nitrate, and chloride ions. Module C filters are analyzed for organic material and light absorbing C. The gravimetric mass of coarse particles (2.5–10.0 μm) is determined by subtracting the module A gravimetric mass from the module D gravimetric mass.

3.5.3.2 Optical Monitoring

Optical monitoring provides a direct quantitative measure of extinction that represents overall visibility conditions. Because water vapor in combination with suspended particles can affect visibility, optical stations also record temperature and relative humidity. Optical monitoring uses ambient long-path transmissometers or ambient nephelometers to collect hourly averaged data. Transmissometers measure the amount of light transmitted through the atmosphere over a known distance (between 0.5 and 10.0 km) between a light source of known intensity (transmitter) and a light measurement device (receiver). The transmission measurements are electronically converted to hourly averaged extinction (scattering plus absorption). Ambient nephelometers draw air into a chamber and measure the scattering component of light extinction. Optical measurements of extinction and scattering include meteorological events such as cloud cover and rain, but the data are filtered by flagging as invalid those data points collected under conditions of high relative humidity (>90%). This filtering process is assumed to remove the largest effects of weather from the dataset. Optical data also provide concurrent, independent measurement of extinction to compare with calculations made using particle data. Transmissometers have gradually been phased out of the IMPROVE program, due to their relatively high expense compared to nephelometers.

3.5.3.3 View Monitoring

View, or scenic, monitoring was formerly accomplished with automated camera systems. Cameras typically took three photographs a day (9:00, 12:00, and 15:00) of selected scenes, which formed a photographic record of characteristic visibility conditions. These photographs reveal how differences in visibility affect vistas in a particular park. Based on April 1995 recommendations of the IMPROVE Steering Committee, view monitoring was discontinued at all National Park Service Class I areas that had a five-year or greater photographic monitoring record. Webcams have been installed at about 20 parks and are linked to air quality monitoring stations (http://www.nature.nps.gov/air/WebCams/index.cfm).

Photos of vistas within many of the Inventory and Monitoring parks were downloaded for this book from http://vista.cira.colostate.edu/IMPROVE/Data/IMPROVE/Data_IMPRPhot.htm. A series of photos representing the best 20%, worst 20%, and annual average visibility were selected and are included as plates associated with some of the case study chapters. If a photo was not available for the exact deciview value needed to correspond with average haze measurements over the period 2004–2008 within a given park, the closest deciview photo was selected. All photos in the series were taken at the same time of day.

3.5.3.4 Regional Haze Rule

In 1977, Congress established a national goal of no human-caused visibility impairment in Class I areas by 2064 and in 1999 promulgated a rule requiring states to develop and implement plans to make continuous progress toward that goal. The EPA provided detailed guidance for assessing regional haze (U.S. EPA 2001a). The assessment of progress toward the national goal requires long-term particle monitoring on which to base estimates of natural* and current visibility conditions. This long-term monitoring is provided by the IMPROVE Program.

The Regional Haze Rule requires states (and tribes who choose to participate) to review how pollution emissions within the state affect visibility in Class I areas. The Regional Haze Rule also requires states to make "reasonable progress" in reducing any effect that air pollution has on visibility conditions in Class I areas and to prevent future impairment of visibility. The states are required by the rule to analyze a linear pathway that takes the Class I areas from current conditions to "natural conditions." The response to these regulations, while aimed at Class I areas, is expected to improve regional visibility conditions throughout the country.

The Regional Haze Rule requires emissions reductions that will reduce the human-caused contributors to regional haze at 156 of the largest national parks, wilderness areas, and national wildlife refuges in the United States. The long-term goal is to achieve visibility that represents natural conditions, unimpacted by human-caused emissions, by the year 2064. In order to comply with requirements of the Regional Haze Rule, multistate

---

* "Natural condition" is a term used in the Clean Air Act, which means that no human-caused pollution can impair visibility.

planning organizations were established to conduct the technical analysis used by state air agencies, as follows:

- Western Regional Air Partnership
- Central Regional Air Planning Association
- Midwest Regional Planning Organization
- Mid-Atlantic/Northeast Visibility Union
- Visibility Improvement State and Tribal Association of the Southeast

Each state must describe ambient visibility conditions, based on the best and worst visibility days for the five baseline years 2000–2004. Because the IMPROVE sampler operates every 3 days, there are data for about 121 days/yr at each site. The 20% highest haze days (approximately 24 haziest days during each year) and the 20% lowest haze days (24 clearest days) are averaged for each baseline year. Each state also must demonstrate that the 20% least impaired days do not degrade from the baseline on days when visibility is mainly affected by natural emissions sources.

Under the Regional Haze Rule, states must estimate the uniform rate of haze reduction that would allow them to progress from the current conditions to the required natural haze level in 2064. Haze conditions for this requirement are expressed as the five-year mean of the annual 20% most impaired visibility conditions at each protected Class I area. States also must document that the five-year mean of the annual 20% clearest days does not deteriorate over time at the protected areas.

In order for states to comply with the Regional Haze Rule, they must estimate haze levels at Class I areas that represent the average of the 20% haziest days so that the 60-year uniform rate of progress can be specified. A default approach was described by the EPA (2001a) to accomplish that. It was based on the original IMPROVE algorithm (Sisler 1996) for estimating extinction. Thus, the EPA provided default estimates of natural background visibility for the mean, clearest, and haziest days for each of the protected areas. States could choose to use these default values or choose to use a refined approach.

The default approach specified by the EPA for estimating natural haze conditions was subsequently reviewed and found to be flawed in multiple respects (Ryan et al. 2005, Pitchford et al. 2007). Problems were identified in the method used to calculate percentiles, and the default approach did not adequately adjust for light scattering by sea salt at coastal locations or the effects of elevation.

The IMPROVE algorithm was revised in 2005 to address some of these problem areas. The use of the revised algorithm for estimating ambient haze also requires the use of the same algorithm for estimating natural conditions in order to avoid biasing the calculation of the required uniform rate of progress glide path needed to accomplish the visibility improvements mandated in the Regional Haze Rule. The revised approach is not without problems, however. For example, it assumes two estimates of natural haze-causing atmospheric species concentrations, one for the East and one for the West. As a consequence, there is a rather steep line of demarcation between the East and the West in the background haze estimates. There may not be sufficient data to justify breaking down the country into smaller units at this time. Additional research may be needed to further improve the approach for specifying natural visibility conditions.

Trend analyses show that over short (2000–2008) and long (1989–2008) time periods, the mass concentrations of major aerosol species have decreased at many IMPROVE sites.

These decreasing trends (improving visibility) were largest for the lowest concentrations and during winter seasons and may not be typical of all sites and species (Hand et al. 2011). However, these results are consistent with 1999–2008 trends at National Park Service sites, which showed larger decreasing trends in deciview on the clearest days compared to the haziest days (NPS 2010).

3.5.3.5 Spatial Patterns in Visibility Impairment

A regional haze dataset was downloaded for the analysis provided in this book from http://vista.cira.colostate.edu/IMPROVE/Default.htm, based on the revised IMPROVE algorithm. Out of this dataset, measured haze values for the years 2004–2008 were extracted. Five-year means of the clearest, haziest, and average 20% visibility days were summarized for the following light extinctions: ammonium sulfate, ammonium nitrate, organic mass, light absorbing C, fine soil, coarse mass, and sea salt.

Relative ranking of haze measurement values, expressed in deciviews, from the IMPROVE sites is provided in Figures 3.7 through 3.9 as five-year averages over the period 2004–2008. Only sites having monitoring data during the period 2004–2008 are included.

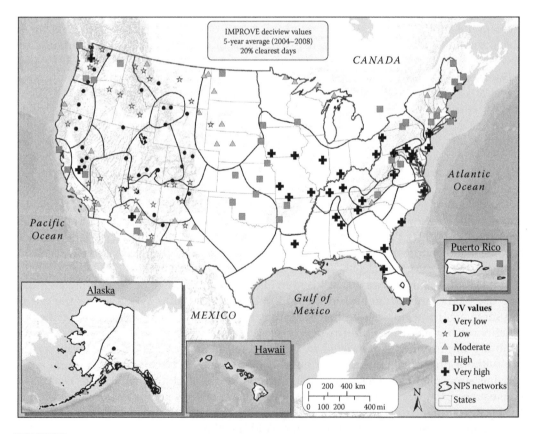

FIGURE 3.7
Relative ranking of haze measurements (in deciviews [dv]) at IMPROVE sites for the 20% clearest days during the five-year period 2004–2008. Low deciview values correspond to better visibility; as deciview increases, visibility decreases.

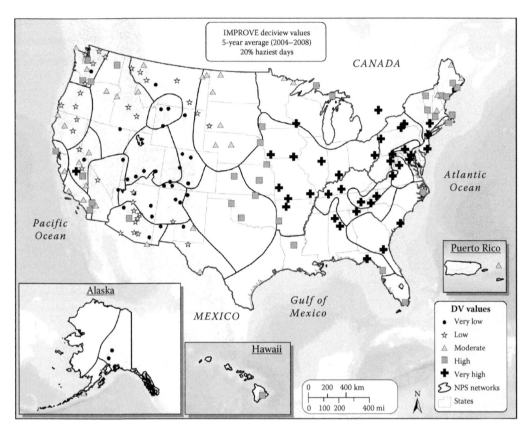

FIGURE 3.8
Relative ranking of haze measurements (in deciviews [dv]) at IMPROVE sites for the 20% haziest days during the five-year period 2004–2008. Low deciview values correspond to better visibility; as deciview increases, visibility decreases.

Results are presented for the 20% of the monitored days having the lowest haze (clearest visibility) at each site (Figure 3.7), the 20% of the days having highest haze (Figure 3.8), and the average 20% days (Figure 3.9). Ambient haze measurements must be interpreted within the context of estimated natural haze levels. In general, parks that exhibited high levels of ambient haze were also estimated to have relatively high natural haze levels.

In order to rank measured haze values into five classes ranging from very low to very high, deciview values for each group were divided using a geometrical interval. This classification scheme is useful for ranking continuous datasets that are not normally distributed. It creates intervals by minimizing the variance among values within each class and results in classes that contain approximately the same number of parks. The ranking for most sites was based on five years of data.

Extinction budgets are provided in the case study chapters of this analysis for each of the park units that contains an IMPROVE monitoring site or has a nearby representative site having data for the time period 2004–2008. The contribution of each aerosol type to extinction is shown relative to particulate light extinction, excluding Rayleigh (blue sky) scattering. This approach was taken because the focus of this book is on the components of haze, rather than on all contributors to extinction.

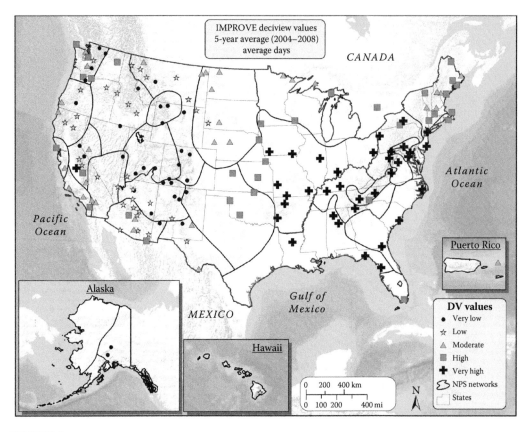

FIGURE 3.9
Relative ranking of haze measurements (in deciviews [dv]) at IMPROVE sites, based on the average days at each site. Low deciview values correspond to better visibility; as deciview increases, visibility decreases.

3.6 Effects of Exposure to Airborne Toxics

Just as for acid precursors and nutrients, atmospheric emissions of toxic substances can affect resources in parks that are downwind of emission sources. Researchers have become increasingly concerned about the potential susceptibility of resources in the national parks to airborne toxic contaminants. These chemical pollutants can be transported by the wind and deposited on land far from where they originate. Relevant toxic chemicals include Hg, various pesticides, and industrial contaminants. Some have the potential to bioaccumulate to relatively high concentrations at upper trophic levels of the food web. Combustion products, including metals and polycyclic aromatic hydrocarbons, may also be important atmospherically deposited toxic and/or carcinogenic (cancer-causing) compounds. Adverse effects on sensitive animal species can include reproductive impairment, reduced growth and health, neurological damage, and behavioral changes. Humans consuming contaminated fish and wildlife may also be affected.

3.6.1 Toxic Substances of Concern for This Book

The most thoroughly studied airborne toxin that affects national parklands is Hg. It is a heavy metal that occurs naturally in several forms. Elemental Hg is released from sources in the earth's crust into the global environment through volcanic and geothermal activity and the weathering of rocks. It is also being emitted by human emissions sources, primarily coal-fired power plants and incinerators. The most important route of entry of Hg into remote national park ecosystems is atmospheric deposition of inorganic Hg (Fitzgerald et al. 1998).

Once released into the atmosphere, Hg can be deposited to the earth's surface and transformed through natural processes into a toxic form, methyl Hg (Ullrich et al. 2001), that can bioaccumulate in food webs. Contamination of aquatic ecosystems by Hg has prompted many states to issue fish consumption advisories and has reduced the benefits provided by fisheries resources in many inland and coastal waters throughout the United States.

A variety of factors influence Hg deposition, fate, and transport. Such factors relate in particular to speciation of the Hg that is emitted, the forms in which it is deposited from the atmosphere, and transformations that occur within the atmosphere and within the aquatic, transitional, and terrestrial compartments of the receiving watersheds (http://www3.epa.gov/mats/actions.html).

Persistent organic pollutants constitute another class of potentially toxic contaminants. Many are semivolatile and therefore are readily emitted into the atmosphere and subsequently deposited from the atmosphere to the ground surface. They are not efficiently degraded in the environment and can be concentrated in the food web toward higher trophic levels. They can accumulate to toxic levels in piscivorous fish and wildlife, wading birds, and humans. Results of persistent organic pollutant biomagnification can include reproductive disruption, neurological disorders, and mortality.

Many persistent organic pollutants have high toxicity, are relatively insoluble in water, and tend to concentrate at each level of the food web (Landers et al. 2010). These compounds include PCBs, dioxins, dichlorodiphenyltrichloroethane (DDT) and its breakdown products, chlordane, dieldrin, endosulfan, polycyclic aromatic hydrocarbons, and others. Although the use of many of these compounds is now restricted, their presence in the environment has human health, environmental, and economic impacts (Clair et al. 2011). Flame retardants (particularly for household goods), developed and deployed to respond to stricter fire regulations, are also of increasing concern. For example, levels of polybrominated diphenylether flame retardants have increased dramatically in smelt in the Great Lakes (Chernyak et al. 2005).

3.6.2 Biomagnification

Biomagnification is the progressive accumulation of chemicals with increasing trophic level (LeBlanc 1995). It describes the process of chemical accumulation in the food web. Pollutants that biomagnify accumulate in body fat or muscle tissue. Organic (methylated) Hg is the most likely metal to biomagnify, in part because organisms can efficiently assimilate methyl Hg and it is slowly eliminated (Reinfelder et al. 1998, Croteau et al. 2005).

Atmospheric emissions of semivolatile organic compounds and metals are of interest primarily because of their detrimental effects on biota, particularly the more sensitive species that can be found in remote national parks. The WACAP study (Landers et al. 2008) showed significant biomagnification of these pollutants in all parks studied; fish showed semivolatile organic compound concentrations that were five to seven orders of magnitude higher than concentrations in abiotic components sampled (air, snow, and water). Concentrations of polycyclic aromatic hydrocarbons, current-use pesticides, and hexachlorocyclohexanes tended to be highest in vegetation, whereas fish tissue accumulated mostly PCBs, chlordanes, DDT, and dieldrin.

The bioavailability of methyl Hg in the environment is also of particular importance to humans who can accumulate high levels of methyl Hg from eating contaminated fish. Avoiding exposure to Hg is particularly important for pregnant women and nursing mothers. Mercury poisoning can cause neurological impairment in children, leading to effects on memory, visual and spatial ability, information processing, and general intelligence (Mahaffey 2005).

Because atmospheric Hg must be methylated in order to bioaccumulate in the food web, significant efforts have been made to understand the bacteria responsible for this chemical transformation. Sulfate-reducing bacteria have been shown to be the main agents of Hg methylation in sediments (Gilmour et al. 1992) and wetlands (St. Louis et al. 1994, Branfireun et al. 1999), although Fe-reducing bacteria can also be important. In ecosystems that are sulfate-poor, the total amount of biologically available S controls the activity of these bacteria and thereby in part the rate of methyl Hg production. Atmospheric sulfate deposition, leading to the stimulation of sulfate-reducing bacteria and increased Hg methylation, may have caused or contributed to the observed postindustrial amplification of methyl Hg concentrations in fish (Jeremiason et al. 2006). Studies have demonstrated increased levels of methyl Hg production with experimental addition of sulfate (Gilmour and Henry 1991, Gilmour et al. 1992, Branfireun et al. 1999, Jeremiason et al. 2006). In addition, fish have shown increased methyl Hg burdens in acidified lakes as compared with nonacidified lakes receiving similar atmospheric Hg deposition (Gilmour and Henry 1991). Therefore, reduced sulfate emissions and deposition may be expected to cause a corresponding reduction in Hg methylation and bioavailability (Jeremiason et al. 2006).

The WACAP analyzed salmonid fish tissue for contaminants and biomarkers to assess fish condition. Vitellogenin concentration was also analyzed as a biomarker indicating exposure to estrogen and estrogen-mimicking chemicals. Many semivolatile organic compound contaminants have been shown to be endocrine-disrupting compounds that interfere with natural hormonal regulation (e.g., dieldrin, PCBs, endosulfan, DDT, polycyclic aromatic hydrocarbons, etc.). The levels of vitellogenin in male fish can constitute a helpful biomarker that indicates exposure to these contaminants.

In order to evaluate possible human health concerns due to biomagnified contaminant concentrations in fish, the WACAP study measured contaminant concentrations in 136 fish from 14 lakes in the eight core parks included in the study and compared their findings with the EPA's *Guidance for Assessing Chemical Contaminant Data for Use in Fish Advisories.* They found that over half of study fish (77) in 11 of the lakes exceeded subsistence fishing human health concentration thresholds for the contaminants dieldrin and/or the DDT breakdown product *p,p*-dichlorodiphenyldichloroethylene (DDE). Nine of the lakes demonstrated mean fish dieldrin and *p,p*-DDE that exceeded human contaminant health thresholds for subsistence fishing. These EPA thresholds were not exceeded by any other semivolatile organic compound contaminant concentrations in fish.

High concentrations of contaminants in fish can be detrimental to piscivorous wildlife. The Hg burdens in fish at many WACAP sites exceeded published health thresholds for mink (*Mustela vison*), river otter (*Lutra canadensis*), and belted kingfisher (*Megaceryle alcyon*). Although the production of DDT was banned in the United States in 1972, concentrations of the DDT and its by-products in fish remained above the health threshold for kingfisher at two sites in Sequoia/Kings Canyon National Park and one in Glacier National Park.

Dieldrin concentrations in fish did not exceed the wildlife health thresholds in any of the parks. Dieldrin production in the United States was banned in 1987, following a 1974 ban of the pesticide for agricultural use. Rocky Mountain National Park contained fish with the highest dieldrin concentrations of any park. This may have resulted from industrial production of dieldrin in Denver between 1952 and 1982 (Landers et al. 2010).

For lakes studied in WACAP, fish in Burial Lake in the arctic Noatak National Preserve bioaccumulated the highest concentration of Hg. The other arctic lake sampled (Matcharak Lake) also contained fish with comparatively high Hg concentrations. Snow, sediments, and vegetation in these parks showed low Hg concentrations, suggesting that the high Hg concentrations found in fish were a result of other characteristics of the watershed, such as efficient transformation of atmospheric Hg into methyl Hg. These two arctic lakes both showed fish with Hg concentrations that exceeded the EPA consumption criteria.

Plant foliage can accumulate elemental Hg over time in response to air and soil exposure (Ericksen et al. 2003, Frescholtz et al. 2003). Although this Hg does not harm the plants, it does affect Hg cycling and bioavailability. Foliar/air Hg exchange has been shown to be dynamic and bidirectional (Millhollen et al. 2006). These investigators compared foliar Hg accumulation over time in three tree species with fluxes measured using a plant gas-exchange system subsequent to soil amendment with mercury chloride. Root tissue Hg concentrations were strongly correlated with soil Hg concentrations, suggesting that below-ground accumulation of Hg by roots may be an important process in the biogeo-chemical cycling of Hg in soil systems. Nevertheless, measured foliar Hg fluxes indicated that the deposition of atmospheric Hg constituted the dominant flux of Hg to the leaf surface (Millhollen et al. 2006).

Persistent organic pollutants can move in successive steps of evaporation, atmospheric transport, and deposition toward colder areas. This is called cold condensation; it can be responsible for high concentrations of persistent organic pollutants in remote arctic and alpine regions (Simonich and Hites 1995, Blais et al. 1998), including in some western national parks. These persistent organic pollutants can then accumulate and biomagnify at higher levels of the food web, eventually potentially impacting fish and wildlife reproduction, behavior, growth, and mortality.

3.6.3 Effects of Atmospherically Deposited Toxic Substances

3.6.3.1 Effects on Fish

Increased body burdens of Hg in fish can lead to decreased reproductive success and behavioral alterations. Yellow perch (*Perca flavescens*) are widely distributed and are often studied as a primary indicator species for Hg contamination in the environment. Ecological effects thresholds for Hg concentration in fish prey have been proposed at levels lower than human health thresholds.

Mercury concentrations in fish were generally lower at WACAP sites in the western United States than in fish from lakes previously sampled in the midwestern and northeastern United States. Nevertheless, concentrations in fish from some WACAP lakes exceeded the EPA thresholds for human consumption.

Histopathological biomarkers provide evidence regarding the effects of toxins on individual organisms. One such biomarker is the macrophage aggregate, a focal accumulation of macrophages in the spleen, kidney, or liver of fish that indicates an immune system response to environmental conditions. The macrophage aggregates are formed in response to tissue damage (Schwindt et al. 2008). Significant correlations have been found between increased macrophage aggregates in fish from polluted waters and fish exposed to Hg or other metals (Handy and Penrice 1993, Meinelt et al. 1997, Manera et al. 2000, Fournie et al. 2001, Khan 2003, Capps et al. 2004). Schwindt et al. (2008) demonstrated an association between Hg and trout kidney and spleen tissue damage, as indicated by increased macrophage aggregate occurrence. This research was conducted on four species of trout collected from 14 lakes in the WACAP national parks or preserves in the western United States. This result suggests that Hg and/or other contaminants might adversely impact fish that inhabit remote and protected lakes in western national parks. Some "intersex" fish (male and female reproductive structures in the same fish) were found in Rocky Mountain National Park and Glacier National Park, but not in any of the other six parks sampled in the WACAP study. The causes and consequences of this reproductive disruption remain to be determined (Schwindt et al. 2009).

3.6.3.2 Effects on Humans

The EPA conducted the National Study of Chemical Residues in Lake Fish Tissue, a screening-level survey of chemical residues in fish throughout the conterminous United States (U.S. EPA 2009b). Sampling sites were selected using a random approach, and results are therefore applicable to numbers and percentages of lakes regionally and nationally. The focus was on identifying chemical concentrations in fish that were above levels of potential concern for humans and piscivorous wildlife (Table 3.4). Fish tissues were analyzed for 268 persistent, bioaccumulative, and toxic chemicals, including Hg, PCB congeners, dioxins and furans, and a variety of pesticides and other semivolatile organic compounds (U.S. EPA 2009b). These pollutants have been delivered to fish in the sampled lakes from both atmospheric and nonatmospheric sources. Over a sampling period of four years, 486 predator fish (fillets) and 395 bottom dwellers (whole body analysis) were analyzed from 500 sampling locations. Results were extrapolated to an estimated 76,559 lakes for predators and 46,190 lakes for bottom dwellers.

Mercury and PCBs were detected in fish collected from all sampled lakes; dioxins and furans were detected in 81% of the predator fish samples and 99% of the bottom-dweller samples. Other contaminants in fish were not as widely distributed. Analyses of predator fish fillets indicated that nearly half (36,422 lakes) of the sampled population had Hg tissue concentrations that exceeded the Hg human health screening value of 0.3 ppm. Nearly 17%

TABLE 3.4

Estimated Numbers and Percentages of Lakes, from among 76,559 Target Lakes in the Conterminous United States That Had Predator Fish Species Having Chemical Concentrations above Screening Values for Protecting Human Health, Based on a Survey by the EPA of Fish in 500 Lakes

		Lakes above Human Health Screening Values	
Chemical	Screening Value	Percent of Population	Number of Lakes
Mercury	0.3 ppm	48.8	36,422
PCBs	12 ppb	16.8	12,886
Dioxins and Furans	0.15 parts per trillion	7.6	5,356
DDT	69 ppb	1.7	1,329
Chlordane	67 ppb	0.3	235

of the sampled population had PCB compounds tissue concentrations above the 12 ppb human health screening value. Smaller percentages of the sampled lakes exceeded screening values of dioxin and furan concentrations, DDT, and chlordane.

Flanagan-Pritz et al. (2014) analyzed fish pesticide data collected in 14 national parks in Alaska and the western United States. The analysis found that concentrations of some historic-use pesticides exceeded the EPA guidelines for human subsistence fish consumers at 13 of 14 parks. Eagles-Smith et al. (2014) examined over 1400 fish across 21 national parks in the western United States. They found that total Hg concentrations in 68% of the fish sampled were above exposure levels recommended by the Great Lakes Advisory Group as being unsuitable for unlimited consumption by humans.

All 50 states issued Hg advisories for human fish consumption in 2008 (Fenn et al. 2011). The U.S. Environmental Protection Agency (2001b) recommended a human health criterion of 0.3 ppm in fish and shellfish tissue to protect the general population. More stringent restrictions may be appropriate for women of child-bearing age and children. Some states, including Maine and Minnesota, use a more restrictive human health standard of 0.2 ppm in fish.

3.6.3.3 *Effects on Piscivorous Wildlife*

Methyl Hg causes damage to the vertebrate central nervous system. In addition, low-level dietary Hg exposures that cause no measureable effects on adult birds can impair egg fertility, hatchling survival, and reproductive success (Scheuhammer 1991). In reproducing females of fish, birds, and piscivorous mammals, methyl Hg passes directly to developing egg or embryo (Evers et al. 2003, Hammerschmidt and Sandheinrich 2005, Heinz et al. 2010). These early life stages are more sensitive than adults to the adverse effects of methyl Hg exposure (Evers et al. 2003, Wiener et al. 2003, Scheuhammer et al. 2007).

Eagles-Smith et al. (2014) found that in 21 western U.S. national parks, total Hg concentrations in 35% of sampled fish exceeded a benchmark for risk to highly sensitive fish-consuming birds. Flanagan-Pritz et al. (2014) found that historic-use pesticides exceeded the EPA's guidelines for wildlife (kingfisher) health thresholds at 13 of 14 parks in the western United States and Alaska.

Reproductive effects on fish-eating birds have been reported at fish Hg levels of 0.16 ppm (Fenn et al. 2011). In piscivorous birds, including loons (*Gavia* spp.) and bald eagles (*Haliaeetus leucocephalus*), Hg poisoning can lead to brain lesions, reduced reproductive success, increased chick mortality, spinal cord collapse, and neuromuscular problems. The common loon (*Gavia immer*) has been widely used as a Hg bioaccumulation indicator for risk to piscivorous birds (Evers et al. 2008, 2011a). It is listed as threatened in Michigan and as a species of special concern in New York and Wisconsin. Because loons feed almost exclusively on fish and crayfish and are relatively long-lived, they can bioaccumulate a substantial amount of Hg (Evers et al. 2011b). Nevertheless, common loons are considered less sensitive to Hg than some other piscivorous birds. They do, however, concentrate Hg in their blood to levels that are high enough to impair the reproduction at some locations (Burgess and Meyer 2008, Evers et al. 2011a). Belted kingfisher has also been used as an indicator of Hg pollution. Kingfishers are piscivorous and widely distributed, occurring in many national parks, including many of the WACAP parks. Lazorchak et al. (2003) developed fish Hg contaminant thresholds for kingfishers; these thresholds were exceeded in nearly all fish in all WACAP lakes (Landers et al. 2010). Average concentrations of Hg in fish in many parks studied by WACAP exceeded risk thresholds for health impacts to fish-eating birds and mammals (Landers et al. 2010).

Mink has been proposed as a sentinel species indicating toxic contaminant exposure (Basu et al. 2007, Martin et al. 2011). This species constitutes a good candidate for biomonitoring due to its wide distribution and abundance, upper level trophic status, and availability of tissue samples from trappers (Mason and Wren 2001). River otter is also widely distributed and a good indicator of Hg contamination. Hg levels in fish from WACAP lakes often exceeded the health thresholds (Lazorchak et al. 2003) for both mink and river otter (Landers et al. 2010).

As more research has been conducted during the last few decades, high concentrations of Hg have been increasingly documented in more species of wildlife, especially across the Great Lakes region and into the northeastern United States. Recent research has also decreased the estimates of effects levels, suggesting sublethal effects on wildlife (including effects on reproduction and biochemical processes) at whole-fish concentrations of 0.2–0.3 ppm (Beckvar et al. 2005, Dillon et al. 2010, Sandheinrich and Wiener 2011).

3.6.4 Spatial Patterns in Exposure to Toxic Substances

3.6.4.1 Mercury

Spatial patterns in Inventory and Monitoring park sensitivity to potential damage from Hg exposure across the United States can be evaluated, in part, using maps of interpolated wet Hg deposition. Wet deposition data are currently available from the Mercury Deposition Network at monitoring sites across the United States. The spatial coverage of the monitoring data is too sparse in some regions to allow rigorous interpolation of site-specific measurements, but the Mercury Deposition Network does provide interpolations for much of the United States. These data reflect only wet deposition. Dry Hg deposition data are not available regionally. To evaluate likely patterns in atmospheric Hg deposition across the Inventory and Monitoring parks, wet Hg deposition estimates are presented in Figure 2.4 for the year 2013 from the Mercury Deposition Network. In general, wet deposition of Hg is highest in Florida, the Gulf States, and north into the Heartland. High deposition also occurs in mountainous areas, including the Rockies, Sierra Nevada, and Cascades, reflecting the higher levels of precipitation in those areas. Wet Hg deposition tends to be lower in the West (at lower elevations), Upper Midwest, Northeast, and Mid-Atlantic regions. Although dry deposition estimates are not available for most areas, limited research indicates that dry Hg deposition can equal or exceed wet deposition, especially in arid regions. In areas of New Mexico, estimated dry Hg deposition was about 60% of total (wet + dry) deposition (Caldwell et al. 2006).

The U.S. Geological Survey (last modified February 20, 2015) predicted surface water methyl Hg concentrations for all hydrologic units included in a National Park Service Inventory and Monitoring Program unit where sufficient empirical data were available to generate a prediction. Predictions were generated using a partial least squares regression model, with input parameters selected based on literature consensus. Partial least squares is a shrinkage method in which a continuous portion of the information contained in linear combinations of the variables (components) is used for regression. Twelve datasets representing water-quality sampling sites from across the conterminous United States were compiled representing various projects having surface water data for pH, sulfate, and organic C together with methyl Hg. Methyl Hg analyses for all data were performed by the Wisconsin Mercury Research Laboratory. Predictions of methyl Hg were made using regression on four selected independent variables: pH, sulfate, total organic C, and watershed percent wetland. Leverage calculations indicated that total organic C was most important relative to the other parameters in describing the variability of methyl Hg concentration. Predicted methyl Hg concentrations were obtained by applying the prediction dataset to the partial least squares model. Total Hg concentrations were not included in the calibration of the model. Therefore, this model represents only the potential sensitivities of ecosystems to surface water methyl Hg contamination providing that sufficient quantities of inorganic Hg are available for methylation.

Predicted methyl Hg concentrations in surface waters varied across the conterminous United States, based on U.S. Geological Survey (2015) estimates for eight-digit hydrologic unit codes containing Inventory and Monitoring parklands (Figure 3.10). Geographic hot spots containing relatively high estimates of methyl Hg occurred throughout the southeast coastal region, upper midwest, intermountain and desert southwest, and northern New England. Estimated methyl Hg concentrations tended to be lower in the interior Mid-Atlantic region, northern Rocky Mountains and Great Plains, and Pacific Northwest. The results of this analysis should be considered a first approximation of risk. Sampling of biota in areas of high predicted methyl Hg can confirm whether bioaccumulation in the food web is occurring.

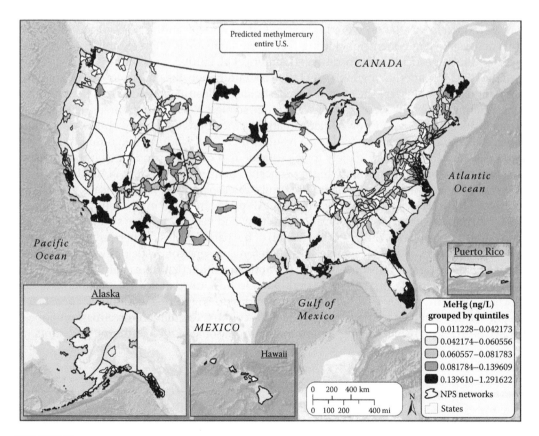

FIGURE 3.10
Predicted methylmercury (MeHg) in nanograms per liter (ng/L) concentrations in surface waters throughout the United States by hydrologic unit codes that contain national parklands. Estimates were generated by the U.S. Geological Survey (last modified February 20, 2015). Rankings are based on quintile distributions across all Inventory and Monitoring parks having estimates by the U.S. Geological Survey.

3.6.4.2 *Semivolatile Organic Compounds*

The WACAP study focused on semivolatile organic compounds and metals, largely because of their capacity for atmospheric transport over long distances. Semivolatile organic compounds can undergo repeated volatilization on surfaces, such as plant foliage, in response to daily changes in temperature. As a consequence, such compounds can be deposited, reemitted, and redeposited multiple times. This behavior can cause these compounds to move large distances in a leap-frog fashion. The semivolatile organic compounds were classified into four groups: current-use pesticides, North American historic-use pesticides, industrial/urban-use compounds, and combustion by-products. WACAP identified and analyzed over 100 semivolatile organic compounds representing various levels of volatility, solubility in water, hydrophobocity, and persistence in the environment. In each of the core parks studied, a variety of ecosystem compartments were sampled, including air, snow, water, sediments, lichens, conifer needles, and fish, to identify concentrations and potential biological impacts of contaminants. Air, lichens, and conifer needles were analyzed in 12 secondary parks. WACAP also used back-trajectory cluster analysis in the core parks to spatially model atmospheric transport of these air pollutants (Landers et al. 2010).

Within the WACAP monitoring effort, the highest pesticide concentrations in snow were found in the Rocky Mountains and the California Sierra Nevada Mountains. Pesticide concentrations in vegetation were highest in California and in the Rocky Mountains; for other ecosystem components, including fish, these same areas often exhibited the highest measured total pesticide concentrations (Landers et al. 2010).

At the WACAP parks having high vegetation contamination, the major semivolatile organic compound pollutants were the current-use pesticides endosulfan and dacthal. Contaminant concentrations generally increased with elevation, so high-elevation areas in parks may be at extra risk for contamination; polycyclic aromatic hydrocarbons were an exception, decreasing at higher elevation. This may have been a result of increased wildfires and human activity at lower elevation. Dieldrin, an acutely toxic pesticide that causes reproductive impairment in fish, was measured at significantly higher concentrations in fish at some WACAP sites than in comparable ecosystems in Canada. Proximity of cropland was a statistically significant indicator of the presence of pesticides in WACAP parks. In parks in the contiguous United States, current-use pesticide concentrations in snow and vegetation were strongly correlated with percent of cropland within 150 km (93 mi) of the park. In Alaska, there are no croplands within 150 km of the WACAP sample sites; therefore, presence of pesticides was assumed to result from long-range atmospheric transport and deposition (Landers et al. 2010).

Historic-use pesticides that are now prohibited in the United States were found at higher concentrations in the lower 48 states than in Alaska, suggesting continued persistence, despite the fact that they are no longer used in this country. Regulations on emissions appear to have been effective in reducing contamination from some of these airborne pollutants.

References

Aber, J.D., K.J. Nadelhoffer, P. Steudler, and J.M. Mellilo. 1989. Nitrogen saturation in northern forest ecosystems. *BioScience* 39(6):378–386.

Achermann, B. and R. Bobbink. 2003. Empirical critical loads for nitrogen. 327. Environmental Documentation No. 164. Swiss Agency for the Environment, Forests, and Landscape (SAEFL), Berne, Switzerland.

Andersen, C.P., W.E. Hogsett, R. Wessling, and M. Plocher. 1991. Ozone decreases spring root growth and root carbohydrate content in ponderosa pine the year following exposure. *Can. J. For. Res.* 21:1288–1291.

Andersen, C.P. and P.T. Rygiewicz. 1991. Stress interactions and mycorrhizal plant response: Understanding carbon allocation priorities. *Environ. Pollut.* 73:217–244.

Andersen, C.P. and P.T. Rygiewicz. 1995. Allocation of carbon in mycorrhizal *Pinus ponderosa* seedlings exposed to ozone. *New Phytol.* 131:471–480.

Andersen, C.P. and C.F. Scagel. 1997. Nutrient availability alters below-ground respiration of ozone-exposed ponderosa pine. *Tree Physiol.* 17:377–387.

Arens, S.J.T., P.F. Sullivan, and J.M. Welker. 2008. Nonlinear responses to nitrogen and strong interactions with nitrogen and phosphorus additions drastically alter the structure and function of a high arctic ecosystem. *J. Geophys. Res.* 113:G03509.

Bailey, S.W., S.B. Horsley, R.P. Long, and R.A. Hallett. 1999. Influence of geologic and pedologic factors on health of sugar maple on the Allegheny Plateau, U.S. *In* S.B. Horsley and R.P. Long (Eds.). *Sugar Maple Ecology and Health: Proceedings of an International Symposium.* USDA Forest Service, Radnor, PA, pp. 63–65.

Baron, J.S. 2006. Hindcasting nitrogen deposition to determine ecological critical load. *Ecol. Appl.* 16(2):433–439.

Baron, J.S., C.T. Driscoll, J.L. Stoddard, and E.E. Richer. 2011. Empirical critical loads of atmospheric nitrogen deposition for nutrient enrichment and acidification of sensitive US lakes. *BioScience* 61(8):602–613.

Basu, N., A.M. Scheuhammer, S. Bursian, K. Rouvinen-Watt, J. Elliott, and H.M. Chan. 2007. Mink as a sentinel in environmental health. *Environ. Res.* 103:130–144.

Beckvar, N., T.M. Dillon, and L.B. Read. 2005. Approaches for linking wholebody fish tissue residues of mercury or DDT to biological effects thresholds. *Environ. Toxicol. Chem.* 24:2094–2105.

Benoit, L.F., J.M. Skelly, L.D. Moore, and L.S. Dochinger. 1982. Radial growth reductions of *Pinus strobus* L. correlated with foliar ozone sensitivity as an indicator of ozone-induced losses in eastern forests. *Can. J. For. Res.* 12:673–678.

Berrang, P., D.F. Karnosky, R.A. Mickler, and J.P. Bennett. 1986. Natural selection for ozone tolerance in *Populus tremuloides. Can. J. For. Res.* 16:1214–1216.

Black, V.J., C.R. Black, J.A. Roberts, and C.A. Stewart. 2000. Impact of ozone on the reproductive development of plants. *New Phytol.* 147:421–447.

Blais, J.M., D.W. Schindler, D.C.G. Muir, L.E. Kimpe, D.B. Donald, and B. Rosenberg. 1998. Accumulation of persistent organochlorine compounds in mountains of western Canada. *Nature* 395:585–588.

Bobbink, R., M. Ashmore, S. Braun, W. Flückiger, and I.J.J. van den Wyngaert. 2003. Empirical nitrogen critical loads for natural and semi-natural ecosystems: 2002 update. *In* B. Achermann and R. Bobbink (Eds.). *Empirical Critical Loads for Nitrogen.* Swiss Agency for Environment, Forest and Landscape SAEFL, Berne, Switzerland, pp. 43–170.

Bobbink, R., K. Hicks, J. Galloway, T. Spranger, R. Alkemade, M. Ashmore, M. Bustamante et al. 2010. Global assessment of nitrogen deposition effects on terrestrial plant diversity: A synthesis. *Ecol. Appl.* 20(1):30–59.

Bowman, W.D. 1994. Accumulation and use of nitrogen and phosphorus following fertilization in two alpine tundra communities. *Oikos* 70:261–270.

Bowman, W.D. and M.C. Fisk. 2001. Primary production. *In* W.D. Bowman and T.R. Seastedt (Eds.). *Structure and Function of an Alpine Ecosystem: Niwot Ridge, Colorado.* Oxford University Press, Oxford, U.K., pp. 177–197.

Bowman, W.D., J.R. Gartner, K. Holland, and M. Wiedermann. 2006. Nitrogen critical loads for alpine vegetation and terrestrial ecosystem response: Are we there yet? *Ecol. Appl.* 16(3):1183–1193.

Bowman, W.D., J. Murgel, T. Blett, and E. Porter. 2012. Nitrogen critical loads for alpine vegetation and soils in Rocky Mountain National Park. *J. Environ. Manage.* 103:165–171.

Bowman, W.D., T.A. Theodose, J.C. Schardt, and R.T. Conant. 1993. Constraints of nutrient availability on primary production in two alpine tundra communities. *Ecology* 74:2085–2097.

Brace, S., D.L. Peterson, and D. Bowers. 1999. A guide to ozone injury in vascular plants of the Pacific Northwest. General Technical Report PNW-GTR-446. USDA Forest Service, Pacific Northwest Research Station, Portland, OR.

Branfireun, B.A., N.T. Roulet, C.A. Kelly, and W.M. Rudd. 1999. *In situ* sulphate stimulation of mercury methylation in a boreal peatland: Toward a link between acid rain and methylmercury contamination in remote environments. *Glob. Biogeochem. Cycles* 13(3):743–750.

Bricker, O.P. and K.C. Rice. 1989. Acidic deposition to streams: A geology-based method predicts their sensitivity. *Environ. Sci. Technol.* 23:379–385.

Britton, A.J. and J.M. Fisher. 2007. Interactive effects of nitrogen deposition, fire and grazing on diversity and composition of low-alpine prostrate *Calluna vulgaris* heathland. *J. Appl. Ecol.* 44:123–135.

Brooks, M.L. 1999. Alien annual grasses and fire in the Mojave Desert. *Madroño* 46:13–19.

Brooks, M.L., C.M. D'Antonio, D.M. Richardson, J.B. Grace, J.E. Keeley, J.M. DiTomaso, R.J. Hobbs, M. Pellant, and D. Pyke. 2004. Effects of invasive alien plants on fire regimes. *BioScience* 54:677–688.

Brooks, M.L. and T.C. Esque. 2002. Alien annual plants and wildfire in desert tortoise habitat: Status, ecological effects, and management. *Chelonian Conserv. Biol.* 4:330–340.

Burgess, N.M. and M.W. Meyer. 2008. Methylmercury exposure associated with reduced productivity in common loons. *Ecotoxicology* 17(2):83–91.

Burkey, K.O., H.S. Neufeld, L. Souza, A.H. Chappelka, and A.W. Davison. 2006. Seasonal profiles of leaf ascorbic acid content and redox state in ozone-sensitive wildflowers. *Environ. Pollut.* 143:427–434.

Burns, D.A., J.A. Lynch, B.J. Cosby, M.E. Fenn, J.S. Baron, and U.S. EPA Clean Air Markets Division. 2011. National acid precipitation assessment program report to congress 2011: An integrated assessment. National Science and Technology Council, Washington, DC.

Bytnerowicz, A. and N.E. Grulke. 1992. Physiological effects of air pollutants on western trees. *In* R.K. Olson, D. Binkley, and M. Böhm (Eds.). *Response of Western Forests to Air Pollution*. Springer-Verlag, New York, pp. 183–233.

Caldwell, C.A., P. Swartzendruber, and E.M. Prestbo. 2006. Concentration and dry deposition of mercury species in arid south central New Mexico (2001–2002). *Environ. Sci. Technol.* 40:7535–7540.

Campbell, S.J., R. Wanek, and J.W. Coulston. 2007. Ozone injury in west coast forests: 6 years of monitoring. General Technical Report PNW-GTR-722. USDA Forest Service, Pacific Northwest Research Station, Portland, OR.

Capps, T., S. Mukhi, J.J. Rinchard, C.W. Theodorakis, V.S. Blazer, and R. Patino. 2004. Exposure to perchlorate induces the formation of macrophage aggregates in the trunk kidney of zebrafish and mosquitofish. *J. Aquat. Anim. Hlth.* 16:145–151.

Chappelka, A.H. 2002. Reproductive development of blackberry (*Rubus cuneifolius*) as influenced by ozone. *New Phytol.* 155:249–255.

Chappelka, A.H., L.J. Samuelson, J.M. Skelly, and A.S. Lefohn. 1996. Effects of ozone on forest trees in the southern Appalachians—An assessment of the current state of knowledge. Prepared for the Southern Appalachian Mountain Initiative (SAMI).

Chernyak, S.M., C.P. Rice, R.T. Quintal, L.J. Begnoche, J.P. Hickey, and B.T. Vinyard. 2005. Time trends (1983–1999) for organochlorines and polybrominated diphenyl ethers in rainbow smelt (*Osmerus mordax*) from Lakes Michigan, Huron, and Superior, U.S.A. *Environ. Toxicol. Chem.* 24:1632–1641.

Clair, T.A., D. Burns, I.R. Pérez, J. Blais, and K. Percy. 2011. Ecosystems. *In* G.M. Hidy, J.R. Brook, K.L. Demerjian, L.T. Molina, W.T. Pennell, and R.D. Scheffe (Eds.). *Technical Challenges of Multipollutant Air Quality Management*. Springer, Dordrecht, the Netherlands, pp. 139–229.

Clark, C.M. and D. Tilman. 2008. Loss of plant species after chronic low-level nitrogen deposition to prairie grasslands. *Nature* 451:712–715.

Cleavitt, N.L., J.W. Hinds, R.L. Poirot, L.H. Geiser, A.C. Dibble, B. Leon, R. Perron, and L.H. Pardo. 2015. Epiphytic macrolichen communities correspond to patterns of sulfur and nitrogen deposition in the northeastern United States. *Bryologist* 118(3):304–324.

Coleman, M.D., R.E. Dickson, J.G. Isebrands, and D.F. Karnosky. 1996. Root growth and physiology of potted and field-grown trembling aspen exposed to tropospheric ozone. *Tree Physiol.* 16:145–152.

Conklin, P.L. and C. Barth. 2004. Ascorbic acid, a familiar small molecule intertwined in the response of plants to ozone, pathogens, and the onset of senescence. *Plant Cell Environ.* 27:959–970.

Cosby, B.J., J.R. Webb, J.N. Galloway, and F.A. Deviney. 2006. Acidic deposition impacts on natural resources in Shenandoah national park. NPS/NER/NRTR-2006/066. U.S. Department of the Interior, National Park Service, Northeast Region, Philadelphia, PA.

Coulston, J.W., G.C. Smith, and W.D. Smith. 2003. Regional assessment of ozone sensitive tree species using bioindicator plants. *Environ. Monitor. Assess.* 83:113–127.

Cowling, E.B. and L.S. Dochinger. 1980. Effects of acidic deposition on health and the productivity of forests. *In Proceedings of Symposium on Effects of Air Pollutants on Mediterranean and Temperate Forest Ecosystems,* Riverside, CA, June 22-27, 1980. USDA Forest Service Technical Report PSW-43. Pacific Southwest Forest and Range Experiment Station, Berkeley, CA, pp. 165–173.

Cronan, C.S., W.A. Reiners, R.C.J. Reynolds, and G.E. Lang. 1978. Forest floor leaching: Contributions from mineral, organic, and carbonic acids in New Hampshire subalpine forests. *Science* 200(4339):309–311.

Cronan, C.S. and C.L. Schofield. 1979. Aluminum leaching response to acid precipitation: Effects on high elevation watersheds in the Northeast. *Science* 204:304–306.

Croteau, M.N., S.N. Luoma, and A.R. Stewart. 2005. Trophic transfer of metals along freshwater food webs: Evidence of cadmium biomagnification in nature. *Limnol. Oceanogr.* 50(5):1511–1519.

D'Antonio, C.M. and P.M. Vitousek. 1992. Biological invasions by exotic grasses: The grass-fire cycle and global change. *Ann. Rev. Ecol. Syst.* 23:63–87.

Day, D.E. and W.C. Malm. 2001. Aerosol light scattering measurements as a function of relative humidity: A comparison between measurements made at three different sites. *Atmos. Environ.* 35:5169–5176.

Day, D.E., W.C. Malm, and S.M. Kreidenweis. 2000. Aerosol light scattering measurements as a function of relative humidity. *J. Air Waste Manage. Assoc.* 50:710–716.

Debell, L.J., K. Gebhart, J.L. Hand, W.C. Malm, M.L. Pitchford, B.A. Schichtel, and W.H. White. 2006. Spatial and seasonal patterns and temporal variability of haze and its constituents in the United States. Report IV. CIRA ISSN 0737-5352-74. Colorado State University, Fort Collins, CO. Available at: https://science.nature.nps.gov/im/units/cakn/Documents/airQuality/visibility%20monitoring/IMPROVE%20monitoring%20network%20report%202006.pdf (accessed October 5, 2016).

Dillon, T., S. Beckvar, and J. Kern. 2010. Residue-based dose–response in fish: An analysis using lethality-equivalent endpoints. *Environ. Toxicol. Chem.* 29:2559–2565.

Dise, N.B. 1984. A synoptic survey of headwater streams in Shenandoah national park, Virginia, to evaluate sensitivity to acidification by acid deposition. MS, Department of Environmental Sciences, University of Virginia, Charlottesville, VA.

Driscoll, C.T., K.M. Driscoll, K.M. Roy, and J. Dukett. 2007. Changes in the chemistry of lakes in the Adirondack region of New York following declines in acidic deposition. *Water Air Soil Pollut.* 22(6):1181–1188.

Driscoll, C.T., G.B. Lawrence, A.J. Bulger, T.J. Butler, C.S. Cronan, C. Eagar, K.F. Lambert, G.E. Likens, J.L. Stoddard, and K.C. Weathers. 2001a. Acid rain revisited: Advances in scientific understanding since the passage of the 1970 and 1990 Clean Air Act Amendments. Volume 1, No. 1. Hubbard Brook Research Foundation, Science Links Publication, Hanover, NH.

Driscoll, C.T., G.B. Lawrence, A.J. Bulger, T.J. Butler, C.S. Cronan, C. Eagar, K.F. Lambert, G.E. Likens, J.L. Stoddard, and K.C. Weathers. 2001b. Acidic deposition in the northeastern United States: Sources and inputs, ecosystem effects, and management strategies. *BioScience* 51(3):180–198.

Driscoll, C.T., K.M. Postek, D. Mateti, K. Sequeira, J.D. Aber, W.J. Kretser, M.J. Mitchell, and D.J. Raynal. 1998. The response of lake water in the Adirondack region of New York to changes in acidic deposition. *Environ. Sci. Policy* 1:185–198.

Eagles-Smith, C.A., J.J. Willacker, Jr., and C.M. Flanagan Pritz. 2014. Mercury in fishes from 21 national parks in the western United States—Inter- and intra-park variation in concentrations and ecological risk. U.S. Geological Survey Open-File Report 2014-1051. U.S. Geological Survey, Reston, VA. Available at: http://dx.doi.org/10.3133/ofr20141051 (accessed October 5, 2016).

Eliason, S.A. and E.B. Allen. 1997. Exotic grass competition in suppressing native shrubland re-establishment. *Restor. Ecol.* 5:245–255.

Elser, J.J., T. Andersen, J.S. Baron, A.-K. Bergström, M. Jansson, M. Kyle, K.R. Nydick, L. Steger, and D.O. Hessen. 2009a. Shifts in lake N:P stoichiometry and nutrient limitation driven by atmospheric nitrogen deposition. *Science* 326:835–837.

Elser, J.J., M. Kyle, L. Steger, K.R. Nydick, and J.S. Baron. 2009b. Nutrient availability and phytoplankton nutrient limitation across a gradient of atmospheric nitrogen deposition. *Ecology* 90(11):3062–3073.

Ericksen, J.A., M.S. Gustin, D.E. Schorran, D.W. Johnson, S.E. Lindberg, and J.S. Coleman. 2003. Accumulation of atmospheric mercury in forest foliage. *Atmos. Environ.* 37(12):1613–1622.

Evans, L.S., J.H. Adamski II, and J.R. Renfro. 1996. Relationships between cellular injury, visible injury of leaves, and ozone exposure levels for several dicotyledonous plant species at Great Smoky Mountains National Park. *Environ. Exp. Bot.* 36(2):229–237.

Evers, D.C., L.J. Savoy, C.R. DeSorbo, D.E. Yates, W. Hanson, K.M. Taylor, L.S. Siegel et al. 2008. Adverse effects from environmental mercury loads on breeding common loons. *Ecotoxicology* 17:69–81.

Evers, D.C., K.M. Taylor, A. Major, R.J. Taylor, R.H. Poppenga, and A.M. Scheuhammer. 2003. Common loon eggs as bioindicators of methylmercury availability in North America. *Ecotoxicology* 12:69–81.

Evers, D.C., J.G. Wiener, C.T. Driscoll, D.A. Gay, N. Basu, B.A. Monson, K.F. Lambert et al. 2011b. Great lakes mercury connections: The extent and effects of mercury pollution in the great lakes region. Report BR1 2011-18. Biodiversity Research Institute, Gorham, ME.

Evers, D.C., K.A. Williams, M.W. Meyer, A.M. Scheuhammer, N. Schoch, A. Gilbert, L. Siegel, R.J. Taylor, R. Poppenga, and C.R. Perkins. 2011a. Spatial gradients of methylmercury for breeding common loons in the Laurentian Great Lakes region. *Ecotoxicology* 20:1609–1625.

Felzer, B., D. Kicklighter, J. Melillo, C. Wang, Q. Zhuang, and R. Prinn. 2004. Effects of ozone on net primary production and carbon sequestration in the conterminous United States using a biogeochemistry model. *Tellus* 56B:230–248.

Fenn, M.E., R. Haeuber, G.S. Tonnesen, J.S. Baron, S. Grossman-Clark, D. Hope, D.A. Jaffe et al. 2003. Nitrogen emissions, deposition, and monitoring in the western United States. *BioScience* 53(4):391–403.

Fenn, M.E., K.F. Lambert, T. Blett, D.A. Burns, L.H. Pardo, G.M. Lovett, R.A. Haeuber, D.C. Evers, C.T. Driscoll, and D.S. Jefferies. 2011. Setting limits: Using air pollution thresholds to protect and restore U.S. ecosystems. Issues in Ecology. Report No. 14. Ecological Society of America. http://www.fs.fed.us/psw/publications/fenn/psw_2011_fenn002.pdf?

Fitzgerald, W.F., D.R. Engstrom, R.P. Mason, and E.A. Nater. 1998. The case for atmospheric mercury contamination in remote areas. *Environ. Sci. Technol.* 32:1–7.

Flanagan Pritz, C.M., J.E. Schrlau, S.L. Massey Simonich, and T.F. Blett. 2014. Contaminants of emerging concern in fish from western U.S. and Alaskan National Parks—Spatial distribution and health thresholds. *J. Am. Water Resour. Assoc.* 50(2):309–323.

Fournie, J.W., J.K. Summers, L.A. Courtney, V.D. Engle, and V.S. Blazer. 2001. Utility of splenic macrophage aggregates as an indicator of fish exposure to degraded environments. *J. Aquat. Anim. Health* 13:105–116.

Fremstad, E., J. Paal, and T. Möls. 2005. Impacts of increased nitrogen supply on Norwegian lichen-rich alpine communities: A 10-year experiment. *J. Ecol.* 93:471–481.

Frescholtz, T.F., M.S. Gustin, D.E. Schorran, and G.C.J. Fernandez. 2003. Assessing the source of mercury in foliar tissue of quaking aspen. *Environ. Toxicol. Chem.* 22:2114–2119.

Geiser, L.H., S.E. Jovan, D.A. Glavich, and M.K. Porter. 2010. Lichen-based critical loads for atmospheric nitrogen deposition in western Oregon and Washington forests, USA. *Environ. Pollut.* 158:2412–2421.

Geiser, L.H. and P.N. Neitlich. 2007. Air pollution and climate gradients in western Oregon and Washington indicated by epiphytic macrolichens. *Environ. Pollut.* 145:203–218.

Geron, C., R. Rasmussen, R.R. Arnts, and A. Guenther. 2000. A review and synthesis of monoterpene speciation from forests in the United States. *Atmos. Environ.* 34:1761–1781.

Gilmour, C.C. and E.A. Henry. 1991. Mercury methylation in aquatic systems affected by acid deposition. *Environ. Pollut.* 71:131–169.

Gilmour, C.C., E.A. Henry, and R. Mitchell. 1992. Sulfate stimulation of mercury methylation in freshwater sediments. *Environ. Sci. Technol.* 26(11):2281–2287.

Glavich, D.A. and L.H. Geiser. 2008. Potential approaches to developing lichen-based critical loads and levels for nitrogen, sulfur, and metal-containing atmospheric pollutants in North America. *Bryologist* 111(4):638–649.

Goodale, C.L. and J.D. Aber. 2001. The long-term effects of land-use history on nitrogen cycling in northern hardwood forests. *Ecol. Appl.* 11(1):253–267.

Grulke, N.E. 1999. Physiological responses of ponderosa pine to gradients of environmental stressors. *In* P.R. Miller and J.R. McBride (Eds.). *Oxidant Air Pollution Impacts in the Montane Forests of Southern California: A Case Study of the San Bernardino Mountains.* Springer, New York, pp. 126–163.

Grulke, N.E., R. Johnson, A. Esperanza, D. Jones, T. Nguyen, S. Posch, and M. Tausz. 2003. Canopy transpiration of Jeffrey pine in mesic and zeric microsites: O_3 uptake and injury response. *Trees* 17:292–298.

Guderian, R. 1977. *Air Pollution—Phytotoxicity of Acidic Gases and Its Significance in Air Pollution Control.* Springer-Verlag, New York.

Hammerschmidt, C.R. and M.B. Sandheinrich. 2005. Maternal diet during oogenesis is the major source of methylmercury in fish embryos. *Environ. Sci. Technol.* 39:3580–3584.

Hand, J.L., S.A. Copeland, D.E. Day, A.M. Dillner, H. Idresand, W.C. Malm, C.E. McDade et al. 2011. IMPROVE (Interagency Monitoring of Protected Visual Environments): Spatial and seasonal patterns and temporal variability of haze and its constituents in the United States. Report V. Cooperative Institute for Research in the Atmosphere, Colorado State University, Fort Collins, CO. Available at: http://vista.cira.colostate.edu/Improve/wp-content/uploads/2016/04/Cover_TOC.pdf (accessed October 5, 2016).

Hand, J.L., B.A. Schichtel, W.C. Malm, S. Copeland, J.V. Molenar, N. Frank, and M. Pitchford. 2014. Reductions in haze across the United States from the early 1990s through 2011. *Atmos. Environ.* 94:671–679.

Handy, R.D. and W.S. Penrice. 1993. The influence of high oral doses of mercuric chloride on organ toxicant concentrations and histopathology in rainbow trout, *Oncorhynchus mykiss. Comp. Biochem. Physiol. C Pharmacol. Toxicol. Endocrinol.* 106:717–724.

Heath, R.L. and G.E. Taylor. 1997. Physiological processes and plant responses to ozone exposure. *In* H. Sandermann, A. R. Wellburn, and R.L. Heath (Eds.). *Forest Decline and Ozone: A Comparison of Controlled Chamber and Field Experiments.* Springer-Verlag, New York, pp. 310–368.

Heck, W.W. and E.B. Cowling. 1997. The need for a long term cumulative secondary ozone standard—An ecological perspective. *Ecol. Manag.* January:22–33.

Heinz, G.H., D. Hoffman, J.D. Klimstra, and K.R. Stebbins. 2010. Predicting mercury concentrations in mallard eggs from mercury in the diet or blood of adult females and from duckling down feathers. *Environ. Toxicol. Chem.* 29:389–392.

Henriksen, A. and M. Posch. 2001. Steady-state models for calculating critical loads of acidity for surface waters. *Water Air Soil Pollut. Focus* 1(1–2):375–398.

Hogsett, W.E., D.T. Tingey, C. Hendricks, and D. Rossi. 1989. Sensitivity of western conifers to SO_2 and season interaction of acid fog and ozone. *In* R.K. Olson and A.S. Lefohn (Eds.). *Transactions, Symposium on the Effects of Air Pollution on Western Forests,* Anaheim, CA, June 1989. Air and Waste Management Association, Pittsburgh, PA, pp. 469–491.

Horsley, S.B., R.P. Long, S.W. Bailey, R.A. Hallet, and T.J. Hall. 1999. Factors contributing to sugar maple decline along topographic gradients on the glaciated and unglaciated Allegheny Plateau. *In* S.B. Horsley and R.P. Long (Eds.). *Sugar Maple Ecology and Health: Proceedings of an International Symposium.* General Technical Report NE-261. U.S. Department of Agriculture, Forest Service, Radnor, PA, pp. 60–62.

Horvath, H. 1996. Spectral extinction coefficients of rural aerosol in southern Italy—A case study of cause and effect of variability of atmospheric aerosol. *J. Aerosol Sci.* 27(3):437–453.

Jeremiason, J.D., D.R. Engstrom, E.B. Swain, E.A. Nater, B.M. Johnson, J.E. Almendinger, B.A. Monson, and R.K. Kolka. 2006. Sulfate addition increases methylmercury production in an experimental wetland. *Environ. Sci. Technol.* 40(12):3800–3806.

Joslin, J.D., J.M. Kelly, and H. Van Miegroet. 1992. Soil chemistry and nutrition of North American spruce-fir stands: Evidence for recent change. *J. Environ. Qual.* 21(1):12–30.

Joslin, J.D. and M.H. Wolfe. 1994. Foliar deficiencies of mature southern Appalachian red spruce determined from fertilizer trials. *Soil Sci. Soc. Am. J.* 58:1572–1579.

Karnosky, D.F., Z.E. Gagnon, R.E. Dickson, M.D. Coleman, E.H. Lee, and J.G. Isebrands. 1996. Changes in growth, leaf abscission, biomass associated with seasonal tropospheric ozone exposures of *Populus tremuloides* clones and seedlings. *Can. J. For. Res.* 26:23–37.

Karnosky, D.F., Z.E. Gagnon, D.D. Reed, and J.A. Witter. 1992. Threshold levels for foliar injury to *Populus tremuloides* by sulfur dioxide and ozone. *Can. J. For. Res.* 6:166–169.

Khan, R.A. 2003. Health of flatfish from localities in Placentia Bay, Newfoundland, contaminated with petroleum and PCBs. *Arch. Environ. Contam. Toxicol.* 44:485–492.

King, J.S., M.E. Kubiske, K.S. Pregitzer, G.R. Hendrey, E.P. McDonald, C.P. Giardina, V.S. Quinn, and D.F. Karnosky. 2005. Tropospheric O_3 compromises net primary production in young stands of trembling aspen, paper birch and sugar maple in response to elevated atmospheric CO_2. *New Phytol.* 168:623–635.

Kline, L.J., D.D. Davis, J.M. Skelly, and D.R. Decoteau. 2009. Variation in ozone sensitivity within Indian hemp and common milkweed selections from the Midwest. *Northeast. Nat.* 16:307–313.

Kline, L.J., D.D. Davis, J.M. Skelly, J.E. Savage, and J. Ferdinand. 2008. Ozone sensitivity of 28 plant selections exposed to ozone under controlled conditions. *Northeast. Nat.* 15:57–66.

Kohut, R. 2007a. Assessing the risk of foliar injury from ozone on vegetation in parks in the U.S. National Park Service's Vital Signs Network. *Environ. Pollut.* 149:348–357.

Kohut, R. 2007b. *Handbook for Assessment of Foliar Ozone Injury on Vegetation in the National Parks,* revised 2nd edn. Boyce Thompson Institute, Cornell University, Ithaca, NY. http://www.nature.nps.gov/air/Permits/ARIS/networks/docs/O3_InjuryAssessmentHandbookD1688.pdf.

Kohut, R.J. 2007c. Ozone risk assessment for vital signs monitoring networks, appalachian national scenic trail, and natchez trace national scenic trail. Natural Resource Technical Report NPS/NRPC/ARD/NRTR—2007/001. National Park Service, Natural Resource Program Center, Fort Collins, CO. Available at: http://www.nature.nps.gov/air/pubs/pdf/03Risk/OzoneRiskAssessmentsCoverRev20070227.pdf.

Kohut, R., C. Flanagan, J. Cheatham, and E. Porter. 2012. Foliar ozone injury on cutleaf coneflower at Rocky Mountain National Park, Colorado. *West. N. Am. Nat.* 72(1):32–42.

Kozlowski, T.T. and H.A. Constantinidou. 1986. Responses of woody plants to environmental pollution. *For. Abstr.* 47:1–51.

Kubiske, M.E., V.S. Quinn, P.E. Marquardt, and D.F. Karnosky. 2007. Effects of elevated atmospheric CO_2 and/or O_3 on intra-and interspecific competitive ability of aspen. *Plant Biol.* 9:342–355.

Landers, D.H., S.L. Simonich, D.A. Jaffe, L.H. Geiser, D.H. Campbell, A.R. Schwindt, C.B. Schreck et al. 2008. The fate, transport, and ecological impacts of airborne contaminants in western national parks (USA). EPA/600/R-07/138. U.S. Environmental Protection Agency, Office of Research and Development, NHEERL, Western Ecology Division, Corvallis, OR.

Landers, D.H., S.M. Simonich, D. Jaffe, L. Geiser, D.H. Campbell, A. Schwindt, C.B. Schreck et al. 2010. The Western Airborne Contaminant Project (WACAP): An interdisciplinary evaluation of the impacts of airborne contaminants in Western U.S. National Parks. *Environ. Sci. Technol.* 44(3):855–859.

Laurence, J.A. 1981. Effects of air pollutants on plant-pathogen interactions. *Z. Pflanzenkr. Pflanzenschutz* 88:156–172.

Laurence, J.A., R.G. Amundson, A.L. Friend, E.J. Pell, and P.J. Temple. 1994. Allocation of carbon in plants under stress: An analysis of the ROPIS experiments. *J. Environ. Qual.* 23:412–417.

Lazorchak, J.M., F.H. McCormick, T.R. Henry, and A.T. Herlihy. 2003. Contamination of fish in streams of the mid-Atlantic region: An approach to regional indicator selection and wildlife assessment. *Environ. Toxicol. Chem.* 22(3):545–553.

LeBlanc, G.A. 1995. Trophic-level differences in the bioconcentration of chemicals: Implications in assessing environmental biomagnification. *Environ. Sci. Technol.* 29:154–160.

Lefohn, A.S. 2011. Comments on the integrated science assessment for ozone and related photo-chemical oxidants: Chapter 9. Environmental effects: Ozone effects on vegetation and ecosystems. Draft Report to National Park Service. A.S.L. & Associates.

Lefohn, A.S., W. Jackson, D.S. Shadwick, and H.P. Knudson. 1997. Effect of surface ozone exposures on vegetation grown in the southern Appalachian Mountains: Identification of possible areas of concern. *Atmos. Environ.* 31(11):1695–1708.

Mahaffey, K.R. 2005. NHANES 1999–2002 update on mercury. *In* Presentation at the *September 2005 EPA Fish Forum*, Washington, DC. http://www.epa.gov/waterscience/fish/forum/2005.

Manera, M., R. Serra, G. Isani, and E. Carpene. 2000. Macrophage aggregates in gilthead sea bream fed copper, iron and zinc enriched diet. *J. Fish. Biol.* 57:457–465.

Martin, P.A., T.V. McDaniel, K.D. Hughes, and B. Hunter. 2011. Mercury and other heavy metals in free-ranging mink of the lower Great Lakes Basin, Canada, 1998–2006. *Ecotoxicology* 20:1701–1712.

Mason, C.F. and C.D. Wren. 2001. Carnivora. *In* R.F. Shore and B.A. Rattner (Eds.). *Ecotoxicology of Wild Mammals.* Wiley, Chichester, U.K., pp. 315–370.

Mason, J. and H.M. Seip. 1985. The current state of knowledge on acidification of surface waters and guidelines for further research. *Ambio* 14:45–51.

Matyssek, R., D. Le Thiec, M. Low, P. Dizengremel, A.J. Nunn, and K.H. Haberle. 2006. Interactions between drought and O_3 stress in forest trees. *Plant Biol.* 8:11–17.

McAinish, M.R., N.H. Evans, L.T. Montgomery, and K.A. North. 2002. Calcium signaling in stomatal responses to pollutants. *New Phytol.* 153:441–447.

McCool, P.M. 1988. Effect of air pollutants on mycorrhizae. *In* S. Schulte-Hostede, N.M. Darrall, L.W. Blank, and A.R. Wellburn (Eds.). *Air Pollution and Plant Metabolism.* Elsevier Applied Science, London, U.K., pp. 356–365.

McDonnell, T.C., S. Belyazid, T.J. Sullivan, H. Sverdrup, W.D. Bowman, and E.M. Porter. 2014. Modeled subalpine plant community response to climate change and atmospheric nitrogen deposition in Rocky Mountain National Park, USA. *Environ. Pollut.* 187:55–64.

McLaughlin, S.B., M. Nosal, S.D. Wullschleger, and G. Sun. 2007a. Interactive effects of ozone and climate on tree growth and water use in a southern Appalachian forest in the USA. *New Phytol.* 174:109–124.

McLaughlin, S.B., S.D. Wullschleger, G. Sun, and M. Nosal. 2007b. Interactive effects of ozone and climate on water use, soil moisture content and streamflow in a southern Appalachian forest in the USA. *New Phytol.* 174:125–136.

Meinelt, T., R. Kruger, M. Pietrock, R. Osten, and C. Steinberg. 1997. Mercury pollution and macrophage centres in pike (*Esox lucius*) tissues. *Environ. Sci. Pollut. Res.* 4:32–36.

Miller, P.R., G.J. Longbotham, and C.R. Longbotham. 1983. Sensitivity of selected western conifers to ozone. *Plant Dis.* 67:1113–1115.

Miller, P.R. and A.A. Millecan. 1971. Extent of air pollution damage to some pines and other conifers in California. *Plant Dis. Rep.* 55:555–559.

Miller, P.R., O.C. Taylor, and R.G. Wilhour. 1982. Oxidant air pollution effects on a western coniferous forest ecosystem. EPA-600/D-82-276. U.S. Environmental Protection Agency, Environmental Research Laboratory, Corvallis, OR.

Millhollen, A.G., M.S. Gustin, and D. Obrist. 2006. Foliar mercury accumulation and exchange for three tree species. *Environ. Sci. Technol.* 40(19):6001–6006.

Minnich, R.A. and R.J. Dezzani. 1998. Historical decline of coastal sage scrub in the Riverside-Perris Plain, California. *West. Birds* 29:366–391.

Musselman, R.C., A.S. Lefohn, W.J. Massman, and R.L. Heath. 2006. A critical review and analysis of the use of exposure- and flux-based ozone indices for predicting vegetation effects. *Atmos. Environ.* 40:1869–1888.

Nanus, L., D.W. Clow, J.E. Saros, V.C. Stephens, and D.H. Campbell. 2012. Mapping critical loads of nitrogen deposition for aquatic ecosystems in the Rocky Mountains, USA. *Environ. Pollut.* 166:125–135.

National Park Service (NPS). 2010. Air quality in national parks: 2009 Annual performance and progress report. Natural Resource Report NPS/NRPC/ARD/NRR-2010/266. National Park Service, Air Resources Division, Denver, CO.

National Park Service. 2011. Rating air quality conditions. National Park Service, Denver, CO. Available at: http://www.nature.nps.gov/air/planning/docs/20111122_Rating-AQ-Conditions.pdf (accessed October 5, 2016)

Nilsson, J. and P. Grennfelt. 1988. Critical loads for sulphur and nitrogen. *Miljorapport* 1988:15.

Nodvin, S.C., H.V. Miegroet, S.E. Lindberg, N.S. Nicholas, and D.W. Johnson. 1995. Acidic deposition, ecosystem processes, and nitrogen saturation in a high elevation southern Appalachian watershed. *Water Air Soil Pollut.* 85:1647–1652.

Norton, S.A., J.J. Akielaszek, T.A. Haines, K.J. Stromborg, and J.R. Longcore. 1982. Bedrock geologic control of sensitivity of aquatic ecosystems in the United States to acidic deposition. National Atmospheric Deposition Program, Fort Collins, CO.

Ollinger, S.V., M.L. Smith, M.E. Martin, R.A. Hallett, C.L. Goodale, and J.D. Aber. 2002. Regional variation in foliar chemistry and N cycling among forests of diverse history and composition. *Ecology* 83(2):339–355.

Padgett, P. and E.B. Allen. 1999. Differential responses to nitrogen fertilization in native shrubs and exotic annuals common to Mediterranean coastal sage scrub of California. *Plant Ecol.* 144:93–101.

Padgett, P.E., E.B. Allen, A. Bytnerowicz, and R.A. Minnich. 1999. Changes in soil inorganic nitrogen as related to atmospheric nitrogenous pollutants in southern California. *Atmos. Environ.* 33:769–781.

Pan, Y.D., R. Birdsey, J. Hom, and K. McCullough. 2009. Separating effects of changes in atmospheric composition, climate and land-use on carbon sequestration of US mid-Atlantic temperate forests. *For. Ecol. Manage.* 259:151–164.

Panek, J.A. 2004. Ozone uptake, water loss and carbon exchange dynamics in annually drought-stressed *Pinus ponderosa* forests: Measured trends and parameters for uptake modeling. *Tree Physiol.* 24:277–290.

Panek, J.A. and A.H. Goldstein. 2001. Response of stomatal conductance to drought in ponderosa pine: Implications for carbon and ozone uptake. *Tree Physiol.* 21:337–344.

Pardo, L.H., M.J. Robin-Abbott, and C.T. Driscoll. 2011. Assessment of nitrogen deposition effects and empirical critical loads of nitrogen for ecoregions of the United States. General Technical Report NRS-80. U.S. Forest Service, Newtown Square, PA.

Parker, D.R., L.W. Zelazny, and T.B. Kinraide. 1989. Chemical speciation and plant toxicity of aqueous aluminum. *In* T.E. Lewis (Ed.). *Environmental Chemistry and Toxicology of Aluminum.* Lewis Publishers, Chelsea, MI. pp. 117–145.

Pell, E.J., N.A. Eckardt, and R.E. Glick. 1994a. Biochemical and molecular basis for impairment of photosynthetic potential. *Photosynth. Res.* 39:453–462.

Pell, E.J., P.J. Temple, A.L. Friend, H.A. Mooney, and W.E. Winner. 1994b. Compensation as a plant response to ozone and associated stresses: An analysis of ROPIS experiments. *J. Environ. Qual.* 23:429–436.

Peterson, D.L. and M.J. Arbaugh. 1992. Coniferous forests of the Colorado front range. Part B: Ponderosa pine second-growth stands. *In* M.J. Mitchell and S.E. Lindberg (Eds.). *Atmospheric Deposition and Forest Nutrient Cycling: A Synthesis of the Integrated Forest Study.* Springer-Verlag, Inc., New York, pp. 433–460.

Peterson, D.L., M.J. Arbaugh, and L.J. Robinson. 1991. Regional growth changes in ozone-stressed ponderosa pine (*Pinus ponderosa*) in the Sierra Nevada, California, USA. *Holocene* 1:50–61.

Peterson, D.L., M.J. Arbaugh, V.A. Wakefield, and P.R. Miller. 1987. Evidence of growth reduction in ozone injured Jeffrey pine in Sequoia and Kings Canyon National Parks. *J. Air Pollut. Control Assoc.* 37:908–912.

Pitchford, M., W. Maim, B. Schichtel, N. Kumar, D. Lowenthal, and J. Hand. 2007. Revised algorithm for estimating light extinction from IMPROVE particle speciation data. *J. Air Waste Water Manage. Assoc.* 57:1326–1336.

Pitchford, M.L. and W.C. Malm. 1994. Development and applications of a standard visual index. *Atmos. Environ.* 28:1049–1055.

Porter, E. 2003. Ozone sensitive plant species on national park service and U.S. fish and wildlife service lands: Results of a June 24–25, 2003 workshop, Baltimore, MD. Natural Resource Report NPS/NRARD/NRR-2003/01.

Porter, E., T. Blett, D.U. Potter, and C. Huber. 2005. Protecting resources on federal lands: Implications of critical loads for atmospheric deposition on nitrogen and sulfur. *BioScience* 55(7):603–612.

Pronos, J., L. Merrill, and D. Dahlsten. 1999. Insects and pathogens in a pollution-stressed forest. *In* P.R. Miller and J.R. McBride (Eds.). *Oxidant Air Pollution Impacts in the Montane Forests of Southern California*. Springer-Verlag, New York, pp. 317–337.

Pronos, J. and D.R. Vogler. 1981. Assessment of ozone injury to pines in the southern Sierra Nevada, 1979/1980. USDA Forest Service Pacific Southwest Region, San Francisco, CA.

Rao, L.E., E.B. Allen, and T. Meixner. 2010. Risk-based determination of critical nitrogen deposition loads for fire spread in southern California deserts. *Ecol. Appl.* 20(5):1320–1335.

Reich, P.B. and R.G. Amundson. 1985. Ambient levels of ozone reduce net photosynthesis in tree and crop species. *Science* 230:566–570.

Reinfelder, J.R., N.S. Fisher, S.N. Luoma, J.W. Nichols, and W.-X. Wang. 1998. Trace element trophic transfer in aquatic organisms: A critique of the kinetic model approach. *Sci. Total Environ.* 219:117–135.

Rood, M.J., D.S. Covert, and T.V. Larson. 1986. Hygroscopic properties of atmospheric aerosol in Riverside, California. *Tellus* 39B:383–397.

Ryan, P.A., D. Lowenthal, and N. Kumar. 2005. Improved light extinction reconstruction in interagency monitoring of protected visual environments. *J. Air Waste Manage. Assoc.* 55:1751–1759.

Rygiewicz, P.T. and C.P. Andersen. 1994. Mycorrhizae alter quality and quantity of carbon allocated below ground. *Nature* 369:58–60.

Samuelson, L. and J.M. Kelly. 2001. Scaling ozone effects from seedlings to forest trees. Tansley review 21. *New Phytol.* 149:21–41.

Sandheinrich, M.B. and J.G. Wiener. 2011. Methylmercury in freshwater fish: Recent advances in assessing toxicity of environmentally relevant exposures. *In* W.N. Beyer and J.P. Meador (Eds.). *Environmental Contaminants in Biota: Interpreting Tissue Concentrations*. CRC Press, Boca Raton, FL, pp. 169–190.

Saros, J.E., D.W. Clow, T. Blett, and A.P. Wolfe. 2011. Critical nitrogen deposition loads in high-elevation lakes of the western US inferred from paleolimnological records. *Water Air Soil Pollut.* 216:193–202.

Saros, J.E., S.J. Interlandi, A.P. Wolfe, and D.R. Engstrom. 2003. Recent changes in the diatom community structure of lakes in the Beartooth Mountain Range, U.S.A. *Arct. Anarct. Alp. Res.* 35(1):18–23.

Scheuhammer, A.M. 1991. Effects of acidification on the availability of toxic metals and calcium to wild birds and mammals. *Environ. Pollut.* 71:329–375.

Scheuhammer, A.M., M.W. Meyer, M.B. Sandheinrich, and M.W. Murray. 2007. Effects of environmental methylmercury on the health of wild birds, mammals, and fish. *Ambio* 36(1):12–18.

Schwindt, A.R., J.W. Fournie, D.H. Landers, C.B. Schreck, and M.L. Kent. 2008. Mercury concentrations in salmonids from western U.S. national parks and relationships with age and macrophage aggregates. *Environ. Sci. Technol.* 42:1365–1370.

Schwindt, A.R., M.L. Kent, L.K. Ackerman, S.L.M. Simonich, D.H. Landers, T. Blett, and C.B. Schreck. 2009. Reproductive abnormalities in trout from western U.S. National Parks. *Trans. Am. Fish. Soc.* 138:522–531.

Simonich, S.L. and R.A. Hites. 1995. Global distribution of persistent organochlorine compounds. *Science* 269:1851–1854.

Singh, H.B., F.L. Ludwig, and W.B. Johnson. 1978. Tropospheric ozone: Concentrations and variabilities in clean remote atmospheres. *Atmos. Environ.* 12:2185–2196.

Singh, H.B., W. Viezee, W.B. Johnson, and F.L. Ludwig. 1980. The impact of stratospheric ozone on tropospheric air quality. *J. Air Pollut. Control Assoc.* 30:1009–1017.

Sisler, J.F. 1996. Spatial and seasonal patterns and long term variability of the composition of the haze in the United States: An analysis of data from the IMPROVE network. Cooperative Institute for Research in the Atmosphere, Colorado State University, Fort Collins, CO.

Smith, G. 2012. Ambient ozone injury to forest plants in northeast and northcentral USA: 16 years of biomonitoring. *Environ. Monitor. Assess.* 184:4049–4065.

Smith, G., J. Coulston, E. Jepsen, and T. Prichard. 2003. A national ozone biomonitoring program: Results from field surveys of ozone sensitive plants in northeastern forests (1994–2000). *Environ. Monitor. Assess.* 87:271–291.

St. Louis, V.L., J.W.M. Rudd, C.A. Kelly, K.G. Beaty, N.S. Bloom, and R.J. Flett. 1994. Importance of wetlands as sources of methylmercury to boreal forest ecosystems. *Can. J. Fish. Aquat. Sci.* 51(5):1065–1076.

Stevens, C.J., N.B. Dise, O.J. Mountford, and D.J. Gowing. 2004. Impact of nitrogen deposition on the species richness of grasslands. *Science* 303:1876–1878.

Stolte, K.W. 1996. Symptomology of ozone injury to pine foliage. *In* P.R. Miller, K.W. Stolte, D.M. Duriscoe, and J. Pronos (tech. coords.). *Evaluating Ozone Air Pollution Effects on Pines in the Western United States.* USDA Forest Service General Technical Report PSW-GTR-155, Albany, CA, pp. 11–18.

Stroo, H.F., P.B. Reich, A.W. Schoettle, and R.G. Amundson. 1988. Effects of ozone and acid rain on white pine (*Pinus strobus*) seedlings grown in five soils. II. Mycorrhizal infection. *Can. J. Bot.* 66:1510–1516.

Sullivan, T.J. 2012. Combining ecosystem service and critical load concepts for resource management and public policy. *Water* 4:905–913.

Sullivan, T.J. 2015. *Air Pollutant Deposition and Its Effects on Natural Resources in New York State.* Cornell University Press, Ithaca, NY, 307pp.

Sullivan, T.J., B.J. Cosby, J.A. Laurence, R.L. Dennis, K. Savig, J.R. Webb, A.J. Bulger et al. 2003. Assessment of air quality and related values in Shenandoah National Park. NPS/NERCHAL/NRTR-03/090. U.S. Department of the Interior, National Park Service, Northeast Region.

Sullivan, T.J. and J. Jenkins. 2014. The science and policy of critical loads of pollutant deposition to protect sensitive ecosystems in NY. *Ann. N. Y. Acad. Sci.* 1313:57–68.

Sullivan, T.J., G.B. Lawrence, S.W. Bailey, T.C. McDonnell, C.M. Beier, K.C. Weathers, G.T. McPherson, and D.A. Bishop. 2013. Effects of acidic deposition and soil acidification on sugar maple in the Adirondack Mountains, New York. *Environ. Sci. Technol.* 47:12687–12694.

Sullivan, T.J. and T.C. McDonnell. 2014. Mapping of nutrient-nitrogen critical loads for selected national parks in the intermountain west and Great Lakes regions. Natural Resource Technical Report NPS/ARD/NRTR—2014/895. National Park Service, Fort Collins, CO.

Sullivan, T.J., J.R. Webb, K.U. Snyder, A.T. Herlihy, and B.J. Cosby. 2007. Spatial distribution of acid-sensitive and acid-impacted streams in relation to watershed features in the southern Appalachian Mountains. *Water Air Soil Pollut.* 182:57–71.

Sverdrup, H., T.C. McDonnell, T.J. Sullivan, B. Nihlgård, S. Belyazid, B. Rihm, E. Porter, W.D. Bowman, and L. Geiser. 2012. Testing the feasibility of using the ForSAFE-VEG model to map the critical load of nitrogen to protect plant biodiversity in the Rocky Mountains region, USA. *Water Air Soil Pollut.* 23:371–387.

Tang, I.N. 1996. Chemical and size effects of hygroscopic aerosols on light scattering coefficients. *J. Geophys. Res.* 101:19245–19250.

Thompson, C.R., D.M. Olszyk, G. Kats, A. Bytnerowicz, P.J. Dawson, and D.C. Wolf. 1984a. Effects of ozone and sulfur dioxide on annual plants of the Mojave Desert. *J. Air Pollut. Control Assoc.* 34:1017–1022.

Thompson, C.R., D.M. Olszyk, G. Kats, A. Bytnerowicz, P.J. Dawson, D.C. Wolf, and C.A. Fox. 1984b. *Air Pollutant Impacts on Plants of the Mojave Desert.* University of California Press, Rosemead, CA.

Treshow, M. and F.K. Anderson. 1989. *Plant Stress from Air Pollution.* John Wiley & Sons, New York, 283pp.

U.S. Environmental Protection Agency. 1996a. Air quality criteria for ozone and related photochemical oxidants. Volume II. EPA/600/P-93/004bF. Washington, DC.

U.S. Environmental Protection Agency. 1996b. Review of national ambient air quality standards for ozone. Assessment of Scientific and Technical Information, OAQPS Staff Paper. EPA-452/R-96-007. Office of Air Quality Planning and Standards, Research Triangle Park, NC.

U.S. Environmental Protection Agency. 2001a. Draft guidance for estimating natural visibility conditions under the regional haze rule. U.S. EPA, Office of Air Quality Planning and Standards, Research Triangle Park, NC.

U.S. Environmental Protection Agency. 2001b. Water quality criterion for the protection of human health—Methylmercury. EPA-823-R-01-001. Office of Science and Technology, Office of Water, U.S. EPA, Washington, DC.

U.S. Environmental Protection Agency. 2004. Air quality criteria for particulate matter. Volumes I and II. EPA/600/P-99/002aF. National Center for Environmental Assessment-RTP, Office of Research and Development, Research Triangle Park, NC.

U.S. Environmental Protection Agency. 2007. Review of the national ambient air quality standards for ozone: Policy assessment of scientific and technical information. OAQPS Staff Paper. EPA 452/R-07-007. Office of Air Quality Planning and Standards, Research Triangle Park, NC.

U.S. Environmental Protection Agency. 2008. Integrated science assessment for oxides of nitrogen and sulfur—Ecological criteria. EPA/600/R-08/082F. National Center for Environmental Assessment, Office of Research and Development, Research Triangle Park, NC.

U.S. Environmental Protection Agency. 2009a. Integrated science assessment for particulate matter. Final report. EPA/600/R-08/139F. U.S. Environmental Protection Agency, Washington, DC. Available at: http://cfpub.epa.gov/ncea/cfm/recordisplay.cfm?deid=216546 (accessed October 5, 2016).

U.S. Environmental Protection Agency. 2009b. The national study of chemical residues in lake fish tissue. EPA-823-R-09-006. U.S. Environmental Protection Agency, Office of Water, Washington, DC.

U.S. Environmental Protection Agency. 2009c. Risk and exposure assessment for review of the secondary national ambient air quality standards for oxides of nitrogen and oxides of sulfur: Final report. EPA-452/R-09-008a. Office of Air Quality Planning and Standards, Health and Environmental Impacts Division, Research Triangle Park, NC.

U.S. Forest Service, National Park Service, and U.S. Fish and Wildlife Service. 2010. Federal land managers' air quality related values work group (FLAG): Phase I report—revised (2010). Natural Resource Report NPS/NRPC/NRR—2010/232. National Park Service, Denver, CO.

U.S. Geological Survey (USGS). Last modified February 20, 2015. Predicted surface water methylmercury concentrations in National Park Service Inventory and Monitoring Program Parks. U.S. Geological Survey. Wisconsin Water Science Center, Middleton, WI. Available at: http://wi.water.usgs.gov/mercury/NPSHgMap.html (accessed February 26, 2015).

Ullrich, S.M., T.W. Tanton, and S.A. Abdrashitova. 2001. Mercury in the aquatic environment: A review of factors affecting methylation. *Crit. Rev. Environ. Sci. Tech.* 31:241–293.

USDA Forest Service—Northern Research Station. 2011. Ozone biomonitoring program. Available at: http://www.nrs.fs.fed.us/fia/topics/ozone/ (accessed September 16, 2011).

van Breemen, N., J. Mulder, and C.T. Driscoll. 1983. Acidification and alkalinization of soils. *Plant Soil* 75:283–308.

Van Miegroet, H., D.W. Cole, and N.W. Foster. 1992b. Nitrogen distribution and cycling. *In* D.W. Johnson and S.E. Lindberg (Eds.). *Atmospheric Deposition and Forest Nutrient Cycling*. Springer-Verlag, New York, pp. 178–196.

Van Miegroet, H., D.W. Johnson, and D.W. Cole. 1992a. Analysis of N cycles in polluted vs unpolluted environment. *In* D.W. Johnson and S.E. Lindberg (Eds.). *Atmospheric Deposition and Forest Nutrient Cycling*. Springer-Verlag, New York, pp. 199–202.

Vitt, D.H., K. Wieder, L.A. Halsey, and M. Turetsky. 2003. Response of *Sphagnum fuscum* to nitrogen deposition: A case study of ombrogenous peatlands in Alberta, Canada. *Bryologist* 106(2):235–245.

Wang, D., D.F. Karnosky, and F.H. Bormann. 1986. Effects of ambient ozone on the productivity of *Populus tremuloides* Michx. grown under field conditions. *Can. J. For. Res.* 16:47–55.

Weinstein, D.A., R.M. Beloin, and R.D. Yanai. 1991. Modeling changes in red spruce carbon balance and allocation in response to interacting ozone and nutrient stress. *Tree Physiol.* 9:127–146.

Weinstein, D.A. and R.D. Yanai. 1994. Integrating the effects of simultaneous multiple stresses on plants using the simulation model TREGRO. *J. Environ. Qual.* 23:418–428.

Wellburn, A.R. 1988. *Air Pollution and Acid Rain: The Biological Impact*. Longman Scientific and Technical, Burnt Mill, U.K.

Wiener, J.G., D.P. Krabbenhoft, G.H. Heinz, and A.M. Scheuhammer. 2003. Ecotoxicology of mercury. *In* D.J. Hoffman, B.A. Rattner, G.A. Burton, and J. Cairns (Eds.). *Handbook of Ecotoxicology*, 2nd edn. CRC Press, Boca Raton, FL, pp. 409–463.

Williams, M.W., J.S. Baron, N. Caine, R. Sommerfeld, and J.R. Sanford. 1996. Nitrogen saturation in the Rocky Mountains. *Environ. Sci. Technol.* 30:640–646.

Wolfe, A.P., J.S. Baron, and R.J. Cornett. 2001. Anthropogenic nitrogen deposition induces rapid ecological changes in alpine lakes of the Colorado Front Range (USA). *J. Paleolimnol.* 25:1–7.

Woodbury, P.B., J.A. Laurence, and G.W. Hudler. 1994. Chronic ozone exposure alters the growth of leaves, stems, and roots of hybrid *Populus*. *Environ. Pollut.* 85:103–108.

Wooldridge, G., K. Zeller, and R. Musselman. 1997. Ozone concentration characteristics at a high-elevation forest site. *Theor. Appl. Climatol.* 56:153–164.

Yoshida, L.C. and E.B. Allen. 2001. Response to ammonium and nitrate by a mycorrhizal annual invasive grass and a native shrub in southern California. *Am. J. Bot.* 88:1430–1436.

Section II

Case Studies

4

Great Smoky Mountains National Park and the Appalachian Highlands Network

4.1 Introduction

The Appalachian Highlands Network contains three national parklands that are larger than 100 mi^2: Great Smoky Mountains National Park (GRSM), Big South Fork National River and Recreation Area (BISO), and Blue Ridge Parkway (BLRI). It also contains one smaller parkland: Obed Wild and Scenic River (OBRI). Larger parks generally have more available data with which to evaluate air pollution sensitivities and effects. In addition, the larger parks generally contain more extensive resources in need of protection against the adverse impacts of air pollution. Since air pollutants can be widespread, reduction of pollution emissions affecting large parks will often result in the protection of smaller parks located nearby as well. Within the Appalachian Highlands Network, effects of air pollutants on air quality–related values have only been well studied in GRSM. Figure 4.1 shows the network boundary, the location of each park, and population centers having more than 10,000 people.

Effects of air pollutants have been well documented in GRSM, including stream acidification, ozone symptoms on plant foliage, and high haze levels. Toxic air contaminants are not addressed in this chapter. This does not imply that metals and persistent organic pollutants are not important in this park, but rather that limited data are available with which to evaluate the levels and impacts of these contaminants.

GRSM is an international biosphere reserve and is the most heavily visited national park in the United States. The park contains the largest remaining area of old-growth spruce-fir forest in the conterminous United States (Shaver et al. 1994). Air pollution constitutes a major threat to the health of natural resources in this park, and elsewhere in the Appalachian Highlands Network. As a consequence, air pollution effects research in GRSM has been substantial. GRSM is designated as Class I, giving it a heightened level of protection under the Clean Air Act against harm caused by new emissions of air pollution.

4.2 Atmospheric Emissions and Deposition

Sulfur dioxide emissions and S deposition near and within the Appalachian Highlands Network have both been high. Malm et al. (2002) analyzed spatial patterns in atmospheric sulfate concentrations throughout the United States from 1988 to 1999 and also atmospheric

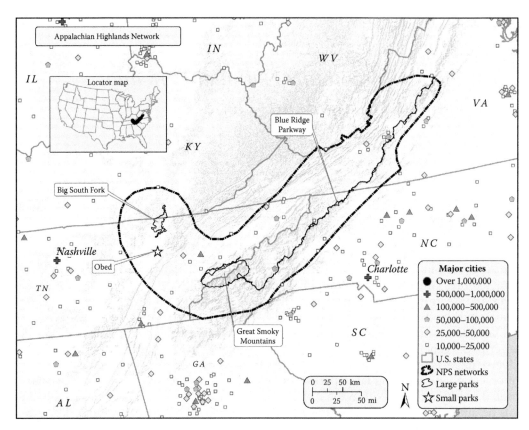

FIGURE 4.1
Appalachian Highlands Network boundary and locations of parks and population centers greater than 10,000 people.

sulfur dioxide emissions from 1990 to 1999. Based upon Interagency Monitoring of Protected Visual Environments (IMPROVE) data, the 90th percentile summer atmospheric sulfate concentrations were highest in the Ohio River Valley and in central Tennessee (generally upwind of GRSM), where emissions of sulfur dioxide have been especially high (Malm et al. 2002). Although S emissions have been substantially reduced over the past two decades by requirements of the Acid Rain Program, the Clean Air Interstate Rule, and other federal and state rules, the southern Appalachian Mountains region still has among the highest emissions levels in the country (Burns et al. 2011). These emissions have the potential to impact sensitive resources in GRSM and elsewhere in the network region. The network is surrounded by many large urban areas and point sources of both S and N emissions.

The atmospheric concentrations of sulfate in the GRSM region during the 1990s were twice as high as concentrations in the northeastern and coastal southeastern United States. However, emissions reductions have been enacted in more recent years in conjunction with federal rules, the North Carolina Clean Smoke Stacks Act, the Georgia Multipollutant Rule, and a legal settlement between North Carolina and the Tennessee Valley Authority. As a result, ambient atmospheric sulfur dioxide concentrations in the Southeast decreased by more than 50% between the periods 1989–1991 and 2007–2009 (Burns et al. 2011). Emissions

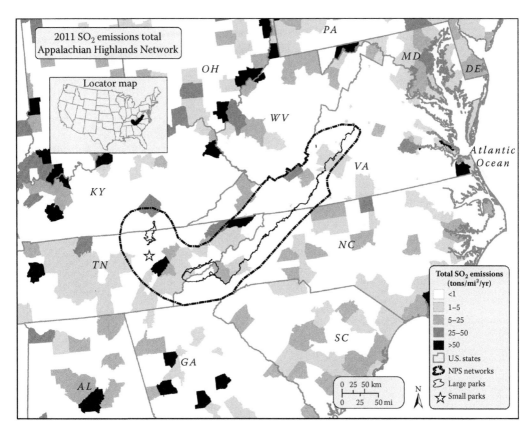

FIGURE 4.2
Total sulfur dioxide (SO$_2$) emissions, by county, near GRSM for the year 2011. (Data from the EPA's National Emissions Inventory, https://www.epa.gov/air-emissions-inventories, accessed January, 2014.)

for 1990–2013 can be found by state at http://www3.epa.gov/airmarkets/progress/reports/emissions_reductions_so2.html#figure2.

County-level emissions near GRSM, based on data from the EPA's National Emissions Inventory during a recent time period (2011), are depicted in Figures 4.2 through 4.4 for sulfur dioxide, oxidized N, and ammonia, respectively. Several counties to the north and west of GRSM had relatively high sulfur dioxide emissions (>50 tons/mi^2/yr; Figure 4.2). Spatial patterns in oxidized N emissions were generally similar, with highest values to the northwest of the park (Figure 4.3). Emissions of ammonia near the park were somewhat lower, with most counties showing emissions levels below 8 tons/mi^2/yr (Figure 4.4).

Total wet plus dry S deposition throughout much of GRSM ranged from about 10 to 15 kg/ha/yr in 2002 (Sullivan et al. 2011) but has since declined. Some locations near the park had estimated wet plus dry S deposition higher than that. Total wet plus dry N deposition within the network ranged from as low as 2 to 5 kg N/ha/yr to as high as 10 to 15 kg N/ha/yr (Sullivan et al. 2011). In addition, it is known that cloud (a type of occult) deposition, which is not represented in the total deposition values depicted on the maps shown here, can be quite high at sites having elevation greater than 1500 m (4921 ft) that occur in and around GRSM. At the highest-elevation locations, the total S and N deposition, including cloud inputs, might each be as much as double the wet plus dry deposition values shown on the maps. However, because of more recent large emissions reductions

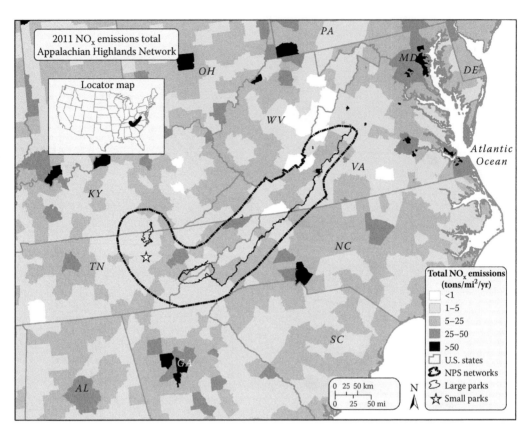

FIGURE 4.3
Total oxidized nitrogen (NO$_x$) emissions, by county, near GRSM for the year 2011. (Data from the EPA's National Emissions Inventory, https://www.epa.gov/air-emissions-inventories, accessed January, 2014.)

achieved by the Acid Rain Program and other air quality improvement programs, total S deposition decreased about 40% in the Southeast between the periods 1989–1991 and 2007–2009 (Burns et al. 2011). Total N deposition decreased about 19%. This more modest decrease in N deposition reflects the fact that air quality management programs for N focused only on oxidized N sources (vehicles, power plants, industry); ammonia sources (agriculture, feedlots), which also contribute to total N deposition, are largely unregulated.

Atmospheric S, and to a lesser extent N, deposition levels have continued to decline at GRSM since 2001, based on Total Deposition project estimates (Table 4.1). Decreases in total S deposition over the previous decade averaged nearly 43%. Estimated total N deposition decreased over that same time period by 0.79 kg N/ha/yr (–7.4%). Oxidized N and ammonium deposition showed opposite patterns, with oxidized N decreasing and ammonium increasing since the monitoring period 2000–2002. Total S deposition in and around GRSM for the period 2010–2012 was generally highest (>10 kg S/ha/yr) to the northwest and lowest (<5 kg S/ha/yr) to the southeast of the network area (Figure 4.5).

Oxidized inorganic N deposition for the period 2010–2012 was in the range of 5–10 kg N/ha/yr throughout much of the park. Most areas received about 2–5 kg N/ha/yr of ammonium from atmospheric deposition during this same period; total N deposition was in the range of 10–15 kg N/ha/yr at most park locations (Figure 4.6). Other areas received 5–10 kg N/ha/yr of total N deposition.

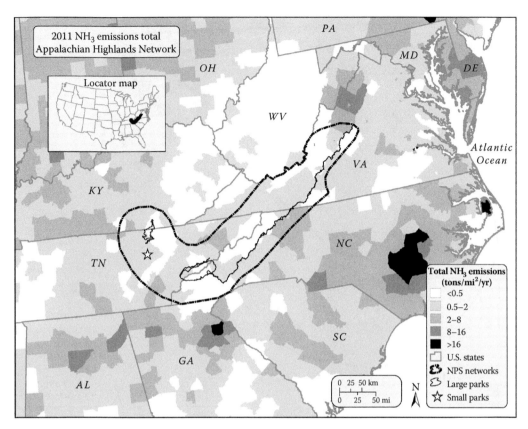

FIGURE 4.4
Total ammonia (NH$_3$) emissions, by county, near GRSM for the year 2011. (Data from the EPA's National Emissions Inventory, https://www.epa.gov/air-emissions-inventories, accessed January, 2014.)

TABLE 4.1

Average Changes in S and N Deposition[a] between 2001 and 2011 across Park Grid Cells at GRSM

Parameter	2001 Average (kg/ha/yr)	2011 Average (kg/ha/yr)	Absolute Change (kg/ha/yr)	Percent Change	2011 Minimum (kg/ha/yr)	2011 Maximum (kg/ha/yr)	2011 Range (kg/ha/yr)
Total S	12.92	7.39	−5.52	−42.8	4.02	9.19	5.17
Total N	10.69	9.89	−0.79	−7.4	6.31	11.68	5.37
Oxidized N	7.64	5.89	−1.75	−22.9	3.62	6.89	3.27
Ammonium	3.04	4.00	0.96	32.1	2.68	4.85	2.17

[a] Deposition estimates were determined by the Total Deposition project, based on three-year averages centered on 2001 and 2011 for all ~4 km grid cells in each park. The minimum, maximum, and range of 2011 S and N deposition within each park are also shown.

Weathers et al. (2006) developed an empirical modeling approach, based on 300–400 throughfall measurements, to estimate the total (wet, dry, cloud) deposition of S and N to complex terrain of GRSM. Throughfall deposition measurements are based on samples of precipitation that have dripped through the canopy and been captured near ground level during summer. The data on water concentration and amount, combined with landscape

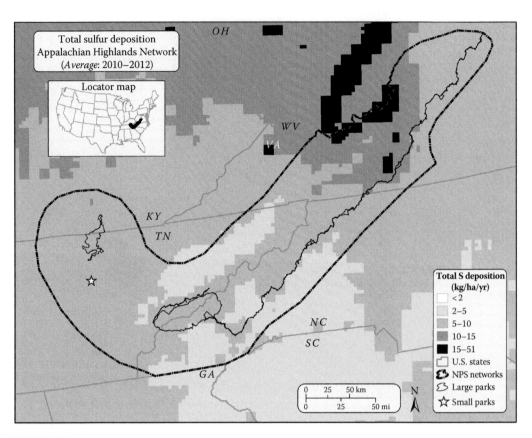

FIGURE 4.5
Total S deposition for the three-year period centered on 2011 in and around GRSM. (From Schwede, D.B. and Lear, G.G., *Atmos. Environ.*, 92. 207, 2014.)

variables, such as elevation, forest type, and slope, explained about 40% of the variation in total deposition estimates. Model estimates were scaled to measured wet and estimated dry deposition values from National Atmospheric Deposition Program and Clean Air Status and Trends Network monitoring sites, respectively. Resulting maps showed substantial spatial variability, with high values of 31 kg N/ha/yr and 42 kg S/ha/yr across the landscapes of this park. Thus, atmospheric loadings of both S and N at high-elevation locations in GRSM have been among the highest of any wildland area in the United States. Much of this deposition has been from dry and cloud deposition processes, which are particularly efficient modes of deposition in spruce-fir forests along the ridge tops. Because of recent emissions reductions, these modeled deposition estimates do not reflect current conditions but are important in assessing the long-term effects of acid deposition in this park.

Cloud water deposition was measured at three high-elevation sites in the Appalachian Mountains in the Mountain Acid Deposition Program. One of those sites was at Clingmans Dome in GRSM. Seasonal cloud water deposition at that site was very high for both sulfate (>50 kg/ha) and N oxides (>25 kg/ha). Baumgardner et al. (2003) concluded that total deposition of acidic anions at high elevation (>1500 m) in GRSM may be 6–20 times higher than the deposition at low elevation, with this difference mainly attributed to cloud deposition at the high elevations.

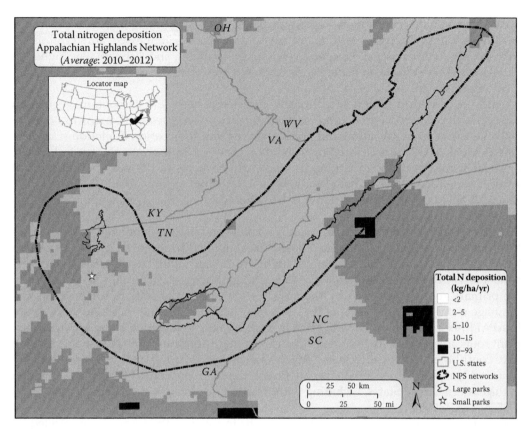

FIGURE 4.6

Total N deposition for the three-year period centered on 2011 in and around GRSM. (From Schwede, D.B. and Lear, G.G., *Atmos. Environ.*, 92. 207, 2014.)

Atmospheric deposition of Hg is also important for resource protection in GRSM. The National Park Service (2010) reported long-term trends in concentrations of Hg in wet deposition during the periods beginning in 1996–2003, and running through 2008, for nine national parks. Three-year means of annual Hg concentration in wet deposition were reported by the National Park Service (2010) for 13 parks that had at least two years of valid data during the period 2006–2008. The highest Hg concentration in precipitation was reported for Everglades National Park in South Florida, followed by GRSM. More recent data are available through the Mercury Deposition Network of the National Atmospheric Deposition Program (http://nadp.sws.uiuc.edu/mdn/). There are two Hg deposition monitoring sites in GRSM, at Elkmont (since 2002) and Clingmans Dome (since 2015).

4.3 Acidification

4.3.1 Acidification of Terrestrial Ecosystems

Some research has been conducted within the GRSM region on the effects of S and N on lichens. McCune et al. (1997) found air pollution–tolerant epiphytic lichen species to be present, along with relatively low epiphytic lichen species richness, in urban and industrial areas within Georgia, North Carolina, South Carolina, Tennessee, and Virginia. In contrast, air pollution–sensitive species were more common, and richness was higher, in rural areas. These effects on lichens may be due to both acidification and nutrient enrichment impacts of enhanced S and N input.

Land cover within GRSM is primarily forested. These forests contain vegetation types that include red spruce and sugar maple, two tree species thought to be especially sensitive to acidification and accompanying soil base cation depletion (U.S. EPA 2008a). Much of GRSM is covered by vegetation types likely to contain either red spruce or sugar maple, with the exception of lands around parts of the park boundary. Red spruce is confined to the higher elevations, generally higher than about 1300 m (4265 ft). Much of this higher elevation land is distributed along the Appalachian Mountain ridgeline.

High-elevation soils in GRSM are mainly shallow and highly weathered, with substantial S adsorption capacity and limited base cation supply (Elwood 1991). Acidic deposition has contributed to a decline in the availability of Ca and other base cations in the soils of acid-sensitive forest ecosystems due to leaching of base cations from foliage and from the primary rooting zone and by the mobilization of Al from soils to soil solution and drainage water (Eagar and Adams 1992, National Acid Precipitation Assessment Program [NAPAP] 1998). Aluminum mobilization from acid soils can impede Ca and Mg uptake by plant roots and potentially induce deficiencies in these nutrients. Both N and S deposition have contributed to these effects in GRSM. As a consequence, foliar Ca levels and soil and root Ca:Al ratios are considered low to deficient over large portions of the spruce-fir region in the eastern United States, including within GRSM (Joslin et al. 1992, Cronan and Grigal 1995, NAPAP 1998).

Researchers have concluded that the spruce-fir forests in GRSM have been affected by atmospheric S and N deposition in at least two important ways (Eager and Adams 1992, Shaver et al. 1994):

1. Inputs of N and other nutrients exceed normal litterfall inputs, thereby altering nutrient cycling and uptake into trees.
2. Pulses of high nitrate and sulfate concentrations in soil water have caused periodic increases in Al concentration in soil water, which could inhibit root growth and plant nutrient base cation uptake.

Calcium must be dissolved in soil water in order to be taken up by roots. Aluminum in soil solution reduces Ca uptake by competing for binding sites in the cortex of fine roots. Tree species may be adversely affected if high Al-to-nutrient ratios limit the uptake of Ca and Mg (Shortle and Smith 1988, Garner 1994). A reduction in Ca uptake suppresses cambial growth, reduces the rate of wood formation, decreases the amount of functional sapwood and live crown, and predisposes trees to disease and injury (Smith 1990).

A variety of factors are known to predispose soils of high-elevation spruce-fir forests in GRSM to potential Al toxicity and Al-induced inhibition of cation uptake. These factors include features of the climate (high precipitation, low temperature), vegetation (coniferous tree litter), bedrock (low base cation production), and soil forming processes such as podzolization (Eagar et al. 1996). Dissolved Al concentrations measured in soil solution at spruce-fir study sites in GRSM frequently exceeded 50 μM and sometimes exceeded 100 μM (Johnson et al. 1991b, Joslin and Wolfe 1992). All studies reviewed by Eagar et al. (1996) showed a strong correlation between Al concentrations and nitrate concentrations in soil solution. They speculated that the occurrence of periodic large pulses of nitrate in soil solution was important in determining soil Al chemistry in this region.

Dendrochronological (tree ring) analyses of tree cores collected from permanent plots in GRSM (37 trees cores from moderate elevation [~1500 m]; 35 tree cores from high-elevation sites [~2000 m]), demonstrated a positive correlation between temporal and spatial trends in red spruce growth and acidic deposition, with a greater response in trees on ridges as compared with trees in draws (Eagar et al. 1996, Webster et al. 2004). In general, ridges are naturally more acidified, receive higher levels of acidic deposition, and have shallower soils with lower base saturation (Webster et al. 2004). However, effects from ozone may be superimposed on acidification effects.

From the 1940s to 1970s, red spruce growth declined in the southern Appalachian Mountains, as upwind emissions of both oxidized N and sulfur dioxide increased (Webster et al. 2004). The growth decline started earlier at higher elevations (around 1940s and 1950s) and was steeper, while the growth decline developed about 20 years later at lower-elevation sites. After the 1980s, growth increased substantially at both the higher- and lower-elevation sites, corresponding to a decrease in sulfur dioxide emissions throughout the eastern United States, while oxidized N emissions held fairly steady. Annual emissions of S oxides plus N oxides explained about 43% of the variability in tree ring growth at high elevation between 1940 and 1998, while climatic variability accounted for about 8% of the growth variation for that period. At low elevation, changes in radial growth could be explained by climatic variables only, and there was no correlation with national S and N oxides emissions trends (Eagar et al. 1996, Webster et al. 2004).

Results of modeling by Sullivan et al. (2002a), using the Nutrient Cycling Model (Liu et al. 1991, Johnson et al. 1993a), together with the results of model simulations published for other watersheds in the southern Appalachian Mountains region, suggested that spruce-fir forests in the region are likely to experience decreased Ca:Al ratios in soil solution under virtually all strategies of reduced future acidic deposition considered in the modeling effort. This was partly because S adsorption on soils is likely to decline in the future, even with dramatically reduced S deposition, allowing increased sulfate leaching. In addition, many spruce-fir forests in the region are at least partially N-saturated, and continued N deposition at moderate or high levels would be expected to contribute to elevated nitrate concentrations in soil water, which could further enhance base cation leaching and mobilization of Al from soils to soil solution (Sullivan et al. 2002a). These processes will be exacerbated by the already naturally low values of base saturation in the soils of many of these forests. Thus, the results of these modeling efforts suggest that changes to forest soils and soil solution will likely continue to occur. It is not clear, however, to what extent these changes in the chemistry of soils and soil solution might actually impact forest growth and health. The state of scientific understanding on this topic suggests that such chemical changes would increase the likelihood that the growth and/or health of spruce-fir forests would be adversely impacted, perhaps making them more susceptible to other stressors associated with such factors as insect pests, pathogens, or extreme climatic conditions. However, the occurrence of low base saturation and low Ca:Al ratio in soil solution will not necessarily be sufficient to cause widespread impacts. Many factors in addition to soil and soil solution acid–base chemistry are important in this regard. Fraser fir (*Abies fraseri*) stands killed by the balsam woolly adelgid (*Adelges piceae*), an exotic insect pest, are largely being replaced by vigorous regrowth of young stands of that species (Van Miegroet et al. 2007). To what extent spruce or fir mortality in the southern Appalachian Mountains will result in a species mix similar to what existed prior to the mortality remains to be seen.

The limited available empirical data suggest that the kinds of changes in soil solution chemistry projected by the Nutrient Cycling Model for spruce-fir stands in the GRSM region are consistent with the kinds of changes that have been associated in the past with reductions in forest growth. The weight of evidence for spruce-fir forests suggests that continued adverse impacts on soil solution chemistry are likely, and adverse impacts on forest growth and health are possible (Sullivan et al. 2002a).

Although scientists are concerned about the potential effects of soil acidification on forest health in GRSM, forest health is a complicated concept. It can be reflected in a variety of physiological indicators, including changes in the growth rate of trees, foliar damage, branch dieback, susceptibility to insects or disease, or tree mortality. Similarly, forest health can be affected by a host of potential stressors, of which air pollution is only one possibility. Climatic conditions, stand competition, outbreak of nonnative pathogens, and forest management (alone or in combination) often contribute greatly to observed forest health problems. Attempts to document, and in particular to quantify, the effects of air pollution on forest health have encountered considerable complexity and uncertainty. Nevertheless, such efforts have produced evidence that suggests that red spruce and sugar maple in some areas in the eastern United States have experienced declining health as a consequence of acidic deposition (Eagar et al. 1996, Sullivan et al. 2002a).

Base saturation values less than about 10% predominate in the soil B horizon in areas where soil and surface water acidification from acidic deposition have been most pronounced, including red spruce forests in the southern Appalachian Mountains (Sullivan et al. 2003). Decreases in concentrations of exchangeable base cations in the Oa and B soil horizons over the past several decades have been common and widespread in the eastern United States. Acidic deposition has been shown to be an important factor causing the observed decreases in concentrations of exchangeable base cations on soils. To some degree, soils can recover their base cation reserves over time in response to reduced future levels of acidic deposition. However, the recovery potential of soil exchangeable base cation concentrations is dependent on resupply by weathering, which is a slow process.

Johnson et al. (1999) explored, through simulation modeling, the potential effects of changes in S, N, and base cation deposition on two contrasting forest ecosystems: a highly acidic red spruce site (1740 m [5709 ft] elevation) at Noland Divide in GRSM and a moderately acidic mixed hardwood site at the Coweeta hydrological research station (720 m [2362 ft] elevation) in North Carolina. The simulations were performed using the Nutrient Cycling Model. Results for the highly acidic Noland Divide site suggested that S and N depositions were the major drivers of soil solution concentrations of Al at this site, and that atmospheric base cation deposition was the major driver of soil solution concentrations of base cations. Results differed at the less acidic Coweeta site, where Al was unimportant, and S and N deposition had a much larger impact than base cation deposition on simulated soil solution chemistry (Johnson et al. 1999).

Soil solution samples collected from spruce-fir forests in GRSM have been shown to exhibit very high concentrations of nitrate and Al, compared with soil solution from similar forests in the northeastern United States (Joslin et al. 1992). At two study sites in GRSM, pulses of nitrate in soil solution caused the Al concentration to exceed levels known to impede the root uptake of Ca and Mg by trees (Johnson et al. 1991a). The molar ratio of soil solution Ca:Al at these sites was consistently below 1.0 and often below 0.5. These values were identified by Cronan and Grigal (1995) as having 50% and 75% risk of adverse impact on tree growth, respectively.

Pardo and Duarte (2007) modeled critical loads for acid deposition effects to forests in GRSM using ecological indicators of forest soil health, including soil base saturation, Al to base cation ratio, and Al concentration. The predicted critical loads at which the indicators were significantly impacted varied by forest type, but for most indicators the critical loads were exceeded by existing deposition levels. Modeling of future deposition scenarios suggested that very large reductions in the atmospheric deposition of oxidized N, sulfur dioxide, and reduced N, well beyond anything currently planned, would be needed to restore forest soil base status.

4.3.2 Acidification of Aquatic Ecosystems

The regions in the Southeast that were identified by Charles (1991) as containing large numbers of low acid neutralizing capacity surface waters included the Appalachian Mountains in and around GRSM. Stream water acid–base chemistry has been extensively studied in this region (e.g., Church et al. 1992, Herlihy et al. 1993, Webb et al. 1994, van Sickle and Church 1995, Sullivan et al. 2002b, 2003). Tennessee has designated 12 park streams as impaired under Section 303(d) of the Clean Water Act for aquatic life, based on having pH less than 6.0. This process has resulted in a multiyear effort (ongoing) to model recovery rates and develop estimates of Total Maximum Daily Loads for these impaired streams.

Elevations in GRSM are among the highest in the eastern United States, with the majority of the parkland being higher than 1000 m (3281 ft). Most streams in the park are first through third order (Sullivan et al. 2011). These low-order* streams are the smallest, coldest streams located at highest elevations. Land slope tends to be fairly steep in GRSM, with most of the park having terrain steeper than 30°. In addition, some land along the BLRI is steeper than 30°. These low-order, high-elevation streams on steep terrain are often particularly sensitive to acidification impacts from both S and N deposition.

One of the most important processes affecting watershed acid neutralization throughout much of the Southeast is S adsorption on soil (Rice et al. 2014). If S adsorption on soil is high, relatively high levels of S deposition have little or no impact on stream acid–base chemistry in the short term. However, this S adsorption capacity can become depleted over time under continued S deposition and cause a delayed acidification response. Because of watershed sulfate retention, soils and surface waters of GRSM region have not yet realized the full effects of elevated S deposition. Sulfur adsorption has been especially high in the soils of the Blue Ridge Province, where typically about half or more of the incoming S has been retained (Herlihy et al. 1993). In general, S adsorption is higher in the southern portions of the Appalachian Mountains region in and around GRSM. Herlihy et al. (1993) concluded that S retention will likely decrease in the future, resulting in further losses of stream acid neutralizing capacity.

Acidic streams and streams with very low acid neutralizing capacity in GRSM are almost all located in small (watershed area < 20 km² [7.7 mi²]), upland, forested catchments in areas of base-poor bedrock. Acidic surface waters in this region are nearly always found in forested watersheds. Localized studies have shown that stream water chemistry is closely related to bedrock mineralogy in the GRSM region (Herlihy et al. 1993). Sullivan et al. (2007) delimited a high-interest area for stream water acidification sensitivity within the southern Appalachian Mountains region (Virginia/West Virginia to Georgia) based on geological classification and elevation. It covered only 28% of the region and yet included almost all known acidic and low acid neutralizing capacity (<20 µeq/L) streams, based on

---

* Low-order streams are the small tributary streams located at relatively high elevation, designated by Strahler Order, generally as first, second, or third order.

the evaluation of about 1000 streams for which water chemistry data were available. They found that the vast majority of low acid neutralizing capacity streams were underlain by the siliceous geologic sensitivity class, which is represented by such lithologies as sandstone and quartzite. Low acid neutralizing capacity stream water throughout the region was also found to be associated with a number of watershed features in addition to lithology and elevation, including ecoregion, physiographic province, soil type, forest type, and watershed area.

Cook et al. (1994) reported that very high nitrate concentrations (~100 µeq/L) in upland streams in GRSM were correlated with elevation and forest stand age. The old-growth sites at higher elevation showed the highest stream nitrate concentrations. This pattern could have been due in part to the higher rates of N deposition and flashier hydrology at high elevation and also to decreased N uptake by trees in older forest stands. High N deposition has likely contributed to both chronic and episodic acidification of stream water in this park (Flum and Nodvin 1995, Nodvin et al. 1995).

Model projections of future changes in acid–base chemistry of streams in the southeastern United States, including the GRSM region, were reported by Sullivan et al. (2002b). They modeled future effects of atmospheric S and N deposition on aquatic resources in the eight-state southern Appalachian Mountains region. Modeling was conducted with the Model of Acidification of Groundwater in Catchments for 40–50 sites within each of three physiographic provinces, stratified by stream water acid neutralizing capacity class. Simulations were based on assumed constant future atmospheric deposition at 1995 levels and on three regional strategies of emissions controls provided by the Southern Appalachian Mountains Initiative, based on simulations performed with the Urban to Regional Multiscale One-Atmosphere model (Odman et al. 2002). The National Stream Survey statistical frame (Kaufmann et al. 1991) was used to estimate the number and percentage of stream reaches in the region that were projected by the model to change their chemistry in response to the emissions control strategies. There was a small decline in the estimated length of projected acidic (acid neutralizing capacity ≤ 0 µeq/L) streams in 2040 from the least to the most restrictive emissions control strategy, but there was little difference in projected stream length in the other acid neutralizing capacity classes as a consequence of adopting one or another emissions control strategy. However, projections of continued future acidification were substantially larger under a scenario in which S and N deposition were held constant into the future at 1995 levels. Thus, emissions controls to date have likely had a large impact on stream acid–base chemistry, but further emissions reductions may have smaller benefits.

Most simulated changes in stream water acid neutralizing capacity from 1995 to 2040 were rather modest, given the very large estimates of decreases in S deposition. Few modeled streams showed projected change in acid neutralizing capacity of more than about 20 µeq/L. Some of the largest changes were simulated for some of the streams that were most acidic in 1995. For such streams, however, even relatively large increases in acid neutralizing capacity would still result in negative acid neutralizing capacity stream water, and therefore, little biological benefit would be expected from the simulated improvement in chemistry. The model results suggested, however, that benefits of emissions controls would continue to accrue well beyond 2040 for all strategies, even if deposition was held constant at 2040 levels into the future (Sullivan et al. 2002b, 2004). Thus, recovery from past stream acidification is expected to be slow and gradual, even with rather large S emissions reductions.

Concentrations of nitrate in two streams in the Noland Divide watershed in GRSM decreased over the period 1991–1998 (Robinson et al. 2004). This trend was attributed in large part to increased N uptake by understory vegetation growth that had been stimulated following the mortality of Fraser fir trees. Robinson et al. (2004) developed, based on monitoring data, multiple regression models for GRSM to predict stream acid neutralizing capacity and concentrations and loads of various ionic constituents. Estimates did not differ depending on whether they were based on weekly to monthly sampling, thereby allowing the National Park Service to redirect resources from the ongoing weekly stream monitoring to other monitoring needs and to focus in the future on monthly monitoring. Despite reduced levels of acidic deposition in recent decades, Robinson et al. (2008) concluded that there has been little chemical recovery of acidified streams in GRSM.

Geology, especially the presence of calcareous or pyritic bedrock, has substantial influence on stream chemistry in GRSM making some streams naturally acidic. The Anakeesta Formation contains pyrite, which weathers when exposed to the air by landslides and road cuts to produce sulfuric acid. It is exposed at many locations in the north-central portion of the park. In the western parts of GRSM, there are deposits of limestone and dolomite that provide substantial amounts of base cations to drainage water. Streams in GRSM that have acid neutralizing capacity >100 µeq/L are mostly influenced by limestone weathering and are located mainly in the western portion of the park. Streams that have acid neutralizing capacity near or below 0 µeq/L are mostly influenced by Anakeesta weathering and are generally located in the north-central portion of the park (Flum and Nodvin 1995). Weathering of the exposed S-bearing rock contributes sulfuric acid to drainage water. Acidification in response to acidic deposition is superimposed on top of these spatial patterns in geology.

4.3.2.1 Episodic Acidification

Temporal variability in surface water and soil solution chemistry, and seasonal patterns in nutrient uptake by terrestrial and aquatic biota, influence acidification processes and pathways. Conditions are constantly changing in response to episodic, seasonal, and interannual cycles and processes. Because of the steep topography and importance of S adsorption on soils in GRSM, episodic acidification of stream water is an important process. In particular, weather and climatic fluctuations that govern the amount and timing of precipitation inputs, snowmelt, vegetative growth, depth to groundwater, and evapoconcentration of solutes influence soil and surface water chemistry and the interactions between pollution stress and sensitive aquatic and terrestrial biological receptors.

Chemical changes during hydrologic episodes are controlled in part by acidic deposition and in part by natural processes, including dilution of base cation concentrations, nitrification, flushing of organic acids from terrestrial to aquatic systems, and the neutral salt effect (acidification of drainage water caused by the addition of a neutral salt, such as NaCl, to acid soil). Episodic acidification pulses may last for hours to weeks and sometimes result in depletion of acid neutralizing capacity in acid-sensitive streams to negative values, with concomitant increases in inorganic Al in solution, often to toxic levels and potentially harming fish and other aquatic life.

The most important factor governing watershed sensitivity to episodic acidification is the pathway followed by snowmelt water and stormflow water through the watershed. These pathways determine the extent of acid neutralization provided by the soils and bedrock in that watershed. High-elevation watersheds with steep topography, extensive areas of exposed bedrock, deep snowpack accumulation, and shallow, base-poor soils tend to be most sensitive to episodic acidification. All of these conditions occur at the high-elevation locations in GRSM.

4.3.2.2 Effects on Aquatic Biota

The southeastern United States is geologically diverse. It was not glaciated during the most recent period of glaciation. As a consequence, a large number of aquatic animal species have evolved and colonized this region over a longer time period than northeastern areas. Streams in the Tennessee-Cumberland aquatic region (Smith et al. 2002) are among the most biologically diverse in the world. This region contains the highest diversity of freshwater mussel and crayfish species and the highest levels of aquatic species endemism (distribution specific to a particular region) in North America (Abell et al. 2000, Smith et al. 2002), including 231 species of fish (67 endemic), 125 species of mussels (20 endemic), and 65 species of crayfish (40 endemic). Regional habitat diversity and intraspecific genetic diversity are also regarded as high. Thus, the Southeast is a unique national biodiversity resource for fish and other aquatic organisms. More than 57 species of fish and 47 species of mussels are considered at risk (Master et al. 1998, Abell et al. 2000) due to a number of stressors, including acidification. The relative sensitivities of the various species to water acidification are largely unknown.

It is likely that aquatic biota in GRSM have been affected by acidification at virtually all levels of the food web. Effects have been best documented for fish and aquatic insects (Bulger et al. 1999, Cosby et al. 2006). Some species, and some life stages within species, are more sensitive than others. In some acidified waters, sensitive species have been eliminated and taxonomic richness (the total number of taxa of a particular type in a stream) has decreased. Surface water acidification typically results in the loss of the most acid-sensitive species first, and there is more loss with higher degrees of acidification, leading to decreasing overall species richness. In some cases, it is likely that all fish species have been eliminated from chronically acidic streams.

Fish communities of high-gradient southern Appalachian streams may contain a variety of species but are often dominated by trout, especially brook trout. Of the 15.1 million ha (37.4 million acres) in the southern Appalachian region (as defined by the Southern Appalachian Man and the Biosphere Program [1996]), 5.9 million ha (14.6 million acres [39%]) are in the range of native brook trout, with up to 53,000 km (33,000 mi) of potential native brook trout streams.

There are clear patterns in species distribution from headwaters to rivers, which can also be seen in community comparisons among reaches at different elevations; the clearest pattern is that species richness increases in a downstream direction. This is thought to result from the rather small number of upstream species, which must tolerate simultaneously highest current velocities and lowest pH values. Fish are absent from the highest headwaters. The highest-elevation fish species is usually brook trout, typically joined downstream by species of dace and darter and by introduced brown (*Salmo trutta*) or rainbow (*Oncorhynchus mykiss*) trout (Wallace et al. 1992). In the context of acidification, the introduced trout are both more acid sensitive than brook trout and will not be present in highly acidified waters. Proceeding downstream, other dace, darters, chubs, shiners, suckers, etc., are often present. In larger downstream reaches, still regarded as high gradient, the important gamefish smallmouth bass (*Micropterus dolomieu*) can be abundant (Wallace et al. 1992). Native brook trout inhabit about 603 mi (970 km) of stream length in GRSM, most of which is above 900 m (2953 ft) elevation (Neff et al. 2009). The trout at higher elevation are more susceptible to acidification because of the lower pH typical of these streams.

In addition to stream chemistry, other factors are related to patterns of distribution and abundance of fish and other aquatic biota, especially temperature, gradient, stream order, and flow regime. Stream gradient can have substantial influence on water velocity, substrate size, number and size of pools, and oxygen content (Moyle and Cech 2000). All of these influence the habitat quality for aquatic biota. At some locations, acidity in the upper reaches and high stream temperatures in the lower reaches limit the availability of suitable habitat for cold-water species (McDonnell et al. 2015).

Effects of stream acidity on benthic invertebrates were investigated by Rosemond et al. (1992) at four sites on Walker Camp Prong and its tributaries (Trout Branch and Cole Creek) in GRSM. Study sites exhibited similar physical stream characteristics but differed in average baseflow pH from 4.5 to 6.4 due to varying influence of pyritic phyllite oxidation associated mainly with road cuts. The varying contributions of geologic acidity would be expected to mimic the effects of mineral acidity from acidic deposition. Species richness declined from 69 species at the highest pH site to 22 species at the lowest pH site. Richness of Ephemoroptera (mayflies) and Tricoptera (caddisflies), both commonly acid sensitive, was positively correlated with pH ($R^2 = 0.96$ for both). Tricoptera species richness was negatively correlated with the concentration of inorganic Al ($R^2 = 0.96$). Differences in benthic invertebrate richness and density were not attributed to differences in food availability but rather were closely associated with stream acid–base chemistry.

When exposed to low-pH and high inorganic Al chemical conditions, fish can die or experience sublethal stress (MacAvoy and Bulger 2004, Baldigo et al. 2007). Acid stress in fish is commonly expressed as ion regulatory disturbance (Booth et al. 1988), interference with gill ion transport, and loss of blood sodium (Na; Grippo and Dunson 1996). Low concentrations of Ca in water increase the vulnerability of fish to acid stress (Cleveland et al. 1991). Neff et al. (2009) conducted *in situ* bioassays in GRSM to evaluate changes in physiological conditions and acid stress in wild southern strain brook trout during episodes of stream acidification. Stress was determined by measuring whole-body Na loss (Grippo and Dunson 1996). Study streams were acidic (Eagle Rocks Prong) or very low in acid neutralizing capacity (<20 µeq/L; Middle Prong, Ramsey Prong; Deyton et al. 2008). Two of the streams had lost brook trout populations during the previous two decades. Episodic acid neutralizing capacity depressions that occurred during the bioassay exposure experiments ranged from 9 to 26 µeq/L. The main contributor to episodic acidification was sulfate; nitrate, organic acids, and base cation dilution also contributed. During episodes, brook trout lost Na in a manner consistent with stress caused by acidification. Whole-body Na concentrations decreased by significant amounts (10%–19% loss) when the 24-hour time-weighted average Al concentration exceeded 200 micrograms per liter (µg/L; 7.4 micromoles [µM]) and pH was less than 5.1. Loss of Na can cause or contribute to mortality (Hesthagen et al. 1999) or sublethal stress (Dussault et al. 2005), stimulate downstream immigration (Gagen et al. 1994), interfere with reproduction (Kaeser and Sharpe 2001), and reduce growth (Cleveland et al. 1991).

4.4 Nutrient Enrichment

Nitrogen deposition poses a potential for nutrient enrichment to temperate forest ecosystems because the growth of trees in these systems is often N-limited (Vitousek and Howarth 1991). Forest canopy structure can enhance the rates of N deposition to terrestrial ecosystems by scavenging dry N deposition from the atmosphere (Lovett 1992). Forests considered sensitive to nutrient enrichment from N deposition include high-elevation forests that receive high rates of N deposition and where effects of N deposition on root allocation or late-season growth may exacerbate other stresses from acidic deposition and harsh climate. The effects of excess nutrient N on sensitive forests can be pronounced in forests that have already experienced soil base cation depletion due to the acidifying effects of acidic deposition.

In some areas, particularly in high-elevation terrestrial ecosystems that have become N-saturated, high levels of atmospheric N deposition have caused elevated levels of nitrate in drainage waters, making these areas potentially more sensitive to added N deposition (Aber et al. 1989, 1998, Stoddard 1994, Campbell et al. 2002, Bowman et al. 2006). In general, sites receiving high N deposition are also subject to other pollutant exposures, including ozone.

There is concern that atmospheric N deposition at GRSM has exceeded biological demand for N, and that excess N will reduce forest growth due to increased frost damage (Friedland et al. 1984) and enhance nitrate leaching, thereby causing nutrient base cation deficiencies and Al mobilization. Stream water acid–base chemistry might also be adversely affected (Stoddard 1994).

The status of red spruce in GRSM is of concern, with evidence for radial growth decline during the previous century at some locations (McLaughlin et al. 1987, Webster et al. 2004). The nature of the relationship between climate, acidic deposition, nutrient loading, and red spruce growth in and near GRSM is complex (Webster et al. 2004). Trees near ridges showed earlier, faster, and more consistent declines in growth than did trees in draws at lower elevation, likely due to both climatic conditions and acidic and nutrient N deposition (Webster et al. 2004). Webster et al. (2004) proposed a conceptual model, supported by field data, to explain the growth differences as follows. Red spruce trees experience more extreme climatic conditions near the ridges due to orographic precipitation, frequent cloud emersion, and enhanced S and N deposition (Mohnen 1992, Lindberg and Owens 1993). The cloud base recorded during the Integrated Forest Study at GRSM was generally at about 1800 m (5906 ft) elevation (Johnson and Lindberg 1992), well within the range of many ridge tops in the park. Soils along ridges in GRSM tend to be shallow, often <20 inches (50 cm), compared to deeper soils downslope (Johnson et al. 1991a). Ridge soils also tend to be less stable, leading to more frequent landslides and transfer via erosion of soil (with associated base cations) to lower elevation positions (Fernandez 1992, White and Cogbill 1992). Soil pools of exchangeable Ca and Mg, and associated base saturation, are small in response to the higher precipitation, enhanced leaching, and shallow depth found at the higher-elevation locations (Johnson and Fernandez 1992). Shallow soils having low base cation supply may lack the capacity to provide sufficient nutrient cations needed for tree growth (Webster et al. 2004). In contrast, the deeper, more base-rich soils in the mountain draws experienced a transient enrichment of base cations in soil solution transported from upslope locations and a delayed negative growth response to acidifying deposition.

Garten and Van Miegroet (1994) demonstrated that naturally occurring N isotope ratios in plant foliage can provide a good index of soil N dynamics in GRSM at sites ranging in elevation from 615 to 1670 m (2018 to 5479 ft) in different forest types. Soils and organic layers from the N-rich high-elevation locations showed greater net N mineralization and nitrification potentials than soils from lower-elevation locations, making them more prone to N saturation. Nitrogen saturation could contribute to adverse effects on forest health. Nevertheless, red spruce has not shown widespread mortality that might be attributed to air pollution (Busing and Wu 1990, Nicholas et al. 1992, Peart et al. 1992, Pauley et al. 1996). There has been some loss of foliage, and secondary wind damage has occurred due to opening of the canopy in response to insect-caused fir mortality (Busing and Pauley 1994). There are complex relationships among insect infestation, wind disturbance, acidic deposition, and nutrient cycling that are not well understood (Pauley et al. 1996).

High-elevation spruce-fir forests in GRSM have been found to lose about 10–20 kg N/ha/ yr to nitrate leaching (Johnson et al. 1993b). This high nitrate leaching reflects high soil N availability and N saturation (Agren and Bosatta 1988, Aber et al. 1989). High atmospheric N deposition, high rate of net nitrification, and low biological demand for N in these primarily old-growth forests further contribute to nitrate leaching losses (Sasser and Binkley 1989, Van Miegroet et al. 1992).

High-elevation spruce-fir forests in GRSM have for the most part not been logged or burned extensively (Harmon et al. 1983, Pyle 1988). As a consequence, soil N pools in spruce-dominated stands tend to be large and the N mineralization potential is high (Van Miegroet et al. 1992, Garten 2000). Because of the high rate of N deposition, these forests are set on a path to N saturation. The rate of change and the time at which the forest becomes N-saturated are influenced by stand vigor, tree age, and disturbance (Creed et al. 2004). Streams in GRSM have been shown to have among the highest nitrate concentrations in the eastern United States (Stoddard 1994), and this is symptomatic of advanced N saturation.

The extensive mortality of Fraser fir (Pauley et al. 1996) has had a large impact on the quantity of coarse woody debris in these forests. An estimated 70% of the standing Fraser fir trees in the Noland Divide watershed were dead at the end of the twentieth century due to the balsam wooly adelgid (Rose 2000).

Barker et al. (2002) estimated spatial variability in tree N uptake in the Noland Divide watershed and assessed the influence of stand and landscape characteristics on forest N uptake. They found that overstory N uptake was highly variable spatially within the small study watershed and depended in large part on the methods used to calculate uptake. It appeared that the overstory component of this high-elevation spruce-fir forest played only a minor role in the retention of atmospherically deposited N (Barker et al. 2002).

Van Miegroet et al. (2001) investigated the magnitude and timing of N fluxes in a small, first-order stream catchment in GRSM. Approximately half of the atmospheric N input was exported in stream water annually. Asynchrony (lack of temporal correspondence) was found for the N inputs and outputs, however. Atmospheric deposition was highest during summer, whereas nitrate leaching losses were largest during the dormant winter season due in part to reduced biotic demand.

Stream nitrate concentrations in GRSM tend to be high in watersheds that receive especially high N deposition (>32 kg N/ha/yr; Weathers et al. 2006, Pardo and Duarte 2007). The critical load for nutrient N addition to forest ecosystems in GRSM has been estimated using the steady-state mass balance modeling approach to range from about 2.5 to 9 kg N/ha/yr by Oja and Arp (1998) and from 3 to 7 kg N/ha/yr by Pardo and Duarte (2007). Gilliam et al. (2011) compiled data on empirical N critical loads for the Eastern Temperate Forests ecoregion. For GRSM, they estimated a critical load of 32 kg N/ha/yr to protect against export of N in stream water. This estimate was based on stream chemistry published by Garten (2000) and Van Miegroet et al. (2001), combined with N deposition estimates of Pardo and Duarte (2007) and Weathers et al. (2006). Ellis et al. (2013) estimated the N critical load for GRSM in the range of 3–8 kg N/ha/yr for the protection of hardwood forests based on data from Pardo et al. (2011).

4.5 Ozone Injury to Vegetation

4.5.1 Ozone Formation

Sources of air pollution that are hundreds of miles from parks in the Appalachian Highlands Network may be linked to ozone exposure at the parks. In particular, these include emissions sources of N oxides and volatile organic compounds in Tennessee, the Ohio Valley, and the Gulf of Mexico regions. The highest ozone exposures in GRSM occur on hot, sunny days having low wind speed, contributing to air stagnation (Tong et al. 2005).

The National Park Service (2010) evaluated the fourth highest 8-hour ozone concentrations for national parks that have consistently been at or above the primary (for protecting human health) 2008 National Ambient Air Quality Standard, based on 11–20 years of monitoring data running through 2008. There were six parks that routinely had ozone concentrations above the standard in place at that time, including GRSM. This park experienced the most rapid rise since the 1990s among eastern national parks in the frequency of occurrence of days during which an 8-hour average ozone concentration exceeded the standard (Tong et al. 2005).

There are recognized differences in the potential to form atmospheric ozone among the southeastern national parks, including GRSM (Kang et al. 2003). Tong et al. (2005) quantified the relative importance of point and nonpoint mobile emissions sources of N oxides, identified the origin of air masses associated with high concentrations of N oxides, and calculated N oxide production and removal budgets for two national parks (GRSM and Mammoth Cave National Park). Air pollution influence areas were identified using cluster analysis (Dorling et al. 1992) of model output generated using the Hybrid Single-Particle Lagrangian Integrated Trajectories model (Draxler 1997) and also emission source identification based on the EPA's National Emissions Inventory (U.S. EPA 2001b). Clusters were labeled according to general direction of movement of the air mass toward the parks.

Diurnal reactive N oxide and sulfur dioxide patterns were identified by Tong et al. (2005) for GRSM, with weak maxima during early morning and minima during late afternoon throughout most of the year. The proportion of reactive N oxide emissions derived from point sources was generally higher during winter and spring and lower during summer and fall.

It is believed that N oxides, rather than volatile organic compounds, constitute the main limiting factor for photochemical ozone production in remote areas of the southeastern United States such as within GRSM (Kang et al. 2003, Tong et al. 2005). On average, a minimum of one-fourth of the reactive N oxides at GRSM are estimated to have been emitted by point sources of N (Tong et al. 2005). Human-caused nonpoint sources are also important. GRSM also has strong biogenic (emitted from natural vegetation) volatile organic compound sources. The most abundant volatile organic compound measured in the air during summer in GRSM during the period 1995–1997 was isoprene (Kang et al. 2001). Upon conversion of all volatile organic compounds into propylene-equivalent concentration units (to assess the relative potential contribution of the various volatile organic compounds to ozone formation), natural biogenic components accounted for 69%–95% of the total atmospheric volatile organic compound concentration in GRSM during all three years of study. Most of the biogenic volatile organic compounds component comprised isoprene; other common volatile organic compounds included α-pinene, β-pinene, and limonene. The major anthropogenic volatile organic compounds present in the air during summer were isopentane, toluene, and propane. These are derived from automobile exhaust (isopentane), solvents (toluene), and natural gas (propane; Kang et al. 2001). Thus, ozone at GRSM is formed from volatile organic compounds that are largely natural and N oxides that are largely human caused.

Episodes of high (>90 ppb) ozone concentrations in GRSM were primarily attributed to atmospheric transport of ozone into the park from outside rather than the formation of ozone within the park (Mueller 1994). Based on model simulations and detailed episode analysis, Mueller et al. (1996) concluded that urban pollution sources of N oxides likely played a large role in determining the episodes of high ozone concentrations in the park. More recent ozone concentration data are available at http://www.nature.nps.gov/air/data/current/index.cfm.

Kang et al. (2003) applied the Multiscale Air Quality Simulation Platform model to predict atmospheric ozone concentration in GRSM and in two national parks in surrounding networks with an overall uncertainty less than 30%. Modeling results suggested that volatile organic compounds were chemically saturated at all three locations, potentially causing decreased ozone production in response to further increases in volatile organic compound emissions. More than half of the ozone at GRSM was simulated to be transported from other areas (Kang et al. 2003). The researchers concluded that, in areas like GRSM where transport from pollutant source areas dominates the local ozone production, a control strategy might focus on the reduction of ozone in the source region(s).

4.5.2 Ozone-Sensitive Plants

GRSM contains more than half of all old-growth forest in the eastern United States and more than three-fourths of the spruce-fir forest in the southern Appalachian Mountains (Burkey et al. 2006). Thus, threats to vegetation in this park are of regional and national importance (Shaver et al. 1994).

The ozone-sensitive plant species that are known or expected to occur within GRSM are listed in Table 4.2. Those considered to be bioindicators display visible and easily recognizable ozone symptoms. They are designated by an asterisk in the table. GRSM contains 36 ozone-sensitive and bioindicator species.

4.5.3 Ozone Exposure Indices and Levels

The southern Appalachian Mountains region has experienced some of the highest ozone concentrations of any physiographic area in the eastern United States (Chappelka et al. 1997, Samuelson and Kelly 1997, Grulke et al. 2007). Visible foliar symptoms suggestive of foliar injury have been observed on over 90 plant species in GRSM, many of which were verified in open-top chamber exposures (Neufeld et al. 1992). Effects are usually induced by cumulative exposures over time; cumulative exposures are expressed as the W126 or the SUM06 metrics, which are calculated by summing up hourly ozone concentrations over a period of time, usually 3 months during the growing season and the hours between 8:00 AM and 8:00 PM. The W126 (a measure of cumulative ozone exposure that preferentially weights higher concentrations) and SUM06 (a measure of cumulative exposure that includes only hourly concentrations over 60 ppb ozone) exposure indices calculated by National Park Service staff are given in Table 4.3, along with Kohut's (2007a) ozone risk ranking that also considered environmental conditions.

The SUM06 index significantly exceeded the threshold for injury at all four monitoring sites in GRSM (Lookout Rock, Cades Cove, Cove Mountain, and Clingmans Dome)

TABLE 4.2

Plant Species That Are Widely Recognized as Ozone-Sensitive or Bioindicator (Asterisk) Plant Species and That Are Known or Thought to Occur in GRSM[a]

Species	Common Name
Ailanthus altissima*	Tree-of-heaven
Apios americana*	Groundnut
Apocynum androsaemifolium*	Spreading dogbane
Apocynum cannabinum	Dogbane, Indian hemp
Asclepias exaltata*	Tall milkweed
Asclepias syriaca*	Common milkweed
Cercis canadensis*	Redbud
Clematis virginiana	Virgin's bower
Corylus americana*	American hazelnut
Fraxinus americana*	White ash
Fraxinus pennsylvanica	Green ash
Gaylussacia baccata*	Black huckleberry
Krigia montana	Mountain dandelion
Liquidambar styraciflua	Sweetgum
Liriodendron tulipifera*	Yellow poplar
Lyonia ligustrina*	Maleberry
Parthenocissus quinquefolia	Virginia creeper
Pinus pungens	Table-mountain pine
Pinus rigida	Pitch pine
Pinus taeda	Loblolly pine
Pinus virginiana	Virginia pine
Platanus occidentalis*	American sycamore
Prunus serotina*	Black cherry
Prunus virginiana	Choke cherry
Robinia pseudoacacia	Black locust
Rubus allegheniensis*	Allegheny blackberry
Rubus canadensis*	Thornless blackberry
Rudbeckia laciniata var. *humilis**	Cutleaf coneflower
Sassafras albidum	Sassafras
Verbesina occidentalis*	Crownbeard
Vitis labrusca*	Northern fox grape

[a] Lists are periodically updated and available at https://irma.nps.gov/NPSpecies/Report.

during the period 1995–1999. In addition, the numbers of hours having ozone concentrations above 100 ppb often met the threshold for injury used in conjunction with the W126 criterion (Kohut 2007b). The highest exposures generally occurred at the Lookout Rock ridge site and the Cove Mountain high-elevation site. In general, when ozone was high, soil moisture was low at both locations, thereby reducing the effectiveness of ozone in causing injury. The probability of injury in GRSM is likely highest when ozone levels exceed thresholds and soil moisture levels are normal (e.g., 1997, 1999) or at least high enough that ozone uptake into foliage is not constrained by stomatal closure (Kohut 2007b).

TABLE 4.3

Ozone Assessment Results for GRSM Based on Estimated Average 3-Month W126 and SUM06 Ozone Exposure Indices for the Period 2005–2009 and Kohut's (2007a) Ozone Risk Ranking for the Period 1995–1999[a,b]

W126		SUM06		Kohut Risk to Vegetation
Value (ppm-hr)	NPS Condition	Value (ppm-hr)	NPS Condition	
12.38	Moderate	16.78	High	High

[a] Parks are classified into one of three ranks (Low, Moderate, High), based on comparison with other Inventory and Monitoring Parks.

[b] Degrees of concern for the W126 and SUM06 indices are based solely on levels of ozone exposure. Kohut's risk to vegetation is based on several factors that contribute to injury in plants, including ozone exposure and environmental variables, and considers the effects of soil moisture on the uptake of ozone.

During the last two decades of the twentieth century, ground-level ozone concentrations in the United States generally decreased about 21% and 12%, based on the previous 1-hour standard and the newer 8-hour standard, respectively (U.S. EPA 2001a). Nevertheless, ozone concentrations did not decrease in all regions. In the southern and north central regions, ozone concentrations generally increased. Seasonal ozone exposure increased dramatically in GRSM between 1990 and 1999 (Chappelka et al. 2003). However, since 2000, both ozone concentrations and ozone exposures decreased substantially at the park.*

The EPA's proposal in 2010 for a new secondary ozone standard to protect vegetation was based on a three-year average of the annual W126-3 month metric (Federal Register Vol. 75, No. 11, 40 CFR Parts 50 and 59, January 2010). The eastern park showing the highest exposure was GRSM. On October 1, 2015, the EPA decided not to set a separate secondary standard to protect the public welfare (including natural vegetation), but the primary and secondary National Ambient Air Quality Standard was revised to 0.070 ppm, based on the annual fourth highest daily maximum 8-hour concentration averaged over three years (http://www3.epa.gov/ozonepollution/actions.html).

4.5.3.1 Field Surveys of Foliar Injury

An intensive ozone effects research effort has been conducted at GRSM since the early 1990s, including field surveys, chamber fumigations, and routine monitoring (Shaver et al. 1994). These studies revealed foliar symptoms, considered to represent adverse physiological effects of ozone on vegetation in the park. Some plant species and populations showed adverse effects on plant growth at ambient ozone exposure levels when compared with plants placed in chambers receiving charcoal-filtered air (Neufeld and Renfro 1993a).

Somers et al. (1998) quantified differences in radial growth of black cherry and yellow poplar (*Liriodendron tulipifera*) among individual trees in GRSM with and without histories of visible foliar symptoms from ozone exposure. Radial growth decreased over both 5- and 10-year periods of measurement for both species, but these decreases were only

---

* See recent data for two monitoring sites as follows:

  Look Rock: http://webcam.srs.fs.fed.us/graphs/o3calc/vegetation.php?state=47&county=009&siteid=01011; Clingmans Dome: http://webcam.srs.fs.fed.us/graphs/o3calc/vegetation.php?state=47&county=155&siteid=01021.

statistically significant ($p \leq 0.05$) for yellow poplar. This study provided empirical evidence, at least for yellow poplar, that visible ozone symptoms are related to growth decline of individual trees growing in a natural setting. Experiments conducted in GRSM further identified black cherry and American sycamore (*Platanus occidentalis*) trees as having foliar symptoms caused by ozone (Neufeld and Renfro 1993a,b). Foliar symptoms attributed to ozone exposure were also documented by Chappelka et al. (1999b) on mature black cherry in GRSM. Trees growing at the highest elevations showed the greatest symptoms during all three years of study. These high-elevation sites also received the highest ozone exposures. The percent of trees injured by ozone was positively correlated with both the SUM06 and W126 indices.

A three-year study in the early 1990s at three locations in GRSM showed that about 11%–12% of all leaves examined of black cherry, sassafras, and yellow poplar showed visible ozone symptoms. Ozone symptoms in the park were greatest at the Cove Mountain site during all three years of study, with less visible symptoms at Look Rock and Twin Creeks (Chappelka et al. 1999a). During the period 1992–1996, the average incidence of ozone-induced foliar symptoms on milkweed in GRSM was 73% (Chappelka et al. 2007). Ozone symptoms can vary with age of leaves. For example, symptoms on tall milkweed in GRSM were most severe in older, lower leaves that had been exposed to ozone for a longer period of time (Souza et al. 2006).

Cutleaf coneflower showed substantial variation in symptoms of ozone injury between and within populations in GRSM. Research reported by Davison et al. (2003) suggested that variation in symptom expression was likely due to variation in light and not to variation in ozone flux. This finding reinforced the importance of evaluation of environmental site conditions in determining visible symptom expression from air pollution injury.

During July and August of most summers, about 25–30 plant species in GRSM, including cutleaf cornflower, develop foliar symptoms of ozone injury (Finkelstein et al. 2004). Cutleaf coneflower is common throughout GRSM at the boundary between forest and meadow. Stands range from about 1 to 20 m (3–65 ft) wide and can run along the forest edge for hundreds of meters (Finkelstein et al. 2004). Roughly half of the plants of this species in GRSM typically show visible symptoms of foliar injury (Finkelstein et al. 2004).

Chappelka et al. (1996) conducted an ozone effects assessment for the southern Appalachian Mountains. They found it difficult to rank tree species' sensitivities to ozone symptoms because of high variability across the study region in environmental conditions, ozone exposure, tree age, and differential sensitivity within a given species. Of the five tree species most studied, however, black cherry and loblolly pine appeared most sensitive and red spruce least sensitive; yellow poplar and northern red oak (*Quercus rubra*) were judged to be intermediate in sensitivity.

There is evidence (cf., McLaughlin et al. 2007b) that ozone can contribute to reduced stream baseflow during fall low-flow periods in GRSM. Trees in GRSM exhibited reduced growth and increased water loss in years that had highest ozone exposure (McLaughlin et al. 2007b). The increased water loss from trees appears to have contributed to reduced soil moisture and decreased fall-season stream flow (McLaughlin et al. 2007b). In response to ozone exposure, there was increased forest water use, perhaps due to increased evapotranspiration, and reduced fall season baseflows in all three watersheds examined near Look Rock. Such changes in streamflow during the critical low-flow period could have substantial effects on water quality and stream ecology, especially in small, low-order streams. Model results further suggested that the cumulative effects of daily peak ozone exposures appeared to be more important than longer-term averages at the scale of the individual tree and the small watershed. Thus, McLaughlin et al. (2007b) suggested that ozone exposure can magnify the effects of drought and exacerbate the effects of climate warming.

McLaughlin et al. (2007a) found growth effects on mature forest trees at three study sites in and near GRSM. Manual measures of stem circumference of 86 trees were recorded at approximately two-week intervals and were linked to electromechanical radius measurements at 30-minute intervals for six trees. The manual measures were of gap changes on tensioned bands and were obtained with digital calipers to an average precision of 0.02 millimeter (mm; 0.00078 inches). High-resolution radius measures were made with electromechanical dendrometers (sensitivity = 0.006 mm [0.00024 in.] radius change). Stem growth reductions were accompanied by changes in water use by trees. They concluded that altered water use could further affect tree growth during drought years and that ozone exposure at the study locations increased water use on an episodic basis and limited tree growth. Episodic increases in ozone exposure during the growing season corresponded with reduced stem increment. Thus, it appears that peak ozone exposure can play a role in regulating water loss from trees via transpiration and can delay the recovery of stem expansion subsequent to episodes of high moisture demand (McLaughlin et al. 2007a).

4.5.3.2 Model Responses

Weinstein et al. (2001) used the tree physiology model, TREGRO, linked to the stand simulation model, ZELIG, to estimate the effects of ozone exposure on yellow poplar forest in the Twin Creeks region of GRSM. Ambient ozone levels measured at low elevation on the west side of GRSM were predicted to cause a 10% reduction in yellow poplar abundance over the next century. In response, other species were simulated to increase in abundance. Reduction of ozone exposure by 50% was projected to be insufficient to prevent a decrease in future yellow poplar tree abundance.

4.6 Visibility Degradation

4.6.1 Natural Background and Ambient Visibility Conditions

Table 4.4 gives the relative park haze rankings on the clearest 20%, haziest 20%, and average days in GRSM. The National Park Service-Air Resources Division uses interpolated estimates of average visibility to assess condition at all parks. Natural background estimates of haze are relatively high in the GRSM region, due to natural biogenic emissions from forest vegetation. However, measured ambient haze for the period 2004–2008 was considerably higher than the estimated natural condition. Visibility data through 2013 are available at http://www.nature.nps.gov/air/data/products/parks/index.cfm. Ambient haze was classified as very high for all visibility days. For example, haze for the average day was more than 13 deciviews higher than the estimated natural haze on the average days. On the haziest 20% of days, ambient haze was more than 18 deciviews higher than the estimated natural condition. A report from the EPA (2008b) on the status and trends in air quality throughout the United States concluded that visibility

TABLE 4.4

Estimated Natural Haze and Measured Ambient Haze at Monitoring Site GRSM1 Averaged over the Period 2004–2008[a]

Estimated Natural Background Haze (Deciview)		
20% Clearest Days	20% Haziest Days	Average Days
4.62	11.24	7.73

Measured Ambient Haze (for Years 2004–2008)					
20% Clearest Days		20% Haziest Days		Average Days	
Deciview	Ranking	Deciview	Ranking	Deciview	Ranking
13.09	Very High	29.79	Very High	21.18	Very High

[a] Parks are classified into one of five haze ranks (Very Low, Low, Moderate, High, or Very High) based on comparison with visibility conditions at all monitored parks.

in GRSM on the 20% haziest days during the period 2000–2004 (30.3 deciviews) was among the poorest in the country.

The Great Smoky Mountains region frequently experiences poor visibility, and this is attributed to anthropogenic aerosols from combustion sources (especially coal), high emissions of biogenic aerosols from forest vegetation, and high humidity (Ames et al. 2000, Day et al. 2000, Malm et al. 2000a,b, Kim and Hopke 2006). Figure 4.7 shows three representative photos of the same vista in GRSM under differing visibility conditions. Photos were selected to correspond with the clearest 20% of visibility conditions, haziest 20% of visibility conditions, and annual average visibility conditions at that location. This series of photos provides a graphic illustration of the visual effect of these differences in haze level in this park. Data from the IMPROVE Program for GRSM show that pollution has reduced average visual range in recent years from about 110 to 25 mi (177 to 40 km). On the haziest days, visual range is reduced from 75 to 10 mi (121 to 16 km). Severe haze episodes occasionally reduce visibility to 6 mi (10 km).

4.6.2 Composition of Haze

Human-caused visibility degradation at GRSM has largely been caused by atmospheric sulfate due to anthropogenic sulfur dioxide emissions (Malm and Sisler 1987, Shaver et al. 1994). Emissions of oxidized N may appreciably contribute to the problem on occasion. Organics are responsible for much of the remaining visibility impairment* (Malm and Sisler 1987). The contribution of sulfate to the current extinction at GRSM is largest on the haziest days (Figure 4.8). On average, 71.5% of the total particulate light extinction was attributable to sulfate, and that percentage increased to 79.0% on the 20% of the days that were haziest. On the clearest 20% of visibility days, the relative importance of sulfate decreased to 58.2% and nitrate extinction became proportionately more important (12.1%) as compared with the haziest days when extinction attributed to nitrate was only 2.3% of the total light extinction.

Annual mean haze levels on the 20% haziest days at 47 monitored park locations during the period 2006–2008 ranged from 1.5 to 21 deciviews higher than the estimated natural

---

* *Visibility impairment* means any humanly perceptible change in visibility (light extinction, visual range, contrast, coloration) from that which would have existed under natural conditions.

FIGURE 4.7
Three representative photos of the same view in GRSM illustrating the 20% clearest visibility days, the 20% haziest visibility days, and the annual average visibility. Extinction is total particulate light extinction. (From http://vista.cira.colostate.edu/improve/Data/IMPROVE/Data_IMPRPhot.htm, accessed December, 2010.)

condition (NPS 2010). The average difference between measured haze and estimated natural condition was 8.3 deciviews. Several eastern parks, including GRSM, had annual mean haze on the haziest days that was substantially higher (more than about 10 deciviews) than estimated natural conditions. Monitoring sites showing the largest differences during the period 2006–2008 between measured visibility on the clearest days and estimated natural conditions included GRSM (NPS 2010).

Kim and Hopke (2006) identified the major sources of fine particulate matter concentrations in the Great Smoky Mountains. The relative contributions to the fine particulate matter total were estimated using a positive matrix factorization source apportionment approach (Paatero 1997). Secondary nitrate particles contributed only 2% of the fine particulate matter mass, some locally produced from ammonium. Gasoline vehicle emissions were tentatively separated from diesel emissions, accounting for 13% and 1%, respectively, of the total fine particulate matter.

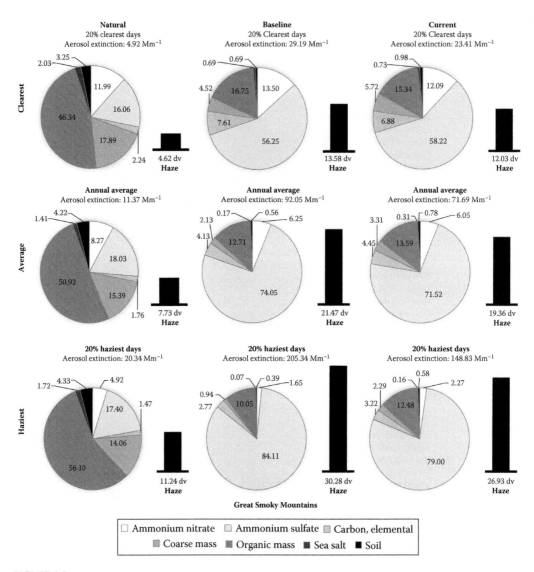

FIGURE 4.8
Estimated natural (preindustrial), baseline (2000–2004), and current (2006–2010) levels of haze (columns) and its composition (pie charts) on the 20% clearest, annual average, and 20% haziest visibility days for GRSM. (From http://views.cira.colostate.edu/fed/Tools/RegionalHazeSummary.aspx, accessed October, 2012.)

Research has been conducted in GRSM to characterize the effects of ambient aerosols on visibility in the park. The Southeastern Aerosol and Visibility Study was conducted during the summer of 1995 (Hand et al. 2002). Measurements were made of aerosol light scattering coefficients, chemical composition, aerosol size distributions, hygroscopicity, and particle refractive index. Cheng and Tanner (2002) reported atmospheric measurements of fine (0.1–0.4 µm diameter) and ultrafine (<0.1 µm diameter) particles and gaseous species in GRSM in 2000. A six-factor model was constructed, which explained nearly 80% of the variation in the atmospheric data. More than one-third of the variability in visibility was attributed to ultrafine particles. Levels of ultrafine particles are high in the atmosphere in

the vicinity of GRSM. This is due to emissions of pollutants from anthropogenic and biogenic sources and to abundant sunshine during summer (Cheng and Tanner 2002).

4.6.3 Trends in Visibility

Although ambient visibility in GRSM is substantially degraded, there is some indication
that visibility conditions are improving in the park. Between 1996 and 2006, visibility in
scenic areas improved on the 20% haziest days at five monitored locations throughout the
United States, including GRSM (U.S. EPA 2008b). A small, but significant, improvement in
visibility demonstrated at GRSM between 1990 and 2004 was largely due to a reduction
in atmospheric sulfate concentration. Improvement has continued in more recent years
(http://www.nature.nps.gov/air/data/products/parks/index.cfm). Model projections suggested that visibility at GRSM is likely to further improve under existing state and federal
emissions regulations (U.S. EPA 2008b).

The Visibility Improvement State and Tribal Association of the Southeast was a collaborative effort among state governments, tribal governments, and federal agencies involved
in the management of visibility and regional haze in the Southeast. The region included
the southeastern United States from Virginia and West Virginia in the north, south to
Florida, and west to Kentucky, Tennessee, and Mississippi. Analyses conducted for this
region included estimation of baseline visibility conditions, calculation of the glide slope
from the baseline necessary to achieve background conditions in 2064, as required by the
Regional Haze Rule, and determination of air pollutant source areas.

Available haze data collected at the GRSM1 IMPROVE site over the period of record are
shown in Figure 4.9. Reductions in haze since 1990 have been most pronounced on the 20%
haziest days, although some reductions in haze also have occurred on the 20% clearest and
average days. To date, there has been progress toward compliance with the Regional Haze

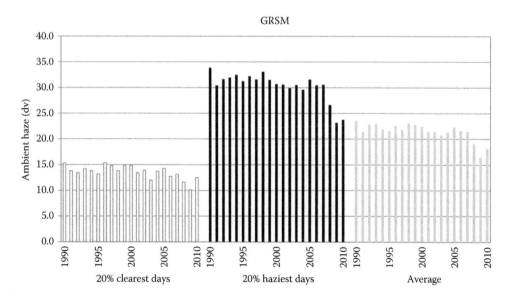

FIGURE 4.9
Trends in ambient haze levels at GRSM, based on IMPROVE measurements on the 20% clearest, 20% haziest,
and annual average visibility days over the monitoring period of record. (From http://vista.cira.colostate.edu/
improve/Data/IMPROVE/summary_data.htm, accessed October, 2012.)

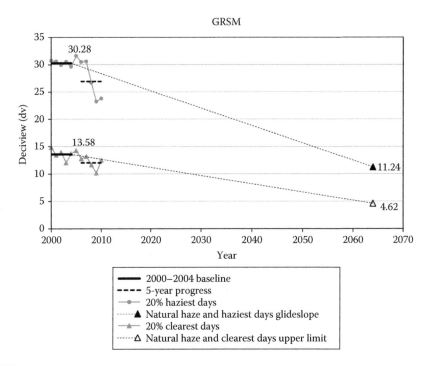

FIGURE 4.10
Glideslopes to achieving natural visibility conditions in 2064 for the 20% haziest (upper line) and the 20% clearest (lower line) days in GRSM. In the Regional Haze Rule, the clearest days do not have a uniform rate of progress glideslope; the rule only requires that the clearest days do not get any worse than the baseline period. Also shown are measured values during the period 2000–2010. (From http://vista.cira.colostate.edu/improve/Data/IMPROVE/summary_data.htm, accessed October, 2012.)

Rule at GRSM (Figure 4.10). Nevertheless, continued progress in reducing haze levels on the haziest days will be needed.

4.7 Summary

This chapter describes the air quality–related values of GRSM in the Appalachian Highlands Network. GRSM contains some of the largest tracts of old-growth forest in the eastern United States. Old-growth spruce-fir forests predominate at elevations above 1500 m (4921 ft) in the park. Mixed hardwood forests cover the lower slopes. Air pollution is an important concern in GRSM, and effects have been well studied. Upwind emissions and deposition of both S and N have been high around the park. In particular, S and N deposition have been high at the upper elevations (especially above about 1500 m), where cloud deposition contributes substantial amounts of S and N to acid-sensitive spruce-fir forests. Wet deposition of Hg is also fairly high.

The most important air pollution concerns in GRSM appear to include visibility, ozone exposure, and acidification of soil and surface waters. Pollution-caused haze levels are very high in the park. Ozone has harmed sensitive plant species. Many streams have acidified. There are known biological effects of acidification on acid-sensitive vegetation and aquatic biota in higher-elevation, low-order stream watersheds. Although acidic deposition has declined substantially over the past two decades in this park, effects of S deposition may intensify in the future even under further reduced S loading. This is because soils in GRSM have adsorbed significant amounts of S over time, limiting acidification under relatively high levels of S deposition. As this adsorption capacity is exhausted, however, continued S input, even at lower levels, will trigger delayed acidification responses in soils and drainage waters.

Although atmospheric deposition of both S and N has been high in the park, deposited N appears to generally have the largest influence on stream chemistry at many locations because watershed retention of atmospherically deposited S has been high (Flum and Nodvin 1995). Streams in GRSM draining old-growth coniferous forests have commonly had nitrate concentrations above 50 μeq/L, substantially higher than at most lower-elevation sites.

Ozone exposures and risk to sensitive vegetation were judged by the National Park Service to be moderate to high in GRSM, depending on the choice of exposure index on which to base the assessment. Based on both the ozone cumulative exposure and Kohut's (2007a) risk ranking index, concern for ozone effects on vegetation was high in GRSM. Exposures have decreased in more recent years.

References

Abell, R., D.M. Olson, E. Dinerstein, P. Hurley, J.T. Diggs, W. Eichbaum, S. Walters et al. 2000. *Freshwater Ecoregions of North America: A Conservation Assessment*. Island Press, Washington, DC.

Aber, J.D., W. McDowell, K. Nadelhoffer, A. Magill, G. Berntson, M. Kamakea, S. McNulty, W. Currie, L. Rustad, and I. Fernandez. 1998. Nitrogen saturation in temperate forest ecosystems; hypotheses revisited. *BioScience* 48(11):921–933.

Aber, J.D., K.J. Nadelhoffer, P. Steudler, and J.M. Mellilo. 1989. Nitrogen saturation in northern forest ecosystems. *BioScience* 39(6):378–386.

Agren, G.I. and E. Bosatta. 1988. Nitrogen saturation in terrestrial ecosystems. *Environ. Pollut.* 54:185–198.

Ames, R.B., J.L. Hand, S.M. Kreidenweis, D.E. Day, and W.C. Malm. 2000. Optical measurements of aerosol size distributions in Great Smoky Mountains National Park: Dry aerosol characterization. *J. Air Waste Manage. Assoc.* 50:665–676.

Baldigo, B.P., G.B. Lawrence, and H.A. Simonin. 2007. Persistent mortality of brook trout in episodically acidified streams of the southwestern Adirondack Mountains, New York. *Trans. Am. Fish. Soc.* 136:121–134.

Barker, M., H. Van Miegroet, N.S. Nicholas, and I.F. Creed. 2002. Variation in overstory nitrogen uptake in a small, high-elevation southern Appalachian spruce-fir watershed. *Can. J. For. Res.* 32:1741–1752.

Baumgardner, R.E., Jr., S.S. Isil, T.F. Lavery, C.M. Rogers, and V.A. Mohnen. 2003. Estimates of cloud water deposition at mountain acid deposition program sites in the Appalachian Mountains. *J. Air Waste Manage. Assoc.* 2003:291–308.

Booth, C.E., D.G. McDonald, B.P. Simons, and C.M. Wood. 1988. Effects of aluminum and low pH on net ion fluxes and ion balance in the brook trout (*Salvelinus fontinalis*). *Can. J. Fish. Aquat. Sci.* 45(9):1563–1574.

Bowman, W.D., J.R. Gartner, K. Holland, and M. Wiedermann. 2006. Nitrogen critical loads for alpine vegetation and terrestrial ecosystem response: Are we there yet? *Ecol. Appl.* 16(3):1183–1193.

Bulger, A.J., B.J. Cosby, C.A. Dolloff, K.N. Eshleman, J.R. Webb, and J.N. Galloway. 1999. SNP:FISH, Shenandoah national park: Fish in sensitive habitats. Project Final Report to National Park Service. University of Virginia, Charlottesville, VA.

Burkey, K.O., H.S. Neufeld, L. Souza, A.H. Chappelka, and A.W. Davison. 2006. Seasonal profiles of leaf ascorbic acid content and redox state in ozone-sensitive wildflowers. *Environ. Pollut.* 143:427–434.

Burns, D.A., J.A. Lynch, B.J. Cosby, M.E. Fenn, J.S. Baron, and U.S. EPA Clean Air Markets Division. 2011. National acid precipitation assessment program report to congress 2011: An integrated assessment. National Science and Technology Council, Washington, DC.

Busing, R.T. and E.F. Pauley. 1994. Mortality trends in a southern Appalachian red spruce population. *For. Ecol. Manage.* 64:41–45.

Busing, R.T. and X. Wu. 1990. Size-specific mortality, growth, and structure of a Great Smoky Mountains red spruce population. *Can. J. For. Res.* 20:206–210.

Campbell, D.H., C. Kendall, C.C.Y. Chang, S.R. Silva, and K.A. Tonnessen. 2002. Pathways for nitrate release from an alpine watershed: Determination using $\delta^{15}N$ and $\delta^{18}O$. *Water Resour. Res.* 31:2811–2821.

Chappelka, A., J. Renfro, G. Somers, and B. Nash. 1997. Evaluation of ozone injury on foliage of black cherry (*Prunus serotina*) and tall milkweed (*Asclepias exaltata*). *Environ. Pollut.* 95(1):13–18.

Chappelka, A., J.M. Skelly, G. Somers, J. Renfro, and E. Hildebrand. 1999b. Mature black cherry used as a bioindicator of ozone injury. *Water Air Soil Pollut.* 116:261–266.

Chappelka, A., G. Somers, and J. Renfro. 1999a. Visible ozone injury on forest trees in Great Smoky Mountains National Park, USA. *Water Air Soil Pollut.* 116:255–260.

Chappelka, A.H., H.S. Neufeld, A.W. Davison, G.L. Somers, and J.R. Renfro. 2003. Ozone injury on cutleaf coneflower (*Rudbeckia laciniata*) and crown-beard (*Verbesina occidentalis*) in Great Smoky Mountains National Park. *Environ. Pollut.* 125:53–59.

Chappelka, A.H., L.J. Samuelson, J.M. Skelly, and A.S. Lefohn. 1996. Effects of ozone on forest trees in the Southern Appalachians—An assessment of the current state of knowledge. Prepared for the Southern Appalachian Mountain Initiative (SAMI).

Chappelka, A.H., G.L. Somers, and J.R. Renfro. 2007. Temporal patterns of foliar ozone symptoms on tall milkweed (*Asclepias exaltata* L.) in Great Smoky Mountains National Park. *Environ. Pollut.* 149:358–365.

Charles, D.F. 1991. *Acidic Deposition and Aquatic Ecosystems: Regional Case Studies.* Springer-Verlag, New York.

Cheng, M.-D. and R.L. Tanner. 2002. Characterization of ultrafine and fine particles at a site near the Great Smoky Mountains National Park. *Atmos. Environ.* 36:5795–5806.

Church, M.R., P.W. Shaffer, K.W. Thornton, D.L. Cassell, C.I. Liff, M.G. Johnson, D.A. Lammers et al. 1992. Direct/delayed response project: Future effects of long-term sulfur deposition on stream chemistry in the mid-Appalachian region of the eastern United States. EPA/600/R-92/186. U.S. Environmental Protection Agency, Corvallis, OR.

Cleveland, L., E.E. Little, C.G. Ingersoll, R.H. Wiedmeyer, and J.B. Hunn. 1991. Sensitivity of brook trout to low pH, low calcium and elevated aluminum concentrations during laboratory pulse exposures. *Aquat. Toxicol.* 19(4):303–318.

Cook, R.B., J.W. Elwood, R.R. Turner, M.A. Bogle, P.J. Mulholland, and A.V. Palumbo. 1994. Acid-base chemistry of high-elevation streams in the Great Smoky Mountains. *Water Air Soil Pollut.* 72:331–356.

Cosby, B.J., J.R. Webb, J.N. Galloway, and F.A. Deviney. 2006. Acidic deposition impacts on natural resources in Shenandoah National Park. NPS/NER/NRTR-2006/066. U.S. Department of the Interior, National Park Service, Northeast Region, Philadelphia, PA.

Creed, I.F., D.L. Morrison, and N.S. Nicholas. 2004. Is coarse woody debris a net sink or source of nitrogen in the red spruce—Fraser fir forest of the southern Appalachians, U.S.A.? *Can. J. For. Res.* 34:716–727.

Cronan, C.S. and D.F. Grigal. 1995. Use of calcium/aluminum ratios as indicators of stress in forest ecosystems. *J. Environ. Qual.* 24:209–226.

Davison, A.W., H.S. Neufeld, A.H. Chappelka, K. Wolff, and P.L. Finkelstein. 2003. Interpreting spatial variation in ozone symptoms shown by cutleaf cone flower, *Rudbeckia laciniata* L. *Environ. Pollut.* 125:61–70.

Day, D.E., W.C. Malm, and S.M. Kreidenweis. 2000. Aerosol light scattering measurements as a function of relative humidity. *J. Air Waste Manage. Assoc.* 50:710–716.

Deyton, E.B., J.S. Schwartz, R.B. Robinson, K.J. Neff, S.E. Moore, and M.A. Kulp. 2008. Characterizing episodic stream acidity during stormflows in the Great Smoky Mountains National Park. *Water Air Soil Pollut.* 196:3–18.

Dorling, S.T., T.D. David, and C.E. Pierce. 1992. Cluster analysis: A technique for estimating the synoptic meteorological controls on air and precipitation chemistry—Method and applications. *Atmos. Environ.* 26A:2575–2581.

Draxler, R.R. 1997. Description of the HYSPLIT_4 modeling system. Technical Memorandum ERL ARL-224. National Oceanic and Atmospheric Administration, Air Resources Laboratory, Silver Spring, MD.

Dussault, E.B., R.C. Playle, D.G. Dixon, and R.S. McKinley. 2005. Effects of chronic aluminum exposure on swimming and cardiac performance in rainbow trout, *Oncorhynchus mykiss. Fish Physiol. Biochem.* 30:137–148.

Eagar, C. and M.B. Adams. 1992. *Ecology and Decline of Red Spruce in the Eastern United States.* Springer-Verlag, New York.

Eagar, C., H. Van Miegroet, S.B. McLaughlin, and N.S. Nicholas. 1996. Evaluation of effects of acidic deposition to terrestrial ecosystems in class I areas of the southern Appalachians. A Technical Report to the Southern Appalachian Mountains Initiative.

Ellis, R.A., D.J. Jacob, M.P. Sulprizio, L. Zhang, C.D. Holmes, B.A. Schichtel, T. Blett, E. Porter, L.H. Pardo, and J.A. Lynch. 2013. Present and future nitrogen deposition to national parks in the United States: Critical load exceedances. *Atmos. Chem. Phys.* 13(17):9083–9095.

Elwood, J.W. 1991. Southeast overview. *In* D.F. Charles (Ed.). *Acidic Deposition and Aquatic Ecosystems: Regional Case Studies.* Springer-Verlag, New York, pp. 291–295.

Fernandez, I.J. 1992. Characterization of eastern U.S. spruce-fir soils. *In* C. Eagar and M.B. Adams (Eds.). *Ecology and Decline of Red Spruce in the Eastern United States.* Springer-Verlag, New York, pp. 40–63.

Finkelstein, P.L., A.W. Davison, H.S. Neufeld, T.P. Meyers, and A.H. Chappelka. 2004. Sub-canopy deposition of ozone in a stand of cutleaf coneflower. *Environ. Pollut.* 131:295–303.

Flum, T. and S.C. Nodvin. 1995. Factors affecting streamwater chemistry in the Great Smoky Mountains, USA. *Water Air Soil Pollut.* 85:1707–1712.

Friedland, A.J., R.A. Gregory, L. Karenlampi, and A.H. Johnson. 1984. Winter damage as a factor in red spruce decline. *Can. J. For. Res.* 14:963–965.

Gagen, C.J., W.E. Sharpe, and R.F. Carline. 1994. Downstream movement and mortality of brook trout (*Salvelinus fontinalis*) exposed to acidic episodes in streams. *Can. J. Fish. Aquat. Sci.* 51(7):1620–1628.

Garner, J.J.B. 1994. Nitrogen oxides, plant metabolism, and forest ecosystem response. *In* R.G. Alscher and A.R. Wellburn (Eds.). *Plant Responses to the Gaseous Environment: Molecular, Metabolic and Physiological Aspects* (*Third International Symposium on Air Pollutants and Plant Metabolism*). Chapman & Hall, Blacksburg, VA, pp. 301–314.

Garten, C.T., Jr. 2000. Nitrogen saturation and soil N availability in a high-elevation spruce and fir forest. *Water Air Soil Pollut.* 120:295–313.

Garten, C.T., Jr. and H. Van Miegroet. 1994. Relationships between soil nitrogen dynamics and natural ^{15}N abundance in plant foliage from Great Smoky Mountains National Park. *Can. J. For. Res.* 24:1636–1645.

Gilliam, F.S., C.L. Goodale, L.H. Pardo, L.H. Geiser, and E.A. Lilleskov. 2011. Eastern temperate forests. *In* L.H. Pardo, M.J. Robin-Abbott, and C.T. Driscoll (Eds.). *Assessment of Nitrogen Deposition Effects and Empirical Critical Loads of Nitrogen for Ecoregions of the United States.* General Technical Report NRS-80. U.S. Forest Service, Newtown Square, PA, pp. 99–116.

Grippo, R.S. and W.A. Dunson. 1996. The body ion loss biomarker. 1. Interactions between trace metals and low pH in reconstructed coal mine-polluted water. *Environ. Toxicol. Chem.* 15(11):1955–1963.

Grulke, N.E., H.S. Neufeld, A.W. Davison, M. Roberts, and A.H. Chappelka. 2007. Stomatal behavior of ozone-sensitive and -insensitive coneflowers (*Rudbeckia laciniata* var. *digitata*) in Great Smoky Mountains National Park. *New Phytol.* 173:100–109.

Hand, J.L., S.M. Kreidenweis, N. Kreisberg, S. Hering, M. Stolzenburg, W. Dick, and P.H. McMurry. 2002. Comparisons of aerosol properties measured by impactors and light scattering from individual particles: Refractive index, number and volume concentrations, and size distributions. *Atmos. Environ.* 36:1853–1861.

Harmon, M.E., S.P. Bratton, and P.S. White. 1983. Disturbance and vegetation response in relation to environmental gradients in Great Smoky Mountains. *Vegetatio* 55:129–139.

Herlihy, A.T., P.R. Kaufmann, M.R. Church, P.J. Wigington, Jr., J.R. Webb, and M.J. Sale. 1993. The effects of acidic deposition on streams in the Appalachian Mountain and Piedmont region of the mid-Atlantic United States. *Water Resour. Res.* 29(8):2687–2703.

Hesthagen, T., J. Heggenes, B.M. Larsen, H.M. Berger, and T. Forseth. 1999. Effects of water chemistry and habitat on the density of young brown trout *Salmo trutta* in acidic streams. *Water Air Soil Pollut.* 112:85–106.

Johnson, D.W., M.S. Cresser, S.I. Milsson, J. Turner, B. Ulrich, D. Binkley, and D.W. Cole. 1991b. Soil changes in forest ecosystems: Evidence for and probable causes. *Proceedings of the Royal Society of Edinburgh Section B: Biological Sciences* 97:81–116.

Johnson, D.W. and I.J. Fernandez. 1992. Soil mediated effects of atmospheric deposition on eastern U.S. spruce fir forests. *In* C. Eagar and M.B. Adams (Eds.). *Ecology and Decline of Red Spruce in the Eastern United States.* Springer-Verlag, New York, pp. 235–270.

Johnson, D.W. and S.E. Lindberg. 1992. *Atmospheric Deposition and Forest Nutrient Cycling: A Synthesis of the Integrated Forest Ecological Series.* Springer, New York, 707pp.

Johnson, D.W., S.E. Lindberg, H. Van Miegroet, G.M. Lovett, D.W. Cole, M.J. Mitchell, and D. Binkley. 1993b. Atmospheric deposition, forest nutrient status, and forest decline: Implications of the Integrated Forest Study. *In* R.F. Huettl and D. Mueller-Dombois (Eds.). *Forest Decline in the Atlantic and Pacific Region.* Springer-Verlag, Berlin, Germany, pp. 66–81.

Johnson, D.W., R.B. Susfalk, P.F. Brewer, and W.T. Swank. 1999. Simulated effects of reduced sulfur, nitrogen, and base cation deposition on soils and solutions in southern Appalachian forests. *J. Environ. Qual.* 28:1336–1346.

Johnson, D.W., W.T. Swank, and J.M. Vose. 1993a. Simulated effects of atmospheric sulfur deposition on nutrient cycling in a mixed deciduous forest. *Biogeochemistry* 23:169–196.

Johnson, D.W., H.V. Van Miegroet, S.E. Lindberg, D.E. Todd, and R.B. Harrison. 1991a. Nutrient cycling in red spruce forests of the Great Smoky Mountains. *Can. J. For. Res.* 21:769–787.

Joslin, J.D., J.M. Kelly, and H. Van Miegroet. 1992. Soil chemistry and nutrition of North American spruce-fir stands: Evidence for recent change. *J. Environ. Qual.* 21(1):12–30.

Joslin, J.D. and M.H. Wolfe. 1992. Red spruce soil solution chemistry and root distribution across a cloud water deposition gradient. *Can. J. For. Res.* 22:893–904.

Kaeser, A.J. and W.E. Sharpe. 2001. The influence of acidic runoff episodes on slimy sculpin reproduction in Stone Run. *Trans. Am. Fish. Soc.* 130(6):1106–1115.

Kang, D., V.P. Aneja, R. Mathur, and J.D. Ray. 2003. Nonmethane hydrocarbons and ozone in three rural southeast United States national parks: A model sensitivity analysis and comparison to measurements. *J. Geophys. Res. Atmos.* 108(D19, 4604):17.

Kang, D., V.P. Aneja, R.G. Zika, C. Farmer, and J.D. Ray. 2001. Nonmethane hydrocarbons in the rural southeast United States national parks. *J. Geophys. Res.* 106(D3):3133–3155.

Kaufmann, P.R., A.T. Herlihy, M.E. Mitch, J.J. Messer, and W.S. Overton. 1991. Stream chemistry in the eastern United States 1. synoptic survey design, acid-base status, and regional patterns. *Water Resour. Res.* 27:611–627.

Kim, E. and P.K. Hopke. 2006. Characterization of fine particle sources in the Great Smoky Mountains area. *Sci. Total Environ.* 368:781–794.

Kohut, R. 2007a. Assessing the risk of foliar injury from ozone on vegetation in parks in the U.S. National Park Service's Vital Signs Network. *Environ. Pollut.* 149:348–357.

Kohut, R.J. 2007b. Ozone risk assessment for vital signs monitoring networks, Appalachian National Scenic Trail, and Natchez Trace National Scenic Trail. Natural Resource Technical Report NPS/NRPC/ARD/NRTR—2007/001. National Park Service, Natural Resource Program Center, Fort Collins, CO.

Lindberg, S.E. and J.G. Owens. 1993. Throughfall studies of deposition to forest edges and gaps in montane ecosystems. *Biogeochemistry* 19:173–194.

Liu, S., R. Munson, D. Johnson, S. Gherini, K. Summers, R. Hudson, K. Wilkinson, and L. Pitelka. 1991. Application of a nutrient cycling model (NuCM) to a northern mixed hardwood and a southern coniferous forest. *Tree Physiol.* 9:173–184.

Lovett, G.M. 1992. Atmospheric deposition and canopy interactions of nitrogen. *In* D.W. Johnson and S.E. Lindberg (Eds.). *Atmospheric Deposition and Forest Nutrient Cycling.* Springer-Verlag, New York, pp. 159–166.

MacAvoy, S.E. and A.J. Bulger. 2004. Sensitivity of blacknose dace (*Rhinichthys atratulus*) to moderate acidification events in Shenandoah National Park, USA. *Water Air Soil Pollut.* 153:125–134.

Malm, W.C., R. Ames, S. Copeland, D. Day, K. Gebhart, M. Pitchford, M. Scruggs, and J. Sisler. 2000b. IMPROVE report, spatial and seasonal patterns and temporal variability of haze and its constituents in the United States: Report III. Cooperative Institute for Research in the Atmosphere (CIRA), Colorado State University, Fort Collins, CO.

Malm, W.C., D.E. Day, and S.M. Kreidenweis. 2000a. Light scattering characteristics of aerosols as a function of relative humidity: Part I—A comparison of measured scattering and aerosol concentrations using the theoretical models. *J. Air Waste Manage. Assoc.* 50:686–700.

Malm, W.C., B.A. Schichtel, R.B. Ames, and K.A. Gebhart. 2002. A 10-year spatial and temporal trend of sulfate across the United States. *J. Geophys. Res.* 107(D22): ACH 11-1–ACH 11-20.

Malm, W.C. and J. Sisler. 1987. Sources of visibility reducing haze at Shenandoah National Park. Paper 87-40A 4. *Proceedings of the 80th Annual Meeting of the Air Pollution Control Association,* Air Pollution Control Association, New York, June 21–26, 1987.

Master, L.L., S.R. Flack, and B.A. Stein. 1998. *Rivers of Life: Critical Watersheds for Protecting Freshwater Biodiversity.* The Nature Conservancy, Arlington, VA.

McCune, B., J. Dey, J. Peck, K. Heiman, and S. Will-Wolf. 1997. Regional gradients in lichen communities of the southeast United States. *Bryologist* 100(2):145–158.

McDonnell, T.C., M.R. Sloat, T.J. Sullivan, C.A. Dolloff, P.F. Hessburg, N.A. Povak, W.A. Jackson, and C. Sams. 2015. Downstream warming and headwater acidity may diminish coldwater habitat in southern Appalachian mountain streams. *PLoS ONE* 10(8):e0134757.

McLaughlin, S.B., D.J. Downing, T.J. Blasing, E.R. Cook, and H.S. Adams. 1987. An analysis of climate and competition as contributors to decline of red spruce in high elevation Appalachian forests of the eastern United States. *Oecologia* 72:487–501.

McLaughlin, S.B., M. Nosal, S.D. Wullschleger, and G. Sun. 2007a. Interactive effects of ozone and climate on tree growth and water use in a southern Appalachian forest in the USA. *New Phytol.* 174:109–124.

McLaughlin, S.B., S.D. Wullschleger, G. Sun, and M. Nosal. 2007b. Interactive effects of ozone and climate on water use, soil moisture content and streamflow in a southern Appalachian forest in the USA. *New Phytol.* 174:125–136.

Mohnen, V.A. 1992. Atmospheric deposition and pollutant exposure of eastern U.S. forests. *In* C. Eagar and M.B. Adams (Eds.). *Ecology and Decline of Red Spruce in the Eastern United States.* Springer-Verlag, New York, pp. 64–124.

Moyle, P.B. and J.J. Cech, Jr. 2000. *Fishes: An Introduction to Ichthyology.* Prentice Hall, Upper Saddle River, NJ, 612pp.

Mueller, S.F. 1994. Characterization of ambient ozone levels in the Great Smoky Mountains National Park. *J. Appl. Meteorol.* 33:465–472.

Mueller, S.F., A. Song, W.B. Norris, S. Gupta, and R.T. McNider. 1996. Modeling pollutant transport during high-ozone episodes in the southern Appalachian Mountains. *J. Appl. Meteorol.* 35:2105–2120.

National Acid Precipitation Assessment Program (NAPAP). 1998. NAPAP biennial report to congress: An integrated assessment. National Acid Precipitation Assessment Program, Silver Spring, MD.

National Park Service (NPS). 2010. Air quality in national parks: 2009 annual performance and progress report. Natural Resource Report NPS/NRPC/ARD/NRR-2010/266. National Park Service, Air Resources Division, Denver, CO.

Neff, K.J., J.S. Schwartz, T.B. Henry, R.B. Robinson, S.E. Moore, and M.A. Kulp. 2009. Physiological stress in native southern brook trout during episodic stream acidification the Great Smoky Mountains National Park. *Arch. Environ. Contam. Toxicol.* 57:366–376.

Neufeld, H.S. and J.R. Renfro. 1993a. Sensitivity of black cherry seedlings (*Prunus serotina* Ehrh.) to ozone in Great Smoky Mountains National Park. The 1989 Seedling Set. Natural Resources Report NPS/NRTR-93/112. U.S. Department of the Interior, National Park Service, Air Quality Division, Denver, CO.

Neufeld, H.S. and J.R. Renfro. 1993b. Sensitivity of sycamore seedlings (*Platanus occidentalis*) to ozone in Great Smoky Mountains National Park. Natural Resources Report NPS/NRTR-93/131. U.S. Department of the Interior, National Park Service, Air Quality Division, Denver, CO.

Neufeld, H.S., J.R. Renfro, W.D. Hacker, and D. Silsbee. 1992. Ozone in Great Smoky Mountains National Park: Dynamics and effects on plants. *In* R.L. Berglung (Ed.). *Tropospheric Ozone and the Environment II*. Air & Waste Management Association, Pittsburgh, PA, pp. 594–617.

Nicholas, N.S., S.M. Zedaker, and C. Eagar. 1992. A comparison of overstory community structure in three southern Appalachian spruce-fir forests. *Bull. Torrey Bot. Club* 119:316–332.

Nodvin, S.C., H.V. Miegroet, S.E. Lindberg, N.S. Nicholas, and D.W. Johnson. 1995. Acidic deposition, ecosystem processes, and nitrogen saturation in a high elevation southern Appalachian watershed. *Water Air Soil Pollut.* 85:1647–1652.

Odman, M.T., J.W. Boylan, J.G. Wilkinson, A.G. Russell, S.F. Mueller, R.E. Imhoff, K.G. Doty, W.B. Norris, and R.T. McNider. 2002. SAMI air quality modeling: Final report. Southern Appalachian Mountains Initiative, Asheville, NC.

Oja, T. and P.A. Arp. 1998. Assessing atmospheric sulfur and nitrogen loads critical to the maintenance of upland forest soils. *In* D.G. Maynard (Ed.). *Sulfur in the Environment*. Marcel Dekker, New York, pp. 337–363.

Paatero, P. 1997. Least square formulation of robust non-negative factor analysis. *Chemometr. Intell. Lab. Syst.* 37:23–35.

Pardo, L.H. and N. Duarte. 2007. Assessment of effects of acidic deposition on forested ecosystems in Great Smoky Mountains National Park using critical loads for sulfur and nitrogen. A report prepared for Tennessee Valley Authority, Knoxville, TN and Great Smoky Mountains National Park, Gatlinburg, TN. USDA Forest Service.

Pardo, L.H., M.J. Robin-Abbott, and C.T. Driscoll. 2011. Assessment of nitrogen deposition effects and empirical critical loads of nitrogen for ecoregions of the United States. General Technical Report NRS-80. U.S. Forest Service, Newtown Square, PA.

Pauley, E.F., S.C. Nodvin, N.S. Nicholas, A.K. Rose, and T.B. Boffey. 1996. Vegetation, biomass, and nitrogen pools in a spruce-fir forest of the Great Smoky Mountains National Park. *Bull. Torrey Bot. Club* 123(4):318–329.

Peart, D.R., N.S. Nicholas, S.M. Zedaker, M.M. Miller-Weeks, and T.G. Siccama. 1992. Condition and recent trends in high elevation red spruce populations. *In* C. Eagar and M.B. Adams (Eds.). *Ecology and Decline of Red Spruce in the Eastern United States*. Springer-Verlag, New York, pp. 125–191.

Pyle, C. 1988. The type and extent of anthropogenic vegetation disturbance in the Great Smoky Mountains before National Park Service acquisition. *Castanea* 53:183–196.

Rice, K.C., T.M. Scanlon, J.A. Lynch, and B.J. Cosby. 2014. Decreased atmospheric sulfur deposition across the southeastern U.S.: When will watersheds release stored sulfate? *Environ. Sci. Technol.* 48(17):10071–10078.

Robinson, R.B., T.W. Barnett, G.R. Harwell, S.E. Moore, M. Kulp, and J.S. Schwartz. 2008. pH and acid anion time trends in different elevation ranges in the Great Smoky Mountains National Park. *J. Environ. Eng.* 134:800–808.

Robinson, R.B., M.S. Wood, J.L. Smoot, and S.E. Moore. 2004. Parametric modeling of water quality and sampling strategy in a high-altitude Appalachian stream. *J. Hydrol.* 287:62–73.

Rose, A.K. 2000. Coarse woody debris and nutrient dynamics in a southern Appalachian spruce–fir forest. Masters, Department of Ecology and Evolutionary Biology, University of Tennessee, Knoxville, TN.

Rosemond, A.D., S.R. Reice, J.W. Elwood, and P.J. Mulholland. 1992. The effects of stream acidity on benthic invertebrate communities in the south-eastern United States. *Freshw. Biol.* 27:193–209.

Samuelson, L.J. and J.M. Kelly. 1997. Ozone uptake in *Prunus serotina, Acer rubrum* and *Quercus rubra* forest trees of different sizes. *New Phytol.* 136:255–264.

Sasser, C.L. and D. Binkley. 1989. Nitrogen mineralization in high-elevation forests of the Appalachians. II. Patterns with stand development in fir waves. *Biogeochemistry* 7:147–156.

Schwede, D.B. and G.G. Lear. 2014. A novel hybrid approach for estimating total deposition in the United States. *Atmos. Environ.* 92:207–220.

Shaver, C.L., K.A. Tonnessen, and T.G. Maniero. 1994. Clearing the air at Great Smoky Mountains National Park. *Ecol. Appl.* 4(4):690–701.

Shortle, W.C. and K.T. Smith. 1988. Aluminum-induced calcium deficiency syndrome in declining red spruce. *Science* 240:1017–1018.

Smith, R.K., P.L. Freeman, J.V. Higgins, K.S. Wheaton, T.W. FitzHugh, A.A. Das, and K.J. Ernstrom. 2002. *Priority Areas for Freshwater Conservation Action: A Biodiversity Assessment of the Southeastern United States.* The Nature Conservancy, Arlington, VA.

Smith, W.H. 1990. *Forests as Sinks for Air Contaminants: Soil Compartment. Air Pollution and Forests: Interactions between Air Contaminants and Forest Ecosystems.* Springer-Verlag, New York, pp. 113–146.

Somers, G.L., A.H. Chappelka, P. Rosseau, and J.R. Renfro. 1998. Empirical evidence of growth decline related to visible ozone injury. *For. Ecol. Manage.* 104:12–137.

Southern Appalachian Man and the Biosphere (SAMAB). 1996. The southern Appalachian assessment aquatics technical report. U.S. Department of Agriculture, Forest Service, Southern Region, Atlanta, GA.

Souza, L., H.S. Neufeld, A.H. Chappelka, K.O. Burkey, and A.W. Davison. 2006. Seasonal development of ozone-induced foliar injury on tall milkweed (*Asclepias exaltata*) in Great Smoky Mountains National Park. *Environ. Pollut.* 141:175–183.

Stoddard, J.L. 1994. Long-term changes in watershed retention of nitrogen: Its causes and aquatic consequences. *In* L.A. Baker (Ed.). *Environmental Chemistry of Lakes and Reservoirs.* American Chemical Society, Washington, DC, pp. 223–284.

Sullivan, T.J., B.J. Cosby, A.T. Herlihy, J.R. Webb, A.J. Bulger, K.U. Snyder, P. Brewer, E.H. Gilbert, and D.L. Moore. 2004. Regional model projections of future effects of sulfur and nitrogen deposition on streams in the southern Appalachian Mountains. *Water Resour. Res.* 40:W02101.

Sullivan, T.J., B.J. Cosby, J.A. Laurence, R.L. Dennis, K. Savig, J.R. Webb, A.J. Bulger et al. 2003. Assessment of air quality and related values in Shenandoah National Park. NPS/NERCHAL/NRTR-03/090. U.S. Department of the Interior, National Park Service, Northeast Region.

Sullivan, T.J., B.J. Cosby, J.R. Webb, K.U. Snyder, A.T. Herlihy, A.J. Bulger, E.H. Gilbert, and D. Moore. 2002b. Assessment of the effects of acidic deposition on aquatic resources in the southern Appalachian Mountains. Report prepared for the Southern Appalachian Mountains Initiative (SAMI). E&S Environmental Chemistry, Inc., Corvallis, OR.

Sullivan, T.J., D.W. Johnson, and R. Munson. 2002a. Assessment of effects of acid deposition on forest resources in the southern Appalachian Mountains. Report prepared for the Southern Appalachian Mountains Initiative (SAMI). E&S Environmental Chemistry, Inc., Corvallis, OR.

Sullivan, T.J., G.T. McPherson, T.C. McDonnell, S.D. Mackey, and D. Moore. 2011. Evaluation of the sensitivity of inventory and monitoring national parks to acidification effects from atmospheric sulfur and nitrogen deposition. U.S. Department of the Interior, National Park Service, Denver, CO.

Sullivan, T.J., J.R. Webb, K.U. Snyder, A.T. Herlihy, and B.J. Cosby. 2007. Spatial distribution of acid-sensitive and acid-impacted streams in relation to watershed features in the southern Appalachian Mountains. *Water Air Soil Pollut.* 182:57–71.

Tong, D.Q., D. Kang, V.P. Aneja, and J.D. Ray. 2005. Reactive nitrogen oxides in the southeast United States National Parks: Source identification, origin, and process budget. *Atmos. Environ.* 39:315–327.

U.S. Environmental Protection Agency. 2001a. Latest finds on national air quality: 2000 status and trends. EPA 454/K-01-002. Office of Air Quality Planning and Standards, Research Triangle Park, NC.

U.S. Environmental Protection Agency. 2001b. National air quality and emissions trends report, 1999. EPA-454/R-01-004. Office of Air Quality Planning and Standards, Emissions Monitoring and Analysis Division, Air Quality Trends Analysis Group, Research Triangle Park, NC.

U.S. Environmental Protection Agency. 2008a. Integrated science assessment for oxides of nitrogen and sulfur—Ecological criteria. EPA/600/R-08/082F. National Center for Environmental Assessment, Office of Research and Development, Research Triangle Park, NC.

U.S. Environmental Protection Agency. 2008b. National air quality status and trends through 2007. EPA-454/R-08-006. U.S. Environmental Protection Agency, Office of Air Quality Planning and Standards, Air Quality Assessment Division, Research Triangle Park, NC.

Van Miegroet, H., D.W. Cole, and N.W. Foster. 1992. Nitrogen distribution and cycling. *In* D.W. Johnson and S.E. Lindberg (Eds.). *Atmospheric Deposition and Forest Nutrient Cycling.* Springer-Verlag, New York, pp. 178–196.

Van Miegroet, H., I.F. Creed, N.S. Nicholas, D.G. Tarboton, K.L. Webster, J. Shubzda, B. Robinson et al. 2001. Is there synchronicity in nitrogen input and output fluxes at the Noland Divide Watershed, a small N-saturated forested catchment in the Great Smoky Mountains National Park. *Sci. World* 1:480–492.

Van Miegroet, H., P.T. Moore, C.E. Tewksbury, and N.S. Nicholas. 2007. Carbon sources and sinks in high-elevation spruce-fir forests of the southeastern US. *For. Ecol. Manage.* 238:249–260.

van Sickle, J. and M.R. Church. 1995. Nitrogen bounding study. Methods for estimating the relative effects of sulfur and nitrogen deposition on surface water chemistry. EPA/600/R-95/172. U.S. Environmental Protection Agency, National Health and Environment Effects Research Laboratory, Corvallis, OR.

Vitousek, P.M. and R.W. Howarth. 1991. Nitrogen limitation on land and in the sea: How can it occur? *Biogeochemistry* 13:87–115.

Wallace, J.B., J.R. Webster, and R.L. Lowe. 1992. High-gradient streams of the Appalachians. *In* C.T. Hackney, S.M. Adams, and W.H. Martin (Eds.). *Biodiversity of Southeastern United States Aquatic Communities.* John Wiley & Sons, New York, pp. 133–192.

Weathers, K.C., S.M. Simkin, G.M. Lovett, and S.E. Lindberg. 2006. Empirical modeling of atmospheric deposition in mountainous landscapes. *Ecol. Appl.* 16(4):1590–1607.

Webb, J.R., F.A. Deviney, J.N. Galloway, C.A. Rinehart, P.A. Thompson, and S. Wilson. 1994. The acid-base status of native brook trout streams in the mountains of Virginia. Department of Environmental Sciences, University of Virginia, Charlottesville, VA.

Webster, K.L., I.F. Creed, N.S. Nicholas, and H.V. Miegroet. 2004. Exploring interactions between pollutant emissions and climatic variability in growth of red spruce in the Great Smoky Mountains National Park. *Water Air Soil Pollut.* 159:225–248.

Weinstein, D.A., B. Gollands, and W.A. Retzlaff. 2001. The effects of ozone on a lower slope forest of the Great Smoky Mountain National Park: Simulations linking an individual tree model to a stand model. *For. Sci.* 47(1):29–42.

White, P.S. and C.V. Cogbill. 1992. Spruce-fir forests of eastern America. *In* C. Eagar and M.B. Adams (Eds.). *Ecology and Decline of Red Spruce in the Eastern United States.* Springer-Verlag, Inc., New York, pp. 3–39.

5

Shenandoah National Park and the Mid-Atlantic Network

5.1 Background

The Mid-Atlantic Network region consists of portions of the Blue Ridge, Ridge and Valley, Piedmont, Coastal Plain, and Northern Piedmont ecoregions, between the Virginia-North Carolina border and southern Pennsylvania. There is only one park in the Mid-Atlantic Network that is larger than 100 mi^2 (259 km^2): Shenandoah National Park (SHEN). There are also nine smaller, mostly historical parks. The focus of this chapter is on SHEN. Since air pollutants can be widespread, reduction of pollution emissions affecting large parks will often result in the protection of smaller parks located nearby as well.

Figure 5.1 shows the network boundary and the location of each national park, along with human population centers around the network that have more than 10,000 people. There are no urban centers within the Mid-Atlantic Network region that are larger than 500,000 people. However, there are many large urban centers within a 300 mi (483 km) radius of the network boundary, including Baltimore; Washington, DC; New York; Richmond; and Philadelphia.

The southern Appalachian area is widely regarded as one of the most diverse landscapes in the Temperate Zone (Southern Appalachian Man and the Biosphere 1996). Fish diversity is quite high. There are about 950 freshwater fish species in North America (Jenkins and Burkhead 1993), of which about 485 species can be found in the Southeast, about 210 species in Virginia (Jenkins and Burkhead 1993), and more than 30 species in SHEN. Regional habitat diversity and intraspecific genetic diversity are also regarded as high. Thus, the Southeast is a unique national biodiversity resource for fish. Unfortunately, the Southern Appalachian Assessment concluded that 70% of sampled stream locations showed moderate to severe fish community degradation and that about 50% of the stream length in West Virginia and Virginia showed habitat impairment (Southern Appalachian Man and the Biosphere 1996).

A great deal of scientific research has been conducted on S and N deposition in SHEN and their effects on biogeochemical processes and aquatic and terrestrial biota. This has largely been due to the proximity of the University of Virginia to the park and the active involvement of numerous University of Virginia scientists in this general field of research. Effects of air pollutants have been well documented in SHEN, including stream acidification, ozone effects on plant foliage, and high haze levels. Toxic air contaminants are not addressed in this chapter. This does not imply that metals and persistent organic pollutants are not important in this park, but rather that limited data are available with which to evaluate the impacts of these contaminants.

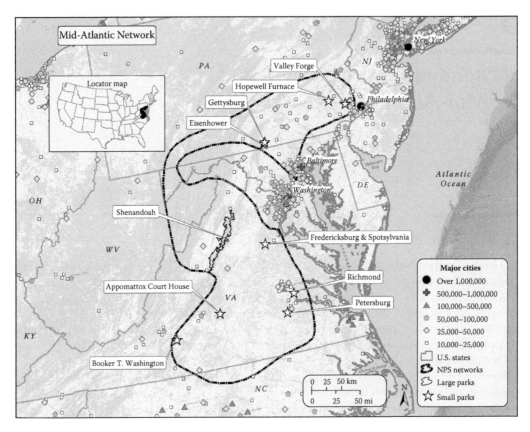

FIGURE 5.1
Network boundary and locations of parks and population centers greater than 10,000 people around the Mid-Atlantic Network region.

5.2 Atmospheric Emissions and Deposition

The major source areas of atmospheric emissions of S and N that impact sensitive resources within SHEN (and presumably many of the other Mid-Atlantic Network parks) have been shown to be located in the Ohio River Valley, northeastern West Virginia, southwestern Pennsylvania, and central and eastern Virginia, based on simulations using the Regional Acid Deposition Model (Sullivan et al. 2003). County-level sulfur dioxide emissions within the Mid-Atlantic Network in 2002, summarized by Sullivan et al. (2011), ranged from less than 1 to greater than 100 tons/mi^2/yr (Sullivan et al. 2011). In most counties within the network region, sulfur dioxide emissions were less than 20 tons/mi^2/yr, with only a few counties exceeding this amount. Although most point sources of S within the Mid-Atlantic Network emitted less than 5,000 tons/yr, there were several sources of greater magnitude, one larger than 40,000 tons/yr. There were also numerous large sulfur dioxide point sources to the northwest of the network. Although emissions of sulfur dioxide upwind from SHEN have been high, they have been gradually decreasing since the early 1980s.

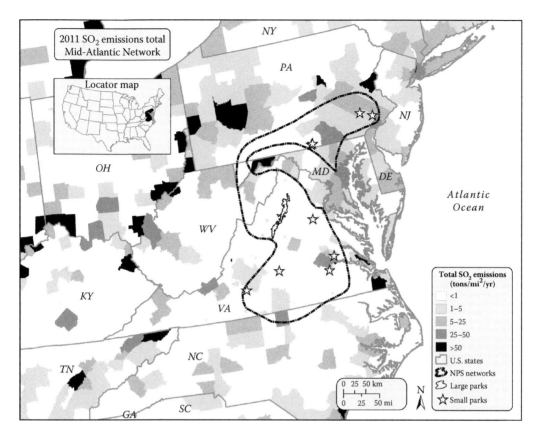

FIGURE 5.2

Total sulfur dioxide (SO$_2$) emissions, by county, near the Mid-Atlantic Network for the year 2011. (Data from the EPA's National Emissions Inventory, https://www.epa.gov/air-emissions-inventories, accessed January, 2014.)

Annual county-level N emissions within the Mid-Atlantic Network in 2002 ranged from less than 1 ton/mi^2/yr to more than 100 tons/mi^2/yr. In general, N emissions were between 1 and 20 tons/mi^2/yr, but there were several counties with higher N emissions, in the range of 20–50 tons/mi^2/yr or more. There were relatively few substantial (larger than 1000 tons/yr) N point sources within this network, but there were a number of point sources of oxidized N that were larger than 5000 tons/yr just to the west of the network boundary (Sullivan et al. 2011). Emissions of N were relatively constant throughout the 1980s and 1990s but began to decline in about 2000.

County-level emissions near the Mid-Atlantic Network, based on data from the EPA's National Emissions Inventory during a recent time period (2011), are depicted in Figures 5.2 through 5.4 for sulfur dioxide, oxidized N, and ammonia, respectively. Several counties to the west of Mid-Atlantic Network parks had relatively high sulfur dioxide emissions (>50 tons/mi^2/yr; Figure 5.2). Spatial patterns in oxidized N emissions near the Mid-Atlantic Network parks were generally higher than sulfur dioxide emissions (Figure 5.3). Emissions of ammonia near Mid-Atlantic Network parks were somewhat lower, with most counties showing emissions levels below 8 tons/mi^2/yr (Figure 5.4).

Total S deposition within the Mid-Atlantic Network region ranged from as low as 5 to 10 kg S/ha/yr to greater than 30 kg S/ha/yr in 2002 (Sullivan et al. 2011). The highest S

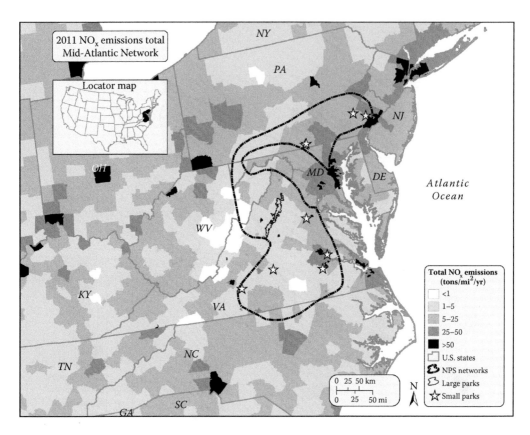

FIGURE 5.3

Total oxidized N (NO$_x$) emissions, by county, near the Mid-Atlantic Network for the year 2011. (Data from the EPA's National Emissions Inventory, https://www.epa.gov/air-emissions-inventories, accessed January, 2014.)

deposition values within the network occurred to the west and north of SHEN. Sulfur deposition in SHEN peaked sometime during the 1970s or early 1980s and has been declining steadily since then (Figure 5.5). Total N deposition within the network in 2002 ranged from as low as 5–10 kg N/ha/yr to greater than 15 kg N/ha/yr. Estimated total N deposition throughout much of the network, including most of SHEN, was in the range of 10–15 kg N/ha/yr.

Atmospheric S, and to a lesser extent N, deposition levels have declined at SHEN (and all Mid-Atlantic Network parks) since 2001, based on Total Deposition project estimates (Table 5.1). Decreases in total S and N deposition over the previous decade for SHEN were –46.7% and –6.2%, respectively. Oxidized N and ammonium showed opposite patterns, with oxidized N deposition decreasing (–38%) and ammonium increasing (+48.1%) since the monitoring period 2000–2002.

Total S deposition in and around the Mid-Atlantic Network for the period 2010–2012 was generally highest (>10 kg S/ha/yr) to the north and lowest (<5 kg S/ha/yr) to the south within the network area (Figure 5.6). Oxidized inorganic N deposition for the period 2010–2012 was in the range of 2–5 kg N/ha/yr at the parklands to the southwest within the Mid-Atlantic Network and 5–10 kg N/ha/yr in the eastern portion of the network. Most areas to the south received less than 5 kg N/ha/yr of reduced inorganic N from atmospheric deposition during this same period; parks to the northeast received higher

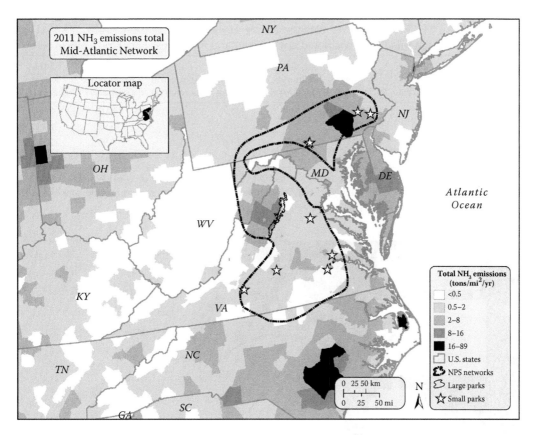

FIGURE 5.4
Total ammonia (NH₃) emissions, by county, near the Mid-Atlantic Network for the year 2011. (Data from the EPA's National Emissions Inventory, https://www.epa.gov/air-emissions-inventories, accessed January, 2014.)

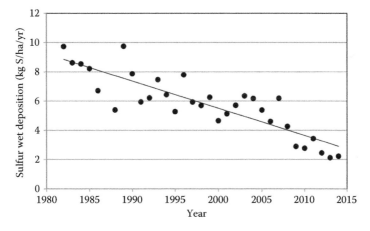

FIGURE 5.5
Wet sulfur deposition for the period of record at the Big Meadows monitoring station in SHEN. Total sulfur deposition, including both wet and dry forms, is probably at least one and a half times these measured wet-only values. Sulfur deposition in the park has decreased by more than 50% over the past three decades in response to emissions control legislation. (From National Atmospheric Deposition Program, http://nadp.sws.uiuc.edu/, accessed December, 2015.)

TABLE 5.1

Average Changes in S and N Deposition[a] between 2001 and 2011 across Park Grid Cells in SHEN

Parameter	2001 Average (kg/ha/yr)	2011 Average (kg/ha/yr)	Absolute Change (kg/ha/yr)	Percent Change	2011 Minimum (kg/ha/yr)	2011 Maximum (kg/ha/yr)	2011 Range (kg/ha/yr)
Total S	13.05	6.98	−6.08	−46.7	4.52	8.02	3.49
Total N	13.60	12.75	−0.85	−6.2	8.74	16.95	8.21
Oxidized N	8.42	5.23	−3.20	−38.0	3.45	5.95	2.50
Reduced N	5.18	7.51	2.35	48.1	3.88	12.99	9.11

[a] Deposition estimates were determined by the Total Deposition project, based on three-year averages centered on 2001 and 2011 for all ~4 km grid cells in each park. The minimum, maximum, and range of 2011 S and N deposition within each park are also shown.

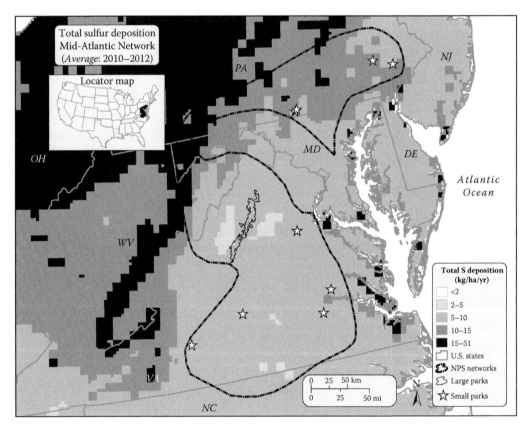

FIGURE 5.6
Total S deposition for the three-year period centered on 2011 in and around the Mid-Atlantic Network region. (From Schwede, D.B. and Lear, G.G., *Atmos. Environ.*, 92, 207, 2014.)

(>5 kg N/ha/yr) amounts. Total N deposition in SHEN was estimated to be 12.8 kg N/ha/yr in 2011 (Table 5.1, Figure 5.7).

Wet deposition collected at 29 national park locations was reported by the National Park Service (2010). Most monitored sites showed statistically significant ($p \leq 0.05$) decreases in concentrations of sulfate in wet deposition between about 1989 and about 2008. One of the

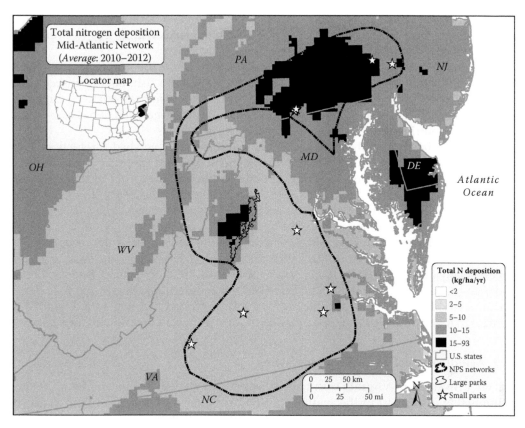

FIGURE 5.7
Total N deposition for the three-year period centered on 2011 in and around the Mid-Atlantic Network region. (From Schwede, D.B. and Lear, G.G., *Atmos. Environ.*, 92, 207, 2014.)

largest decreasing trends was reported for SHEN (−0.77 µeq/L/yr; NPS 2010), second only to Indiana Dunes in the Great Lakes Network (−1.4 µeq/L/yr). There was not, however, a statistically significant change in nitrate concentration in wet deposition at SHEN during that period.

The National Park Service (2010) also reported long-term trends in concentrations of Hg in wet deposition in SHEN and eight other national parks during the period 2003–2008. Statistically significant ($p \leq 0.05$) decreases in wet Hg deposition were reported for two parks, but Hg deposition remained unchanged at SHEN. Three-year means of annual Hg concentration in wet deposition were reported by the National Park Service (2010) for 13 parks that had at least two years of valid data during the period 2006–2008. SHEN was among the five monitored parks having highest Hg concentrations in precipitation.

5.3 Acidification

The network rankings developed by Sullivan et al. (2011) in a coarse screening assessment for acid pollutant exposure, ecosystem sensitivity to acidification, and park protection yielded an overall network acidification summary risk ranking for the Mid-Atlantic

Network that was the highest among all networks. The overall level of concern for acidification effects on Inventory and Monitoring program parks within this network was judged by Sullivan et al. (2011) to be very high.

Sulfur deposition has been high enough at SHEN to cause the acidification of many streams within the park, with associated harmful impacts to fish, aquatic insects, and other life forms (Cosby et al. 2006). Nitrogen deposition has also been relatively high but, because deposited N has been strongly retained in soils and vegetation, it has constituted only a minor contributor to stream acidification to date.

Sulfate mass balance analyses indicated that, because of watershed S retention, soils and surface waters of the Mid-Atlantic Network and southward throughout the southern Appalachian Mountain region have not yet realized the full effects of elevated S deposition. On average, based on National Stream Survey data from the 1980s, sites in the Blue Ridge Mountains retained 35% of incoming S from atmospheric deposition. Sulfur adsorption varies by physiographic province in the Appalachian Mountain region. It is highest in the soils of the southern Blue Ridge, where typically about half of the incoming S is retained. Adsorption is lower in the Valley and Ridge watersheds to the west of SHEN and especially in the Appalachian Plateau (Herlihy et al. 1993). In general, S adsorption is moderate in the vicinity of SHEN and higher in the southern portions of the Blue Ridge Province. Herlihy et al. (1996) and Rice et al. (2014) concluded that S retention will likely decrease in the future, resulting in further losses of stream acid neutralizing capacity.

5.3.1 Acidification of Terrestrial Ecosystems

In 2000, University of Virginia scientists collected and analyzed soil samples at 79 sites within SHEN in support of the acidic deposition component of the Sullivan et al. (2003) assessment. The samples were collected from 14 different watersheds, 5 each on primarily siliciclastic and basaltic bedrock and 4 primarily on granitic bedrock. Between four and eight soil pits were excavated within each watershed, distributed to account for differences in slope, aspect, land-use history, fire history, and forest cover type.

The soils within watersheds situated primarily on siliciclastic bedrock generally showed the lowest soil pH (median 4.4–4.5), cation exchange capacity (median 3.5–7.5 centimoles per kilogram [cmol/kg]), and base saturation (median 8%–12%). Values for watersheds having soils primarily on granitic bedrock were generally intermediate (base saturation less than about 14%), and basaltic watersheds were higher in all three parameters (Table 5.2). The range of base saturation values obtained for soils associated with both siliciclastic and granitic bedrock in SHEN (Welsch et al. 2001) included values in the 10%–20% range cited as possible threshold values for incomplete acid neutralization and increased leaching of Al (Reuss and Johnson 1986, Binkley et al. 1989, Cronan and Schofield 1990). Also, the measured pH of soils located over both siliciclastic and granitic bedrock included values in the highly acidic range (pH < 4.5) in which the more toxic forms of Al often predominate (Binkley et al. 1989). The low base cation availability in soils of watersheds underlain by siliciclastic or granitic bedrock was probably due to a combination of low base cation content of the parent bedrock and Ca depletion. Soils may stop changing their base cation status before deficiencies develop in vegetation as acidic deposition levels decline, or weathering rates may be sufficient to maintain adequate cation nutrient supplies.

Soil acidification, base cation depletion, and Al toxicity have collectively contributed to decline in red spruce, sugar maple, and perhaps other tree species in portions of the northeastern United States and Appalachian Mountains (likely including within SHEN)

TABLE 5.2

Interquartile Distribution of pH, Cation Exchange Capacity (CEC), and Percent Base Saturation for Soil Samples[a] Collected in SHEN Study Watersheds during the 2000 Soil Survey

Site ID	Watershed	n	pH 25th	Med	75th	CEC (cmol/kg) 25th	Med	75th	Percent Base Saturation 25th	Med	75th
Siliciclastic bedrock class[b]											
VT35	Paine Run	6	4.4	4.5	4.7	3.7	5.7	5.7	7.1	10.0	24.9
WOR1	White Oak Run	6	4.3	4.4	4.4	4.8	7.5	7.8	5.3	7.5	8.5
DR01	Deep Run	5	4.3	4.4	4.5	3.9	5.0	5.8	7.2	8.9	10.8
VT36	Meadow Run	6	4.4	4.4	4.5	3.1	3.5	7.6	7.8	8.7	11.3
VT53	Two Mile Run	5	4.3	4.5	4.5	4.6	6.0	6.9	11.7	12.3	13.6
Granitic bedrock class											
VT59	Staunton River	6	4.7	4.8	4.9	6.5	7.5	9.2	9.1	13.9	29.5
NFDR	NF of Dry Run	5	4.4	4.5	4.7	7.3	8.0	9.2	7.5	10.8	12.4
VT58	Brokenback Run	5	4.6	4.7	4.7	7.3	8.4	9.6	6.0	6.7	9.7
VT62	Hazel River	4	4.5	4.7	4.8	5.3	5.3	6.5	12.3	12.8	21.6
Basaltic bedrock class											
VT60	Piney River	6	4.7	5.0	5.3	7.3	7.7	10.0	17.0	24.0	57.0
VT66	Rose River	8	4.8	5.0	5.3	7.3	10.1	10.7	19.1	38.0	63.5
VT75	White Oak Canyon	6	4.9	5.1	5.5	7.1	7.5	9.3	15.6	32.8	43.4
VT61	NF of Thornton River	7	5.1	5.2	5.3	7.7	9.6	10.8	35.6	54.4	71.2
VT51	Jeremy's Run	4	4.7	5.0	5.3	6.3	7.6	7.7	15.0	22.8	46.1

Source: Reprinted from Sullivan, T.J. et al., *Environ. Monit. Assess.*, 137, 85, 2008, Table 3. With permission.

[a] Samples collected from mineral soil >20 cm depth.

[b] Watersheds are stratified according to the predominant bedrock class present in each watershed.

that have experienced soil acidification as a consequence of S and N deposition. Effects have included reduced growth and increased stress to overstory trees, and likely changes in the species distributions of understory plants. Nearly all of SHEN is covered by vegetation types that contain sugar maple. Sugar maple is known to be sensitive to acidification (Sullivan et al. 2013). Potentially sensitive spruce-fir forest resources are found at high elevation (generally above about 1370 m [4495 ft]) in the southern Appalachian Mountains. There are small (~18 ha [45 ac]) relic populations of red spruce in SHEN at Limberlost and Hawksbill Peak (NPS 1981). These populations are likely under considerable natural stress and may be more susceptible to the effects of high S and/or N deposition than other forest types in SHEN.

There is no clear evidence to demonstrate that the levels of N deposition observed in SHEN have caused forest decline in the deciduous forests of the park. Because the forests in SHEN are almost exclusively second growth, subsequent to large-scale timber harvesting and localized agricultural land use prior to park creation, the N demand of the regrowing forest is likely to be high. This precludes nitrate leaching at current N deposition levels in the absence of significant disturbance.

Nevertheless, the complex relationships between atmospheric inputs and forest health remain a concern in SHEN, especially in view of the known effects of acidic deposition on soil acidity, nutrient supply, metal toxicity, and tree growth. It is not known whether forest ecosystems at SHEN are currently experiencing subtle effects of elevated N deposition, and whether, for example, some plant species are favored at the expense of other species by the relatively high N environment.

Effects of insect-caused defoliation on the N cycle at SHEN can be pronounced. The foliar N consumed by insects is deposited on the forest floor as insect feces (frass), greenfall, and insect biomass. Some of this deposited N is subsequently taken up by tree roots and soil microbes, with little effect on the nutritional condition of the trees or the site. Where a sizable component of this N is leached in drainage water, the nutritional consequences can be more significant. There are also various feedback mechanisms. For example, low N supply can slow the population growth of defoliating insects (Mason et al. 1992) and enhance the tree's chemical defenses against insects (Hunter and Schultz 1995). The amount of N leaching loss is generally small, relative to atmospheric deposition inputs and relative to the amount of N transferred to the forest floor with the defoliation (Lovett and Ruesink 1995, Lovett et al. 2002). Nevertheless, it can be high enough to contribute to base cation depletion of soils and effects on downstream receiving waters. The extent of nitrate leaching may be partly related to the extent of defoliation and tree mortality that occurs and also to the amount of precipitation that occurs immediately after the defoliation (Lovett et al. 2002).

5.3.2 Acidification of Aquatic Ecosystems

Acidic streams and streams with very low acid neutralizing capacity in the Mid-Atlantic Network are almost all located in small (watershed area < 20 km^2), upland, forested catchments in areas of base-poor bedrock. Thin soils and steep slopes make these watersheds generally unsuitable for agriculture and other development and contribute to their sensitivity to acidic deposition (Figure 5.8; Baker et al. 1990c).

In the subpopulation of upland forested streams, which comprises about half of the total stream population in the mid-Appalachian area, data from various local surveys in the 1980s and 1990s showed that 5%–20% of the streams were acidic, and about 25%–50% had acid neutralizing capacity <50 µeq/L (Herlihy et al. 1993). The National Stream Survey estimates for the whole region showed that there were 2330 km (1448 mi) of acidic streams and 7500 km (4660 mi) of streams with acid neutralizing capacity <50 µeq/L. In the forested reaches, 12% of the upstream reaches were acidic and 17% had pH ≤ 5.5. Sulfur has been the primary determinant of precipitation acidity and sulfate has been the dominant acid anion associated with acidic streams, both in the central Appalachian Mountains region and within SHEN. Sulfur deposition has acidified park streams and affected in-stream biota, particularly in watersheds dominated by siliciclastic bedrock types that give rise to soils with low base saturation and relatively low sulfur adsorption and to streams with low acid neutralizing capacity. Nitrate concentrations in SHEN streams are negligible in the absence of insect infestation.

Aquatic effects research at SHEN has contributed significantly to the development of scientific understanding of watershed processes that control aquatic effects of acidic deposition (cf., Galloway et al. 1983). This research has also contributed to understanding of relationships between geology and sensitivity to acidification (Lynch and Dise 1985, Bricker and Rice 1989, Sullivan et al. 2007) and of the effects of forest insect infestation on episodic chemical processes (Webb et al. 1995, Eshleman et al. 2001).

Stream water acid neutralizing capacity is closely related to bedrock mineralogy in the Mid-Atlantic Network and surrounding areas (Herlihy et al. 1993). In fact, geological type, soils conditions that developed from underlying geology, and water chemistry conditions are all closely interrelated within the network. This is partly because rock and soils

FIGURE 5.8
Many small streams in SHEN are sensitive to acidification. (National Park Service photo.)

materials in the Mid-Atlantic Network and elsewhere in the Southeast were not transported from place to place (and thereby mixed) by the process of glaciation. Sullivan et al. (2007) delimited a high-interest area for stream water acidification sensitivity within the southern Appalachian Mountains region (Virginia/West Virginia to Georgia) based on geological classification and elevation. It covered only 28% of the region and yet included almost all known acidic and low acid neutralizing capacity (<20 µeq/L) streams, based on the evaluation of about 1000 streams for which water chemistry data were available. They found that the vast majority of low acid neutralizing capacity sampled streams were underlain by the siliceous geologic sensitivity class, which is represented by such lithologies as sandstone and quartzite. Low acid neutralizing capacity stream water throughout the region was also found to be associated with a number of watershed features in addition to lithology and elevation, including ecoregion, physiographic province, soil type, forest type, and watershed area.

The SHEN landscape includes three major geologic sensitivity types: siliciclastic, granitic, and basaltic. Each of these bedrock types influences about one-third of the stream length in the park. The observed patterns in stream water chemistry in SHEN are strongly related to patterns in bedrock geology (Sullivan et al. 2003). As observed by Sullivan et al. (2007) throughout the region, siliciclastic sites have the greatest sensitivity to acidification, whereas granitic sites have moderate sensitivity and basaltic sites have low sensitivity.

There are many streams on siliciclastic bedrock in SHEN that have chronically low acid neutralizing capacity, in the range where adverse effects are likely to occur on sensitive aquatic biota and where episodic acidification to acid neutralizing capacity values near or below zero frequently occur during hydrological events. The streams that are most susceptible to adverse chronic or episodic effects on in-stream biota are those having chronic acid neutralizing capacity less than about 50 μeq/L, especially those having chronic acid neutralizing capacity less than about 20 μeq/L. These are primarily situated on siliciclastic bedrock.

The Shenandoah Watershed Study was initiated by the University of Virginia in 1979, with the establishment of water quality monitoring on two streams (Webb et al. 1993). The ongoing watershed data collection involves 14 primary study watersheds, including a combination of discharge gauging, routine quarterly and weekly water quality sampling, and high-frequency episodic, or storm-flow, sampling (Galloway et al. 1999, Cosby et al. 2006). In addition, a number of extensive stream chemistry surveys, fish population surveys, and other watershed data collection efforts have been conducted throughout the park in support of various research efforts. The Shenandoah Watershed Study is coordinated with the Virginia Trout Stream Sensitivity Study, which extends quarterly sampling to an additional 51 native brook trout streams located on public lands throughout western Virginia (primarily in the George Washington and Jefferson national forests).

Since the 1980s, the Shenandoah Watershed Study has developed a uniquely comprehensive watershed database for SHEN, while making major contributions to scientific understanding of surface water acidification and the biogeochemistry of forested mountain watersheds. Data and analyses provided through the Shenandoah Watershed Study have contributed significantly to regional and national assessments of acidic deposition effects, including several that led to enactment of the Clean Air Act Amendments of 1990 (e.g., Baker et al. 1990a, National Acid Precipitation Assessment Program [NAPAP] 1991). The Shenandoah Watershed Study and the Virginia Trout Stream Sensitivity Study have provided some of the most comprehensive data available for use in aquatic effects assessment. The combined Shenandoah Watershed Study and Virginia Trout Stream Sensitivity Study contributed data to the EPA for use in Congressionally mandated the evaluation of air pollution control program benefits relative to acidic deposition effects on sensitive surface waters. As a consequence of this extensive monitoring, research, and assessment activity, SHEN is a leader among the national parks with respect to park-specific knowledge of acidic deposition effects and watershed ecosystem conditions in general.

A clear relationship was found between stream water acid neutralizing capacity and measured soil base saturation among the Shenandoah Watershed Study watersheds (Figure 5.9). All watersheds that were characterized by soil base saturation less than 15% had average stream water acid neutralizing capacity <100 μeq/L. Watersheds that had higher soil base saturation (all of which were >22%) were dominated by the basaltic bedrock type and had average stream water acid neutralizing capacity >100 μeq/L. Lowest base saturation values (7%–14%) were found in the siliciclastic and granitic watersheds, with much higher values in the basaltic watersheds.

Although N is thought to play an important role in the chronic acidification of streams in some areas (cf., Sullivan et al. 1997), N is generally tightly cycled in SHEN and does not usually contribute significantly to stream water acidification. However, forest insect infestations can provide a dramatic exception: the gypsy moth (*Lymantria dispar*) infestation traversed SHEN in the 1990s and affected all of the Shenandoah Watershed Study

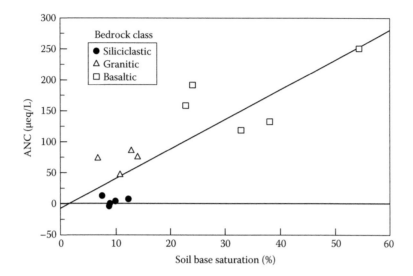

FIGURE 5.9
Median spring acid neutralizing capacity of streams in Shenandoah Watershed Study watersheds during the period 1988–1999 versus median base saturation of watershed soils. Soils data were collected by Welsch et al. (2001). (Reprinted from Sullivan, T.J. et al., *Environ. Monit. Assess.*, 137, 85, 2008, Figure 1. With permission.)

watersheds (Webb 1999). The repeated consumption and processing of foliage by the gypsy moth larvae disrupted the ordinarily tight cycling of N in SHEN forests. Some areas of the park were heavily defoliated two to three years in a row. The White Oak Run watershed, for example, was more than 90% defoliated in both 1991 and 1992. This insect infestation of forest ecosystems in SHEN resulted in substantial effects on stream water chemistry. The most notable effects of the defoliation on park streams were dramatic increases in the concentration and export of nitrate and base cations in stream water. The elevated concentrations of nitrate in SHEN streams following defoliation did not appear to contribute to chronic baseflow acidification in White Oak Run. This was due to a con-current increase in concentrations of base cations in stream water (Webb et al. 1995). Both nitrate and base cation concentrations increased during high-runoff conditions, although the increase in base cations did not fully compensate for the episodic increase in nitrate. Episodic acidification following defoliation thus became more frequent and more extreme in terms of observed minimum acid neutralizing capacity (Webb et al. 1995). Following defoliation, nitrate export increased to previously unobserved levels and remained high for over six years before returning to predefoliation levels. Eshleman et al. (2001) estimated that park-wide export of nitrate in 1992, the year of peak defoliation, increased 1700% from a predefoliation baseline of about 0.1 kg N/ha/yr. The very low predefoliation levels of nitrate export in park streams were consistent with expectations for N-limited, regenerat-ing forests (Aber et al. 1989, Stoddard 1994). Release of nitrate to surface waters following defoliation was likewise consistent with previous observations of increased N export due to forest disturbance (Likens et al. 1970, Swank 1988). Webb et al. (1994) compared pre- and postdefoliation stream water chemistry for 23 Virginia Trout Stream Sensitivity Study watersheds. Nitrate concentrations, measured quarterly, increased in most of the streams in response to defoliation, typically by 10–20 µeq/L or more.

Jastram et al. (2013) synthesized data on surface water quality and aquatic biota in SHEN during the period 1979–2009. The effort focused on stream chemistry, benthic macroinvertebrates, and fish. Long-term trends in stream acid neutralizing capacity suggested some improvements, but such increases in stream acid neutralizing capacity were minimal for the most sensitive and impacted streams, which are located on siliciclastic bedrock. The watersheds having greater proportions of siliciclastic and granitic geology, with smaller size and higher minimum elevation, were generally experiencing continued degradation in stream acid–base chemistry. Trends in benthic macroinvertebrates generally suggested declines in stream condition over time. The streams on siliciclastic bedrock showed the lowest fish species richness and exhibited no evidence of recent recovery.

Research at SHEN has also contributed significantly to the development of predictive models of watershed response to future conditions. The Model of Acidification of Groundwater in Catchments (Cosby et al. 1985a–c) of watershed response was initially developed largely using data collected within SHEN. It was the principal model used by the National Acid Precipitation Assessment Program to estimate future damage to lakes and streams in the eastern United States (Thornton et al. 1990, NAPAP 1991) and has been the most widely used dynamic acid–base chemistry model in the United States and Europe (Sullivan 2000).

The EPA's National Stream Survey statistical frame (Kaufmann et al. 1991) was used to estimate the number and percentage of stream reaches in the Southern Appalachian Mountains Initiative region that were simulated by the Model of Acidification of Groundwater in Catchments to change their chemistry in response to various emissions control strategies under consideration by the Southern Appalachian Mountains Initiative (Sullivan et al. 2002, 2004). Simulations were based on assumed constant future atmospheric deposition at 1995 levels and on three regional strategies of emissions controls, based on the Urban to Regional Multiscale One-Atmosphere model (Odman et al. 2002). There was a small decline in the estimated length of projected acidic streams by 2040 from the least to the most restrictive emissions control strategy, but there was little difference in projected stream length in the other acid neutralizing capacity classes as a consequence of adopting one or another emissions control strategy. However, projections of continued future acidification were substantially larger under a scenario in which S and N deposition were held constant into the future at the relatively high 1995 levels.

Measured stream chemistry in SHEN showed relatively minor recovery in response to reductions in S emissions and deposition to date, in agreement with the Southern Appalachian Mountains Initiative simulation results. Sulfur deposition declined by about 40% between 1985 and 2000 in western Virginia, yet the acid–base chemistry of 65 monitoring stream sites showed no significant regional trends in stream acid neutralizing capacity. For the subset of 14 monitoring sites in SHEN, Webb et al. (2004) found median decreases in stream sulfate concentration and acid neutralizing capacity of −0.229 and −0.168 µeq/L/yr, respectively. These recovery trends were very small relative to regional deposition reductions, estimates of historical acidification, and concurrent recovery at acid-sensitive surface water sites in the northeastern United States (Webb et al. 2004).

The steady-state critical load for protecting stream water acid neutralizing capacity to a level of 50 µeq/L in SHEN was modeled by Reynolds et al. (2012) and McDonnell et al. (2014) in the Ecosystem Management Decision Support study. Calculations were based on the Steady State Water Chemistry model, with the weathering parameter extrapolated from Model of Acidification of Groundwater in Catchments calibrations to watersheds in the full study area. Results are given in Figure 5.10 and Table 5.3. Model-predicted critical loads to protect and restore the acid neutralizing capacity of streams in SHEN to 50 µeq/L varied from largely in the range of ≤5 kg S/ha/yr in the southwestern portion of SHEN to the range of 5–10 kg S/ha/yr in the northern sections of the park. The median critical load

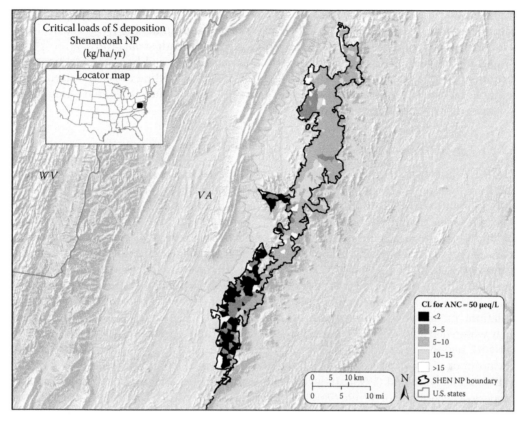

FIGURE 5.10

Steady-state critical loads of atmospheric S deposition for maintaining stream water acid neutralizing capacity at or above 50 µeq/L in SHEN. Estimates were generated using data from within the Ecosystems Management Decision Support modeling effort of Reynolds et al. (2012) and McDonnell et al. (2014).

TABLE 5.3

Distribution of Steady-State Critical Loads of Sulfur (in kg/ha/yr) to Protect Stream Acid Neutralizing Capacity to 50 µeq/L for SHEN as Modeled in the Ecosystem Management Decision Support Project within the Mid-Atlantic Network

# of CLs[a]	Minimum	25th Percentile	Median	25th Percentile	Maximum
1072	0	4	7	9	High

Source: McDonnell, T.C. et al., *J. Environ. Manage.*, 146, 407, 2014.

Notes: 0 signifies that the target acid neutralizing capacity cannot be attained, even if S deposition is reduced to zero; "High" signifies that the critical load is significantly higher than ambient deposition; Critical loads of S in units of kg/ha/yr can be converted to meq/m²/yr by multiplying by 6.25.

[a] Number of small watersheds (generally approximately 1 km²), for which critical loads (CLs) were calculated that are wholly or partly within the park.

in SHEN was 7 kg S/ha/yr. Much of the central portion of the park was estimated to have critical load less than 5 kg S/ha/yr, and some areas were estimated to have S critical load below 2 kg S/ha/yr, which is likely near the background preindustrial level of atmospheric S input to this park. This result suggests that Ca and other soil base cations may have become sufficiently depleted in portions of SHEN and surrounding areas that stream acid neutralizing capacity will not increase to the target level of 50 µeq/L, even if S deposition is decreased to near-background levels.

5.3.2.1 Episodic Acidification

The degree to which acidic deposition results in chronic loss of acid neutralizing capacity in streams within SHEN depends mainly upon two watershed processes associated with acid–base status: (1) acid anion retention in watershed soils and (2) base cation release from watershed soils and rocks (Elwood et al. 1991, Church et al. 1992). The degree to which acidic deposition results in episodic loss of acid neutralizing capacity in surface water depends largely upon the hydrologic flow paths associated with high-runoff conditions (Turner et al. 1990, Wigington et al. 1990). These pathways largely determine the extent of acid neutralization provided by the soils and bedrock in that watershed. High-elevation watersheds with steep topography, extensive areas of exposed bedrock, snowpack accumulation, and shallow, base-poor soils tend to be most sensitive to episodic acidification. Such watershed features are common within SHEN and elsewhere in the Mid-Atlantic Network region.

A number of studies of episodic acidification have been conducted in streams within SHEN. Miller-Marshall (1993) analyzed data from the Shenandoah Watershed Study for the period 1988–1991 for White Oak Run, North Fork Dry Run, Deep Run, and Madison Run and also conducted a field experiment in 1992 at White Oak Run and North Fork Dry Run. Acid anion flushing was identified as the predominant acidification mechanism during hydrological episodes. Base cation dilution frequently also played a large role, depending on the underlying bedrock geology and baseflow acid neutralizing capacity. The ratio of change in the sum of base cation concentration to the change in acid neutralizing capacity varied from 0.5 at Madison Run (median spring baseflow acid neutralizing capacity = 63 µeq/L) to −1.4 at Deep Run (median spring baseflow acid neutralizing capacity = 1 µeq/L). Ratios were intermediate (0.2 and 0.3, respectively) at North Fork Dry Run (median spring baseflow acid neutralizing capacity = 40 µeq/L) and White Oak Run (median spring baseflow acid neutralizing capacity = 16 µeq/L). At the site exhibiting the lowest baseflow acid neutralizing capacity (Deep Run), base cations increased during episodes. At the other sites, base cation concentrations were diluted during episodes, with the greatest dilution occurring in the streams that were highest in baseflow acid neutralizing capacity (Miller-Marshall 1993).

Data regarding episodic variability in stream water acid neutralizing capacity for six intensively studied sites within SHEN for the period 1993–1999 are presented in Figure 5.11. The minimum measured acid neutralizing capacity each year at each site (which generally is recorded during a large rain or snowmelt episode) is plotted against the median spring acid neutralizing capacity for that year at that site. Sites that had median spring acid neutralizing capacity below about 20 µeq/L (Paine Run, White Oak Run, Deep Run) generally had minimum measured acid neutralizing capacity about 10 µeq/L lower than median spring acid neutralizing capacity. In contrast, at the high acid neutralizing capacity Piney River site, the minimum measured acid neutralizing capacity was generally more than about 40 µeq/L lower than the respective median spring acid neutralizing capacity.

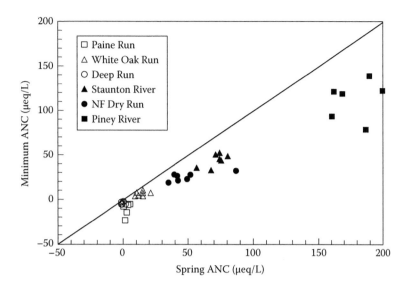

FIGURE 5.11

Minimum stream water acid neutralizing capacity sampled at each site during each year versus median spring acid neutralizing capacity for all samples collected at that site during that spring season. Data are provided for all intensively studied streams within SHEN during the period 1993–1999. A 1:1 line is provided for reference. The vertical distance from each sample point upward to the 1:1 line indicates the acid neutralizing capacity difference between the median spring value and the lowest sample value for each site and year. (From Sullivan, T.J. et al., Assessment of air quality and related values in shenandoah national park, NPS/NERCHAL/NRTR-03/090, U.S. Department of the Interior, National Park Service, Northeast Region, 2003, available at: http://www.nps.gov/nero/science/FINAL/shen_air_quality/shen_airquality.html.)

At sites having intermediate acid neutralizing capacity values, the minimum acid neutralizing capacity measured each year was generally about 20–30 µeq/L lower than the respective median spring acid neutralizing capacity. This reflects a clear pattern in SHEN of larger episodic acid neutralizing capacity depressions in streams having higher median acid neutralizing capacity and smaller episodic acid neutralizing capacity depressions in streams having lower median acid neutralizing capacity.

Eshleman and Hyer (2000) estimated the contribution of each major ion to observed episodic acid neutralizing capacity depressions in Paine Run, Staunton River, and Piney River during a three-year period. During the study, 33 discrete storm events were sampled and water chemistry values were compared between antecedent baseflow and the point of minimum measured acid neutralizing capacity, which occurred near peak discharge. The relative contribution of each ion to the acid neutralizing capacity depressions was estimated using the method of Molot et al. (1989), which normalized the change in ion concentration by the overall change in acid neutralizing capacity during the episode. At the low acid neutralizing capacity (~0 µeq/L) Paine Run site, increases in nitrate and sulfate, and to a lesser extent organic acid anions, were the primary causes of episodic acidification. Base cations tended to compensate for most of the increases in acid anion concentration. Acid neutralizing capacity declined by 3–21 µeq/L (median 7 µeq/L) during the episodes studied.

At the intermediate acid neutralizing capacity (~60 to 120 µeq/L) Staunton River site, increases in sulfate and organic acid anions, and to a lesser extent nitrate, were the primary causes of episodic acidification. Base cation increases compensated these changes to a large degree, and acid neutralizing capacity declined by 2–68 µeq/L during the episodes (median decrease in acid neutralizing capacity was 21 µeq/L).

At the high acid neutralizing capacity (~150–200 µeq/L) Piney River site, base cation concentrations declined during episodes, in contrast with the other two sites where base cation concentrations increased. Sulfate and nitrate usually increased. The change in acid neutralizing capacity during the studied episodes ranged from 9 to 163 µeq/L (median 57 µeq/L; Eshleman and Hyer 2000). Changes in base cation concentrations during episodes contributed to the acid neutralizing capacity of Paine Run, had little effect in Staunton River, and consumed acid neutralizing capacity in Piney River (Hyer 1997).

In general, pre-episode acid neutralizing capacity is a good predictor of minimum episodic acid neutralizing capacity and also a reasonable predictor of episodic change in acid neutralizing capacity in SHEN streams. Higher values of pre-episodic acid neutralizing capacity contributed to larger changes during episodes, but minimum values of such streams were generally not especially low. Lowest minimum acid neutralizing capacity values were reached in streams that had low pre-episode acid neutralizing capacity, but the changes during episodes for such streams were generally relatively small.

Webb et al. (1994) developed an approach to calibration of an episodic acidification model for Virginia Trout Stream Sensitivity Study long-term monitoring streams in western Virginia that was based on the regression method described by Eshleman (1988). Median spring quarter acid neutralizing capacity concentrations in stream water for the period 1988–1993 were used to represent chronic acid neutralizing capacity, from which episodic acid neutralizing capacity was predicted. Regression results were very similar for the four lowest acid neutralizing capacity watershed classes, and they were therefore combined to yield a single regression model to predict the minimum measured acid neutralizing capacity from the chronic acid neutralizing capacity. Extreme acid neutralizing capacity values were about 20% lower than chronic values, based on the regression equation:

$$\text{Acid neutralizing capacity}_{min} = 0.79 \text{ acid neutralizing capacity}_{chronic}$$

(5.1)

$$(r^2 = 0.97; \quad \text{SE of slope} = 0.02, \quad p \le 0.001)$$

Thus, episodic acidification of streams in SHEN can be attributed to a number of causes, including dilution of base cations and flushing of nitrate, sulfate, and/or organic acids from forest soils to drainage water (Kahl et al. 1992, Wigington et al. 1996, Wigington 1999, Lawrence 2002). For streams having low pre-episodic acid neutralizing capacity, episodic decreases in pH and acid neutralizing capacity and increases in toxic Al concentrations can have adverse impacts on fish populations. Not all of the causes of episodic acidification are related to acidic deposition. Base cation dilution and increase in organic acid anions during high-flow conditions are natural processes. The contribution of nitric acid, indicated by increased nitrate concentrations, has evidently been (at least for streams in the park) mainly related to forest defoliation by the gypsy moth (Webb et al. 1995, Eshleman et al. 1998). However, significant contributions of sulfuric acid, indicated by increased sulfate concentrations during episodes in some streams, are an effect of atmospheric deposition and the dynamics of S adsorption on soils (Eshleman and Hyer 2000).

5.3.2.2 Effects on Aquatic Biota

Aquatic biota in the Mid-Atlantic Network have been affected by acidification at virtually all levels of the food web. Effects have been clearly documented for fish, aquatic insects, other invertebrates, and algae. Some species, and some life stages within species, are more sensitive than others. In some acidified waters, sensitive species have been eliminated and taxonomic richness has decreased. In some cases, all fish species have been eliminated from acidified waters.

Relatively little is known about sublethal changes in the condition of fish or other aquatic biota resulting from acidification in the Mid-Atlantic Network region. It is expected that sublethal effects will occur in acid-sensitive species well before the species is eliminated from a particular habitat. For that reason, loss of an acid-sensitive species is not necessarily an ideal indicator of acid stress. Clearly, stress begins to occur prior to species elimination. Sublethal effects are more difficult to quantify but are nevertheless important.

Condition factor is one measure of sublethal effect that has been used to quantify the impacts of acidification on fish. Condition factor is an index to describe the relationship between fish weight and length. Expressed as fish weight/length, multiplied by a scaling constant, this index reflects potential depletion of stored energy reserves (Everhart and Youngs 1981, Goede and Barton 1990, Dennis and Bulger 1995). Fish with higher condition factor are more robust than fish having lower condition factor.

Differences in condition factor may occur because maintenance of internal chemistry in the more acidic streams would require energy that otherwise would be available for growth and weight gain (Dennis and Bulger 1999, Webb 2003). The energy costs to fish for active iono-osmoregulation can be substantial (Farmer and Beamish 1969, Bulger 1986). Because of the steep gradient in Na and chloride concentrations between fish blood and freshwater, there is constant diffusional loss of these ions, which must be replaced by energy-requiring active transport to move these ions back into the fish. Low pH increases the rate of passive loss of blood electrolytes (especially Na and chloride); dissolved Al elevates losses of Na and chloride above the levels that occur due to acid stress alone (Wood 1989).

It is possible that the loss of sensitive individuals or early life stages within a species may reduce competition for food among the survivors, resulting in better growth rates, survival, or condition. Similarly, competitive release (increase in growth or abundance subsequent to removal of a competitor) may result from the loss of a sensitive species, with positive effects on the density, growth, or survival of competitor population(s) of other species (Baker et al. 1990b). However, in some cases where acidification continued, transient positive effects on the size of surviving fish were shortly followed by extirpation (Bulger et al. 1993).

Field studies in the Mid-Atlantic Network have shown lower condition factor in blacknose dace (*Rhinichthys atratulus*) found in the more acidic streams (Dennis and Bulger 1995, Webb 2003). This species is widely distributed in Appalachian Mountain streams and is moderately tolerant of low pH and acid neutralizing capacity, relative to other fish species in the region. Bulger et al. (1999) observed a positive relationship between dace condition factor and pH in streams in SHEN. Dennis and Bulger (1995) found a reduction in the condition factor for blacknose dace in waters near pH 6.0. The mean length-adjusted condition factor of fish from the study stream having lowest acid neutralizing capacity was about 20% lower than that of the fish in best condition.

Acid stress is at least partly responsible for the low condition factor of many blacknose dace populations in SHEN. Reduced access to food or lower food quality (Baker et al. 1990b), either resulting from the nature of soft water streams or exacerbated by acidification, may also be important.

A direct outcome of population loss caused by acidification is a decline in taxonomic richness (the total number of taxa of a particular type in a stream). Taxonomic richness is a metric that is commonly used to quantify the effects of an environmental stress such as acidification. The richness metric can be applied at various taxonomic levels. For example, the number of fish species can be used as an index of acidification (cf., Bulger et al. 1999). Similarly, acidification effects on aquatic insects can be evaluated on the basis of the number of families or genera of mayflies (order Ephemeroptera; Sullivan et al. 2003).

Acidification is a conspicuous threat to three trout species in the Mid-Atlantic Network region: brook trout, brown trout, and rainbow trout. Although brown and rainbow trout are found within SHEN, brook trout predominate. Of the three, native brook trout is the most acid tolerant, brown trout introduced from Europe is intermediate in acid tolerance, and rainbow trout introduced from the western United States is most sensitive. The Southern Appalachian Man and the Biosphere (1996) survey concluded that trout populations are regarded by residents of the southern Appalachian Mountains region as among the region's most valuable aquatic natural resources. Trout populations and trout habitat are major concerns to the public in the southern Appalachians. Sources of concern generally fall into three categories: (1) fisheries for native brook trout and introduced rainbow and brown trout; (2) "existence value" for brook trout, regarded as a beautiful and intrinsically valuable native species; and (3) the presence of trout as indicators of high water quality.

The three-year Fish in Sensitive Habitats study of stream acidification in SHEN demonstrated negative effects on fish from both chronic and episodic acidification (Bulger et al. 1999). Biological effects in low versus high acid neutralizing capacity streams included changes in species richness, population density, condition factor, age, size, and field bioassay survival. As an element of the Fish in Sensitive Habitats project (Bulger et al. 1999), numbers of fish species were compared among 13 SHEN streams spanning a range of pH and acid neutralizing capacity conditions. There was a highly significant ($p < 0.0001$) relationship between stream acid–base status and fish species richness among the 13 streams. The streams with the lowest acid neutralizing capacity hosted the fewest species (Figure 5.12). The best fit regression line suggested, on average, a loss of one species for every 21 µeq/L decline in annual minimum recorded acid neutralizing capacity value.

Median stream acid neutralizing capacity values and watershed areas are shown in Table 5.4 for the streams used by Bulger et al. (1999) to develop the relationship between acid neutralizing capacity and fish species richness shown in Figure 5.12. Despite the overall similarities, these study streams vary in watershed area by a factor of 10, and this can affect the fish species richness. It appears that fish species richness is controlled by both acidification and watershed area. Smaller watersheds may contain smaller streams having less diversity of habitat, more pronounced effects on fish from high-flow conditions, or lower food availability. The streams that have larger watershed area generally have more fish species than the streams having smaller watershed area. All of the rivers have watersheds larger than 10 km² (4 mi²) and acid neutralizing capacity higher than 75 µeq/L. In contrast, most of the runs have watershed area smaller than 10 km² and acid neutralizing capacity less than 20 µeq/L. All of the streams that have watershed areas smaller than 10 km² have three or fewer known species of fish present. All of the streams having larger watersheds (>10 km²) have three or more known fish species; seven of nine have five or more species; and the average number of fish species is six. There is no clear distinction between river and run. As small streams in SHEN combine and flow into larger streams and eventually to rivers, however, acid sensitivity generally declines and the habitat becomes suitable for additional fish species (Sullivan et al. 2003).

Sullivan et al. (2003) combined dynamic water chemistry model projections with biological dose-response relationships to estimate declines in fish species richness with

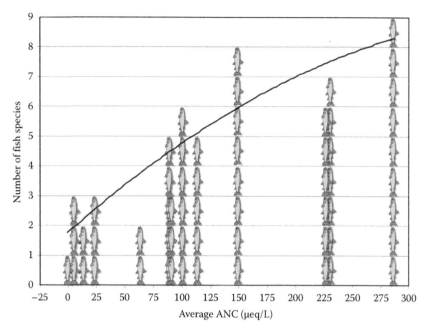

FIGURE 5.12

Number of fish species among 13 streams in SHEN. Values of acid neutralizing capacity are means based on quarterly measurements, 1987–1994. The regression analysis showed a highly significant relationship (p < 0.0001) between mean stream acid neutralizing capacity and number of fish species. Streams having acid neutralizing capacity consistently <75 µeq/L had three or fewer species. (Redrawn from Bulger, A.J. et al., SNP:FISH, Shenandoah National Park: fish in sensitive habitats, Project Final Report to National Park Service, University of Virginia, Charlottesville, VA, 1999.)

acidification. A relationship derived from the data in Figure 5.12 was used with stream acid neutralizing capacity values predicted by the Model of Acidification of Groundwater in Catchments to provide estimates of the expected number of fish species in each of the modeled streams for the past, present, and future chemical conditions simulated for each stream. The coupled geochemical and biological model predictions were evaluated by comparing the predicted species richness in each of the 13 streams with the observed number of species that occurred in each stream. The agreement between predicted and observed species numbers was good, with a root mean squared error of 1.2 species across the 13 streams. The average error was 0.3 species, indicating that the coupled models were unbiased in their predictions. Model reconstructions of past species richness in the streams suggested that historical loss of species had been largest in the streams located on the most sensitive geological class. The average number of species lost from streams on the three bedrock types examined were estimated as follows: 1.6 species on siliciclastic bedrock; 0.4 species on granitic bedrock; and 0.4 species on basaltic bedrock (Sullivan et al. 2003). In the case of the siliciclastic streams, the projected past changes were much larger than the average error and root mean squared error of the coupled models.

Bulger et al. (1999) concluded that the most important cause of the observed decline in fish species richness with decreasing acid neutralizing capacity in SHEN was acid stress associated with acidification. However, an additional causal factor may have been the decrease in the number of available aquatic niches when moving from downstream

TABLE 5.4

Median Stream Water Acid Neutralizing Capacity and Watershed Area of Streams in SHEN That Have Water Chemistry and Fish Species Richness Data

Site ID	Watershed Area (km^2)	Median Acid Neutralizing Capacity (μeq/L)	Number of Fish Species
Smaller watersheds ($\leq$10 km^2)			
North Fork Dry Run	2.3	48.7	2
Deep Run	3.6	0.3	N.D.[a]
White Oak Run	4.9	16.2	3
Two Mile Run	5.4	10.0	2
Meadow Run	8.8	−3.1	1
Larger watersheds ($\geq$10 km^2)			
Brokenback Run	10.1	74.4	3
Staunton River	10.6	76.8	5
Piney River	12.4	191.9	7
Paine Run	12.7	3.7	3
Hazel River	13.2	86.8	6
White Oak Canyon	14.0	119.3	7
N. Fork Thornton River	18.9	249.1	9
Jeremy's Run	22.0	158.5	6
Rose River	23.6	133.6	8

Source: Sullivan, T.J. et al., Assessment of air quality and related values in Shenandoah National Park, NPS/NERCHAL/NRTR-03/090, U.S. Department of the Interior, National Park Service, Northeast Region, 2003, available at: http://www.nps.gov/nero/science/FINAL/shen_air_quality/shen_airquality.html.

[a] No data were available regarding the number of fish species in Deep Run.

locations (which are seldom low in pH and acid neutralizing capacity) to upstream locations (which are often low in pH and acid neutralizing capacity in this region; Sullivan et al. 2003). The relative importance of this latter factor, compared with the importance of acid stress, in determining this relationship is not known.

Effects on biota can be assessed as effects on a particular sensitive species or a species perceived to be important, or as effects on the richness or diversity of fish or other potentially sensitive life form. Acid neutralizing capacity criteria have been used for the evaluation of potential acidification effects on fish communities. The utility of these criteria lies in the association between acid neutralizing capacity and the surface water constituents that directly contribute to or ameliorate acidity-related stress, mainly pH, Ca, and Al. Bulger et al. (2000) developed acid neutralizing capacity thresholds for brook trout response to acidification in forested headwater catchments in western Virginia (Table 5.5). Note that because brook trout are comparatively acid tolerant, adverse effects on many other fish species should be expected at relatively higher acid neutralizing capacity values. These values given in the table were based on annual average stream water chemistry and therefore represent chronic exposure conditions. The likelihood of additional episodic stress is incorporated into the response categories in the manner in which they are interpreted. For example, the episodically acidic response category, that has chronic acid neutralizing capacity in the range of 0–20 μeq/L, represents streams that are expected to acidify to acid neutralizing capacity near or below 0 during rainfall or snowmelt episodes. In such streams, sublethal and/or lethal effects on brook trout are possible (Bulger et al. 2000, Sullivan et al. 2003).

TABLE 5.5

Brook Trout Acidification Response Categories Developed by Bulger et al. (2000) for Streams in Virginia

Response Category	Chronic Acid Neutralizing Capacity Range (µeq/L)	Expected Response
Suitable	>50	Reproducing brook trout expected if other habitat features are also suitable
Indeterminate	20–50	Brook trout response expected to be variable
Episodically acidic	0–20	Sublethal and/or lethal effects on brook trout are possible
Chronically acidic	<0	Lethal effects on brook trout probable

Source: Reproduced from Bulger, A.J. et al., *Can. J. Fish. Aquat. Sci.*, 57(7), 1515, 2000. Copyright 2008 Canadian Science Publishing or its licensors. With permission.

Streams with chronic acid neutralizing capacity greater than about 50 µeq/L are generally considered suitable for brook trout in the southeastern United States. Reproducing brook trout populations are expected if the habitat is otherwise suitable (Bulger et al. 2000), although some streams may periodically experience episodic chemistry that affects species more sensitive than brook trout. Streams having annual average acid neutralizing capacity from 20 to 50 µeq/L may or may not experience episodic acidification during storms that can be lethal to juvenile brook trout, as well as other fish. Streams that are designated as episodically acidic (chronic acid neutralizing capacity between 20 and 50 µeq/L) are considered marginal for brook trout (Hyer et al. 1995), although the frequency and magnitude of episodes vary. Streams that are chronically acidic (chronic acid neutralizing capacity less than 0 µeq/L) are generally not expected to support healthy brook trout populations (Bulger et al. 2000).

Acidification also affects aquatic life forms other than fish. Within stream systems, macroinvertebrate communities are among the most sensitive life forms to disturbances, including those associated with atmospheric deposition (Cairns and Pratt 1993). In addition, they are relatively easy to sample in the field (Plafkin et al. 1989, Resh et al. 1995, Karr and Chu 1999, Potyondy et al. 2006). Considerable data are available for aquatic insects in SHEN.

Low stream water pH can be associated with reductions in invertebrate species richness or diversity (Townsend et al. 1983, Raddum and Fjellheim 1984, Burton et al. 1985, Kimmel et al. 1985, Hall and Ide 1987, Peterson and Van Eechhaute 1992, Rosemond et al. 1992, Sullivan et al. 2003), and sometimes density (Hall et al. 1980, Townsend et al. 1983, Aston et al. 1985, Burton et al. 1985, Kimmel et al. 1985). A decrease in species richness with decreasing pH has been found in almost all such studies (Rosemond et al. 1992); decreases in pH of one unit or more typically result in species loss. Invertebrate taxa that are most sensitive to acidification include mayflies, amphipods, snails, and clams. Porak (1981) found that all species of mayfly were intolerant of the acid condition in the streams containing acidic Anakeesta leachates. Caddisflies are also highly sensitive. At low levels of acidification (pH 5.5–6.0), acid-sensitive species are replaced by more acid-tolerant species, yielding little or no change in total community species richness, diversity, density, or biomass. If pH decreases are larger, more species will be lost without replacement, resulting in decreased richness and diversity.

The Ephemeroptera-Plecoptera-Tricoptera (EPT) Index is a common measure of stream macroinvertebrate community integrity. The EPT metric is the total number of families present in those three insect orders (mayflies, stoneflies, and caddisflies, respectively). The total number of families is generally lower at acidified sites because species within those families tend to exhibit varying acid sensitivity (cf., SAMAB 1996). Mayflies tend to be most sensitive of the three, and stoneflies tend to be least sensitive (Peterson and Van Eechhaute 1992).

5.4 Nutrient Nitrogen Enrichment

Nitrogen deposition can affect nutrient dynamics in northern hardwood forests in SHEN. Hardwood forests found throughout the northeastern United States and south to Virginia appear to have experienced increased tree growth with increasing N deposition across a depositional gradient from about 3 to 11 kg N/ha/yr (Thomas et al. 2010).

In addition to growth effects on trees, eastern hardwood forests can respond to increased N deposition with a decrease in biodiversity of understory species (Gilliam et al. 2006a,b). The response of the herbaceous layer of the forest can be pronounced, with an initial increase in herbaceous plant cover, followed by decrease in species richness and decrease in species evenness. The response time is typically shorter if ambient N deposition is relatively high (Gilliam et al. 2006a,b, Fraterrigo et al. 2009, Royo et al. 2010, Gilliam et al. 2011). Because the plants in the herbaceous layer of eastern hardwood forests tend to have foliage with relatively high nutrient content, herbaceous plants influence N cycling to a level that is disproportionate to their biomass (Muller 2003, Moore et al. 2007, Welch et al. 2007, Gilliam et al. 2011).

There is evidence that air pollution may have impacted lichen species distributions in the Mid-Atlantic Network region. McCune et al. (1997) found some air pollution–tolerant epiphytic lichen species, but relatively low richness of epiphytic lichen species, in urban and industrial areas within Virginia and states to the south. In rural areas, in contrast, air pollution–sensitive lichen species were more common, and richness was higher.

Ellis et al. (2013) estimated the critical load for nutrient N deposition to protect the most sensitive ecosystem receptors in 45 national parks, based on the data of Pardo et al. (2011). They estimated the N critical load for SHEN in the range of 3–8 kg N/ha/yr for the protection of hardwood forests.

5.5 Ozone Injury to Vegetation

Of all national parks where ozone is monitored, SHEN has had among the highest measured atmospheric concentrations. During the period 1997–2001, the park's air quality did not meet the EPA's ozone standard to protect human health and welfare (Sullivan et al. 2003), but more recently ozone levels have decreased and the area now meets the standard, even as the standard has decreased. Scientists have found that some sensitive plants in natural ecosystems may be injured at SUM06 levels of 8–12 ppm-h (Heck and Cowling 1997). The average cumulative ozone exposure at Big Meadows, a monitoring station in SHEN, during the late 1990s was several times higher than these suggested injury thresholds. During the 1990s, ozone levels in SHEN generally showed an increasing trend. Over the following several years, ozone concentrations stabilized (Sullivan et al. 2003) and subsequently declined.

5.5.1 Ozone Exposure Indices and Levels

Kohut (2007b) assessed the risk of foliar symptoms from ozone exposure during the period 1995–1999 in 10 parks in the Mid-Atlantic Network, based on in-park monitoring of ozone concentration in SHEN and kriging of data from surrounding monitoring stations in the other 9 parks. The SUM06 index exceeded the threshold for foliar injury in all 10 parks investigated. The W126 index exceeded the threshold in eight parks. In SHEN, the 1-hour concentration of ozone fulfilled the W126 threshold in only two of five years evaluated. However, the ozone concentrations in SHEN exceeded 100 ppb for a few hours during most years and for many hours during one year. It is likely that these higher exposures could injure plant foliage (Kohut 2007b). In general, in each of the parks studied, when ozone concentration was high, soil moisture was low, thereby reducing the uptake of ozone into foliage and its effectiveness in producing foliar injury (Kohut 2007b). The probability of injury at the parks was judged to be greatest during years when ozone levels exceeded thresholds and soil moisture was high enough so that it was unlikely that there were long-term constraints on ozone uptake.

The W126 and SUM06 exposure indices calculated by National Park Service staff for SHEN are given in Table 5.6, along with Kohut's (2007a) ozone risk ranking. Ozone condition in SHEN, as rated by the National Park Service, was moderate. Kohut's evaluation of risk to plants was also moderate.

5.5.2 Ozone Formation

Ozone formation is governed by the atmospheric levels of both oxidized N and volatile organic compounds. Some information is available regarding the relative importance of these two ozone precursors in SHEN. The most abundant volatile organic compound measured in the air during summer in SHEN during the period 1995–1997 was isoprene, a natural biogenic compound (Kang et al. 2001). Upon conversion of all volatile organic compounds into propylene-equivalent concentration units (to assess the relative potential contribution of volatile organic compounds to ozone formation), biogenic components accounted for more than two-thirds of the total atmospheric volatile organic compound

TABLE 5.6

Ozone Assessment Results for SHEN Based on Estimated Average 3-Month W126 and SUM06 Ozone Exposure Indices for the Period 2005–2009 and Kohut's (2007a) Ozone Risk Ranking for the Period 1995–1999[a,b]

W126		SUM06		Kohut Risk to Vegetation
Value (ppm-h)	NPS Condition	Value (ppm-hr)	NPS Condition	
10.05	Moderate	12.88	Moderate	Moderate

[a] Parks are classified into one of three ranks (Low, Moderate, High) based on comparison with other Inventory and Monitoring parks.

[b] Degrees of concern for the W126 and SUM06 indices are based solely on levels of ozone exposure. Kohut's risk to vegetation is based on several factors that contribute to injury in plants, including ozone exposure and environmental variables, and considers the effects of soil moisture on the uptake of ozone.

concentration during all three study years. About 80% of the biogenic volatile organic compounds comprised isoprene; other common volatile organic compounds included α-pinene, β-pinene, and limonene. The major human-caused volatile organic compounds during summer were isopentane, toluene, and propane. These are derived from automobile exhaust (isopentane), solvents (toluene), and natural gas (propane; Kang et al. 2001).

Kang et al. (2003) applied the Multiscale Air Quality Simulation Platform model to predict atmospheric ozone concentration in SHEN and two other parks. Modeling results suggested that volatile organic compounds were chemically saturated, potentially causing decreased ozone production in response to further increases in volatile organic compound emissions. More than half of the ozone at SHEN was simulated to be transported from other areas (Kang et al. 2003). The researchers concluded that, in areas like SHEN where transport from pollutant source areas dominates the local ozone production, a control strategy might focus on the reduction of ozone in the source region(s).

5.5.3 Ozone Exposure Effects

Ground-level ozone constitutes a stress to ozone-sensitive vegetation in SHEN that can interact with other, potentially exacerbating stresses such as drought, insects, and diseases. Ozone also causes visible foliar symptoms on several plant species in SHEN, including but not limited to milkweed, black cherry, yellow poplar, and white ash (*Fraxinus americana*; Chappelka et al. 1999, Sullivan et al. 2003). However, little is known about the relationships among ozone exposure, visible foliar symptoms, and the growth or vitality of sensitive plant species.

Foliar injury due to ozone exposure was documented by Chappelka et al. (1999) on mature black cherry in SHEN. Trees growing at the highest elevations showed the most severe symptoms during all three years of study. These high-elevation sites also received the highest ozone exposures. The percent of trees with symptoms attributed to ozone was positively correlated with both the SUM06 and W126 indices.

Responses of eight tree species to the isolated effects of ozone exposures were simulated by Sullivan et al. (2003) using the TREGRO model and ranked in order from most to least sensitive to growth and species composition impacts:

> White ash > Basswood = Chestnut oak > Red maple > Yellow poplar >
> Black cherry = Red oak > Sugar maple.

Simulations suggested that white ash is more sensitive to growth and species composition from ozone impacts than other tree species in SHEN, both as an individual and as a component of a forest stand. Ambient ozone concentrations caused an estimated 1% decrease in total growth of white ash, a long-lived species, over the three-year simulation period. Over a 100-year simulation period, ambient ozone exposures were expected to cause a 50% decrease in white ash species composition in chestnut oak (*Quercus prinus*) forests. Ground-level ozone exposures greater than ambient levels were projected to cause a less than 10% decrease in white ash and yellow poplar species composition in cove hardwood forests over 100 years.

Foliar injury surveys were conducted on five native plant species in SHEN in 1982. Three of the five species (virgin's bower [*Clematis virginiana*], black locust (*Robinia pseudoacacia*), and wild grape [*Vitis* sp.]) displayed increased injury with increased elevation (Winner et al. 1989). In 1991, trend plots of the ozone-sensitive hardwoods yellow poplar, black cherry, and white ash were established near the ambient ozone monitors at Dickey

Ridge, Big Meadows, and Sawmill Run in the park. Marked trees in each plot were evaluated for foliar ozone symptoms in 1991, 1992, and 1993 (Hildebrand et al. 1996). Black cherry and white ash exhibited increased foliar symptoms with increased ozone exposure across all three sites and at each site across all years of study. Whereas the amount of foliar symptoms on yellow poplar at Dickey Ridge corresponded well with ozone exposure, there was no correlation for this species at the Big Meadows or Sawmill Run sites. The researchers speculated the lack of correlation for yellow poplar at Big Meadows and Sawmill Run may have been due to extremes in moisture availability at the two sites. Hildebrand et al. (1996) concluded that cumulative ozone statistics, such as SUM06 and W126, best represented foliar symptom observations, particularly for black cherry, during the period of study.

5.6 Visibility Degradation

5.6.1 Natural Background and Ambient Visibility Conditions

Visibility is degraded throughout SHEN, detracting from visitor enjoyment of scenic vistas. This is especially problematic along Skyline Drive, which is designated a Virginia State Scenic Highway. These scenic vistas are also enjoyed by visitors hiking the Appalachian National Scenic Trail through the park. There is an Interagency Monitoring of Protected Visual Environments (IMPROVE) site at SHEN. Natural background and ambient visibility conditions have been estimated for SHEN on the 20% clearest visibility days, the 20% haziest visibility days, and average days (Table 5.7). Haze was very high at SHEN on the average and haziest days and high on the clearest days. Sullivan et al. (2003) concluded that the ambient annual average visual range was only about 20% of the estimated natural visual range of about 115 mi (185 km). Even the mean of the clearest 20% of days, which occur mostly in winter, are degraded by human-made particles in the air in SHEN.

Annual mean haze levels on the haziest days at 47 monitored park locations during the period 2006–2008 ranged from 1.5 to 21 deciviews higher than the estimated natural

TABLE 5.7

Estimated Natural Haze and Measured Ambient Haze in SHEN Averaged over the Period 2004–2008[a]

Estimated Natural Background Haze (Deciview)		
20% Clearest Days	20% Haziest Days	Average Days
3.14	6.90	11.35
Measured Ambient Haze (for Years 2004–2008)					
20% Clearest Days		20% Haziest Days		Average Days	
Deciview	Ranking	Deciview	Ranking	Deciview	Ranking
9.95	High	28.76	Very High	18.92	Very High

[a] Parks are classified into one of five haze ranks (Very Low, Low, Moderate, High, or Very High) based on comparison with other monitored parks.

condition (NPS 2010). The average difference between measured haze and estimated natural condition was 8.3 deciviews. Several eastern parks had annual mean deciviews on the haziest days that were substantially higher (more than about 10 deciviews) than estimated natural conditions; these included SHEN. Monitoring sites showing the largest differences during the period 2006–2008 between measured visibility and estimated natural conditions on the clearest days also included SHEN (NPS 2010). On average days during the period 2004–2008, ambient haze at SHEN was nearly 22 deciviews higher than the estimated average natural haze.

Representative photos of a selected vista in SHEN under three different visibility conditions are shown in Figure 5.13. Photos were selected to correspond with the clearest 20% of visibility conditions, haziest 20% of visibility conditions, and annual average visibility conditions at that location during the period 2004–2008. This series of photos provides a graphic illustration of the visual effect of these differences in haze level in this park.

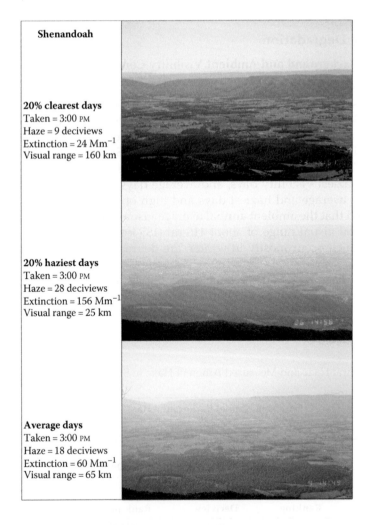

FIGURE 5.13
Three representative photos of the same view in SHEN illustrating the 20% clearest visibility, the 20% haziest visibility, and the annual average visibility. Extinction is total particulate light extinction. (From http://vista. cira.colostate.edu/improve/Data/IMPROVE/Data_IMPRPhot.htm, accessed December, 2010.)

5.6.2 Composition of Haze

Figure 5.14 shows estimated natural (preindustrial), baseline (2000–2004), and recent (2006–2010) levels of haze and its composition for SHEN. The figure illustrates that sulfate is the primary component of ambient haze on the 20% clearest, annual average, and 20% haziest visibility days. Organics and nitrates also contribute significantly to ambient haze. Under estimated natural conditions, in the absence of anthropogenic air pollution, organics dominated the light extinction budget.

On an annual average basis, 71% of total particulate light extinction was attributable to sulfate, while 9% was contributed by nitrate, and 11% was from organics. On the 20%

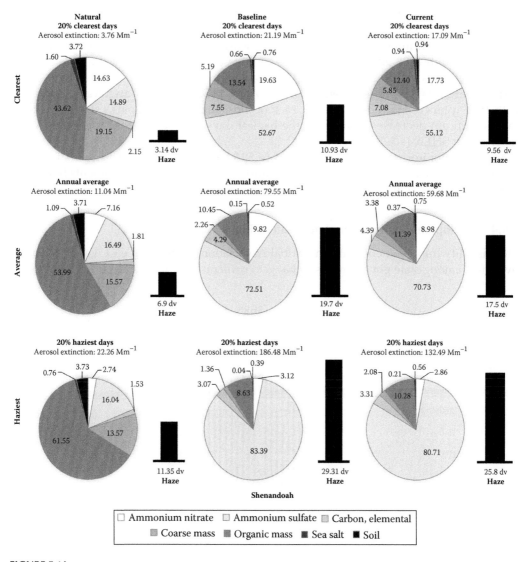

FIGURE 5.14

Estimated natural (preindustrial), baseline (2000–2004), and current (2006–2010) levels of haze (columns) and its composition (pie charts) on the 20% clearest, annual average, and 20% haziest visibility days for SHEN. (From http://views.cira.colostate.edu/fed/Tools/RegionalHazeSummary.aspx, accessed October, 2012.)

haziest days, total particulate light extinction was mostly attributable to sulfate, which accounted for 81% of total particulate light extinction; nitrate contributed only 3%, and organics contributed 10%. On the 20% clearest visibility days, the sulfate contribution decreased to 55% while nitrate and organics increased to 18% and 12% of total particulate light extinction, respectively.

5.6.3 Trends in Visibility

Changes in visibility over time in the Mid-Atlantic Network are mainly caused by changes in the concentration of particles of ammonium sulfate in the atmosphere. Atmospheric ammonium sulfate is primarily derived from human-caused sources of air pollution.

The National Park Service (2010) reported long-term trends in haze on the clearest and haziest 20% of days at monitoring sites in 29 national parks. All 27 parks that showed statistically significant ($p \leq 0.05$) trends on the clearest days for the 11–20-year monitoring periods through 2008 exhibited decreases in haze over time. None of the sites showed increasing trends on the clearest days. The steepest declines (−0.18 to −0.20 deciviews/yr) on the clearest days were reported for SHEN, Acadia National Park, and Washington DC, with 18–19 years of monitoring data at each of those locations. Ten parks showed statistically significant decreases in haze on the haziest days, with the second steepest decline reported for SHEN (−0.27 deciviews/yr).

Available monitoring data shown in Figure 5.15 suggest some improvements in recent years on the 20% haziest, 20% clearest, and annual average days. Improvements have been most pronounced on the haziest days.

5.6.4 Development of State Implementation Plans

The Visibility Improvement State and Tribal Association of the Southeast was a collaborative effort among state governments, tribal governments, and federal agencies involved in

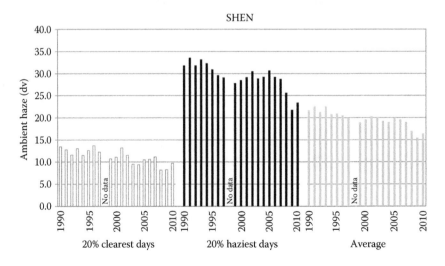

FIGURE 5.15
Trends in ambient haze levels at SHEN, based on IMPROVE measurements on the 20% clearest, 20% haziest, and annual average visibility days over the monitoring period of record. (From http://vista.cira.colostate.edu/improve/Data/IMPROVE/summary_data.htm, accessed October, 2012.)

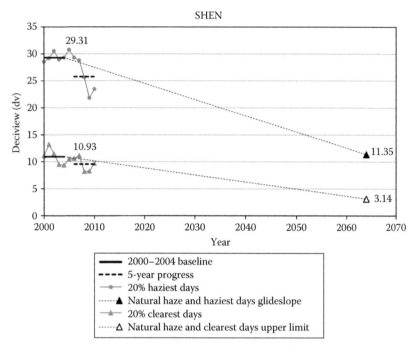

FIGURE 5.16

Glideslopes to achieving natural visibility conditions in 2064 for the 20% haziest (upper line) and the 20% clearest (lower line) days in SHEN. In the Regional Haze Rule, the clearest days do not have a uniform rate of progress glideslope; the rule only requires that the clearest days do not get any worse than the baseline period. Also shown are measured values during the period 2000–2010. (From http://vista.cira.colostate.edu/improve/Data/IMPROVE/summary_data.htm, accessed October, 2012.)

conducting the technical analysis of visibility and regional haze in the Southeast. The group conducted analyses that included determination of baseline visibility conditions, calculation of the glideslope from the baseline necessary to achieve background conditions in 2064, and determination of air pollutant source areas. The Virginia Department of Environmental Quality used this information to develop its Regional Haze State Implementation Plan. The glideslope analysis depicted in Figure 5.16 shows that recent improvements in visibility on the haziest days are generally following a trajectory in compliance with the Regional Haze Rule requirement for attaining natural visibility conditions in 2064.

5.7 Summary

SHEN receives air pollutants from regional and local emissions sources, including large power plants, urban areas, agriculture, and oil and gas development. Air pollution is an important stressor to natural resources in the park. SHEN is one of the national parks that has been most intensively studied regarding air pollution and its effects on park resources.

The major source areas of atmospheric emissions of S and N that impact sensitive resources within SHEN and surrounding areas within the Mid-Atlantic Network are located in the Ohio River Valley, northeastern West Virginia, southwestern Pennsylvania, and central and eastern Virginia (Sullivan et al. 2003), based on simulations using the Regional Acid Deposition atmospheric transport model. There are many large urban centers within a 300 mi (483 km) radius of the network boundary.

Sulfur and to a lesser extent N air pollutants cause acidification of streams, lakes, and soils in SHEN. Sulfur deposition in SHEN peaked sometime during the 1970s or early 1980s and has been declining steadily since then. Nitrogen pollutants can also cause undesirable nutrient enrichment of natural ecosystems, leading to changes in plant species diversity and soil nutrient cycling. Nitrogen deposition in the Mid-Atlantic Network region was high throughout the 1980s and 1990s but has declined since 2000. Sulfur and N oxide deposition are expected to continue to decrease as the Acid Rain Program and other air quality management programs are fully implemented (Burns et al. 2011).

SHEN has received some of the highest levels of air pollution (including S, N, and ozone) of any national park studied. Over 20 years of scientific research and monitoring in SHEN have shown that, despite some recent improvements, the park's sensitive scenic, vegetative, and aquatic resources remain degraded by human-made air pollution. Wet Hg deposition is also relatively high in the Mid-Atlantic Network. Three-year means of annual Hg concentration in rain and snow were reported by the National Park Service (2010) for 13 national parks that had at least two years of valid data during the period 2006–2008. SHEN was among the five parks having highest Hg concentration in precipitation.

The most thoroughly studied air pollution impact in SHEN has probably been stream acidification. Aquatic biota have been affected by acidification at virtually all levels of the food web. Effects have been clearly documented for fish, aquatic insects, other invertebrates, and algae.

The vast majority of low acid neutralizing capacity sampled streams were underlain by the silica-based geologic sensitivity class. Modeled streams on siliciclastic bedrock showed critical S deposition loads to protect against stream acidification to acid neutralizing capacity of 50 µeq/L in the year 2100 ranging from less than 0 kg S/ha/yr (target acid neutralizing capacity value not attainable, even with no future acidic deposition) to 6 kg S/ha/yr. Of the 14 modeled park streams, none had simulated preindustrial stream water acid neutralizing capacity less than 50 µeq/L, suggesting that these streams may have supported a greater variety of aquatic fauna prior to the Industrial Revolution.

Low acid neutralizing capacity park streams generally have lower fish species richness, lower population density, poorer blacknose dace condition, fewer age classes, smaller sizes, and lower field bioassay survival than higher acid neutralizing capacity streams. Modeling results suggested that park streams that occur on siliciclastic bedrock have generally lost one or two species, and some streams may have lost up to four species, of fish in response to acidic deposition. Higher acid neutralizing capacity streams generally have greater numbers of families and numbers of individuals than low acid neutralizing capacity streams in each of the important benthic insect orders: mayflies, stoneflies, and caddisflies.

Of all monitored eastern national parks, the air in SHEN has had among the highest measured concentrations of ozone. The probability of injury to plant foliage at SHEN was judged to be greatest during years when ozone levels exceeded thresholds and soil moisture was high enough so that it was unlikely that there were long-term constraints on ozone uptake into plant foliage. Ground-level ozone causes visible foliar symptoms on several plant species in SHEN, including but not limited to milkweed, black cherry, yellow poplar, and white ash. From 2000 to 2009, concentrations of ozone in the air decreased significantly at SHEN (NPS 2013).

Visibility is degraded by haze in SHEN, detracting from visitor enjoyment of scenic vistas, especially along Skyline Drive, which is designated as a Virginia State Scenic Highway, and the Appalachian National Scenic Trail, which passes through the park. Changes in visibility over time in this park are mainly caused by changes in the concentration of particles of ammonium sulfate in the atmosphere.

The National Park Service (2010) reported long-term trends in annual average haze on the clearest and haziest 20% of days at monitoring sites in 29 national parks. The steepest declines (−0.18 to −0.20 deciviews per year) on the clearest days were reported for SHEN, Acadia National Park in Maine, and Washington DC. Ten parks showed statistically significant decreases in haze on the haziest days, with the second steepest decline reported for SHEN (−0.27 deciviews/yr), based on 18 years of data. More recently (2000–2009), haze measurements continued to decrease significantly in SHEN on the clearest days, but there was no significant trend on the haziest days (NPS 2013).

References

Aber, J.D., K.J. Nadelhoffer, P. Steudler, and J.M. Mellilo. 1989. Nitrogen saturation in northern forest ecosystems. *BioScience* 39(6):378–386.

Aston, R.J., K. Sadler, A.G.P. Milner, and S. Lynam. 1985. The effects of pH and related factors on stream invertebrates. Central Electric Generating Board, Surrey, U.K.

Baker, J.P., D.P. Bernard, S.W. Christensen, and M.J. Sale. 1990a. Biological effects of changes in surface water acid-base chemistry. State of Science/Technology Report 13. National Acid Precipitation Assessment Program, Washington, DC.

Baker, J.P., S.A. Gherini, S.W. Christensen, C.T. Driscoll, J. Gallagher, R.K. Munson, R.M. Newton, K.H. Reckhow, and C.L. Schofield. 1990b. *Adirondack Lakes Survey: An Interpretive Analysis of Fish Communities and Water Chemistry, 1984–1987.* Adirondack Lakes Survey Corporation, Ray Brook, NY.

Baker, L.A., P.R. Kauffman, A.T. Herlihy, and J.M. Eilers. 1990c. Current status of surface water acid-base chemistry. State of Science/Technology Report 9. National Acid Precipitation Assessment Program, Washington, DC.

Binkley, D., C. Driscoll, H.L. Allen, P. Schoeneberger, and D.C. McAvoy. 1989. *Acidic Deposition and Forest Soils: Context and Case Studies in the Southeastern U.S.* Springer-Verlag, New York, 149pp.

Bricker, O.P. and K.C. Rice. 1989. Acidic deposition to streams: A geology-based method predicts their sensitivity. *Environ. Sci. Technol.* 23:379–385.

Bulger, A.J. 1986. Coincident peaks in serum osmolality and heat-tolerance rhythms in seawater-acclimated killifish (*Fundulus heteroclitus*). *Physiol. Zool.* 59(2):169–174.

Bulger, A.J., B.J. Cosby, C.A. Dolloff, K.N. Eshleman, J.R. Webb, and J.N. Galloway. 1999. SNP:FISH, Shenandoah National Park: Fish in sensitive habitats. Project final report to National Park Service. University of Virginia, Charlottesville, VA.

Bulger, A.J., B.J. Cosby, and J.R. Webb. 2000. Current, reconstructed past, and projected future status of brook trout (*Salvelinus fontinalis*) streams in Virginia. *Can. J. Fish. Aquat. Sci.* 57(7):1515–1523.

Bulger, A.J., L. Lien, B.J. Cosby, and A. Henriksen. 1993. Trout status and chemistry from the Norwegian thousand lake survey: Statistical analysis. *Can. J. Fish. Aquat. Sci.* 50:575–585.

Burns, D.A., J.A. Lynch, B.J. Cosby, M.E. Fenn, J.S. Baron, and U.S. EPA Clean Air Markets Division. 2011. National acid precipitation assessment program report to Congress 2011: An Integrated assessment. National Science and Technology Council, Washington, DC.

Burton, T.M., R.M. Stanford, and J.W. Allan. 1985. Acidification effects on stream biota and organic matter processing. *Can. J. Fish. Aquat. Sci.* 42:669–675.

Cairns, J., Jr. and J.R. Pratt. 1993. A history of biological monitoring using benthic macroinvertebrates. *In* D.M. Rosenberg and V.H. Resh (Eds.). *Freshwater Biomonitoring and Benthic Macroinvertebrates.* Chapman and Hall, New York, pp. 10–28.

Chappelka, A., J.M. Skelly, G. Somers, J. Renfro, and E. Hildebrand. 1999. Mature black cherry used as a bioindicator of ozone injury. *Water Air Soil Pollut.* 116:261–266.

Church, M.R., P.W. Shaffer, K.W. Thornton, D.L. Cassell, C.I. Liff, M.G. Johnson, D.A. Lammers et al. 1992. Direct/delayed response project: Future effects of long-term sulfur deposition on stream chemistry in the mid-Appalachian region of the eastern United States. EPA/600/R-92/186. U.S. Environmental Protection Agency, Corvallis, OR.

Cosby, B.J., G.M. Hornberger, J.N. Galloway, and R.F. Wright. 1985a. Modelling the effects of acid deposition: Assessment of a lumped parameter model of soil water and streamwater chemistry. *Water Resour. Res.* 21(1):51–63.

Cosby, B.J., G.M. Hornberger, J.N. Galloway, and R.F. Wright. 1985b. Time scales of catchment acidification: A quantitative model for estimating freshwater acidification. *Environ. Sci. Technol.* 19:1144–1149.

Cosby, B.J., J.R. Webb, J.N. Galloway, and F.A. Deviney. 2006. Acidic deposition impacts on natural resources in Shenandoah National Park. NPS/NER/NRTR-2006/066. U.S. Department of the Interior, National Park Service, Northeast Region, Philadelphia, PA.

Cosby, B.J., R.F. Wright, G.M. Hornberger, and J.N. Galloway. 1985c. Modelling the effects of acid deposition: Estimation of long-term water quality responses in a small forested catchment. *Water Resour. Res.* 21(11):1591–1601.

Cronan, C.S. and C.L. Schofield. 1990. Relationships between aqueous aluminum and acidic deposition in forested watersheds of North America and northern Europe. *Environ. Sci. Technol.* 24:1100–1105.

Dennis, T.E. and A.J. Bulger. 1995. Condition factor and whole-body sodium concentrations in a freshwater fish: Evidence for acidification stress and possible ionoregulatory over-compensation. *Water Air Soil Pollut.* 85:377–382.

Dennis, T.E. and A.J. Bulger. 1999. The susceptibility of blacknose dace (*Rhinichthys atratulus*) to acidification in Shenandoah National Park. *In* A.J. Bulger, B.J. Cosby, C.A. Dolloff, K.N. Eshleman, J.N. Galloway, and J.R. Webb. (Eds.). *Shenandoah National Park: Fish in Sensitive Habitats.* Project Final Report. Volume IV: Stream Bioassays, Aluminum Toxicity, Species Richness and Stream Chemistry, and Models of Susceptibility to Acidification. Department of Environmental Sciences, University of Virginia, Charlottesville, VA.

Ellis, R.A., D.J. Jacob, M.P. Sulprizio, L. Zhang, C.D. Holmes, B.A. Schichtel, T. Blett, E. Porter, L.H. Pardo, and J.A. Lynch. 2013. Present and future nitrogen deposition to national parks in the United States: Critical load exceedances. *Atmos. Chem. Phys.* 13(17):9083–9095.

Elwood, J.W., M.J. Sale, P.R. Kaufmann, and G.F. Cada. 1991. The southern blue ridge province. *In* D.F. Charles (Ed.). *Acidic Deposition and Aquatic Ecosystems: Regional Case Studies.* Springer-Verlag, New York, pp. 319–364.

Eshleman, K.N. 1988. Predicting regional episodic acidification of surface waters using empirical techniques. *Water Resour. Res.* 24:1118–1126.

Eshleman, K.N., D.A. Fiscus, N.M. Castro, J.R. Webb, and F.A. Deviney, Jr. 2001. Computation and visualization of regional-scale forest disturbance and associated dissolved nitrogen export from Shenandoah National Park, Virginia. *Sci. World* 1(Suppl. 2):539–547.

Eshleman, K.N. and K.E. Hyer. 2000. Discharge and water chemistry at the three intensive sites. *In* A.J. Bulger, B.J. Cosby, C.A. Dolloff, K.N. Eshleman, J.R. Webb, and J.N. Galloway (Eds.). *Shenandoah National Park: Fish in Sensitive Habitats.* Project Final Report. Volume II: Stream Water Chemistry and Discharge, and Synoptic Water Quality Surveys. Department of Environmental Sciences, University of Virginia, Charlottesville, VA, pp. 51–92.

Eshleman, K.N., R.P. Morgan II, J.R. Webb, F.A. Deviney, and J.N. Galloway. 1998. Temporal patterns of nitrogen leakage from mid-Appalachian forested watersheds: Role of insect defoliation. *Water Resour. Res.* 34(8):2005–2116.

Everhart, W.H. and W.D. Youngs. 1981. *Principles of Fishery Science*. Cornell University Press, Ithaca, NY.

Farmer, G.J. and F.W.H. Beamish. 1969. Oxygen consumption of *Tilapia nilotica* in relation to swimming speed and salinity. *J. Fish. Res. Board Can.* 26:2807–2821.

Fraterrigo, J.M., S.M. Pearson, and M.G. Turner. 2009. The response of understory herbaceous plants to nitrogen fertilization in forests of different land use history. *For. Ecol. Manage.* 257:2182–2188.

Galloway, J.N., F.A. Deviney, Jr., and J.R. Webb. 1999. Shenandoah watershed study data assessment: 1980–1993. Department of Environmental Sciences, University of Virginia, Charlottesville, VA.

Galloway, J.N., S.A. Norton, and M.R. Church. 1983. Freshwater acidification from atmospheric deposition of sulfuric acid: A conceptual model. *Environ. Sci. Technol.* 17:541A–545A.

Gilliam, F.S., C.L. Goodale, L.H. Pardo, L.H. Geiser, and E.A. Lilleskov. 2011. Eastern temperate forests. *In* L.H. Pardo, M.J. Robin-Abbott, and C.T. Driscoll (Eds.). *Assessment of Nitrogen Deposition Effects and Empirical Critical Loads of Nitrogen for Ecoregions of the United States*. General Technical Report NRS-80. U.S. Forest Service, Newtown Square, PA, pp. 99–116.

Gilliam, F.S., A.W. Hockenberry, and M.B. Adams. 2006a. Effects of atmospheric nitrogen deposition on the herbaceous layer of a central Appalachian hardwood forest. *J. Torrey Bot. Soc.* 133:240–254.

Gilliam, F.S., W.J. Platt, and R.K. Peet. 2006b. Natural disturbances and the physiognomy of pine savannas: A phenomenological model. *Appl. Veg. Sci.* 9:83–96.

Goede, R.W. and B.A. Barton. 1990. Organismic indices and an autopsy-based assessment as indicators of health and condition of fish. *Am. Fish. Soc. Symp.* 8:80–93.

Hall, R.J. and F.P. Ide. 1987. Evidence of acidification effects on stream insect communities in central Ontario between 1937 and 1985. *Can. J. Fish. Aquat. Sci.* 44:1652–1657.

Hall, R.J., G.E. Likens, S.B. Fiance, and G.R. Hendrey. 1980. Experimental acidification of a stream in the Hubbard Brook Experimental Forest, New Hampshire. *Ecology* 61:976–989.

Heck, W.W. and E.B. Cowling. 1997. The need for a long term cumulative secondary ozone standard— An ecological perspective. *Ecol. Manag.* January:22–33.

Herlihy, A.T., P.R. Kaufmann, M.R. Church, P.J. Wigington, Jr., J.R. Webb, and M.J. Sale. 1993. The effects of acidic deposition on streams in the Appalachian Mountain and Piedmont region of the mid-Atlantic United States. *Water Resour. Res.* 29(8):2687–2703.

Herlihy, A.T., P.R. Kaufmann, J.L. Stoddard, K.N. Eshleman, and A.J. Bulger. 1996. Effects of acidic deposition on aquatic resources in the southern Appalachians with a special focus on class I wilderness areas. Prepared for the Southern Appalachian Mountains Initiative (SAMI), Asheville, NC.

Hildebrand, E., J.M. Skelly, and T.S. Fredericksen. 1996. Foliar response of ozone-sensitive hardwood trees species from 1991 to1993 in Shenandoah National Park, Virginia. *Can. J. For. Res.* 26:658–669.

Hunter, M.D. and J.C. Schultz. 1995. Fertilization mitigates chemical induction and herbivore responses within damaged trees. *Ecology* 76:1226–1232.

Hyer, K.E. 1997. Episodic acidification of streams in Shenandoah National Park. Masters, University of Virginia, Charlottesville, VA.

Hyer, K.E., J.R. Webb, and K.N. Eshleman. 1995. Episodic acidification of three streams in Shenandoah National Park, Virginia, USA. *Water Air Soil Pollut.* 85:523–528.

Jastram, J.D., C.D. Snyder, N.P. Hitt, and K.C. Rice. 2013. Synthesis and interpretation of surface-water quality and aquatic biota data collected in Shenandoah National Park, Virginia, 1979–2009. U.S. Geological Survey Scientific Investigations Report 2013–5157, Reston, VA.

Jenkins, R.E. and N.M. Burkhead. 1993. *Freshwater Fishes of Virginia*. American Fisheries Society, Bethesda, MD.

Kahl, J.S., S.A. Norton, T.A. Haines, E.A. Rochette, R.C. Heath, and S.C. Novdin. 1992. Mechanisms of episodic acidification in low-order streams in Maine, USA. *Environ. Pollut.* 78:37–44.

Kang, D., V.P. Aneja, R. Mathur, and J.D. Ray. 2003. Nonmethane hydrocarbons and ozone in three rural southeast United States national parks: A model sensitivity analysis and comparison to measurements. *J. Geophys. Res. Atmos.* 108(D19, 4604):17.

Kang, D., V.P. Aneja, R.G. Zika, C. Farmer, and J.D. Ray. 2001. Nonmethane hydrocarbons in the rural southeast United States national parks. *J. Geophys. Res.* 106(D3):3133–3155.

Karr, J.R. and L.W. Chu. 1999. *Restoring Life in Running Rivers: Better Biological Monitoring.* Island Press, Washington, DC, 206pp.

Kaufmann, P.R., A.T. Herlihy, M.E. Mitch, J.J. Messer, and W.S. Overton. 1991. Stream chemistry in the eastern United States 1. Synoptic survey design, acid-base status, and regional patterns. *Water Resour. Res.* 27:611–627.

Kimmel, W.G., D.J. Murphy, W.E. Sharpe, and D.R. DeWalle. 1985. Macroinvertebrate community structure and detritus processing rates in two southwestern Pennsylvania streams acidified by atmospheric deposition. *Hydrobiologia* 124:97–102.

Kohut, R. 2007a. Assessing the risk of foliar injury from ozone on vegetation in parks in the U.S. National Park Service's Vital Signs Network. *Environ. Pollut.* 149:348–357.

Kohut, R.J. 2007b. Ozone risk assessment for vital signs monitoring networks, Appalachian National Scenic Trail, and Natchez Trace National Scenic Trail. Natural Resource Technical Report NPS/NRPC/ARD/NRTR—2007/001. National Park Service, Natural Resource Program Center, Fort Collins, CO.

Lawrence, G.B. 2002. Persistent episodic acidification of streams linked to acid rain effects on soil. *Atmos. Environ.* 36:1589–1598.

Likens, G.E., N.M.J. F.H. Bormann, D.W. Fisher, and R.S. Pierce. 1970. Effects of forest cutting and herbicide treatment on nutrient budgets in the Hubbard Brook watershed-ecosystem. *Ecol. Monogr.* 40:23–47.

Lovett, G.M., L.M. Christensen, P.M. Groffman, C.G. Jones, J.E. Hart, and M.J. Mitchell. 2002. Insect defoliation and nitrogen cycling in forests. *BioScience* 52(4):335–341.

Lovett, G.M. and A.E. Ruesink. 1995. Carbon and nitrogen assimilation in red oaks (*Quercus rubra* L.) subject to defoliation and nitrogen stress. *Tree Physiol.* 12:259–269.

Lynch, D.D. and N.B. Dise. 1985. Water resources investigations report 85-4115. U.S. Geological Survey, Richmond, VA.

Mason, R.R., B.E. Wickman, R.C. Beckwith, and H.G. Paul. 1992. Thinning and nitrogen fertilization in a grand fir stand infested with spruce budworm. Part I: Insect response. *For. Sci.* 38:235–251.

McCune, B., J. Dey, J. Peck, K. Heiman, and S. Will-Wolf. 1997. Regional gradients in lichen communities of the southeast United States. *Bryologist* 100(2):145–158.

McDonnell, T.C., T.J. Sullivan, P.F. Hessburg, K.M. Reynolds, N.A. Povak, B.J. Cosby, W. Jackson, and R.B. Salter. 2014. Steady-state sulfur critical loads and exceedances for protection of aquatic ecosystems in the U.S. southern Appalachian Mountains. *J. Environ. Manage.* 146:407–419.

Miller-Marshall, L.M. 1993. Mechanisms controlling variation in stream chemical composition during hydrologic episodes in the Shenandoah National Park, Virginia. Masters, Department of Environmental Sciences, University of Virginia, Charlottesville, VA.

Molot, L.A., P.J. Dillon, and B.D. LaZerte. 1989. Changes in ionic composition of streamwater during snowmelt in central Ontario. *Can. J. Fish. Aquat. Sci.* 46:1658–1666.

Moore, P.T., H. Van Miegroet, and N.S. Nicholas. 2007. Relative role of understory and overstory carbon and nitrogen cycling in a southern Appalachian spruce-fir forest. *Can. J. For. Res.* 37:2689–2700.

Muller, R.N. 2003. Nutrient relations of the herbaceous layer in deciduous forest ecosystems. *In* F.S. Gilliam and M.R. Roberts (Eds.). *The Herbaceous Layer in Forests of Eastern North America.* Oxford University Press, New York, pp. 15–37.

National Acid Precipitation Assessment Program (NAPAP). 1991. Integrated assessment report. National Acid Precipitation Assessment Program, Washington, DC.

National Park Service. 1981. Biogeography of red spruce in Shenandoah National Park. National Park Service, Denver, CO.

National Park Service (NPS). 2010. Air quality in national parks: 2009 annual performance and progress report. Natural Resource Report NPS/NRPC/ARD/NRR-2010/266. National Park Service, Air Resources Division, Denver, CO.

Odman, M.T., J.W. Boylan, J.G. Wilkinson, A.G. Russell, S.F. Mueller, R.E. Imhoff, K.G. Doty, W.B. Norris, and R.T. McNider. 2002. SAMI air quality modeling: Final report. Southern Appalachian Mountains Initiative, Asheville, NC.

Pardo, L.H., M.J. Robin-Abbott, and C.T. Driscoll. 2011. Assessment of nitrogen deposition effects and empirical critical loads of nitrogen for ecoregions of the United States. General Technical Report NRS-80. U.S. Forest Service, Newtown Square, PA.

Peterson, R.H. and L. Van Eechhaute. 1992. Distributions of ephemeroptera, plecoptera, and trichoptera of three maritime catchments differing in pH. *Freshw. Biol.* 27:65–78.

Plafkin, J.L., M.T. Barbour, K.D. Porter, S.K. Gross, and R.M. Hughes. 1989. Rapid bioassessment protocols for use in streams and rivers: Benthic macroinvertebrates and fish. EPA 440-89-001. U.S. Environmental Protection Agency, Office of Water Regulations and Standards, Washington, DC.

Porak, W.F. 1981. The effects of acid drainage mitigation upon fish, benthic macroinvertebrates, and water quality in streams of the Cherokee National Forest, Tennessee. Masters, Tennessee Technological University, Cookeville, TN.

Potyondy, J.P., B.B. Roper, S.E. Hixson, R.L. Leiby, R.L. Lorenz, and C.M. Knopp. 2006. Aquatic ecological unit inventory technical guide: Valley segment and river reach. General Technical Report. USDA Forest Service, Ecosystem Management Coordination Staff, Washington, DC.

Raddum, G.G. and A. Fjellheim. 1984. Acidification and early warning organisms in freshwater in western Norway. *Int. Ver. Theor. Angew. Limnol. Verh.* 22:1973–1980.

Resh, V.H., R.H. Norris, and M.T. Barbour. 1995. Design and implementation of rapid assessment approaches for water resource monitoring using benthic macroinvertebrates. *Aust. J. Ecol.* 20:108–121.

Reuss, J.O. and D.W. Johnson. 1986. *Acid Deposition and the Acidification of Soil and Water.* Springer-Verlag, New York.

Reynolds, K.M., P.F. Hessburg, T. Sullivan, N. Povak, T. McDonnell, B. Cosby, and W. Jackson. 2012. Spatial decision support for assessing impacts of atmospheric sulfur deposition on aquatic ecosystems in the southern Appalachian region. *Proceedings of the 45th Hawaiian International Conference on System Sciences*, Maui, HI, January 4–7, 2012.

Rice, K.C., T.M. Scanlon, J.A. Lynch, and B.J. Cosby. 2014. Decreased atmospheric sulfur deposition across the southeastern U.S.: When will watersheds release stored sulfate? *Environ. Sci. Technol.* 48(17):10071–10078.

Rosemond, A.D., S.R. Reice, J.W. Elwood, and P.J. Mulholland. 1992. The effects of stream acidity on benthic invertebrate communities in the south-eastern United States. *Freshw. Biol.* 27:193–209.

Royo, A.A., R. Collins, M.B. Adams, C. Kirschbaum, and W.P. Carson. 2010. Pervasive interactions between ungulate browsers and disturbance regimes promote temperate forest herbaceous diversity. *Ecology* 91(1):93–105.

Schwede, D.B. and G.G. Lear. 2014. A novel hybrid approach for estimating total deposition in the United States. *Atmos. Environ.* 92:207–220.

Southern Appalachian Man and the Biosphere (SAMAB). 1996. The Southern Appalachian assessment aquatics technical report. U.S. Department of Agriculture, Forest Service, Southern Region, Atlanta, GA.

Stoddard, J.L. 1994. Long-term changes in watershed retention of nitrogen: Its causes and aquatic consequences. *In* L.A. Baker (Ed.). *Environmental Chemistry of Lakes and Reservoirs.* American Chemical Society, Washington, DC, pp. 223–284.

Sullivan, T.J. 2000. *Aquatic Effects of Acidic Deposition.* CRC Press, Boca Raton, FL.

Sullivan, T.J., B.J. Cosby, A.T. Herlihy, J.R. Webb, A.J. Bulger, K.U. Snyder, P. Brewer, E.H. Gilbert, and D.L. Moore. 2004. Regional model projections of future effects of sulfur and nitrogen deposition on streams in the southern Appalachian Mountains. *Water Resour. Res.* 40:W02101.

Sullivan, T.J., B.J. Cosby, J.A. Laurence, R.L. Dennis, K. Savig, J.R. Webb, A.J. Bulger et al. 2003. Assessment of air quality and related values in Shenandoah National Park. NPS/NERCHAL/NRTR-03/090. U.S. Department of the Interior, National Park Service, Northeast Region, Philadelphia, PA.

Sullivan, T.J., B. J. Cosby, J.R. Webb, R.L. Dennis, A.J. Bulger, and F.A. Deviney, Jr. 2008. Streamwater acid-base chemistry and critical loads of atmospheric sulfur deposition in Shenandoah National Park, Virginia. *Environ. Monit. Assess.* 137:85–99.

Sullivan, T.J., B.J. Cosby, J.R. Webb, K.U. Snyder, A.T. Herlihy, A.J. Bulger, E.H. Gilbert, and D. Moore. 2002. Assessment of the effects of acidic deposition on aquatic resources in the southern Appalachian Mountains. Report prepared for the Southern Appalachian Mountains Initiative (SAMI). E&S Environmental Chemistry, Inc., Corvallis, OR.

Sullivan, T.J., J.M. Eilers, B.J. Cosby, and K.B. Vaché. 1997. Increasing role of nitrogen in the acidification of surface waters in the Adirondack Mountains, New York. *Water Air Soil Pollut.* 95(1–4):313–336.

Sullivan, T.J., G.B. Lawrence, S.W. Bailey, T.C. McDonnell, C.M. Beier, K.C. Weathers, G.T. McPherson, and D.A. Bishop. 2013. Effects of acidic deposition and soil acidification on sugar maple in the Adirondack Mountains, New York. *Environ. Sci. Technol.* 47:12687–12694.

Sullivan, T.J., G.T. McPherson, T.C. McDonnell, S.D. Mackey, and D. Moore. 2011. Evaluation of the sensitivity of inventory and monitoring national parks to acidification effects from atmospheric sulfur and nitrogen deposition. U.S. Department of the Interior, National Park Service, Denver, CO.

Sullivan, T.J., J.R. Webb, K.U. Snyder, A.T. Herlihy, and B.J. Cosby. 2007. Spatial distribution of acid-sensitive and acid-impacted streams in relation to watershed features in the southern Appalachian Mountains. *Water Air Soil Pollut.* 182:57–71.

Swank, W.T. 1988. Stream chemistry responses to disturbance. *In* W.T. Swank and D.S. Crossley (Eds.). *Forest hydrology at Coweeta.* Springer-Verlag, New York, pp. 339–358.

Thomas, R.Q., C.D. Canham, K.C. Weathers, and C.L. Goodale. 2010. Increased tree carbon storage in response to nitrogen deposition in the US. *Nat. Geosci.* 3:13–17.

Thornton, K., D. Marmorek, and P. Ryan. 1990. Methods for projecting future changes in surface water acid-base chemistry. National Acid Precipitation Assessment Program, Washington, DC.

Townsend, C.R., A.G. Hildrew, and J. Francis. 1983. Community structure in some southern English streams: The influence of physiochemical factors. *Freshw. Biol.* 13:521–544.

Turner, R.S., R.B. Cook, H. van Miegroet, D.W. Johnson, J.W. Elwood, O.P. Bricker, S.E. Lindberg, and G.M. Hornberger. 1990. Watershed and lake processes affecting chronic surface water acid-base chemistry. State of the Science, SOS/T 10. National Acid Precipitation Assessment Program, Washington, DC.

Webb, J.R. 1999. Synoptic stream water chemistry. *In* A.J. Bulger, B.J. Cosby, C.A. Dolloff, K.N. Eshleman, J.R. Webb and J.N. Galloway (Eds.). *Shenandoah National Park: Fish in Sensitive Habitats.* Project Final Report, Volume II. University of Virginia, Charlottesville, VA. pp. 1–50.

Webb, J.R. 2003. Effects of acidic resources of the Central Appalachian Mountain region—Significance of the W.H. Sammis Electric Power Generating Facility. Monterey, VA.

Webb, J.R., B.J. Cosby, F.A. Deviney, Jr., K.N. Eshleman, and J.N. Galloway. 1995. Change in the acid-base status of an Appalachian Mountain catchment following forest defoliation by the gypsy moth. *Water Air Soil Pollut.* 85:535–540.

Webb, J.R., B.J. Cosby, F.A. Deviney, J.N. Galloway, S.W. Maben, and A.J. Bulger. 2004. Are brook trout streams in western Virginia and Shenandoah National Park recovering from acidification? *Environ. Sci. Technol.* 38(15):4091–4096.

Webb, J.R., F.A. Deviney, and J.N. Galloway. 1993. Shenandoah watershed study: Program evaluation. Report to U.S. National Park Service. University of Virginia, Charlottesville, VA.

Webb, J.R., F.A. Deviney, J.N. Galloway, C.A. Rinehart, P.A. Thompson, and S. Wilson. 1994. The acid-base status of native brook trout streams in the mountains of Virginia. Department of Environmental Sciences, University of Virginia, Charlottesville, VA.

Welch, N.T., J.M. Belmont, and J.C. Randolph. 2007. Summer ground layer biomass and nutrient contribution to aboveground litter in an Indiana temperate deciduous forest. *Am. Midl. Nat.* 157:11–26.

Welsch, D.L., J.R. Webb, and B.J. Cosby. 2001. Description of summer 2000 field work. Collection of soil samples and tree cores in the Shenandoah National Park with summary soils data. Report submitted to National Park Service. Department of Environmental Science, University of Virginia, Charlottesville, VA.

Wigington, P.J. 1999. Episodic acidification: Causes, occurrence and significance to aquatic resources. *In* J.R. Drohan (Ed.). *The Effects of Acidic Deposition on Aquatic Ecosystems in Pennsylvania. 1998 PA Acidic Deposition Conference.* Environmental Resources Research Institute, University Park, PA, pp. 1–5.

Wigington, P.J., Jr., T.D. Davies, M. Tranter, and K.N. Eshleman. 1990. Episodic acidification of surface waters due to acidic deposition. State of Science/Technology Report 12. National Acid Precipitation Assessment Program, Washington, DC.

Wigington, P.J., Jr., D.R. DeWalle, P.S. Murdoch, W.A. Kretser, H.A. Simonin, J. Van Sickle, and J.P. Baker. 1996. Episodic acidification of small streams in the northeastern United States: Ionic controls of episodes. *Ecol. Appl.* 6(2):389–407.

Winner, M.W., J.S. Baron, I.S. Cotter, C.S. Greitner, J. Nellessen, L.R. McEvoy Jr., R.L. Olson, C.J. Atkinson, and L.D. Moore. 1989. Plant responses to elevational gradients of ozone exposures in Virginia. *Proc. Natl. Acad. Sci.* 86:8828–8832.

Wood, C.M. 1989. The physiological problems of fish in acid waters. *In* R. Morris, E.W. Taylor, D.J.A. Brown, and J.A. Brown (Eds.). *Acid Toxicity and Aquatic Animals.* Cambridge University Press, Cambridge, U.K., pp. 125–152.

6

Acadia National Park and the Northeast Temperate Network

6.1 Background

Acadia National Park (ACAD) is the largest park in the Northeast Temperate Network. It is also the oldest established national park in the eastern United States. There are many medium to large urban population centers in the Northeast, especially along the coastal corridor between Boston and Washington, DC. Elsewhere in the network region, there are few population centers larger than 50,000 people. Figure 6.1 shows the network boundary, the location of ACAD and other parks, and population centers having more than 10,000 people.

6.2 Atmospheric Emissions and Deposition

Portions of the network region away from the most intensive urban development (including around ACAD) generally had low sulfur dioxide emissions, less than 1 ton/mi²/yr in most counties. Although there were many point sources of sulfur dioxide within the region, predominantly in the heavily urbanized areas, most emitted less than 5000 tons/yr.

County-level N emissions within the network region ranged from less than 1 ton/mi²/yr, mainly in the north, to greater than 100 tons/mi²/yr in 2002 in some of the densely urbanized areas in and around Boston and near New York City. However, annual county N emissions were generally less than 20 tons/mi²/yr throughout all but the most heavily urbanized areas. There were few N point sources larger than 500–1000 tons/yr within this network region. Emissions of N were at relatively constant and high levels during the 1980s and 1990s but have declined since about 2000 (Sullivan et al. 2011).

County-level emissions near the Northeast Temperate Network, based on data from the EPA's National Emissions Inventory during a more recent time period (2011), are depicted in Figures 6.2 through 6.4 for sulfur dioxide, oxidized N, and ammonia, respectively. Many counties to the southwest of network parks had relatively high sulfur dioxide emissions (>25 tons/mi²/yr; Figure 6.2). Spatial patterns in N oxide emissions were generally similar, with highest values generally to the south and southwest of network parks (Figure 6.3). Emissions of ammonia were somewhat lower, with most counties showing emissions levels below 8 tons/mi²/yr (Figure 6.4).

Some states have adopted emissions controls for toxic substances that go well beyond federal requirements (Smith and Trip 2005). In 1998, the New England Governors and

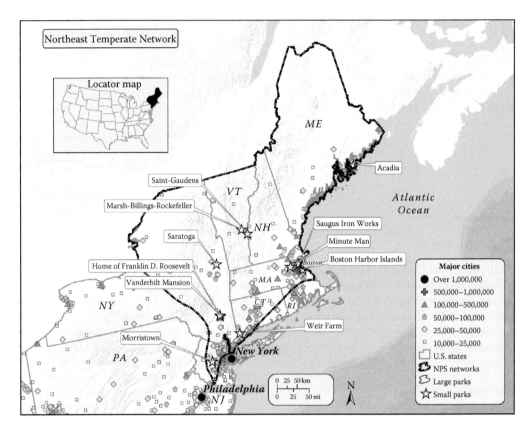

FIGURE 6.1
Network boundary and locations of parks and population centers greater than 10,000 people within the Northeast Temperate Network region.

Eastern Canadian Premiers Mercury Action Plan was adopted to regulate Hg emissions and to control the use of Hg-containing products in the region. Aggressive emissions reduction goals were adopted by the New England states, Atlantic provinces, and Quebec. New York and New Jersey have also participated. This effort resulted in a reduction of Hg emissions by 55% in 2003 as compared with mid-1990 levels, and 2010 Hg emissions levels likely reflected reductions of over 80% (U.S. EPA 2010). These values are in line with the 1998 goals of the Mercury Action Plan of at least 50% and 75% reductions by 2003 and 2010, respectively (Smith and Trip 2005).

Total S and total N deposition within the Northeast Temperate Network region in 2002 were both highly variable from north to south, ranging from between 2 and 5 kg S or N/ha/yr in northern Maine to between 10 and 15 kg S or N/ha/yr in much of the southern portion of the network region (Sullivan et al. 2011).

Patterns of deposition have been extensively studied at ACAD. This park typically receives 20%–40% more wet deposition of S than inland sites in Maine. During winter, some air masses that reach ACAD track over the ocean before reaching the park and incorporate marine air into their circulation and precipitation. Partly as a consequence of this pattern, Hg and S wet deposition at ACAD are highest in summer and lowest in winter. Wet chloride deposition shows the opposite pattern (Kahl et al. 2007a).

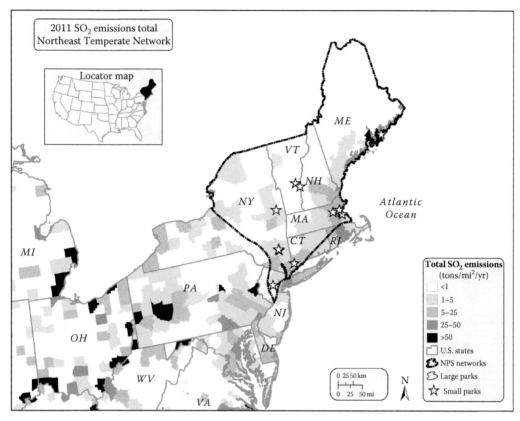

FIGURE 6.2
Total sulfur dioxide (SO₂) emissions, by county, near Northeast Temperate Network for the year 2011. (Data from the EPA's National Emissions Inventory, https://www.epa.gov/air-emissions-inventories, accessed January, 2014.)

Dry deposition at ACAD varies with elevation and vegetation cover. The undisturbed coniferous forested landscape at ACAD facilitates high dry deposition of S, N, and Hg (Weathers et al. 2000, 2006). Highest total S deposition occurs at high-elevation sites. Weathers et al. (2006) simulated total S deposition hot spots in ACAD up to 25 kg S/ha/yr, and Kahl et al. (2007a) estimated maxima near 13 kg S/ha/yr. Results of both studies indicate that localized S deposition in parts of ACAD can be much higher than park or regional averages. Occult (fog and cloud) deposition can also be an important vehicle for depositing acidifying substances to this park and other coastal and high-elevation locations within the region. Fog pH below 3.5 has been documented in the past (Weathers et al. 1988a,b, Jagels et al. 1989).

Dry deposition of Hg at ACAD equals or exceeds wet deposition of Hg during both the growing season and the winter (Miller et al. 2005, Vaux et al. 2008). Snow Hg deposition can be substantial (Nelson et al. 2007), but a portion of the Hg deposited via snowfall is subsequently volatilized and reemitted back into the atmosphere (Vaux et al. 2008), limiting in-park environmental impacts.

Weathers et al. (2006) developed an empirical modeling approach, based on 300–400 throughfall measurements, to estimate the total (wet, dry, plus cloud) deposition of S and N to the complex terrain of ACAD. Throughfall deposition measurements taken

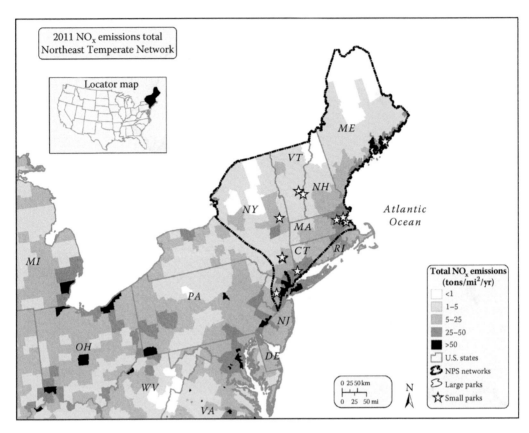

FIGURE 6.3

Total nitrogen oxide (NO_x) emissions, by county, near Northeast Temperate Network for the year 2011. (Data from the EPA's National Emissions Inventory, https://www.epa.gov/air-emissions-inventories, accessed January, 2014.)

during summer, combined with landscape variables such as elevation, forest type, and slope, explained about 40% of the variation in total deposition estimates. Model estimates were scaled to wet and dry deposition measurements and estimated values from National Atmospheric Deposition Program and Clean Air Status and Trends Network monitoring sites. Resulting maps showed substantial spatial variability in S and N deposition across the landscape of this park. Results of throughfall monitoring at 21 locations within ACAD indicated that canopy openness (which could also be represented as canopy height or vegetation type) and elevation were the most important landscape variables influencing spatial variability in total atmospheric deposition of S and Hg. The litterfall Hg flux was just as important as precipitation to the total input of Hg to the soil at the monitored sites. Elevated Hg deposition in ACAD has also been inferred from lake sediment accumulation rates of Hg and other contaminants (Kahl et al. 1985, Norton et al. 1997).

During the period 1985–2002, wet deposition of S decreased substantially at ACAD (Kahl et al. 2004), but there was not a significant change in N oxide or ammonium wet deposition (Lehmann et al. 2005). Because of the lack of decrease in the levels of wet N deposition, Tonnessen and Manski (2007) recommended continued monitoring of ecosystems in ACAD to assess potential changes caused by fertilization and eutrophication.

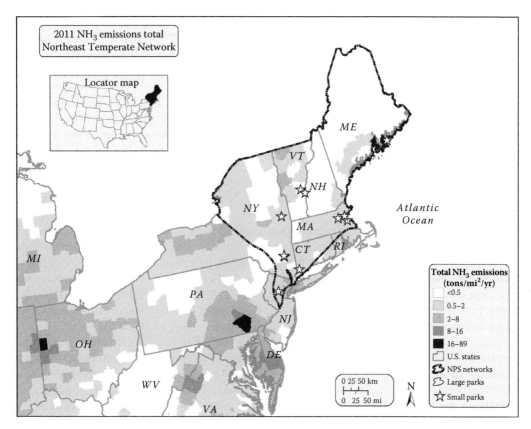

FIGURE 6.4

Total ammonia (NH_3) emissions, by county, near Northeast Temperate Network for the year 2011. (Data from the EPA's National Emissions Inventory, https://www.epa.gov/air-emissions-inventories, accessed January, 2014.)

TABLE 6.1

Average Changes in S and N Deposition[a] between 2001 and 2011 across Park Grid Cells at ACAD

Parameter	2001 Average (kg/ha/yr)	2011 Average (kg/ha/yr)	Absolute Change (kg/ha/yr)	Percent Change	2011 Minimum (kg/ha/yr)	2011 Maximum (kg/ha/yr)	2011 Range (kg/ha/yr)
Total S	6.55	3.28	−3.28	−50.0	3.14	5.66	2.52
Total N	4.39	3.08	−1.31	−29.8	2.97	5.17	2.20
Oxidized N	3.06	1.82	−1.24	−40.4	1.76	3.38	1.62
Reduced N	1.33	1.26	−0.07	−5.5	1.20	1.79	0.58

[a] Deposition estimates were determined by the Total Deposition project, based on three-year averages centered on 2001 and 2011 for all ~4 km grid cells in the park. The minimum, maximum, and range of 2011 S and N deposition are also shown.

Atmospheric S deposition levels have further decreased since 2001, based on Total Deposition project estimates (Schwede and Lear 2014; Table 6.1). Total S deposition decreased by 50% over the previous decade at ACAD and now averages only about 3 kg S/ha/yr. Estimated total N and oxidized N deposition over that same time period also decreased substantially. Reduced N deposition changed little.

Total S deposition for the period 2010–2012 was generally in the range of 3–5 kg S/ha/yr at ACAD. Oxidized inorganic N deposition for the period 2010–2012 was lower. ACAD received less than 2 kg N/ha/yr of ammonium from atmospheric deposition during this same period. Total N deposition was generally in the range of 3–5 kg N/ha/yr, with higher values to the southwest of the park.

A study of two forested watersheds in ACAD by Johnson et al. (2007) investigated the effects of landscape factors, such as watershed aspect and forest type, on atmospheric Hg deposition. The coniferous forest study site received higher levels of total Hg (but not methyl Hg) deposition compared with the site covered with hardwood vegetation (Johnson et al. 2007). The researchers interpreted this difference as likely due to the more efficient dry deposition scavenging of the conifer canopy, which has waxy cuticle, greater surface roughness, higher foliar surface area, and denser canopy than a deciduous forest canopy (Johnson et al. 2007).

Litterfall can constitute an important depositional source of Hg to soils in ACAD. Sheehan et al. (2006) investigated the importance of Hg inputs to soil from litterfall in vegetated landscapes in ACAD. Results showed that Hg concentrations were significantly different among vegetation classes sampled. Highest concentrations were found in litter from coniferous forests (58.8 ± 3.3 nanograms Hg per gram [ng Hg/g]), followed by mixed (41.7 ± 2.8 ng Hg/g) and scrub (40.6 ± 2.7 ng Hg/g) vegetation types. Lowest Hg concentrations were found in litter from hardwoods (31.6 ± 2.6 ng Hg/g; Sheehan et al. 2006). There were no significant differences, however, among vegetation classes in litter Hg flux (including consideration of both concentration and amount of litter). This may have been due to the high variability in the data and differences in litterfall mass between hardwoods and softwoods. Hardwood litter had higher autumnal Hg flux, whereas softwood litter had higher Hg concentrations and higher litterfall flux throughout the winter and spring (Sheehan et al. 2006). Results suggested that the litterfall Hg flux can be just as large as the wet deposition flux.

Snow is a factor to consider when determining atmospheric Hg deposition to soil but has not been well studied. To address this gap in understanding, a study by Nelson et al. (2008) in ACAD examined Hg concentrations in snow and compared concentrations at sites with and without substantial conifer forest canopies, using differing collection methods. Results showed that Hg deposition at sites with no canopy, including the Mercury Deposition Network site, was approximately 3.4 times lower than deposition at forested sites, as measured after snowfall "events" (snowfall of >8 cm [3.2 in.]). Mercury concentrations were lowest when measured as season-long throughfall Hg flux (1.8 $\mu g/m^2$), slightly higher in the bulk snowpack (2.38 ± 0.68 $\mu g/m^2$), and highest in measurements made after snowfall events (5.63 ± 0.38 $\mu g/m^2$; Nelson et al. 2008). These results illustrate the importance of dry deposition during the cold season, and subsequent reemission of Hg from the snowpack back into the atmosphere, to the annual Hg flux from the atmosphere to the soil. Additionally, the researchers conducted an Hg tracer study and found evidence for the movement of Hg from the soil into the snowpack (Nelson et al. 2008).

6.3 Acidification

Previous research on acidification in northeastern national parks has focused on ACAD. A pair of gauged watershed research sites was established at ACAD in 1990, as part of the Park Research and Intensive Monitoring of Ecosystems Network, a joint program of the EPA and the National Park Service. Research focused on quantifying the influence of landscape features on atmospheric S, N, and Hg deposition; quantifying throughfall fluxes of Hg; investigating seasonal patterns of deposition; and documenting relationships between atmospheric deposition and stream chemistry (Kahl et al. 2007a).

Most of ACAD is underlain by granite. The low buffering capacity of the overlying soils, together with areas having steep slopes, and in some areas also relatively high elevation, influences susceptibility of soil and drainage water to acidification (Kahl et al. 2000). Rainfall in ACAD is about 140 cm/yr (55 inches/yr), which is about 20 cm/yr (8 inches/yr) higher than most of the rest of Maine (Nielsen and Kahl 2007). These features enhance the sensitivity of streams and soils in the park to acidification. The predominant sensitive vegetation type in ACAD is red spruce. Sugar maple is also common. These species are well known to be acid sensitive.

6.3.1 Acidification of Terrestrial Ecosystems

The Northeast Temperate Network has conducted forest health monitoring in ACAD since 2006. Nearly all of the 176 forest health monitoring plots have been sampled twice through 2013 (Miller et al. 2014). Some forest soils in ACAD have high organic content, low pH, and low base saturation (Miller et al. 2014). Overall forest health in ACAD is good, although the forests are subjected to multiple stressors, including soil acidification.

Acid-sensitive forest soils have been acidified by S oxides and N oxide deposition throughout the Northeast Temperate Network region. Acidic deposition has been shown to be an important factor causing decreases in concentrations of exchangeable base cations in soils, which were naturally low historically. Forest soils in ACAD, which have high organic content, low pH, and low base saturation, are highly vulnerable to acidification. Base saturation values less than 10% predominate in the soil B horizon in areas where soil and surface water acidification from acidic deposition have been most pronounced (David and Lawrence 1996, Bailey et al. 2004, Sullivan et al. 2006b). At Hubbard Brook Experimental Forest in New Hampshire, model hindcast simulations suggested that soil percent base saturation has decreased due to acidic deposition and accumulation of nutrient cations by growing forest vegetation. Sulfur and N emissions controls put into place in the United States (and neighboring Canada) since passage of the Clean Air Act and its Amendments have slowed the pace of damage to sensitive ecosystems in the Northeast Temperate Network region.

To some degree, previously acidified soils can recover their base cation reserves over time in response to reduced levels of acidic deposition. Some degree of soil and/or vegetation recovery may be beginning to occur in some places in the Northeast Temperate Network as air pollution levels decline, but scientists have only been able to confirm that response to a limited extent. It is unlikely that recovery from soil damage has occurred to any significant extent in the Northeast in response to decreased levels of air pollution. More likely, soil conditions and plant health are continuing to decline, albeit more slowly, at most acid-sensitive locations. The recovery potential of soil exchangeable base cation concentrations is dependent on weathering rates, which can be very slow. Additionally, there is evidence of a delayed watershed acidification recovery response. In the past, most of the S deposited in watersheds in the Northeast Temperate Network region moved more or less directly through soils and into water as sulfate, with the capacity to acidify soil and water along the way. Nevertheless, a fraction of the S deposited each year was stored in the soil. Now that S deposition inputs to the watersheds have decreased due to air pollution regulations, some of that stored S is being released to drainage water, making recovery slower as a consequence of the cumulative damage to the soil from past air pollution. Current scientific understanding suggests that additional cuts in emissions, beyond those required as of about 2003, might enable the most sensitive ecosystems in the Northeast to continue to recover and prevent renewed acidification in response to base cation loss from soils under continued, but relatively lower, levels of atmospheric S and N deposition (Sullivan et al. 2006a).

Acidic deposition and soil acidification can have detrimental effects on multiple forest plant species. It is difficult to quantify these effects, in part because plants are simultaneously being affected by multiple stressors; besides air pollution, plants are particularly affected by changing climatic conditions, insect pests, competition with nonnative species, and disease. Nevertheless, it is clear that plant species throughout portions of the Northeast Temperate Network region have been damaged by air pollution. At sensitive locations, acidic deposition has resulted in base cation depletion and decreased Ca:Al ratio in soil solution. Aber et al. (2003) concluded that N deposition is altering the N status of northeastern forests; the study reported a decrease in C:N ratio from about 35 to about 25 along an increasing atmospheric N deposition gradient of 3–12 kg N/ha/yr across the Northeast.

Soil acidification, Al toxicity, and exposure of foliage to acidic deposition have collectively contributed to decline in some tree species in portions of the northeastern United States that have experienced soil acidification as a consequence of sulfur dioxide and N oxide deposition. Effects have included reduced growth and increased stress to overstory trees, and likely changes in the species distributions of understory plants. Acidic deposition has been implicated as a causal factor for the decline of red spruce at high elevation throughout the Northeast Temperate Network region (DeHayes et al. 1999). Spruce dieback has been observed and has been most severe at high elevations in the Adirondack and Green Mountains, where more than 50% of the canopy trees died during the 1970s and 1980s. In the White Mountains, about 25% of the canopy spruce died during that same period (Craig and Friedland 1991). Dieback of red spruce trees was also observed in mixed hardwood-conifer stands at relatively low elevations in the western Adirondack Mountains (Shortle et al. 1997). This rapid dieback was linked with two aspects of air pollution: exposure of foliage to acidic cloud water and an increase in the amount of dissolved Al compared with dissolved Ca in soil water. Such effects may have occurred in ACAD and other national parks of the Northeast Temperate Network, but at a lower level.

The study of DeHayes et al. (1999) suggested that the direct deposition of acidic compounds on red spruce needles preferentially removes membrane-associated Ca. More recently, a strong link has also been established between availability of soil Ca and winter injury (Hawley et al. 2006). Acidification can also affect base saturation and the availability and leaching loss of important base cation nutrients in hardwood forests. The deciduous tree species most commonly associated with adverse acidification-related effects of S and N deposition is sugar maple, which occurs widely in ACAD. Sugar maple is a key species of the northern hardwood forest. Several studies, mainly in Pennsylvania (Bailey et al. 2004) and the Adirondack Mountains (Sullivan et al. 2013), have indicated that sugar maple decline is linked to the occurrence of relatively high levels of acidic deposition and base-poor soils.

Along an increasing N deposition gradient in the northeastern United States, from 4.2 to 11.1 kg N/ha/yr, Lovett and Rueth (1999) found a twofold increase in N mineralization in soils of sugar maple stands but no significant relationship between increased deposition and mineralization in American beech (*Fagus grandifolia*) stands. This difference might be attributable to the lower litter quality in beech stands. Sugar maple appears to be more susceptible to effects of increasing N deposition and concomitant soil acidification from either direct leaching of nitrate or enhanced nitrification. For northeastern hardwoods, Aber et al. (2003) found a decrease in soil C:N ratio from 24 to 17 over a deposition gradient of 3 to 12 kg N/ha/yr. This decrease was similar to, but less steep than, the decrease seen in soil under conifers.

The health of sugar maple trees is strongly influenced by the availability of Ca in soil. Calcium is depleted by acidic deposition. Trees that grow on soils having low base cation supply are stressed and consequently often become more susceptible to damage from defoliating insects, drought, and extreme weather. The overall response can include death of mature trees and poor regeneration of seedlings.

6.3.2 Acidification of Aquatic Ecosystems

6.3.2.1 Status

The relatively high rates of atmospheric deposition of S and N to the watersheds of lakes and streams in the Northeast Temperate Network region, combined with naturally low contributions from some rock types of base cations that serve to neutralize acidity, are among the most important causes of acidity in many lakes and streams within this region, including within ACAD. Sulfur deposition has contributed to chronic surface water acidification in the northeastern United States to a greater extent than has N deposition. Nitrate concentrations in acid-sensitive drainage waters are generally much lower than sulfate concentrations. However, at the peak of snowmelt, the influence of N deposition becomes proportionately more important at some locations and in some waters is just as important as S acidity. The seasonal shift in the relative importance of S- and N-caused acidity is related to the dynamics of plant and microbial growth cycles and snowpack accumulation and melting.

Most lakes in ACAD are circumneutral, with average pH between 6.5 and 7.5. Only two are known to have had pH < 5.0: Duck Pond is naturally acidic from organic acids; Sargent Mountain Pond has limited soil in its watershed and has likely been acidified by acidic deposition (Kahl et al. 2000, Vaux et al. 2008). The relatively recent monitoring has not revealed any acidic lakes in ACAD. Most surveyed streams in the park have acid neutralizing capacity <100 µeq/L, and seven surveyed streams were acidic at the time of sampling (Vaux et al. 2008).

Kahl et al. (2007a) surveyed 28 streams in ACAD. Stream water sulfate concentrations followed the observed spatial patterns in S deposition. Deposition and stream water sulfate concentration were both highest in the western and southern sections of the park.

6.3.2.2 Past Trends

Over the past 25 years, S deposition to acid-sensitive watersheds in the Northeast Temperate Network has decreased by about half. The pH of rainfall has increased by about a fourth of a pH unit, and N deposition has decreased slightly.

The EPA's Long Term Monitoring Program has been collecting monitoring data since the early 1980s for many lakes and streams in acid-sensitive areas of the United States, including within the Northeast Temperate Network region. These data allow evaluation of trends and variability in key components of lake and stream water chemistry prior to, during, and subsequent to Clean Air Act Amendment Title IV implementation. Throughout the Northeast Temperate Network, the concentration of sulfate in surface waters has decreased substantially, often by a third or more, subsequent to the Clean Air Act Amendments. Monitoring data from the Long Term Monitoring and the Temporally Integrated Monitoring of Ecosystems projects indicated that most regions included in the monitoring efforts showed large declines in sulfate concentrations in surface waters over the period of monitoring (1990s) analyzed by Stoddard et al. (2003), with rates of change for individual lakes ranging from about –1.5 to –3 µeq/L/yr. These declines in lake and stream sulfate concentrations were considered consistent with observed declines in S wet deposition and were corroborated by other studies that showed that sulfate concentrations in Adirondack lakes have decreased steadily since at least 1978 (e.g., Driscoll et al. 1995, Stoddard et al. 2003).

Trend analysis results for the period 1982–1994 were reported by Stoddard et al. (1998) for 36 lakes in the northeastern United States having acid neutralizing capacity ≤100 µeq/L. Trend statistics among sites were combined through a meta-analytical technique to determine whether the combined results from multiple sites had more significance than the individual Seasonal Kendall Test statistics. All lakes showed significant declining trends in sulfate concentration (change in sulfate = –1.7 µeq/L/yr; $p \leq 0.001$). Lake water acid neutralizing capacity responses were regionally variable. Lakes in New England showed evidence of acid neutralizing capacity recovery (change in acid neutralizing capacity = 0.8 µeq/L/yr; $p \leq 0.001$).

There were no significant changes in the acid neutralizing capacity or sulfate concentration in streams in ACAD during the period 1982–1990 (Heath et al. 1993). During the period 1990–2000, however, the concentrations of sulfate and base cations in lakes in ACAD decreased by significant amounts (10% decrease for sulfate), but the pH did not increase (Kahl et al. 2004).

The observed changes in the concentration of nitrate in some surface waters in the Northeast Temperate Network region have likely been due to a variety of factors, including changes in N deposition and climate. During the 1980s, nitrate concentration increased in many surface waters (Driscoll and Van Dreason 1993, Murdoch and Stoddard 1993). There was concern that some forests were becoming N-saturated, leading to increased nitrate leaching from forest soils throughout the region. Such a response could partially negate the benefits of decreased sulfate concentrations in lake and stream waters. However, this trend was reversed in about 1990, and the reversal could not be attributed to a change in N deposition. Nitrate leaching through soils to drainage waters is the result of a complex set of biological and hydrological processes. Key components include N uptake by plants and microbes, transformations between the various forms of inorganic and organic N, and local precipitation patterns. Most of the major processes are influenced by climatic factors, including temperature, moisture, and snowpack development. Therefore, nitrate concentrations in surface waters respond to many factors in addition to N deposition and can be difficult to predict. It is likely that monitoring programs of several decades or longer will be needed to separate trends in nitrate leaching from climatic variability in forested watersheds in the Northeast Temperate Network.

Lake water sulfate concentrations in the most acid-sensitive Maine lakes declined by about 12%–22% during the period 1982–1998 (Kahl 1999). Only in the seepage lakes, however, was there evidence of a small decline in lake water acidity during that period. The high-elevation lakes in Maine showed small declines in lake water acidity during the 1980s, but that trend slowed or reversed in the 1990s (Kahl 1999). Whereas nitrate concentrations decreased during the 1990s in many lakes (cf., Stoddard et al. 2003), the high-elevation lakes in Maine continued to show relatively high nitrate concentrations. Both seepage and high-elevation drainage lakes in Maine showed increased dissolved organic

C concentrations of 10%–20%, generally about 0.5–1.0 mg/L. The increase in dissolved organic matter would be expected to limit the extent of acid neutralizing capacity and pH recovery that would otherwise accompany the observed decreases in sulfate concentration. This has been attributed to both decreased S deposition and climate change (Saros 2014).

The Northeast Temperate Network monitors water quality in nine parks, including ACAD. Data are summarized at https://irma.nps.gov/App/Portal/Home. Sampling is conducted annually from May through October at multiple sites in each park (Gawley et al. 2014). Surface water acid neutralizing capacity is measured at all sites. Nearly all acid neutralizing capacity measurements in ACAD from 2006 to 2013 were <100 μeq/L, suggesting some sensitivity to acidification. Although upwind acidifying emissions have been reduced since the 1990 Clean Air Act Amendments, there has been little corresponding change in the acid neutralizing capacity or pH of surface waters in ACAD (Bank et al. 2006).

6.3.2.3 Critical Loads for Acidification

The Conference of the New England Governors and Eastern Canadian Provinces sponsored the development of steady-state critical loads for the protection of forest soils and lakes against acidification in the Northeast Temperate Network region and in eastern Canada (Ouimet et al. 2001, 2006, Dupont et al. 2005, Miller 2006). Dupont et al. (2005) used the Steady-State Water Chemistry aquatic critical loads model to calculate critical loads of acidity and associated exceedances for lakes in the northeastern United States and eastern Canada. Atmospheric acid loads were assessed based on atmospheric deposition of both S and N, using a critical limit of pH = 6 to protect aquatic biota. This pH level approximately corresponds with acid neutralizing capacity = 40 μeq/L in this region (Small and Sutton 1986). Lakes having lowest calculated critical loads included many in eastern and northern Maine. Exceedances of critical loads, based on estimated acidic deposition in 2002, were high in portions of Maine. Eastern Maine was a notable "hot spot" where ambient S + N deposition exceeded critical loads by more than 10 meq/m^2/yr (Dupont et al. 2005).

Data from over 12,500 streams and lakes were used by the Critical Loads of Atmospheric Deposition Science Committee of the National Atmospheric Deposition Program (http://nadp.sws.uiuc.edu/committees/clad/) to develop steady-state critical loads for acidity of surface waters based on multiple approaches for estimating base cation weathering. Water quality data were obtained from a variety of sources, including EPA's Long Term Monitoring sites, lake surveys, Environmental Monitoring and Assessment Program, and National Stream Surveys; U.S. Geological Survey; National Park Service Vital Signs program; and U.S. Forest Service air program. The average water quality measurements from the most recent five years of data were used for sites with long-term water monitoring records. The Critical Loads of Atmospheric Deposition Science Committee database included 85 sites in ACAD. Some streams in ACAD had modeled critical load near zero, suggesting limited ability to neutralize additional S deposition acidity.

6.3.2.4 Temporal Variability in Water Chemistry

Episodic acidification of stream water by nitrate has been documented in ACAD by Kahl et al. (1992). Stream pH values in the park have declined during episodes to values as low as 4.7 (Kahl et al. 1992, Heath et al. 1993). The mechanisms that produce acidic episodes can include dilution of base cations and flushing of nitrate, sulfate, chloride, and/or organic acids from forest soils to drainage water (Sullivan et al. 1986, Kahl et al. 1992, Wigington et al. 1996, Wigington 1999, Lawrence 2002). Acidic deposition can contribute to episodic acidification of surface water by supplying N, which can produce pulses of nitrate during high flow periods, contributing hydrologically mobile sulfate, and by lowering baseline pH and acid neutralizing capacity, so that episodes are sufficient to produce short-term biologically harmful conditions (Stoddard et al. 2003).

Decreases in pH with increases in flow have been well documented in the Northeast Temperate Network region. The most severe acidification of surface waters in this network generally occurs during spring snowmelt (Charles 1991). Stoddard et al. (2003) found that on average, spring acid neutralizing capacity values in New England, the Adirondacks, and the Northern Appalachian Plateau were about 30 µeq/L lower than summer values during the period 1990–2000. This implies that lakes and streams in these regions would need to recover to chronic acid neutralizing capacity values above about 30 µeq/L before they could be expected to not experience acidic episodes (Stoddard et al. 2003).

The transient nature of high flows and remote location of some sensitive waters makes episodic acidification difficult to measure in the Northeast Temperate Network. Therefore, assessments have generally estimated the number of lakes and streams prone to episodic acidification by combining episode information from a few sites with base flow values of acid neutralizing capacity determined in large surveys or modeling studies (Eshleman et al. 1995, Bulger et al. 2000, Driscoll et al. 2001). Inclusion of episodically acidified water bodies in regional assessments substantially increases estimates of the extent of surface water acidification.

There is evidence of episodic acidification of headwater streams in ACAD to pH < 5.0, likely in part a result of marine salt and fog deposition to thin, acidic soils (Heath et al. 1992). Salt inputs at ACAD can occur due to marine aerosol deposition and/or road salt application. Vaux et al. (2008) reported higher salt concentrations at sites below roads, as compared with above roads, in six watersheds in ACAD.

A study by Heath et al. (1992) attributed episodic acidification of low-order streams in ACAD primarily to wet and dry atmospheric deposition of sea salts and stream water dilution, rather than acidic deposition of S and N. Short-term pH depressions of up to 2 pH units and decreases in acid neutralizing capacity of up to 130 µeq/L were noted. This appears to have been the first documented example of the "neutral salt" acidification effect in North America. This effect requires acidic soils and inputs of sodium chloride or magnesium chloride that facilitate ion exchange of Na or Mg for H in soil solution. Storm events and resulting acid episodes occur in ACAD most often in the spring and fall (Bank et al. 2006).

6.4 Nutrient Nitrogen Enrichment

A study by Sievering et al. (2000) in central Maine investigated the effects of atmospheric N deposition on forest C storage. Net N uptake by the spruce-fir forest canopy was in the range of 1–5 kg N/ha/yr, while recycled root N uptake was between 10 and 30 kg N/ha/yr. Because N availability is usually limiting for photosynthesis, higher N uptake by the canopy can stimulate enhanced C storage in these and other eastern conifer forests (Sievering et al. 2000).

Nitrogen deposition increases the potential for nutrient enrichment to temperate forest ecosystems in the Northeast Temperate Network region because the growth of trees in these systems is often N-limited (Vitousek and Howarth 1991). Atmospheric deposition of N has decreased C:N ratios in soils and contributed to an increase in net nitrification and associated production of acidity in soils.

Thomas et al. (2010) analyzed Forest Inventory Analysis data in the Northeast Temperate Network region to determine tree growth enhancement across a gradient of atmospheric N deposition from about 3 to 11 kg N/ha/yr. Some tree species showed increased growth across the N input gradient (yellow poplar, black cherry, white ash). Some showed highest growth at intermediate levels of N deposition (quaking aspen and scarlet oak [*Quercus coccinea*]). Red pine (*Pinus resinosa*) exhibited growth decline across the gradient of increasing N deposition (Thomas et al. 2010). Thus, N deposition at ambient levels can have both positive and negative effects on tree growth, depending in part on species and deposition level.

Baron et al. (2011) estimated that 34% of 4361 New England lakes represented in the Eastern Lakes Survey were likely N-limited, based on having dissolved inorganic N:total P ratio (by weight) less than 4. In an analysis of data collected during the mid- to late 1990s from lakes and streams throughout the northeastern United States, Aber et al. (2003) suggested that nearly all N deposition is retained or denitrified in northeastern watersheds that receive less than about 8–10 kg N/ha/yr. An analysis of N deposition to forest land in the northeastern United States suggested that approximately 36% of the forests in the region received 8 kg N/ha/yr or more and may therefore be susceptible to elevated nitrate leaching (Driscoll et al. 2003).

Most lakes and ponds in and near ACAD are nutrient poor (Kahl et al. 2000). Total P is positively correlated with chlorophyll concentration in Maine lakes and also within the subset that occurs in ACAD, suggesting that P is generally limiting to lakes (Vaux et al. 2008). Although eutrophication is a concern in many Maine lakes (Nieratko 1992), there is little evidence that this is occurring to any appreciable extent in ACAD (Vaux et al. 2008).

Researchers at the University of Maine and the U.S. Geological Survey conducted a paired watershed study in ACAD to investigate the effects on N retention of the fires that burned about a third of the park in 1947. The unburned study watershed exported 10–20 times more inorganic N than the burned watershed (Nelson et al. 2007). Retention of inorganic N was 96% in the burned watershed compared with 72% in the unburned watershed (Campbell et al. 2004).

Nutrient loading to coastal ecosystems in the Northeast Temperate Network is an important concern (Nielsen and Kahl 2007), and the issue of nutrient loading to coastal waters was identified by ACAD as a high-priority resource management issue for the park (Kahl et al. 2000). Many estuaries in this network region have become increasingly eutrophic, largely in response to N loading from nonpoint pollution sources (Howes et al. 1996, Nixon 1996, Kinney and Roman 1998). Data from ACAD are important in this context because there have been few studies of coastal New England streams that have little human influence other than atmospheric input to serve as benchmarks (Nielsen and Kahl 2007). Watershed export of N and P to coastal waters around Mt. Desert Island in Maine, in and around ACAD, may be greatly affected by land-use history and human influence. Total N and total P exported by watersheds entirely within ACAD were significantly lower than exports by watersheds that were partly or completely outside of the park (Nielsen and Kahl 2007).

Pardo et al. (2011) compiled data on empirical critical load for protecting sensitive resources in Level I ecoregions across the conterminous United States against nutrient enrichment effects caused by atmospheric N deposition. Available data on empirical critical load of nutrient N in the Northeast Temperate Network suggested that the lower end of estimates of the critical load for resource protection was largely between about 3 and 8 kg N/ha/yr. This level of loading that was considered to be potentially problematic pertained to the protection of mycorrhizal fungi, lichens, and forest vegetation and also prevention of nitrate leaching into drainage water. Pardo et al. (2011) estimated that ambient N deposition was consistently higher than these critical load values. Thus, these empirical critical load data suggest the possibility of widespread exceedance of nutrient N critical load within parks in the Northeast Temperate Network.

6.5 Ozone Injury to Vegetation

Ozone levels are high throughout much of the Northeast. More than a dozen ozone-sensitive plant species are known or thought to occur within ACAD. In the past, polluted air masses from the industrialized northeastern corridor of the United States created elevated ozone exposure conditions in the Northeast Temperate Network during the growing season that could potentially damage sensitive plant species. Forests in ACAD have experienced periodic episodes of ozone above 80 ppb by volume throughout the summer months, and foliar symptoms were observed in several plant species in the 1990s (Bartholomay et al. 1997). Ozone has been shown to reduce photosynthesis in white pine, even in the absence of visible foliar injury (Spence et al. 1990, Kozlowski et al. 1991). Bartholomay et al. (1997) used dendroclimatic techniques to investigate the relationship between ozone exposure in ACAD and white pine growth rates. Regression analysis indicated that ozone exposure explained more variation in tree growth than did climate. This may have been because elevated ozone concentrations can disrupt photosynthetic rates and reduce carbohydrate availability (Bartholomay et al. 1997).

From 1983 to 2005, ozone levels at the McFarland Hill monitoring site in ACAD exceeded the previous standard of 80 ppb (based on an 8-hour average; Vaux et al. 2008) one or more times each year. An area is found to be in exceedance of the standard if the 8-hour average is surpassed four times during the year. In 2004, the EPA designated Hancock County, Maine, where ACAD is located, as a nonattainment area for ozone. Between 1995 and 2005, the mean annual number of 8-hour ozone exceedances was 4.6 at the Cadillac Mountain monitoring site (Vaux et al. 2008). Because of various air quality improvement programs stemming from the Clean Air Act and its amendments, ozone levels decreased at ACAD and elsewhere in the Northeast. Hancock County was redesignated as a Maintenance Area in 2007. Since that time, ozone levels have continued to decrease (NPS-ARD 2013) and Hancock County is more recently in compliance with the more stringent 2015 ozone standard (70 ppb).

The W126 and SUM06 exposure indices calculated by National Park Service staff for ACAD parks are given in Table 6.2, along with Kohut's (2007) ozone risk ranking. Ozone condition, as rated by National Park Service, was low in ACAD. This park was ranked as having moderate risk by Kohut. The results of both ranking systems should be considered when evaluating the potential for ozone injury to park vegetation.

In the eastern United States, ground-level ozone formation is controlled to a greater extent by N oxide emissions than by volatile organic compounds (NRC 1992, Ryerson et al. 2001). Indirect effects of N oxide emissions through ground-level ozone production

TABLE 6.2

Ozone Assessment Results for ACAD Based on Estimated Average 3-Month W126 and SUM06 Ozone Exposure Indices for the Period 2005–2009 and Kohut's (2007) Ozone Risk Ranking for the Period 1995–1999[a,b]

W126		SUM06		Kohut Risk to Vegetation
Value (ppm-h)	NPS Condition	Value (ppm-h)	NPS Condition	
4.40	Low	4.78	Low	Moderate

[a] Parks are classified into one of three ranks (Low, Moderate, High) based on comparison with other Inventory and Monitoring parks.

[b] Degrees of concern for the W126 and SUM06 indices are based solely on levels of ozone exposure. Kohut's risk to vegetation is based on several factors that contribute to injury in plants, including ozone exposure and environmental variables, and considers the effects of soil moisture on the uptake of ozone.

include reduction in net photosynthetic capacity (Reich 1987) and associated changes in biomass production and C allocation (Laurence et al. 1994). Ozone-related decreases in aboveground forest growth in the Northeast Temperate Network in the 1990s appeared to be in the range of negligible to 10% per year, based on analyses reported by Chappelka and Samuelson (1998). In more recent years, decreases in ozone exposure levels would be expected to alleviate these previous growth effects.

6.6 Visibility Degradation

6.6.1 Natural Background and Ambient Visibility Conditions

Table 6.3 gives the park haze rankings for ACAD on the 20% clearest, 20% haziest, and average days. The relative ranking for measured ambient haze for the period 2004–2008 was high for all groups. Ambient haze was substantially higher (about 3.5 deciviews on the clearest days to 10 deciviews on the haziest days) than natural haze values (Table 6.3).

TABLE 6.3

Estimated Natural Haze and Measured Ambient Haze in ACAD Averaged over the Period 2004–2008[a]

Estimated Natural Haze (Deciviews)		
20% Clearest Days	20% Haziest Days	Average Days
4.66	12.43	7.88

Measured Ambient Haze (for Years 2004–2008)					
20% Clearest Days		20% Haziest Days		Average Days	
dv	Ranking	dv	Ranking	dv	Ranking
8.06	High	22.06	High	14.01	High

[a] Parks are classified into one of five haze ranks (Very Low, Low, Moderate, High, or Very High) based on comparison with other monitored parks.

Interagency Monitoring of Protected Visual Environments (IMPROVE) data allow estimation of visual range. Data from ACAD indicate that air pollution has reduced average visual range in the park from 100 to 50 mi (161 to 80 km). On the haziest days, visual range has been reduced from 70 to 20 mi (113 to 24 km). Severe haze episodes occasionally reduce visibility to 6 mi (10 km).

6.6.2 Composition of Haze

Sulfate is the most important cause of fine particle pollution and visibility impairment across the Mid-Atlantic/Northeast Visibility Union states, including within the Northeast Temperate Network region. Sulfate impairment is exacerbated by high humidity. At the Class I sites in this region, sulfate accounts for about one-half to two-thirds of total fine particle mass on the 20% haziest days (Northeast States for Coordinated Air Use Management [NESCAUM] 2006). On the 20% clearest days, sulfate typically accounts for 40% or more of total fine particle mass. Regional sulfate emissions generally control visibility in the Mid-Atlantic/Northeast Visibility Union region during summer. During winter, visibility depends more on both regional and local S sources and also on local meteorological conditions (including temperature inversions). Organic C is, after sulfate, the next most important contributor to regional haze within the Mid-Atlantic/Northeast Visibility Union region (NESCAUM 2006).

The majority of total particulate light extinction in ACAD was attributable to sulfate and organics (Figure 6.5). On average, sulfate contributed 62.3% of total particulate light extinction, and organics contributed 14.4%. On the 20% haziest days, the contribution of sulfate increased to 69.2% of total particulate light extinction, and organics decreased to 12.4%. On the clearest 20% visibility days, the contribution of sulfate decreased to 51.3%, and organics increased to 19.6% of total particulate light extinction.

Visibility at ACAD varies with wind direction. Days having clearest visibility tend to be those when air masses derive from the north. Haziest visibility occurs on days when air masses derive from the south and southwest (Vaux et al. 2008), where many of the largest nearby emissions sources are located.

6.6.3 Trends in Visibility

The National Park Service (2010) reported long-term trends in annual haze on the clearest and haziest 20% of days at monitoring sites in 29 national parks. All 27 parks that showed statistically significant ($p \leq 0.05$) trends on the 20% clearest days for the 11–20 year monitoring periods through 2008 exhibited decreases in haze over time. None of the sites showed increasing trends on the clearest days. The steepest declines (−0.18 to −0.20 deciviews per year) on the clearest days were reported for Shenandoah National Park, ACAD, and Washington, DC, with 18–19 years of monitoring data at each of those locations. Ten parks showed statistically significant decreases in deciviews on the haziest days, with the steepest declines reported for Mount Rainier National Park (−0.38 deciviews/yr), Shenandoah National Park (−0.27 deciviews/yr), ACAD (−0.26 deciviews/yr), and Washington, DC (−0.25 deciviews/yr), with 18–19 years of data for each of those locations.

Annual mean haze levels on the haziest days at 47 monitored park locations during the period 2006–2008 ranged from 1.5 to 21 deciviews higher than the estimated natural condition (NPS 2010). The average difference between measured haze and estimated natural condition was 8.3 deciviews. Several eastern parks, including ACAD, had annual

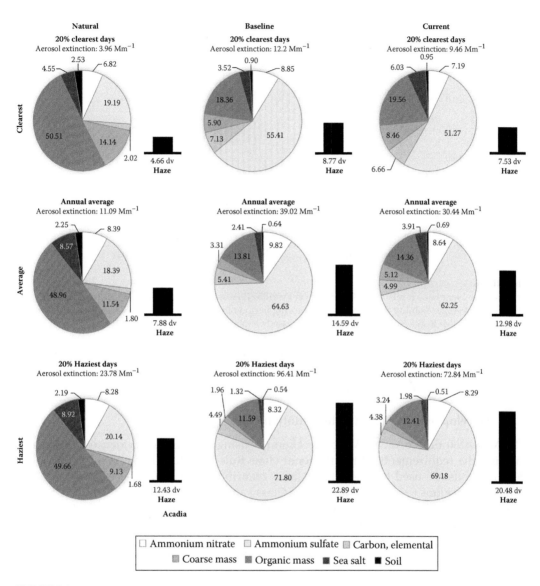

FIGURE 6.5
Estimated natural (preindustrial), baseline (2000–2004), and current (2006–2010) levels of haze (columns) and its composition (pie charts) on the 20% clearest, annual average, and 20% haziest visibility days for ACAD. (From http://views.cira.colostate.edu/fed/Tools/RegionalHazeSummary.aspx, accessed October, 2012.)

mean haze levels on the haziest days that were substantially higher (more than about 10 deciviews) than estimated natural conditions.

Within the northeastern United States, visibility improvement during recent years has been most pronounced in the western and southern portions of the Mid-Atlantic/ Northeast Visibility Union region, in and near the Northeast Temperate Network. The area of greatest visibility improvement has been in proximity to large power plant sources of sulfur dioxide emissions in the Ohio River and Tennessee Valleys (NESCAUM 2006).

Visibility at ACAD on the 20% clearest, 20% haziest, and average days has been improving throughout the period of monitoring record since 1990 (Figure 6.6). The steepest declines

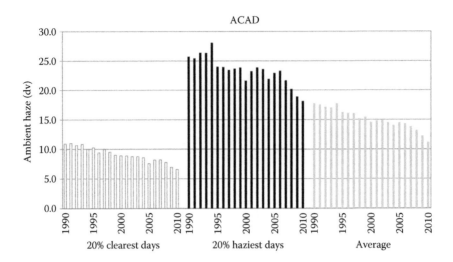

FIGURE 6.6
Trends in ambient haze levels at ACAD, based on IMPROVE measurements on the 20% clearest, 20% haziest, and annual average visibility days over the monitoring period of record. (From http://vista.cira.colostate.edu/improve/Data/IMPROVE/summary_data.htm, accessed October, 2012.)

in ambient haze at ACAD have occurred on the 20% haziest days (~8 deciviews improvement). Progress has also been substantial at this park on the 20% clearest (~5 deciviews) and average days (~7 deciviews).

6.6.4 Development of State Implementation Plans

The Mid-Atlantic/Northeast Visibility Union is using a weight of evidence approach to meeting the requirements of the Regional Haze Rule to reduce haze. Multiple independent methods are used to assess the relative contribution of different emissions sources and source regions to regional haze in the Class I areas within and near the Mid-Atlantic/Northeast Visibility Union region (NESCAUM 2006). These approaches include application of Eulerian (grid-based) source models, Lagrangian (air parcel–based) source dispersion models, back trajectory calculations, and analysis of monitoring data.

A variety of model estimates suggested that emissions from the Mid-Atlantic/Northeast Visibility Union states account for about 25%–30% of the fine particulate sulfate observed in ACAD and other Class I sites within the Mid-Atlantic/Northeast Visibility Union region. The Midwest Regional Planning Organization and the Visibility Improvement State and Tribal Association of the Southeast states each accounted for another approximately 15% of total fine particulate sulfate at ACAD and about 25% each at other Class I areas within the Mid-Atlantic/Northeast Visibility Union region (NESCAUM 2006). Although the Union is focusing primarily on fine particle sulfate abatement in its initial efforts toward compliance with the Regional Haze Rule (NESCAUM 2006), controls on other haze-producing constituents, such as organic C and both urban and mobile sources of N oxide during winter, will also be important.

Progress to date in meeting the national visibility goal is illustrated in Figure 6.7 using a uniform rate of progress glideslope. Improvements in visibility on the 20% haziest days at all of the monitored parks in the Northeast Temperate Network have so far been sufficient to comply with the glideslope requirements of the Regional Haze Rule. Additional monitoring will be required to assure continued compliance.

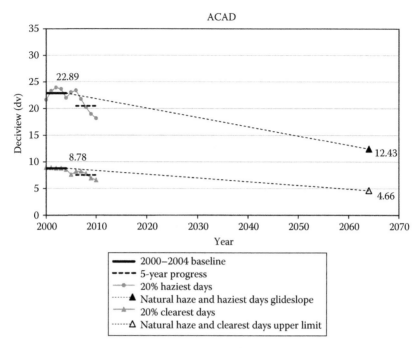

FIGURE 6.7

Glideslopes to achieving natural visibility conditions in 2064 for the 20% haziest (upper line) and the 20% clearest days (lower line) in ACAD. In the Regional Haze Rule, the clearest days do not have a uniform rate of progress glideslope; the rule only requires that the clearest days do not get any worse than the baseline period. Also shown are measured values during the period 2000–2010. (From http://vista.cira.colostate.edu/improve/Data/IMPROVE/summary_data.htm, accessed October, 2012.)

6.7 Toxic Airborne Contaminants

Estimates of Hg methylation potential generated by the U.S. Geological Survey (Last modified February 20, 2015) for watershed boundaries (based on eight-digit hydrologic unit codes) containing national parklands in the northern portion of the Northeast Temperate Network region suggested high methylation potential at many parks (Figure 6.8). Estimates were highest in and around ACAD. This result is likely driven mainly by relatively high concentrations of total organic C in surface waters in and near this park.

Much of the research in the United States that has been conducted on Hg methylation, the influence of sulfate on methylation rates, and controls on Hg transport within watersheds has been conducted in three park network areas, one of which is the Northeast Temperate Network (cf., Chen et al. 2005, Evers 2005, Evers et al. 2005, 2007, 2008, Kamman et al. 2005, Driscoll et al. 2007). Myers et al. (2007) proposed a national map showing the relative sensitivity of aquatic ecosystems to Hg contamination. Areas showing the highest estimated sensitivity included portions of the Northeast Temperate Network region. ACAD has been one area of substantial research focus on Hg biogeochemistry.

The ACAD region lies within a coastal zone of generally low levels of Hg in surface water compared with the rest of Maine (Peckenham et al. 2007). However, Mt. Desert Island represents a hot spot of locally high Hg within the regional context (Vaux et al. 2008).

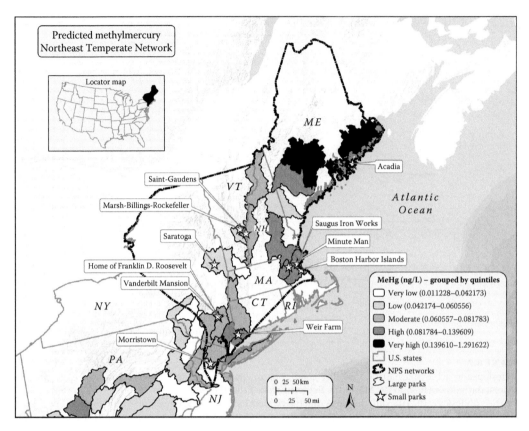

FIGURE 6.8
Predicted methylmercury (MeHg; in ng/L) concentrations in surface waters by hydrologic unit codes that contain national parklands in the Northeast Temperate Network. Estimates were generated by the U.S. Geological Survey (last modified February 20, 2015). Rankings are based on quintile distributions across all Inventory and Monitoring parks having estimates by the U.S. Geological Survey.

Total dissolved Hg was measured in streams across Mt. Desert Island by Peckenham et al. (2007). Highest concentrations were recorded in Squid Cove Brook, Oak Hill Stream, Hodgdon Brook, and Whalesback Brook. Hodgdon Brook is a tributary to Hodgdon Pond, where the highest Hg concentration in a Maine fish was found in 1995 (Vaux et al. 2008).

Research conducted in ACAD suggested that fire may play a role in Hg cycling within forested watersheds in the park (Amirbahman et al. 2004). Nelson et al. (2007) investigated the effects of fire on Hg and N dynamics at two watersheds in ACAD. The reference watershed, Hadlock Brook, had been largely undisturbed for the previous 300 years and was dominated by red spruce and balsam fir (*Abies balsamea*; Schauffler et al. 2007). In contrast, the Cadillac Brook watershed was the site of a 1947 stand-replacing wildfire that decreased soil organic matter. The burned watershed supports a heterogeneous mix of hardwood forest and softwood stands (Schauffler et al. 2007).

Analysis of stream water and precipitation samples collected from 1999 to 2000 showed that Hg transport was lower in the watershed recovering from fire damage (0.4 µg/m²/yr at Cadillac Brook versus 1.3 µg/m²/yr at Hadlock Brook). This was also true for dissolved inorganic N export, which was 11.5 eq/ha/yr from Cadillac Brook and 92.5 eq/ha/yr from Hadlock Brook (Nelson et al. 2007). Higher Hg deposition occurred in the

Hadlock Brook watershed, where mature conifer stands are more efficient at scavenging Hg from the atmosphere relative to the bare soil, exposed bedrock, and regenerating hardwood vegetation that occurs throughout much of the Cadillac Brook watershed (Nelson et al. 2007).

Peckenham et al. (2007) found that total Hg concentrations in stream water in ACAD were statistically correlated with both the amount of upstream wetland in the watershed and the total dissolved organic C in stream water. Mercury concentrations were highest in lower-elevation portions of streams, likely due to cumulative inputs from wetlands and riparian areas (Peckenham et al. 2007). This study also reported that headwater forested watersheds are more efficient in transporting Hg per unit dissolved organic C, which to some extent counterbalances the higher total dissolved organic C generated in wetlands.

The National Park Service sampled surface waters in ACAD to distinguish water quality effects from local organic and metal pollutant sources (in particular, traffic-related emissions) versus long-range atmospheric transport and deposition (Peckenham et al. 2006). Samples were collected across a gradient from immediately below high-use roads to remote areas of the park. Differences between busy roads and remote areas were used to estimate the importance of local sources of pollutants. They did not detect any polynuclear aromatic hydrocarbon or volatile organic compounds above 1 µg/L. Low concentrations of metals that are known to be associated with motor fuels were detected at all sample locations. The locations of sites showing the highest metal detection occurrences suggested a likely important contribution from vehicles to localized metal loading (Peckenham et al. 2006).

In the northeastern United States, high concentrations of Hg in yellow perch and common loon have been shown to be significantly correlated with water chemistry: total P < 30 µg/L, dissolved organic C > 4 mg/L, pH < 6.0, acid neutralizing capacity < 100 µeq/L (Chen et al. 2005, Driscoll et al. 2007). High Hg concentrations have also been shown to be correlated with several landscape characteristics: wetland abundance, low lake:watershed area ratio, high percent forest cover, and high atmospheric Hg deposition (Roué-LeGall et al. 2005, Driscoll et al. 2007). These characteristics affect Hg transport, methylation, and trophic transfer (St. Louis et al. 1996, Wiener et al. 2006, Driscoll et al. 2007, Turnquist et al. 2011).

Webber and Haines (2003) investigated the biological and behavioral effects of methyl Hg exposure on fish (golden shiner [*Notemigonus crysoleucas*]) collected from Brewer, Maine. Results indicated that, although growth rate and mortality did not differ with varying methyl Hg loads, fish with increased methyl Hg exposure displayed less efficient predator-avoidance behaviors. Webber and Haines (2003) concluded that Hg exposure at levels found in the Northeast Temperate Network region can alter fish predator avoidance behavior and may increase vulnerability to predation. This has implications regarding food chain transfer of Hg to piscivorous wildlife.

Elevated methyl Hg accumulation in fish-eating birds in the northeastern United States has been linked to lake acidification (Evers et al. 2007). Methylation is critical to the effects of Hg on aquatic biota; this form of Hg is toxic, bioavailable, and accumulates in top predators to levels of concern for both human health and the environment (Evers et al. 2007).

Methylation is correlated with multiple aspects of water acid–base chemistry (Wiener et al. 2006, Driscoll et al. 2007); methylating bacteria generally require sulfate to carry on their metabolic activities, and therefore, increased S deposition has been shown to increase rates of Hg methylation in freshwater wetlands (Galloway and Branfireun 2004, ICF

International 2006, Jeremiason et al. 2006). Wetlands act as important sources of methyl Hg to freshwater ecosystems. This is likely due in large part to two characteristics of wetlands: (1) high availability of dissolved organic C and (2) anaerobic conditions in sediments. Both enhance methylation rates. Mercury binds to organic matter. As a consequence, dissolved organic C is an important parameter affecting Hg bioavailability and transport through watersheds (Grigal 2002). The abundance of dissolved organic C enhances the transport of methyl Hg to downstream receiving waters. Many lakes, especially small lakes in the Northeast Temperate Network region, have extensive wetlands in their watersheds and contain relatively high concentrations of dissolved organic C (cf., Linthurst et al. 1986, Kretser et al. 1989). As a consequence of wetland influences on Hg methylation and transport, the percentage of wetland area within watersheds is commonly correlated with methyl Hg flux (Grigal 2002).

The concentration of Hg in fish tissue in the Northeast Temperate Network is often positively correlated with lake and/or watershed area and negatively correlated with pH, acid neutralizing capacity, and zooplankton density (Chen et al. 2005, Driscoll et al. 2007). Lake types that are generally associated with the most Hg bioaccumulation are poorly buffered, low in pH and productivity, and have forested watersheds and little human development within the watershed (Chen et al. 2005). The review of Evers (2005) on Hg pollution in the Northeast classifies Hg-sensitive surface waters as those having high sulfate concentrations, low pH and acid neutralizing capacity, extensive wetlands, large watershed area relative to lake area, fluctuating water level, and low nutrient concentration.

Several piscivorous bird and mammal species have been suggested as biomonitors of Hg bioaccumulation in the Northeast Temperate Network (Wolfe et al. 2007). In particular, the common loon and bald eagle are good indicators of Hg risk to wildlife (Evers 2006). Increased concentrations of Hg have been found to be associated with behavioral, physiological, and reproductive effects on these species (Burgess and Meyer 2008, Evers et al. 2008). Accumulation of methyl Hg in fish-eating birds can result in damage to nervous, excretory, and reproductive systems (Wolfe et al. 1998). Reproduction is considered one of the most sensitive endpoints for chronic low-level methyl Hg exposure of fish-eating birds (Wolfe et al. 1998). Reduced clutch size, increased number of eggs laid outside the nest, eggshell thinning, and increased embryo mortality have all been documented (Wolfe et al. 1998). Kramar et al. (2005) determined that the extent of wetland located in close proximity (less than 150 m [492 ft]) to loon territory was positively correlated with Hg concentrations in loon blood.

The lowest observed adverse effect level provides a benchmark for quantifying potential injury to wildlife from Hg exposure. For example, the lowest observed adverse effect level for the common loon has been established as 3.0 micrograms per gram (μg/g) of Hg in adult loon blood (Evers et al. 2007, 2008). This level of Hg is associated with reproductive effects such as reduced fledgling success (Burgess and Meyer 2008). However, loon Hg exposure data can include either sex; different ages, locations, and time periods; and can be derived from the analysis of blood, tissue, or eggs. It has therefore been difficult to standardize the data regarding methyl Hg exposure. Evers et al. (2011) developed linkages among loon Hg measurements in eggs, blood, and fish prey in the Great Lakes region, including the western portion of the Northeast Temperate Network region. Data were normalized into standard loon tissue units. The use of a standard unit of measure that combines multiple tissues of a high profile species and its principal prey items facilitates the examination of spatial gradients in pollution effects (Evers et al. 2011). Based on an analysis of over 8000 male loon units, seven biological Hg hot spots were identified in the region. The average male loon unit concentration across the region was 1.8 µg/g; 82% were above 1 µg/g and 9.8% were above the lowest observed adverse effect level of 3 µg/g.

Relatively little information is available on the effects of Hg on amphibians. However, Bank et al. (2005) reported concentrations of total and methyl Hg in several northern two-lined salamanders (*Eurycea bislineata bislineata*) collected from streams in ACAD, the Bear Brook watershed in Maine, and Shenandoah National Park. Total Hg levels in salamander larvae were significantly higher at sites in Maine, as compared with sites in Shenandoah National Park. At the Bear Brook watershed, total Hg levels in salamanders were higher in the ammonium sulfate treated subwatershed than in the reference subwatershed. Study results suggested that watershed features, including fire, ammonium sulfate addition, wetland extent, and forest cover type, affected Hg bioaccumulation in salamanders (Bank et al. 2005).

Four out of nine ponds in ACAD (44%) sampled by Bank et al. (2007a) had Hg methylation efficiencies of more than 10%, suggesting that palustrine food webs in ACAD are highly susceptible to bioaccumulation of methyl Hg. Total Hg in water was a strong indicator of both methyl Hg concentrations in the pond and total Hg in green frog (*Rana clamitans*) and bullfrog (*Rana catesbeiana*) tadpoles. One-third of sampled ponds had sediment Hg concentrations that approximated or exceeded the established effect thresholds for fresh-water sediments. Mercury toxicity is of particular concern to amphibians, which cannot shunt toxics away from vital organs to body parts such as feathers, hair, or carapaces (as in the case of birds, mammals, and reptiles, respectively; Bank et al. 2006). Mercury concentrations in the two-lined salamander in ACAD were considered elevated, and tadpoles collected from streams in ACAD had Hg body burdens approximately threefold higher than green frog and bullfrog tadpoles. Because the salamander tadpoles are predators, whereas frog tadpoles are grazers, differences in Hg burdens may be due to differences in diet (Bank et al. 2007a). This puts two-lined salamanders and other higher-trophic level predators in ACAD at potential risk due to Hg toxicity. Bank et al. (2006) suggested that the effects of toxic Hg and Al may be causes of the dramatic and perhaps irreversible decline in ACAD of the northern dusky salamander (*Desmognathus fuscus fuscus*), whose diet includes two-lined salamanders and other Hg-contaminated prey.

Davis (2013) investigated Hg methylation in vernal pools in Northeast Temperate Network parks. These pools provide important breeding habitat for amphibians. Environmental conditions in the pools, including anoxic conditions at the sediment/water interface, low pH, and high dissolved organic C, contribute to enhanced methylation. He found high methylation efficiency (mean 43%; maximum 58%) at the eight study sites, suggesting high bioavailability of Hg to vernal pool biota.

Evers et al. (2005) compiled a database of over 4700 records of avian Hg levels in the northeastern United States and eastern Canada. Using the belted kingfisher and bald eagle as indicators, they found increased Hg bioavailability from marine, to estuarine, to riverine systems. Bioavailability was highest in lakes. Differences in Hg exposure among species were correlated mainly with trophic position and availability of methyl Hg (Evers et al. 2005).

Studies of the transfer of Hg within food webs have mainly focused on freshwater aquatic ecosystems, which are considered to be at greatest risk of Hg biomagnification. Some studies have also been conducted on terrestrial upland ecosystems, including studies of Hg bioaccumulation in passerine birds. For example, Rimmer et al. (2005) documented methyl Hg availability in insectivorous passerines at 21 mountaintop locations in the northeastern United States. Mean blood Hg concentration at breeding site locations varied from 0.08 to 0.38 µg/g (wet weight) and was highest in southern portions of the study area. Older male

Bicknell's thrush (*Catharus bicknelli*) that breed in New England were judged to be at highest risk (Rimmer et al. 2005).

Hames et al. (2002) demonstrated a strong negative relationship between acidic deposition and wood thrush breeding, after accounting for several habitat variables. Breeding populations of wood thrush appeared to be most strongly affected in areas of low soil pH at high elevation where forest habitat was fragmented. It has been hypothesized that Hg is incorporated from leaf litter into invertebrates that feed on leaf tissue, and by insect predators, which are in turn preyed upon by song birds (Evers et al. 2009).

A study by Longcore et al. (2007b) compared Hg contamination in tree swallows in ACAD with tree swallows at an Hg-contaminated EPA Superfund site in Ayer, Massachusetts. They determined that methyl Hg concentrations in feathers of birds living at Aunt Betty Pond in ACAD were higher than in birds at the Superfund site. Eggs from both locations exceeded the embryotoxicity threshold for Hg. Swallows with high Hg burdens in their feathers may expose raptors and other high trophic-level predators to concentrated and potentially toxic levels of Hg. For example, peregrine falcons (*Falco peregrines*) nesting in ACAD may feed on Hg-contaminated songbirds (Longcore et al. 2007a). Additionally, the researchers suggested that Hg residues in tree swallow tissues in the northeastern United States were higher than those of birds from the Midwest, perhaps due to greater local sources of Hg, higher Hg atmospheric deposition rates, and more prevalent hydrological conditions that promote the conversion of inorganic Hg to methyl Hg (Longcore et al. 2007b).

Although the fauna in ACAD show elevated tissue Hg concentrations, the physiological and ecological implications of this Hg contamination are unclear (Vaux et al. 2008). Exposure of park fauna to Hg pollution likely represents a moderate to high risk (Bank et al. 2007b). Continued study of potential adverse impacts is needed. Identified priority research needs at ACAD include continuation of long-term air monitoring, research to establish critical loads, studies to identify biological and ecological endpoints associated with pollutant exposure, and process-based Hg experiments (Maniero and Breen 2004, Tonnessen and Manski 2007).

6.8 Summary

Substantial research on air pollution effects has been conducted in ACAD, but not in other parks within the Northeast Temperate Network. ACAD is designated as Class I, giving it a heightened level of protection against harm caused by poor air quality under the Clean Air Act. Steep slopes, thin soils, high mountains, wetlands, and exposure to coastal fog in ACAD increase the park's susceptibility to deposition of long-range atmospheric pollutants, including S, N, Hg, and other contaminants. Most of ACAD is underlain by granite. The relatively low buffering capacity of the overlying soils, together with areas having steep slopes, and in some areas also relatively high elevation, influence susceptibility of soil and drainage water to acidification.

Although upwind acidifying emissions have been reduced since passage of the 1990 Clean Air Act Amendments, there has been little corresponding change in the acid neutralizing capacity and pH of surface waters in ACAD. Some episodic acidification of headwater streams likely occurs in ACAD as a result of S, N, and marine salt and fog deposition to thin, acidic soils.

Nutrient enrichment in response to atmospheric N deposition is also an important research and management issue. In some areas that have moved toward a condition of N saturation, high levels of N deposition have contributed to elevated concentrations of nitrate in drainage waters. Elevated nitrate leaching causes depletion of base cations from forest soils, with adverse effects on sensitive tree species and acidification of drainage waters in watersheds that have base-poor soils. Several streams in ACAD have been shown to have chronically elevated nitrate concentrations (Johnson et al. 2007, Nelson et al. 2008), which suggests possible N saturation of some ACAD forests (Kahl et al. 2007b). Forests considered especially sensitive to N deposition include high-elevation red spruce and sugar maple forests. Effects of N deposition on root allocation or late-season growth may exacerbate other stresses from acidic deposition and harsh climate. Base cation depletion of soil may be associated with an increased likelihood of Al toxicity to plants and the decline of red spruce and/or sugar maple trees. The health of sugar maple trees is strongly influenced by the acid–base chemistry, especially the availability of Ca, in soil. Trees that grow on soils having low base cation supply are stressed and consequently can become more susceptible to damage from defoliating insects, drought, and extreme weather.

Ozone pollution can harm human health, reduce plant growth, and cause visible injury to foliage. Urban-derived plumes of ozone have been tracked from the Boston and New York City areas to remote parks, including ACAD. Ozone impacts to vegetation have been studied in ACAD, which has experienced periodic episodes of elevated ground-level ozone throughout the summer months. Foliar symptoms of ozone injury were observed in several plant species in this park, although ozone levels have declined in recent years.

Particulate pollution can cause haze, reducing visibility. Levels of ambient haze in ACAD were high for all groups (clearest, haziest, average) and were about 10 deciviews higher than natural haze on the 20% haziest days. Sulfate is the most important cause of fine particle pollution and visibility impairment across the Mid-Atlantic/Northeast Visibility Union states. At the Class I sites (including ACAD) in this region, sulfate accounts for about one-half to two-thirds of atmospheric total fine particle mass on the 20% haziest days. Although the Mid-Atlantic/Northeast Visibility Union is focusing primarily on fine particle sulfate abatement in its initial efforts toward compliance with the Regional Haze Rule, controls on other haze-producing constituents, such as organic C and both urban and mobile sources of N oxides during winter, will also be important.

Airborne contaminants, including Hg, can accumulate in food webs, reaching toxic levels in top predators. Substantial research has been conducted in ACAD on Hg methylation, the influence of sulfate on methylation rates, and controls on Hg transport within watersheds. Ecosystems in ACAD are especially sensitive to Hg bioaccumulation, due in part to relatively high Hg deposition and in particular due to watershed and lake characteristics that enhance Hg transport, methylation, and bioaccumulation.

In the northeastern United States, high concentrations of Hg in yellow perch and common loon have been shown to be significantly correlated with aspects of water chemistry, including total P, dissolved organic C, and acid neutralizing capacity. High Hg concentrations in fish have also been shown to be correlated with several landscape characteristics: wetland abundance, low lake:watershed area ratio, high percent forest cover, and high atmospheric Hg deposition. These characteristics affect Hg transport, methylation, and trophic transfer. Methylation is critical to the effects of Hg on aquatic biota and piscivorous wildlife; methyl Hg is toxic, bioavailable, and accumulates in top predators to levels of concern for both human health and the environment.

Accumulation of methylated Hg in fish-eating birds can result in damage to nervous, excretory, and reproductive systems. Reduced clutch size, increased number of eggs laid outside the nest, eggshell thinning, and increased embryo mortality have all been documented. Wetland and associated watershed food webs in ACAD are highly susceptible to bioaccumulation of methyl Hg. Effects of toxic Hg and Al are thought to be causes of the dramatic and perhaps irreversible decline in ACAD of the northern dusky salamander, whose diet includes Hg-contaminated prey.

References

Aber, J.D., C.L. Goodale, S.V. Ollinger, M.-L. Smith, A.H. Magill, M.E. Martin, R.A. Hallett, and J.L. Stoddard. 2003. Is nitrogen deposition altering the nitrogen status of northeastern forests? *BioScience* 53(4):375–389.

Amirbahman, A., P.L. Ruck, I.J. Fernandez, T.A. Haines, and J.S. Kahl. 2004. The effect of fire on mercury cycling in the soils of forested watersheds: Acadia National Park, Maine, U.S.A. *Water Air Soil Pollut.* 152:313–331.

Bailey, S.W., S.B. Horsley, R.P. Long, and R.A. Hallett. 2004. Influence of edaphic factors on sugar maple nutrition and health on the Allegheny Plateau. *Soil Sci. Soc. Am. J.* 68:243–252.

Bank, M.S., J.R. Burgess, D.C. Evers, and C.S. Loftin. 2007b. Mercury contamination of biota from Acadia National Park, Maine: A review. *Environ. Monitor. Assess.* 126:105–115.

Bank, M.S., J. Crocker, B. Connery, and A. Amirbahman. 2007a. Mercury bioaccumulation in green frog (*Rana clamitans*) and bullfrog (*Rana catesbeiana*) tadpoles from Acadia National Park, Maine, USA. *Environ. Toxicol. Chem.* 26(1):118–128.

Bank, M.S., J.B. Crocker, S. Davis, D.K. Brotherton, R. Cook, J. Behler, and B. Connery. 2006. Population decline of northern dusky salamanders at Acadia National Park, Maine, USA. *Biol. Conserv.* 130:230–238.

Bank, M.S., C.S. Loftin, and R.E. Jung. 2005. Mercury bioaccumulation in northern two-lined salamanders from streams in the northeastern United States. *Ecotoxicology* 14:181–191.

Baron, J.S., C.T. Driscoll, and J.L. Stoddard. 2011. Inland surface water. *In* L.H. Pardo, M.J. Robin-Abbott, and C.T. Driscoll (Eds.). *Assessment of Nitrogen Deposition Effects and Empirical Critical Loads of Nitrogen for Ecoregions of the United States.* General Technical Report NRS-80. U.S. Forest Service, Newtown Square, PA, pp. 209–228.

Bartholomay, G.A., R.T. Eckert, and K.T. Smith. 1997. Reductions in tree-ring widths of white pine following ozone exposure at Acadia National Park, Maine, U.S.A. *Can. J. For. Res.* 27:361–368.

Bulger, A.J., B.J. Cosby, and J.R. Webb. 2000. Current, reconstructed past, and projected future status of brook trout (*Salvelinus fontinalis*) streams in Virginia. *Can. J. Fish. Aquat. Sci.* 57(7):1515–1523.

Burgess, N.M. and M.W. Meyer. 2008. Methylmercury exposure associated with reduced productivity in common loons. *Ecotoxicology* 17(2):83–91.

Campbell, J.L., J.W. Hornbeck, M.J. Mitchell, M.B. Adams, M.S. Castro, C.T. Driscoll, J.S. Kahl et al. 2004. A synthesis of nitrogen budgets from forested watersheds in the northeastern United States. *Water Air Soil Pollut.* 151:373–396.

Chappelka, A.H. and L.J. Samuelson. 1998. Ambient ozone effects on forest trees of the eastern United States: A review. *New Phytol.* 139:91–108.

Charles, D.F. 1991. *Acidic Deposition and Aquatic Ecosystems: Regional Case Studies.* Springer-Verlag, New York.

Chen, C.Y., R.S. Stemberger, N.C. Kamman, B.M. Mayes, and C.L. Folt. 2005. Patterns of Hg bioaccumulation and transfer in aquatic food webs across multi-lake studies in the Northeast US. *Ecotoxicology* 14:135–147.

Craig, B.W. and A.J. Friedland. 1991. Spatial patterns in forest composition and standing dead red spruce in montane forests of the Adirondacks and northern Appalachians. *Environ. Monitor. Assess.* 18:129–140.

David, M.B. and G.B. Lawrence. 1996. Soil and soil solution chemistry under red spruce stands across the northeastern United States. *Soil Sci.* 161(5):314–328.

Davis, E. 2013. Seasonal changes in mercury stocks and methylation ratios in vernal pools in the northeastern United States. Masters, University of Vermont, Burlington, VT.

DeHayes, D.H., P.G. Schaberg, G.J. Hawley, and G.R. Strimbeck. 1999. Acid rain impacts on calcium nutrition and forest health. *BioScience* 49(1):789–800.

Driscoll, C.T., Y.-J. Han, C.Y. Chen, D.C. Evers, K.F. Lambert, T.M. Holsen, N.C. Kamman, and R.K. Munson. 2007. Mercury contamination in forest and freshwater ecosystems in the northeastern United States. *BioScience* 57(1):17–28.

Driscoll, C.T., G.B. Lawrence, A.J. Bulger, T.J. Butler, C.S. Cronan, C. Eagar, K.F. Lambert, G.E. Likens, J.L. Stoddard, and K.C. Weathers. 2001. Acidic deposition in the northeastern United States: Sources and inputs, ecosystem effects, and management strategies. *BioScience* 51(3):180–198.

Driscoll, C.T., K.M. Postek, W. Kretser, and D.J. Raynal. 1995. Long-term trends in the chemistry of precipitation and lake water in the Adirondack region of New York, USA. *Water Air Soil Pollut.* 85:583–588.

Driscoll, C.T. and R. Van Dreason. 1993. Seasonal and long-term temporal patterns in the chemistry of Adirondack lakes. *Water Air Soil Pollut.* 67:319–344.

Driscoll, C.T., D. Whitall, J. Aber, E. Boyer, M. Castro, C. Cronan, C. Goodale et al. 2003. Nitrogen pollution: Sources and consequences in the U.S. *Northeast Environ.* 45(7):9–22.

Dupont, J., T.A. Clair, C. Gagnon, D.S. Jeffries, J.S. Kahl, S.J. Nelson, and J.M. Peckenham. 2005. Estimation of critical loads of acidity for lakes in northeastern United States and eastern Canada. *Environ. Monitor. Assess.* 109:275–291.

Eshleman, K.N., T.D. Davies, M. Tranter, and P.J. Wigington, Jr. 1995. A two-component mixing model for predicting regional episodic acidification of surface waters during spring snowmelt periods. *Water Resour. Res.* 31(4):1011–1021.

Evers, D., M. Duron, D.E. Yates, and N. Schoch. 2009. An exploratory study of methylmercury availability in terrestrial wildlife of New York and Pennsylvania, 2005–2006. New York State Energy Research and Development Authority, Albany, NY.

Evers, D.C. 2005. *Mercury Connections: The Extent and Effects of Mercury Pollution in Northeastern North America.* BioDiversity Research Institute, Gorham, ME.

Evers, D.C. 2006. Loons as biosentinels of aquatic integrity. *Environ. Bioindicators* 1:18–21.

Evers, D.C., N.M. Burgess, L. Champoux, B. Hoskins, A. Major, W.M. Goodale, R.J. Taylor, and R. Poppenga. 2005. Patterns and interpretation of mercury exposure in freshwater avian communities in northeastern North America. *Ecotoxicology* 14:193–222.

Evers, D.C., Y. Han, C.T. Driscoll, N.C. Kamman, M.W. Goodale, K.F. Lambert, T.M. Holsen, C.Y. Chen, T.A. Clair, and T. Butler. 2007. Biological mercury hotspots in the northeastern United States and southeastern Canada. *BioScience* 57(1):29–43.

Evers, D.C., L.J. Savoy, C.R. DeSorbo, D.E. Yates, W. Hanson, K.M. Taylor, L.S. Siegel et al. 2008. Adverse effects from environmental mercury loads on breeding common loons. *Ecotoxicology* 17:69–81.

Evers, D.C., K.A. Williams, M.W. Meyer, A.M. Scheuhammer, N. Schoch, A. Gilbert, L. Siegel, R.J. Taylor, R. Poppenga, and C.R. Perkins. 2011. Spatial gradients of methylmercury for breeding common loons in the Laurentian Great Lakes region. *Ecotoxicology* 20:1609–1625.

Galloway, M.E. and B.A. Branfireun. 2004. Mercury dynamics of a temperate forested wetland. *Sci. Total Environ.* 325:239–254.

Gawley, W.G., B.R. Mitchell, and E.A. Arsenault. 2014. Northeast temperate network lakes, ponds, and streams monitoring protocol: 2014 revision. Natural Resource Report NPS/NETN/NRR—2014/770. National Park Service, Fort Collins, CO.

Grigal, D.F. 2002. Inputs and outputs of mercury from terrestrial watersheds: A review. *Environ. Rev.* 10:1–39.

Hames, R.S., K.V. Rosenberg, J.D. Lowe, S.E. Barker, and A.A. Dhondt. 2002. Adverse effects of acid rain on the distribution of the wood thrush *Hylocichla mustelina* in North America. *Proc. Natl. Acad. Sci. USA* 99:11235–11240.

Hawley, G.J., P.G. Schabery, C. Eager, and C.H. Borer. 2006. Calcium addition at the Hubbard Brook Experimental Forest reduced winter injury to red spruce in a high-injury year. *Can. J. For. Res.* 36:2544–2549.

Heath, R.H., J.S. Kahl, and S.A. Norton. 1992. Episodic stream acidification caused by atmospheric deposition of sea salts at Acadia National Park, Maine, United States. *Water Resour. Bull.* 28(4):1081–1088.

Heath, R.H., J.S. Kahl, S.A. Norton, and W.R. Brutsaert. 1993. Elemental mass balances and episodic and ten-year changes in the chemistry of surface water, Acadia National Park, Maine: Final report. Technical Report NPS/NAROSS/NRTR-93/16. National Park Service, North Atlantic Region, Boston, MA.

Howes, B.L., P.K. Wieskel, D.D. Goehringer, and J.M. Teal. 1996. Interception of freshwater and nitrogen transport from uplands to coastal waters: The role of saltmarshes. *In* K. Nordstrom and C.T. Roman (Eds.). *Estuarine Shores, Evolution, Environments, and Human Alterations.* John Wiley & Sons, New York, pp. 287–310.

ICF International. 2006. Mercury transport and fate through a watershed. synthesis report of research from EPA's Science to Achieve Results (STAR) grant program. Prepared for US EPA, Office of Research and Development, Washington, DC.

Jagels, R., J. Carlisle, R. Cunningham, S. Serreze, and P. Tsai. 1989. Impact of fog and ozone in coastal red spruce. *Water Air Soil Pollut.* 48:193–208.

Jeremiason, J.D., D.R. Engstrom, E.B. Swain, E.A. Nater, B.M. Johnson, J.E. Almendinger, B.A. Monson, and R.K. Kolka. 2006. Sulfate addition increases methylmercury production in an experimental wetland. *Environ. Sci. Technol.* 40(12):3800–3806.

Johnson, K.B., T.A. Haines, J.S. Kahl, S.A. Norton, A. Amirbahman, and K.D. Sheehan. 2007. Controls on mercury and methylmercury deposition for two watersheds in Acadia National Park, Maine. *Environ. Monitor. Assess.* 126:55–67.

Kahl, J.S., J.L. Andersen, and S.A. Norton. 1985. Water resource baseline data and assessment of impacts from acidic precipitation, Acadia National Park, Maine. Technical Report #16, National Park Service, North Atlantic Region Water Resources Program, Orono, ME. 123 pp.

Kahl, J.S., D. Manski, M. Flora, and N. Houtman. 2000. Water resources management plan. Acadia National Park, Mount Desert Island, Maine.

Kahl, J.S., S.J. Nelson, I. Fernandez, T. Haines, S. Norton, G.B. Wiersma, G. Jacobson, Jr. et al. 2007b. Watershed nitrogen and mercury geochemical fluxes integrate landscape factors in long-term research watersheds at Acadia National Park, Maine, USA. *Environ. Monit. Assess.* 126:9–25.

Kahl, J.S., S.J. Nelson, I.J. Fernandez, and K.C. Weathers. 2007a. Understanding atmospheric deposition to complex landscapes at Acadia National Park, Maine 2002–2005. Technical Report NPS/NER/NRTR-2007/080. National Park Service, Northeast Region, Boston, MA.

Kahl, J.S., S.A. Norton, T.A. Haines, E.A. Rochette, R.C. Heath, and S.C. Nodvin. 1992. Mechanisms of episodic acidification in low-order streams in Maine, USA. *Environ. Pollut.* 78:37–44.

Kahl, J.S., J.L. Stoddard, R. Haeuber, S.G. Paulsen, R. Birnbaum, F.A. Deviney, J.R. Webb et al. 2004. Have U.S. surface waters responded to the 1990 Clean Air Act Amendments? *Environ. Sci. Technol.* 38:485A–490A.

Kahl, S. 1999. Responses of Maine surface waters to the Clean Air Act Amendments of 1990. Final report. EPA project CX826563-01-0. Water Research Institute, University of Maine, Orono, ME.

Kamman, N.C., N.M. Burgess, C.T. Driscoll, H.A. Simonin, W. Goodale, J. Linehan, R. Estabrook et al. 2005. Mercury in freshwater fish of northeast North America—A geographic perspective based on fish tissue monitoring databases. *Ecotoxicology* 14:163–180.

Kinney, E.H. and C.T. Roman. 1998. Response of primary producers to nutrient enrichment in a shallow estuary. *Mar. Ecol. Prog. Ser.* 163:89–98.

Kohut, R. 2007. Assessing the risk of foliar injury from ozone on vegetation in parks in the U.S. National Park Service's Vital Signs Network. *Environ. Pollut.* 149:348–357.

Kozlowski, T.T., P.J. Kramer, and S.G. Pallardy. 1991. *The Physiological Ecology of Woody Plants.* Academic Press, New York.

Kramar, D., W.M. Goodale, L.M. Kennedy, L.W. Carstensen, and T. Kaur. 2005. Relating land cover characteristics and common loon mercury levels using geographic information systems. *Ecotoxicology* 14:253–262.

Kretser, W., J. Gallagher, and J. Nicolette. 1989. Adirondack lakes study, 1984–1987: An evaluation of fish communities and water chemistry. Adirondack Lakes Survey Corporation, Ray Brook, NY.

Laurence, J.A., R.G. Amundson, A.L. Friend, E.J. Pell, and P.J. Temple. 1994. Allocation of carbon in plants under stress: An analysis of the ROPIS experiments. *J. Environ. Qual.* 23:412–417.

Lawrence, G.B. 2002. Persistent episodic acidification of streams linked to acid rain effects on soil. *Atmos. Environ.* 36:1589–1598.

Lehmann, J., Z. Lan, C. Hyland, S. Sato, D. Solomon, and Q.M. Ketterings. 2005. Long-term dynamics of phosphorus forms and retention in manure-amended soils. *Environ. Sci. Technol.* 39:6672–6680.

Linthurst, R.A., D.H. Landers, J.M. Eilers, D.F. Brakke, W.S. Overton, E.P. Meier, and R.E. Crowe. 1986. Characteristics of lakes in the Eastern United States. Volume I: Population Descriptions and Physico-chemical Relationships. EPA/600/4-86/007a. U.S. Environmental Protection Agency, Washington, DC.

Longcore, J.R., R. Dineli, and T.A. Haines. 2007a. Mercury and growth of tree swallows at Acadia National Park, and at Orono, Maine, USA. *Environ. Monitor. Assess.* 126:117–127.

Longcore, J.R., T.A. Haines, and W.A. Halteman. 2007b. Mercury in tree swallow food, eggs, bodies, and feathers at Acadia National Park, Maine, and an EPA superfund site, Ayer, Massachusetts. *Environ. Monitor. Assess.* 126:129–143.

Lovett, G.M. and H. Rueth. 1999. Soil nitrogen transformation in beech and maple stands along a nitrogen deposition gradient. *Ecol. Appl.* 9(4):1330–1344.

Maniero, T. and B. Breen. 2004. Quality programmatic, monitoring, and research needs. Natural Resources Report NPS/NER/NRR-2004/002. National Park Service, Northeast Region, Boston, MA.

Miller, E., A. Vanarsdale, G. Keeler, A. Chalmers, L. Poissant, N. Kamman, and R. Brulotte. 2005. Estimation and mapping of wet and dry mercury deposition across northeastern North America. *Ecotoxicology* 14:53–70.

Miller, E.K. 2006. Assessment of forest sensitivity to nitrogen and sulfur deposition in Maine. Report to the Maine Department of Environmental Protection. Ecosystems Research Group, Ltd., Norwich, VT.

Miller, K.M., B.R. Mitchell, P.J. Curtin, and J.S. Wheeler. 2014. Forest health monitoring in Acadia National Park. Northeast Temperate Network 2006–2013 Summary Report. Natural Resource Report NPS/NETN/NRR-2014/777. U.S. Department of the Interior, National Park Service, Fort Collins, CO.

Murdoch, P.S. and J.L. Stoddard. 1993. Chemical characteristics and temporal trends in eight streams of the Catskill Mountains. *Water Air Soil Pollut.* 67:367–395.

Myers, M.D., M.A. Ayers, J.S. Baron, P.R. Beauchemin, K.T. Gallagher, M.B. Goldhaber, D.R. Hutchinson et al. 2007. USGS goals for the coming decade. *Science* 318:200–201.

National Park Service (NPS). 2010. Air quality in national parks: 2009 annual performance and progress report. Natural Resource Report NPS/NRPC/ARD/NRR-2010/266. National Park Service, Air Resources Division, Denver, CO.

National Park Service–Air Resources Division (NPS–ARD). 2013. Air quality in national parks: Annual trends (2000–2009) and conditions (2005–2009) report. Natural Resource Report NPS/NRSS/ARD/NRR—2013/683. National Park Service, Denver, CO.

National Research Council (NRC). 1992. *Rethinking the Ozone Problem in Urban and Regional Air Pollution.* National Academy Press, Washington, DC, 524pp.

Nelson, S.J., K.B. Johnson, J.S. Kahl, T.A. Haines, and I.J. Fernandez. 2007. Mass balances of mercury and nitrogen in burned and unburned forested watersheds at Acadia National Park, Maine, USA. *Environ. Monitor. Assess.* 126:69–80.

Nelson, S.J., K.B. Johnson, K.C. Weathers, C.S. Loftin, I.J. Fernandez, J.S. Kahl, and D.P. Krabbenhoft. 2008. A comparison of winter mercury accumulation at forested and no-canopy sites measured with different snow sampling techniques. *Appl. Geochem.* 23:384–398.

Nielsen, M.G. and J.S. Kahl. 2007. Nutrient export from watersheds on Mt. Desert Island, Maine, as a function of land use and fire history. *Environ. Monitor. Assess.* 126:81–96.

Nieratko, D.P. 1992. Factors controlling phosphorus loading to lakes in Maine: A statistical analysis. MS thesis, University of Maine, Orono, ME.

Nixon, S.W. 1996. The fate of nitrogen and phosphorus at the land–sea margin of the North Atlantic Ocean. *Biogeochemistry* 35:141–180.

Northeast States for Coordinated Air Use Management (NESCAUM). 2006. Contributions to regional haze in the northeast and Mid-Atlantic United States. Mid-Atlantic/Northeast Visibility Union (MANE-VU) Contribution Assessment. Report prepared by NESCAUM for the Mid-Atlantic/Northeast Visibility Union. Updated July, 2012.

Norton, S.A., G.C. Evans, and J.S. Kahl. 1997. Comparison of Hg and Pb fluxes to hummocks and hollows of ombrotrophic Big Heath bog and to nearby Sargent Mt. Pond, Maine, USA. *Water Air Soil Pollut.* 100:271.276.

Ouimet, R., P.A. Arp, S.A. Watmough, J. Aherne, and I. Demerchant. 2006. Determination and mapping critical loads of acidity and exceedances for upland forest soils in eastern Canada. *Water Air Soil Pollut.* 172:57–66.

Ouimet, R., L. Duchesne, D. Houle, and P.A. Arp. 2001. Critical loads of atmospheric S and N deposition and current exceedances for northern temperate and boreal forests in Quebec. *Water Air Soil Pollut.* 1:119–134.

Pardo, L.H., M.J. Robin-Abbott, and C.T. Driscoll. 2011. Assessment of nitrogen deposition effects and empirical critical loads of nitrogen for ecoregions of the United States. General Technical Report NRS-80. U.S. Forest Service, Newtown Square, PA.

Peckenham, J.M., J.S. Kahl, and A. Amirbahman. 2006. The impact of vehicle traffic on water quality in Acadia National Park. Technical Report NPS/NER/NRTR-2006/035. National Park Service, Northeast Region, Boston, MA.

Peckenham, J.M., J.S. Kahl, K.B. Johnson, and T.A. Haines. 2007. Landscape controls on mercury in streamwater at Acadia National Park, USA. *Environ. Monitor. Assess.* 126:97–104.

Reich, P.B. 1987. Quantifying plant response to ozone: A unifying theory. *Tree Physiol.* 3:63–91.

Rimmer, C.C., K.P. McFarland, D.C. Evers, E. Miller, K. Y. Aubry, D. Busby, and R.J. Taylor. 2005. Mercury concentrations in Bicknell's thrush and other insectivorous passerines in montane forests of northeastern North America. *Ecotoxicology* 14:223–240.

Roué-LeGall, A., M. Lucotte, J. Carreau, R. Canuel, and E. Garcia. 2005. Development of an ecosystem sensitivity model regarding mercury levels in fish using a preference modeling methodology: Application to the Canadian boreal system. *Environ. Sci. Technol.* 39(24):9412–9423.

Ryerson, T.B., M. Trainer, J.S. Holloway, D.D. Parrish, L.G. Huey, D.T. Sueper, G.J. Frost et al. 2001. Observation of ozone formation in power plant plumes and implications of ozone control strategies. *Science* 292(5517):719–723.

Saros, J.E. 2014. Determining critical nitrogen loads to boreal lake ecosystems. The response of phytoplankton at Acadia and Isle Royale National Parks. National Resources Technical Report NPS/ACAD/NRTR-2014/862. U.S. Department of the Interior, National Park Service, Fort Collins, CO.

Schauffler, M., S.J. Nelson, G.L. Jacobson, Jr., T.A. Haines, W.A. Patterson III, and K.B. Johnson. 2007. Paleoecological assessment of watershed history in PRIMENet watersheds at Acadia National Park. USA. *Environ. Monitor. Assess.* 126:39–53.

Schwede, D.B. and G.G. Lear. 2014. A novel hybrid approach for estimating total deposition in the United States. *Atmos. Environ.* 92:207–220.

Sheehan, K.D., I.J. Fernandez, J.S. Kahl, and A. Amirbahman. 2006. Litterfall mercury in two forested watersheds at Acadia National Park, Maine, USA. *Water Air Soil Pollut.* 170:249–265.

Shortle, W.C., K.T. Smith, R. Minocha, G.B. Lawrence, and M.B. David. 1997. Acidic deposition, cation mobilization, and biochemical indicators of stress in healthy red spruce. *J. Environ. Qual.* 26(3):871–876.

Sievering, H., I. Fernandez, J. Lee, J. Hom, and L. Rustad. 2000. Forest canopy uptake of atmospheric nitrogen deposition at eastern U.S. conifer sites: Carbon storage implications? *Glob. Biogeochem. Cycles* 14(4):1153–1159.

Small, M.J. and M.C. Sutton. 1986. A regional pH–alkalinity relationship. *Water Res.* 20:335–343.

Smith, C.M. and L.J. Trip. 2005. Mercury policy and science in northeastern North America: The Mercury Action Plan of the New England Governors and Eastern Canadian Premiers. *Ecotoxicology* 14:19–35.

Spence, R.D., E.J. Rykiel, and P.J.H. Sharpe. 1990. Ozone alters carbon allocation in loblolly pine: Assessment with carbon-11 labeling. *Environ. Pollut.* 64:93–106.

St. Louis, V.L., J.W.M. Rudd, C.A. Kelly, K.G. Beaty, R.J. Flett, and N.T. Roulet. 1996. Production and loss of methylmercury and loss of total mercury from boreal forest catchments containing different types of wetlands. *Environ. Sci. Technol.* 30(9):2719–2729.

Stoddard, J., J.S. Kahl, F.A. Deviney, D.R. DeWalle, C.T. Driscoll, A.T. Herlihy, J.H. Kellogg, P.S. Murdoch, J.R. Webb, and K.E. Webster. 2003. Response of surface water chemistry to the clean air act amendments of 1990. EPA 620/R-03/001. U.S. Environmental Protection Agency, Office of Research and Development, National Health and Environmental Effects Research Laboratory, Research Triangle Park, NC.

Stoddard, J.L., C.T. Driscoll, J.S. Kahl, and J.H. Kellogg. 1998. A regional analysis of lake acidification trends for the northeastern U.S., 1982–1994. *Environ. Monitor. Assess.* 51:399–413.

Sullivan, T.J., N. Christophersen, I.P. Muniz, H.M. Seip, and P.D. Sullivan. 1986. Aqueous aluminum chemistry response to episodic increases in discharge. *Nature* 323:324–327.

Sullivan, T.J., C.T. Driscoll, B.J. Cosby, I.J. Fernandez, A.T. Herlihy, J. Zhai, R. Stemberger et al. 2006b. Assessment of the extent to which intensively-studied lakes are representative of the Adirondack Mountain Region. Final report 06-17. New York State Energy Research and Development Authority, Albany, NY.

Sullivan, T.J., I.J. Fernandez, A.T. Herlihy, C.T. Driscoll, T.C. McDonnell, N.A. Nowicki, K.U. Snyder, and J.W. Sutherland. 2006a. Acid-base characteristics of soils in the Adirondack Mountains, New York. *Soil Sci. Soc. Am. J.* 70:141–152.

Sullivan, T.J., G.B. Lawrence, S.W. Bailey, T.C. McDonnell, C.M. Beier, K.C. Weathers, G.T. McPherson, and D.A. Bishop. 2013. Effects of acidic deposition and soil acidification on sugar maple in the Adirondack Mountains, New York. *Environ. Sci. Technol.* 47:12687–12694.

Sullivan, T.J., G.T. McPherson, T.C. McDonnell, S.D. Mackey, and D. Moore. 2011. Evaluation of the sensitivity of inventory and monitoring national parks to acidification effects from atmospheric sulfur and nitrogen deposition. Natural Resource Report NPS/NRPC/ARD/NRR—2011/349. U.S. Department of the Interior, National Park Service, Denver, CO.

Thomas, R.Q., C.D. Canham, K.C. Weathers, and C.L. Goodale. 2010. Increased tree carbon storage in response to nitrogen deposition in the US. *Nat. Geosci.* 3:13–17.

Tonnessen, K. and D. Manski. 2007. The contribution of Acadia PRIMENet research to science and resource management in the National Park Service. *Environ. Monitor. Assess.* 126:3–8.

Turnquist, M.A., C.T. Driscoll, K.L. Schulz, and M.A. Schlaepfer. 2011. Mercury concentrations in snapping turtles (*Chelydra serpentina*) correlate with environmental and landscape characteristics. *Ecotoxicology* 20:1599–1608.

U.S. Environmental Protection Agency. 2010. Clean water act section 319 (g) mercury conference: Mercury summary for northeast states. Available at: https://www.epa.gov/sites/production/files/2015-09/documents/mercury_summaries_northeast_states_06-18update.pdf (accessed January, 2013)

U.S. Geological Survey (USGS). Last modified February 20, 2015. Predicted surface water methylmercury concentrations in National Park Service Inventory and Monitoring Program Parks. U.S. Geological Survey. Wisconsin Water Science Center, Middleton, WI. Available at: http://wi.water.usgs.gov/mercury/NPSHgMap.html (accessed February 26, 2015).

Vaux, P.D., S.J. Nelson, N. Rajakarune, G. Mittelhauser, K. Bell, B.S. Kopp, J. Peckenham, and G. Longsworth. 2008. Assessment of natural resource conditions in and adjacent to Acadia National Park, Maine. National Park Service, Fort Collins, CO.

Vitousek, P.M. and R.W. Howarth. 1991. Nitrogen limitation on land and in the sea: How can it occur? *Biogeochemistry* 13:87–115.

Weathers, K.C., G.E. Likens, F.H. Bormann, S.H. Bicknell, B.T. Bormann, J.S. Eaton, J.N. Galloway et al. 1988b. Cloudwater chemistry from ten sites in North America. *Environ. Sci. Technol.* 22:1018–1026.

Weathers, K.C., G.E. Likens, F.H. Bormann, J.S. Eaton, K.D. Kimball, J.N. Galloway, T.G. Siccama, and D. Smiley. 1988a. Chemical concentrations in cloud water from four sites in the eastern United States. *In* M.H. Unsworth and D. Fowler (Eds.). *Acid Deposition at High Elevation Sites.* Kluwer Academic Publishers, Dordrecht, the Netherlands, pp. 345–357.

Weathers, K.C., G.M. Lovett, G.E. Likens, and R. Lathrop. 2000. The effect of landscape features on deposition to Hunter Mountain, Catskill Mountains, New York. *Ecol. Appl.* 10:528–540.

Weathers, K.C., S.M. Simkin, G.M. Lovett, and S.E. Lindberg. 2006. Empirical modeling of atmospheric deposition in mountainous landscapes. *Ecol. Appl.* 16(4):1590–1607.

Webber, H.M. and T.A. Haines. 2003. Mercury effects on predator avoidance behavior of a forage fish, golden shiner (*Notemigonus crysoleucas*). *Environ. Toxicol. Chem.* 22(7):1556–1561.

Wiener, J.G., B.C. Knights, M.B. Sandheinrich, J.D. Jeremiason, M.E. Brigham, D.R. Engstrom, L.G. Woodruff, W.F. Cannon, and S.J. Balogh. 2006. Mercury in soils, lakes, and fish in Voyageurs National Park (Minnesota): Importance of atmospheric deposition and ecosystem factors. *Environ. Sci. Technol.* 40(20):6261–6268.

Wigington, P.J. 1999. Episodic acidification: Causes, occurrence and significance to aquatic resources. *In* J.R. Drohan (Ed.). *The Effects of Acidic Deposition on Aquatic Ecosystems in Pennsylvania. 1998 PA Acidic Deposition Conference.* Environmental Resources Research Institute, University Park, PA, pp. 1–5.

Wigington, P.J., Jr., D.R. DeWalle, P.S. Murdoch, W.A. Kretser, H.A. Simonin, J. Van Sickle, and J.P. Baker. 1996. Episodic acidification of small streams in the northeastern United States: Ionic controls of episodes. *Ecol. Appl.* 6(2):389–407.

Wolfe, M.F., T. Atkeson, W. Bowerman, K. Burger, D.C. Evers, M.W. Murray, and E. Zillioux. 2007. Wildlife indicators. *In* R.C. Harris, D.P. Krabbenhoft, R.P. Mason, M.W. Murray, R.J. Reash, and T. Saltman (Eds.). *Ecosystem Responses to Mercury Contamination: Indicators of Change.* SETAC, CRC Press, Taylor & Francis Group, Boca Raton, FL, pp. 123–189.

Wolfe, M.F., S. Schwarzbach, and R.A. Sulaiman. 1998. Effects of mercury on wildlife: A comprehensive review. *Environ. Toxicol. Chem.* 17(2):146–160.

7

South Florida/Caribbean Network

7.1 Background

In the conterminous United States, the tropical and subtropical humid forest ecoregion occurs only in southern Florida. Biodiversity is very high in this ecoregion, which contains many species of plants, including epiphytes. The tropical humid forest zone of southern Florida is biologically one of the most diverse vegetation zones in the United States. Wetlands are widespread, both saltwater and freshwater, including mangrove swamps and tropical tree islands.

There are many human population centers in the range of 50,000–500,000 people in the vicinity of the South Florida/Caribbean Network, scattered along the south Florida coastline, but the Virgin Islands National Park (VIIS) is more remote. Figure 7.1 shows the network boundaries along with locations of the parks and the population centers having more than 10,000 people. Both Everglades National Park (EVER) and VIIS are Class I air quality areas. These two parks are the focus of this chapter.

7.2 Atmospheric Emissions and Deposition

Annual county-level S emissions in 2002 generally ranged from less than 1 ton of sulfur dioxide per square mile per year to 20 tons/mi^2/yr in the South Florida/Caribbean Network (Sullivan et al. 2011a). Point source emissions of sulfur dioxide were mostly sources that emitted less than 5000 tons of sulfur dioxide per year. There were a few larger sulfur dioxide point sources that emitted between 5,000 and 20,000 tons/yr.

In general, annual county N emissions in 2002 were between 1 and 20 tons/mi^2/yr throughout most of the network region, with higher and lower values in a few areas. There were many relatively large (larger than 2000 tons/yr) point sources of N. The largest point sources emitted oxidized N, although there were also some moderate size sources of ammonia emissions.

County-level emissions near the South Florida/Caribbean Network, based on data from the EPA's National Emissions Inventory during a recent time period (2011), are depicted in Figures 7.2 through 7.4 for sulfur dioxide, oxidized N, and ammonia, respectively. The counties near the South Florida/Caribbean Network parks had relatively low sulfur dioxide

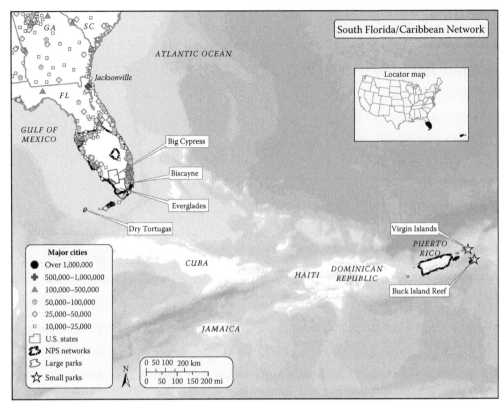

FIGURE 7.1
Network boundary and locations of national parks and human population centers near the South Florida/Caribbean Network.

emissions (<5 tons/mi^2/yr; Figure 7.2). Oxidized N emissions were generally higher, with highest values in the range of 5–50 tons/mi^2/yr (Figure 7.3). Emissions of ammonia near South Florida/Caribbean Network parks were lower, with most counties showing emissions levels below 2 tons/mi^2/yr (Figure 7.4).

Total estimated S deposition within the network was generally between 5 and 10 kg S/ha/yr in 2002 (Sullivan et al. 2011a). There were areas of lower and higher estimated S deposition, ranging from 2 to 5 kg S/ha/yr in the south to more than 30 kg S/ha/yr in the northwest. Total estimated N deposition in 2002 ranged from as low as 5 to 10 kg N/ha/yr to as high as 10 to 15 kg N/ha/yr across broad areas of the network. Smaller areas receiving both lower (2–5 kg/ha/yr) and higher (more than 15 kg/ha/yr) estimated atmospheric N deposition also occurred in limited portions of the network region.

Total S deposition in and around the South Florida/Caribbean Network for the period 2010–2012 was generally in the range of 2–5 kg S/ha/yr at park locations within the network area (Figure 7.5). Oxidized inorganic N deposition for the period 2010–2012 was less than 5 kg N/ha/yr throughout most of the parklands within the network (Figure 7.6). Most areas also received less than 5 kg N/ha/yr of ammonium from atmospheric deposition

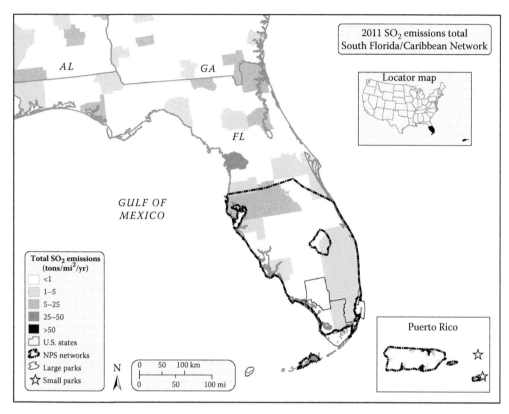

FIGURE 7.2
Total sulfur dioxide (SO$_2$) emissions, by county, near the South Florida/Caribbean Network for the year 2011. (Data from EPA's National Emissions Inventory, https://www.epa.gov/air-emissions-inventories, accessed January, 2014.)

during this same period (Figure 7.7). Total N deposition was less than 10 kg N/ha/yr at most park locations (Figure 7.8).

Atmospheric S and N deposition levels have declined at South Florida/Caribbean Network parks since 2001, based on Total Deposition project estimates (Table 7.1 for EVER). Some of the decreases have been sizeable (>20% change). Oxidized N and ammonium showed opposite patterns, with oxidized N decreasing and reduced N increasing at EVER since the monitoring period 2000–2002.

The National Park Service (2010) reported long-term trends in concentrations of Hg in wet deposition in EVER during the period beginning in 1996 and running through 2008. Three-year means of annual Hg concentration in wet deposition were reported for 13 national parks that had at least two years of valid data during the period 2006–2008. The highest Hg concentration in precipitation was reported for EVER. In fact, South Florida has one of the nation's highest rates of atmospheric Hg deposition, originating from both local and international sources (Gustin et al. 1999 and material provided by Jed Redwine, National Park Service, from the National Resource Condition Assessment under development, July 2014).

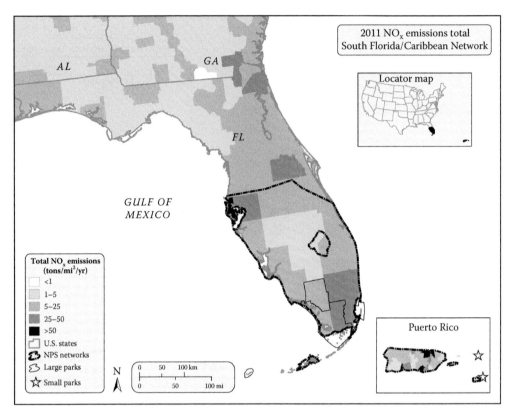

FIGURE 7.3
Total oxidized N (NO_x) emissions, by county, near the South Florida/Caribbean Network for the year 2011. (Data from EPA's National Emissions Inventory, https://www.epa.gov/air-emissions-inventories, accessed January, 2014.)

7.3 Acidification

I am not aware of any data documenting acid sensitivity of either aquatic or terrestrial resources in southern Florida. Based on the geology and low topographic relief of this region, resources are likely not highly sensitive to, or affected by, acidification attributable to acidic deposition. Alkaline soils in southern Florida can effectively buffer acidic inputs from the atmosphere (Hall 2011). However, there are also peat soils that may under certain conditions be more sensitive to acidification from S or N deposition. The Ca supply to network wetlands is high in many places due to the prevalence of calcium carbonate geology (Jed Redwine, National Park Service, personal communication, July 2014).

7.4 Nutrient Nitrogen Enrichment

The predominant vegetation types in the South Florida/Caribbean Network that are thought to be especially sensitive to eutrophication effects from nutrient N addition are

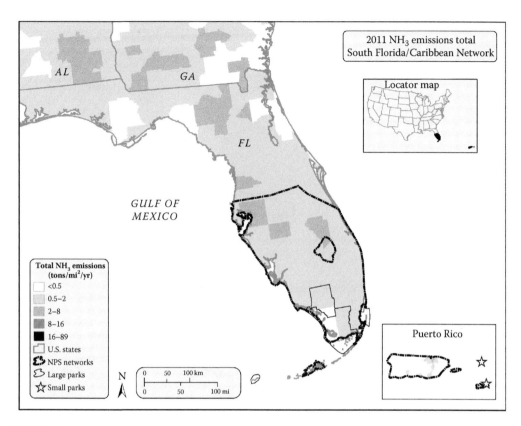

FIGURE 7.4

Total ammonia (NH$_3$) emissions, by county, near the South Florida/Caribbean Network for the year 2011. (Data from EPA's National Emissions Inventory, https://www.epa.gov/air-emissions-inventories, accessed January, 2014.)

wetland and seagrass. Wetlands covered much of South Florida prior to human alterations of the hydrologic system, which began more than a century ago. The quantity, timing, location, and quality of freshwater flows to estuaries and wetlands have been dramatically modified (Science Subgroup 1996). These hydrologic changes interact in complex ways with nutrient inputs, most of which are derived from nonatmospheric sources such as agriculture. Harmful effects on Everglades ecosystem structure and function from water diversion and nutrient loading (mostly P) have been documented (Chiang et al. 2000, Gaiser et al. 2006, Wright et al. 2008, Richardson 2009).

Based on a coarse screening analysis by Sullivan et al. (2011b), the network rankings for nutrient N pollutant exposure, ecosystem sensitivity to nutrient N enrichment, and park protection yielded an overall network nutrient N enrichment summary risk ranking for the South Florida/Caribbean Network that was in the highest quintile (upper 20%) among all networks. The overall level of concern for nutrient N enrichment effects on parks within this network was judged by Sullivan et al. (2011b) to be very high. Although rankings provide a rough indication of risk, park-specific data, particularly regarding nutrient-enrichment sensitivity, are needed to fully evaluate risk from nutrient N addition.

Tropical and subtropical forests are often relatively rich in N, and other nutrients are more often limiting (Chadwick et al. 1999). Such ecosystems commonly show high rates of N leaching and denitrification, irrespective of atmospheric deposition (Lewis et al. 1999, Davidson et al. 2007, Hall 2011). On tropical or subtropical sites where N is not limiting,

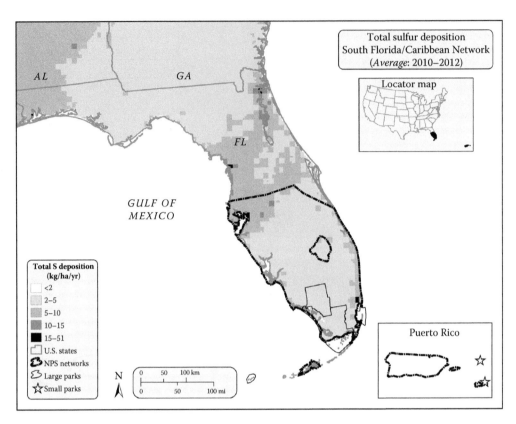

FIGURE 7.5

Total S deposition for the three-year period centered on 2011 in and around the Florida section of the South Florida/Caribbean Network. (From Schwede, D.B. and Lear, G.G., *Atmos. Environ.*, 92, 207, 2014.)

atmospheric N deposition would not be expected to alter productivity or plant species composition. However, loss of N to the atmosphere as gases produced by denitrification or to drainage water as nitrate may be stimulated by increased N deposition (Herbert and Fownes 1995, Hall and Matson 1999, Lohse and Matson 2005, Templer et al. 2008, Hall 2011). On sites where the N supply is limiting, added N would be expected to increase plant growth and perhaps change plant community composition, eventually leading to N saturation (Hall and Matson 1999, Erickson et al. 2001, Feller et al. 2007).

Nitrogen retention in tropical and subtropical forests differs from retention in temperate forests. Evergreen tropical forests have high leaf area throughout the year and therefore retain more N in their canopies. This prevents much of the added N from reaching the microbial and plant communities that develop on the soil surface (Bakwin et al. 1990, Sparks et al. 2001, Hall 2011).

Data are not available with which to evaluate the extent to which wetlands in the South Florida/Caribbean Network have been affected by nutrient enrichment from N deposition. The levels of N deposition have been relatively high and may or may not have been sufficiently high as to cause species shifts in wetland plants. The risk of species composition change is important, in part because wetland ecosystems often contain large numbers of rare plant species.

One of the adverse impacts of eutrophication on estuarine and coastal ecosystems is a decrease in the extent of seagrasses such as turtle grass (*Thalassia testudinum*) and other

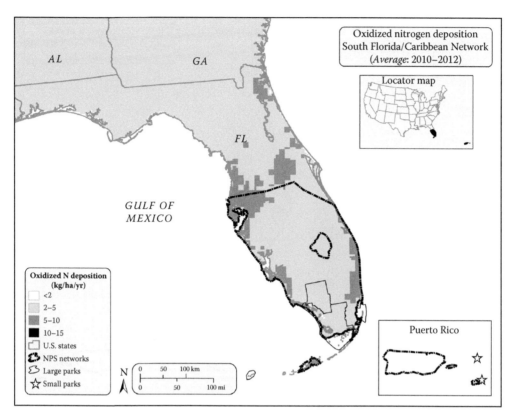

FIGURE 7.6

Total oxidized inorganic N deposition for the three-year period centered on 2011 in and around the Florida section of the South Florida/Caribbean Network. (From Schwede, D.B. and Lear, G.G., *Atmos. Environ.*, 92, 207, 2014.)

submerged aquatic vegetation that provide habitat for a wide range of estuarine and marine species. Investigators in some parks are delineating the extent of submerged aquatic vegetation in park estuaries to provide data needed for more effective resource management. The U.S. Geological Survey and National Oceanic and Atmospheric Administration have partnered with National Park Service to map submerged resources in coastal parks at several locations, including VIIS (Cross and Curdts 2011).

Florida Bay has experienced major seagrass die-offs and noxious algal blooms attributable to nutrient inputs and hydrological changes (Science Subgroup 1996, Philips et al. 1999). Long-term management plans are aimed at restoring, to the extent feasible, the natural hydrological conditions of this system (Gilbert et al. 2004).

7.5 Ozone Injury to Vegetation

Using the National Park Service (2010) and Kohut (2007) criteria, ozone levels at EVER were rated low (Table 7.2). Monitoring was discontinued at VIIS in 2003. Kohut's (2007) ranking was low across both parks.

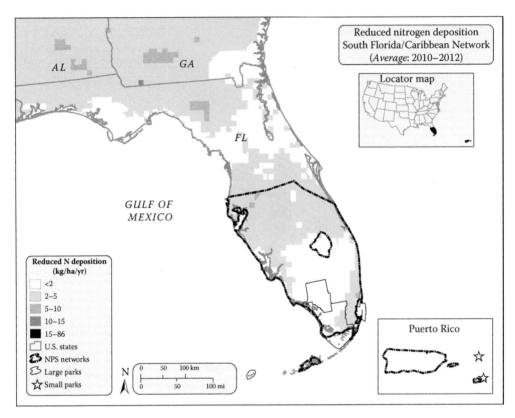

FIGURE 7.7
Reduced inorganic N (ammonium) deposition for the three-year period centered on 2011 in and around the Florida section of the South Florida/Caribbean Network. (From Schwede, D.B. and Lear, G.G., *Atmos. Environ.*, 92, 207, 2014.)

7.6 Visibility Degradation

7.6.1 Natural Background and Ambient Visibility Conditions

Haze is monitored by the Interagency Monitoring of Protected Visual Environments (IMPROVE) network in EVER and VIIS. Because of their proximity to the ocean, these parks have relatively high levels of natural haze (Table 7.3) caused by sea salt and, to some extent, by marine sources of sulfate and high humidity. Wildfire also contributes to haze in EVER. Current haze levels are substantially elevated above the estimated natural levels for the parks and are considered impaired at times due to anthropogenic pollution.

7.6.2 Composition of Haze

Figure 7.9 shows estimated natural (preindustrial), baseline (2000–2004), and current (2006–2010) levels of haze and its composition for EVER and VIIS. The largest contributor to total particulate light extinction in EVER was sulfate, followed by organics (Figure 7.9).

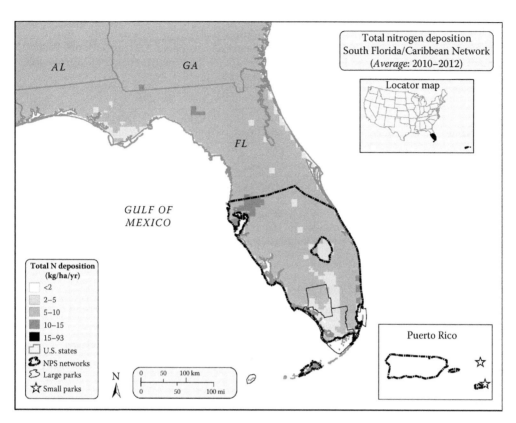

FIGURE 7.8
Total N deposition for the three-year period centered on 2011 in and around the Florida section of the South Florida/Caribbean Network. (From Schwede, D.B. and Lear, G.G., *Atmos. Environ.*, 92, 207, 2014.)

TABLE 7.1

Average Changes in S and N Deposition[a] between 2001 and 2011 across Park Grid Cells at EVER

Parameter	2001 Average (kg/ha/yr)	2011 Average (kg/ha/yr)	Absolute Change (kg/ha/yr)	Percent Change	2011 Minimum (kg/ha/yr)	2011 Maximum (kg/ha/yr)	2011 Range (kg/ha/yr)
Total S	5.00	3.84	−1.15	−22.7	3.37	5.05	1.68
Total N	5.49	4.94	−0.55	−7.7	3.32	6.85	3.53
Oxidized N	4.06	3.02	−1.04	−23.0	2.03	4.74	2.71
Reduced N	1.43	1.92	0.49	35.2	1.29	2.60	1.31

[a] Deposition estimates were determined by the Total Deposition project based on three-year averages centered on 2001 and 2011 for all ~4 km grid cells. The minimum, maximum, and range of 2011 S and N deposition are also shown.

The contribution of sulfate was highest on the 20% haziest days (57.4%). The contribution of organics was about 14% on the 20% clearest and 20% haziest days. Nitrates, coarse mass, and sea salt also contributed to haze.

In VIIS, the majority of total particulate light extinction was attributable to sulfate, sea salt, and coarse mass (Figure 7.9). On an annual average basis, sulfate contributed 29.5% of

TABLE 7.2

Ozone Assessment Results for EVER and VIIS Based on Estimated Average 3-Month W126 and SUM06 Ozone Exposure Indices for the Period 2005–2009 and Kohut's (2007) Ozone Risk Ranking for the Period 1995–1999[a,b]

Park Name	Park Code	W126 Value (ppm-hr)	W126 NPS Condition	SUM06 Value (ppm-hr)	SUM06 NPS Condition	Kohut Ozone Risk Ranking
Everglades	EVER	6.50	Low	7.50	Low	Low
Virgin Islands	VIIS	No data	No rank	No data	No rank	Low

[a] Parks are classified into one of three ranks (Low, Moderate, High) based on comparison with other Inventory and Monitoring parks.

[b] Degrees of concern for the W126 and SUM06 indices are based solely on levels of ozone exposure. Kohut's risk to vegetation is based on several factors that contribute to injury in plants, including ozone exposure and environmental variables, and considers the effects of soil moisture on the uptake of ozone.

TABLE 7.3

Estimated Natural Haze and Measured Ambient Haze in EVER and VIIS Averaged over the Period 2004–2008[a]

Park Name	Park Code	Site ID	Estimated Natural Haze (in Deciviews) 20% Clearest Days	Estimated Natural Haze (in Deciviews) 20% Haziest Days	Estimated Natural Haze (in Deciviews) Average Days
Everglades	EVER	EVER1	5.22	12.15	7.77
Virgin Islands	VIIS	VIIS1	4.41	10.68	7.10

Park Name	Park Code	Site ID	Measured Ambient Haze (for Years 2004–2008) (in Deciviews) 20% Clearest Days	Measured Ambient Haze (for Years 2004–2008) (in Deciviews) 20% Haziest Days	Measured Ambient Haze (for Years 2004–2008) (in Deciviews) Average Days
Everglades	EVER	EVER1	12.14	21.43	16.17
Virgin Islands	VIIS	VIIS1	9.16	18.08	13.23

[a] Parks are classified into one of five haze ranks (Very Low, Low, Moderate, High, or Very High) based on comparison with other monitored parks.

total particulate light extinction, sea salt 26.5%, and coarse mass 24.1%. On the 20% haziest days, sulfate accounted for 25.3% of total particulate light extinction, sea salt 19.3%, and the contribution of coarse mass increased slightly to 27.2%. On the 20% clearest visibility days, the contribution of sulfate increased to 35.2% of total particulate light extinction, sea salt contributed 27.8%, and the contribution of coarse mass decreased to 21.6%.

7.6.3 Trends in Visibility

The EPA monitors visibility in 155 national parks through the IMPROVE program. Over the period 1996–2006, visibility on the 20% clearest days improved (decreased haze) or remained constant at all monitored sites in the conterminous United States except EVER (U.S. EPA 2008). Available IMPROVE data suggest, however, that haze may be decreasing in more recent years at EVER, especially on the 20% haziest days (Figure 7.10). In marked contrast, haze at VIIS has generally increased since monitoring began in 2001, especially on the 20% haziest days.

7.6.4 Development of State Implementation Plans

The Visibility Improvement State and Tribal Association of the Southeast was a collaborative effort among state governments, tribal governments, and federal agencies involved in the management of visibility and regional haze in the Southeast. The region included the southeastern United States from Virginia and West Virginia in the north, south to Florida,

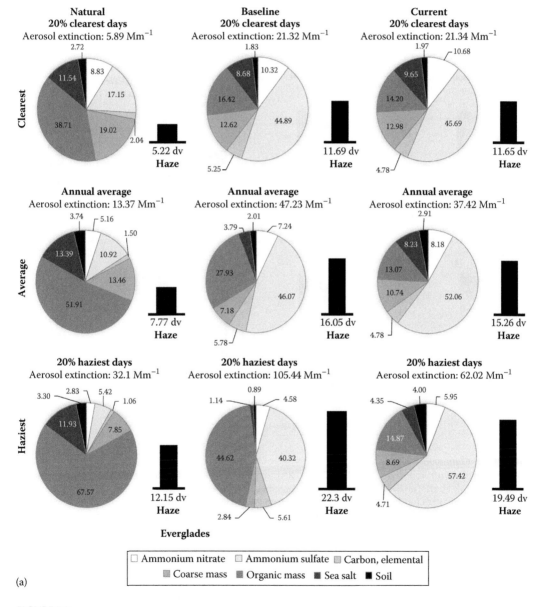

(a)

FIGURE 7.9

Estimated natural (preindustrial), baseline (2000–2004), and current (2006–2010) levels of haze (columns) and its composition (pie charts) on the 20% clearest, annual average, and 20% haziest visibility days for (a) EVER and (b) VIIS. EVER has no data for the year 2000. VIIS has no data for the years 2000 and 2007. Ammonium sulfate is the most important nonnatural substance that causes haze in southern Florida's National Park Service units. (From http://views.cira.colostate.edu/fed/Tools/RegionalHazeSummary.aspx, accessed October, 2012.)   (*Continued*)

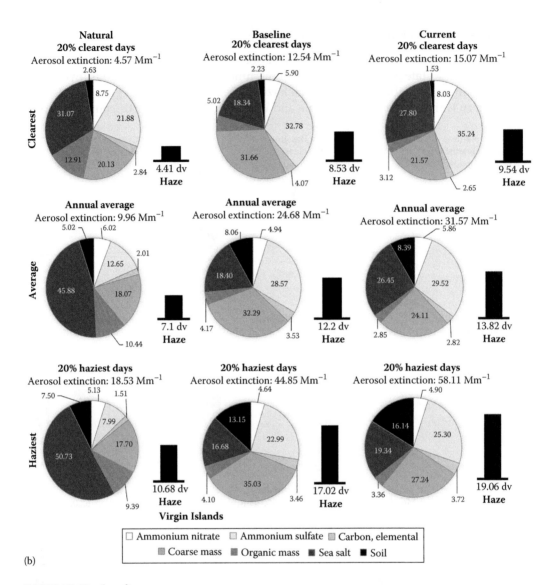

(b)

FIGURE 7.9 (Continued)
Estimated natural (preindustrial), baseline (2000–2004), and current (2006–2010) levels of haze (columns) and its composition (pie charts) on the 20% clearest, annual average, and 20% haziest visibility days for (a) EVER and (b) VIIS. EVER has no data for the year 2000. VIIS has no data for the years 2000 and 2007. Ammonium sulfate is the most important nonnatural substance that causes haze in southern Florida's National Park Service units. (From http://views.cira.colostate.edu/fed/Tools/RegionalHazeSummary.aspx, accessed October, 2012.)

and west to Kentucky, Tennessee, and Mississippi. Analyses have included determination of baseline visibility conditions, calculation of the glideslope from the baseline necessary to achieve background conditions in 2064, and determination of air pollutant source areas. Progress in meeting the national visibility goal is illustrated in Figure 7.11. Results of this analysis suggest that improvements to date at EVER in visibility on the 20% haziest days exceeded the glideslope required for Regional Haze Rule compliance. This has clearly not been the case at VIIS. On the clearest 20% of days, ambient haze appears to be increasing at VIIS.

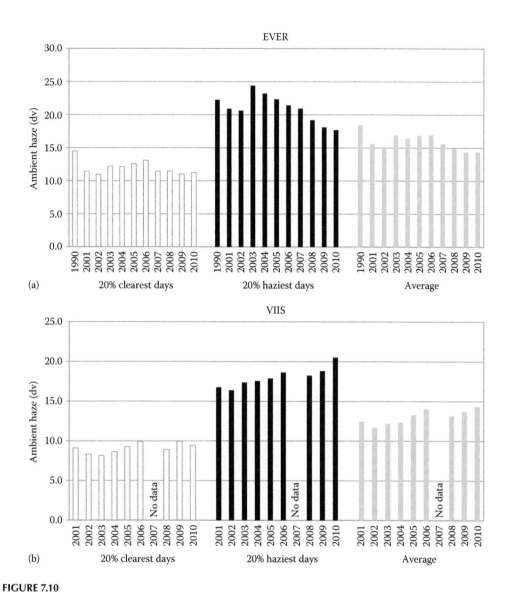

FIGURE 7.10
Trends in ambient haze levels at (a) EVER and (b) VIIS, based on Interagency Monitoring of Protected Visual Environments measurements on the 20% clearest, 20% haziest, and annual average visibility days over the monitoring period of record. (From http://vista.cira.colostate.edu/improve/Data/IMPROVE/summary_data.htm, accessed October, 2012.)

7.7 Toxic Airborne Contaminants

Airborne toxics include various semivolatile organics (e.g., pesticides, industrial by-products) and heavy metals (e.g., Hg). Pesticide residues attributable to the deposition of airborne contaminants likely contribute to adverse impacts on sensitive species in the South Florida/Caribbean Network. However, there are no available data to suggest that atmospheric contributions of pesticides constitute an important part of the total pesticide

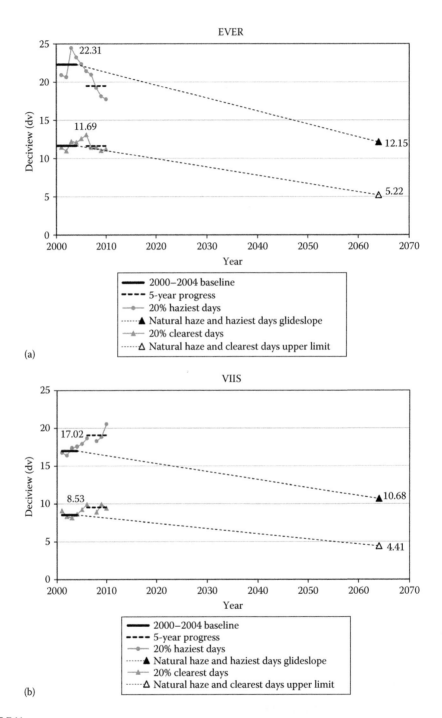

FIGURE 7.11
Glideslopes to achieving natural visibility conditions in 2064 for the 20% haziest (upper line) and the 20% clearest (lower line) days in (a) EVER and (b) VIIS. In the Regional Haze Rule, the clearest days do not have a uniform rate of progress glideslope; the rule only requires that the clearest days do not get any worse than the baseline period. Also shown are measured values during the period 2001–2010. EVER has no data for the year 2000. VIIS has no data for the years 2000 and 2007. (From http://vista.cira.colostate.edu/improve/Data/IMPROVE/summary_data.htm, accessed October, 2012.)

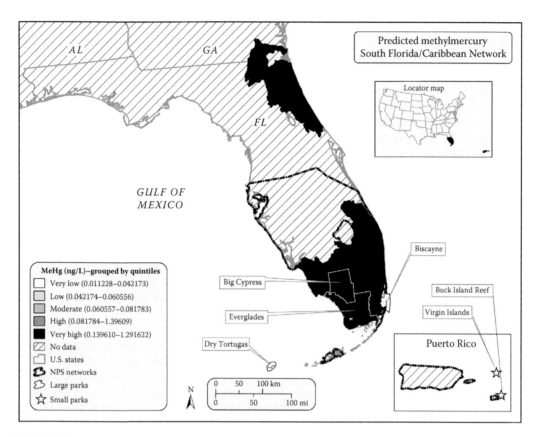

FIGURE 7.12

Predicted methylmercury (MeHg; in ng/L) concentrations in surface waters by hydrologic unit codes that contain national parklands in the South Florida/Caribbean Network. Estimates were generated by U.S. Geological Survey (last modified February 20, 2015). Rankings are based on quintile distributions across all parks having estimates by the U.S. Geological Survey.

loading to sensitive surface waters in this network. There is some evidence of high levels of atmospheric deposition of endosulfan (Potter et al. 2014). Estimates of Hg methylation potential generated by the U.S. Geological Survey (last modified February 20, 2015) for watershed boundaries (based on eight-digit hydrologic unit codes containing national parklands in the South Florida/Caribbean Network) suggested high methylation potential at the parks considered in this analysis (Figure 7.12). This result is likely driven mainly by relatively high concentrations of total organic C in surface waters in this network, coupled with relatively high atmospheric deposition of Hg and S.

Wetlands, which are common in parts of the South Florida/Caribbean Network, act as important sources of biologically available methyl Hg to freshwater ecosystems. This is likely due in large part to three characteristics of south Florida wetlands: (1) high availability of dissolved organic C, (2) common occurrence of anaerobic conditions in sediments, and (3) high atmospheric S deposition. All enhance Hg methylation rates. The abundance of dissolved organic C also enhances the transport of methyl Hg to downstream receiving waters. As a consequence of these wetland influences on Hg methylation and transport, the percentage of wetland areas within watersheds is commonly correlated with methyl Hg flux (Grigal 2002). EVER has extensive wetland coverage. It is likely that S enrichment of wetlands in the Everglades, in response to both atmospheric S deposition and agricultural inputs (Bates et al. 2001), has increased Hg methylation in wetland ecosystems in EVER (McCormick et al. 2011).

High precipitation, coupled with elevated concentrations of dissolved organic C in surface waters, availability of S, and the high chemical reducing capacity of the EVER wetlands, support rapid transformation of Hg to the biologically available methyl Hg (Chen et al. 2012), which contributes to high Hg bioaccumulation in fish. Mercury can biomagnify, especially at higher trophic levels, to concentrations that can potentially damage the nervous system of sensitive species of biota. This issue is of particular concern to people and wildlife that consume large quantities of fish. Body burdens of Hg in sunfish (*Lepomis* spp.), largemouth bass, and bluegill (*Lepomis macrochirus*) in northern EVER exceeded the wildlife criteria levels established by the EPA (U.S. EPA 2007, Chen et al. 2012). Exposure to methyl Hg above the levels that cause adverse effects have been estimated for great egret (*Ardea alba*), bald eagle, and wood stork (*Mycteria americana*) in the northern part of the park. There is also concern for how Hg exposure may impact the reproductive success of the endangered Florida panther (*Puma concolor coryi*; U.S. EPA 2007). To protect humans from adverse effects associated with Hg contamination, the state of Florida has issued fish consumption advisories that ban or limit consumption by humans of nine fish species over two million acres in EVER (U.S. EPA 2007).

In the Everglades Water Conservation areas adjacent to EVER, methyl Hg concentrations in fish have been declining since the early 1990s. Nevertheless, they still generally exceed the EPA thresholds for human consumption and for wildlife (Axelrad et al. 2007).

In 1999, a survey was conducted of 28 American alligators (*Alligator mississippiensis*) along a transect through the Everglades by a multiagency team including researchers from the U.S. Geological Survey, the U.S. Fish and Wildlife Service, and the Florida Fish and Wildlife Conservation Commission. Results showed that alligators in EVER had Hg body burdens that were roughly twice as high as the Everglades-wide average (10.4 and 1.2 mg/kg versus 4.0 and 0.64 mg/kg in the liver and tail, respectively; Rumbold et al. 2002). Nevertheless, Hg levels in alligators in EVER appear to have declined since 1994 when compared with a survey by Yanochko et al. (1997).

Monitoring of Hg levels in pig frogs (*Rana grylio*) in EVER is important because this amphibian is an abundant midlevel component of the Everglades food web and because the frogs are harvested for human consumption. In a survey by Ugarte et al. (2005), the highest Hg concentrations in frog legs were found in EVER, where harvesting is prohibited; however, frog legs collected at some sites in the Everglades outside EVER had Hg levels that exceeded the EPA 0.3 mg/kg fish tissue residue criterion. The role played by frogs in the transfer of Hg through the wetland system in EVER may be significant, as Hg levels measured in some frogs were higher than the threshold values for piscivorous wildlife. Spatial patterns in Hg concentrations in the frog samples generally corresponded with results for other wildlife species (Ugarte et al. 2005).

Wading birds (order Ciconiiformes) include herons, egrets, ibis, and spoonbill. These are upper trophic level aquatic feeders that may be at risk from high levels of Hg exposure in portions of the South Florida/Caribbean Network having high ambient Hg in surface water. Sundlof et al. (1994) compared Hg concentrations in livers of young Ciconiiforme birds collected across southern Florida, including within and adjacent to EVER. Mercury was measured in the livers of 144 birds. Liver Hg concentration varied by location, age, diet, and body fat content. Birds collected from the central Everglades and eastern Florida Bay had significantly higher liver Hg concentration than birds collected in other areas. Species that had a prey base of larger fish had about four times the liver Hg concentration compared with species that consumed small fish or crustaceans. Four great blue herons (*Ardea herodias*) collected in the central Everglades had livers containing Hg concentrations commonly associated with neurologic symptoms (≥30 μg/g). Between 30% and 80% of potential breeding-age birds collected from the central Everglades had liver Hg concentrations that have previously been associated with reproduction impairment in ducks and pheasants (Sundlof et al. 1994).

The wading bird populations in Florida have declined dramatically (Runde 1991, Ogden 1994). Habitat loss, altered regional hydrology, and P pollution have been important contributing factors to the observed declines. Hg contamination of their food supply may also be an important factor (Sundlof et al. 1994). Toxicity to birds from methyl Hg exposure can be expressed as damage to nervous, excretory, immune, or reproductive systems (Wiener et al. 2003). Embryos and hatchlings are especially vulnerable (Heinz and Hoffman 2003).

Julian et al. (2014) provided an assessment of the Hg and S status of the Everglades Protection Area during water year 2013. The reported total Hg concentration in large-mouth bass ranged from 0.02 to 2.0 mg/kg, with a median concentration of 0.4 mg/kg. In the trophic level 3 sunfish species surveyed, total Hg concentration ranged from 0.1 to 0.25 mg/kg, which exceeded the federal methyl Hg criterion of 0.077 mg/kg for trophic level 3 fish in order to protect piscivorous wildlife.

7.8 Summary

Parks in the South Florida/Caribbean Network have been heavily influenced by human activities. Much of the historical Everglades ecosystem has been lost or degraded (McCormick et al. 2009). Atmospheric deposition is only a small part of that complicated story. VIIS is more remote from major human-caused pollutant emissions sources. There are many population centers in the range of 50,000–500,000 people scattered along the south Florida coastline. Miami, Fort Lauderdale, Key West, and Naples are all within about 50 mi (80 km) of EVER. Emissions of S and N are high in some areas.

Total estimated S deposition within the network was relatively high in 2002, generally between 5 and 10 kg S/ha/yr. There were areas of lower and higher estimated S deposition, ranging from 2 to 5 kg S/ha/yr in the south to more than 30 kg S/ha/yr in the northwest. Total estimated N deposition in 2002 ranged from as low as 5 to 10 kg N/ha/yr to as high as 10 to 15 kg N/ha/yr across broad areas of the network. Both S and N deposition decreased substantially between 2001 and 2011 at most South Florida/Caribbean Network parks. Nevertheless, ammonium deposition actually increased at all parks, in all cases by more than 10%. These increases in ammonium deposition partly counteracted concurrent decreases in oxidized N deposition.

Atmospheric S and N can cause acidification of streams, lakes, and soils. Parks in this network are relatively insensitive to freshwater acidification, however, because of generally low relief, strong buffering capacity of some soils, and ample contact between drainage water and weatherable soils and geologic materials.

Nitrogen can cause undesirable nutrient enrichment of natural ecosystems, leading to changes in plant species distribution and diversity. The overall level of concern for nutrient N enrichment effects on parks within this network was judged by Sullivan et al. (2011b) in a coarse screening assessment to be very high. Wetland is the predominant vegetation type found in EVER that is thought to be especially sensitive to eutrophication effects from nutrient N addition. Ambient levels of N deposition in some areas are comparable to those found to cause species shifts in wetlands elsewhere in the United States (Greaver et al. 2011). However, P limitation is thought to be common in South Florida wetlands, in part because many tropical wetland plants fix N through symbiotic relationships with bacteria. Thus, N addition may have limited impact on the nutrient status of wetlands in the South Florida/Caribbean Network except where there are also elevated P inputs. Estuaries are also highly sensitive to N addition from both atmospheric and land-based sources. Seagrass communities are especially vulnerable.

Haze has been monitored by the IMPROVE network at EVER and VIIS. Visibility is impaired in those parks. Although some of the haze is natural (caused largely by sea salt and marine sulfate), a substantial portion is caused by anthropogenic emissions of air pollutants. The largest contributor to total particulate light extinction in EVER is sulfate, followed by organics. Atmospheric sulfate in the South Florida/Caribbean Network parks is derived from both human-caused and natural (marine) sources. On the 20% haziest days at EVER, atmospheric sulfate at this park contributed more than half of the light extinction. In VIIS, the majority of the light extinction was attributable to a combination of sulfate, sea salt, and coarse mass. On the 20% haziest days in VIIS, sulfate accounted for 24.1% of light extinction.

Airborne toxics, including Hg and other heavy metals, can accumulate in, and is evident at, all levels of the food web. These contaminants can reach toxic levels in midlevel and top predators. South Florida is considered to be a region of high Hg methylation potential. The issue has been well studied in EVER. Atmospheric deposition accounts for an estimated 95% of all Hg inputs in EVER (Landing et al. 1995). South Florida, including the parks and preserve, has some of the highest wet deposition levels of Hg in the United States (NADP 2008). This is due to a combination of high concentrations of Hg in precipitation and high amounts of precipitation in the region.

High Hg and S deposition rates, coupled with elevated concentrations of dissolved organic C and the high reducing capacity of soils in the EVER wetlands, support rapid transformation of Hg to the biologically available and significantly more toxic methyl Hg form (Chen et al. 2012), which contributes to high Hg bioaccumulation in many species, including fish, panthers, alligators, and wading birds. Mercury can biomagnify at higher trophic levels to concentrations that can potentially damage the nervous system of sensitive species. This issue is of particular concern to people and wildlife that consume large quantities of Hg-contaminated fish and shellfish. Body burdens of Hg in sunfish, largemouth bass, and bluegill in northern EVER exceeded the human and wildlife criteria levels established by the EPA (U.S. EPA 2007, Chen et al. 2012). Exposure to methyl Hg above the levels that cause adverse effects has been estimated for great egret, bald eagle, and wood stork in the northern part of the park. There is also concern regarding how Hg exposure may impact the reproductive success of the endangered Florida panther (U.S. EPA 2007).

The wading bird population in Florida has declined dramatically compared to its original size (Runde 1991). Habitat loss has been an important contributor to that decline. However, Hg contamination of their food supplies may also be an important factor (Sundlof et al. 1994). Toxicity to birds from methyl Hg exposure can be expressed as damage to nervous, excretory, immune, or reproductive systems (Wiener et al. 2003). Embryos and hatchlings are especially vulnerable (Heinz and Hoffman 2003).

References

Axelrad, D., T. Atkeson, T. Lange, C. Pollman, C. Gilmour, W. Orem, I. Mendelssohn et al. 2007. Mercury monitoring, research and environmental assessment in South Florida. Chapter 3B. 2007 South Florida Environment Report. South Florida Water Management District and Florida Department of Environmental Protection, West Palm Beach, FL.

Bakwin, P.S., S.C. Wofsy, and S. Fan. 1990. Measurements of reactive nitrogen oxides (NO_y) within and above a tropical forest canopy in the wet season. *J. Geophys. Res.* 95(D10):16765–16772.

Bates, A.L., W.H. Orem, J.W. Harvey, and E.C. Spiker. 2001. Geochemistry of sulfur in the Florida Everglades, 1994–1999. Open file report 01-007. U.S. Geological Survey, Reston, VA.

Chadwick, O.A., L.A. Derry, P.M. Vitousek, B.J. Huebert, and L.O. Hedin. 1999. Changing sources of nutrients during four million years of ecosystem development. *Nature* 397(6719):491–497.

Chen, C.Y., C.T. Driscoll, and N.C. Kamman. 2012. Mercury hotspots in freshwater ecosystems: Drivers, processes, and patterns. *In* M.S. Bank (Ed.). *Mercury in the Environment: Pattern and Process.* University of California Press, Berkeley, CA, pp. 143–166.

Chiang, C., C.B. Craft, D.W. Rogers, and C.J. Richardson. 2000. Effects of 4 years of nitrogen and phosphorus additions on Everglades plant communities. *Aquat. Bot.* 68(1):61–78.

Cross, J. and T. Curdts. 2011. Mapping submerged resources in ocean, coastal, and Great Lakes parks. Abstract. *George Wright Society Conference on Parks, Protected Areas, & Cultural Sites*, New Orleans, LA.

Davidson, E.A., C.J.R. de Carvalho, A.M. Figueira, F.Y. Ishida, J. Ometto, G.B. Nardoto, R.T. Saba et al. 2007. Recuperation of nitrogen cycling in Amazonian forests following agricultural abandonment. *Nature* 447(7147):995–998.

Erickson, H., M. Keller, and E.A. Davidson. 2001. Nitrogen oxide fluxes and nitrogen cycling during postagricultural succession and forest fertilization in the humid tropics. *Ecosystems* 4(1):67–84.

Feller, I.C., C.E. Lovelock, and K.L. McKee. 2007. Nutrient addition differentially affects ecological processes of *Avicennia germinans* in nitrogen versus phosphorus limited mangrove ecosystems. *Ecosystems* 10(3):347–359.

Gaiser, E.E., D.L. Childers, R.D. Jones, J.H. Richards, L.J. Scinto, and J.C. Trexler. 2006. Periphyton responses to eutrophication in the Florida Everglades: Cross-system patterns of structural and compositional change. *Limnol. Oceanogr.* 51:617–630.

Gilbert, P.M., C.A. Heil, D. Hollander, M. Revilla, A. Hoare, J. Alexander, and S. Murasko. 2004. Evidence for dissolved organic nitrogen and phosphorus uptake during a cyanobacterial bloom in Florida Bay. *Mar. Ecol. Prog. Ser.* 280:73–83.

Greaver, T., L. Liu, and R. Bobbink. 2011. Wetlands. *In* L.H. Pardo, M.J. Robin-Abbott, and C.T. Driscoll (Eds.). *Assessment of Nitrogen Deposition Effects and Empirical Critical Loads of Nitrogen for Ecoregions of the United States.* General Technical Report NRS-80. U.S. Forest Service, Newtown Square, PA.

Grigal, D.F. 2002. Inputs and outputs of mercury from terrestrial watersheds: A review. *Environ. Rev.* 10:1–39.

Gustin, M.S., S. Lindberg, F. Marsik, A. Casimir, R. Ebinghaus, G. Edwards, C. Hubble-Fitzgerald et al. 1999. Nevada STORMS project: Measurement of mercury emissions from naturally enriched surfaces. *J. Geophys. Res. Atmos.* 104(D17):21831–21844.

Hall, S.J. 2011. Tropical and subtropical humid forests. *In* L.H. Pardo, M.J. Robin-Abbott, and C.T. Driscoll (Eds.). *Assessment of Nitrogen Deposition Effects and Empirical Critical Loads of Nitrogen for Ecoregions of the United States.* General Technical Report NRS-80. U.S. Forest Service, Newtown Square, PA, pp. 181–192.

Hall, S.J. and P.A. Matson. 1999. Nitrogen oxide emissions after nitrogen additions in tropical forests. *Nature* 400(6740):152–155.

Heinz, G.H. and D.J. Hoffman. 2003. Embryonic thresholds of mercury: Estimates from individual mallard ducks. *Arch. Environ. Contam. Toxicol.* 44:257–264.

Herbert, D.A. and J.H. Fownes. 1995. Phosphorus limitation of forest leaf area and net primary pro-
    duction on a highly weathered soil. *Biogeochemistry* 29:223–235.
Julian, P., II, B. Gu, R. Frydenborg, T. Lange, A.L. Wright, and J.M. McCray. 2014. Mercury and sulfur
    environmental assessment for the Everglades. Chapter 3B. 2014 South Florida Environmental
    Report. South Florida Environment Report. South Florida Water Management District and
    Florida Department of Environmental Protection, West Palm Beach, FL.
Kohut, R. 2007. Assessing the risk of foliar injury from ozone on vegetation in parks in the U.S.
    National Park Service's Vital Signs Network. *Environ. Pollut.* 149:348–357.
Landing, W.M., J.J. Perry, J.L. Guentzel, G.A. Gill, and C.D. Pollman. 1995. Relationships between
    the atmospheric deposition of trace-elements, major ions, and mercury in Florida—The FAMS
    Project (1992–1993). *Water Air Soil Pollut.* 80:343–352.
Lewis, W.M., J.M. Melack, W.H. McDowell, M. McClain, and J.E. Richey. 1999. Nitrogen yields from
    undisturbed watersheds in the Americas. *Biogeochemistry* 46(1–3):149–162.
Lohse, K.A. and P.A. Matson. 2005. Consequences of nitrogen additions for soil losses from wet
    tropical forests. *Ecol. Appl.* 15(5):1629–1648.
McCormick, P., S. Newman, and L. Vilchek. 2009. Landscape responses to wetland eutrophication:
    Loss of slough habitat in the Florida Everglades, USA. *Hydrobiologia* 621(1):105–114.
McCormick, P.V., J.W. Harvey, and E.S. Crawford. 2011. Influence of changing water sources and
    mineral chemistry on the Everglades ecosystem. *Crit. Rev. Environ. Sci. Tech.* 41(Suppl. 1):
    28–63.
National Acid Deposition Program (NADP). 2008. Mercury deposition network. Available at: http://
    nadp.sws.uiuc.edu/mdn/.
National Park Service (NPS). 2010. Air quality in national parks: 2009 annual performance and prog-
    ress report. Natural Resource Report NPS/NRPC/ARD/NRR-2010/266. National Park Service,
    Air Resources Division, Denver, CO.
Ogden, J.C. 1994. A comparison of wading bird nesting dynamics 1931–1946 and 1974–1989, as an
    indication of ecosystem conditions in the southern Everglades. *In* S. Davis and O.J. C. (Eds.).
    Everglades, Spatial and Temporal Patterns in Guidelines for Ecosystem Restoration. University of
    Florida Press, Gainesville, FL, pp. 533–570.
Philips, E.J., S. Badylak, and T.C. Lynch. 1999. Blooms of the picoplanktonic cyanobacterium
    Synechococcus in Florida Bay, a subtropical inner-shelf lagoon. *Limnol. Oceanogr.* 44:1166–1175.
Potter, T.L., C.J. Hapeman, L.L. McConnell, J.A. Harman-Fetcho, W.F. Schmidt, C.P. Rice, and
    B. Schaffer. 2014. Endosulfan wet deposition in southern Florida (USA). *Sci. Total Environ.*
    468–469(0):505–513.
Richardson, C.J. 2009. The Everglades: North America's subtropical wetland. *Wetl. Ecol. Manage.*
    18(5):517–542.
Rumbold, D.G., L.E. Fink, K.A. Laine, S.L. Niemczyk, T. Chandrasekhar, S.D. Wankel, and C.
    Kendall. 2002. Levels of mercury in alligators (*Alligator mississippiensis*) collected along a tran-
    sect through the Florida Everglades. *Sci. Tot. Environ.* 297:239–252.
Runde, D.E. 1991. Trends in wading bird nesting populations in Florida 1976–1978 and 1986–1989.
    Final performance report. Survey #7612, Tallahassee, FL. Florida Game and Fresh Water
    Commission, Nongame Program, Tallahassee, FL.
Schwede, D.B. and G.G. Lear. 2014. A novel hybrid approach for estimating total deposition in the
    United States. *Atmos. Environ.* 92:207–220.
Science Subgroup. 1996. South Florida ecosystem restoration: Scientific information needs. Report to
    the Working Group of the South Florida Ecosystem Restoration Task Force.
Sparks, J.P., R.K. Monson, K.L. Sparks, and M. Lerdau. 2001. Leaf uptake of nitrogen dioxide (NO_2) in
    a tropical wet forest: Implications for tropospheric chemistry. *Oecologia* 127(2):214–221.
Sullivan, T.J., T.C. McDonnell, G.T. McPherson, S.D. Mackey, and D. Moore. 2011b. Evaluation of the
    sensitivity of inventory and monitoring national parks to nutrient enrichment effects from
    atmospheric nitrogen deposition. Natural Resource Report NPS/NRPC/ARD/NRR—2011/313.
    U.S. Department of the Interior, National Park Service, Denver, CO.

Sullivan, T.J., G.T. McPherson, T.C. McDonnell, S.D. Mackey, and D. Moore. 2011a. Evaluation of the sensitivity of inventory and monitoring national parks to acidification effects from atmospheric sulfur and nitrogen deposition. Natural Resource Report NPS/NRPC/ARD/NRR—2011/349. U.S. Department of the Interior, National Park Service, Denver, CO.

Sundlof, S.F., M.G. Spalding, J.D. Wentworth, and C.K. Steible. 1994. Mercury in livers of wading birds (Ciconiiformes) in southern Florida. *Arch. Environ. Contam. Toxicol.* 27:299–305.

Templer, P.H., W.L. Silver, J. Pett-Ridge, K.M. DeAngelis, and M.K. Firestone. 2008. Plant and microbial controls on nitrogen retention and loss in a humid tropical forest. *Ecology* 89(11):3030–3040.

U.S. Environmental Protection Agency. 2007. Everglades ecosystem assessment: Water management and quality, eutrophication, mercury contamination, soils and habitat. monitoring for adaptive management. A R-EMAP Status Report. EPA 904-R-07-001. Support Division and Water Region 4 Science & Ecosystem Management Division, Athens, GA.

U.S. Environmental Protection Agency. 2008. National air quality status and trends through 2007. EPA-454/R-08-006. U.S. Environmental Protection Agency, Office of Air Quality Planning and Standards, Air Quality Assessment Division, Research Triangle Park, NC.

U.S. Geological Survey (USGS). Last modified February 20, 2015. Predicted surface water methylmercury concentrations in National Park Service Inventory and Monitoring Program Parks. U.S. Geological Survey. Wisconsin Water Science Center, Middleton, WI. Available at: http://wi.water.usgs.gov/mercury/NPSHgMap.html (accessed February 26, 2015).

Ugarte, C.A., K.G. Rice, and M.A. Donnelly. 2005. Variation of total mercury concentrations in pig frogs (*Rana grylio*) across the Florida Everglades, USA. *Sci. Total Environ.* 345:51–59.

Wiener, J.G., D.P. Krabbenhoft, G.H. Heinz, and A.M. Scheuhammer. 2003. Ecotoxicology of mercury. *In* D.J. Hoffman, B.A. Rattner, G.A. Burton, and J. Cairns (Eds.). *Handbook of Ecotoxicology*, 2nd edn. CRC Press, Boca Raton, FL, pp. 409–463.

Wright, A.L., K.R. Reddy, and S. Newman. 2008. Biogeochemical response of the Everglades landscape to eutrophication. *Glob. J. Environ. Res.* 2(3):102–109.

Yanochko, G.M., C.H. Jagoe, and I.L. Brisbin. 1997. Tissue mercury concentrations in alligators (*Alligator mississippiensis*) from the Florida Everglades and the Savannah River Site, South Carolina. *Arch. Environ. Contam. Toxicol.* 32:323–328.

8

Great Lakes Network

8.1 Background

The Great Lakes Network contains nine national parks. Isle Royal (ISRO) and Voyageurs (VOYA) national parks are discussed here. Figure 8.1 shows the network boundary and the location of each national park, along with human population centers with more than 10,000 people. The largest urban centers in the network region are Chicago, Detroit, Minneapolis/St. Paul, and Milwaukee. Columbus and Indianapolis also have large populations and lie within a 300 mi (483 km) radius of the network. Air pollution effects have been well studied in the Great Lakes region, especially Hg biomagnification and the sensitivity of lakes to acidification.

8.2 Atmospheric Emissions and Deposition

Sullivan et al. (2011) identified sulfur dioxide point sources within the Great Lakes Network region with emissions higher than 20,000 tons of sulfur dioxide per year. In addition, sulfur dioxide point sources of greater magnitude were found to the southeast, outside of the network boundary. Relatively large N point sources (greater than 2500 tons/yr) also occur in the Great Lakes Network region. Many of the larger N point sources within the network region are located along the perimeter of the southern section of Lake Michigan. Most have been sources of oxidized, rather than reduced, N (Sullivan et al. 2011). There were also many smaller sources of ammonia near the network, mainly in the southern half of Minnesota.

Regional air pollutants are transported in the atmosphere to VOYA and ISRO. Pulp and paper plants in Minnesota and Ontario constitute local sources of oxidized N. Power plants, mining, and ore processing are important regional sources of sulfur dioxide, particulate matter, and oxidized N (Swackhamer and Hornbuckle 2004). ISRO and VOYA are located in the northernmost portion of the network, where emissions and deposition of S and N are generally lowest.

County-level emissions near the Great Lakes Network, based on data from the EPA's National Emissions Inventory during a recent time period (2011), are depicted in Figures 8.2 through 8.4 for sulfur dioxide, oxidized N, and ammonia, respectively. Many counties in the region had relatively high sulfur dioxide emissions (>5 tons/mi^2/yr; Figure 8.2). Spatial patterns in oxidized N emissions were generally similar, with highest values above 50 tons/mi^2/yr (Figure 8.3). Emissions of ammonia near Great Lakes Network parks

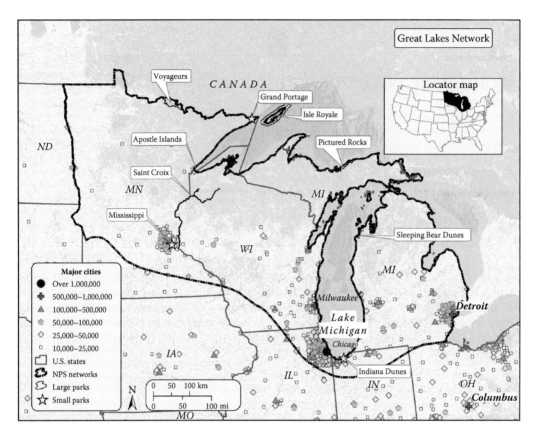

FIGURE 8.1
The Great Lakes Network boundary, locations of the nine network parks, and population centers with more than 10,000 people.

were somewhat lower, with most counties showing emissions levels below 8 tons/mi^2/yr (Figure 8.4).

Emissions of N from snowmobiles and boats are potentially important in VOYA, although the impacts are poorly understood. Snowmobile use at VOYA is second only to Yellowstone National Park among national parks (Swackhamer and Hornbuckle 2004).

Estimated total S deposition within the Great Lakes Network in 2002 ranged from less than 2 kg S/ha/yr in the northwestern portion of the network region, near the United States–Canadian border, to greater than 30 kg S/ha/yr in the south-southeastern portion, near Chicago and Lake Michigan (Sullivan et al. 2011). In general, the estimated S deposition within the network region ranged from 2 to 15 kg S/ha/yr; higher S deposition occurred to the south and lower S deposition occurred to the north. Total N deposition ranged from 2 to 5 kg N/ha/yr in the north, 15 to 20 kg N/ha/yr near heavily urbanized areas to the south, and up to 20 to 30 kg N/ha/yr at some locations. Throughout much of the network region, total N deposition ranged from 10 to 15 kg N/ha/yr in the south and 5 to 10 kg N/ha/yr in the north (Sullivan et al. 2011).

Total S deposition in and around the Great Lakes Network for the period 2010–2012 was generally highest (>5 kg S/ha/yr) to the east and lowest (<5 kg S/ha/yr) to the west of the network area. Oxidized inorganic N deposition for the period 2010–2012 was in

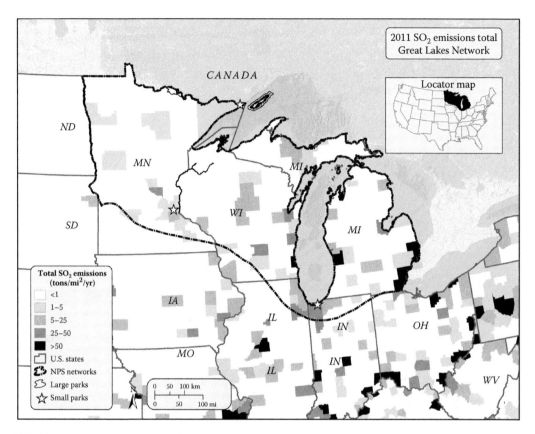

FIGURE 8.2

Total sulfur dioxide (SO$_2$) emissions, by county, near the Great Lakes Network for the year 2011. (Data from EPA's National Emissions Inventory, https://www.epa.gov/air-emissions-inventories, accessed January, 2014.)

the range of 5–10 kg N/ha/yr throughout some of the parklands within the Great Lakes Network. Most parklands received less than 10 kg N/ha/yr of reduced inorganic N from atmospheric deposition during this same period; a few areas received higher amounts. Total N deposition was variable, with generally lowest values to the north (Figure 8.5).

Total wet plus dry S deposition decreased by a relatively small amount at both study parks during the period 2001–2011 (Table 8.1). During that same time period, oxidized N deposition decreased at both parks by about one-fourth; results for reduced N were mixed.

Hall et al. (2005) reported total and methyl Hg concentrations in precipitation at five sites in the Great Lakes region, including one site at ISRO. Monitoring was conducted from 1997 to 2003. The highest methyl Hg concentrations were measured during small rain events, suggesting washout of methyl Hg from the atmosphere. This methyl Hg may have formed in conjunction with lake-effect clouds and fog, emitted from wetlands around Lake Superior, and/or released into the atmosphere by Lake Superior sediments (Hall et al. 2005).

Risch et al. (2012) reported wet Hg deposition throughout the Great Lakes region during the period 2002–2008, based on data from three Hg and precipitation monitoring networks. Areas in southern Indiana and Illinois, eastern Pennsylvania, and central Michigan had highest estimated wet Hg deposition, ranging from about 10 to 14 µg/m². Such high wet deposition levels were caused by high Hg concentration in precipitation (>12 ng/L)

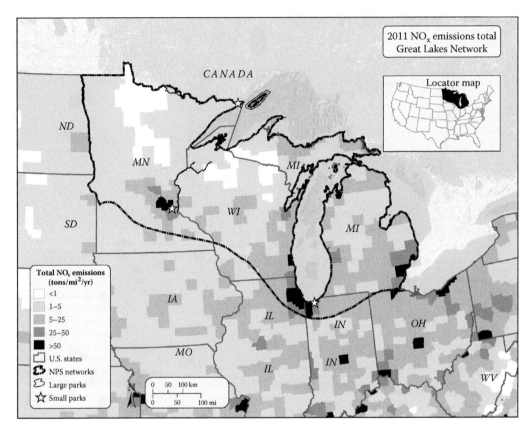

FIGURE 8.3
Total oxidized N (NO$_x$) emissions, by county, near the Great Lakes Network for the year 2011. (Data from EPA's National Emissions Inventory, https://www.epa.gov/air-emissions-inventories, accessed January, 2014.)

and/or high precipitation amount (120–140 cm/yr [47–55 inches/yr]). There was no trend in deposition over the course of the monitoring period; any small decrease that may have occurred in Hg concentration was apparently offset by increased precipitation (Risch et al. 2012). More recent analyses by Weiss-Penzias et al. (2016) show increases in Hg air concentrations in this region.

Initial scientific focus on Hg contamination in the Great Lakes Network region was directed at industrial point sources such as pulp and paper mills and chlor-alkali plants (Turner and Southworth 1999, Wiener et al. 2003, 2012a). However, point source release of Hg directly into surface waters decreased substantially in the 1980s, and atmospheric deposition is now the main source of Hg to watersheds in the Great Lakes region. The region contains a number of anthropogenic atmospheric emissions sources that collectively emit more than 50 kg/yr of Hg (Wiener et al. 2012a).

Dry Hg deposition estimates, derived by Zhang et al. (2012) using the Community Multiscale Air Quality and the Global/Regional Atmospheric Heavy Metals models, suggested that total annual dry Hg deposition was generally <5 µg/m^2 to the water surface of the Great Lakes. Estimates were higher, between 5 and 40 µg/m^2, to land surfaces. This result was attributed to higher dry deposition velocities to vegetative surfaces as compared with open water.

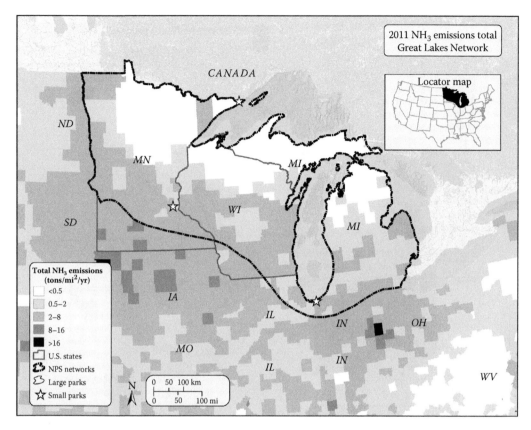

FIGURE 8.4

Total ammonia (NH_3) emissions, by county, near the Great Lakes Network for the year 2011. (Data from EPA's National Emissions Inventory, https://www.epa.gov/air-emissions-inventories, accessed January, 2014.)

8.3 Acidification

8.3.1 Acidification of Terrestrial Ecosystems

Site-specific data are generally not available regarding the sensitivity of terrestrial ecosystems to acidification in the Great Lakes Network or the extent to which terrestrial acidification has occurred in the parks in this network. There have been no local or regional studies documenting that terrestrial acidification is an important concern in this network. Nevertheless, sugar maple occurs and is known to be sensitive to acidification on base-poor soils. In northern New York, sugar maple is not regenerating on sites that have soil B-horizon base saturation less than about 12% (Sullivan et al. 2013). It is not known whether soil acidification has occurred sufficiently within the Great Lakes Network to cause damage to this sensitive tree species.

In response to decreased S deposition across the Great Lakes region since the 1980s, there have been large decreases in Michigan in the concentrations of various elements in sugar maple foliage: S (−16%), Ca (−17%), and Al (−42%; Talhelm et al. 2011). Thus, the likelihood of acidification effects on sugar maple may be declining in the Great Lakes

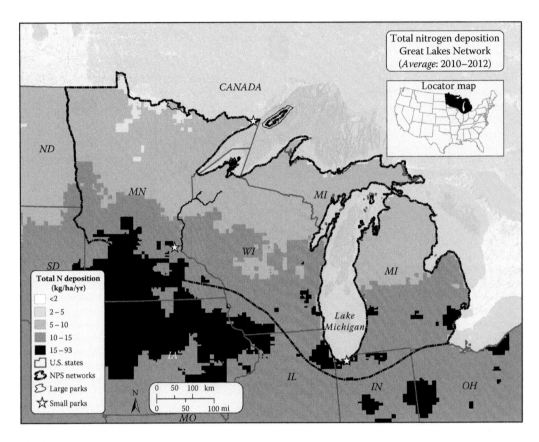

FIGURE 8.5
Total N deposition for the three-year period centered on 2011 in and around Great Lakes Network. (From Schwede, D.B. and Lear, G.G., *Atmos. Environ.*, 92, 207, 2014.)

TABLE 8.1

Average Changes in S and N Deposition[a] between 2001 and 2011 across Park Grid Cells at ISRO and VOYA Parks

Park Name	Parameter	2001 Average (kg/ha/yr)	2011 Average (kg/ha/yr)	Absolute Change (kg/ha/yr)	Percent Change	2011 Minimum (kg/ha/yr)	2011 Maximum (kg/ha/yr)	2011 Range (kg/ha/yr)
ISRO	Total S	3.55	3.26	−0.29	−8.4	2.51	5.17	2.66
	Total N	5.47	4.59	−0.88	−16.2	3.83	6.07	2.24
	Oxidized N	2.84	2.20	−0.65	−23.1	1.70	3.26	1.57
	Reduced N	2.62	2.39	−0.23	−8.8	2.12	2.85	0.73
VOYA	Total S	2.69	2.63	−0.06	−2.4	1.41	3.14	1.73
	Total N	5.61	5.05	−0.56	−9.8	3.67	5.45	1.78
	Oxidized N	3.13	2.32	−0.81	−25.7	1.61	2.52	0.91
	Reduced N	2.48	2.73	0.25	10.4	2.05	2.93	0.88

[a] Deposition estimates were determined by the Total Deposition project, based on three-year averages centered on 2001 and 2011 for all ~4 km grid cells in each park. The minimum, maximum, and range of 2011 S and N deposition within each park are also shown.

Network in response to decreases over time in S deposition and associated manifestations in tree foliage.

Bennett and Wetmore (1997) sampled and analyzed over a period of three years the chemical contents of four species of lichen (*Cladina rangiferina, Evernia mesomorpha, Hypogymnia physodes,* and *Parmelia sulcata*) at 10 sites along a 24 km (15 mi) transect between International Falls, Minnesota, and the park boundary of VOYA. There was not a clear pattern of decreasing metal and other element concentrations with increasing distance from S and metal emissions sources in International Falls in any of the lichens studied. However, the researchers concluded that there was sufficient evidence to suggest a high probability of physiological impairment to the lichens collected near VOYA from relatively high atmospheric inputs of S and heavy metals.

8.3.2 Acidification of Aquatic Ecosystems

The Upper Midwest contains numerous lakes created by glaciation. They have been the subject of extensive research on lake acidification (cf., Charles 1991, NAPAP 1991). Acid-sensitive lakes are common within the region, but not specifically within the Great Lakes Network parks. The region has little topographic relief. Acid-sensitive surface waters in the region are mainly groundwater recharge seepage lakes (Eilers et al. 1983). Most drainage lakes and some of the seepage lakes (the flow-through type) in the Upper Midwest receive substantial inflow from groundwater, which is generally high in base cation concentrations from dissolution of carbonate and silicate minerals. Relatively high concentrations of base cations in most of these lakes make them insensitive to acidification from acidic deposition. The seepage lakes that have low base cation concentrations, and that are therefore acid sensitive, generally receive most of their water input from precipitation directly on the lake surface (Baker et al. 1991). These groundwater recharge seepage lakes can be identified on the basis of having Si concentration less than about 1 mg/L (Baker et al. 1991). The acid sensitivity of streams within the region has not been well studied but is not expected to be especially high.

8.3.2.1 Status

Based on the EPA's Eastern Lakes Survey, the Great Lakes region had a large population of low acid neutralizing capacity lakes, but relatively few chronically acidic (acid neutralizing capacity ≤0 μeq/L) lakes in the 1980s (Linthurst et al. 1986a,b). Acidic lakes in this region are primarily small, shallow seepage lakes that are hydrologically isolated from the surrounding terrain and that have low concentrations of base cations and Al and moderate sulfate concentrations. Organic anions, estimated by both the Oliver et al. (1983) method and the anion deficit method, tended to be less than half the measured sulfate concentrations in the acidic lakes (Eilers et al. 1988). Organic anion concentrations were much higher in many of the drainage lakes in the region that are less sensitive to acidification from acidic deposition. These relatively insensitive, high dissolved organic C drainage lakes occur mainly in Minnesota, in the western portion of the Great Lakes Network region.

Lake water concentrations of inorganic N reported by the Eastern Lakes Survey were low throughout the Upper Midwest. In addition, snowmelt would not be expected to provide any significant nitrate influx to lakes in the Upper Midwest because most snowmelt in this region infiltrates the soil before reaching the drainage lakes and because snowmelt input of N into seepage lakes would be limited mainly to the snow on the lake surface and immediate nearshore environment.

Aluminum concentrations are far lower in the Great Lakes Network region than in lakes of similar pH in the northeastern United States. Base cation production is the dominant ion enrichment process in most lakes in the Great Lakes Network region. Even in groundwater-recharge seepage lakes having low acid neutralizing capacity, base cation production accounts for an estimated 72%–86% of total acid neutralizing capacity production (Cook and Jager 1991).

8.3.2.2 Trends

Regional trends analyses for long-term monitoring lakes in the Great Lakes Network region during the period 1990–2000 suggested that sulfate declined in lake water by −3.63 µeq/L/yr, whereas lake water nitrate concentrations were relatively constant (Stoddard et al. 2003). The large decrease in sulfate concentration in lake water was mainly balanced by a combination of a large decrease in base cation concentrations (−1.42 µeq/L/yr) and an increase in acid neutralizing capacity (+1.07 µeq/L/yr). All of these trends were significant at $p < 0.01$ (Stoddard et al. 2003). In the Great Lakes Network region, an estimated 80 of 251 lakes that were acidic in the mid-1980s were no longer acidic in 2000. This change was probably caused mainly by decreased S deposition (Stoddard et al. 2003).

8.4 Nutrient Enrichment

8.4.1 Terrestrial Ecosystems

Reduced species richness and biodiversity in response to N addition are of concern in the Great Lakes Network, in large part because of the potential interactions among diversity and ecosystem processes and functioning. It is more likely that a species-poor plant community may exhibit compromised ecosystem functioning as compared with a species-rich community when exposed to elevated N input.

Reich et al. (2001) investigated the influence of plant species composition and diversity on plant biomass enhancement and C acquisition in response to nutrient addition in a grassland in Minnesota. Exposure to added carbon dioxide, N, or both, enhanced plant biomass accumulation. The effect was greater in species-rich plant communities, as compared with species-poor communities.

Experimental studies in successional fields in Cedar Creek Natural History Area in Minnesota have shown that N is the major limiting resource in grasslands, given sufficient precipitation. Terrestrial plant biomass increased and community composition changed after N addition to grasslands (Tilman 1987, Tilman and Wedin 1991). Increased availability of N to grasses can also affect herbivores that feed on grasses by altering food quality, quantity, and phenology and also perhaps by changing the relationships between herbivores and their predators (Throop and Lerdau 2004). Nitrogen fertilization at rates of 54 and 170 kg N/ha/yr (as ammonium nitrate) led to a decline in plant species composition in an oak savanna site (Avis et al. 2003). These fertilization levels are much higher than ambient N deposition in the Great Lakes Network region. In the control plots, five species collectively accounted for more than 40% of the plant cover versus four plant species in the lower N addition plots. In the higher N addition plots, a single plant species accounted for more than 40% of the plant cover.

Nitrogen additions of 120 kg N /ha/yr in Michigan old fields had a significant positive growth effect on annual dicot biomass but no significant growth effect on annual grass biomass (Huberty et al. 1998). In tallgrass prairie, C_3 grasses (*Elymus virginicus, E. canadensis* L.) showed a greater positive growth response to N additions than C_4 grasses (*Andropogon geradii, Schizachyrium scoparium*) and forbs (*Solidago nemoralis, S. rigida*; Lane and BassiriRad 2002). Species with smaller initial biomass exhibited the greatest increase in biomass, with a seven- to eightfold increase for *S. nemoralis* and *E. canadensis* and only a threefold increase for *S. rigida* (Lane and BassiriRad 2002). In experiments where N fertilization was applied to common ragweed (*Ambrosia artemisifolia*), vegetative and seed biomass increased and root:shoot ratios decreased (Throop 2005).

Experimental N addition at >50 kg N/ha/yr to tallgrass prairie increased productivity and decreased species richness, probably as a consequence of reduced light penetration and increased litter biomass (Baer et al. 2003). Wedin and Tilman (1996) presented results of 12 years of experimental N addition to 162 grassland plots in Minnesota. The N loading dramatically changed plant species composition, decreased species diversity, and increased aboveground productivity in experimental plots. Species richness declined by more than 50% across the N application gradient, with the greatest losses at levels of N input of 10 to 50 kg N/ha/yr. This loss of species diversity was accompanied by large changes in plant species composition, with C_4 grasses declining and the weedy Eurasian C_3 grass quackgrass (*Agropyron repens*) becoming dominant at high N addition rates (Wedin and Tilman 1996). The authors concluded that very high N loading is a major threat to grassland ecosystems and causes loss of diversity, increased abundance of nonnative species, and the disruption of ecosystem functioning. A major uncertainty, however, is the rate of N loading at which such changes may be manifested. Total N loading to Great Lakes Network ecosystems is lower than the loading rates used in the experimental treatments of Wedin and Tilman (1996) and the other terrestrial effects studies discussed earlier.

Effects of N fertilization, in turn, can influence faunal populations. For example, fertilization in a Minnesota old field at modest rates (≤20 kg N/ha/yr) for 14 years resulted in decreased species diversity of insects in response to decreased diversity of plant food sources. Plant species richness decreased, and quackgrass and Kentucky bluegrass (*Poa pratensis*) became dominant in response to the N addition (Haddad et al. 2000). Changes in the abundance of insect functional groups were also observed. Herbivores (especially the dominant species) increased in numbers; parasitoid insect species decreased. Over the long term, changes in plant species composition would be expected to either increase or decrease insect herbivore activity, depending on whether there is a shift toward or away from herbivore-preferred plant species (Throop et al. 2004, Clark 2011).

In another study, experimental N addition for 12 years to two Minnesota old fields at levels of 10 and 20 kg N/ha/yr decreased the abundance of four native grasses by almost 20%, and increased the abundance of invasive Kentucky bluegrass and quackgrass (Knops and Reinhart 2000). Losses of grass species and changes in species composition generally increased over time with experimental N addition, especially at lower addition levels (Clark 2011). After 22 years of N addition at a level of 10 kg N/ha/yr, the number of species declined by 17% (Clark and Tilman 2008). Tilman (1993) hypothesized that species loss was caused mainly by increased litter mass, although light limitation from competing plants was also likely important (Hautier et al. 2009).

Much of the work on eutrophication effects of N deposition on grasslands has been conducted in Europe. Nitrophilous plant species have been shown to increase in abundance, and N-sensitive species to decline in European grasslands, over approximately the last half of the twentieth century (Bobbink et al. 1998). The clearest evidence that atmospheric N deposition over a large region has actually impacted terrestrial biodiversity was provided by Stevens et al. (2004), who reported the results of 68 plot surveys (4 m² [43 ft²] quadrants) across a range of atmospheric N deposition from 5 to 35 kg N/ha/yr. Species richness was found to decline as a linear function of N deposition, with a reduction on average of one plant species for every 2.5 kg N/ha/yr of N input.

Forests can also be responsive to nutrient addition. Northern hardwood forests that predominate in the northern portions of Minnesota, Wisconsin, and Michigan are expected to show growth enhancement by some, but not all, tree species in response to N addition. Eastern hardwood forests from Wisconsin to Maine and south to Virginia were shown to have experienced increased growth with increasing N deposition across a depositional gradient from about 3 to 11 kg N/ha/yr (Thomas et al. 2010). Nevertheless, responses were species specific. Growth increases were most pronounced for red maple (*Acer rubrum*), sugar maple, white ash, yellow poplar, black cherry, balsam fir, pignut hickory (*Carya glabra*), eastern white pine, trembling aspen, northern red oak, and scarlet oak. Other tree species showed negative response to N addition. Growth decreased by a statistically significant level with increasing deposition for red pine, red spruce, and northern white cedar (*Thuja occidentalis*). Mortality increased with increasing deposition for yellow birch (*Betula alleghaniensis*), eastern white pine, basswood (*Tilia* sp.), trembling aspen, bigtooth aspen (*Populus grandidentata*), scarlet oak, chestnut oak, and northern red oak. At higher rates of N supply on nutrient-poor sites, tree growth might be expected to decline (Pardo et al. 2011b).

Eastern hardwood forests can also respond to increased N deposition with a decrease in herbaceous plant biodiversity (Gilliam et al. 2006). The response of the herbaceous layer of the forest can be pronounced, with initial increases in herbaceous plant cover, followed by decreases in species richness and species evenness, two components of biodiversity. The response time is typically shorter if ambient N deposition is relatively high (Gilliam et al. 2006, Fraterrigo et al. 2009, Royo et al. 2010, Gilliam et al. 2011). Because the plants in the herbaceous layer of eastern hardwood forests tend to have foliage with relatively high nutrient content, herbaceous plants in these forests influence N cycling to a level that is disproportionate to their biomass (Muller 2003, Moore et al. 2007, Welch et al. 2007, Gilliam et al. 2011). Lichens have also been heavily impacted throughout the eastern hardwood forests. High levels of air pollution, including both S and N, has played an important role in causing such impacts (McCune 1988, Wetmore 1988).

Pardo et al. (2011a) compiled data on empirical critical loads for protecting sensitive resources in Omernik Level I ecoregions across the conterminous United States against nutrient enrichment effects caused by atmospheric N deposition. Available data on empirical critical loads of nutrient N suggested that the lower end of the critical load range for resource protection was about 3–5 kg N/ha/yr in ISRO and VOYA for the protection of mycorrhizal fungi, lichens, and forests (Table 8.2). Ambient N deposition reported by Pardo et al. (2011a) was higher than the empirical critical load at all of the parks in the Great Lakes Network that were evaluated. At some parks, including ISRO and VOYA, empirical critical loads were also exceeded for protecting herbaceous plants and preventing nitrate leaching in drainage water. Thus, there appears to be widespread exceedance of these nutrient empirical critical loads throughout the Great Lakes Network parks. This pattern roughly reflects the north–south human population gradient.

A critical load exceedance map for VOYA is shown in Figure 8.6. Much of the park is covered by vegetation thought to be sensitive to nutrient N enrichment. Essentially all of the identified N-sensitive plant community types in this park are in exceedance, based

TABLE 8.2

Empirical Critical Loads for Nitrogen in the Great Lakes Network, by Ecoregion and Receptor, from Pardo et al. (2011a)[a]

National Park Service Unit	Ecoregion	N Deposition (kg N/ha/yr)	Critical Load (kg N/ha/yr)				
			Mycorrhizal Fungi	Lichen	Herbaceous Plant	Forest	Nitrate Leaching
Voyageurs NP	Northern Forests	4.7	5–7	4–6	7–21	3–26	8
Isle Royale NP	Northern Forests	5.0	5–7	4–6	7–21	3–26	8

[a] Ambient N deposition reported by Pardo et al. (2011a) is compared to the lowest critical load for a receptor to identify potential exceedance, indicated by gray cells. A critical load exceedance suggests that the receptor is at increased risk for harmful effects.

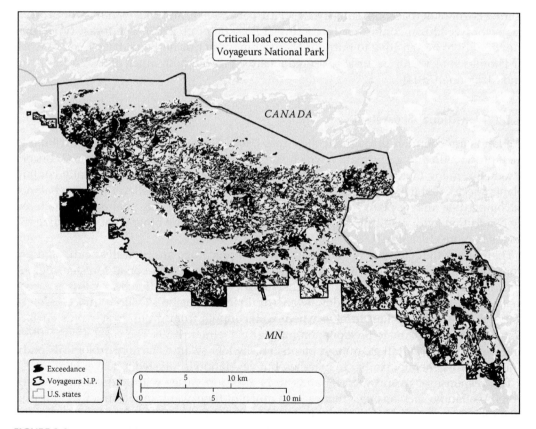

FIGURE 8.6

Locations of vegetation types in Voyageurs National Park that exhibit critical load (CL) exceedance due to nutrient N enrichment. The area of sensitive vegetation is shaded if total N deposition in the year 2008 is in exceedance of the lower limit of the critical load (3 kg N/ha/yr) estimated by Pardo et al. (2011b) for the sensitive vegetation types found in the park. (From Sullivan, T.J. and McDonnell, T.C., Mapping of nutrient-nitrogen critical loads for selected national parks in the intermountain west and Great Lakes regions, Report prepared for National Park Service, Natural Resource Technical Report NPS/ARD/NRTR—2014/895, National Park Service, Fort Collins, CO, 2014.)

on the Pardo et al. (2011b) lower limit of the empirical critical load of 3 kg N/ha/yr and atmospheric deposition model projections of 4–6 kg N/ha/yr of ambient deposition. An analysis of exceedance based on Greaver et al. (2011) estimate of peatland critical load of 2.7 kg N/ha/yr would yield similar results.

Ellis et al. (2013) estimated the critical load for nutrient N deposition to protect the most sensitive ecosystem receptors in 45 national parks based on the critical load data of Pardo et al. (2011a). The lowest terrestrial critical load of N is generally estimated for the protection of lichens (Geiser et al. 2010). Changes to lichen communities may signal the beginning of other changes to the ecosystem that might affect structure and function (Pardo et al. 2011c). Ellis et al. (2013) estimated the N critical load for ISRO and VOYA in the range of 3–8 kg N/ha/yr for the protection of forests.

Common vegetation types within VOYA that are considered potentially sensitive to nutrient N enrichment effects attributed to atmospheric N deposition include forests, wetlands, and mycorrhizal fungi (Pardo et al. 2011b). Based on the gradient analysis of Thomas et al. (2010), the critical loads for protecting forest ecosystems in VOYA were assumed to be as low as 3 kg N/ha/yr for protecting forest trees, 7 kg N/ha/yr for protecting cover of herbaceous plant species (Hurd et al. 1998), and 5 kg N/ha/yr for protecting ectomycorrhizal community structure (Lilleskov et al. 2008). The forest types that are thought to be sensitive to nutrient N enrichment in this network, based on the work of Thomas et al. (2010), occur at scattered locations throughout much of VOYA (Sullivan and McDonnell 2014).

8.4.2 Transitional Ecosystems

Wetlands are common throughout the Great Lakes Network region. They contribute to high production of organic matter, which is reflected in high dissolved organic C concentrations in many lakes. Wetlands considered sensitive to N deposition typically contain plants species that have evolved under N-limited conditions. It is believed that the balance of competition among plant species in some sensitive wetland ecosystems can be altered by N addition, with resulting displacement of some species by others that can utilize the excess N more efficiently (U.S. EPA, 1993).

There are different types of wetlands. Ombotrophic bogs are typically acidic and are dominated by mosses. They are especially common in northern boreal forested regions and develop where precipitation is higher than evapotranspiration water loss in areas that exhibit an impediment to downward drainage in the soil (Mitsch and Gosselink 2000). Freshwater marshes develop where water inputs from groundwater plus surface water inflow approximate precipitation input (Koerselman 1989). The vegetation in these marshes is primarily tall graminoid plants. Freshwater swamps have hydrological conditions that are generally similar to marshes, but vegetation is forested (Greaver et al. 2011).

The response of freshwater wetlands to N addition varies with hydrological conditions. Wetlands, such as bogs, that receive much of their water input directly from precipitation are most sensitive to the effects of N input (Morris 1991). At the high end of the N deposition responsiveness threshold are the ombotrophic bogs. Most of their nutrient supply comes from precipitation and as a consequence plants are adapted to low inputs of N and other nutrients (Shaver and Melillo 1984, Bridgham et al. 1995). Atmospheric N deposition generally has greater impact on the bog systems as compared with other kinds of wetlands.

Peat-forming bog ecosystems are among the transitional ecosystems most sensitive to the effects of N deposition. In the conterminous United States, peat-forming bogs are most common in areas that were glaciated, especially in portions of the Northeast and Upper Midwest (U.S. EPA, 1993). Bogs can host several federally listed rare and endangered plant species. These include multiple species of quillworts (*Isoetes* spp.), sphagnum mosses (*Sphagnum* spp.), and the green pitcher plant (*Sarracenia oreophila*; Greaver et al. 2011).

Nutrient concentrations in wetland waters associated with the Laurentian Great Lakes suggest that algal productivity in coastal Great Lakes wetlands is N-limited. Historically, freshwater ecosystems, including the open waters of the Great Lakes themselves, have been considered P limited (Hutchinson 1971, Rose and Axler 1998). However, Hill et al. (2006) found that more coastal wetlands were N than P limited at each of the five Laurentian Great Lakes, and Morrice et al. (2004) measured a low ratio of N to P in Lake Superior coastal wetlands, also suggesting N limitation. These results are consistent with the apparent general N limitation of North American marsh lands (Bedford et al. 1999) and may be due to differences in nutrient cycling in wetlands as compared to open-water ecosystems (Morrice et al. 2004). Wetlands sediments tend to be oxygen poor and organic C oxidation by primary producers in anoxic environments can be limited by the availability of nitrate for use as an electron acceptor (Sundareshwar et al. 2003).

8.4.3 Great Lakes Ecosystems

Nutrient loading to coastal wetlands is a concern throughout the lower Great Lakes (Lakes Erie and Ontario, and the southern part of Lake Michigan) and in some localized areas of the upper lakes (Hill et al. 2006). Both agricultural and atmospheric sources of nutrients contribute to this stress. Atmospheric N deposition has been shown to increase from Lake Superior (north) to Lake Ontario (south), with N deposition in the upper lakes classified by the National Atmospheric Deposition Program as "moderately high," as compared to "very high" in the lower lakes (NADP 2004, Hill et al. 2006). Agricultural inputs of nutrients to wetlands also follow this north–south trend, with the highest chemical loading to Lake Erie and the lowest to Lake Superior (Hill et al. 2006).

Hill et al. (2006) showed that total N dissolved in surface waters of Great Lakes wetlands is directly correlated with atmospheric N deposition and the degree of nearby agricultural activity. Consistent with this, they also found lower N concentrations in wetland sediments in upper lakes (Lakes Superior and Huron) as compared with lower lakes. This can result in greater increases in eutrophication, turbidity, sedimentation, and algal blooms in watersheds associated with the lower lakes.

The extent to which wetlands in the Great Lakes Network have been affected by nutrient enrichment from N deposition is not clear. Wetlands are widely distributed, including within parks that receive moderate levels of N deposition. The moderate levels of N deposition commonly found in areas that contain appreciable areas of wetland may or may not be sufficiently high to cause shifts in wetland plant species. If such effects do occur, they are most likely in wetlands such as bogs and poor fens that normally receive most of their nutrients from atmospheric inputs. Such ecosystems in the Great Lakes Network tend to occur in areas that receive much lower levels of N deposition than do the affected wetlands in Europe. It is not clear to what extent such effects occur under ambient N deposition levels. The risk of species composition change is important, in part because wetland ecosystems often contain relatively large numbers of rare plant species.

Greaver et al. (2011) estimated that the critical load of atmospheric N deposition to protect peatlands from increased productivity, with possible change in species composition, was in the range of 2.7–13 kg N/ha/yr. This estimate was based on the consideration of the results of the studies of Aldous (2002), Moore et al. (2004), Rochefort et al. (1990), and Vitt et al. (2003).

Some studies on aquatic nutrient limitation have been conducted on the Great Lakes. It is generally believed that the Laurentian Great Lakes are P limited (Schelske 1991, Downing and McCauley 1992, Rose and Axler 1998). Water quality in the open waters of these lakes has improved in response to controls on point sources of P (Nicholls et al. 2001). Work by Levine et al. (1997), however, suggested a more complicated pattern of response to nutrient addition for Lake Champlain. They added nutrients to *in situ* enclosures and measured indicators of P status, including alkaline phosphatase activity and orthophosphate turnover time. Although P appeared to be the principal limiting nutrient during summer, N addition also resulted in algal growth stimulation. Phosphorus sufficiency appeared to be as common as P deficiency. During spring, phytoplankton growth was not limited by P, N, or Si, but perhaps by light or temperature (Levine et al. 1997).

In Lake Erie, N input was found to be limiting to phytoplankton on a seasonal basis (Moon and Carrick 2007, North et al. 2007). Lake nitrate concentrations in Lake Superior increased over the twentieth century, causing strong P limitation (Finlay et al. 2007, Sterner et al. 2007).

Silica depletion due to nutrient enrichment has been reported for the Great Lakes (Conley et al. 1993). Increased growth of silicate-utilizing diatoms as a result of nitrate- and phosphate-induced eutrophication, and subsequent removal of fixed biogenic Si via sedimentation, has the potential to change the ratios of nutrient elements Si, N, and P. In turn, such changes might cause shifts from diatoms to nonsiliceous phytoplankton (Ittekot 2003).

A diatom biomonitoring program is conducted by the Great Lakes Network as a complement to water quality monitoring efforts. In 2011 and 2012, it included monitoring of lakes, lagoons, and flowages in ISRO (Edlund et al. 2013). Recent data collected in this monitoring effort suggest that many Great Lakes Network lakes are undergoing biological change. Continued monitoring will be needed to determine the cause(s) and effects.

Recent changes in water chemistry in the Great Lakes could also result from the 1989 and 1991 introduction of two exotic dreissenid mussel species, whose filtration capacity can drastically deplete phytoplankton nutrient levels and lake production (Mills et al. 1993). Mills et al. (2006) found declines in open-water P concentrations from 1995 to 2005 in Lake Ontario that are thought to be detrimental to healthy levels of ecosystem productivity. This result is consistent with observed increases in water transparency (Barbiero et al. 2006) and declining phytoplankton biomass (Munawar et al. 2006) in the lake. Specifically, diatom abundances in Lake Ontario have decreased since the invasion of dreissenid mussels, leading to a corresponding increase in Si concentrations in lake water in response to reduced Si uptake by diatoms (Millard et al. 2003).

Dove (2009) found that the proportion of total filtered P has increased since 1992, particularly in nearshore waters where mussels are most abundant. Excretion by invasive mussels may also explain the elevated proportion of soluble reactive P, a measure of biologically available P, observed in nearshore waters as compared with low soluble reactive P levels in open waters (Dove 2009). Thus, the increasing incidence of nuisance algal blooms along Lake Ontario's shore may be, at least in part, a result of mussel activity rather than further nutrient addition (Dove 2009).

Nitrate appears to have accumulated in Lake Ontario (Dove 2009) and in Lake Superior (Finlay et al. 2007). This was mostly likely due to nitrification of N derived from external sources such as agriculture, human waste, and atmospheric deposition (Bennett 1986, Finlay et al. 2007). As a result, N:P ratios in these lakes are relatively high; this relative N abundance and P depletion may help protect lake waters against cyanobacteria blooms, which tend to dominate at low N:P (Smith 1983, Dove 2009).

The EPA Great Lakes National Program has set trophic state goals for each of the Laurentian Great Lakes. Most recent data show that the upper lakes (Lakes Superior, Huron, and Michigan) are currently meeting their oligotrophy goals. Surface water quality in these lakes is considered "excellent," and water quality in Lakes Huron and Michigan shows improvement since monitoring began in the 1980s (http://www.epa. gov/glindicators/water/trophicb.html). Lake Ontario is also currently meeting its trophic state goal of oligomesotrophy. However, Lake Erie shows signs of concern. Historically, it was the first lake to show evidence of nuisance algal blooms and oxygen-depletion problems associated with eutrophication. Lake Erie is also most affected by proximate agriculture and urban areas. Surface waters in Lake Erie's eastern, central, and western basins are currently meeting their EPA tropic state goals of oligotrophy, oligomesotrophy, and mesotrophy, respectively. However, the central basin has experienced elevated P concentrations since the 1990s and shows a corresponding summer oxygen depletion in its deeper waters, jeopardizing the health of fish and other aquatic biota (EPA; http://www.epa.gov/ glindicators/water/trophicb.html).

8.5 Ozone Injury to Vegetation

The ozone-sensitive plant species that are known or thought to occur within ISRO and VOYA are listed in Table 8.3. Those considered to be bioindicators, because they exhibit distinctive symptoms when injured by ozone (e.g., dark stipple), are designated in the table by an asterisk. Both parks support a number of ozone-sensitive and bioindicator plant species.

The W126 and SUM06 exposure indices calculated by National Park Service staff are given in Table 8.4, along with Kohut's (2007a) ozone risk ranking. Kohut's approach considers both ozone exposure and environmental conditions (soil moisture). The risk to plants in ISRO and VOYA was judged to be low. Neither of the parks showed a clear association between ozone exposure and soil moisture conditions. Soil moisture varies throughout the growing season and locally from one area to another within a park. During periods of water stress, the likelihood of injury may be reduced. Under mild water stress, however, the soil moisture level may not be low enough to significantly constrain the uptake of ozone into foliage and consequent injury to vegetation (Kohut 2007b). The criteria for injury under the SUM06 and W126 indices were generally not satisfied. Similarly, Swackhamer and Hornbuckle (2004) reported that there was no evidence of symptoms attributed to ozone exposures in ISRO or VOYA.

However, Percy et al. (2007) developed dose-response regression relationships to predict growth decreases in trembling aspen in the Great Lakes region based on measured ozone, expressed as the fourth highest daily maximum 8-hour average ozone concentration, growing degree days, and wind speed. The models had high statistical significance. They estimated appreciable growth reductions in aspen trees throughout substantial portions of the Great Lakes region due to ambient ozone exposure.

TABLE 8.3

Ozone-Sensitive and Bioindicator (Asterisk) Plant Species Known or Thought to Occur in ISRO and VOYA[a]

Species	Common Name	ISRO	VOYA
		Park	
Alnus incana spp. *rugosa**	Speckled alder	×	
Amelanchier alnifolia	Saskatoon serviceberry		×
*Apocynum androsaemifolium**	Spreading dogbane	×	×
Apocynum cannabinum	Dogbane, Indian hemp		×
*Artemisia ludoviciana**	Silver wormwood		×
Asclepias incarnata	Swamp milkweed	×	×
*Asclepias syriaca**	Common milkweed	×	×
Clematis virginiana	Virgin's bower	×	
*Corylus americana**	American hazelnut		×
*Eurybia macrophylla**	Big-leaf aster	×	×
Fraxinus pennsylvanica	Green ash	×	×
Parthenocissus quinquefolia	Virginia creeper		×
Pinus banksiana	Jack pine	×	×
*Populus tremuloides**	Trembling aspen	×	×
*Prunus serotina**	Black cherry		×
Prunus virginiana	Chokecherry	×	×
*Rubus allegheniensis**	Allegheny blackberry	×	×
*Rubus canadensis**	Thornless blackberry	×	×
*Rubus parviflorus**	Thimbleberry	×	×
*Rudbeckia laciniata**	Cutleaf coneflower		×
Sambucus racemosa var. *racemosa**	Red elderberry	×	
Solidago canadensis	Goldenrod	×	
*Symphoricarpos albus**	Common snowberry	×	×

[a] Lists are periodically updated and available at https://irma.nps.gov/NPSpecies/Report.

TABLE 8.4

Ozone Assessment Results for ISRO and VOYA Based on Estimated Average 3-Month W126 and SUM06 Ozone Exposure Indices for the Period 2005–2009 and Kohut's (2007a) Ozone Risk Ranking for the Period 1995–1999[a,b]

Park Name	Park Code	W126		SUM06		Kohut Ozone Risk Ranking
		Value (ppm-h)	NPS Condition	Value (ppm-h)	NPS Condition	
Isle Royale	ISRO	4.95	Low	4.97	Low	Low
Voyageurs	VOYA	4.38	Low	3.78	Low	Low

[a] Parks are classified into one of three ranks (Low, Moderate, High) based on comparison with other Inventory and Monitoring parks.

[b] Degrees of concern for the W126 and SUM06 indices are based solely on levels of ozone exposure. Kohut's risk to vegetation is based on several factors that contribute to injury in plants, including ozone exposure and environmental variables, and considers the effects of soil moisture on the uptake of ozone.

8.6 Visibility Degradation

8.6.1 Natural Background and Ambient Visibility Conditions

Current visibility estimates reflect recent pollution levels and were used to rank conditions at parks in order to provide park managers with information on spatial differences in visibility and air pollution. Rankings range from very low haze (very good visibility) to very high haze (very poor visibility). Table 8.5 gives the relative park haze rankings for ISRO and VOYA on the 20% clearest, 20% haziest, and average days.

Natural background visibility, the goal of the Clean Air Act and the Regional Haze Rule, assumes no human-caused pollution. At both ISRO and VOYA, natural haze is relatively high in comparison with other parks for the 20% clearest natural haze conditions and for the average of all natural haze conditions (Table 8.5). Natural haze levels are particularly high at these parks on the 20% haziest days. This may be partly due to the influence of wildfires. The measured haze for the period 2004–2008 was considerably higher than the estimated natural condition largely due to organic mass. Current haze in ISRO was classified as moderate for the 20% clearest and average days and high for the 20% haziest days. Current haze in VOYA was ranked as moderate for all groups.

Interagency Monitoring of Protected Visual Environments (IMPROVE) data allow estimation of visual range. Data indicate that air pollution at ISRO has reduced average visual range from 110 to 60 mi (177 to 97 km). On the haziest days, visual range has been reduced from 65 to 30 mi (105 to 48 km). Severe haze episodes occasionally reduce visibility to 7 mi (11 km). At VOYA, pollution has reduced average visual range from 110 to 60 mi (177 to 97 km). On the haziest days, visual range has been reduced from 70 to 35 mi (113 to 56 km). Severe haze episodes occasionally reduce visibility to 11 mi (18 km).

8.6.2 Composition of Haze

The majority of the current total particulate light extinction in both of the parks is attributable to sulfate (Figure 8.7). On average, about half of particulate light extinction was

TABLE 8.5

Estimated Natural Haze and Measured Ambient Haze in ISRO and VOYA Averaged over the Period 2004–2008[a]

Park Name	Site ID	Estimated Natural Haze (in Deciviews)		
		20% Clearest Days	20% Haziest Days	Average Days
Isle Royale	ISLE1	3.72	12.37	7.33
Voyageurs	VOYA2	4.26	12.06	7.58

		Measured Ambient Haze (for Years 2004–2008)					
		20% Clearest Days		20% Haziest Days		Average Days	
Park Name	Site ID	dv	Ranking	dv	Ranking	dv	Ranking
Isle Royale	ISLE1	6.31	Moderate	21.55	High	12.54	Moderate
Voyageurs	VOYA2	6.54	Moderate	19.43	Moderate	11.96	Moderate

[a] Parks are classified into one of five haze ranks (Very Low, Low, Moderate, High, or Very High) based on comparison with other monitored parks.

attributable to sulfate, and nitrate contributed about 20%. On the 20% clearest visibility days, sulfate accounted for more than 50% of extinction, followed by organics, coarse mass, and nitrate. On the 20% haziest days, sulfate again was the dominant source of light extinction in both parks, followed by nitrate and organics (Figure 8.7). The largest difference between clear days and hazy days is that coarse mass contributes more extinction on clear days, and nitrate contributes more on hazy days.

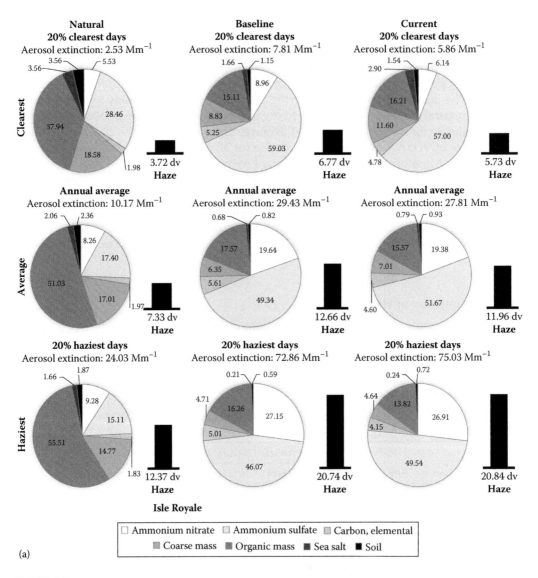

FIGURE 8.7
Estimated natural (preindustrial), baseline (2000–2004), and current (2006–2010) levels of haze (columns) and its composition (pie charts) on the 20% clearest, annual average, and 20% haziest visibility days for (a) ISRO and (b) VOYA. Note that there were no data available for natural conditions at VOYA. (From http://views.cira. colostate.edu/fed/Tools/RegionalHazeSummary.aspx, accessed October, 2012.)   *(Continued)*

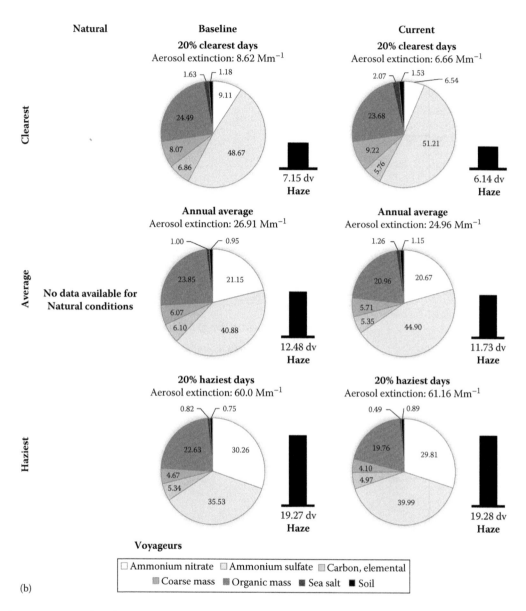

FIGURE 8.7 (*Continued*)
Estimated natural (preindustrial), baseline (2000–2004), and current (2006–2010) levels of haze (columns) and its composition (pie charts) on the 20% clearest, annual average, and 20% haziest visibility days for (a) ISRO and (b) VOYA. Note that there were no data available for natural conditions at VOYA. (From http://views.cira. colostate.edu/fed/Tools/RegionalHazeSummary.aspx, accessed October, 2012.)

8.6.3 Development of State Implementation Plans

The measurements made during the most recent years of monitoring at each park suggest progress toward compliance with the Regional Haze Rule (Figure 8.8). Visibility improvement in the Great Lakes Network may be on track to attain natural visibility conditions (~12 deciviews) by 2064. This will require a reduction of approximately 50% in haze on the 20% haziest days (Figure 8.8).

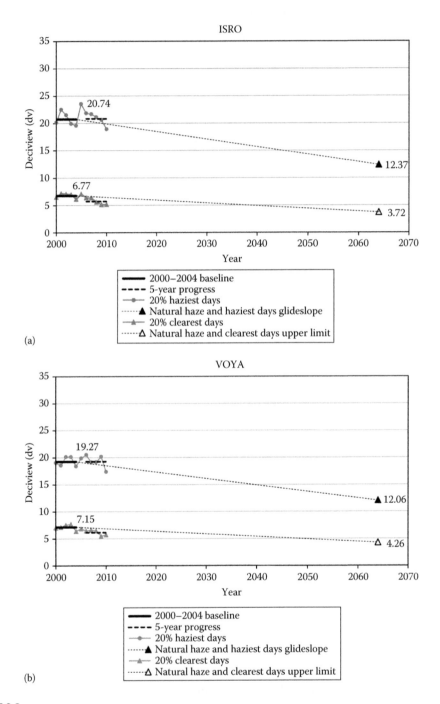

FIGURE 8.8
Glideslopes to achieving natural visibility conditions in 2064 for the 20% haziest (upper line) and the 20% clearest (lower line) days in (a) ISRO and (b) VOYA. In the Regional Haze Rule, the clearest days do not have a uniform rate of progress glideslope; the rule only requires that the clearest days do not get any worse than the baseline period. Also shown are measured values during the period 2000–2010. (From http://vista.cira.colostate.edu/improve/Data/IMPROVE/summary_data.htm, accessed October, 2012.)

8.7 Toxic Airborne Contaminants

In large part because of the common occurrence of wetlands, Hg methylation and bio-accumulation are important concerns for ISRO and VOYA. Estimates of Hg methylation potential generated by the U.S. Geological Survey (last modified February 20, 2015) for watershed boundaries (based on eight-digit hydrologic unit codes) containing national parklands in the Great Lakes Network suggested high methylation potential at most parks (Figure 8.9). This result is likely driven mainly by relatively high concentrations of total organic carbon in surface waters.

Swackhamer and Hornbuckle (2004) assessed air quality and air pollutant impacts in ISRO and VOYA in a report to the National Park Service. They concluded that the air pollutants of most concern in these two parks were Hg and persistent organic pollutants, such as dioxins, PCBs, chlorinated pesticides, and brominated flame retardants. Exposure of wildlife to these chemicals occurs largely through fish consumption and may be linked to reproductive, neurological, and developmental problems. Inputs of these toxic substances to park ecosystems are primarily atmospheric, attributable mostly to long-range transport.

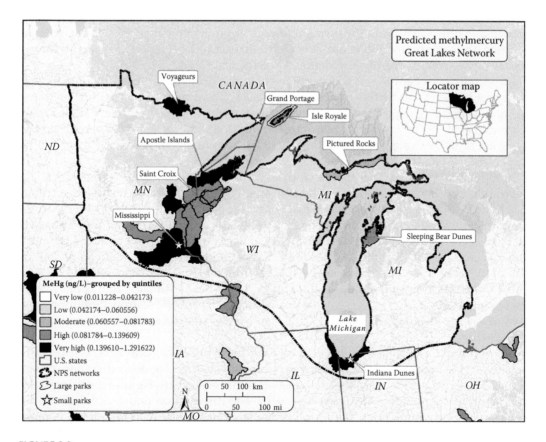

FIGURE 8.9
Predicted methylmercury (MeHg; in ng/L) concentrations in surface waters by hydrologic unit codes that contain national parklands in the Great Lakes Network. Estimates were generated by the U.S. Geological Survey (last modified February 20, 2015). Rankings are based on quintile distributions across all parks having estimates by the U.S. Geological Survey.

Possible effects of atmospherically deposited toxic metals have been the focus of research interest at ISRO for several decades (cf., Bennett 1995). Subtle patterns of slightly elevated elements in two lichen species collected in ISRO suggested potential impacts from atmospheric deposition. Lichens collected from areas having higher exposures were found to have higher element concentrations of S, Mn, selenium (Se), and other heavy metals (Bennett 1995).

The National Park Service and the U.S. Geological Survey have monitored lichen species in parks of the Great Lakes Network for decades (Bennett 2007). The parks south of 46°N latitude average 180 lichen species per park. Parks north of 46°N latitude average more than twice as many lichen species per park (Bennett 2007). Elements associated with human activities (i.e., Cu, Pb, S, Zn) increase in lichen tissue from west to east, with increasing proximity to human population centers.

Trends in contaminant concentrations in bald eagles along the southern shore of Lake Superior decreased substantially between 1989 and 2008 (Dykstra et al. 2010). The concentrations of DDE (a breakdown product of DDT) in eagle nestling blood plasma collected along Lake Superior decreased an average of 3% annually over the 10-year period. Total PCBs decreased 4% annually. Mercury in nestling feathers decreased from 1991 to 2008 by 2.4% per year. With the possible exception of Hg concentration in the upper St. Croix River, mean concentrations in 2006–2008 of all three of these contaminants were below levels expected to cause significant impairment of eagle reproduction at all monitoring sites. In addition, the eagle reproduction rate averaged more than 1.2 young per occupied territory, a level suggestive of a healthy population (Dykstra et al. 2010).

The importance of atmospheric deposition of Hg as a source of aquatic contamination in the upper Midwest was further established when a study of lake sediments illustrated that wet atmospheric Hg deposition could explain most of the observed sediment Hg loadings (Sorensen et al. 1990, Glass and Sorensen 1999). Analyses of lake sediment cores reported by Swain et al. (1992) suggested that the rate of Hg accumulation in lake sediments in the Upper Midwest had nearly tripled since preindustrial time.

Stratigraphic data from eight lakes in rural Minnesota indicated that atmospheric Hg deposition in this region peaked in the 1960s and 1970s, followed by decreases attributable to decreased emissions from regional sources (Engstrom and Swain 1997). Based on comparison with data from remote lakes in southeastern Alaska, the authors estimated that global Hg emissions in the Northern Hemisphere could account for about 7.4 µg/m^2 of annual Hg deposition in rural Minnesota. The remaining deposition (about 5.1 µg/m^2, or 40% of the total) was estimated to derive mainly from regional anthropogenic Hg sources in the upper Midwest and eastern United States. It has been estimated that about 60% of the Hg in lake sediments in VOYA was derived from atmospheric sources (Sorensen et al. 2005, Wiener et al. 2006). Some of that deposited Hg probably originated in Asia (Pittman et al. 2011).

The northern Great Lakes region, and eastward into northern New York, is especially sensitive to Hg bioaccumulation, due in part to relatively high Hg deposition and in particular due to watershed and lake characteristics that exacerbate Hg transport, methylation, and bioaccumulation (Evers et al. 2011b). The distribution and magnitude of the impacts of Hg contamination of natural ecosystems in the Great Lakes region is more substantial than previously recognized, with concentrations of Hg exceeding risk thresholds in many species of fish and wildlife across the region (Evers et al. 2011b). Many species of fish in the inland waters of the Great Lakes region have average Hg concentration in fillets above the general risk threshold for piscivorous wildlife (0.27 ppm; Sandheinrich and Wiener 2011).

The Hg that has accumulated historically in lake sediments provides an index of changes over time in Hg inputs from atmospheric deposition and other sources. Drevnick et al. (2012) analyzed 91 sediment cores collected from relatively undisturbed inland lakes in the Great Lakes region. Inferred rates of Hg accumulation in these lakes increased about sevenfold from preindustrial times to a peak in about the 1980s. Over the past two decades, inferred inputs declined about 20%. This recent decrease in sediment Hg is consistent with documented trends in Hg deposition (Drevnick et al. 2012).

8.7.1 Hg in Fish

Wiener et al. (2012b) synthesized data on 6400 yellow perch from the Great Lakes and associated inland lakes and reservoirs in and around the Great Lakes Network. Yellow perch from 6.5% of the waters examined had average whole fish concentrations of Hg high enough to cause adverse effects on piscivorous fish. Concentrations of Hg were higher in fish at higher trophic levels.

Gorski et al. (2003) investigated the bioaccumulation of Hg in northern pike (*Esox lucius*) in two inland lakes in ISRO: one Hg advisory lake (Sargent Lake) and one nonadvisory lake (Lake Richie). Concentrations of total and methyl Hg were analyzed in water, zooplankton, macroinvertebrates, and fish. Although concentrations of total Hg were significantly higher in pike collected from Sargent Lake, concentrations in open water were slightly higher in Lake Richie. Analyses of stable isotopes of C and N in biota indicated that pike from the two lakes were positioned at approximately the same trophic level (4.2 and 4.3). However, the food web in Sargent Lake was more pelagic based, whereas in Lake Richie it was more benthic based. Gorski et al. (2003) speculated that the pelagic food web in Sargent Lake may enhance bioaccumulation. For example, Campbell et al. (2000) showed by ^{13}C isotope analysis that organochlorines can bioaccumulate to a greater extent in pelagic food webs.

The concentration of Hg in fish tissue is often positively correlated with lake and/ or watershed area and negatively correlated with pH, acid neutralizing capacity, and zooplankton density (Chen et al. 2005, Driscoll et al. 2007). Lake types that are generally associated with the most Hg bioaccumulation are poorly buffered, low in pH and productivity, and have forested watersheds with little human development (Chen et al. 2005). Such lakes and watersheds are common in the Great Lakes Network. Lake pH, sulfate concentration, and extent of wetland connectivity were the major factors identified by Wiener et al. (2006) controlling Hg bioaccumulation in predatory fish within VOYA. Low acid neutralizing capacity, low-pH lakes are common in VOYA. Such lakes are prone to having high levels of Hg in fish (Swain and Helwig 1989, Wiener and Spry 1996).

The U.S. Geological Survey sampled 20 lakes in VOYA during the period 2000–2002 and analyzed the water and fish for Hg. Near-surface water samples had methyl Hg concentrations ranging from below the method detection level of 0.04 ng/L to concentrations that were more than an order of magnitude higher. In nearby Ontario, Scheuhammer and Blancher (1994) estimated that 30% of studied lakes had prey-size fish that had Hg concentrations sufficiently high to cause reproductive impairment in common loons that consumed those fish.

Changes in atmospheric Hg deposition have been shown to have an important influence on Hg concentrations in fish. Hrabik and Watras (2002) found approximately a 30% reduction in fish Hg concentrations between 1994 and 2000 in a Wisconsin seepage lake in response to decreased atmospheric Hg deposition of about 10% per year between 1995 and 1999 (Watras et al. 2000). It must be noted, however, that seepage lakes, such as the lake studied by Hrabik and Watras (2002), might be expected to respond more markedly and quickly to decreased atmospheric Hg loading (Harris et al. 2007), compared with drainage lakes that typically receive a large portion of their Hg input via the watershed.

VOYA contains many lakes that vary in size, productivity, and fish Hg content (Sorensen et al. 2005, Wiener et al. 2006). Within VOYA, lakes with extremely low and extremely high fish Hg content are situated within a few kilometers of each other. Fish collected from Ryan Lake and Mukooda Lake in VOYA had some of the highest and lowest Hg concentrations recorded in Minnesota for northern pike (Rolfhus et al. 2011). These two lakes receive nearly identical Hg inputs in atmospheric deposition (Wiener et al. 2006, Rolfhus et al. 2011). Thus, lake- and watershed-specific factors have large influence on Hg bioaccumulation.

Within the Great Lakes region, walleye and largemouth bass exhibited increasing Hg concentrations from south to north and from west to east. Portions of the region having highest concentrations were forested and contained substantial wetland area (Evers et al. 2011b).

Sorensen et al. (2005) reported the results of a monitoring study of Hg concentrations in young-of-the-year yellow perch in the context of water level fluctuations in 14 lakes in northeastern Minnesota, including 6 lakes in or adjacent to VOYA. Twelve years of monitoring data were collected for Sand Point Lake, and three years for each of the other lakes. Over the three-year period of record across all lakes, mean fish Hg concentrations varied in each lake by nearly a factor of 2, on average. For the 12-year monitoring period at Sand Point Lake, values ranged from 38 ng/g (wet weight) in 1998 to 200 ng/g (wet weight) in 2001. Annual water level fluctuation was correlated with Hg concentrations in young-of-the-year perch. Because the Hg concentration measurements were made on young-of-the-year fish at the end of each growing season, they represented water conditions and bioaccumulation for each year individually. The researchers concluded that annual water level fluctuations in the reservoir systems in VOYA have significant influence on Hg bioaccumulation in perch. This relationship provides a management opportunity for controlling Hg bioaccumulation in VOYA.

Bioaccumulation of Hg and the factors that control it have been topics of considerable research in VOYA. Results of a key study were reported by Wiener et al. (2006), who studied Hg concentrations and bioaccumulation in 17 lakes in the park. They assessed the importance of atmospheric and geologic sources of Hg to interior lakes and their watersheds and identified factors associated with lake-to-lake variation in methyl Hg contamination of lacustrine food webs. Geologic sources were judged to be small. Most Hg found in the upper O and A soil horizons was of atmospheric origin, primarily from anthropogenic sources. The most important factors influencing methyl Hg concentrations in lake water and fish were identified as lake pH, sulfate concentration in lake water, and total organic C in lake water, which largely reflects wetland influence. The area of connected wetlands adjoining the 17 study lake watersheds represented from 3.5 to 20% of total watershed area (Wiener et al. 2006).

Northern pike from some lakes in VOYA have contained the highest concentrations of Hg in pike reported in Minnesota (Wiener et al. 2006). Concentrations varied almost 10-fold among the small lakes in the park (Wiener et al. 2006). This reflected the importance of lake and watershed characteristics, in addition to atmospheric deposition, as controlling factors in the cycling and bioaccumulation of Hg. There was no indication in the deposition measurements or distribution of local Hg emissions sources that the atmospheric loading of Hg would explain these spatial differences.

Mean concentrations of total Hg in one-year-old yellow perch (n = 612) varied more than fivefold across the 17 study lakes of Wiener et al. (2006) in Mukooda Lake to 942 ng/g [dry weight] in Ryan Lake). A model with pH and lake sulfate concentration explained 53% of the variation in total Hg concentration in yellow perch. A model with sulfate concentration, total organic C, and (total organic C)^2 explained 60% of the variation in total Hg. The model judged best, based on the Akaike Information Criterion, for predicting methyl Hg concentration in lake water included sulfate, total organic C, and (total organic C)^2 and explained 79% of the variation in methyl Hg concentration in water (Wiener et al. 2006).

Ryan and Tooth lakes had both the highest concentrations of Hg in yellow perch and the highest concentrations of lake sulfate (up to 74 µeq/L) among the studied lakes. Fish Hg concentrations were lowest in the three lakes (Little Trout, Mukooda, O'Leary) that had pH > 7.0 (Wiener et al. 2006), a pattern observed previously (Grieb et al. 1990, Spry and Wiener 1991).

Analyses of soil, bedrock, and lake sediments indicated that atmospheric deposition was the primary source of Hg to 17 study lakes in VOYA (Wiener et al. 2006). Bedrock and lower soil horizon material contained much lower Hg concentrations than upper A and O soil horizons. Within the A horizon, Hg concentration was positively correlated with organic content ($r^2 = 0.82$). Lake sediment cores showed patterns of low background concentration in sediments deposited before 1860 (47–167 ng/g [dry weight]) to maximum values in the late twentieth century (102–364 ng/g [dry weight]). Sediment Hg concentrations were lowest in the lake cores that exhibited the highest sedimentation rates, suggesting dilution of sediment Hg concentration by sediment flux. The ratio of recent Hg concentration in sediment divided by preindustrial Hg concentration was relatively consistent among lakes (mean ± standard deviation: 2.1 ± 0.1). Measured Hg fluxes represented in lake sediment cores collected from five lakes were not controlled mainly by atmospheric Hg deposition. Rather, they were strongly influenced by sedimentation patterns within the lakes and watershed characteristics that control runoff, including watershed-to-lake area ratio and the percent of the watershed covered by wetlands. Analyses conducted by Wiener et al. (2006) confirmed that both watershed and in-lake processes affect the concentrations of Hg in lacustrine food webs in the Great Lakes Network. The concentration of sulfate, total organic C, and pH in water are robust indicators of the sensitivity of lake watersheds to Hg inputs in landscapes affected by atmospheric Hg deposition.

8.7.2 Piscivorous Wildlife

Relatively high levels of Hg have been documented in many species of bird, mammal, and fish at multiple trophic positions in a variety of habitats across the Great Lakes region (Evers et al. 2011b). Fish constitute an important pathway for transferring methyl Hg to wild mammals and birds. Monitoring of fish for Hg in the Great Lakes Network has often focused on both small prey fish and large, older piscivorous fish. The former are expected to respond quickly to changes in Hg exposure and reflect trophic transport of contaminants in food webs (Wiener et al. 2007). The latter respond more gradually to Hg bioavailability and are influenced by such factors as fish age, fish size, nutrient input, and interspecies competition (Mason et al. 2005).

In the southern portions of the Great Lakes region, agricultural land uses predominate and nutrient inputs from fertilizer use and livestock waste have contributed to increased algal populations. The abundant algae biodilute methyl Hg in the food web, resulting in lower concentrations in fish and wildlife at higher trophic levels (Evers et al. 2011b).

Several piscivorous bird and mammal species have been suggested as biomonitors of Hg bioaccumulation in the Great Lakes Network (Wolfe et al. 2007). The common loon and bald eagle are especially good indicators of Hg risk to wildlife (Evers 2006). Increased concentrations of Hg have been found to be associated with behavioral, physiological, and reproductive effects on these bird species (Burgess and Meyer 2008, Evers et al. 2008).

Acidic deposition might contribute to Hg toxicity in fish-eating birds in the Great Lakes Network because sulfate addition to wetland environments could stimulate the production of methyl Hg, thereby increasing lake water concentrations of methyl Hg (Jeremiason et al. 2006). Kramar et al. (2005) determined that the extent of wetland located in close proximity (<150 m [492 ft]) to loon territory was positively correlated with Hg concentrations in loon blood.

Evers et al. (2011a) identified two biological Hg hot spots within the Great Lakes Network for which rationales were well described for the observed high concentrations of Hg in loons: northeastern Minnesota and northern Wisconsin/upper Michigan. Both were characterized by mixed deciduous and coniferous forest (79% and 61%, respectively) interspersed with scrub–shrub and emergent wetland (12% and 20%, respectively). In the northeastern Minnesota hot spot, high Hg levels were further associated with reservoir impoundments characterized by fluctuating water levels and with small natural lakes having low pH and high sulfate and dissolved organic C concentrations. Some of these lakes were located in VOYA (Wiener et al. 2006) and nearby Superior National Forest. In northern Wisconsin/upper Michigan, high Hg levels were associated with low lake acid neutralizing capacity (Cope et al. 1990, Meyer et al. 1995, Meyer et al. 1998) and interactions with sulfate and dissolved organic C (Watras and Morrison 2008).

The loon Hg biological hot spot in northeastern Minnesota identified by Evers et al. (2011a) included reservoirs as an important landscape feature. Water-level fluctuations are likely important for increasing methyl Hg bioavailability. Within the northern Wisconsin/upper Michigan loon Hg hot spot, reservoirs were less common, but nevertheless were often the water bodies having high Hg (Evers et al. 2011a).

Myers et al. (2007) proposed a national map showing the relative sensitivity of aquatic ecosystems to Hg contamination. Areas showing the highest estimated sensitivity were mainly confined to the East Coast, Gulf Coast, and the region in and around the Great Lakes Network.

Total Hg concentrations in mink in the Great Lakes region have been in many cases high enough to suggest the likelihood of subclinical effects (Basu et al. 2007). In one survey, Hg concentrations in mink were highest in wetlands along impounded rivers that had fluctuating water levels and that were downstream of large historical point sources (Hamilton et al. 2011). Similarly, bald eagles in this region accumulated sufficient Hg as to suggest subclinical neurological damage. An estimated 14%–27% of the eagles studied had Hg tissue concentration above the proposed risk threshold for liver toxicity of 16.7 ppm (Zillioux et al. 1993, Rutkiewicz et al. 2011). Dykstra et al (2010) showed declines in bald eagle nestlings in Great Lakes Network study areas.

Rolfhus et al. (2011) evaluated the results of 10 studies in the western Great Lakes region to document the trophic transfer efficiency of methyl Hg in the pelagic food webs of lakes. The largest increases in methyl Hg (largest biomagnification) was found at the base of the food web between water and suspended particles. The observed similarity in the efficiency of trophic transfer suggested that the aqueous supply of methyl Hg controls much of the bioaccumulation in pelagic food webs in this region.

Bald eagles in the United States have commonly been found to contain substantial amounts of Hg in feathers, eggs, liver, and brain (cf., Wood et al. 1996, Bechard et al. 2007, Scheuhammer et al. 2008). Impacts of Hg toxicity on eagle reproduction and survival can affect individual and population condition. It is therefore important to identify markers of subclinical effects of Hg on the brain of eagles and other piscivorous bird species as early warning signals. Rutkiewicz et al. (2011) evaluated Hg exposure to bald eagles collected in Iowa, Michigan, Minnesota, Ohio, and Wisconsin. Levels of Hg in eagle brains were associated with neurochemical receptors and enzymes. Results suggested that bald eagles in this region are exposed to Hg at levels high enough to cause subclinical neurological damage (Rutkiewicz et al. 2011). The concentrations of total Hg in bald eagle brain and liver tissues were lowest in Ohio and Wisconsin and highest in Michigan, Iowa, and Minnesota.

Feathers from nestling bald eagles in VOYA have been collected over the past two decades and analyzed for Hg content. This monitoring has focused on nesting sites near three impoundments: Rainy, Kabetogama, and Namakan (included together with Crane and Sand Pointe) lakes. The VOYA environment has high potential for Hg methylation because wetlands are abundant (Grim and Kallemeyn 1995). Since 1999, the water levels of these impoundments have been stabilized to more closely match natural conditions. The lake-level stabilization was ordered by the International Joint Commission because research had shown that large changes in lake stage can increase methyl Hg concentrations in lake water (Rudd 1995, St. Louis et al. 2004, Sorensen et al. 2005). In apparent response to this control order, the annual geometric mean of Hg concentrations in bald eagle nestling feathers decreased 74.4% from 1989 to 2010 (Pittman et al. 2011).

Wiener et al. (2013) investigated contaminant bioaccumulation in fish and larval dragonflies (Insecta, Odonata) in six national park units in the Great Lakes Network, including ISRO and VOYA, during 2008 and 2009. The objectives of the study were to assess the spatial patterns of contamination of aquatic biota to identify locations where contaminant concentrations may pose a risk to aquatic biota and to document temporal trends. The investigated contaminants included total and methyl Hg, Pb, the insecticide DDT and its metabolites, PCBs, perfluorinated chemicals, and polybrominated diphenyl ethers. Adult piscivorous fish, prey fish, and larval dragonflies were sampled and analyzed. The concentration of Hg in adult piscivorous fish from several water bodies exceeded the EPA tissue residue criterion for methyl Hg (0.3 ppm), which was established to protect human health. In some predatory fish, the concentrations of Hg were high enough to adversely impact fish health and reproduction. One or more fish collected in ISRO and VOYA exceeded 0.5 ppm (wet weight), a level associated with impaired fish reproduction and other health effects. Concentrations of Hg in prey fish were also considered high enough to impact piscivorous fish or birds.

8.7.3 Persistent Organic Pollutants

In addition to the substantial Hg research conducted within the Great Lakes Network region, there has also been considerable work in this region on atmospheric inputs of other toxic compounds. Hites (2006) provided a comprehensive summary, with chapters covering PCBs, pesticides, polycyclic aromatic hydrocarbons, perfluorinated chemicals, and others. Polychlorinated dibenzo-p-dioxins and polychlorinated dibenzofurans are lipophylic environmental contaminants that are emitted into the atmosphere via combustion of municipal and industrial waste (Czuczwa et al. 1984, Czuczwa and Hites 1986). They deposit readily from the atmosphere and accumulate in lake sediments. Cores of undisturbed lake sediment can produce a record of changes in atmospheric inputs of these contaminants over time for lakes lacking substantial nonatmospheric inputs. This approach has been applied to sediment cores collected from Siskiwit Lake, a remote lake on Isle Royale (Czuczwa and Hites 1986, Baker and Hites 2000). Sediment inferred lake concentrations of polychlorinated dibenzofurans were near zero prior to about 1930, increasing to peak concentrations about 1970, and subsequently declining by about 50% (Baker and Hites 2000).

The Integrated Atmospheric Deposition Network has monitored persistent organic pollutants in air and precipitation at each of the five Great Lakes since 1990 (Clair et al. 2011). Concentrations of many chlorinated persistent organic pollutants have declined over the Great Lakes region in recent decades (Simcik et al. 2000). In particular, decreases have been observed for PCBs and organochlorinated pesticides (including dieldrin, chlordanes, and endosulfans; Sun et al. 2006, 2007). Monitoring of brominated flame retardants was initiated in recent years and shows some evidence of decline since controls were initiated (Sun et al. 2007, Venier and Hites 2008, Clair et al. 2011).

Brominated flame retardants, including polybrominated diphenyl ethers, are added to various commercial products to reduce flammability. Their usage increased markedly after about 1980 and are now ubiquitous in the environment (Hites 2004). Venier et al. (2010) reported the concentrations of polybrominated diphenyl ethers in the plasma of nestling bald eagles sampled throughout the Great Lakes region. The effects of these atmospherically deposited contaminants on the health and reproduction of bald eagles, or other wildlife, is not known.

Levels of legacy pollutants (DDT and its metabolites, PCBs, Pb) and emerging contaminants (polybrominated diphenyl ethers and perfluorinated chemicals) were generally low or below the detection limit in ISRO and VOYA (Wiener et al. 2013). Measured levels of these contaminants were not considered harmful. The average Hg concentrations in predatory fish were closely correlated with the average concentrations in larval dragonflies of the genus *Gompus* in lakes of ISRO and VOYA (http://science.nature.nps.gov/im/units/GLKN/monitor/contaminants/contaminants.cfm).

Control programs for persistent organic pollutants have shown evidence of effectiveness, based on monitoring contaminant concentrations in wildlife. Measurements of persistent organic pollutants in archived herring gull eggs from the Great Lakes showed 75%–80% decreases in PCB concentrations between 1971 and 1982 (Hebert et al. 1999).

8.8 Summary

Air pollution is an important stressor to natural resources within national park units in the Great Lakes Network. Effects have generally been well studied in this region (Charles 1991, Kohut 2007a,b, Evers et al. 2011b, Greaver et al. 2011). Atmospheric emissions and deposition of S and N vary dramatically across the network region, with relatively low values to the northwest and high values to the southeast.

The air pollution effects of greatest concern to Great Lakes Network parks appear to include Hg methylation and nutrient enrichment of wetlands. Airborne contaminants, including Hg, can accumulate in food webs, reaching toxic levels in top predators. The major source of Hg in the watersheds of this region is atmospheric deposition (Mason and Sullivan 1997, Rolfhus et al. 2003, Wiener et al. 2006), most of which originates from human-caused sources (Swain et al. 1992, Lorey and Driscoll 1999, Lamborg et al. 2002). Within the Great Lakes region, including the northern portions of the Great Lakes Network, methyl Hg is known to constitute a substantial risk to the health of wildlife. Much of the landscape is conducive to transformation from elemental Hg deposited from the atmosphere to methyl Hg. Tooth, Ryan, and Mukooda Lakes in VOYA have among the highest concentrations of Hg in fish among all studied lakes in Minnesota (Goldstein et al. 2003, Rolfhus et al. 2011).

Sulfur and N pollutants can cause acidification of streams, lakes, and soils. Lake acidification is mostly confined to groundwater recharge seepage lakes in the northeastern portion of the network region. Acidification in response to human-caused S and N deposition has not been documented in surface waters in network parks.

Nitrogen pollutants can cause undesirable enrichment of natural ecosystems, leading to changes in plant species diversity and soil nutrient cycling. Wetlands are especially sensitive to N enrichment effects, stimulating the growth of certain plant species, with the subsequent loss of some rare species. Nitrogen-limited lakes are also vulnerable to excess N enrichment that can cause unwanted increases in productivity and changes in natural aquatic plant and phytoplankton species composition.

Ozone pollution can harm human health, reduce plant growth, and cause visible injury to foliage. Ozone is an important stressor to park vegetation in this network, primarily in the more southerly parks that are in proximity to major N oxide emissions sources.

Particulate pollution can cause haze, reducing visibility. Ambient haze, including haze caused by both natural and anthropogenic particles and aerosols, is high on the haziest days in ISRO. Levels of haze are also high at times in other Great Lakes Network parks.

References

Aldous, A.R. 2002. Nitrogen retention by Sphagnum mosses: Responses to atmospheric nitrogen deposition and drought. *Can. J. Bot.* 80(7):721–731.

Avis, P.G., D.J. McLaughlin, B.C. Dentinger, and P.B. Reich. 2003. Long-term increases in nitrogen supply alters above and belowground ectomycorrhizal communities and increases dominance of *Russula* species. *New Phytol.* 160(1):239–253.

Baer, S.G., J.M. Blair, S.L. Collins, and A.K. Knapp. 2003. Soil resources regulate productivity and diversity in newly established tallgrass prairie. *Ecology* 84:724–735.

Baker, J.I. and R.A. Hites. 2000. Siskiwit Lake revisited: Time trends of polychlorinated dibenzo-p-dioxin and dibenzofuran deposition at Isle Royale, Michigan. *Environ. Sci. Technol.* 34(14):2887–2891.

Baker, L.A., A.T. Herlihy, P.R. Kaufmann, and J.M. Eilers. 1991. Acidic lakes and streams in the United States: The role of acidic deposition. *Science* 252:1151–1154.

Barbiero, R.P., M.L. Tuchman, and E.S. Millard. 2006. Post-dreissenid increases in transparency during summer stratification in the offshore waters of Lake Ontario; is a reduction in whiting events the cause? *J. Great Lakes Res.* 32:131–141.

Basu, N., A.M. Scheuhammer, K. Rouvinen-Watt, N. Grochowina, R.D. Evans, M. O'Brien, and H.M. Chan. 2007. Decreased N-methyl-dasparctic acid (NMDA) receptor levels are associated with mercury exposure in wild and captive mink. *Neurotoxicology* 28:587–593.

Bechard, M.J., D.N. Perkins, G.S. Kaltenecker, and S. Alsup. 2007. Mercury contamination in Idaho bald eagles, (*Haliaeetus leucocephalus*). *Bull. Environ. Contam. Toxicol.* 83:698–702.

Bedford, B.L., M.R. Walbridge, and A. Aldous. 1999. Patterns in nutrient availability and plant diversity of temperate North American wetlands. *Ecol. Soc. Am.* 80(7):2151–2169.

Bennett, E.B. 1986. The nitrifying of Lake Superior. *Ambio* 15(5):272–275.

Bennett, J.P. 1995. Abnormal chemical element concentrations in lichens of Isle Royale National Park. *Environ. Exp. Bot.* 35(3):259–277.

Bennett, J.P. 2007. Twenty-four years of Great Lakes lichen studies provide park biomonitoring base-lines. *In* J. Selleck (Ed.). *Natural Resource Year in Review: 2006*. Publication D-1859. National Park Service, Denver, CO, pp. 39–40.

Bennett, J.P. and C.M. Wetmore. 1997. Chemical element concentrations in four lichens on a transect entering Voyageurs National Park. *Environ. Exp. Bot.* 37:173–185.

Bobbink, R., M. Hornung, and J.G.M. Roelofs. 1998. The effects of air-borne nitrogen pollutants on species diversity in natural and semi-natural European vegetation. *J. Ecol.* 86:717–738.

Bridgham, S.D., J. Pastor, C.A. McClaugherty, and C.J. Richardson. 1995. Nutrient-use efficiency: A litterfall index, a model, and a test along a nutrient-availability gradient in North Carolina peatlands. *Am. Nat.* 145:1–21.

Burgess, N.M. and M.W. Meyer. 2008. Methylmercury exposure associated with reduced productiv-ity in common loons. *Ecotoxicology* 17(2):83–91.

Campbell, L.M., D.W. Schindler, D.C.G. Muir, D.B. Donald, and K.A. Kidd. 2000. Organochlorine transfer in the food web of subalpine Bow Lake, Banff National Park. *Can. J. Fish. Aquat. Sci.* 57:1258–1269.

Charles, D.F. 1991. *Acidic Deposition and Aquatic Ecosystems: Regional Case Studies*. Springer-Verlag, New York.

Chen, C.Y., R.S. Stemberger, N.C. Kamman, B.M. Mayes, and C.L. Folt. 2005. Patterns of Hg bioac-cumulation and transfer in aquatic food webs across multi-lake studies in the Northeast US. *Ecotoxicology* 14:135–147.

Clair, T.A., D. Burns, I.R. Pérez, J. Blais, and K. Percy. 2011. Ecosystems. *In* G.M. Hidy, J.R. Brook, K.L. Demerjian, L.T. Molina, W.T. Pennell, and R.D. Scheffe (Eds.). *Technical Challenges of Multipollutant Air Quality Management*. Springer, Dordrecht, the Netherlands, pp. 139–229.

Clark, C.M. 2011. Great plains. *In* L.H. Pardo, M.J. Robin-Abbott, and C.T. Driscoll (Eds.). *Assessment of Nitrogen Deposition Effects and Empirical Critical Loads of Nitrogen for Ecoregions of the United States*. General Technical Report NRS-80. U.S. Forest Service, Newtown Square, PA, pp. 117–132.

Clark, C.M. and D. Tilman. 2008. Loss of plant species after chronic low-level nitrogen deposition to prairie grasslands. *Nature* 451:712–715.

Conley, D.J., C.L. Schelske, and E.F. Stoermer. 1993. Modification of the biogeochemical cycle of silica with eutrophication. *Mar. Ecol. Prog. Ser.* 101:179–192.

Cook, R.B. and H.I. Jager. 1991. Upper Midwest: The effects of acidic deposition on lakes. *In* D.F. Charles (Ed.). *Acidic Deposition and Aquatic Ecosystems: Regional Case Studies*. Springer-Verlag, New York.

Cope, W.G., J.G. Wiener, and R.G. Rada. 1990. Mercury accumulation in yellow perch in Wisconsin seepage lakes: Relation to lake characteristics. *Environ. Toxicol. Chem.* 9:931–940.

Czuczwa, J.M. and R.A. Hites. 1986. Airborne dioxins and dibenzofurans: Sources and fates. *Environ. Sci. Technol.* 20:195–200.

Czuczwa, J.M., B.D. McVeety, and R.A. Hites. 1984. Polychlorinated dibenzo-p-dioxins and dibenzo-furans in sediments from Siskiwit Lake, Isle Royale. *Science* 226(4674):568–569.

Dove, A. 2009. Long-term trends in major ions and nutrients in Lake Ontario. *Aquat. Ecosyst. Health Manage.* 12(3):281–295.

Downing, J.A. and E. McCauley. 1992. The nitrogen-phosphorus relationship in lakes. *Limnol. Oceanogr.* 37:936–945.

Drevnick, P.E., D.R. Engstrom, C.T. Driscoll, E.B. Swain, S.J. Balogh, N.C. Kamman, D.T. Long et al. 2012. Spatial and temporal patterns of mercury accumulation in lacustrine sediments across the Laurentian Great Lakes region. *Environ. Pollut.* 161:252–260.

Driscoll, C.T., Y.-J. Han, C.Y. Chen, D.C. Evers, K.F. Lambert, T.M. Holsen, N.C. Kamman, and R.K. Munson. 2007. Mercury contamination in forest and freshwater ecosystems in the northeast-ern United States. *BioScience* 57(1):17–28.

Dykstra, C.R., W.T. Route, M.W. Meyer, and P.W. Rasmussen. 2010. Contaminant concentrations in bald eagles nesting on Lake Superior, the Upper Mississippi River, and the St. Croix River. *J. Great Lakes Res.* 36(3):561–569.

Edlund, M.B., J.R. Hobbs, and D.R. Engstrom. 2013. Biomonitoring using diatoms and paleolimnology in the western Great Lakes national parks. NPS/GLKN/NRTR-2013/814. National Park Service, For Collins, CO.

Eilers, J.M., D.F. Brakke, and D.H. Landers. 1988. Chemical and physical characteristics of lakes in the Upper Midwest, United States. *Environ. Sci. Technol.* 22:164–172.

Eilers, J.M., G.E. Glass, K.E. Webster, and J.A. Rogalla. 1983. Hydrologic control of lake susceptibility to acidification. *Can. J. Fish. Aquat. Sci.* 40:1896–1904.

Ellis, R.A., D.J. Jacob, M.P. Sulprizio, L. Zhang, C.D. Holmes, B.A. Schichtel, T. Blett, E. Porter, L.H. Pardo, and J.A. Lynch. 2013. Present and future nitrogen deposition to national parks in the United States: Critical load exceedances. *Atmos. Chem. Phys.* 13(17):9083–9095.

Engstrom, D.R. and E.B. Swain. 1997. Recent declines in atmospheric mercury deposition in the Upper Midwest. *Environ. Sci. Technol.* 31:960–967.

Evers, D.C. 2006. Loons as biosentinels of aquatic integrity. *Environ. Bioindicators* 1:18–21.

Evers, D.C., L.J. Savoy, C.R. DeSorbo, D.E. Yates, W. Hanson, K.M. Taylor, L.S. Siegel et al. 2008. Adverse effects from environmental mercury loads on breeding common loons. *Ecotoxicology* 17:69–81.

Evers, D.C., J.G. Wiener, C.T. Driscoll, D.A. Gay, N. Basu, B.A. Monson, K.F. Lambert et al. 2011b. Great Lakes mercury connections: The extent and effects of mercury pollution in the Great Lakes region. Report BR1 2011–18. Biodiversity Research Institute, Gorham, ME.

Evers, D.C., K.A. Williams, M.W. Meyer, A.M. Scheuhammer, N. Schoch, A. Gilbert, L. Siegel, R.J. Taylor, R. Poppenga, and C.R. Perkins. 2011a. Spatial gradients of methylmercury for breeding common loons in the Laurentian Great Lakes region. *Ecotoxicology* 20:1609–1625.

Finlay, J.C., R.W. Sterner, and S. Kumar. 2007. Isotopic evidence for in-lake production of accumulating nitrate in Lake Superior. *Ecol. Appl.* 17:2323–2332.

Fraterrigo, J.M., S.M. Pearson, and M.G. Turner. 2009. The response of understory herbaceous plants to nitrogen fertilization in forests of different landuse history. *For. Ecol. Manage.* 257:2182–2188.

Geiser, L.H., S.E. Jovan, D.A. Glavich, and M.K. Porter. 2010. Lichen-based critical loads for atmospheric nitrogen deposition in western Oregon and Washington forests, USA. *Environ. Pollut.* 158:2412–2421.

Gilliam, F.S., C.L. Goodale, L.H. Pardo, L.H. Geiser, and E.A. Lilleskov. 2011. Eastern temperate forests. *In* L.H. Pardo, M.J. Robin-Abbott, and C.T. Driscoll (Eds.). *Assessment of Nitrogen Deposition Effects and Empirical Critical Loads of Nitrogen for Ecoregions of the United States.* General Technical Report NRS-80. U.S. Forest Service, Newtown Square, PA, pp. 99–116.

Gilliam, F.S., A.W. Hockenberry, and M.B. Adams. 2006. Effects of atmospheric nitrogen deposition on the herbaceous layer of a central Appalachian hardwood forest. *J. Torrey Bot. Soc.* 133:240–254.

Glass, G.E. and J.A. Sorensen. 1999. Six-year trend (1990–1995) of wet mercury deposition in the Upper Midwest, U.S.A. *Environ. Sci. Technol.* 33:3303–3312.

Goldstein, R.M., M.E. Brigham, L. Steuwe, and M.A. Menheer. 2003. Mercury data from small lakes in Voyageurs National Park, Northern Minnesota, 2000–02. U.S. Department of the Interior, U.S. Geological Survey, Mounds View, MN.

Gorski, P.R., L.B. Cleckner, J.P. Hurley, M.E. Sierszen, and D.E. Armstrong. 2003. Factors affecting enhanced mercury bioaccumulation in inland lakes of Isle Royale National Park, USA. *Sci. Tot. Environ.* 304:327–348.

Greaver, T., L. Liu, and R. Bobbink. 2011. Wetlands. *In* L.H. Pardo, M.J. Robin-Abbott, and C.T. Driscoll (Eds.). *Assessment of Nitrogen Deposition Effects and Empirical Critical Loads of Nitrogen for Ecoregions of the United States.* General Technical Report NRS-80. U.S. Forest Service, Newtown Square, PA.

Grieb, T.M., C.T. Driscoll, S.P. Gloss, C.L. Schofield, G.L. Bowie, and D.B. Porcella. 1990. Factors affecting mercury accumulation in fish in the Upper Michigan Peninsula. *Environ. Toxicol. Chem.* 9:919–930.

Grim, L.H. and L.W. Kallemeyn. 1995. Reproduction and distribution of bald eagles in Voyageurs National Park, Minnesota, 1973–1993. Biological Report 1. U.S. Department of the Interior, National Biological Service, Washington, DC.

Haddad, N.M., J. Haarstad, and D. Tilman. 2000. The effects of long-term nitrogen loading on grassland insect communities. *Oecologia* 124:73–84.

Hall, B.D., H. Manolopoulos, J.P. Hurley, J.J. Schauer, V.L. St. Louis, D. Kenski, J. Graydon, C.L. Babiarz, L.B. Cleckner, and G.J. Keeler. 2005. Methyl and total mercury in precipitation in the Great Lakes region. *Atmos. Environ.* 39:7557–7569.

Hamilton, M., A. Scheuhammer, and N. Basu. 2011. Mercury, selenium, and neurochemical biomarkers in different brain regions of migrating common loons from Lake Erie, Canada. *Ecotoxicology* 20(7):1677–1683.

Harris, R.C., J.W.M. Rudd, M. Amyot, C.L. Babiarz, K.G. Beaty, P.J. Blanchfield, R.A. Bodaly et al. 2007. Whole-ecosystem study shows rapid fish-mercury response to changes in mercury deposition. *Proc. Nat. Acad. Sci.* 104(42):16586–16591.

Hautier, Y.R., P.A. Niklaus, and A. Hector. 2009. Competition for light causes plant biodiversity loss after eutrophication. *Science* 324:636–638.

Hebert, C.E., R.J. Norstrom, J.P. Zhu, and C.R. Macdonald. 1999. Historical changes in PCB patterns I n Lake Ontario and Green Bay, Lake Michigan, 1971 to 1982, from herring gull egg monitoring data. *J. Great Lakes Res.* 25:220–233.

Hill, B.H., C.M. Elonen, T.M. Jicha, A.M. Cotter, A.S. Trebitz, and N.P. Danz. 2006. Sediment microbial enzyme activity as an indicator of nutrient limitation in Great Lakes coastal wetlands. *Freshw. Biol.* 51:1670–1683.

Hites, R.A. 2004. Polybrominated diphenyl ethers in the environment and in people: A meta-analysis of concentrations. *Environ. Sci. Technol.* 38(4):945–956.

Hites, R.A. 2006. *Persistent Organic Pollutants in the Great Lakes*. Springer-Verlag, Berlin, Germany.

Hrabik, T.R. and C.J. Watras. 2002. Recent declines in mercury concentration in a freshwater fishery: Isolating the effects of de-acidification and decreased atmospheric mercury deposition in Little Rock Lake. *Sci. Total Environ.* 297:229–237.

Huberty, L.E., K.L. Gross, and C.J. Miller. 1998. Effects of nitrogen addition on successional dynamics and species diversity in Michigan old-fields. *J. Ecol.* 86:794–803.

Hurd, T.M., A.R. Brach, and D.J. Raynal. 1998. Response of understory vegetation of Adirondack forests to nitrogen additions. *Can. J. For. Res.* 28:799–807.

Hutchinson, G.E. 1971. *A Treatise on Limnology*. Volume 1: Geography, Physics, and Chemistry. John Wiley & Sons, New York.

Ittekot, V. 2003. Carbon-silicon interactions. *In* J.M. Melillo, C.B. Field and B. Moldan (Eds.). *Interactions of the Major Biogeochemical Cycles*. Island Press, Washington, DC, pp. 311–322.

Jeremiason, J.D., D.R. Engstrom, E.B. Swain, E.A. Nater, B.M. Johnson, J.E. Almendinger, B.A. Monson, and R.K. Kolka. 2006. Sulfate addition increases methylmercury production in an experimental wetland. *Environ. Sci. Technol.* 40(12):3800–3806.

Knops, J.M.H. and K. Reinhart. 2000. Specific leaf area along a nitrogen fertilization gradient. *Am. Midl. Nat.* 144:265–272.

Koerselman, W. 1989. Groundwater and surface water hydrology of a small groundwater-fed fen. *Wetl. Ecol. Manage.* 1:31–43.

Kohut, R. 2007a. Assessing the risk of foliar injury from ozone on vegetation in parks in the U.S. National Park Service's Vital Signs Network. *Environ. Pollut.* 149:348–357.

Kohut, R.J. 2007b. Ozone risk assessment for vital signs monitoring networks, Appalachian national scenic trail, and natchez trace national scenic trail. Natural Resource Technical Report NPS/NRPC/ARD/NRTR—2007/001. National Park Service, Natural Resource Program Center, Fort Collins, CO.

Kramar, D., W.M. Goodale, L.M. Kennedy, L.W. Carstensen, and T. Kaur. 2005. Relating land cover characteristics and common loon mercury levels using geographic information systems. *Ecotoxicology* 14:253–262.

Lamborg, C.H., W.F. Fitzgerald, A.W.H. Damman, J.M. Benoit, P.H. Balcom, and D.R. Engstrom. 2002. Modern and historic atmospheric mercury fluxes in both hemispheres: Global and regional mercury cycling implications. *Global Biogeochem. Cycles* 16:1104–1114.

Lane, D.R. and H. BassiriRad. 2002. Differential responses of tallgrass prairie species to nitrogen loading and varying ratios of NO_3^- to NH_4^+. *Funct. Plant Biol.* 29:1227–1235.

Levine, S.N., A.D. Shambaugh, S.E. Pomeroy, and M. Braner. 1997. Phosphorus, nitrogen, and silica as controls on phytoplankton biomass and species composition in Lake Champlain (USA-Canada). *Intern. Assoc. for Great Lakes Res.* 23(2):131–148.

Lilleskov, E.A., P.M. Wargo, K.A. Vogt, and D.J. Vogt. 2008. Mycorrhizal fungal community relationship to root nitrogen concentration over a regional atmospheric nitrogen deposition gradient in the northeastern US. *Can. J. For. Res.* 38:1260–1266.

Linthurst, R.A., D.H. Landers, J.M. Eilers, D.F. Brakke, W.S. Overton, E.P. Meier, and R.E. Crowe. 1986a. Characteristics of lakes in the eastern united states. Volume I: Population Descriptions and Physico-Chemical Relationships. EPA/600/4–86/007a. U.S. Environmental Protection Agency, Washington, DC.

Linthurst, R.A., D.H. Landers, J.M. Eilers, P.E. Kellar, D.F. Brakke, W.S. Overton, R. Crowe, E.P. Meier, P. Kanciruk, and D.S. Jeffries. 1986b. Regional chemical characteristics of lakes in North America Part II: Eastern United States. *Water Air Soil Pollut.* 31:577–591.

Lorey, P. and C.T. Driscoll. 1999. Historical trends of mercury deposition in Adirondack lakes. *Environ. Sci. Technol.* 33(5):718–722.

Mason, R.P., M.L. Abbott, R.A. Bodaly, O.R. Bullock, Jr., C.T. Driscoll, D. Evers, S.B. Lindberg, M. Murray, and E.B. Swain. 2005. Monitoring the response to changing mercury deposition. *Environ. Sci. Technol.* 39:14A-22A.

Mason, R.P. and K.A. Sullivan. 1997. Mercury in Lake Michigan. *Environ. Sci. Technol.* 31:942–947.

McCune, B. 1988. Lichen communities along O_3 and SO_2 gradients in Indianapolis. *Bryologist* 91(3):223–228.

Meyer, M.W., D.C. Evers, T. Daulton, and W.E. Braselton. 1995. Common loons (*Gavia immer*) nesting on low pH lakes in northern Wisconsin have elevated blood mercury content. *Water Air Soil Pollut.* 80:871–880.

Meyer, M.W., D.C. Evers, J.J. Hartigan, and P.S. Rasmussen. 1998. Patterns of common loon (*Gavia immer*) mercury exposure reproduction and survival in Wisconsin, USA. *Environ. Toxicol. Chem.* 17:184–190.

Millard, E.S., O.E. Johannsson, M.A. Neilson, and A.H. El-Shaarawi. 2003. Long-term, seasonal and spatial trends in nutrients, chlorophyll *a* and light attenuation in Lake Ontario. *In* M. Munawar (Ed.). *State of Lake Ontario (SOLO): Past, Present, and Future.* Ecovision World Monograph Series, Aquatic Ecosystem Health and Management Society, Burlington, Ontario, Canada, pp. 97–132.

Mills, E.L., R.M. Dermott, E.F. Roseman, D. Dustin, E. Mellina, D.B. Conn, and A.P. Spindle. 1993. Colonization, ecology, and population structure of the "quagga" mussel (bivalivia: Dreissenidae) in the lower Great Lakes. *Can. J. Fish. Aquat. Sci.* 50:2304–2314.

Mills, E.L., C.E. Hoffman, J.P. Gillette, L.G. Rudstam, R. McCullough, D. Bishop, W. Pearsall et al. 2006. 2005 Status of the Lake Ontario ecosystem: A biomonitoring approach. Section 20. NYSDEC Lake Ontario Annual Report 2005. New York State Department of Environmental Conservation, Albany, NY.

Mitsch, W.J. and J.G. Gosselink. 2000. *Wetlands.* John Wiley & Sons, New York, 936pp.

Moon, J.B. and H.J. Carrick. 2007. Seasonal variation of phytoplankton nutrient limitation in Lake Erie. *Aquat. Microb. Ecol.* 48:61–71.

Moore, P.T., H. Van Miegroet, and N.S. Nicholas. 2007. Relative role of understory and overstory carbon and nitrogen cycling in a southern Appalachian spruce-fir forest. *Can. J. For. Res.* 37:2689–2700.

Moore, T., C. Blodau, J. Turenen, N. Roulet, and P.J.H. Richard. 2004. Patterns of nitrogen and sulfur accumulation and retention in ombrotrophic bogs, eastern Canada. *Glob. Change Biol.* 11:356–367.

Morrice, J.A., J.R. Kelly, A.S. Trebitz, A.M. Cotter, and M.L. Knuth. 2004. Temporal dynamics of nutrients (N and P) and hydrology in a Lake Superior coastal wetland. *J. Great Lakes Res.* 30(Suppl. 1):82–96.

Morris, J.T. 1991. Effects of nitrogen loading on wetland ecosystems with particular reference to atmospheric deposition. *Ann. Rev. Ecol. Syst.* 22:257–279.

Muller, R.N. 2003. Nutrient relations of the herbaceous layer in deciduous forest ecosystems. *In* F.S. Gilliam and M.R. Roberts (Eds.). *The Herbaceous Layer in Forests of Eastern North America.* Oxford University Press, New York, pp. 15–37.

Munawar, M., I.F. Munawar, R. Dermott, M. Fitzpatrick, and H. Niblock. 2006. The threat of exotic species to the food web in Lake Ontario. *Verh. Int. Verein. Limnol.* 29:1194–1198.

Myers, M.D., M.A. Ayers, J.S. Baron, P.R. Beauchemin, K.T. Gallagher, M.B. Goldhaber, D.R. Hutchinson et al. 2007. USGS goals for the coming decade. *Science* 318:200–201.

National Acid Precipitation Assessment Program (NAPAP). 1991. Integrated assessment report. National Acid Precipitation Assessment Program, Washington, DC.

National Atmospheric Deposition Program (NADP). 2004. National atmospheric deposition program 2003 annual summary. NADP Data Report 2004-1. Illinois State Water Survey, Champaign, IL.

Nicholls, K.H., G.J. Hopkins, S.J. Standke, and L. Nakamoto. 2001. Trends in total phosphorus in Canadian nearshore waters of Laurentian Great Lakes: 1976–1999. *Great Lakes Res.* 27:402–422.

North, R.L., S.J. Guidford, R.E.H. Smith, S.M. Havens, and M.R. Twiss. 2007. Evidence for phosphorus, nitrogen, and iron colimitation of phytoplankton communities in Lake Erie. *Limnol. Oceanogr.* 52:315–328.

Oliver, B.G., E.M. Thurman, and R.L. Malcolm. 1983. The contribution of humic substances to the acidity of colored natural waters. *Geochim. Cosmochim. Acta* 47:2031–2035.

Pardo, L.H., M.E. Fenn, C.L. Goodale, L.H. Geiser, C.T. Driscoll, E.B. Allen, J.S. Baron et al. 2011c. Effects of nitrogen deposition and empirical nitrogen critical loads for ecoregions of the United States. *Ecol. Appl.* 21(8):3049–3082.

Pardo, L.H., C.L. Goodale, E.A. Lilleskov, and L.H. Geiser. 2011b. Northern forests. *In* L.H. Pardo, M.J. Robin-Abbott, and C.T. Driscoll (Eds.). *Assessment of Nitrogen Deposition Effects and Empirical Critical Loads of Nitrogen for Ecoregions of the United States.* General Technical Report NRS-80. U.S. Forest Service, Newtown Square, PA, pp. 61–74.

Pardo, L.H., M.J. Robin-Abbott, and C.T. Driscoll. 2011a. Assessment of nitrogen deposition effects and empirical critical loads of nitrogen for ecoregions of the United States. General Technical Report NRS-80. U.S. Forest Service, Newtown Square, PA.

Percy, K.E., M. Nosal, W. Heilman, T. Dann, J. Sober, A.H. Legge, and D.F. Karnosky. 2007. New exposure-based metric approach for evaluating O_3 risk to North American aspen forests. *Environ. Pollut.* 147:554–566.

Pittman, H.T., W.W. Bowerman, L.H. Grim, T.G. Grubb, and W.C. Bridges. 2011. Using nestling feathers to assess spatial and temporal concentrations of mercury in bald eagles at Voyageurs National Park, Minnesota, USA. *Ecotoxicology* 20:1626–1635.

Reich, P.B., J. Knops, D. Tilman, J. Craine, D. Ellsworth, M. Tjoelker, T. Lee et al. 2001. Plant diversity enhances ecosystem responses to elevated CO_2 and nitrogen deposition. *Nature* 410:809–812.

Risch, M.R., J.F. DeWild, D.P. Krabbenhoft, R.K. Kolka, and L. Zhang. 2012. Litterfall mercury dry deposition in the eastern USA. *Environ. Pollut.* 161:264–290.

Rochefort, L., D.H. Vitt, and S.E. Bayley. 1990. Growth, production and decomposition dynamics of *Sphagnum* under natural and experimentally acidified conditions. *Ecology* 71(5):1986–2000.

Rolfhus, K.R., B.D. Hall, B.A. Monson, A.M. Paterson, and J.D. Jeremiason. 2011. Assessment of mercury bioaccumulation within the pelagic food web of lakes in the western Great Lakes region. *Ecotoxicology* 20:1520–1529.

Rolfhus, K.R., H.E. Sakamoto, L.B. Cleckner, R.W. Stoor, C.L. Babiarz, R.C. Back, H. Manolopoulos, and J.P. Hurley. 2003. The distribution and fluxes of total and methylmercury in Lake Superior. *Environ. Sci. Technol.* 37:865–872.

Rose, C. and R.P. Axler. 1998. Uses of alkaline phosphatase activity in evaluating phytoplankton community phosphorous deficiency. *Hydrobiologia* 361:145–156.

Royo, A.A., R. Collins, M.B. Adams, C. Kirschbaum, and W.P. Carson. 2010. Pervasive interactions between ungulate browsers and disturbance regimes promote temperate forest herbaceous diversity. *Ecology* 91(1):93–105.

Rudd, J.W.M. 1995. Sources of methyl mercury to freshwater ecosystems: A review. *Water Air Soil Pollut.* 80:697–713.

Rutkiewicz, J., D.-H. Nam, T. Cooley, K. Neumann, I.B. Padilla, W. Route, S. Strom, and N. Basu. 2011. Mercury exposure and neurochemical impacts in bald eagles across several Great Lakes states. *Ecotoxicology* 20:1669–1676.

Sandheinrich, M.B. and J.G. Wiener. 2011. Methylmercury in freshwater fish: Recent advances in assessing toxicity of environmentally relevant exposures. *In* W.N. Beyer and J.P. Meador (Eds.). *Environmental Contaminants in Biota: Interpreting Tissue Concentrations.* CRC Press, Boca Raton, FL, pp. 169–190.

Schelske, C.L. 1991. Historical nutrient enrichment of Lake Ontario: Paleolimnological evidence. *Can. J. Fish. Aquat. Sci.* 48:1529–1538.

Scheuhammer, A.M., N. Basu, N.M. Burgess, J.E. Elliott, G.D. Campbell, M. Wayland, L. Champoux, and J. Rodrigue. 2008. Relationships among mercury, selenium, and neurochemical parameters in common loons (*Gavia immer*) and bald eagles (*Haliaeetus leucocephalus*). *Ecotoxicology* 17:93–101.

Scheuhammer, A.M. and P.J. Blancher. 1994. Potential risk to common loons (*Gavia immer*) from methylmercury exposure in acidified lakes. *Hydrobiologia* 279/280:445–455.

Schwede, D.B. and G.G. Lear. 2014. A novel hybrid approach for estimating total deposition in the United States. *Atmos. Environ.* 92:207–220.

Shaver, G.R. and J.M. Melillo. 1984. Nutrient budgets of marsh plants—Efficiency concepts and relation to availability. *Ecology* 655:1491–1510.

Simcik, M.F., R.M. Hoff, W.M.J. Strachan, C.W. Sweet, I. Basu, and R.A. Hites. 2000. Temporal trends of semivolatile organic contaminants in Great Lakes precipitation. *Environ. Sci. Technol.* 34:361–367.

Smith, V.H. 1983. Low nitrogen to phosphorus ratios favor dominance by blue-green algae in lake phytoplankton. *Science* 221:669–671.

Sorensen, J.A., G.E. Glass, K.W. Schmidt, J.K. Huber, and G.R.J. Rapp. 1990. Airborne mercury deposition and watershed characteristics in relation to mercury concentrations in water, sediments, plankton, and fish of eighty northern Minnesota lakes. *Environ. Sci. Technol.* 24(11):1716–1727.

Sorensen, J.A., L.W. Kallemeyn, and M. Sydor. 2005. Relationship between mercury accumulation in young-of-the-year yellow perch and water-level fluctuations. *Environ. Sci. Technol.* 39:9237–9243.

Spry, D.J. and J.G. Wiener. 1991. Metal bioavailability and toxicity to fish in low-alkalinity lakes: A critical review. *Environ. Pollut.* 71:243–304.

St. Louis, V.L., J.W.M. Rudd, C.A. Kelly, R.A. Bodaly, J.M. Paterson, K.G. Beaty, R.H. Hesslein, A. Heyes, and A. Majewski. 2004. The rise and fall of mercury methylation in an experimental reservoir. *Environ. Sci. Technol.* 38:1348–1358.

Sterner, R.W., E. Anagnostou, S. Brovold, G.S. Bullerjahn, J.C. Finlay, S. Kumar, R.M.L. McKay, and R.M. Sherrell. 2007. Increasing stoichiometric imbalance in North America's largest lake: Nitrification in Lake Superior. *Geophys. Res. Lett.* 34:L10406.

Stevens, C.J., N.B. Dise, O.J. Mountford, and D.J. Gowing. 2004. Impact of nitrogen deposition on the species richness of grasslands. *Science* 303:1876–1878.

Stoddard, J., J.S. Kahl, F.A. Deviney, D.R. DeWalle, C.T. Driscoll, A.T. Herlihy, J.H. Kellogg, P.S. Murdoch, J.R. Webb, and K.E. Webster. 2003. Response of surface water chemistry to the clean air act amendments of 1990. EPA 620/R-03/001. U.S. Environmental Protection Agency, Office of Research and Development, National Health and Environmental Effects Research Laboratory, Research Triangle Park, NC.

Sullivan, T.J., G.B. Lawrence, S.W. Bailey, T.C. McDonnell, and G.T. McPherson. 2013. Effects of acidic deposition and soil acidification on sugar maple trees in the Adirondack Mountains, New York. NYSERDA Report No. 13-04. New York State Energy Research and Development Authority, Albany, NY.

Sullivan, T.J. and T.C. McDonnell. 2014. Mapping of nutrient-nitrogen critical loads for selected national parks in the intermountain west and Great Lakes regions. Natural Resource Technical Report NPS/ARD/NRTR—2014/895. National Park Service, Fort Collins, CO.

Sullivan, T.J., G.T. McPherson, T.C. McDonnell, S.D. Mackey, and D. Moore. 2011. Evaluation of the sensitivity of inventory and monitoring national parks to acidification effects from atmospheric sulfur and nitrogen deposition. Natural Resource Report NPS/NRPC/ARD/NRR—2011/349. U.S. Department of the Interior, National Park Service, Denver, CO.

Sun, P., S.I. Basu, P. Blanchard, K.A. Brice, and R.A. Hites. 2007. Temporal and spatial trends of concentrations near the Great Lakes. *Environ. Sci. Technol.* 41:1131–1136.

Sun, P., P. Blanchard, K. Brice, and R.A. Hites. 2006. Atmospheric organochlorine pesticide concentrations near the Great Lakes: Temporal and spatial trends. *Environ. Sci. Technol.* 40:6587–6593.

Sundareshwar, P.V., J.T. Morris, E. Koepfler, and B. Fornwalt. 2003. Phosphorus limitation of coastal ecosystem processes. *Science* 299:563–565.

Swackhamer, D.L. and K.C. Hornbuckle. 2004. Assessment of air quality and air pollutant impacts in Isle Royale National Park and Voyageurs National Park. Report prepared for the U.S. National Park Service.

Swain, E.B., D.R. Engstrom, M.E. Brigham, T.A. Henning, and P.L. Brezonik. 1992. Increasing rates of atmospheric mercury deposition in midcontinental North America. *Science* 257:784–787.

Swain, E.B. and D.D. Helwig. 1989. Mercury in fish from northeastern Minnesota lakes: Historical trends, environmental correlates, and potential sources. *J. Minn. Acad. Sci.* 55(1):103–109.

Talhelm, A.F., K.S. Pregitzer, A.J. Burton, and D.R. Zak. 2011. Air pollution and the changing biogeochemistry of northern forests. *Front. Ecol. Environ.* 10(4):181–185.

Thomas, R.Q., C.D. Canham, K.C. Weathers, and C.L. Goodale. 2010. Increased tree carbon storage in response to nitrogen deposition in the US. *Nat. Geosci.* 3:13–17.

Throop, H.L. 2005. Nitrogen deposition and herbivory affect biomass production and allocation in an annual plant. *Oikos* 111:91–100.

Throop, H.L., E.A. Holland, W.J. Parton, D.S. Ojima, and C.A. Keough. 2004. Effects of nitrogen deposition and insect herbivory on patterns of ecosystem-level carbon and nitrogen dynamics: Results from the CENTURY model. *Glob. Change Biol.* 2004(10):1092–1105.

Throop, H.L. and M.T. Lerdau. 2004. Effects of nitrogen deposition on insect herbivory: Implications for community and ecosystem processes. *Ecosystems* 7:109–133.

Tilman, D. 1987. Secondary succession and the pattern of plant dominance along experimental nitrogen gradients. *Ecol. Monogr.* 57(3):189–214.

Tilman, D. 1993. Species richness of experimental productivity gradients: How important is colonization limitation. *Ecology* 74:2179–2191.

Tilman, D. and D. Wedin. 1991. Dynamics of nitrogen competition between successional grasses. *Ecology* 72:1038–1049.

Turner, R.R. and G.W. Southworth. 1999. Mercury contaminated industrial and mining sites in North America: An overview with selected case studies. *In* R. Ebinghaus, R.R. Turner, L.D. Lacerda, O. Vasiliev, and W. Salomons (Eds.). *Mercury Contaminated Sites.* Springer-Verlag, Berlin, Germany.

U.S. Environmental Protection Agency. 1993. Air quality criteria for oxides of nitrogen. Volumes I–III. EPA600/8-91/049af. U.S. Environmental Protection Agency, Research Triangle Park, NC.

U.S. Geological Survey (USGS). Last modified February 20, 2015. Predicted surface water methylmercury concentrations in National Park Service Inventory and Monitoring Program Parks. U.S. Geological Survey. Wisconsin Water Science Center, Middleton, WI. Available at: http://wi.water.usgs.gov/mercury/NPSHgMap.html (accessed February 26, 2015).

Venier, M. and R.A. Hites. 2008. Flame retardants in the atmosphere near the Great Lakes. *Environ. Sci. Technol.* 42:4745–4751.

Venier, M., M. Wierda, W.W. Bowerman, and R.A. Hites. 2010. Flame retardants and organochlorine pollutants in bald eagle plasma from the Great Lakes region. *Chemosphere* 80:1234–1240.

Vitt, D.H., K. Wieder, L.A. Halsey, and M. Turetsky. 2003. Response of *Sphagnum fuscum* to nitrogen deposition: A case study of ombrogenous peatlands in Alberta, Canada. *Bryologist* 106(2):235–245.

Watras, C.J. and K.A. Morrison. 2008. The response of two remote, temperate lakes to changes in atmospheric mercury deposition, sulfate, and the water cycle. *Can. J. Fish. Aquat. Sci.* 65:100–116.

Watras, C.J., K.A. Morrison, R.J.M. Hudson, T.M. Frost, and T.K. Kratz. 2000. Decreasing mercury in northern Wisconsin: Temporal patterns in bulk precipitation and a precipitation-dominated lake. *Environ. Sci. Technol.* 34(19):4051–4057.

Wedin, D. and D. Tilman. 1996. Influence of nitrogen loading and species composition on the carbon balance of grasslands. *Science* 274:1720–1723.

Weiss-Penzias, P.S., D.A. Gay, M.E. Brigham, M.T. Parsons, M.S. Gustin, and A. ter Schure. 2016. Trends in mercury wet deposition and mercury air concentrations across the U.S. and Canada. *Sci. Total Environ.* 568:546–556.

Welch, N.T., J.M. Belmont, and J.C. Randolph. 2007. Summer ground layer biomass and nutrient contribution to aboveground litter in an Indiana temperate deciduous forest. *Am. Midl. Nat.* 157:11–26.

Wetmore, C.M. 1988. Lichens and air quality in Indian Dunes National Lakeshore. *Mycotaxon* 33:25–39.

Wiener, J.G., R.A. Bodaly, S.S. Brown, M. Lucotte, M.C. Newman, D.B. Porcella, R.J. Reash, and E.B. Swain. 2007. Monitoring and evaluating trends in methylmercury accumulation in aquatic biota. *In* R.C. Harris, D.P. Krabbenhoft, R.P. Mason, M.W. Murray, R.J. Reash, and T. Saltman (Eds.). *Ecosystem Responses to Mercury Contamination: Indicators of Change.* SETAC, CRC Press, Taylor & Francis Group, Boca Raton, FL, pp. 87–122.

Wiener, J.G., D.C. Evers, D.A. Gay, H.A. Morrison, and K.A. Williams. 2012a. Mercury contamination in the Laurentian Great Lakes region: Introduction and overview. *Environ. Pollut.* 161:243–251.

Wiener, J.G., R.J. Haro, K.R. Rolfhus, M.B. Sandheinrich, S.W. Bailey, and R.M. Northwick. 2013. Bioaccumulation of contaminants in fish and larval dragonflies in six national park units of the western Great Lakes region, 2008–2009. Natural Resource Data Series NPS/GLKN/NRDS—2013/427. National Park Service, Fort Collins, CO.

Wiener, J.G., B.C. Knights, M.B. Sandheinrich, J.D. Jeremiason, M.E. Brigham, D.R. Engstrom, L.G. Woodruff, W.F. Cannon, and S.J. Balogh. 2006. Mercury in soils, lakes, and fish in Voyageurs National Park (Minnesota): Importance of atmospheric deposition and ecosystem factors. *Environ. Sci. Technol.* 40(20):6261–6268.

Wiener, J.G., D.P. Krabbenhoft, G.H. Heinz, and A.M. Scheuhammer. 2003. Ecotoxicology of mercury. *In* D.J. Hoffman, B.A. Rattner, G.A. Burton, and J. Cairns (Eds.). *Handbook of Ecotoxicology,* 2nd edn. CRC Press, Boca Raton, FL, pp. 409–463.

Wiener, J.G., M.B. Sandheinrich, S.P. Bhavsar, J.R. Bohr, D.C. Evers, B.A. Monson, and C.S. Schrank. 2012b. Toxicological significance of mercury in yellow perch in the Laurentian Great Lakes region. *Environ. Pollut.* 161:350–357.

Wiener, J.G. and D.J. Spry. 1996. Toxicological significance of mercury in freshwater fish. *In* W.N. Beyer, G.H. Heinz, and A.W. Redmon (Eds.). *Environmental Contaminants in Wildlife: Interpreting Tissue Concentrations.* Lewis Publishers, Boca Raton, FL, pp. 297–339.

Wolfe, M.F., T. Atkeson, W. Bowerman, K. Burger, D.C. Evers, M.W. Murray, and E. Zillioux. 2007. Wildlife indicators. *In* R.C. Harris, D.P. Krabbenhoft, R.P. Mason, M.W. Murray, R.J. Reash, and T. Saltman (Eds.). *Ecosystem Responses to Mercury Contamination: Indicators of Change.* SETAC, CRC Press, Taylor & Francis Group, Boca Raton, FL, pp. 123–189.

Wood, P.B., J.H. White, A. Steffer, J.M. Wood, C.F. Facemire, and F. Percival. 1996. Mercury concentrations in tissues of Florida bald eagles. *J. Wildl. Manage.* 60:178–185.

Zhang, L., P. Blanchard, D. Johnson, A. Dastoor, A. Ryzhkov, C.J. Lin, K. Vijayaraghavan et al. 2012. Assessment of modeled mercury dry deposition over the Great Lakes region. *Environ. Pollut.* 161:272–283.

Zillioux, E.J., D.B.B. Porcella, and B.J. M. 1993. Mercury cycling and effects in freshwater wetland ecosystems. *Environ. Toxicol. Chem.* 12:2245–2264.

9

Mammoth Cave National Park and the Cumberland Piedmont Network

9.1 Background

The Cumberland Piedmont Network contains 14 national parks. Most are historic and military parks. Mammoth Cave National Park (MACA) is the main network park considered here. There are many large population centers (more than 500,000 people) near MACA and the network region, including Atlanta, Charlotte, Nashville, and Memphis. Figure 9.1 shows the network boundary along with locations of each park and population centers with more than 10,000 people. Land cover in and around the Cumberland Piedmont Network is variable. The predominant cover types within this network region are generally pasture/hay, forest, and row crop, based on the National Land Cover Dataset. The agricultural and urban land uses may be important source areas for atmospheric N emissions.

The southeastern and south central United States is geologically diverse and was not glaciated during the most recent period of glaciation. As a consequence, a large number of aquatic animal species have evolved and colonized this region. Streams in the Tennessee-Cumberland aquatic region are among the most biologically diverse in the world (Smith et al. 2002). This region contains the highest diversity of freshwater mussel and crayfish species and the highest levels of aquatic species endemism in North America (Abell et al. 2000, Smith et al. 2002). The region contains 231 species of fish (67 endemic), 125 species of mussels (20 endemic), and 65 species of crayfish (40 endemic). More than 57 species of fish and 47 species of mussels are considered at risk (Master et al. 1998, Abell et al. 2000). Therefore, potential effects of acidic deposition on acid-sensitive streams in this region are of great concern.

9.2 Atmospheric Emissions and Deposition

County-level emissions near the Cumberland Piedmont Network, based on data from the EPA's National Emissions Inventory during a recent time period (2011), are depicted in Figures 9.2 through 9.4 for sulfur dioxide, oxidized N, and ammonia, respectively. Many counties to the north and west of Cumberland Piedmont Network parks had relatively high sulfur dioxide emissions (>50 tons/mi^2/yr; Figure 9.2). Spatial patterns in oxidized N emissions were generally similar, with highest values to the northwest of Cumberland

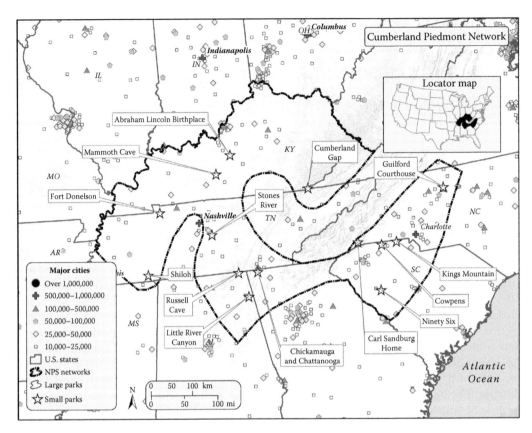

FIGURE 9.1
Network boundary and locations of parks and population centers having more than 10,000 people near the Cumberland Piedmont Network.

Piedmont Network parks (Figure 9.3). Emissions of ammonia near Cumberland Piedmont Network parks were somewhat lower, with most counties showing emissions levels below 8 tons/mi^2/yr (Figure 9.4).

Total S deposition throughout most of the Cumberland Piedmont Network region generally ranged from about 10 to 15 kg S/ha/yr in 2002, based on Community Multiscale Air Quality model estimates (Sullivan et al. 2011). Some areas of the network, predominately in the northern part, were estimated to receive as much as 20–30 kg S/ha/yr. Total N deposition within the network ranged from as low as 5 to 10 kg N/ha/yr to as high as 15 to 20 kg N/ha/yr. Throughout most of the network region, total N deposition was relatively high compared to other networks, in the range of 10–15 kg N/ha/yr (Sullivan et al. 2011).

Total S deposition in and around the Cumberland Piedmont Network for the period 2010–2012 was generally highest (>10 kg S/ha/yr) to the north and lowest (<5 kg S/ha/yr) in the southeastern portion of the network area (Figure 9.5). Oxidized inorganic N deposition for the period 2010–2012 was less than 10 kg N/ha/yr throughout the parklands within the Cumberland Piedmont Network. Some areas received less than 5 kg N/ha/yr. Reduced inorganic N (ammonium) deposition during this same period was typically in

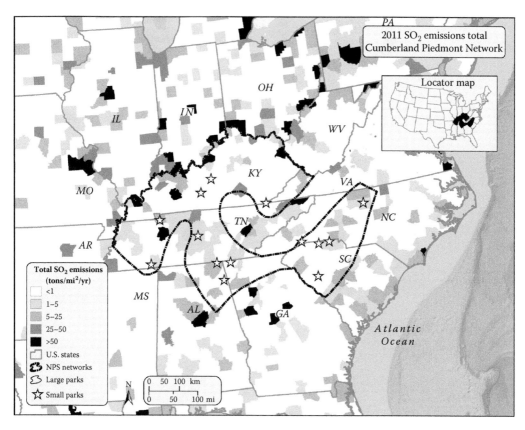

FIGURE 9.2

Total sulfur dioxide (SO_2) emissions, by county, near the Cumberland Piedmont Network for the year 2011. (Data from EPA's National Emissions Inventory, https://www.epa.gov/air-emissions-inventories, accessed January, 2014.)

the range of 5–10 kg N/ha/yr. Total N deposition was 10–15 kg N/ha/yr at most park locations (Figure 9.6).

Atmospheric S and oxidized N deposition levels have declined at MACA since 2001, based on Total Deposition project estimates (Table 9.1). Sulfur deposition at MACA over the past decade decreased 30%. Estimated total N deposition was unchanged. Ammonium deposition increased 49% while oxidized N deposition decreased 25%.

The National Park Service (2010) reported long-term trends in concentrations of Hg in wet deposition during the period beginning in 1996–2003, and running through 2008, for nine national parks located throughout the conterminous United States. Most of the parks, including MACA, did not show statistically significant ($p \leq 0.05$) changes in wet Hg deposition. Three-year mean values of annual Hg concentration in wet deposition were reported by the National Park Service (2010) for 13 parks that had at least two years of valid data during the period 2006–2008. MACA had among the highest levels of wet Hg deposition, as did Great Smoky Mountains and Shenandoah national parks and Indiana Dunes National Lakeshore in nearby networks. These results suggest the possibility of relatively high Hg concentrations in the biota of Cumberland Piedmont Network parks.

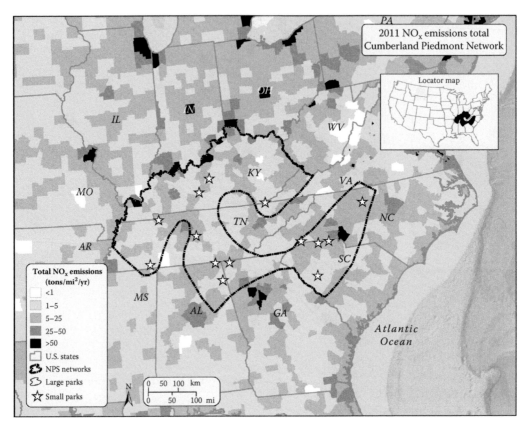

FIGURE 9.3
Total oxidized N (NO_x) emissions, by county, near the Cumberland Piedmont Network for the year 2011. (Data from EPA's National Emissions Inventory, https://www.epa.gov/air-emissions-inventories, accessed January, 2014.)

9.3 Acidification

The network rankings determined in a coarse screening assessment by Sullivan et al. (2011) for acid pollutant exposure, ecosystem sensitivity to acidification, and park protection yielded an overall network acidification Summary Risk ranking for the Cumberland Piedmont Network that was relatively high among networks.

9.3.1 Acidification of Terrestrial Ecosystems

Local data are not available regarding the effects of acidic deposition on terrestrial resources within the Cumberland Piedmont Network parks. However, sugar maple, a species known to be sensitive to acidification, occurs in several of the Cumberland Piedmont Network parks. Acidification of soils has been linked to sugar maple decline in eastern forests (Long et al. 2009, Sullivan et al. 2013). Given the relatively high levels of acidic deposition in the Cumberland Piedmont Network, it is likely that some soil acidification has occurred, with potential effects on acid-sensitive plant species such as sugar maple.

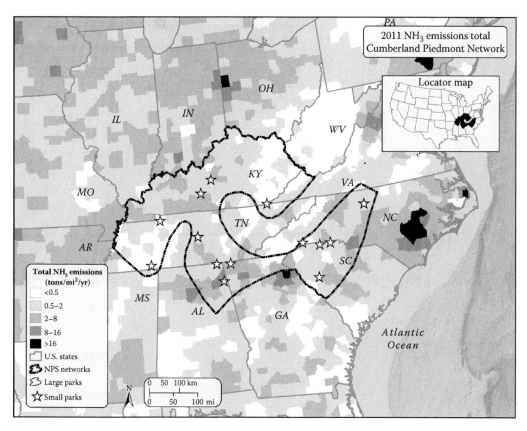

FIGURE 9.4

Total ammonia (NH_3) emissions, by county, near the Cumberland Piedmont Network for the year 2011. (Data from EPA's National Emissions Inventory, https://www.epa.gov/air-emissions-inventories, accessed January, 2014.)

9.3.2 Acidification of Aquatic Ecosystems

The Cumberland Piedmont Network has an ongoing water quality sampling program at the network parks. Monitoring results suggest that many park streams have naturally low pH and acid neutralizing capacity because they are located on bedrock types (especially sandstones) that provide limited acid-buffering and base cation supply. These streams are likely sensitive to atmospheric acid inputs. Given the high levels of acidic deposition that the region has experienced for over 40 years, it is likely that pH and acid neutralizing capacity have been depressed by human-caused pollution. Eight parks in the Cumberland Piedmont Network contain streams that have naturally low pH (<6.0) and acid neutralizing capacity (<50 μeq/L): Carl Sandburg Home National Historic Site (CARL), Chickamauga and Chattanooga National Military Park (CHCH), Cowpens National Battlefield (COWP), Cumberland Gap National Historical Park (CUGA), Guilford Courthouse National Military Park (GUCO), Kings Mountain National Military Park (KIMO), Little River Canyon National Preserve (LIRI), and Shiloh National Military Park (SHIL; Meiman 2005, 2007a–e, 2009a,b). These pH and acid neutralizing capacity values have likely been further depressed by acidic deposition.

Reynolds et al. (2012) and McDonnell et al. (2014) reported S critical load estimates generated within the Ecosystem Management Decision Support aquatic acidification assessment

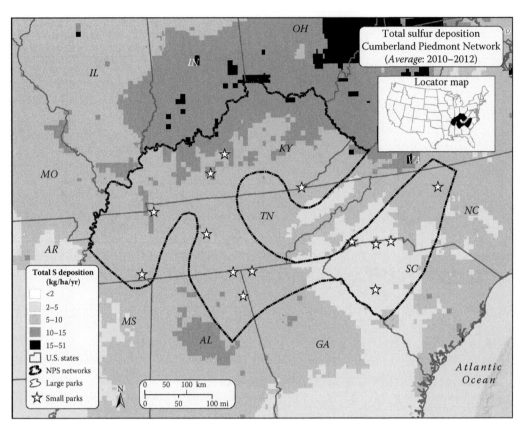

FIGURE 9.5
Total S deposition for the three-year period centered on 2011 in and around the Cumberland Piedmont Network. (From Schwede, D.B. and Lear, G.G., *Atmos. Environ.*, 92, 207, 2014.)

for the southern Appalachian Mountains region (Table 9.2). Steady-state model critical load estimates were given for two parks in the Cumberland Piedmont Network: CARL and CUGA. The estimated critical load values for seven streams in CARL were fairly low. In CUGA, 146 streams were modeled; the median stream had a high critical load, but 25% of the streams had critical load ≤6 kg S/ha/yr, which is lower than the current deposition of S in some areas. The most acid-sensitive stream in CUGA had a modeled critical load equal to 0 kg S/ha/yr, suggesting that even if S deposition was reduced to zero, the stream would still not recover to acid neutralizing capacity above 50 μeq/L.

9.4 Nutrient Nitrogen Enrichment

It appears likely that some nutrient enrichment impacts on terrestrial and aquatic ecosystems have occurred in the Cumberland Piedmont Network region in response to N deposition. For example, McCune et al. (1997) found N-tolerant epiphytic lichen species to be present, along with relatively low epiphytic lichen species richness in urban and industrial areas within Georgia, North Carolina, South Carolina, Tennessee, and Virginia.

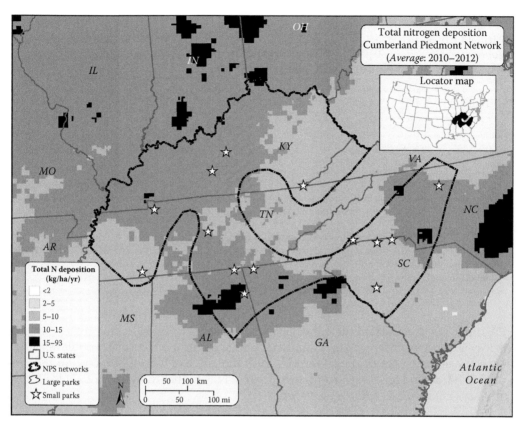

FIGURE 9.6
Total N deposition for the three-year period centered on 2011 in and around the Cumberland Piedmont Network. (From Schwede, D.B. and Lear, G.G., *Atmos. Environ.*, 92, 207, 2014.)

TABLE 9.1

Average Changes in S and N Deposition[a] between 2001 and 2011 across Park Grid Cells at MACA

Parameter	2001 Average (kg/ha/yr)	2011 Average (kg/ha/yr)	Absolute Change (kg/ha/yr)	Percent Change	2011 Minimum (kg/ha/yr)	2011 Maximum (kg/ha/yr)	2011 Range (kg/ha/yr)
Total S	14.78	10.38	−4.40	−29.8	10.06	10.97	0.91
Total N	12.73	12.67	−0.06	−0.4	12.23	13.21	0.98
Oxidized N	8.48	6.34	−2.14	−25.2	6.20	6.53	0.34
Reduced N	4.25	6.33	2.08	49.1	5.79	6.82	1.03

[a] Deposition estimates were determined by the Total Deposition project, based on three-year averages centered on 2001 and 2011 for all ~4 km grid cells for the park. The minimum, maximum, and range of 2011 S and N deposition are also shown.

In contrast, sensitive species were more common, and richness was higher, in rural areas. Brookshire et al. (2007) documented increased stream nitrate concentrations across North Carolina and West Virginia with increasing N deposition ranging from 5 to 32 kg N/ha/yr. This finding suggests that watersheds in this region may be experiencing some level of N saturation.

TABLE 9.2

Distribution of Steady-State Critical Loads of Sulfur (in kg/ha/yr) to Protect Stream Acid Neutralizing Capacity = 50 μeq/L for Parks Modeled in the Ecosystem Management Decision Support Project within the Cumberland Piedmont Network

Park Name	Park Code	# of Critical Loads[a]	Minimum	25th Percentile	Median	25th Percentile	Maximum
Carl Sandburg Home	CARL	7	5	6	6	7	8
Cumberland Gap	CUGA	146	0	6	High	High	High

Sources: Reynolds, K.M. et al., Spatial decision support for assessing impacts of atmospheric sulfur deposition on aquatic ecosystems in the Southern Appalachian Region, *Proceedings of the 45th Hawaiian International Conference on System Sciences*, Maui, HI, January 4–7, 2012; McDonnell, T.C. et al., *J. Environ. Manage.*, 146, 407, 2014.

Notes: 0 signifies that the target acid neutralizing capacity cannot be attained, even if S deposition is reduced to zero; "High" signifies that the critical load is significantly higher than ambient deposition; Critical loads of S in units of kg/ha/yr can be converted to meq/m²/yr by multiplying by 6.25.

[a] Number of small watersheds (generally approximately 1 km²), for which critical load were calculated by Reynolds et al. (2012), that are wholly or partly within the park.

Ellis et al. (2013) estimated the critical load for nutrient N deposition to protect the most sensitive ecosystem receptors in 45 national parks, using the data of Pardo et al. (2011). They estimated the N critical load for MACA in the range of 3–8 kg N/ha/yr for the protection of hardwood forests. Current deposition of N is higher than this critical load, suggesting critical load exceedance (Table 9.1).

9.5 Ozone Injury to Vegetation

9.5.1 Ozone Exposure Indices and Concentrations

The W126 (a measure of cumulative ozone exposure that preferentially weights higher concentrations) and SUM06 (a measure of cumulative ozone exposure that includes only hourly concentrations over 60 ppb) exposure indices calculated by National Park Service staff for parks in the Cumberland Piedmont Network during the period 2005–2009 are given in Table 9.3, along with Kohut's (2007) ozone risk ranking. Ozone condition, as rated by the National Park Service, was moderate or high in the Cumberland Piedmont Network parks. Kohut's evaluation of risk to plants was moderate or high for all parks except Ninety Six (NISI). In the five years analyzed, drought conditions at NISI were judged likely to constrain ozone uptake by plants in the park.

The Cumberland Piedmont Network has an ongoing monitoring program to evaluate ozone symptoms on sensitive bioindicator plants. Ozone symptoms have been documented at CUGA (Jernigan et al. 2009); COWP, MACA, and SHIL (Jernigan et al. 2010); CHCH (Jernigan et al. 2011); LIRI and RUCA (Jernigan et al. 2012); and GUCO (Howie Neufeld, National Park Service, personal communication, July 2014). Carl Sandburg Home NHS and Fort Donelson National Battlefield (FODO) also had confirmed ozone injury (Johnathan Jernigan, National Park Service, personal communication, July 2014).

TABLE 9.3

Ozone Assessment Results for Parks in the Cumberland Piedmont Network Based on Estimated
Average 3-Month W126 and SUM06 Ozone Exposure Indices for the Period 2005–2009 and Kohut's
(2007) Ozone Risk Ranking for the Period 1995–1999[a,b]

| Park Name | Park Code | W126 | | SUM06 | | Kohut Ozone Risk Ranking |
		Value (ppm-hr)	NPS Condition	Value (ppm-hr)	NPS Condition	
Abraham Lincoln Birthplace	ABLI	12.60	Moderate	17.32	High	Moderate
Carl Sandburg Home	CARL	11.88	Moderate	15.99	High	Moderate
Chickamauga and Chattanooga	CHCH	14.09	High	19.06	High	High
Cowpens	COWP	13.45	High	18.40	High	High
Cumberland Gap	CUGA	12.70	Moderate	16.44	High	High
Fort Donelson	FODO	13.59	High	17.41	High	High
Guilford Courthouse	GUCO	14.01	High	18.95	High	High
Kings Mountain	KIMO	14.41	High	19.66	High	High
Little River Canyon	LIRI	13.10	High	18.08	High	High
Mammoth Cave	MACA	12.77	Moderate	17.60	High	High
Ninety Six	NISI	12.44	Moderate	17.38	High	Low
Russell Cave	RUCA	13.75	High	18.77	High	Moderate
Shiloh	SHIL	12.33	Moderate	16.49	High	High
Stones River	STRI	13.20	High	17.79	High	High

[a] Parks are classified into one of three ranks (Low, Moderate, High) based on comparison with other Inventory and Monitoring parks.
[b] Degrees of concern for the W126 and SUM06 indices are based solely on levels of ozone exposure. Kohut's risk to vegetation is based on several factors that contribute to injury in plants, including ozone exposure and environmental variables, and considers the effects of soil moisture on the uptake of ozone.

9.5.2 Ozone Formation

There are recognized differences in the potential to form atmospheric ozone among the prominent southeastern national parks, including MACA in the Cumberland Piedmont Network (Kang et al. 2003). In areas like MACA, where ozone is mainly produced locally, reducing oxidized N and/or volatile organic compound emissions locally should reduce the capacity for ozone production. Tong et al. (2005) quantified the relative importance of point and nonpoint mobile emissions sources of oxidized N, identified the origin of air masses associated with high concentrations of oxidized N, and calculated oxidized N production and removal budgets for MACA. Air pollution influence areas were identified using cluster analysis (Dorling et al. 1992) of model output generated using the Hybrid Single-Particle Lagrangian Integrated Trajectories model (Draxler 1997) and also emission source identification based on the EPA's National Emissions Inventory (U.S. EPA 2001). Clusters were labeled according to the general direction of movement of the air mass toward the parks. MACA experienced a larger contribution of oxidized N from point sources, with the maximum occurring during midday and the minimum just before sunrise. This pattern was attributed by Tong et al. (2005) to the formation of a stable nighttime boundary layer over MACA that prevents air masses having high pollutant concentrations from reaching the surface at night. This process is coupled with enhanced nighttime removal of pollutants by dry deposition due to the lower boundary layer and higher relative humidity at night (Finkelstein et al. 2000). Breakup of the nighttime boundary layer at MACA is triggered by sunrise, allowing air masses containing high concentrations of transported S and N to mix downward toward the surface during the day (Tong et al. 2005).

At MACA, the proportion of oxidized N emissions derived from point sources was generally higher during winter and spring and lower during summer and fall. In midsummer, point sources sometimes constituted an estimated 70% of total oxidized N at MACA during episodes of high air pollutant levels; more commonly, summer point source contributions at MACA were much lower than mobile sources. On average, a minimum of one-fourth of reactive oxidized N at MACA is estimated to have been emitted by point sources (Tong et al. 2005). Point sources tend to be episodic at MACA and occur mainly during daytime hours. However, MACA has exhibited high concentrations of atmospheric ozone, in excess of the 2008 8-hour standard (75 ppb), several times each year (Kang et al. 2001, 2003, Ryerson et al. 2001). It is generally assumed that oxidized N is the main limiting factor for photochemical ozone production in remote areas of the southeastern United States, including at MACA (Kang et al. 2003, Tong et al. 2005).

MACA also has strong biogenic volatile organic compound sources. The most abundant volatile organic compound measured in the air during summer in MACA and nearby parks during the period 1995–1997 was isoprene, a natural biogenic compound produced by vegetation (Kang et al. 2001). Upon conversion of all volatile organic compound concentrations into propylene-equivalent concentration units (to assess the relative potential contribution of volatile organic compounds to ozone formation), biogenic components accounted for 69%–95% of the total atmospheric volatile organic compound concentration during all three years investigated. About 80% of the total comprised isoprene; other common volatile organic compounds included α-pinene, β-pinene, and limonene. The major anthropogenic volatile organic compounds during summer were isopentane, toluene, and propane. These are derived from automobile exhaust (isopentane), solvents (toluene), and natural gas (propane; Kang et al. 2001).

Kang et al. (2003) applied the Multiscale Air Quality Simulation Platform model to predict atmospheric ozone concentration with an overall uncertainty less than 30% in three eastern parks, including MACA. The largest ozone production due to local chemistry was simulated for the low elevation (219 m [719 ft]) site at MACA, nearly twice as high as for two nearby high-elevation sites. Modeling results suggested that volatile organic compounds were chemically saturated at all three locations, potentially causing decreased ozone production in response to further increases in volatile organic compound emissions. More than half of the ozone at the high-elevation sites was simulated to be transported from other areas, as compared with only 20% at MACA (Kang et al. 2003).

9.6 Visibility Degradation

9.6.1 Natural Background and Ambient Visibility Conditions

Within the Cumberland Piedmont Network, haze has been estimated by the Interagency Monitoring of Protected Visual Environments (IMPROVE) program for the monitoring site (MACA1) located in MACA for the 20% clearest visibility days, the 20% haziest visibility days, and average days. Data are also available that are considered to be representative of visibility conditions in two Class II areas (Abraham Lincoln Birthplace [ABLI] and FODO), based on measured values taken from nearby IMPROVE sites (MACA1 and CADI1, respectively; CADI1 is in Cadiz, Kentucky).

Ambient visibility estimates reflect the pollution level at the time of measurement and were used to rank conditions at parks in order to provide information on spatial differences in visibility and air pollution. Rankings range from very low haze (very good visibility) to very high haze (very poor visibility). Table 9.4 gives the relative park haze rankings on the 20% clearest, 20% haziest, and average days. Measured haze values in all of the parks studied for the period 2004 through 2008 in this network were considerably higher than the estimated natural conditions; the haze rankings for all three groups were very high. MACA exhibits some of the highest visibility degradation from air pollutants among all parks in the United States.

Annual mean haze levels on the haziest days at 47 monitored park locations during the period 2006–2008 ranged from 1.5 to 21 deciviews higher than the estimated natural condition (NPS 2010). The average difference between measured haze and estimated natural condition was 8.3 deciviews. Several eastern parks had annual mean deciviews on the haziest days that were substantially higher (more than about 10 deciviews) than estimated natural conditions; these included MACA and two parks close to the Cumberland Piedmont Network boundary: GRSM and SHEN. Monitoring sites showing the largest differences during the period 2006–2008 between measured visibility on the clearest days and estimated natural conditions included MACA (NPS 2010).

IMPROVE data allow estimation of visual range. Data indicate that at the MACA1 monitoring site, air pollution has reduced average visual range from 110 to 20 mi (177 to 32 km). On the haziest days, visual range has been reduced from 80 to 10 mi (129 to 16 km). Severe haze episodes occasionally reduce visibility to 4 mi (6 km). At the CADI1 monitor, representative of FODO, pollution has reduced average visual range from 110 to 25 mi (177 to 40 km). On the haziest days, visual range has been reduced from 80 to 10 mi (129 to 16 km). Severe haze episodes occasionally reduce visibility to 5 mi (8 km).

Figure 9.7 shows estimated natural (preindustrial), baseline (2000–2004), and current (2006–2010) levels of haze and its composition for ABLI, FODO, and MACA. Data for ABLI and FODO are from nearby sites. The majority of total particulate light extinction in MACA is attributable to sulfate extinction (Figure 9.7). On average, sulfate contributed 67.9% of annual light extinction, nitrate 12.7%, and organics 11.4%. For the other two parks, sulfate was also the largest contributor to ambient haze. Thus, sulfate has a disproportionately large impact on light extinction in this network, especially on the 20% haziest days. Nitrate is also important. Even on the 20% clearest visibility days, sulfate plus nitrate contributes almost three-fourths of the total light extinction.

Zhao and Hopke (2006) analyzed data from the IMPROVE monitoring program in order to distinguish the chemical composition and sources of fine aerosols at MACA and ultimately inform pollution control strategies. The data resolved nine sources of ambient aerosols at this park and determined the average contribution of each source to total particulate matter mass, as follows: gasoline emission (6.7%), diesel emission (3.1%), summer secondary sulfate (49.0%), winter secondary sulfate (0.6%), organic C-rich secondary sulfate (16.2%), secondary nitrate (2.8%), intercontinental dust plus soil (4.9%), wood smoke (13.6%), and aged sea salt (3.2%; Zhao and Hopke 2006). The researchers suggested that the intercontinental dust plus soil source (4.9% of the total) may be primarily composed of Saharan dust from summer sandstorms in Africa (Zhao and Hopke 2006).

TABLE 9.4

Estimated Natural Haze and Measured Ambient Haze in Parks in the Cumberland Piedmont Network Averaged over the Period 2004–2008[a]

Park Name	Park Code	Site ID	Estimated Natural Haze (Deciviews)		
			20% Clearest Days	20% Haziest Days	Average Days
Abraham Lincoln Birthplace[b]	ABLI	MACA1	4.99	11.08	7.77
Fort Donelson[b]	FODO	CADI1	5.17	10.84	7.77
Mammoth Cave	MACA	MACA1	4.99	11.08	7.77

Park Name	Park Code	Site ID	Measured Ambient Haze (for Years 2004–2008)					
			20% Clearest Days		20% Haziest Days		Average Days	
			Deciviews	Ranking	Deciviews	Ranking	Deciviews	Ranking
Abraham Lincoln Birthplace[b]	ABLI	MACA1	16.19	Very High	31.38	Very High	23.20	Very High
Fort Donelson[b]	FODO	CADI1	15.11	Very High	29.86	Very High	22.04	Very High
Mammoth Cave	MACA	MACA1	16.19	Very High	31.38	Very High	23.20	Very High

[a] Parks are classified into one of five haze ranks (Very Low, Low, Moderate, High, or Very High) based on comparison with other monitored parks.

[b] Data are borrowed from a nearby site. A monitoring site is considered by IMPROVE to be representative of an area if it is within 60 mi (100 km) and 425 ft (130 m) in elevation of that area. MACA1 is in MACA; CADI1 is in Cadiz, Kentucky.

9.6.2 Trends in Visibility

Only modest long-term changes in haze levels are apparent in the available monitoring data collected at MACA and at the IMPROVE site located near FODO (Figure 9.8). No long-term trend in deciviews was observed on the clearest or haziest 20% of days for the visibility monitoring site in MACA during the period 1992–2008 (NPS 2010). However, haze levels have decreased over the last approximately six years of monitoring data.

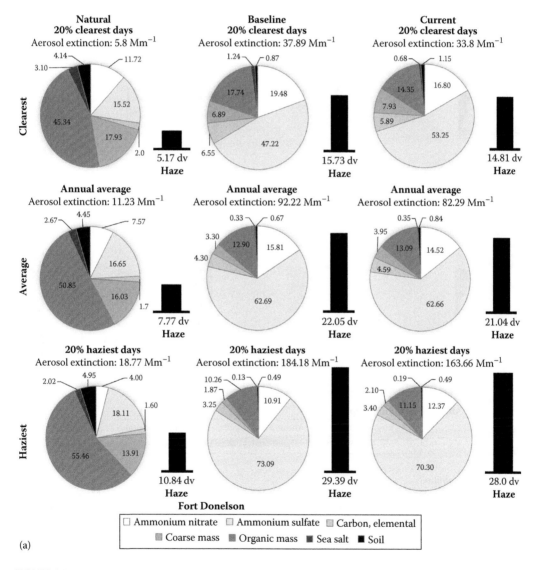

FIGURE 9.7

Estimated natural (preindustrial), baseline (2000–2004; FODO has no data for the years 2000 and 2001), and current (2006–2010) levels of haze (columns) and its composition (pie charts) on the 20% clearest, annual average, and 20% haziest visibility days for (a) FODO and (b) MACA and ABLI. Data for ABLI and FODO were taken from nearby sites (MACA1 and CADI1, respectively). (From http://views.cira.colostate.edu/fed/Tools/RegionalHazeSummary.aspx, accessed October, 2012.)   (*Continued*)

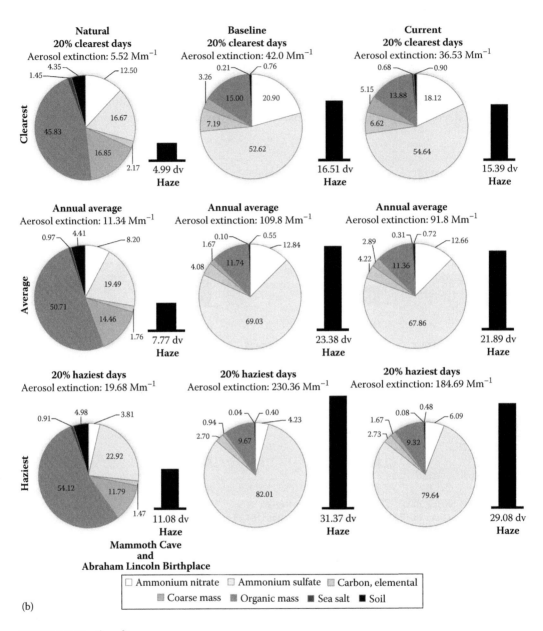

FIGURE 9.7 (Continued)
Estimated natural (preindustrial), baseline (2000–2004; FODO has no data for the years 2000 and 2001), and current (2006–2010) levels of haze (columns) and its composition (pie charts) on the 20% clearest, annual average, and 20% haziest visibility days for (a) FODO and (b) MACA and ABLI. Data for ABLI and FODO were taken from nearby sites (MACA1 and CADI1, respectively). (From http://views.cira.colostate.edu/fed/Tools/RegionalHazeSummary.aspx, accessed October, 2012.)

9.6.3 Development of State Implementation Plans

The Visibility Improvement State and Tribal Association of the Southeast was a collaborative effort among state governments, tribal governments, and federal agencies involved in the management of visibility and regional haze in the Southeast. The region included the

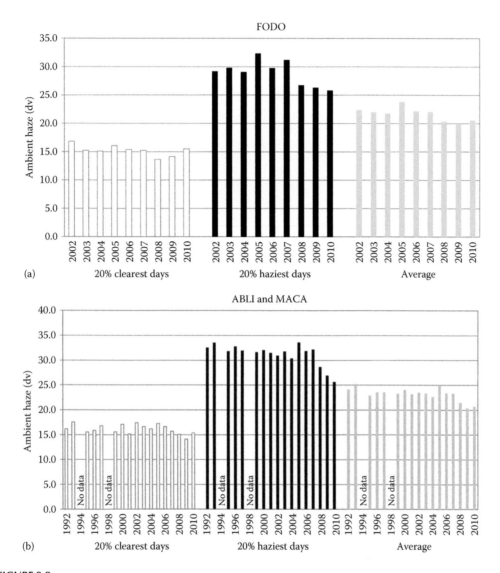

FIGURE 9.8
Trends in ambient haze levels at (a) FODO and (b) MACA and ABLI, based on IMPROVE measurements on the 20% clearest, 20% haziest, and annual average visibility days over the monitoring period of record. Data for ABLI and FODO were taken from nearby sites (MACA1 and CADI1, respectively). (From http://vista.cira. colostate.edu/improve/Data/IMPROVE/summary_data.htm, accessed October, 2012.)

southeastern United States from Virginia and West Virginia in the north, south to Florida, and west to Kentucky, Tennessee, and Mississippi, including parts of the Cumberland Piedmont Network. Analyses conducted on behalf of this organization have included determination of baseline visibility conditions, calculation of the glideslope from the baseline necessary to achieve background conditions in 2064, and determination of air pollutant source areas. The apparent recent improvement in visibility in the monitored parks in the Cumberland Piedmont Network is evident in the glideslope analysis shown in Figure 9.9. Recent measured haze values on the 20% haziest days are decreasing roughly in accordance with the glideslope needed to attain natural visibility by 2064.

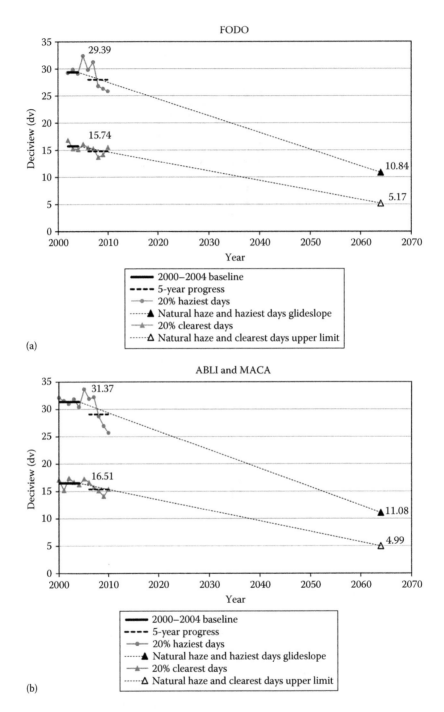

FIGURE 9.9
Glideslopes to achieving natural visibility conditions in 2064 for the 20% haziest (upper line) and the 20% clearest (lower line) days in (a) FODO and (b) MACA and ABLI. In the Regional Haze Rule, the clearest days do not have a uniform rate of progress glideslope; the rule only requires that the clearest days do not get any worse than the baseline period. Also shown are measured values during the period 2000–2010. Data for ABLI and FODO were taken from nearby sites (MACA1 and CADI1, respectively). (From http://vista.cira.colostate.edu/improve/Data/IMPROVE/summary_data.htm, accessed October, 2012.)

9.7 Toxic Airborne Contaminants

Mercury emissions and deposition have likely been high in the southcentral and southeastern United States because of the prevalence of coal-burning power plants. Biota have been evaluated for Hg contamination in some of the parks. The National Park Service has used dragonfly larvae to evaluate Hg in some parks (http://www.nature.nps.gov/air/Studies/air_toxics/dragonfly/index.cfm). Three sites were sampled at MACA, two of which had total Hg concentrations in dragonfly larvae higher than the median across all 22 study parks (112 ppb dry wright). The potential impacts of these Hg concentrations are not known.

9.8 Summary

Emissions from large power plants and urban, industrial, and agricultural sources affect air quality in MACA and other Cumberland Piedmont Network parks. There are several large population centers near the network area, including Nashville, TN; Chattanooga, TN; Charlotte, NC; Memphis, TN; and Atlanta, GA. Levels of atmospheric S and N deposition are high across the network parks. Many streams in some of these parks have low pH and acid neutralizing capacity, suggesting acid sensitivity and/or human-caused acidification. Some of the acidification sensitivity may be attributable to the fact that underlying rocks are low in buffering capacity. Given the relatively high levels of acidic deposition over the last half century in this region, it is likely that S and N depositions have also caused some depletion of acid neutralizing capacity and pH in some streams. This acidification may have already affected acid-sensitive fish and aquatic invertebrates. Terrestrial species may also be affected by soil acidification. Sugar maple, a species known to be sensitive to soil acidification, occurs in several of the Cumberland Piedmont Network parks. Acidification of soils has been linked to sugar maple decline in northeastern forests, but to date not in the Cumberland Piedmont Network parks (Jernigan et al. 2012).

In addition to contributing to acidification, N deposition in the Cumberland Piedmont Network can also cause undesirable nutrient enrichment of natural ecosystems, leading to changes in plant species diversity and soil nutrient cycling. Although eastern forests generally have the capacity to assimilate some level of increased N from deposition, sensitive species may be affected at current levels of N deposition in this region.

Ozone pollution can harm human health, reduce plant growth, and cause visible injury to foliage. Ozone concentrations and seasonal exposures are high in the Cumberland Piedmont Network region, and visible injury to a number of plant species has been documented in Cumberland Piedmont Network parks (Jernigan et al. 2012).

Particulate pollution can cause haze, reducing visibility. Haze levels are very high in the Cumberland Piedmont Network region, primarily as the result of sulfates forming in the atmosphere from sulfur dioxide emitted by coal-burning power plants. Nitrates and organic aerosols also contribute to haze in this network. MACA exhibits some of the highest visibility degradation from air pollutants among all parks in the United States.

Airborne contaminants, including Hg and other heavy metals, can accumulate in food webs, reaching toxic levels in top predators. The National Park Service (2010) reported long-term trends in concentrations of Hg in wet deposition during the period beginning in 1996–2003, and running through 2008, for nine national parks. MACA had among the highest wet Hg deposition. Mercury has been detected in dragonfly larvae and other biota in Cumberland Piedmont Network parks.

References

Abell, R., D.M. Olson, E. Dinerstein, P. Hurley, J.T. Diggs, W. Eichbaum, S. Walters et al. 2000. *Freshwater Ecoregions of North America: A Conservation Assessment*. Island Press, Washington, DC.

Brookshire, E.N.J., H.M. Valett, S.A. Thomas, and J.R. Webster. 2007. Atmospheric N deposition increases organic N loss from temperate forests. *Ecosystems* 10:252–262.

Dorling, S.T., T.D. David, and C.E. Pierce. 1992. Cluster analysis: A technique for estimating the synoptic meteorological controls on air and precipitation chemistry—Method and applications. *Atmos. Environ.* 26A:2575–2581.

Draxler, R.R. 1997. Description of the HYSPLIT_4 modeling system. Technical Memorandum ERL ARL-224. National Oceanic and Atmospheric Administration, Air Resources Laboratory, Silver Spring, MD.

Ellis, R.A., D.J. Jacob, M.P. Sulprizio, L. Zhang, C.D. Holmes, B.A. Schichtel, T. Blett, E. Porter, L.H. Pardo, and J.A. Lynch. 2013. Present and future nitrogen deposition to national parks in the United States: Critical load exceedances. *Atmos. Chem. Phys.* 13(17):9083–9095.

Finkelstein, P.L., T.G. Ellestad, J.F. Clarke, T.P. Meyers, D.B. Schwede, E.O. Hebert, and J.A. Neal. 2000. Ozone and sulfur dioxide dry deposition to forests: Observations and model evaluation. *J. Geophys. Res.* 105:15365–15377.

Jernigan, J.W., B. Carson, and T. Leibfreid. 2009. Ozone and foliar injury report for Cumberland Piedmont Network parks: Annual report 2008. Natural Resource Data Series NPS/CUPN/NRDS—2009/012. National Park Service, Fort Collins, CO.

Jernigan, J.W., B. Carson, and T. Leibfreid. 2010. Ozone and foliar injury report for the Cumberland Piedmont Network Parks consisting of Cowpens NB, Fort Donelson NB, Mammoth Cave NP and Shiloh NMP: Annual report 2009. Natural Resource Data Series NPS/CUPN/NRDS—2010/110. National Park Service, Fort Collins, CO.

Jernigan, J.W., B. Carson, and T. Leibfreid. 2011. Cumberland Piedmont Network ozone and foliar injury report—Chickamauga and Chattanooga NMP, Mammoth Cave NP and Stones River NB: Annual report 2010. Natural Resource Data Series NPS/CUPN/NRDS—2011/219. National Park Service, Fort Collins, CO.

Jernigan, J.W., B. Carson, and T. Leibfreid. 2012. Cumberland Piedmont Network ozone and foliar injury report—Little River Canyon National Preserve, Mammoth Cave NP and Russell Cave NM: Annual report 2011. Natural Resource Data Series NPS/CUPN/NRDS—2012/xxx. National Park Service, Fort Collins, CO.

Kang, D., V.P. Aneja, R. Mathur, and J.D. Ray. 2003. Nonmethane hydrocarbons and ozone in three rural southeast United States national parks: A model sensitivity analysis and comparison to measurements. *J. Geophys. Res. Atmos.* 108(D19, 4604):17.

Kang, D., V.P. Aneja, R.G. Zika, C. Farmer, and J.D. Ray. 2001. Nonmethane hydrocarbons in the rural southeast United States national parks. *J. Geophys. Res.* 106(D3):3133–3155.

Kohut, R. 2007. Assessing the risk of foliar injury from ozone on vegetation in parks in the U.S. National Park Service's Vital Signs Network. *Environ. Pollut.* 149:348–357.

Long, R.P., S.B. Horsley, R.A. Hallett, and S.W. Bailey. 2009. Sugar maple growth in relation to nutrition and stress in the northeastern United States. *Ecol. Appl.* 19(6):1454–1466.

Master, L.L., S.R. Flack, and B.A. Stein. 1998. *Rivers of Life: Critical Watersheds for Protecting Freshwater Biodiversity.* The Nature Conservancy, Arlington, VA.

McCune, B., J. Dey, J. Peck, K. Heiman, and S. Will-Wolf. 1997. Regional gradients in lichen communities of the southeast United States. *Bryologist* 100(2):145–158.

McDonnell, T.C., T.J. Sullivan, P.F. Hessburg, K.M. Reynolds, N.A. Povak, B.J. Cosby, W. Jackson, and R.B. Salter. 2014. Steady-state sulfur critical loads and exceedances for protection of aquatic ecosystems in the U.S. southern Appalachian Mountains. *J. Environ. Manage.* 146:407–419.

Meiman, J. 2005. Cumberland Piedmont water quality report: Chickamagua and Chattanooga National Military Park. National Park Service, Atlanta, GA.

Meiman, J. 2007a. Cumberland Piedmont Network water quality report: Cowpens national battlefield. National Park Service, Atlanta, GA.

Meiman, J. 2007b. Cumberland Piedmont Network water quality report: Guilford Courthouse National Military Park. National Park Service, Atlanta, GA.

Meiman, J. 2007c. Cumberland Piedmont Network water quality report: Shiloh National Military Park. National Park Service, Atlanta, GA.

Meiman, J. 2007d. Cumberland Piedmont water quality report: Carl Sandburg Home National Historic Site. National Park Service, Atlanta, GA.

Meiman, J. 2007e. Cumberland Piedmont water quality report: Kings Mountain National Military Park. National Park Service, Atlanta, GA.

Meiman, J. 2009a. Cumberland Piedmont Network water quality report: Cowpens National Battlefield. National Park Service, Atlanta, GA.

Meiman, J. 2009b. Cumberland Piedmont Network water quality report: Little River Canyon National Preserve. National Park Service, Atlanta, GA.

National Park Service (NPS). 2010. Air quality in national parks: 2009 annual performance and progress report. Natural Resource Report NPS/NRPC/ARD/NRR-2010/266. National Park Service, Air Resources Division, Denver, CO.

Pardo, L.H., M.J. Robin-Abbott, and C.T. Driscoll. 2011. Assessment of nitrogen deposition effects and empirical critical loads of nitrogen for ecoregions of the United States. General Technical Report NRS-80. U.S. Forest Service, Newtown Square, PA.

Reynolds, K.M., P.F. Hessburg, T. Sullivan, N. Povak, T. McDonnell, B. Cosby, and W. Jackson. 2012. Spatial decision support for assessing impacts of atmospheric sulfur deposition on aquatic ecosystems in the southern Appalachian Region. *Proceedings of the 45th Hawaiian International Conference on System Sciences*, Maui, HI, January 4–7, 2012.

Ryerson, T.B., M. Trainer, J.S. Holloway, D.D. Parrish, L.G. Huey, D.T. Sueper, G.J. Frost et al. 2001. Observation of ozone formation in power plant plumes and implications of ozone control strategies. *Science* 292(5517):719–723.

Schwede, D.B. and G.G. Lear. 2014. A novel hybrid approach for estimating total deposition in the United States. *Atmos. Environ.* 92:207–220.

Smith, R.K., P.L. Freeman, J.V. Higgins, K.S. Wheaton, T.W. FitzHugh, A.A. Das, and K.J. Ernstrom. 2002. Priority areas for freshwater conservation action: A biodiversity assessment of the southeastern United States. The Nature Conservancy, Arlington, VA.

Sullivan, T.J., G.T. McPherson, T.C. McDonnell, S.D. Mackey, and D. Moore. 2011. Evaluation of the sensitivity of inventory and monitoring national parks to acidification effects from atmospheric sulfur and nitrogen deposition. Natural Resource Report NPS/NRPC/ARD/NRR—2011/349. U.S. Department of the Interior, National Park Service, Denver, CO.

Sullivan, T.J., G.B. Lawrence, S.W. Bailey, T.C. McDonnell, C.M. Beier, K.C. Weathers, G.T. McPherson, and D.A. Bishop. 2013. Effects of acidic deposition and soil acidification on sugar maple in the Adirondack Mountains, New York. *Environ. Sci. Technol.* 47:12687–12694.

Tong, D.Q., D. Kang, V.P. Aneja, and J.D. Ray. 2005. Reactive nitrogen oxides in the southeast United States National Parks: Source identification, origin, and process budget. *Atmos. Environ.* 39:315–327.

U.S. Environmental Protection Agency. 2001. National air quality and emissions trends report, 1999. EPA-454/R-01-004. Office of Air Quality Planning and Standards, Emissions Monitoring and Analysis Division, Air Quality Trends Analysis Group, Research Triangle Park, NC.

Zhao, W. and P.K. Hopke. 2006. Source identification for fine aerosols in Mammoth Cave National Park. *Atmos. Res.* 80:309–322.

10

Northern Colorado Plateau Network

10.1 Background

There are five parks in the Northern Colorado Plateau Network that are larger than 100 mi^2 (259 km^2): Arches National Park (ARCH), Canyonlands National Park (CANY), Capitol Reef National Park (CARE), Dinosaur National Monument (DINO), and Zion National Park (ZION). Larger parks generally have more available data with which to evaluate air pollution sensitivities and effects and generally contain more extensive resources in need of protection against the adverse impacts of air pollution. They are the focus of this evaluation. Since air pollutants can be widespread, reduction of pollution emissions affecting large parks will often result in the protection of smaller parks located nearby as well.

All population centers of considerable magnitude (larger than 25,000 people) in this network region are found near the Great Salt Lake. The remainder of the network region is sparsely populated. Figure 10.1 shows the network boundary, the location of each park, and population centers around the network having more than 10,000 people.

10.2 Atmospheric Emissions and Deposition

Annual county-level S emissions in the Northern Colorado Plateau Network region ranged from less than 1 ton/mi^2 of sulfur dioxide to between 1 and 5 tons/mi^2 in 2002. In general, county-level S and N emissions were both less than 5 tons/mi^2/yr at most locations within the network region. There were scattered sulfur dioxide point sources in the network; most were relatively small and emitted less than 5000 tons/yr.

Annual county-level N emissions within the network ranged from less than 1 ton/mi^2 to between 5 and 20 tons/mi^2 in 2002. There were several relatively large (larger than about 4000 tons/yr) point sources of oxidized N in this network; point source emissions of ammonia were much lower.

Oil and gas emissions of N per unit land area in 2008 were relatively high near several of the Northern Colorado Plateau Network parks (Figure 10.2). Furthermore, there was an increase in oil and gas development at many locations subsequent to 2008. In some counties in the Northern Colorado Plateau Network, oil and gas N emissions represented a substantial percentage (>20%) of the overall N emissions from all sources in 2008 (Figure 10.3).

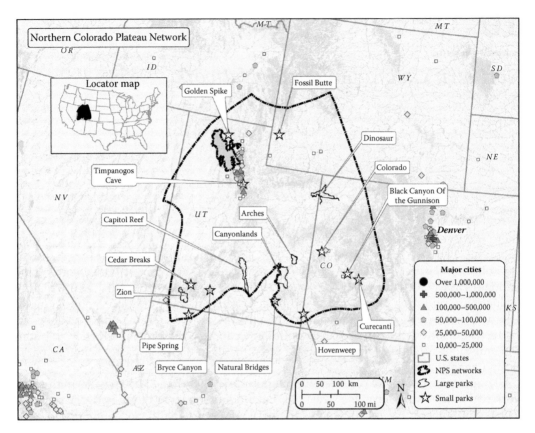

FIGURE 10.1
Network boundary and locations of parks and population centers with more than 10,000 people near the
Northern Colorado Plateau Network.

County-level emissions near the Northern Colorado Plateau Network, based on data
from the EPA's National Emissions Inventory during a recent time period (2011), are
depicted in Figures 10.4 through 10.6 for sulfur dioxide, oxidized N, and ammonia, respectively. Many counties to the north and west of Northern Colorado Plateau Network parks
had relatively high sulfur dioxide emissions (>5 tons/mi^2/yr; Figure 10.4). Spatial patterns in oxidized N emissions were generally similar, with highest values to the northwest of Northern Colorado Plateau Network parks (Figure 10.5). Emissions of ammonia
near Northern Colorado Plateau Network parks were somewhat lower, with most counties
showing emissions levels below 2 tons/mi^2/yr (Figure 10.6).

Total S deposition in and around the Northern Colorado Plateau Network for the
period 2010–2012 was generally less than 1.5 kg S/ha/yr. Oxidized inorganic N deposition for the period 2010–2012 was less than 3 kg N/ha/yr throughout the parkland
within the network. Total N deposition was in the range of 2–5 kg N/ha/yr at most park
locations (Figure 10.7).

Atmospheric S deposition levels have declined at most Northern Colorado Plateau
Network parks since 2001, based on Total Deposition project estimates (Table 10.1).
Estimated total N deposition showed variable responses. At almost all parks in the

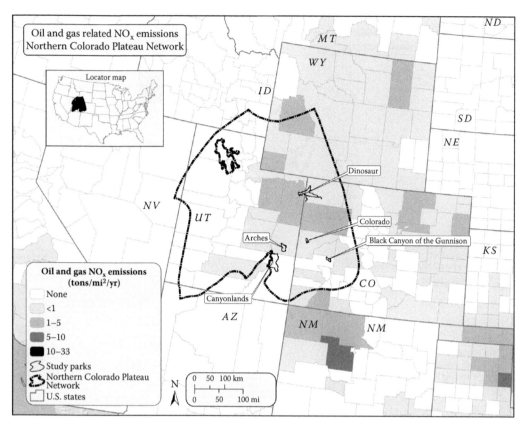

FIGURE 10.2
Estimated oil and gas oxidized N emissions per unit land area (tons/mi^2/year) in the Northern Colorado Plateau Network for the year 2008. (From Sullivan, T.J. and McDonnell, T.C., Mapping of nutrient-nitrogen critical loads for selected national parks in the intermountain west and Great Lakes regions, Report prepared for National Park Service, Natural Resource Technical Report NPS/ARD/NRTR—2014/895, National Park Service, Fort Collins, CO, 2014.)

Northern Colorado Plateau Network, ammonia deposition increased in one case (CARE) by 47%. Oxidized N deposition decreased at all parks considered here.

10.3 Acidification

In general, surface waters of the Northern Colorado Plateau Network are considered resistant to acidification due to the nature of the hydrogeology of the region (Binkley et al. 1997). Data are not available that suggest substantial sensitivity of aquatic or terrestrial resources to acidification (Figure 3.2). Given the relatively low levels (below 1 kg/ha/yr) of acidic deposition (especially of S), acidification effects are unlikely in this network at the present time.

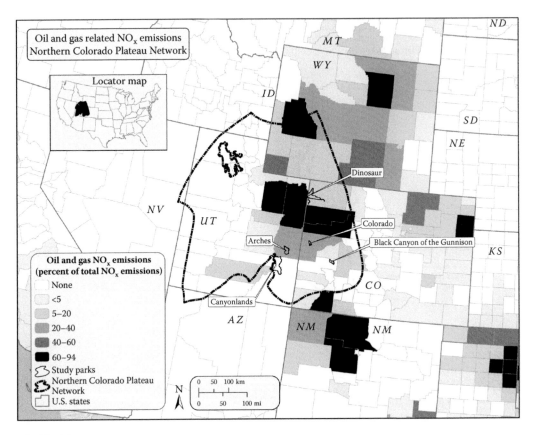

FIGURE 10.3
Estimated oil and gas oxidized N emissions in the Northern Colorado Plateau Network as a percentage of total oxidized N emissions for the year 2008. (From Sullivan, T.J. and McDonnell, T.C., Mapping of nutrient-nitrogen critical loads for selected national parks in the intermountain west and Great Lakes regions, Report prepared for National Park Service, Natural Resource Technical Report NPS/ARD/NRTR—2014/895, National Park Service, Fort Collins, CO, 2014.)

10.4 Nutrient Nitrogen Enrichment

Historically, ecosystems in the arid Southwest derived much of their N from N fixation in biological soil crusts (Schwinning et al. 2005). Atmospheric N deposition now constitutes an additional important source of N input to these systems. Biological crusts are active mainly during the cooler months, from autumn to spring (Belnap 2003), when the native plant N requirement is highest. While active, these crusts release N to the soil as ammonium, which is then transformed to nitrate. This close coupling between N release to the soil by crusts and peak N demand by native plants may have been important in maintaining native grasslands. The simultaneous decline of soil crust cover and increase in atmospheric N deposition may have decreased ecosystem resilience to environmental change and increased the risk of invasion by weedy summer annuals (Schwinning et al. 2005). For example, a fertilization experiment in an arid grassland on the Colorado Plateau demonstrated a large difference in the response of native and nonnative plants to added N (Schwinning et al. 2005).

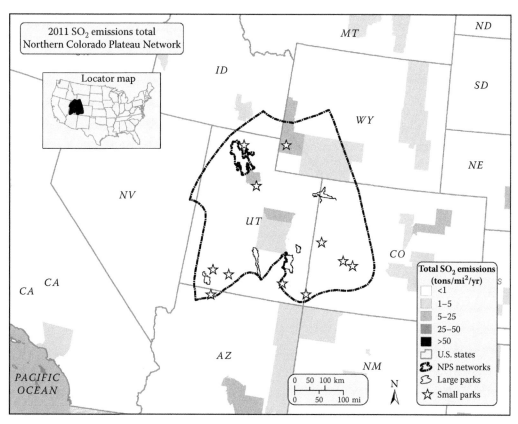

FIGURE 10.4

Total sulfur dioxide (SO_2) emissions, by county, near the Northern Colorado Plateau Network for the year 2011. (Data from EPA's National Emissions Inventory, https://www.epa.gov/air-emissions-inventories, accessed January, 2014.)

For two years, plots were treated with 0, 10, 20, or 40 kg N/ha/yr as a potassium nitrate solution. Galleta (*Hilaria jamesii*) and Indian ricegrass (*Oryzopis hymenoides*) showed no increase in leaf photosynthesis or tiller size, but ricegrass showed a 50% increase in tiller density in the second year at the very high 20 and 40 kg N/ha/yr application levels. For both species, the increased N application hastened the onset of water stress. Unexpectedly, a nonnative species, Russian thistle (*Salsola iberica*), showed a rapid growth response to the highest fertilization rate in the first summer, when rainfall was above average. Schwinning et al. (2005) suggested that the timing and amount of N deposition could facilitate noxious weed invasion and thus change community composition in arid grasslands.

Sullivan and McDonnell (2014) mapped empirical nutrient N critical loads for selected national parks in the intermountain West, including CANY, DINO, and ARCH. This assessment was based on plant communities within Level I ecoregions for which empirical critical loads were compiled by Pardo et al. (2011). These parks are largely covered by vegetation types that were found by Pardo et al. (2011) to be highly sensitive to potential nutrient N enrichment. There were no areas of nutrient N exceedance in ARCH, DINO, or CANY. However, virtually all of DINO and the southern half of ARCH received N deposition in 2008 that was below, but within 1 kg N/ha/yr, of exceedance. Thus, any appreciable increase in N deposition from further oil and gas, agricultural,

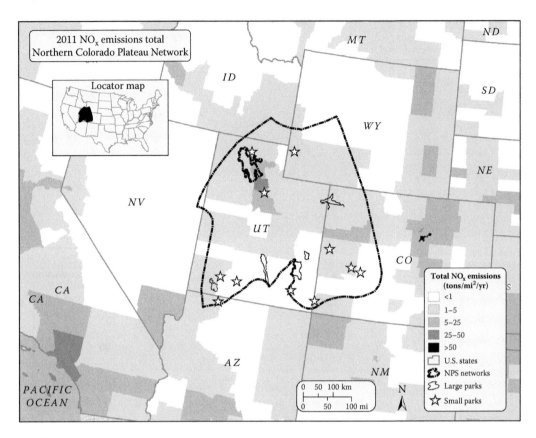

FIGURE 10.5
Total oxidized N (NO$_x$) emissions, by county, near the Northern Colorado Plateau Network for the year 2011. (Data from EPA's National Emissions Inventory, https://www.epa.gov/air-emissions-inventories, accessed January, 2014.)

or other N emissions sources may increase the risk of excess nutrient effects throughout most of the parklands in the Northern Colorado Plateau Network (Sullivan and McDonnell 2014).

10.5 Ozone Injury to Vegetation

Ozone assessment results for the parks considered here are given in Table 10.2. The W126 and SUM06 results were high in four of the parks. Due to generally low soil moisture, however, Kohut (2007) classified parks in this network as having low risk of foliar injury due to ozone exposure. Low soil moisture can limit ozone uptake during drier years.

The National Park Service (2010) evaluated the fourth highest 8-hour ozone concentrations for national parks that have consistently been at or above the 2008 standard of 75 ppb, based on 11–20 years of monitoring data running through 2008. There were six parks that routinely had such high concentrations, and six additional parks had ozone levels that were typically below the 75 ppb standard, but were within the EPA's proposed range for

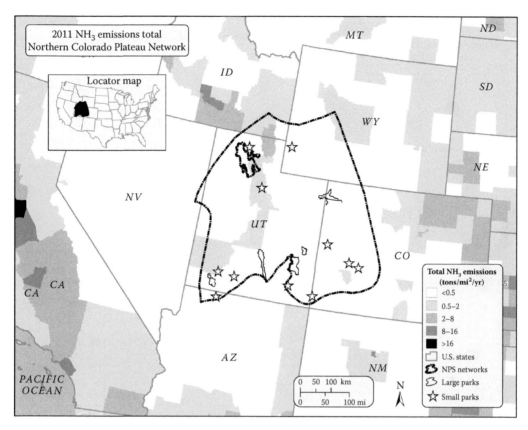

FIGURE 10.6

Total ammonia (NH$_3$) emissions, by county, near the Northern Colorado Plateau Network for the year 2011. (Data from EPA's National Emissions Inventory, https://www.epa.gov/air-emissions-inventories, accessed January, 2014.)

a possible new standard (60–70 ppb; NPS 2010). These latter parks included CANY. The standard was revised to 70 ppb in 2015.

10.6 Visibility Impairment

10.6.1 Natural Background and Ambient Visibility Conditions

Visibility is a source of concern for all parks in the northern Colorado Plateau because of the magnificent vistas viewed from within and looking into and out of the parks. Although some of the best visibility in the conterminous United States occurs in this region, haze is an important air quality issue. Regional and local pollution sources that cause haze and contribute to visibility impairment throughout the Northern Colorado Plateau Network include automobiles, coal- and oil-fired power plants, smelters, wildfires, urban emissions (Peterson et al. 1998), and recently oil and gas development (Sullivan and McDonnell 2014).

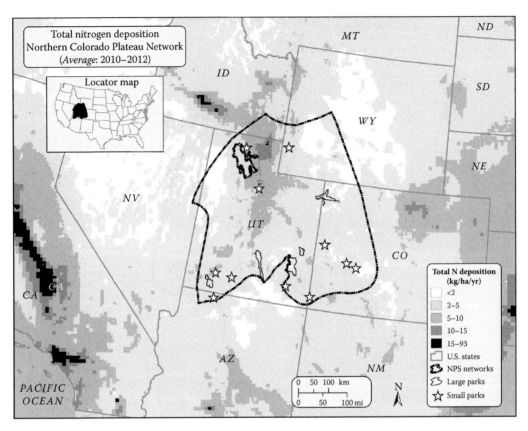

FIGURE 10.7
Total N deposition for the three-year period centered on 2011 in and around the Northern Colorado Plateau Network. (From Schwede, D.B. and Lear, G.G., *Atmos. Environ.*, 92, 207, 2014.)

Interagency Monitoring of Protected Visual Environments (IMPROVE) monitors are located in CANY (CANY1), CARE (CAPI1), and ZION (ZICA1; ZION1 operated from 2001 to 2003, at which time the sampler was moved to a new location and the site was renamed ZICA1). Data are also available that are considered to be representative of visibility conditions in ARCH. A monitoring site is considered by IMPROVE to be representative of an area if it is within 60 mi (100 km) and 425 ft (130 m) in elevation of that area.

Table 10.3 gives the estimated natural (i.e., no man-made impairment) conditions and degree of impairment on the average days and the 20% clearest and 20% haziest days. In general, ambient haze levels at these parks are only slightly higher than estimated background haze levels.

IMPROVE data allow estimation of visual range, a measurement more easily understood by many park visitors than deciview or light extinction. At the CANY site (CANY1; also representative of ARCH), air pollution has reduced average visual range from 170 to 110 mi (274 to 177 km). On the haziest days, visual range has been reduced from 130 to 75 mi (209 to 121 km). Severe haze episodes occasionally reduce visibility to 26 mi (42 km). At the CARE site (CAPI1), air pollution has reduced average visual range from 170 to 110 mi (274 to 177 km). On the haziest days, visual range has been reduced from 130 to 80 mi (209 to 129 km). Severe haze episodes occasionally reduce visibility to 21 mi (34 km). At the ZION site (ZICA1), air pollution has reduced average visual range from 160 to 100 mi

TABLE 10.1

Average Changes in S and N Deposition[a] between 2001 and 2011 across Park Grid Cells at Selected Northern Colorado Plateau Network Parks

Park Code	Parameter	2001 Average (kg/ha/yr)	2011 Average (kg/ha/yr)	Absolute Change (kg/ha/yr)	Percent Change	2011 Minimum (kg/ha/yr)	2011 Maximum (kg/ha/yr)	2011 Range (kg/ha/yr)
ARCH	Total S	1.01	0.68	−0.33	−32.3	0.62	0.78	0.16
	Total N	3.11	2.29	−0.82	−26.2	2.13	2.65	0.52
	Oxidized N	2.31	1.51	−0.79	−34.4	1.40	1.71	0.31
	Reduced N	0.80	0.78	−0.02	−2.6	0.70	0.94	0.24
CANY	Total S	0.97	0.68	−0.29	−29.7	0.59	0.92	0.33
	Total N	2.85	2.22	−0.63	−22.2	1.94	3.10	1.16
	Oxidized N	2.20	1.50	−0.70	−31.7	1.31	2.09	0.78
	Reduced N	0.65	0.72	0.06	9.7	0.63	1.01	0.38
CARE	Total S	0.82	0.70	−0.11	−14.0	0.48	1.19	0.71
	Total N	2.75	2.69	−0.06	−2.7	1.70	4.58	2.88
	Oxidized N	2.02	1.61	−0.41	−20.5	1.15	2.54	1.38
	Reduced N	0.73	1.08	0.35	47.1	0.55	2.04	1.49
DINO	Total S	0.66	0.81	0.15	0.23	0.58	1.02	0.43
	Total N	2.20	2.39	0.19	0.09	1.86	2.74	0.87
	Oxidized N	1.61	1.56	−0.05	−0.03	1.17	1.81	0.63
	Reduced N	0.59	0.82	0.23	0.41	0.65	1.34	0.69
ZION	Total S	1.10	0.98	−0.12	−0.11	0.74	1.35	0.62
	Total N	3.71	3.71	0.00	0.00	2.89	5.16	2.27
	Oxidized N	2.73	2.11	−0.62	−0.23	1.72	2.61	0.90
	Reduced N	0.98	1.60	0.61	0.62	1.16	2.62	0.00

[a] Deposition estimates were determined by the Total Deposition project, based on three-year averages centered on 2001 and 2011 for all ~4 km grid cells in each park. The minimum, maximum, and range of 2011 S and N deposition within each park are also shown.

TABLE 10.2

Ozone Assessment Results for Selected Parks in the Northern Colorado Plateau Network Based on Estimated Average 3-Month W126 and SUM06 Ozone Exposure Indices for the Period 2005–2009 and Kohut's (2007) Ozone Risk Ranking for the Period 1995–1999[a,b]

Park Name	Park Code	W126 Value (ppm-hr)	W126 NPS Condition	SUM06 Value (ppm-hr)	SUM06 NPS Condition	Kohut Ozone Risk Ranking
Arches	ARCH	13.97	High	18.98	High	Low
Canyonlands	CANY	14.70	High	20.35	High	Low
Capitol Reef	CARE	16.16	High	22.82	High	Low
Dinosaur	DINO	10.81	Moderate	12.62	Moderate	Low
Zion	ZION	18.16	High	27.62	High	Low

[a] Parks are classified into one of three ranks (Low, Moderate, High) based on comparison with other Inventory and Monitoring parks.

[b] Degrees of concern for the W126 and SUM06 indices are based solely on levels of ozone exposure. Kohut's risk to vegetation is based on several factors that contribute to injury in plants, including ozone exposure and environmental variables, and considers the effects of soil moisture on the uptake of ozone.

TABLE 10.3

Estimated Natural Haze and Measured Ambient Haze in Selected Parks Averaged over the Period 2004–2008[a]

Park Name	Park Code	Site ID	Estimated Natural Haze (in Deciviews)		
			20% Clearest Days	20% Haziest Days	Average Days
Arches[b]	ARCH	CANY1	1.05	6.43	3.37
Canyonlands	CANY	CANY1	1.05	6.43	3.37
Capitol Reef	CARE	CAPI1	1.25	6.03	3.31
Zion	ZION	ZICA1	1.80	6.70	4.06
Park Name	Park Code	Site ID[b]	Measured Ambient Haze (for Years 2004–2008) (in Deciviews)		
			20% Clearest Days	20% Haziest Days	Average Days
Arches	ARCH	CANY1	2.86	10.64	6.64
Canyonlands	CANY	CANY1	2.86	10.64	6.64
Capitol Reef	CARE	CAPI1	3.08	11.37	6.95
Zion	ZION	ZICA1	4.39	12.60	8.20

[a] Parks are classified into one of five haze ranks (Very Low, Low, Moderate, High, or Very High) based on comparison with other monitored parks.

[b] For ARCH, data are borrowed from a nearby site (CANY1). A monitoring site is considered by IMPROVE to be representative of an area if it is within 60 mi (100 km) and 425 ft (130 m) in elevation of that area.

(258 to 161 km). On the haziest days, visual range has been reduced from 120 to 70 mi (193 to 113 km). Severe haze episodes occasionally reduce visibility to 21 mi (34 km).

10.6.2 Composition of Haze and Sources of Visibility Impairment

Figure 10.8 shows estimated natural (preindustrial), baseline (2000–2004), and more recent (2006–2010) levels of haze and its composition for the monitored parks in the Northern Colorado Plateau Network. For more information on these monitoring data, see http://vista.cira.colostate.edu/improve/Data/IMPROVE/summary_data.htm.

Analyses conducted by the Western Regional Air Partnership indicated that a variety of sources contribute to visibility impairment throughout the western United States, including within the Northern Colorado Plateau Network. The Western Regional Air Partnership provided technical assistance for western states in developing their plans to meet the requirements of the Regional Haze Rule. As part of this, the Partnership conducted regional modeling to identify what type of pollution sources are affecting visibility in Class I areas and where that pollution is coming from.

10.6.3 Development of State Implementation Plans

Because of the Best Available Retrofit Technology requirement of the Regional Haze Rule, major emissions reductions have been made by power plants (electric generating units), and many more are anticipated from sources impacting the Northern Colorado Plateau Network region. From 2000 to 2013, large emissions reductions occurred in Colorado and Utah. These reductions will result in visibility improvements if other visibility-reducing pollutants do not increase. However, organics and carbon fine particles also are major contributors to visibility impairment in the region.

10.7 Toxic Airborne Contaminants

Atmospheric deposition of Hg from coal-burning power plants has been identified as a major source of Hg to remote ecosystems (Landers et al. 2008). Because of the size and number of coal-burning power plants in the Northern Colorado Plateau Network region, Hg is a potential problem for ecosystems in the parks. Although Hg is not monitored in the Northern Colorado Plateau Network parks, data from Hg monitors in the region, including

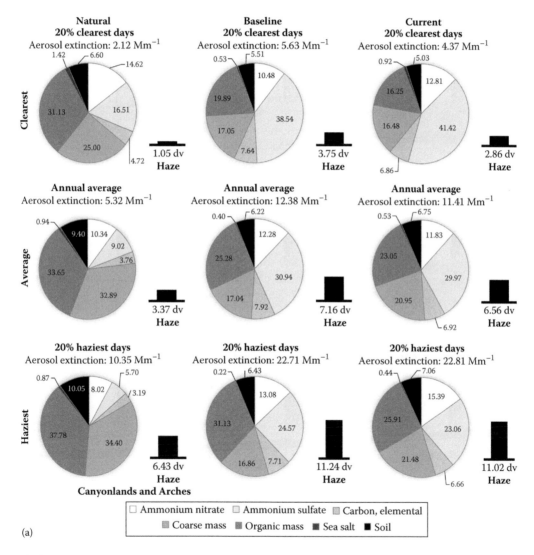

(a)

FIGURE 10.8

Estimated natural (preindustrial), baseline (2000–2004), and more recent (2006–2010) levels of haze in deciviews (columns) and light extinction (aerosol extinction) in inverse megameters (Mm⁻¹) and its composition (pie charts) on the 20% clearest, annual average, and 20% haziest visibility days for the parks in the Northern Colorado Plateau Network. (a) Canyonlands and Arches. (b) Capitol Reef. (c) Zion. Data for ARCH were taken from a representative site (see Table 10.3). (From http://views.cira.colostate.edu/fed/Tools/RegionalHazeSummary.aspx, accessed October, 2012.) *(Continued)* a monitor near Salt Lake City, UT, suggest that concentrations of Hg in rainfall and snow are relatively high compared with elsewhere in the West.

Eagles-Smith et al. (2014) examined fish from 21 national parks in the western U.S., including 2 Northern Colorado Plateau Network parks, CARE (from the Fremont River) and ZION (Virgin River). In both parks, only speckled dace, a small, invertebrate-feeding species, were sampled. Because of their small size, they are not consumed by humans, and therefore, no conclusions can be drawn regarding toxicity related to human consumption.

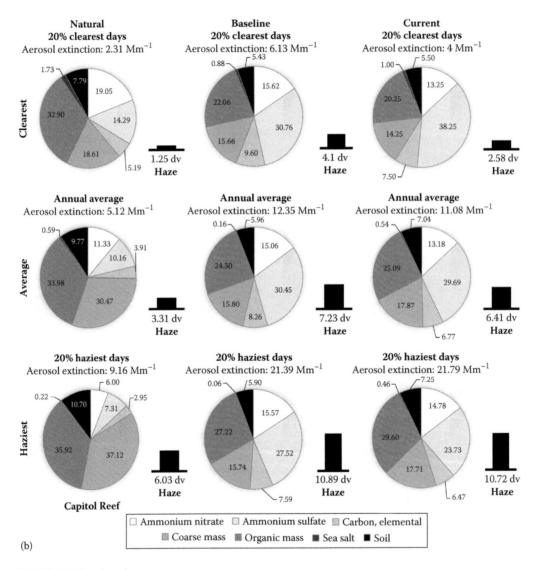

(b)

FIGURE 10.8 (*Continued*)
Estimated natural (preindustrial), baseline (2000–2004), and more recent (2006–2010) levels of haze in deciviews (columns) and light extinction (aerosol extinction) in inverse megameters (Mm⁻¹) and its composition (pie charts) on the 20% clearest, annual average, and 20% haziest visibility days for the parks in the Northern Colorado Plateau Network. (a) Canyonlands and Arches. (b) Capitol Reef. (c) Zion. Data for ARCH were taken from a representative site (see Table 10.3). (From http://views.cira.colostate.edu/fed/Tools/RegionalHazeSummary. aspx, accessed October, 2012.) (*Continued*)

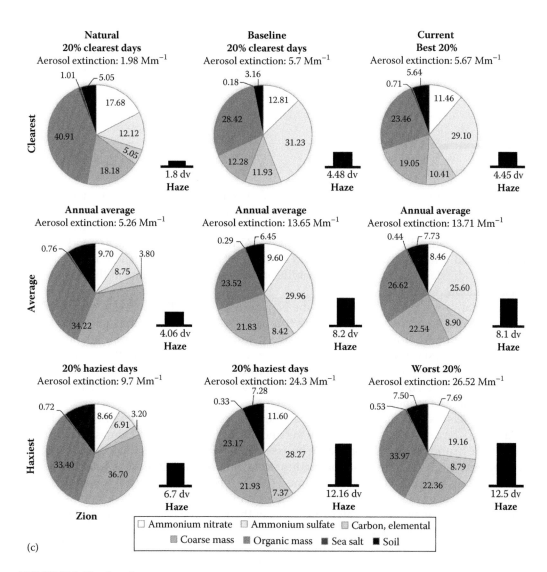

(c)

FIGURE 10.8 (*Continued*)
Estimated natural (preindustrial), baseline (2000–2004), and more recent (2006–2010) levels of haze in deciviews (columns) and light extinction (aerosol extinction) in inverse megameters (Mm^{-1}) and its composition (pie charts) on the 20% clearest, annual average, and 20% haziest visibility days for the parks in the Northern Colorado Plateau Network. (a) Canyonlands and Arches. (b) Capitol Reef. (c) Zion. Data for ARCH were taken from a representative site (see Table 10.3). (From http://views.cira.colostate.edu/fed/Tools/RegionalHazeSummary.aspx, accessed October, 2012.)

However, Hg concentrations were as high as or higher than those in the largest, long-lived predatory species, such as lake trout, sampled in other national parks. Speckled dace serve as potential prey items for predatory fishes and piscivorous birds, and concentrations in dace from both parks exceeded levels associated with biochemical and reproductive effects in fish and reproductive impairment in birds (Eagles-Smith et al. 2014).

Other airborne contaminants pose a risk to biota in the Northern Colorado Plateau Network parks. These include persistent organic pollutants and other semivolatile organic compounds such as pesticides, industrial- and urban-use compounds, and polycyclic aromatic hydrocarbon combustion products (Landers et al. 2010). Limited information is available regarding the levels of semivolatile organic compound contaminants in the Northern Colorado Plateau Network parks (http://science.nature.nps.gov/im/units/ncpn/publications.cfm).

10.8 Summary

Emissions and deposition of S and N in and around the Northern Colorado Plateau Network region are low. In general, county-level S and N emissions were both less than 5 tons/mi^2/yr in most counties within the network region. Total S deposition in both 2001 and 2011 was generally less than 1.5 kg S/ha/yr. Throughout most of the network region, the estimated total N deposition was less than 5 kg N/ha/yr. Given these relatively low levels of acidic deposition, the risk of acidification to soils and streams in this network is considered low.

Terrestrial resources in the parks in the Northern Colorado Plateau Network are likely highly sensitive to nutrient N enrichment from atmospheric N deposition, given the widespread distribution of arid and semiarid plant communities. In several network parks, N deposition is exceeding or close to exceeding levels known to affect the natural diversity of plant and lichen communities. Potential future increases in local N emissions from agriculture and oil and gas development near some of the parks may be cause for concern (Sullivan and McDonnell 2014).

Ozone levels are relatively low in the Northern Colorado Plateau Network region but in recent years have increased in some areas, including DINO. Recent increases in wintertime ozone in and near DINO appear to be associated with emissions related to oil and gas development in the Uintah Basin of northeastern Utah and might pose a potential health risk to visitors and park staff. Summertime ozone levels are generally not considered high enough to harm plants, and ozone uptake is often limited by dry soil conditions. Plants in riparian areas may be at greater risk because their uptake of ozone is less likely to be limited by dry soil conditions.

Ozone precursors are emitted within and upwind from the Northern Colorado Plateau Network region from a variety of oxidized N source types, including power plants, industry, motor vehicles, oil and gas development, and others. Forest fires contribute to these emissions. They emit considerable quantities of oxidized N and hydrocarbons and therefore can contribute to ozone formation. It appears that the increase in fires in the western United States has largely been responsible for the observed increase in summer ozone concentrations reported by Jaffe and Ray (2007). Increasing temperature in the future will likely further influence the effects of fire on ozone formation, mainly during summer and fall months. High wintertime ozone episodes are associated with elevated levels of emissions from other source types and occur when the ground is covered with snow and weather conditions promote the formation of a strong temperature inversion.

Visibility is a source of concern for all parks in the northern Colorado Plateau because of the magnificent vistas viewed from within and looking into and out of these parks. However, some of the best visibility in the conterminous United States typically occurs in the Colorado Plateau, and effects of haze are limited and mostly stable or improving.

Other airborne contaminants, including Hg and other heavy metals and also semivolatile organic compounds, can accumulate in food webs. Some fish at CARE have been found to have elevated levels of Hg in their tissues; whether the source of the Hg is atmospheric or runoff from mine contamination is undetermined. Information on semivolatile organic compounds in the Northern Colorado Plateau Network is generally not available.

References

Belnap, J. 2003. The worlds at your feet: Desert biological crusts. *Front. Ecol. Environ.* 1:181–189.

Binkley, D., C. Giardina, D. Morse, M. Scruggs, and K. Tonnessen. 1997. Status of air quality and related values in Class I National Parks and Monuments of the Colorado Plateau, Chapter 9. Grand Canyon National Park, National Park Service, Air Resources Division, Denver, CO.

Eagles-Smith, C.A., J.J. Willacker, Jr., and C.M. Flanagan Pritz. 2014. Mercury in fishes from 21 national parks in the western United States—Inter- and intra-park variation in concentrations and ecological risk. U.S. Geological Survey Open-File Report 2014-1051. U.S. Geological Survey in cooperation with National Park Service. Air Resources Division. Reston, VA.

Jaffe, D.A. and J. Ray. 2007. Increase in ozone at rural sites in the western U.S. *Atmos. Environ.* 41(26):5452–5463.

Kohut, R. 2007. Assessing the risk of foliar injury from ozone on vegetation in parks in the U.S. National Park Service's Vital Signs Network. *Environ. Pollut.* 149:348–357.

Landers, D.H., S.L. Simonich, D.A. Jaffe, L.H. Geiser, D.H. Campbell, A.R. Schwindt, C.B. Schreck, M.L. Kent, W.D. Hafner, H.E. Taylor, K.J. Hageman, S. Usenko, L.K. Ackerman, J.E. Schrlau, N.L. Rose, T.F. Blett, and M.M. Erway. 2008. The fate, transport, and ecological impacts of airborne contaminants in western National Parks (USA). EPA/600/R-07/138. U.S. Environmental Protection Agency, Office of Research and Development, NHEERL, Western Ecology Division, Corvallis, Oregon.

Landers, D.H., S.M. Simonich, D. Jaffe, L. Geiser, D.H. Campbell, A. Schwindt, C.B. Schreck et al. 2010. The Western Airborne Contaminant Project (WACAP): An interdisciplinary evaluation of the impacts of airborne contaminants in western U.S. National Parks. *Environ. Sci. Technol.* 44(3):855–859.

National Park Service (NPS). 2010. Air quality in national parks: 2009 annual performance and progress report. Natural Resource Report NPS/NRPC/ARD/NRR-2010/266. National Park Service, Air Resources Division, Denver, CO.

Pardo, L.H., M.J. Robin-Abbott, and C.T. Driscoll. 2011. Assessment of nitrogen deposition effects and empirical critical loads of nitrogen for ecoregions of the United States. General Technical Report NRS-80. U.S. Forest Service, Newtown Square, PA.

Peterson, D.L., T.J. Sullivan, J.M. Eilers, S. Brace, D. Horner, K. Savig, and D. Morse. 1998. Assessment of air quality and air pollutant impacts in national parks of the Rocky Mountains and northern Great Plains. NPS D-657. U.S. Department of the Interior, National Park Service, Air Resources Division, Denver, CO.

Schwede, D.B. and G.G. Lear. 2014. A novel hybrid approach for estimating total deposition in the United States. *Atmos. Environ.* 92:207–220.

Schwinning, S., B.I. Starr, N.J. Wojcik, M.E. Miller, J.E. Ehleringer, and R.S. Sanford. 2005. Effects of nitrogen deposition on an arid grassland in the Colorado Plateau cold desert. *Rangeland Ecol. Manage.* 58:565–574.

Sullivan, T.J. and T.C. McDonnell. 2014. Mapping of nutrient-nitrogen critical loads for selected national parks in the intermountain west and Great Lakes regions. Natural Resource Technical Report NPS/ARD/NRTR—2014/895. National Park Service, Fort Collins, CO.

11

The Grand Canyon and the Southern Colorado Plateau Network

11.1 Background

There are five national parks in the Southern Colorado Plateau Network that are larger than 100 mi^2: Canyon de Chelly National Monument (CACH), El Malpais National Monument (ELMA), Glen Canyon National Recreation Area (GLCA), Grand Canyon National Park (GRCA), and Petrified Forest National Park (PEFO). In addition, there are 14 smaller parks, including Mesa Verde National Park (MEVE) and Bandelier National Monument (BAND). Larger parks generally have more available data with which to evaluate air pollution sensitivities and effects. In addition, the larger parks generally contain more extensive resources in need of protection against the adverse impacts of air pollution. This chapter focuses mainly on GRCA, MEVE, and BAND, which have the most complete data regarding air pollution effects within the network. Since air pollutants can be widespread, reduction of pollution emissions affecting large parks will often result in the protection of smaller parks located nearby as well.

There are few human population centers within the network; only Albuquerque is larger than 100,000 people. Figure 11.1 shows the network boundary and the location of each national park along with human population centers that have more than 10,000 people.

11.2 Atmospheric Emissions and Deposition

County-level emissions near the Southern Colorado Plateau Network, based on data from the EPA's National Emissions Inventory during a recent time period (2011), are depicted in Figures 11.2 through 11.4 for sulfur dioxide, oxidized N, and ammonia, respectively. Most counties near Southern Colorado Plateau Network parks had relatively low sulfur dioxide emissions (<5 tons/mi^2/yr; Figure 11.2). Spatial patterns in oxidized N emissions showed many counties in the range of 1–5 tons/mi^2/yr or higher (Figure 11.3). Emissions of ammonia near Southern Colorado Plateau Network parks were somewhat lower, with most counties showing emissions levels below 2 tons/mi^2/yr (Figure 11.4).

Deposition estimates from the Total Deposition project were used to evaluate changes in the deposition of S and N at park locations over a recent 10-year period. From 2001 to 2011, total S deposition in the Southern Colorado Plateau Network decreased slightly (≤0.33 kg S/ha/yr). Total oxidized N deposition decreased by a larger amount (≤0.61 kg N/ha/yr),

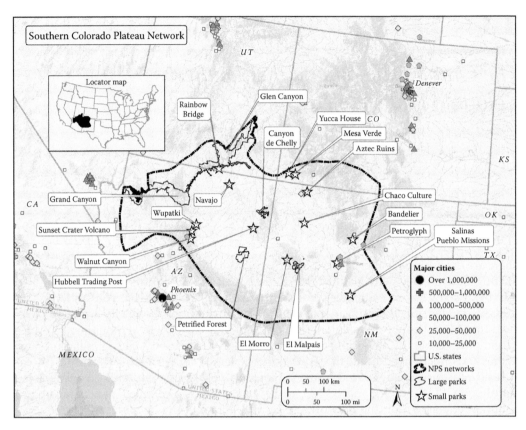

FIGURE 11.1
Network boundary and locations of national parks and human population centers within the Southern Colorado Plateau Network region.

but that was balanced by an increase in total ammonium deposition of a larger magnitude. Table 11.1 gives three-year average changes in the atmospheric deposition of S and N components between 2001 and 2011 in each Southern Colorado Plateau Network park considered here.

Total S deposition in and around the Southern Colorado Plateau Network for the period 2010–2012 was generally less than 2 kg S/ha/yr at park locations within the network area. Oxidized inorganic N deposition for the period 2010–2012 was less than about 3 kg N/ha/yr throughout the parklands within Southern Colorado Plateau Network. Most areas received less than 2 kg N/ha/yr of ammonium from atmospheric deposition during this same period; a few areas received higher amounts. Total N deposition was less than 5 kg N/ha/yr at most park locations (Figure 11.5); some areas received higher amounts.

Rainfall in the arid southwest is infrequent and most N deposition occurs as dry inputs (Fenn et al. 2003). Dry deposition is difficult to quantify, in large part because the deposition flux is influenced by a multitude of factors, including the mix of air pollutants present, surface characteristics of soil and vegetation, and meteorological conditions (Weathers et al. 2006). Because the vegetation is sparse in arid ecosystems, soil surfaces are potentially exposed to substantial direct dry deposition of atmospheric pollutants. Padgett et al. (1999) found that inorganic N can accumulate on the soil surface of coastal sage scrub vegetation

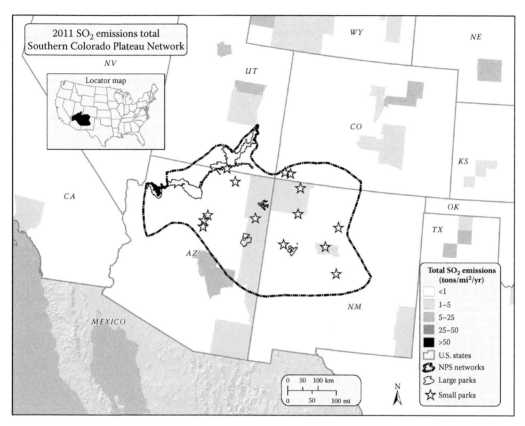

FIGURE 11.2

Total sulfur dioxide (SO_2) emissions, by county, near the Southern Colorado Plateau Network for the year 2011. (Data from EPA's National Emissions Inventory, https://www.epa.gov/air-emissions-inventories, accessed January, 2014.)

communities in southern California during dry smoggy periods. Upon onset of the rainy season, the concentration of nitrate in the surface soil decreased and nitrate was leached down the soil profile (Padgett et al. 1999, Padgett and Bytnerowicz 2001).

From 2002 to 2007, WACAP studied airborne contaminants in a number of national parks of the western United States from Alaska to southern California, including BAND in the Southern Colorado Plateau Network. Semivolatile organic compounds in the air at BAND were mainly the current-use pesticides endosulfan, dacthal, and trifluralin and the historic-use pesticides hexachlorobenzene, hexachlorocyclohexane-α, and chlordanes.

11.3 Acidification

11.3.1 Effects on Natural Ecosystems

Given the low levels of N, and especially S, deposition in this network and the low precipitation, it is very unlikely that any appreciable soil acidification has occurred to date. Large rivers, such as the Colorado River that flows through the Grand Canyon, typically

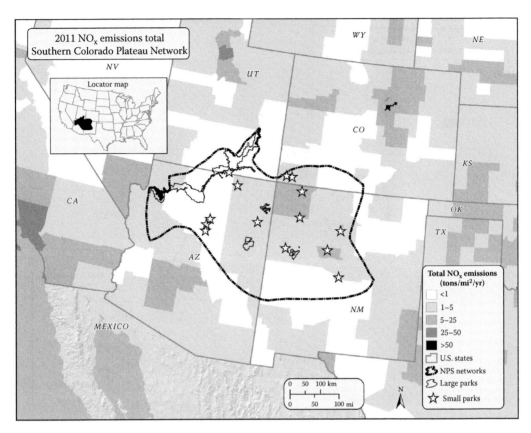

FIGURE 11.3
Total oxidized N (NO_x) emissions, by county, near the Southern Colorado Plateau Network for the year 2011. (Data from EPA's National Emissions Inventory, https://www.epa.gov/air-emissions-inventories, accessed January, 2014.)

contain high solute concentrations and are not acid sensitive. The arid conditions that prevail throughout much of the Southern Colorado Plateau Network also do not contribute to acid sensitivity, which tends to be enhanced by high precipitation amounts. At the higher elevations within parklands in the Southern Colorado Plateau Network that receive greater amounts of precipitation, it is possible that there are acid-sensitive surface waters. Data are not available that would suggest that is the case, however (Figure 3.2).

There are a number of amphibian species living in GRCA and other parks within the Southern Colorado Plateau Network, but they do not appear to be threatened by acidification caused by air pollution. Many of these species, including the Great Basin spadefoot toad (*Scaphiopus intermontanus*), the red-spotted frog (*Bufo punctatus*), and the canyon treefrog (*Hyla arenicolor*), breed in small pools in sandstone. The tiger salamander (*Ambystoma tigrinum*) and the Chiricahua leopard frog (*Rana chiricahuensis*), both found in GRCA, are on the State of Arizona threatened list. However, Binkley et al. (1997) reported that the main threat to these species is habitat loss and that there is no evidence indicating risk from acidic deposition.

Binkley et al. (1997) additionally concluded that it is unlikely that fish in the Colorado River in GRCA are affected by air pollution; they again cite habitat elimination as the greatest risk to native fish species in the river. The Colorado River was transformed from

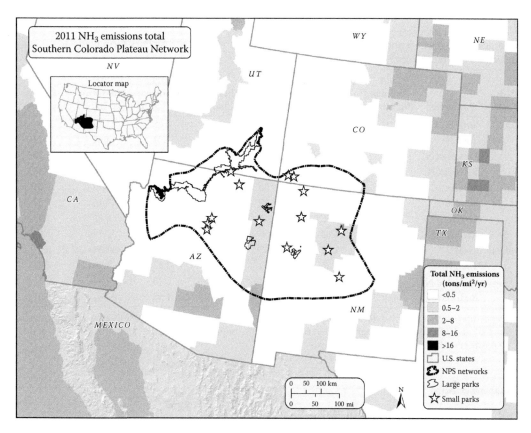

FIGURE 11.4

Total ammonia (NH$_3$) emissions, by county, near the Southern Colorado Plateau Network for the year 2011. (Data from EPA's National Emissions Inventory, https://www.epa.gov/air-emissions-inventories, accessed January, 2014.)

a warm, muddy river to a cold, relatively clear river as a result of construction of the Glen Canyon Dam. Operation of the dam resulted in the loss of habitat for native species, such as the Colorado squawfish (*Ptychocheilus lucius*), bonytail chub (*Gila elegans*), humpback chub (*Gila cypha*), razorback sucker (*Xyrauchen texanus*), roundtail chub (*Gila robusta*), and flannelmouth sucker (*Catostomus insignis*; Binkley et al. 1997).

11.3.2 Effects on Historic Structures

In the 1980s, the NPS became increasingly concerned that archeological resources, specifically sandstone masonry structures, in the Four Corners region were at risk from anthropogenic stressors, including air pollution. To evaluate potential air pollution impacts, sandstone masonry deterioration was monitored at MEVE over three years. Investigators found that, during this time, the environment did not seem to be sufficiently wet or acidic to accelerate sandstone deterioration beyond what would be expected from natural erosion processes. However, they noted that in more exposed areas (e.g., mesa top sites), chemical erosion due to acidic deposition could be more pronounced. Increases in acidic deposition inputs in the future might further increase erosion (Dolske et al. 1991).

TABLE 11.1

Average Changes in S and N Deposition between 2001 and 2011 across Park Grid Cells at Selected Southern Colorado Plateau Network Parks

Park Code	Parameter	2001 Average (kg/ha/yr)	2011 Average (kg/ha/yr)	Absolute Change (kg/ha/yr)	Percent Change	2011 Minimum (kg/ha/yr)	2011 Maximum (kg/ha/yr)	2011 Range (kg/ha/yr)
BAND	Total S	1.11	0.95	−0.15	−14.5	0.68	1.48	0.80
	Total N	3.33	3.28	−0.05	−2.2	2.51	4.65	2.13
	Oxidized N	2.44	2.17	−0.27	−11.6	1.67	2.94	1.28
	Reduced N	0.89	1.11	0.22	23.6	0.84	1.70	0.86
GRCA	Total S	0.84	0.75	−0.08	−8.9	0.39	1.70	1.31
	Total N	2.91	2.65	−0.25	−8.9	1.54	5.16	3.62
	Oxidized N	2.24	1.63	−0.61	−27.5	1.05	3.02	1.97
	Reduced N	0.67	1.03	0.36	55.8	0.49	2.14	1.64
MEVE	Total S	1.76	1.42	−0.33	−17.6	1.25	1.54	0.29
	Total N	3.75	4.28	0.53	15.7	3.87	4.90	1.03
	Oxidized N	2.72	2.59	−0.13	−4.0	2.35	2.88	0.53
	Reduced N	1.02	1.69	0.66	69.9	1.50	2.01	0.51

[a] Deposition estimates were determined by the Total Deposition project based on three-year averages centered on 2001 and 2011 for all ~4 km grid cells in each park. The minimum, maximum, and range of 2011 S and N deposition within each park are also shown.

11.4 Nutrient Nitrogen Enrichment

Response of desert vegetation communities to N addition has been shown to include change in plant species composition, increase in biomass of nonnative invasive species, and decrease in relative abundance and richness of native herbaceous species (Allen and Geiser 2011). The abundance of invasive nonnative plant species has increased in desert ecosystems, likely due in part to N deposition at some locations (Brooks 2003, Allen et al. 2009). Increases in N have been found to promote the invasions of fast-growing exotic annual grasses (e.g., cheatgrass [*Bromus tectorum*]) and forbs (e.g., Russian thistle [*Salsola kali*]) at the expense of native species (Brooks 2003, Schwinning et al. 2005, Allen et al. 2009). Nitrogen addition favors buffelgrass (*Pennisetum ciliare*) over native species in some arid ecosystems, as buffelgrass is able to rapidly absorb and assimilate N (Lyons et al. 2013). Increased cover of exotic grasses can increase fire risk and frequency (Rao et al. 2010, Balch et al. 2013) and affect plant biodiversity. Inputs of N may also decrease water-use efficiency in big sagebrush (Inouye 2006).

Limited experimental data are available and this, in turn, limits our understanding of what levels of N deposition cause changes. Biomass of native blue grama grass (*Bouteloua gracilis*) in desert grassland at the Sevilleta Long-Term Ecological Research site in New Mexico increased in response to N addition at a rate of 20 kg N/ha/yr (Báez et al. 2007). In contrast, a change in ambient N deposition at the control plot during the 16-year experiment from 1.7 to 2.4 kg N/ha/yr did not cause a change in blue grama biomass.

Experimental N addition reduced the soil moisture in sagebrush steppe. This would be expected to decrease herbaceous plant productivity (Inouye 2006). Experimental fertilization with N at very high levels (72 kg N/ha/yr) at the Jornada Long-Term Ecological

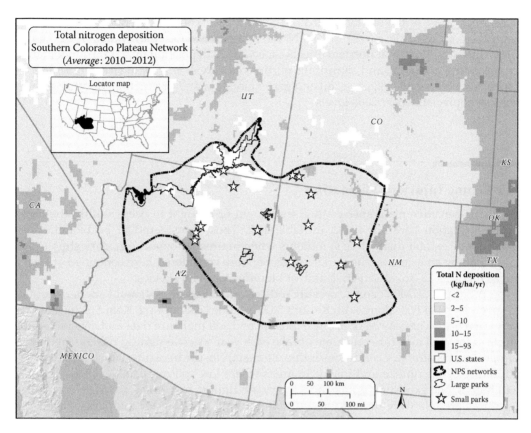

FIGURE 11.5
Total N deposition for the three-year period centered on 2011 in and around the Southern Colorado Plateau Network. (From Schwede, D.B. and Lear, G.G., *Atmos. Environ.*, 92, 207, 2014.)

Research site in New Mexico caused increased growth of native annual forbs during winter, but decreases during summer (Gutierrez and Whitford 1987). This variable response may have been due, at least in part, to water limitation during summer. Barker et al. (2006) determined that water availability, rather than N supply, was the most important limiting factor for the productivity of creosote bush (*Larrea tridentata*).

Ellis et al. (2013) estimated the critical load for nutrient N deposition to protect the most sensitive ecosystem receptors in 45 national parks, based on the data of Pardo et al. (2011). They estimated the N critical load for GRCA and MEVE in the range of 3–8.4 kg N/ha/yr for the protection of herbaceous plants and 3 kg N/ha/yr for the protection of lichens.

Sullivan and McDonnell (2014) analyzed nutrient N deposition and critical loads at selected national parks in the Intermountain West, including MEVE and GRCA in the Southern Colorado Plateau Network. Most of MEVE received ambient N deposition in excess of the lower estimate of the nutrient N critical load reported by Allen and Geiser (2011). None of GRCA was reported to be in exceedance, although much of the park was within 1 kg N/ha/yr of exceedance (Sullivan and McDonnell 2014).

Most of MEVE received ambient N deposition estimated using the Comprehensive Air Quality Model with Extensions that was in exceedance of the lower estimate of the nutrient N empirical critical load reported by Allen and Geiser (2011); the remainder was largely below, but within 1 kg N/ha/yr of, exceedance. Although none of GRCA was in exceedance, most of the western third of the park and scattered portions of the remainder of the park were close (within 1 kg N/ha/yr) to exceedance (Sullivan and McDonnell 2014). Thus, it is likely that nutrient effects caused by N deposition may occur under ambient N loading at MEVE, and significant portions of GRCA may be at risk if N emissions from electricity generating units, oil and gas exploration and production, or other sources increase appreciably over 2008 levels.

11.5 Ozone Injury to Vegetation

The ozone-sensitive plant species that are known or thought to occur within the parks under evaluation here are listed in Table 11.2. Those considered to be bioindicators, because they exhibit distinctive symptoms when injured by ozone (e.g., dark stipple), are designated by an asterisk. The parks in this network considered here contained at least 4 and as many as 12 ozone-sensitive and/or bioindicator plant species.

The W126 and SUM06 ozone exposure indices calculated by National Park Service staff are given in Table 11.3, along with Kohut's (2007) ozone risk ranking. Kohut examined five individual years (1995–1999) of ozone exposure and soil moisture data and considered the effects of low soil moisture on ozone uptake each year when assigning risk. Soil moisture is important because dry conditions induce stomatal closure in plants, which has the effect of limiting ozone uptake and injury. In areas where low soil moisture levels correspond with high ozone exposure, uptake and injury are limited by stomatal closure even when exposures are relatively high. Kohut's (2007) ranking was low for the parks considered here.

TABLE 11.2

Ozone-Sensitive and Bioindicator (Asterisk) Plant Species Known or Thought to Occur in Selected Parks of the Southern Colorado Plateau Network[a]

		Park		
Species	Common Name	BAND	GRCA	MEVE
Ailanthus altissima*	Tree-of-heaven	×	×	
Amelanchier alnifolia	Saskatoon serviceberry			×
Apocynum androsaemifolium*	Spreading dogbane	×	×	×
Apocynum cannabinum	Dogbane, Indian hemp	×	×	×
Artemisia ludoviciana*	Silver wormwood		×	×
Oenothera elata*	Evening primrose		×	×
Pinus ponderosa*	Ponderosa pine		×	×
Populus tremuloides*	Quaking aspen		×	×
Prunus virginiana	Choke cherry		×	×
Rhus trilobata*	Skunkbush		×	×
Robinia pseudoacacia	Black locust	×		
Salix gooddingii	Goodding's willow		×	
Salix scouleriana*	Scouler's willow		×	
Solidago canadensis var. *scabra*	Goldenrod		×	

[a] Lists are available and periodically updated at https://irma.nps.gov/NPSpecies/Report.

TABLE 11.3

Ozone Assessment Results for Selected Parks in the Southern Colorado Plateau Network Based on Estimated Average 3-Month W126 and SUM06 Ozone Exposure Indices for the Period 2005–2009 and Kohut's (2007) Ozone Risk Ranking for the Period 1995–1999[a,b]

Park Name	Park Code	W126 Value (ppm-hr)	W126 NPS Condition	SUM06 Value (ppm-hr)	SUM06 NPS Condition	Kohut Ozone Risk Ranking
Bandelier	BAND	11.94	Moderate	15.16	High	Low
Grand Canyon	GRCA	17.78	High	26.22	High	Low
Mesa Verde	MEVE	13.96	High	19.62	High	Low

[a] Parks are classified into one of three ranks (Low, Moderate, High) based on comparison with other Inventory and Monitoring parks.

[b] Degrees of concern for the W126 and SUM06 indices are based solely on levels of ozone exposure. Kohut's risk to vegetation is based on several factors that contribute to injury in plants, including ozone exposure and environmental variables, and considers the effects of soil moisture on the uptake of ozone.

Between 1983 and 1994, average hourly ozone concentrations in GRCA were measured at around 25–50 ppb, with a peak 1-hour concentration of up to 80 ppb. Although there were no reports of ozone damage to park plants, these values may be high enough to cause visible foliar or growth effects on sensitive plant species (Binkley et al. 1997).

The National Park Service (2010) evaluated the fourth highest 8-hour ozone concentrations for national parks that have consistently been at or above the 2008 standard of 75 ppb, based on 11–20 years of monitoring data running through 2008. There were six parks that routinely had such high concentrations. Six additional parks had ozone levels that were typically below the 75 ppb standard but were within the EPA's proposed range for a possible new standard (60–70 ppb; NPS 2010). These parks included GRCA. The standard was revised to 70 ppb in 2015. The National Park Service (2010) reported long-term trends in the annual fourth highest 8-hour daily maximum ozone concentration for 31 monitoring sites in 27 national parks having more than 10 years of data through 2008. Statistically significant increases were reported for four parks, including MEVE in the Southern Colorado Plateau Network.

11.6 Visibility Impairment

The southern Colorado Plateau landscape includes massive landforms, unique geology, and vivid rock colors (Grand Canyon Visibility Transport Commission [GCVTC] 1996). The panoramic views evident at many locations in parklands within this network provide exceptional visual experiences for visitors, and high air quality is central to full enjoyment of those experiences (Figure 11.6). Some of the best visibility in the conterminous United States occurs in the Colorado Plateau region (Savig and Morse 1998, Hand et al. 2011). As a consequence, reduced air quality that might go relatively unnoticed elsewhere is acutely evident on the Plateau (GCVTC 1996).

Air pollution causes haze and visibility impairment in all of the Southern Colorado Plateau Network parks at times. Regional and local pollution sources that appear to

FIGURE 11.6
Visibility is an integral part of the visitor experience in GRCA and many other national parks. (National Park Service photo by Michael Quinn.)

contribute to visibility impairment include automobiles, coal- and oil-fired power plants, smelters, oil and gas production, wildfires, and urban emissions.

11.6.1 Natural Background and Ambient Visibility Conditions

Interagency Monitoring of Protected Visual Environments (IMPROVE) monitors are located at BAND, GRCA, and MEVE. Table 11.4 gives the estimated natural (i.e., no human-caused impairment) conditions and degree of impairment during the baseline period 2000–2004 on the average of the 20% clearest and 20% haziest days. Measured ambient haze values for the Regional Haze Rule baseline period 2000–2004 were substantially higher than the estimated natural condition on both the clearest and haziest days.

TABLE 11.4

Estimated Natural Haze and Measured Current Baseline Haze in Selected Parks in the Southern Colorado Plateau Network Averaged over the Regional Haze Rule Baseline Period 2000–2004

			Measured Baseline and Estimated Natural Haze			
			20% Clearest Days (Deciviews)		20% Haziest Days (Deciviews)	
Park Name	Park Code	Site ID	Baseline	Natural	Baseline	Natural
Bandelier	BAND	BAND1	4.95	1.29	12.22	6.26
Grand Canyon	GRCA	GRCA2	2.16	0.31	11.66	7.04
Mesa Verde	MEVE	MEVE1	4.32	1.01	13.03	6.81

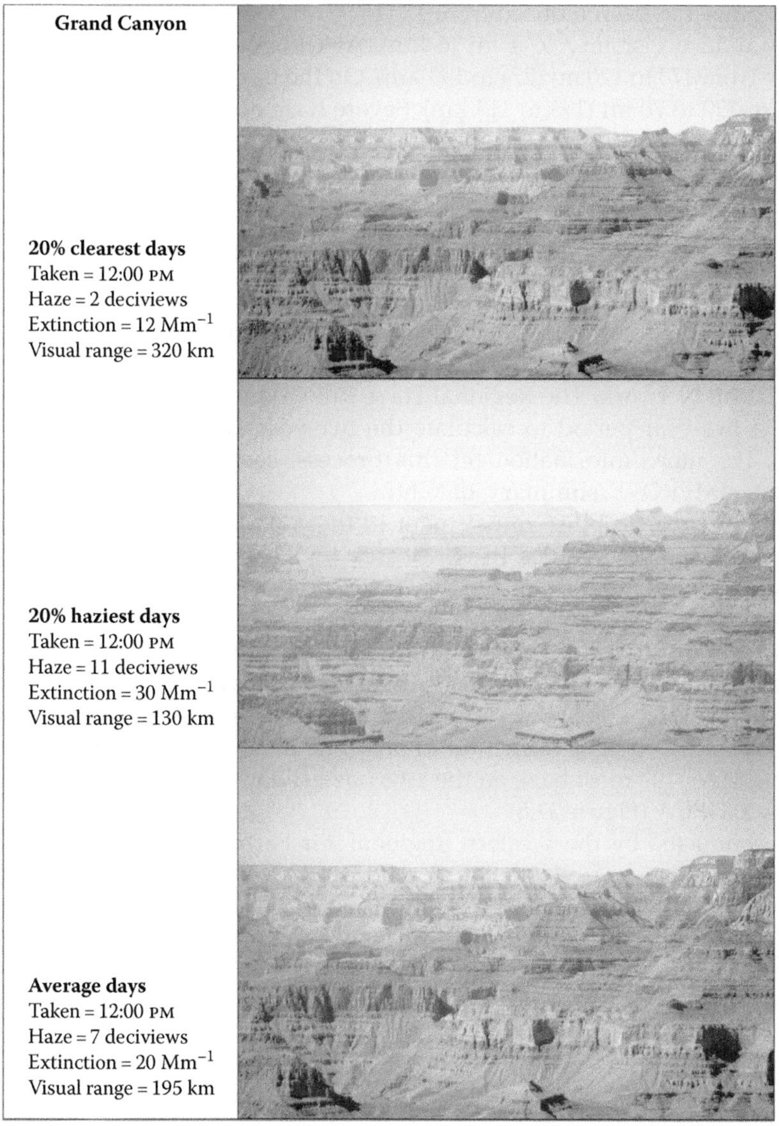

FIGURE 11.7
Three representative photos of the same view in GRCA illustrating the 20% clearest visibility, the 20% haziest visibility, and the annual average visibility. Extinction is total particulate light extinction. (From http://vista. cira.colostate.edu/improve/Data/IMPROVE/Data_IMPRPhot.htm, accessed December, 2010.)

In addition to the IMPROVE samplers, cameras have been used to illustrate visibility conditions. Representative photos of selected vistas in GRCA under three different visibility conditions are shown in Figure 11.7. Photos were selected to correspond with the clearest 20% of visibility conditions, the haziest 20% of visibility conditions, and annual average visibility conditions. This series of photos provides a graphic illustration of the visual effect of these differences in haze level on a representative vista in this park.

IMPROVE data allow estimation of visual range. Data indicate that in BAND, air pollution has reduced average visual range from 170 to 95 mi (274 to 153 km). On the haziest days, visual range has been reduced from 120 to 60 mi (193 to 97 km). Severe haze episodes occasionally reduce visibility to 4 mi (6 km). At GRCA, pollution has reduced average visual range from 170 to 120 mi (274 to 193 km). On the haziest days, visual range has been reduced from 120 to 70 mi (193 to 113 km). Severe haze episodes occasionally reduce visibility to 14 mi (23 km). At MEVE, air pollution has reduced average visual range from 170 to 100 mi (274 to 161 km). On the haziest days, visual range has been reduced from 120 to 60 mi (193 to 97 km). Severe haze episodes occasionally reduce visibility to 10 mi (16 km).

11.6.2 Composition of Haze

Figure 11.8 shows the composition and levels of estimated natural (preindustrial), baseline (2000–2004), and more recent (2006–2010) haze for the monitored parks in the Southern Colorado Plateau Network. The Regional Haze Rule requires at least three years of valid data out of a five-year period to calculate the five-year average for the baseline period (2000–2004). For more information on this process, see http://vista.cira.colostate.edu/improve/Data/IMPROVE/summary_data.htm.

The majority of the visibility impairment in these parks was typically attributable to sulfate, coarse mass, and organics. In some cases, light absorbing carbon, soils, and nitrate were also quantitatively important. On the 20% clearest days in all of the monitored parks, sulfate was the largest contributor to visibility impairment, followed by organics. Sulfate particles can remain suspended in the atmosphere for many days to weeks and be transported long distances (Hall 1981). For the annual average days, sulfates, followed by organics, were responsible for the largest contributions to haze. On the 20% haziest days, sulfates contributed the most to light extinction in only one park, BAND. In MEVE, coarse mass was responsible for the most haze on the 20% haziest days, and organics were the largest contributors in GRCA (Figure 11.8).

Analyses conducted by the Western Regional Air Partnership indicated that organics from natural emissions sources, including wildfire and biogenic sources (vegetation), contribute to substantial visibility impairment throughout the western United States, including within the Southern Colorado Plateau Network. In addition, air pollution sources outside the Western Regional Air Partnership domain, including international offshore shipping and sources from Mexico, Canada, and Asia, can in some cases be substantial contributors to haze (Suarez-Murias et al. 2009).

A study of air composition was conducted by Eatough et al. (2006) during summer at Meadview, AZ, just west of GRCA. They determined that reduced visibility was most strongly influenced by suspended fine particulate organic material, which contributed 19% of light extinction, on average, followed by fine particulate sulfate, which contributed an average of 14%. When light extinction was highest, these species, along with light-absorbing carbon, were the dominant contributors to light extinction. During periods of highest light extinction, sulfate and associated water were responsible for 33% of light extinction (Eatough et al. 2006). These were also the periods of highest humidity. Coarse particles (e.g., dust) were also important contributors to visibility impairment and were judged to be responsible for 18% of the light extinction on average and 19% during periods of the highest light extinction (Eatough et al. 2006). In a study of coarse particle species composition at nine sites chosen to be representative of the continental United States, Malm et al. (2007) found that crustal materials accounted for 76% of coarse material at GRCA, which was the highest percentage among all study sites. Contribution to coarse material from organic mass was 8.6% at GRCA, which was the lowest among the study sites (along with San Gorgonio wilderness area in southern California; Malm et al. 2007).

11.6.3 Sources of Visibility Impairment in the Southern Colorado Plateau Network

The Western Regional Air Partnership has provided technical assistance for western states in developing plans to meet the requirements of the Regional Haze Rule. As part of this effort, the partnership conducted regional modeling to identify pollution sources that were affecting visibility in Class I areas and where these sources were located. Source regions included the Western Regional Air Partnership region (AZ, CA, CO, ID, MT, ND, NM, NV, OR, SD, UT, WA, and WY), Pacific Offshore (PO), CENRAP (AL, AR, IA, KS, MN, MO, NE, OK, and TX), Eastern U.S., Canada, Mexico, and outside domain (outside the modeling

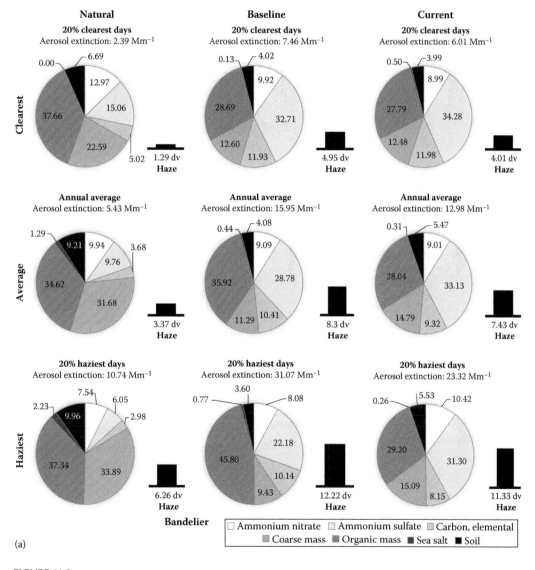

FIGURE 11.8
Estimated natural (preindustrial), baseline (2000–2004), and current (2006–2010) levels of haze (columns) and its composition (pie charts) on the 20% clearest, annual average, and the 20% haziest visibility days for (a) BAND, (b) GRCA, and (c) MEVE. (From http://views.cira.colostate.edu/fed/Tools/RegionalHazeSummary. aspx, accessed October, 2012.) *(Continued)*

domain). Source types identified include outside domain (undefined), point (e.g., power plants, smelters), area (e.g., oil and gas development), mobile (e.g., vehicles), prescribed burns, and natural fires and biogenic sources.

The West-Wide Jump-Start Air Quality Modeling Study (WestJump) updated the previous Western Regional Air Partnership work with emissions inventories for the year 2008 (ENVIRON et al. 2013). The WestJump results for nitrate and sulfate are described and compared to the WRAP data in the following text. WestJump data have some of the

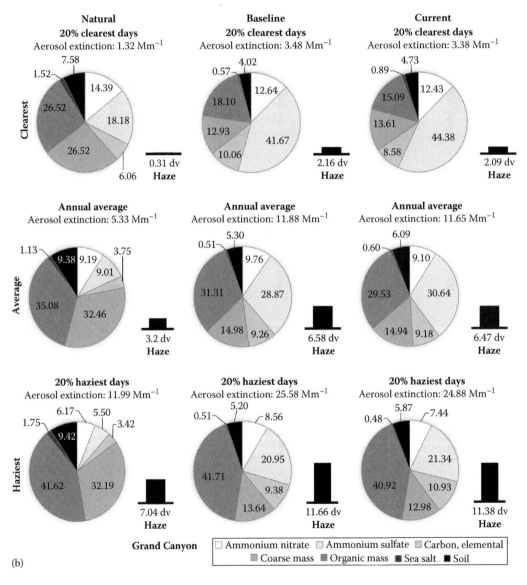

(b)

FIGURE 11.8 (*Continued*)
Estimated natural (preindustrial), baseline (2000–2004), and current (2006–2010) levels of haze (columns) and its composition (pie charts) on the 20% clearest, annual average, and the 20% haziest visibility days for (a) BAND, (b) GRCA, and (c) MEVE. (From http://views.cira.colostate.edu/fed/Tools/RegionalHazeSummary.aspx, accessed October, 2012.)   (*Continued*)

same limitations as the Western Regional Air Partnership data in that they do not include recent emissions reductions and increases occurring since 2008. The WestJump data agree with the Western Regional Air Partnership data in that they also show that organic and elemental carbon from prescribed burns and other anthropogenic sources are at times significant, and in some cases major, contributors to haze and visibility impairment in the Southern Colorado Plateau Network parks. Ongoing work continues to improve emissions inventories, particularly for fires (both wildfires and prescribed burns) through the Western Regional Air Partnership and the Joint Fire Science Program (2014). Carbon from

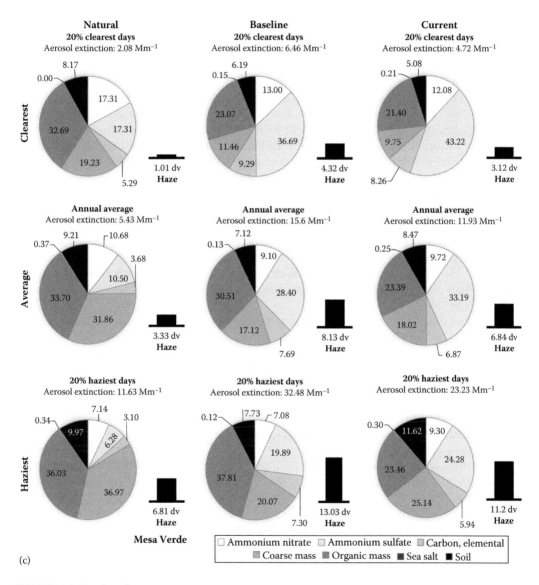

(c)

FIGURE 11.8 (Continued)
Estimated natural (preindustrial), baseline (2000–2004), and current (2006–2010) levels of haze (columns) and its composition (pie charts) on the 20% clearest, annual average, and the 20% haziest visibility days for (a) BAND, (b) GRCA, and (c) MEVE. (From http://views.cira.colostate.edu/fed/Tools/RegionalHazeSummary. aspx, accessed October, 2012.)

prescribed burns is, at present, considered controllable by the EPA. This can be problematic for States and Federal Land Managers because prescribed burns, while causing visibility impairment, are needed to improve and maintain forest health and to avoid more catastrophic wildfires in the future.

Data from the WestJump emissions inventory found that most of the pollutants that reduced visibility in BAND in 2008 came from sources within the WRAP region, mainly New Mexico. Nitrate was a somewhat larger contributor to haze than sulfate, but organic and elemental carbon from prescribed fires and other sources, primarily in New Mexico, were the major controllable pollutants reducing visibility in BAND (ENVIRON et al. 2013).

11.6.4 Visibility Research at GRCA

Because visibility is such an important air quality–related value in GRCA, this park became the focus of considerable visibility research several decades ago. It was recognized that the most important canyon levels for visibility are located at and slightly below the canyon rim. Visibility at these locations is most strongly affected by atmospheric conditions between the viewer's vantage point and the geological formations on the opposite side of the canyon (Banta et al. 1999). As a consequence, effectiveness of pollutant vertical transport down into and within the canyon is a key component of visibility degradation central to the visitor experience in this park. These issues were studied by Banta et al. (1999) using Doppler lidar. They focused on three wind flow patterns that occurred during winter: flow reversal, coupled flow inside and above the canyon, and thermally forced up-and-down-canyon winds. Other studies have focused on how the vertical transport of haze-producing aerosols is affected by the flow structure within the canyon as well as the dynamics of advections of clean air into the canyon that can flush out pollutants (e.g., Steinacker 1984, McKee and O'Neal 1989, Whiteman et al. 1999a,b).

Mazurek et al. (1997) used gas chromatography to examine the composition of aerosol organics collected in GRCA in August 1989 and compared them with the components of diluted, aged air samples from the Los Angeles area. The researchers found that certain characteristics of the samples were very different based on sample origin. They therefore concluded that significant organic aerosols in GRCA likely originated from sources other than Los Angeles. Nevertheless, atmospheric tracer studies have demonstrated some summer pollutant transport from the Los Angeles basin to GRCA (Hering et al. 1981, Macias et al. 1981, Pitchford et al. 1981). Summer long-range pollutant transport from the Los Angeles basin initiates with prevailing westerlies that are enhanced during summer by the formation of thermal lows over the interior California deserts. Pollutants are ejected vertically by passage of the air mass over the San Gabriel Mountains and are then transported to the southwest toward the Southern Colorado Plateau Network with winds from the Pacific High (Chen et al. 1999). Southwesterly flows have also been shown to advect pollution from southern Arizona smelters into the GRCA region (Chen et al. 1999).

Xu et al. (2006) identified major source regions of the atmospheric sulfate aerosols measured at Class I areas throughout the western United States. Airmass back-trajectories were calculated using the NOAA HYSPLIT v4.6 model for each of 84 Class I areas every 3 hours over the period 2000–2002. Results for a site on the south rim of the Grand Canyon suggested that air flows that reached GRCA most commonly followed a southeasterly path along the Pacific coast before turning inland and approaching GRCA from the southwest. A secondary flow regime included air flows to GRCA from the southeast.

The identified top five source regions that contributed sulfate to GRCA were the Pacific Ocean, Pacific Coast, southwestern Arizona, Mexico, and California (Xu et al. 2006). The sulfate originating from southwestern Arizona and Mexico was likely from coal-burning power plants. The sulfate originating from the Pacific Ocean and Pacific Coast source regions was from multiple sources, including ocean shipping and aerosol transport from Asia.

A substantial portion of winter anthropogenic visibility impairment at GRCA has been attributed to emissions from the coal-burning units at the Navajo Generating Station power plant (NRC 1990), located at the eastern end of the Grand Canyon, about 6 mi (10 km) east of Page, Arizona. The Navajo Generating Station impacts on GRCA visibility were generally highest from December to February. This was likely a result of sulfur dioxide interactions with humidity, which is most prevalent at GRCA during the winter (Green et al. 2005). This finding was further supported by modeling results based on the 1990 winter Navajo Generating Station visibility field study (Gray et al. 1991). Chen et al. (1999) used data from that study as input to several models to analyze the transport of the Navajo Generating Station plume during a four-day period. Results revealed that complex terrain features dramatically affected wind flow patterns during winter in the GRCA area.

Chen et al. (1999) determined the path taken by the Navajo Generating Station emissions plume to GRCA and evaluated the resulting effects on winter visibility in the park. The analysis of data collected during the 1990 Navajo Generating Station Visibility Field Study showed that terrain complexity influences wind flow patterns during winter and creates such local effects as channeling, decoupled canyon winds, and slope and valley flows. Both direct and indirect plume transport occurred from the Navajo Generating Station to GRCA, resulting from northeasterly winds or from wind direction shifts caused by passing synoptic weather systems, respectively (Chen et al. 1999).

A field program associated with the project SCENES air quality and visibility assessment around GRCA evaluated the contribution of emissions from the coal-fired power plant at the Navajo Generating Station to winter visibility impairment in GRCA, Canyonlands National Park in the Northern Colorado Plateau Network, and GLCA (Malm et al. 1988, 1989). During the 1990 Navajo Generating Station winter visibility study, surface and upper air meteorological measurement systems were operated in and around GRCA from January 10 to March 31, 1990. The network included surface monitoring stations, balloon sounding stations, Doppler radar wind profilers, and Doppler sodars (Lindsey et al. 1999). Results suggested the atmospheric transport of Navajo Generating Station emissions to GRCA, followed by buildup over multiple days, and then advection out of the area by frontal passages.

Lindsey et al. (1999) summarized the knowledge gained from the analyses of the 1990 Navajo Generating Station data, highlighting the mechanisms responsible for emissions transport to GRCA. They determined that the primary mode of transport was synoptic-scale winds driven by pressure gradients associated with passing storms, creating a ridge of high pressure following the passage of a low-pressure system. The topographical features of the region channel these winds and also influence mesoscale circulation, another mechanism for emissions transport that forms during light wind conditions. Short-distance transport of emissions can also occur due to slope and valley flows, which are directed by the regional topography but driven by the diurnal heating cycle. However, these flows are disrupted by passing storms and may not play a significant role in overall emissions transport from the Navajo Generating Station to GRCA (Lindsey et al. 1999).

In 1990, the Navajo Generating Station agreed to reduce sulfur dioxide emissions by 90%. Three scrubbers were installed in 1997–1999, and the IMPROVE Aerosol

Network monitored air quality in GRCA to evaluate the effects of emissions reductions. Previously, the Navajo Generating Station emitted about 200 tons/day of sulfur dioxide. Post-scrubber concentrations of particulate sulfate were significantly reduced (by ~33%) during the winter as compared with concentrations before scrubbers were installed. Reducing emissions had a beneficial effect by decreasing winter haze and improving visibility in GRCA. Visibility on the haziest days having the highest particulate sulfate improved markedly following the reduction in Navajo Generating Station emissions (Green et al. 2005). During summer, however, when Navajo Generating Station impacts on GRCA were minimal, there was only a small decrease in particulate sulfate (Green et al. 2005).

Installation of combustion controls to reduce oxidized N at the Navajo Generating Station began in 2009 and has resulted in a decline from 35,000 tons/yr to around 16,000 tons of N oxides per year. Remaining plant-wide emissions of sulfur dioxide and oxidized N are expected to drop by one-third in 2019 due to decisions by the Los Angeles Department of Water and Power and by Nevada Energy to eliminate coal-fired electricity from their generation portfolios. In 2014, the EPA finalized a rule to reduce air pollutants from the Navajo Generating Station. When fully implemented by 2030, the plan will reduce oxidized N emissions by about 80% and the visual impairment caused by the Navajo Generating Station by about 73%.

Project MOHAVE (Measurement of Haze and Visual Effects) was a large EPA air quality and visibility research program conducted between 1990 and 1999 to evaluate causes of visibility impairment in and surrounding GRCA. The MOHAVE project was particularly comprehensive and included tracer/receptor field experiments in 1992, source emissions simulations, and extensive statistical analyses (Pitchford et al. 1999). The Mohave Power Project was identified by the MOHAVE study as the single emissions source of haze-producing aerosols most likely to decrease visibility in GRCA (Pitchford et al. 1999). The Mohave Power Project was a large coal-fired power plant located 90 mi southeast of Las Vegas in Laughlin, Nevada. The link between the Mohave Power Project emissions and air quality in GRCA had been subject of debate and study for over two decades (cf., Pitchford et al. 1999).

The Project MOHAVE final report (Pitchford et al. 1999) concluded that field measurements did not directly link elevated sulfate concentrations in GRCA with Mohave Power Project emissions. Rather, concentrations of all visibility impairing species were affected by regional sources and regional meteorology. The dominant sources of GRCA visibility impairment were area sources in Southern California, Arizona, northern Mexico and, to some extent, the Las Vegas area.

Eatough et al. (2001) found that, although Mohave Power Project was the major source of sulfate at Meadview, AZ, only 4.3%–5.5% of total atmospheric sulfate in GRCA appeared to be derived from Mohave Power Project emissions. They identified Las Vegas and urban areas in the Baja California and Imperial Valley regions as likely primary sources of atmospheric sulfate in the park (Eatough et al. 2001). They identified northwestern Mexico as a significant source of haze-producing particulates during an extreme event of unusually high sulfate concentrations in the Grand Canyon in 1992 and as the largest contributor of sulfate aerosols to the sample site at Hopi Point in GRCA during Project MOHAVE. Subsequent analyses of Project MOHAVE extinction data and mass balance analyses of sulfur oxide sources by Eatough et al. (2006) also suggested that Los Angeles, CA, and Las Vegas, NV, were likely the largest anthropogenic contributors to light extinction in GRCA.

As a result of Project MOHAVE, the company and stakeholders entered into a consent decree in 1999 that required the Mohave Power Project to install pollution control equipment by 2006 in order to address concerns about the impacts of plant emissions on GRCA visibility and ongoing violations of the plant's opacity emission limits (67 FR 6130; February 8, 2002). The Mohave Power Project opted to not install the equipment and, in 2005, elected to close the plant. The permanent closure of the Mohave Power Project facility provided an opportunity to ascertain the level of impact on GRCA visibility that had been due to its operation. In advance of and following this permanent power plant closure, analyses were conducted using the IMPROVE Aerosol Network database to predict the resulting changes in atmospheric sulfur dioxide concentrations and visibility in GRCA. Project MOHAVE's analysis of these data and tracer studies concluded that although low levels of particulate sulfate originating in Mohave Power Project emissions were reaching GRCA, there were negligible correlations between tracer concentration and measurements of light extinction in the park, suggesting that other aerosol sources contributed substantially to haze (Pitchford et al. 1999). Terhorst and Berkman (2010) analyzed pre- and post-closure visibility in GRCA in order to evaluate whether visibility improved after the closure of the plants and to identify the level of visibility degradation that could be attributed solely to the Mohave Power Project. Their analyses found a 3%–10% drop in atmospheric fine sulfate concentrations in GRCA following the Mohave Power Project closure, but no statistically significant change in visibility as measured by the haze index. Based on this finding, they concluded that the closure of the plant did not improve visibility in the Grand Canyon. White et al. (2012) found that conclusion to be misleading, noting that the analysis of Terhorst and Berkman actually confirmed the results of Project MOHAVE, which predicted that fine sulfate concentrations in GRCA would decrease if Mohave Power Project sulfur emissions were controlled. White et al. (2012) also noted that Terhorst and Berkman based their conclusion of no improvement in visibility on haze estimates derived directly from the IMPROVE filter concentrations of sulfate and other chemical species, an accounting framework in which lower sulfate concentrations necessarily imply better visibility if other species are held constant. However, other chemical species fluctuated during the time period examined, likely masking the effect of the relatively small decreases in sulfate. White et al. (2012) commented that scientists had recognized for some time that visibility in GRCA is impaired by a combination of emissions from diverse sources over a large region and that imperceptible increments, such as from the Mohave Power Project, can add up to perceptible haze.

11.6.5 Trends in Visibility

Visibility at 12 locations in the GRCA area declined from the 1950s to the 1970s by an estimated 10% to 30% (Trijonis 1979). IMPROVE long-term trend data are available for BAND (1992–2012), GRCA (1999–2012), MEVE (1990–2012), and PEFO (1990–2012). Trends analyses indicated that visibility has significantly improved in all four parks on the 20% clearest days, but only in BAND has visibility improved significantly on the 20% haziest days.

Table 11.5 gives estimates of visibility at four Southern Colorado Plateau Network Class I areas on the 20% clearest and haziest days. Haze is significantly elevated over natural conditions on both clear and hazy days for both the baseline (2000–2004) period and 2018. Natural conditions on the haziest days are not expected to be achieved by 2064. The EPA estimated that natural conditions will not be reached in these parks for many years (101–261 years).

TABLE 11.5

Estimates of Natural and Baseline (2000–2004) and Projected 2018 Visibility on the 20% Clearest and Haziest Days at BAND, GRCA, MEVE, and PEFO, as well as the Estimated Years to Natural Conditions on the Haziest Days Given Actual and Anticipated Emissions Reductions in the Region

| Park Name | Measured Baseline, Projected 2018, and Estimated Natural Haze | | | | | | Years to Natural Conditions |
| | 20% Clearest Days (dv) | | | 20% Haziest Days (dv) | | | |
	Baseline	2018	Natural	Baseline	2018	Natural	
Bandelier[a]	4.95	4.89	1.29	12.22	11.90	6.26	261
Grand Canyon[b]	2.16	2.02	0.31	11.66	10.58	7.04	101
Mesa Verde[c]	4.32	4.10	1.01	13.03	12.50	6.81	164
Petrified Forest[b]	5.02	4.62	1.07	13.21	12.64	6.49	165

[a] EPA, 40 CFR Part 52, [EPA–R06–OAR–2009–0050; FRL–9683–9], Approval and Promulgation of State Implementation Plans; New Mexico; Regional Haze Rule Requirements for Mandatory Class I Areas Agency (EPA). ACTION: Proposed rule. June 15, 2012.

[b] EPA, 40 CFR Part 52, [EPA–R09–OAR–2013–0588; FRL–9912–97–OAR], Promulgation of Air Quality Implementation Plans; Arizona; Regional Haze and Interstate Visibility Transport Federal Implementation Plan, ACTION: Final rule. September 3, 2014.

[c] EPA, 40 CFR Part 52, [EPA–R08–OAR–2011–0770, FRL–9650–7], Approval and Promulgation of Implementation Plans; State of Colorado; Regional Haze State Implementation Plan, ACTION: Proposed rule. March 26, 2012.

11.7 Toxic Airborne Contaminants

Because of the size and number of coal-burning power plants in the Southern Colorado Plateau Network region, Hg is a potential problem for ecosystems in the network parks. Burning coal releases Hg, which can be transported long distances in the atmosphere. It is routinely monitored in rain at MEVE and levels at the park are the highest among western U.S. sites (NADP 2010). A pilot study at the park did not find elevated mercury levels in songbirds, invertebrates, fish, or crayfish (Nydick and Williams 2010). Sample sizes were very small, however, and the study did not preclude the possibility of Hg accumulation in park wildlife.

Eagles-Smith et al. (2014) examined fish from 21 national parks in the western United States, including 2 Southern Colorado Plateau Network parks, GRCA and MEVE. In GRCA, brown trout and rainbow trout were sampled (n = 42), and mean total Hg concentrations (76 ng/g wet weight) were similar to the average across all fish in the study (77.7 ng/g wet weight). However, size-adjusted total Hg concentrations were among the highest in the size class (101.2 ng/g wet weight) across all parks. Moreover, size-adjusted total Hg concentrations in Shinumo Creek (178.8 ng/g wet weight) were more than twofold higher than those in the other two sites (71.5 and 81.0 ng/g wet weight) in GRCA, suggesting that Hg exposure within the park is variable. None of the fish exceeded the EPA's criterion for human consumption (300 ng/g wet weight), but 10% of the fish exceeded toxicity thresholds for birds considered to be highly sensitive to Hg, including osprey (*Pandion haliaetus*) and snowy egret (*Egretta thula*). In MEVE, speckled dace were sampled (n = 10). Their total mean Hg concentration (74.9 ng/g wet weight) was comparable to the study-wide mean value (77.7 ng/g wet weight) but much lower than concentrations measured in dace from nearby Capitol Reef and Zion National Parks. These differences may reflect differences in environmental processes occurring in the parks. None of the fish in MEVE exceeded toxicity thresholds for birds. Human health thresholds were not applied, as speckled dace is a very small fish, not consumed by humans (Eagles-Smith et al. 2014).

Data on other airborne toxics in Southern Colorado Plateau Network ecosystems are very limited. Pesticide levels in air samples in BAND were near the median for all WACAP parks. Relative to other WACAP parks, BAND vegetation had medium to high concentrations of current-use pesticides (endosulfans, dacthal, and chlorpyrifos) and medium to low concentrations of historic-use pesticides (hexachlorobenzene and hexachlorocyclohexane-α).

11.8 Summary

Although the Southern Colorado Plateau Network parks are located in relatively remote areas, they experience air pollutants from regional and local sources, including large power plants, urban areas, agriculture, oil and gas development, and fires. There are few human population centers within the Southern Colorado Plateau Network; only Albuquerque is larger than 100,000 people.

Total S deposition at the locations of Southern Colorado Plateau Network parks during the period 2010–2012 was generally less than 2 kg S/ha/yr. Total N deposition was less than 5 kg N/ha/yr at most park locations.

The arid conditions that prevail throughout much of the Southern Colorado Plateau Network do not contribute to acid sensitivity, which tends to be enhanced by high precipitation due to the leaching by the precipitation of base cations from watershed soils. At the higher elevations within parklands in the Southern Colorado Plateau Network that receive greater amounts of precipitation, it is possible that there are some acid-sensitive surface waters. Ecosystem sensitivity to acidification is also influenced by land slope, which often affects the degree of acid neutralization provided by soils and bedrock within the watershed. Land slope varies considerably among the parks in the Southern Colorado Plateau Network but is most pronounced in GRCA, suggesting increased potential for acid sensitivity of aquatic ecosystems. However, stream acidification has not been documented in any of the Southern Colorado Plateau Network parks and is considered unlikely in this network.

The predominant vegetation types within these parks that are known to be sensitive to nutrient N enrichment are arid and semi-arid plant communities. Nitrogen inputs in the form of atmospheric deposition can promote the growth of annual grasses, such as buffelgrass (*Pennisetum ciliare*) and cheatgrass (*Bromus tectorum*), with gradual loss of native forbs and shrubs. Increases in annual grasses in these parks can also increase fuel loading, leading to greater fire risk.

Although ozone concentrations and long-term exposures are somewhat elevated at times in the Southern Colorado Plateau Network region, risk to plants has been rated as low to moderate. Relatively dry soil conditions that predominate in the region tend to limit gas exchange in plants, reducing ozone uptake. However, plants growing in well-watered riparian areas may be more vulnerable to ozone damage.

Clean, clear skies are typical of the Southern Colorado Plateau Network parks, although all suffer some impairment of visibility. Most parks in the network afford magnificent vistas, with massive landforms, picturesque geology, and varied rock colors. The panoramic views evident at many locations in parklands within this network provide unique visual experiences for visitors, and high air quality is central to full enjoyment of those experiences. Because the Colorado Plateau has some of the best visual air quality in the conterminous United States, reduced air quality that might go relatively unnoticed elsewhere is acutely evident on the Plateau.

The park in the Southern Colorado Plateau Network in which specific visibility concerns are best understood is GRCA. Because of the importance of visibility to visitors at GRCA, the sources of atmospheric aerosols that impair that visibility, even by a relatively minor amount, have been a subject of considerable research. The most important canyon locations for visibility are located at and slightly below the canyon rim. Visibility at these locations is most strongly affected by atmospheric conditions between the viewer's vantage point and the geological formations on the opposite side of the canyon.

Visibility in GRCA varies with the seasons. It is worse in the spring and summer than in the winter. Regardless of season, visibility is typically worse inside the canyon than on the rim. Average atmospheric concentrations of sulfate, organics, and elemental C are highest in summer, whereas nitrate concentrations are generally highest in winter and spring, reflecting seasonal differences in atmospheric chemistry and pollutant emissions.

Much of the work on the sources of visibility degradation in GRCA has focused on long-range transport from the Pacific coast and local transport from two large point sources, the Navajo Generating Station and the Mohave Power Project. In 1990, the Navajo Generating Station agreed to reduce sulfur dioxide emissions by 90%. Three scrubbers were installed in 1997–1999. Reducing emissions had a beneficial effect by decreasing winter haze and improving visibility in GRCA. Visibility on the haziest days, having highest fine particulate sulfate, improved markedly following the reduction in Navajo Generating Station emissions. During summer, however, when Navajo Generating Station impacts on GRCA were minimal, there was only a small decrease in atmospheric fine particulate sulfate.

The Mohave Power Project elected to shut down operations in 2005. As predicted by NPS studies, the closure of the Mohave Power Project improved air quality in the Grand Canyon but not by a substantial amount. Mean haze and extinction levels did not respond to the closure in a statistically significant way. Aerosol sulfate concentrations did decrease, but not by enough to induce a measureable improvement in light extinction or haze.

Tracer experiments conducted during both winter and summer have demonstrated the importance of air flow channeling within valleys and canyons along the Colorado River to visibility in GRCA. Three situations have most commonly been associated with poor air quality in GRCA: (1) regional haze from southern California, (2) high humidity and cloud cover from summer monsoons, and (3) slow moving anticyclones over the Great Basin that trap pollutants in the planetary boundary layer. Other studies have identified Las Vegas and urban areas in the Baja California and Imperial Valley regions as likely primary sources of atmospheric sulfate in the park.

Several reservoirs in southwestern Colorado near MEVE have fish consumption advisories because of elevated Hg levels. Concentrations of Hg in rainfall are relatively high in the region, but Hg deposition is relatively low because precipitation amounts are low (deposition is a function of the concentration of Hg in the precipitation and the precipitation amount). A pilot study examining Hg bioaccumulation in birds and aquatic biota in MEVE and vicinity did not find high levels of Hg, but recommended additional study focused on the top predators in aquatic ecosystems that would be most likely to accumulate toxic levels of Hg.

References

Allen, E.B. and L.H. Geiser. 2011. North American deserts. *In* L.H. Pardo, M.J. Robin-Abbott, and C.T. Driscoll (Eds.). *Assessment of Nitrogen Deposition Effects and Empirical Critical Loads of Nitrogen for Ecoregions of the United States*. General Technical Report NRS-80. U.S. Forest Service, Newtown Square, PA, pp. 133–142.

Allen, E.B., L.E. Rao, and R.J. Steers. 2009. Impacts of atmospheric nitrogen deposition on vegetation and soils at Joshua Tree National Park. *In* R.H. Webb, J.S. Fenstermaker, J.S. Heaton, D.L. Hughson, E.V. McDonald, and D.M. Miller (Eds.). *The Mojave Desert: Ecosystem Processes and Sustainability*. University of Nevada Press, Las Vegas, NV, pp. 78–100.

Báez, S., J. Fargione, D.I. Moore, S.L. Collins, and J.R. Gosz. 2007. Atmospheric nitrogen deposition in the northern Chihuahuan Desert: Temporal trends and potential consequences. *J. Arid Environ.* 68:640–651.

Balch, J.K., B.A. Bradley, C.M. D'Antonio, and J. Gómez-Dans. 2013. Introduced annual grass increases regional fire activity across the arid western USA (1980–2009). *Glob. Change Biol.* 19(1):173–183.

Banta, R.M., L.S. Darby, P. Kaufmann, D.H. Levinson, and C.-J. Zhu. 1999. Wind-flow patterns in the Grand Canyon as revealed by Doppler Lidar. *J. Appl. Meteorol.* 38:1069–1083.

Barker, D.H., C. Vanier, E. Naumburg, T.N. Charlet, K.M. Nielsen, B.A. Newingham, and S.D. Smith. 2006. Enhanced monsoon precipitation and nitrogen deposition affect leaf traits and photosynthesis differently in spring and summer in the desert shrub *Larrea tridentata*. *New Phytol.* 169:799–808.

Binkley, D., C. Giardina, D. Morse, M. Scruggs, and K. Tonnessen. 1997. Status of air quality and related values in Class I National Parks and Monuments of the Colorado Plateau. Chapter 9. Grand Canyon National Park. National Park Service, Air Resources Division, Denver, CO.

Brooks, M.L. 2003. Effects of increased soil nitrogen on the dominance of alien annual plants in the Mojave Desert. *J. Appl. Ecol.* 40:344–353.

Chen, J., R. Bornstein, and C.G. Lindsey. 1999. Transport of a power plant tracer plume over Grand Canyon National Park. *J. Appl. Meteorol.* 38:1049–1068.

Dolske, D.A., W.T. Petuskey, and D.A. Richardson. 1991. Impacts of microclimate and air quality on sandstone masonry of Anasazi dwelling ruins at Mesa Verde National Park. *In* A. Hutchinson and S.J. E. (Eds.). *Proceedings of the Anasazi Symposium 1991*. Pub. LC93-080585. Mesa Verde Museum Association., Mesa Verde National Park, CO.

Eagles-Smith, C.A., J.J. Willacker, Jr., and C.M. Flanagan Pritz. 2014. Mercury in fishes from 21 national parks in the western United States—Inter- and intra-park variation in concentrations and ecological risk. U.S. Geological Survey Open-File Report 2014-1051, Reston, VA.

Eatough, D.J., W. Cui, and J. Hull. 2006. Fine particulate chemical composition and light extinction at Meadview, AZ. *J. Air Waste Manage. Assoc.* 56:1694–1706.

Eatough, D.J., M. Green, W. Moran, and R. Farber. 2001. Potential particulate impacts at the Grand Canyon from northwestern Mexico. *Sci. Tot. Environ.* 276:69–82.

Ellis, R.A., D.J. Jacob, M.P. Sulprizio, L. Zhang, C.D. Holmes, B.A. Schichtel, T. Blett, E. Porter, L.H. Pardo, and J.A. Lynch. 2013. Present and future nitrogen deposition to national parks in the United States: Critical load exceedances. *Atmos. Chem. Phys.* 13(17):9083–9095.

ENVIRON, Alpine Geophysics LLC, and University of North Carolina. 2013. Western Regional Air Partnership West-wide Jump-start Air Quality Modeling Study (WestJumpAQMS). Report prepared for Western Regional Air Partnership, Fort Collins, CO.

Fenn, M.E., R. Haeuber, G.S. Tonnesen, J.S. Baron, S. Grossman-Clark, D. Hope, D.A. Jaffe et al. 2003. Nitrogen emissions, deposition, and monitoring in the western United States. *BioScience* 53(4):391–403.

Grand Canyon Visibility Transport Commission (GCVTC). 1996. Recommendations for improving western vistas. Report to the U.S. Environmental Protection Agency. Available at: http://www.westgov.org/images/dmdocuments/GCVTCFinal.PDF (accessed January, 2014).

Gray, H.A., M.P. Ligocki, G.E. Moore, C.A. Emery, R.C. Kessler, J.P. Cohen, C.C. Chang et al. 1991. Deterministic modeling in the Navajo generating station visibility study. SYSAPP-91/045. Systems Applications International, San Rafael, CA.

Green, M., R. Farber, N. Lien, K. Gebhart, J. Molenar, H. Iyer, and D. Eatough. 2005. The effects of scrubber installation at the Navajo Generation Station on particulate sulfur and visibility levels in the Grand Canyon. *J. Air Waste Manage. Assoc.* 55:1675–1682.

Gutierrez, J.R. and W.G. Whitford. 1987. Chihuahuan Desert annuals—Importance of water and nitrogen. *Ecology* 68:2032–2045.

Hall, F.F. 1981. Visibility reductions from soil dust in the southwestern United States. *Atmos. Environ.* 15:1929–1933.

Hand, J.L., S.A. Copeland, D.E. Day, A.M. Dillner, H. Idresand, W.C. Malm, C.E. McDade et al. 2011. IMPROVE (Interagency Monitoring of Protected Visual Environments): Spatial and seasonal patterns and temporal variability of haze and its constituents in the United States. Cooperative Institute for Research in the Atmosphere Report, Fort Collins, CO.

Hering, S.V., J.L. Bowen, J.G. Wengert, and L.W. Richards. 1981. Characterization of the regional haze in the southwestern United States. *Atmos. Environ.* 15:1999–2009.

Inouye, R.S. 2006. Effects of shrub removal and nitrogen addition on soil moisture in sagebrush steppe. *J. Arid Environ.* 65:604–618.

Joint Fire Science Program (JFSP). 2014. Progress report 2014. BLM/FA/GI-14/002+9270. National Interagency Fire Center, Boise, ID.

Kohut, R. 2007. Assessing the risk of foliar injury from ozone on vegetation in parks in the U.S. National Park Service's Vital Signs Network. *Environ. Pollut.* 149:348–357.

Lindsey, C.G., J. Chen, T.S. Dye, L.W. Richards, and D.L. Blumenthal. 1999. Meteorological processes affecting the transport of emissions from the Navajo generating station to Grand Canyon National Park. *J. Appl. Meteorol.* 38:1031–1048.

Lyons, K.G., B.G. Maldonado-Leal, and G. Owen. 2013. Community and ecosystem effects of buffelgrass (*Pennisetum ciliare*) and nitrogen deposition in the Sonoran Desert. *Invasive Plant Sci. Manage.* 6(1):65–78.

Macias, E.S., J.O. Zwicker, and W.H. White. 1981. Regional haze case studies in the southwestern U.S.: II. Source contributions. *Atmos. Environ.* 15:1987–1997.

Malm, W., K. Gebhart, D. Latimer, T. Cahill, R. Eldred, A. Pielke, R. Stocker, and J. Watson. 1989. Winter Haze Intensive Tracer Experiment (WHITEX). Final report prepared for the National Park Service. Fort Collins, CO.

Malm, W., M.L. Pitchford, C. McDade, and L.L. Ashbaugh. 2007. Coarse particle speciation at selected locations in the rural continental United States. *Atmos. Environ.* 41:2225–2239.

Malm, W.C., M. Pitchford, and H.K. Iyer. 1988. Design and implementation of the winter haze intensive tracer experiment-WHITEX. Paper 88-52.1. *81st Annual Meeting of the Air Pollution Control Association*, Dallas, TX.

Mazurek, M., M.C. Masonjones, H.D. Masonjones, L.G. Salmon, G.R. Cass, K.A. Hallock, and M. Leach. 1997. Visibility-reducing organic aerosols in the vicinity of Grand Canyon National Park: Properties observed by high resolution gas chromatography. *J. Geophys. Res.* 102(D3): 3779–3793.

McKee, T.B. and R.D. O'Neal. 1989. The role of valley geometry and energy budget in the formation of nocturnal valley winds. *J. Appl. Meteorol.* 28:445–456.

National Atmospheric Deposition Program (NADP). 2010. National atmospheric deposition program 2009 annual summary. NADP Data Report 2010-01. Illinois State Water Survey, University of Illinois at Urbana–Champaign, Urbana–Champaign, IL.

National Park Service (NPS). 2010. Air quality in national parks: 2009 annual performance and progress report. Natural Resource Report NPS/NRPC/ARD/NRR-2010/266. National Park Service, Air Resources Division, Denver, CO.

National Research Council (NRC). 1990. *Haze in the Grand Canyon: An Evaluation of the WHITEX*. National Academy Press, Washington, DC, 224pp.

Nydick, K. and K. Williams. 2010. Final report. Pilot study of the ecological effects of mercury deposition in Mesa Verde National Park, Colorado. Report 2010-01. Mountain Studies Institute, Silverton, CO.

Padgett, P.E., E.B. Allen, A. Bytnerowicz, and R.A. Minnich. 1999. Changes in soil inorganic nitrogen as related to atmospheric nitrogenous pollutants in southern California. *Atmos. Environ.* 33:769–781.

Padgett, P.E. and A. Bytnerowicz. 2001. Deposition and adsorption of the air pollutant HNO_3 vapor to soil surfaces. *Atmos. Environ.* 35:2405–2415.

Pardo, L.H., M.J. Robin-Abbott, and C.T. Driscoll. 2011. Assessment of nitrogen deposition effects and empirical critical loads of nitrogen for ecoregions of the United States. General Technical Report NRS-80. U.S. Forest Service, Newtown Square, PA.

Pitchford, A., M. Pitchford, W. Malm, R. Floccini, T. Cahill, and E. Walther. 1981. Regional analysis of factors affecting visual air quality. *Atmos. Environ.* 15:2043–2054.

Pitchford, M., M. Green, H. Kuhns, I. Tombach, W. Malm, M. Scruggs, R. Farber, and V. Mirabella. 1999. Project MOHAVE final report. Available at: http://vista.cira.colostate.edu/Improve/?s=mohave (accessed August 22, 2011).

Rao, L.E., E.B. Allen, and T. Meixner. 2010. Risk-based determination of critical nitrogen deposition loads for fire spread in southern California deserts. *Ecol. Appl.* 20(5):1320–1335.

Savig, K. and D. Morse. 1998. Visibility sections. *In* D.L. Peterson, T.J. Sullivan, J.M. Eilers, S. Brace and D. Horner. *Assessment of Air Quality and Air Pollutant Impacts in National Parks of the Rocky Mountains and Northern Great Plains*. NPS D-657. U.S. Department of the Interior, National Park Service, Air Resources Division, Denver, CO.

Schwede, D.B. and G.G. Lear. 2014. A novel hybrid approach for estimating total deposition in the United States. *Atmos. Environ.* 92:207–220.

Schwinning, S., B.I. Starr, N.J. Wojcik, M.E. Miller, J.E. Ehleringer, and R.S. Sanford. 2005. Effects of nitrogen deposition on an arid grassland in the Colorado Plateau cold desert. *Rangeland Ecol. Manage.* 58:565–574.

Steinacker, R. 1984. Air mass stability associated with winter haziness at Grand Canyon National Park. *Proceedings of the 84th Annual Meeting*, Air & Waste Management Association, Vancouver, British Columbia, Canada, Vol. 2, pp. 1–8.

Suarez-Murias, T., J. Glass, E. Kim, L. Melgoza, and T. Najita. 2009. California regional haze plan. California Environmental Protection Agency, Air Resources Board, Sacramento, CA.

Sullivan, T.J. and T.C. McDonnell. 2014. Mapping of nutrient-nitrogen critical loads for selected national parks in the intermountain west and Great Lakes regions. Natural Resource Technical Report NPS/ARD/NRTR—2014/895. National Park Service, Fort Collins, CO.

Terhorst, J. and M. Berkman. 2010. Effect of coal-fired power generation on visibility in a nearby national park. *Atmos. Environ.* 44:2524–2531.

Trijonis, J. 1979. Visibility in the Southwest—An exploration of the historical data base. *Atmos. Environ.* 13:833–843.

Weathers, K.C., S.M. Simkin, G.M. Lovett, and S.E. Lindberg. 2006. Empirical modeling of atmospheric deposition in mountainous landscapes. *Ecol. Appl.* 16(4):1590–1607.

White, W.H., R.J. Farber, W.C. Malm, M. Nuttall, M.L. Pitchford, and B.A. Schichtel. 2012. Comment on "Effect of coal-fired power generation on visibility in a nearby National Park (Terhorst and Berkman, 2010)". *Atmos. Environ.* 55(0):173–178.

Whiteman, C.D., X. Bian, and J.L. Sutherland. 1999b. Wintertime surface wind patterns in the Colorado River Valley. *J. Appl. Meteorol.* 38:1118–1130.

Whiteman, C.D., S. Zhong, and X. Bian. 1999a. Wintertime boundary layer structure in the Grand Canyon. *J. Appl. Meteorol.* 38:1084–1102.

Xu, J., D. DuBois, M. Pitchford, M. Green, and V. Etyemezian. 2006. Attribution of sulfate aerosols in Federal Class I areas of the western United States based on trajectory regression analysis. *Atmos. Environ.* 40:3433–3447.

12

Coast and Cascades Network

12.1 Background

There are three national parks larger than 100 mi^2 (259 km^2) in the North Coast and Cascades Network: Mount Rainier (MORA), North Cascades (NOCA), and Olympic (OLYM) national parks. There are also four historical parks that are smaller than 100 mi^2. Larger parks generally have more available data with which to evaluate air pollution sensitivities and effects. In addition, the larger parks generally contain more extensive resources in need of protection against the adverse impacts of air pollution. The focus here is on the three larger parks.

The major urban centers (larger than 500,000 people) in the region are Seattle, Washington, and Portland, Oregon. Vancouver, British Columbia, lies just to the north of the network boundary. Figure 12.1 shows the network boundary and the location of each national park along with human population centers that have more than 10,000 people. MORA, NOCA, and OLYM are within about 60 mi (100 km) of emissions sources in the Seattle/Tacoma metropolitan area. OLYM and NOCA are in fairly close proximity to large urban centers in Vancouver and Victoria, British Columbia, Canada. OLYM is adjacent to busy marine shipping channels.

Weather influences atmospheric deposition processes and ecosystem acidification. The seaward slopes of the Olympic Mountains receive more precipitation than any other place in the contiguous United States. Between 120 and 160 inches (300 and 400 cm) of precipitation falls annually in the Hoh and Queets valleys of OLYM. Elevation in OLYM ranges from sea level to 7965 ft (2428 m) at Mt. Olympus. Eastern portions of the park are drier. The western slopes of the Cascade Mountains in Washington also receive substantial precipitation, exceeding 80–120 inches (200–300 cm) at many locations. At the higher elevations, much of this precipitation falls as snow. Mount Rainier is the highest of the Cascade Mountain volcanoes, at 14,410 ft (4,392 m), and the fifth tallest mountain in the contiguous United States. The mountain occupies more than one-fourth of the area of MORA. Its glaciers comprise the largest mass of permanent ice in the United States outside Alaska.

The network parks contain alpine peaks, glaciers, and large rivers. OLYM also includes a coastal zone. The dominant aquatic features of OLYM are the 13 major rivers flowing from the Olympic Mountains in all directions and the 648 lakes and ponds found in the park. These are generally well-buffered aquatic systems draining sedimentary bedrock and glaciers with high silt loads. There are many high-elevation lakes, some of which have been sampled for water chemistry, and several large low-elevation lakes that have been studied

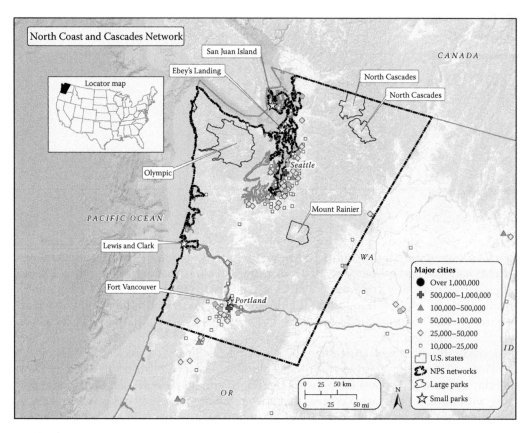

FIGURE 12.1
Network boundary and locations of national parks and population centers with more than 10,000 people in the North Coast and Cascades Network region.

in some detail (Eilers et al. 1994). OLYM has sedimentary and basaltic rocks derived from uplifted marine strata.

Terrain in NOCA is steep and rugged and includes numerous alpine peaks, glaciers, and rivers, with strongly dissected deep forested valleys. Many of the 245 natural lakes in NOCA are in subalpine and alpine settings and are accessible only on foot. The natural lakes and stream valleys were formed by glacial action that is evident throughout the park. Glaciers cover about 42 mi² (109 km²) in NOCA, more than any other park unit in the conterminous United States. Igneus (granitic) and metamorphic rocks form a central core in the park. The granitic rocks are mainly granodiorite and quartz diorite, which are resistant to weathering and contribute little buffering in the form of base cations to drainage waters, making these water resources sensitive to acidification.

MORA has an extensive network of streams and rivers radiating from the summit and high-elevation glaciers of Mt. Rainier. The park contains 35 mi² (91 km²) of snowfields and glaciers. Glacial activity has created nearly 200 lakes and ponds. The glaciers that remain on the mountain feed the streams and some of the lakes with meltwater. The lakes are distributed around the mountain and extend from montane to alpine settings. Many of the lakes are located at higher elevations and may remain ice-free only three to four months of the year.

12.2 Atmospheric Emissions and Deposition

Local and regional sources of air pollutants in the North Coast and Cascades Network region include mainly motor vehicles, power plants, and industrial facilities in the Puget Sound area, the Portland metropolitan area, and the eastern end of the Columbia River Gorge (Clow and Campbell 2008). Marine vessels are also a significant source of emissions that are transported to OLYM. Sulfur is generally emitted as sulfur dioxide from coal-burning power plants, smelting, and diesel engines, depositing eventually as sulfate. Nitrogen is emitted as oxidized N from vehicles, power plants, industry, and fires, or as reduced N (ammonia) from agriculture and fertilizer use, eventually depositing mainly as nitrate or ammonium. Toxic air pollutants include a wide variety of organic compounds and heavy metals that may be emitted from power plants, industry, and agriculture.

12.2.1 Emissions

In general, annual county average sulfur dioxide emissions in 2002 were estimated at less than 5 tons per square mile (tons/mi^2/yr) throughout most of the North Coast and Cascades Network region, with only one county in the 5–20 tons/mi^2/yr range (Sullivan et al. 2011a). Most individual point sources of sulfur dioxide were below 5,000 tons per year with only one point source, a coal-fired power plant, in the range of 5,000–20,000 tons/yr. Only one N point source larger than 2500 tons per year exists in the North Coast and Cascades Network region, but N emissions from this source are expected to decrease substantially, and the coal-fired boilers will be replaced by a natural gas–fired power plant by 2025 (T. Cummings, National Park Service, personal communication, August 2014). Even before the shutdown of the coal-fired boilers, contributions to total human-caused oxidized N emissions in the Cascadia region, including the North Coast and Cascades Network, were much higher from mobile sources than from area sources or point sources (70% vs. 12% and 18%, respectively; Barna et al. 2000).

Local emissions sources of pesticides, Hg, polycyclic aromatic hydrocarbons, and other toxic materials also influence North Coast and Cascades Network parks. Such emissions derive from agriculture, power plants, fire, and other point and nonpoint sources, including some in Asia. To date, many of these sources have not been well quantified.

County-level emissions near the North Coast and Cascades Network, based on data from the EPA's National Emissions Inventory during a recent time period (2011), are depicted in Figures 12.2 through 12.4 for sulfur dioxide, oxidized N, and ammonia, respectively. Many counties near the parks had sulfur dioxide emissions in the range of 1–5 tons/mi^2/yr (Figure 12.2). Patterns in oxidized N emissions were generally higher, at many locations higher than 5 tons/mi^2/yr (Figure 12.3). Emissions of ammonia near network parks were mostly below 2 tons/mi^2/yr (Figure 12.4).

12.2.2 Deposition

Wet deposition appears to be higher than dry deposition in North Coast and Cascades Network parks, in part due to high precipitation. It was estimated to make up more than 70% of total wet plus dry N and S deposition, based on Community Multiscale Air Quality atmospheric deposition model simulations (Sullivan et al. 2011a). The contributions from occult deposition are unknown.

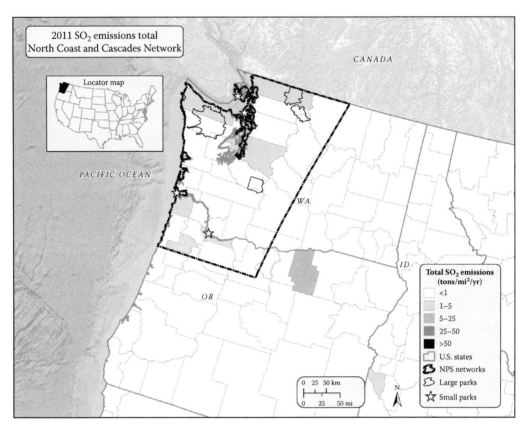

FIGURE 12.2
Total sulfur dioxide (SO_2) emissions, by county, near the North Coast and Cascades Network for the year 2011. (Data from EPA's National Emissions Inventory, https://www.epa.gov/air-emissions-inventories, accessed January, 2014.)

Atmospheric S deposition levels have declined at the network parks since 2001, based on Total Deposition project estimates (Table 12.1). Decreases in total S deposition over the previous decade for the parks in this network have been larger than 25%. Estimated total N deposition over that same time period decreased at MORA and NOCA and increased at OLYM. Oxidized N decreased at all three parks. Reduced N deposition increased substantially at all three network parks.

Total S deposition in and around the North Coast and Cascades Network for the period 2010–2012 was generally <5 kg S/ha/yr at park locations. Oxidized inorganic N deposition for the period 2010–2012 was less than 5 kg N/ha/yr throughout the parklands within the network. Most areas received less 2 kg N/ha/yr of reduced inorganic N from atmospheric deposition during this same period; a few areas received higher amounts. Total N deposition was less than 5 kg N/ha/yr at most park locations (Figure 12.5); some portions of the region were estimated to have received higher amounts, up to 10 kg N/ha/yr.

Data from a U.S. Geological Survey study in MORA and NOCA suggested that, during the 1986–2005 period, wet S deposition in rain and snow declined significantly while wet N deposition remained relatively stable (Clow and Campbell 2008). Decreases in S deposition were more pronounced after 2001 when emissions controls were added to TransAlta's

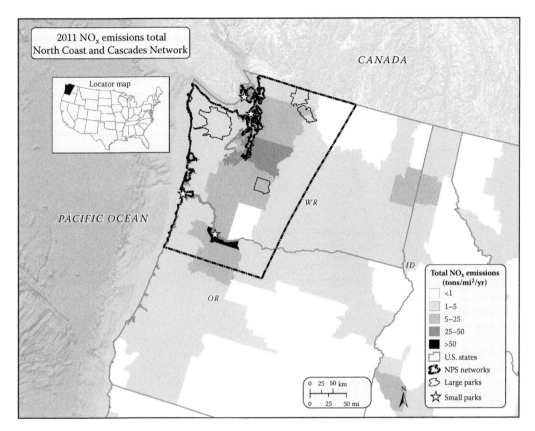

FIGURE 12.3
Total oxidized N (NO$_x$) emissions, by county, near the North Coast and Cascades Network for the year 2011. (Data from EPA's National Emissions Inventory, https://www.epa.gov/air-emissions-inventories, accessed January, 2014.)

Centralia power plant (Clow and Campbell 2008). Sulfur deposition in the North Coast and Cascades Network region has never been high, compared to areas in the eastern United States, for example, and these relatively low deposition levels have been further reduced over the past two decades in response to emissions controls.

Clouds and fog are relatively common throughout the network region. Cloudwater chemistry sampled by Basabe et al. (1994) in western Washington was relatively acidic and similar to acidity values reported in California, the Midwest, and the Northeast. Fog at low elevation contained higher ammonium concentrations than fog at high elevation, reducing its acidity. High-elevation clouds on the Olympic Peninsula and in the North Cascades Mountains were less acidic than high-elevation clouds nearer the greater Seattle area, in closer proximity to the major regional S and N emissions sources (Basabe et al. 1994). Unquantified cloud and fog deposition of S and N to sensitive high-elevation and coastal receptors in the network region increases impacts caused by wet- and dry-deposited materials by an unknown amount.

Atmospheric deposition of air toxics is an important air pollution concern in North Coast and Cascades Network parks. The WACAP studied airborne contaminants in a number of national parks of the western United States, including OLYM, MORA, and NOCA from 2002 to 2007. The dominant semivolatile organic compounds detected in environmental

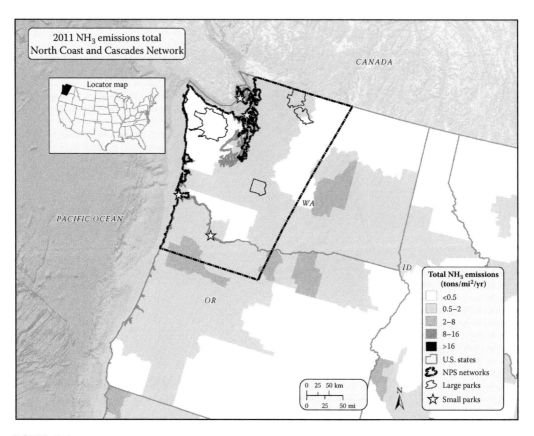

FIGURE 12.4
Total ammonia (NH_3) emissions, by county, near the North Coast and Cascades Network for the year 2011. (Data from EPA's National Emissions Inventory, https://www.epa.gov/air-emissions-inventories, accessed January, 2014.)

samples collected in these parks included current and historic-use (banned) pesticides, fungicides, and combustion by-products (Landers et al. 2010).

Measurements of atmospheric ozone, carbon monoxide, hydrocarbons, and aerosol chemistry in air arriving in the Pacific Northwest from Eurasia have shown that a variety of contaminants can be transported across the Pacific Ocean (Jaffe et al. 2003). There is a high degree of variability, however, in the chemistry of the trans-Pacific episodes of atmospheric pollution transport.

12.3 Acidification

Both N and S deposition can contribute to acidification of soils, lakes, and streams in the North Coast and Cascades Network, with possible effects on biota. If deposition is sufficiently high to acid-sensitive resources, these effects can include loss of acid-sensitive fish and macroinvertebrates; disruption of soil nutrient cycling; and loss of aquatic, terrestrial, and wetland ecosystem biodiversity.

TABLE 12.1

Average Changes in S and N Deposition[a] between 2001 and 2011 across Park Grid Cells at Selected North Coast and Cascades Network Parks

Park Code	Parameter	2001 Average (kg/ha/yr)	2011 Average (kg/ha/yr)	Absolute Change (kg/ha/yr)	Percent Change	2011 Minimum (kg/ha/yr)	2011 Maximum (kg/ha/yr)	2011 Range (kg/ha/yr)
MORA	Total S	4.19	1.86	−2.33	−55.1	1.34	2.63	1.29
	Total N	4.08	3.16	−0.91	−21.7	2.61	4.27	1.65
	Oxidized N	3.06	1.83	−1.23	−40.0	1.52	2.43	0.91
	Reduced N	1.02	1.34	0.32	34.3	1.08	1.84	0.75
NOCA	Total S	2.21	1.56	−0.65	−29.4	0.84	2.32	1.47
	Total N	3.07	2.92	−0.16	−5.2	1.84	4.33	2.49
	Oxidized N	2.41	1.99	−0.42	−17.6	1.23	2.74	1.51
	Reduced N	0.67	0.93	0.27	39.9	0.57	1.85	1.28
OLYM	Total S	6.46	2.98	−3.48	−53.6	1.03	5.19	4.16
	Total N	2.92	2.99	0.07	3.9	1.49	5.23	3.74
	Oxidized N	2.24	1.87	−0.38	−15.0	1.03	3.09	2.06
	Reduced N	0.68	1.12	0.45	67.6	0.46	2.37	1.91

[a] Deposition estimates were determined by the Total Deposition project, based on three-year averages centered on 2001 and 2011 for all ~4 km grid cells in each park. The minimum, maximum, and range of 2011 S and N deposition within each park are also shown.

12.3.1 Acidification Risk Analysis Results

In a coarse screening analysis by Sullivan et al. (2011a), all parks and networks in the National Park Service Inventory and Monitoring Program were ranked according to their relative risk of acidification from atmospheric deposition. The analysis considered three major factors: pollutant exposure, ecosystem sensitivity, and park protection mandates. The summary risk ranking for the North Coast and Cascades Network was among the highest of all networks (Sullivan et al. 2011a). Although estimated relative risk varied by park, the overall level of concern for acidification effects on Inventory and Monitoring parks within this network was judged by Sullivan et al. (2011a) to be very high.

Ecosystem sensitivity was rated high, in large part because high-elevation lakes and low-order streams that do not receive substantial glacial meltwater inputs are common within these parks. High-elevation lakes that do not have glacial meltwater inputs tend to be more acid sensitive than lower-elevation lakes because they typically receive low contributions of base cations in runoff that influences lake chemistry. Each of these parks also contains an extensive network of potentially acid-sensitive high-elevation and low-order streams. Ecosystem sensitivity to acidification rankings also take into account land slope, which often influences the degree of acid neutralization provided by soils and bedrock within the watershed. Terrain in NOCA, MORA, and OLYM is steep, with large portions of all three parks having an average slope above 40°. These steep slopes suggest areas with high potential for acid sensitivity of aquatic ecosystems. Parts of OLYM along the coast are much less steep, with average slopes below 20°. Although soil acidification sensitivity may be high in some areas, soil acidification is not expected to be an important concern in the North Coast and Cascades Network, given the relatively low levels of acidic deposition.

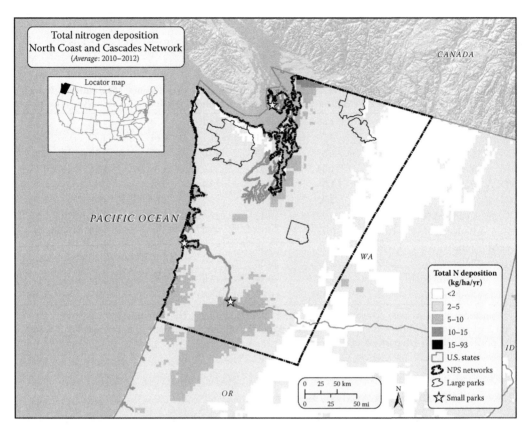

FIGURE 12.5
Total N deposition for the three-year period centered on 2011 in and around the North Coast and Cascades Network. (From Schwede, D.B. and Lear, G.G., *Atmos. Environ.*, 92, 207, 2014.)

12.3.2 Chronic Surface Acid–Base Chemistry

The EPA's Western Lakes Survey in 1985 indicated that the Cascade Mountains, including within the North Coast and Cascades Network region, are among the western mountain ranges with the greatest numbers of acid-sensitive streams and lakes (Landers et al. 1987). Lakes in this mountain range are especially sensitive to acidic deposition because of the predominance of base-poor bedrock and thin acidic soils, large amounts of precipitation, prevalence of coniferous vegetation, and the dilute nature of the lakes (Nelson 1991). Substantial portions of the North Coast and Cascades Network contain areas of exposed bedrock, with little soil cover to neutralize acidic inputs. However, the percentage of exposed bedrock in a watershed does not always indicate acid sensitivity. If the bedrock contains even small deposits of calcareous minerals or if physical weathering such as that caused by glaciers causes a high production of base cations within the watershed (Drever and Hurcomb 1986), surface waters may be alkaline and not sensitive to acidification from acidic deposition.

Highest nitrate concentrations in surface waters in this network are generally found at high elevation, where biological activity, and therefore nitrate uptake, by terrestrial and aquatic biota is lowest (Gibson et al. 1983). The most sensitive (to N-driven acidification) aquatic resources likely include high-elevation lakes that are ice covered more than eight or nine months of the year. These lakes may not achieve water temperatures >10°C during the brief summer, and biological activity is limited. Therefore, they may be sensitive to accumulating N throughout the growing season (Eilers et al. 1994). Lakes and streams receiving glacial meltwater are not likely to be acid-sensitive receptors because of the high physical weathering rates associated with glacial action. Weathering contributes calcium and other base cations that buffer acidity.

Results from the Western Lakes Survey (Landers et al. 1987) concluded that other than lakes affected by geothermal activity or acid mine drainage, there were virtually no acidic (acid neutralizing capacity ≤0 μeq/L) lakes in the West. Both MORA and NOCA and the surrounding Cascade Mountains contain lakes and streams that have acid neutralizing capacity ≤50 μeq/L (Figure 3.2; Larson et al. 1994, 1999); acid neutralizing capacity below 50 to 100 μeq/L indicates acid sensitivity. These low acid neutralizing capacity waters are considered to be highly sensitive to acidification (Figure 12.6; Clow and Campbell 2008).

Sampling of high-elevation lakes by Brakke (1984, 1985), Landers et al. (1987), and Liss et al. (1991) showed that very low acid neutralizing capacity (~10 μeq/L) lakes were present and presumably highly sensitive to acidic deposition. Eilers et al. (1994) stated that headwater streams in the high-elevation areas were also presumed to be acid sensitive, although data were not available to confirm this. Relatively high nitrate concentrations found in some streams in NOCA (Clow and Campbell 2008) could be due to the presence of natural N-fixing bacteria associated with red alder (*Alnus rubra*) trees (Edmonds 1980). Clow and Campbell (2008) suggested that monitoring and assessment of effects of atmospheric N deposition in NOCA should focus on high-elevation ecosystems to avoid confounding effects of N-fixation associated with alder trees.

Some lakes throughout MORA, especially those on the west side, are considered highly sensitive to acidification from atmospheric deposition (Eilers et al. 1994). Eunice Lake and Lake Louise have been well studied (cf., Clow and Campbell 2008). These are small,

FIGURE 12.6
High-elevation ecosystems in NOCA are sensitive to nutrient enrichment and acidification impacts. (National Park Service photo by D. Astudillo.)

shallow, high-elevation, acid-sensitive lakes with fast hydrologic flushing, steep slopes, and thin coarse soils. Volume-weighted average acid neutralizing capacity was determined by Clow and Campbell (2008) to be between 25 and 50 µeq/L in both lakes.

The North Coast and Cascades Network Mountain Lakes Monitoring Protocol includes monitoring the chemistry of 20 reference lakes in MORA, NOCA, and OLYM. The reference lakes are small (most smaller than about 3 ha in surface area). Study watersheds differ in their ability to buffer acidic atmospheric inputs. None of the eight monitoring lakes in OLYM had acid neutralizing capacity <100 µeq/L. The other two parks contained two (MORA) and three (NOCA) such lakes (Fradkin et al. 2012).

Summit Lake, located at 5413 ft (1650 m) in the western Washington Cascades, was sampled by Landers et al. (1987) and again in 1993 by the Mt. Baker-Snoqualmie National Forest. It had lake acid neutralizing capacity about 1 µeq/L, with nonmarine sulfate concentration of 7 µeq/L. Because the nonmarine base cation concentrations (those not attributable to marine aerosol contributions) were very low (about 10 µeq/L) and could largely be attributed to atmospheric deposition, there appeared to be little watershed contribution of S to the lake. Therefore, the lake water sulfate concentration was assumed to be of atmospheric origin. Summit Lake is close to local and regional sources of atmospheric S emissions, and it is conceivable that the lake has lost up to 7 µeq/L of acid neutralizing capacity in response to human-caused acidic deposition (Sullivan and Eilers 1994). It is possible that some of the other high-elevation lakes in North Coast and Cascades Network parks have been similarly affected.

12.3.3 Temporal Variability in Water Chemistry

The acid–base chemistry of Cascade Mountain lakes varies seasonally, with lowest acid neutralizing capacity during the spring snowmelt when accumulated acid anions are released from the melting snowpack, increasing to higher values during fall. For example, the acid neutralizing capacity of Gertrude Lake, located at 5702 ft (1738 m) elevation in the Goat Rocks wilderness in Washington, varied seasonally from a minimum of about 10 µeq/L to a maximum of 91 µeq/L (Eilers and Vaché 1998).

Episodic acidification is an especially important concern for surface waters throughout mountainous areas in the North Coast and Cascades Network. This network contains widespread high-elevation watersheds with steep topography, extensive areas of exposed bedrock, deep snowpack accumulation, and shallow, base-poor soils. These conditions help to determine the pathway followed by snowmelt and stormflow water through the watershed and, therefore, the extent of acid neutralization provided by the soils and bedrock. Conditions predispose the surface water to be highly sensitive to episodic acidification because they receive an elevated proportion of flow that is derived from water that has moved laterally through the surface soil without infiltration to deeper soil horizons (Wigington et al. 1990, Sullivan 2000); in general, more subsurface contact contributes to higher surface water acid neutralizing capacity (Turner et al. 1990).

The North Coast and Cascades Network has not demonstrated any clear pattern of change in lake or stream acid–base chemistry over time. Water chemistry in acid-sensitive watersheds, such as are commonly found in these parks, changes on both intra-annual and interannual time scales in response to changes in environmental conditions. Because of this variability, many years of data are required to establish the existence of trends in surface water chemistry. Assignment of causality to changes that are found to occur is even more difficult. Synoptic lake surveys, such as those conducted in the North Coast and Cascades Network by the Western Lakes Survey (Landers et al. 1987) and some more localized studies, have typically been conducted during the summer or autumn "index period," during which time lake water chemistry generally exhibits low temporal variability.

Most research in the United States on episodic processes has been conducted on stream systems; data from intensive study sites suggest that episodic depression of acid neutralizing capacity and pH may be more pronounced for streams than for lakes. However, there are no available systematic stream chemistry data within the North Coast and Cascades Network region with which to assess the sensitivity of streams to acidic deposition. Most acid–base chemistry data collected in this region have been collected from lakes. Spatial variability can be considerable in lakes, and this complicates efforts to quantify the magnitude of episodic effects (Gubala et al. 1991).

Although many Cascade Mountain lakes are highly sensitive to potential acidic deposition effects (Nelson 1991), it does not appear that chronic acidification has occurred to any significant degree. Episodic acidification has likely occurred in some surface waters but has not been quantified. The data that would be needed for determining the extent and magnitude of episodic acidification have not been collected to a sufficient degree in acid-sensitive waters in the North Coast and Cascades Network to support regional assessment of episodic acidification. Nevertheless, episodic acidification likely occurs in this network, and it remains an important potential stressor at high elevation in the parks in this network. Some high-elevation lakes may receive N deposition sufficiently high to cause some limited chronic nitrate leaching. What little information has been collected suggests that acidification from nitrate is probably not occurring to any substantial degree (cf., Eilers et al. 1994), and any biological effects that have occurred would be expected to have been minimal.

Eilers et al. (1994) offered a preliminary evaluation of the critical load of atmospheric S deposition in the North Coast and Cascades Network, below which sensitive resources will be protected from adverse effects. For the protection of sensitive lakes and streams in MORA and NOCA, they recommended an interim nonmarine S deposition guideline of 20 meq/m^2/yr (equal to about 3 kg S/ha/yr; 9 kg sulfate/ha/yr). They recommended that the critical load for S should probably be adjusted upward, however, to account for precipitation greater than 39 inches (1 m), especially in OLYM and portions of the west slope of the Cascade Mountains. For example, areas that receive 78 inches (2 m) of annual precipitation might require double the annual S load to experience surface water concentrations comparable to similar areas receiving only 39 inches (1 m) of precipitation (Eilers et al. 1994).

12.4 Nutrient Nitrogen Enrichment

In a coarse screening study, Sullivan et al. (2011b) ranked all Inventory and Monitoring networks nationwide for relative risk of nutrient N enrichment. The network rankings for nutrient N pollutant exposure, ecosystem sensitivity to nutrient N enrichment, and park protection yielded an overall network nutrient enrichment summary risk ranking for the North Coast and Cascades Network that was among the highest of all National Park Service networks (Sullivan et al. 2011b). The overall level of concern for nutrient N enrichment effects on parks within this network was judged by Sullivan et al. (2011b) to be very high.

12.4.1 Terrestrial Effects of Nutrient Enrichment

Lichens are known to be highly sensitive to N input (Geiser et al. 2010). Lichen communities in many areas in the Pacific Northwest are exposed to sufficient levels of air pollution such that they show signs of air pollution damage (Fenn et al. 2003, Geiser and Neitlich 2007). Although the Pacific Northwest retains widespread populations of air pollution–sensitive lichens in some areas (Fenn et al. 2003), there are few sensitive lichen species in urban areas, intensive agricultural zones, or downwind of major urban and industrial centers. Various factors influence lichen community composition, but research has found strong correlations between elevated N deposition and changes in lichen diversity. Indicators of clean sites and polluted sites were used by Geiser and Neitlich (2007) to identify six lichen zones of air quality within the region, from worst (all sensitive species absent) to best (all sensitive species present). Geiser et al. (2010) developed empirical N critical loads for forests in western Washington and Oregon, based on observed relationships between epiphytic lichens and total N deposition estimates derived from the Community Multiscale Air Quality model. Estimated total N deposition equal to 3–9 kg N/ha/yr was associated with declines in sensitive lichen species of 20%–40%. Sensitivity varied with precipitation amount. Geiser et al. (2010) concluded that no single lichen critical load is appropriate across land areas that receive varying amounts of precipitation. A model was developed with which to predict N critical loads at any desired lichen community composition, expressed as the percent composition by oligotrophic (low nutrient) versus eutrophic (high-nutrient) lichen species. In addition, high N concentrations have been measured in lichen tissue collected in areas of relatively high N deposition within the North Coast and Cascades Network region (Fenn et al. 2003). With N enrichment, especially around urban and agricultural areas, there is a shift toward weedy, nitrophilous (N-loving) lichen species (Fenn et al. 2003). Replacement of sensitive lichens by nitrophilous species has undesirable ecological consequences. In late-successional, naturally N-limited forests of the Coast Range and western Cascade Mountains, for example, lichens make important contributions to mineral cycling and soil fertility (Pike 1978, Sollins et al. 1980). Large, pollution-sensitive lichens comprise integral parts of the food web for mammals, insects, and birds (McCune and Geiser 1997, U.S. EPA 2005). Sensitive lichen species appear to be negatively affected by N inputs as low as 3–8 kg/ha/yr (Fenn et al. 2003).

Geiser and Neitlich (2007) found that air pollution was associated more with effects on community composition of lichens rather than species richness (number of species). The most widely observed effects included paucity of sensitive, endemic species, and enhancement of nitrophilous and nonnative species (Geiser and Neitlich 2007). The strongest relationship was found with wet ammonium deposition, consistent with findings in California (Jovan and McCune 2005) and Europe (van Dobben et al. 2001). The zone of worst air quality was associated with an absence of sensitive lichen species, enhancement of nitrophilous lichens, mean wet ammonium deposition >0.06 mg N/L in precipitation, and lichen tissue N concentration >0.6%.

Ellis et al. (2013) estimated the N critical load in the range of 2.5–7.1 kg N/ha/yr for the protection of lichens in MORA and NOCA, based on Pardo et al. (2011). For the protection of lichens in OLYM, the estimated N critical load was in the range of 2.7–9.2 kg N/ha/yr.

12.4.2 Aquatic Effects of Nutrient Enrichment

Surveys of the literature on fertilization experiments and lake studies found that remote oligotrophic waters are commonly N-limited, especially undisturbed northern temperate or boreal lakes that receive relatively low levels of atmospheric N deposition (Bergström et al. 2005, Elser et al. 2009a,b). Based on current understanding, eutrophication effects on freshwater ecosystems from the atmospheric deposition of N are of greatest concern in lakes and streams that have very low productivity and nutrient levels and that are located in remote areas, such as in the national park units in the North Coast and Cascades Network (Figure 12.6). In more productive and less remote fresh waters, nutrient enrichment from N deposition does not necessarily stimulate productivity or community changes because P is more commonly the limiting nutrient. Also, in many places with even minor levels of human disturbance, nutrient enrichment with both N and P from nonatmospheric human-caused sources is common.

High-elevation lakes in NOCA, MORA, and OLYM might be prone to N limitation and therefore potentially more susceptible to eutrophication in response to atmospheric N input. Examination of Western Lakes Survey data (Eilers et al. 1987) found enhanced nitrate concentrations in high-elevation lakes in this region adjacent to and downwind of urban centers (Fenn et al. 2003).

Sheibley et al. (2014) investigated N deposition critical loads to protect lakes in NOCA, MORA, and OLYM against changes in the diatom community caused by nutrient N enrichment. They collected lake sediment cores using gravity corers from 12 high-elevation lakes (above the treeline), four in each park. All study lakes were dilute and oligotrophic. Calculated dissolved inorganic N to total P ratios suggested N limitation or N and P colimitation at all of the lakes. Nine study lakes showed a lack of response of the diatom community to changes in atmospheric N input. This result suggests that the nutrient N critical load for these lakes is not exceeded under existing levels of N deposition of about 3 kg N/ha/yr.

At Hoh Lake in OLYM, there was a shift in the relative abundances of *Asterionella formosa* and *Fragilaria tenera*, two diatom species that are suggestive of N enrichment, beginning during the time period 1969–1975. This information was used together with estimates of past N deposition to derive an empirical critical load of 1.0–1.2 kg N/ha/yr for this lake. There was no information suggesting that the shift in diatom relative abundance was caused by disturbance such as fire or by fish stocking.

12.5 Ozone Injury to Vegetation

Ozone is of concern with respect to air pollution effects on vegetation in North Coast and Cascades Network parks, mainly due to expanding urban centers and vehicular traffic. Among the parks in the network, MORA probably faces the highest risk. Ozone concentrations in the Pacific Northwest have occasionally exceeded National Ambient Air Quality Standards intended to protect human health and welfare (including ecosystems). Nevertheless, ozone exposures are generally low in network parks (W126 less than about 1.5 ppm-hr in NOCA to about 3.0 ppm-hr in MORA; Table 12.2; http://www.nature.nps.gov/air/Maps/AirAtlas/IM_materials.cfm), and vegetation surveys have not found ozone injury symptoms (Campbell et al. 2007). Kohut (2007) rated vegetation in all North Coast and Cascades Network parks at low risk from ozone injury.

Maximum atmospheric ozone concentrations observed at park locations have been downwind of urban centers in the Portland and Puget Sound areas (Barna et al. 2000) and at higher elevation (Peterson et al. 1999). Brace and Peterson (1998) measured highest ozone

TABLE 12.2

Ozone Assessment Results for Selected Inventory and Monitoring Parks in the North Coast and Cascades Network Based on Estimated Average 3-Month W126 and SUM06 Ozone Exposure Indices for the Period 2005–2009 and Kohut's (2007) Ozone Risk Ranking for the Period 1995–1999[a,b]

Park Name	Park Code	W126		SUM06		Kohut Ozone Risk Ranking
		Value (ppm-hr)	NPS Condition	Value (ppm-hr)	NPS Condition	
Mount Rainier	MORA	3.02	Low	3.10	Low	Low
North Cascades	NOCA	1.42	Low	1.16	Low	Low
Olympic	OLYM	1.57	Low	1.71	Low	Low

[a] Parks are classified into one of three ranks (low, moderate, high), based on comparison with other Inventory and Monitoring parks.

[b] Degrees of concern for the W126 and SUM06 indices are based solely on levels of ozone exposure. Kohut's risk to vegetation is based on several factors that contribute to injury in plants, including ozone exposure and environmental variables, and considers the effects of soil moisture on the uptake of ozone.

concentrations in MORA at high elevation, and concentrations were higher in the western portion of the park, as compared with the eastern portion. The highest weekly average ozone concentrations in western Washington were measured at the two highest elevation sites in MORA (Peterson et al. 1999).

The W126 (a measure of cumulative ozone exposure that preferentially weights higher concentrations) and SUM06 (a measure of cumulative exposure that includes only hourly concentrations over 60 ppb ozone) exposure indices calculated by NPS staff are given in Table 12.2, along with Kohut's (2007) ozone risk ranking. The W126 and SUM06 indices, as well as Kohut's (2007) risk ranking, were ranked low across all parks within the North Coast and Cascades Network.

Studies by Hogsett et al. (1989) suggested that, among the tree species studied in the Pacific Northwest, ponderosa pine is the most sensitive to damage from ozone exposure. Once weakened by ozone stress, ponderosa pine trees are more susceptible to root rot (*Fomes annosus*) and western pine beetle infestation (Cobb and Stark 1970). These stresses can increase tree mortality. Within the network region, ponderosa pine is at risk of injury if ozone concentrations in the atmosphere increase substantially in this region in the future (Schoettle et al. 1999).

On the dry sites of the lower eastern foothills of NOCA (a small percentage of the overall park), ozone-sensitive ponderosa pine predominates. This portion of the park is further away, however, from the principal sources of ozone precursor emissions, which are located to the west of the park. In addition, dry conditions on the east side of the park tend to inhibit ozone uptake by plant stomata. The climate at moderate elevations on the eastern slope of the Cascade Mountains, with its typical summer drought, limits foliar uptake of C and ozone in ecosystems that contain ponderosa pine. For example, Panek and Goldstein (2001) compared diurnal foliar patterns in young ponderosa pine trees in mixed conifer watered and control (natural drought) plots at moderate elevation 4265–4920 ft (1300–1500 m) in the Sierra Nevada. When plant stomata are closed, ozone flux into the leaf is reduced, even if ambient ozone exposure remains high. Thus, increased rainfall enhances vulnerability to ozone damage (Panek and Goldstein 2001). Estimated ozone uptake by trees in the ambient drought stress plot was about 40% less than in the watered plot. Panek and Goldstein (2001) speculated that an important implication of this research is that increased soil water availability, which is predicted to occur in the Pacific Northwest with ongoing climate change, could make ponderosa pine trees vulnerable to ozone damage at exposure levels that previously did not cause damage.

12.6 Visibility Degradation

12.6.1 Natural Background Visibility and Monitored Visibility Conditions

Within the North Coast and Cascades Network, natural background and ambient visibility conditions have been estimated by Interagency Monitoring of Protected Visual Environments (IMPROVE) for MORA, NOCA, and OLYM for the 20% clearest visibility days, the 20% haziest visibility days, and average days (Table 12.3). Haze levels ranged from very low (NOCA average days) to moderate. Visibility impairment was generally lowest at NOCA.

12.6.2 Composition of Haze

Figure 12.7 shows estimated natural (preindustrial), baseline (2000–2004), and current (most recent available five-year period; 2006–2010) levels of haze and its composition for MORA, NOCA, and OLYM. Sulfates are the primary components of haze at all three parks on the 20% clearest, annual average, and the 20% haziest visibility days. Some of the sulfate is derived from natural marine sources. Organics and nitrates also contribute significantly to haze at these parks.

TABLE 12.3

Estimated Natural Haze and Measured Ambient Haze in Selected North Coast and Cascades Network Parks Averaged over the Period 2004–2008[a]

Park Name	Park Code	Site ID	Estimated Natural Background Visibility (in Deciviews)		
			20% Clearest Days	20% Haziest Days	Average Days
Mount Rainier	MORA	MORA1	2.56	8.54	5.46
North Cascades	NOCA	NOCA1	1.93	8.39	4.44
Olympic	OLYM	OLYM1	2.70	8.44	5.34

Park Name	Park Code	Site ID	Baseline Visibility (for Years 2004–2008)					
			20% Clearest Days		20% Haziest Days		Average Days	
			Deciview	Ranking	Deciview	Ranking	Deciview	Ranking
Mount Rainier	MORA	MORA1	5.04	Low	16.81	Moderate	10.95	Moderate
North Cascades	NOCA	NOCA1	3.19	Low	13.20	Low	7.88	Very Low
Olympic	OLYM	OLYM1	5.54	Moderate	15.44	Low	10.81	Moderate

[a] Parks are classified into one of five haze ranks (very low, low, moderate, high, or very high haze) based on comparison with other monitored parks.

Analyses conducted by the Western Regional Air Partnership indicated that organics from natural emissions sources, including wildfire and biogenic sources (vegetation), contribute to substantial (18%–29% of total) visibility impairment throughout the western United States, including within the North Coast and Cascades Network region. In addition, air pollution sources outside the Western Regional Air Partnership domain, including international offshore shipping and sources from Mexico, Canada, and Asia, can in some cases be substantial contributors to haze in this network (Suarez-Murias et al. 2009).

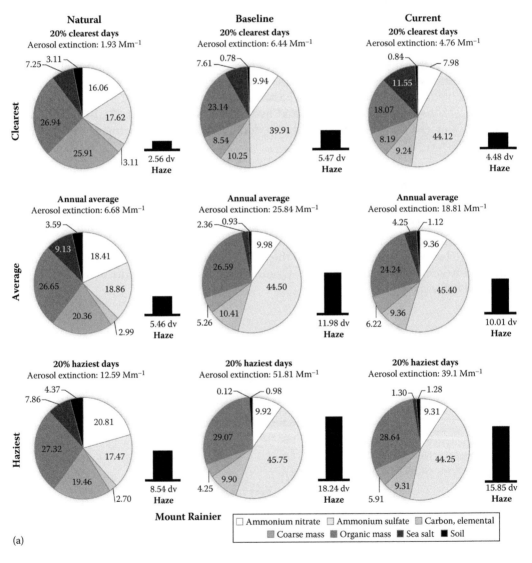

FIGURE 12.7
Estimated natural (preindustrial), baseline (2000–2004), and current (2006–2010) levels of haze (columns) and its composition (pie charts) on the 20% clearest, annual average, and 20% haziest visibility days for (a) MORA, (b) NOCA, and (c) OLYM. The baseline figures for NOCA use the RHR2_VS method of calculation since the years 2003 and 2004 were not valid years for this park. (From http://views.cira.colostate.edu/fed/Tools/RegionalHazeSummary.aspx, accessed October, 2012.) *(Continued)*

12.6.3 Trends in Visibility

The National Park Service (2010) reported long-term trends in visibility on the clearest and haziest 20% of days at monitoring sites in 29 national parks. Twenty-seven parks showed statistically significant ($p \leq 0.05$) visibility improvement on the clearest days for 11–20-year monitoring periods through 2008. Ten parks showed statistically significant

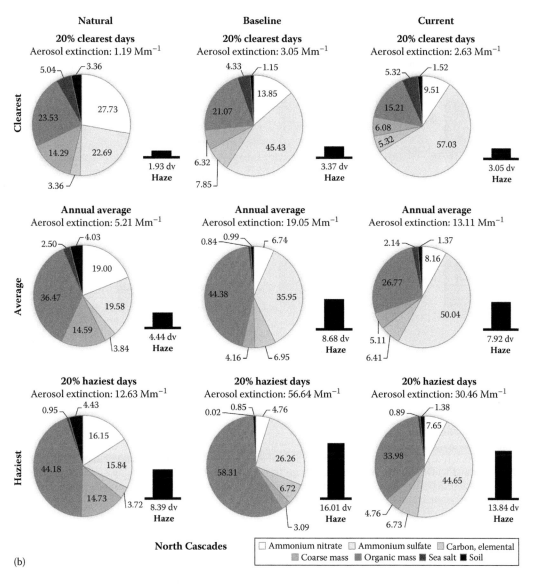

North Cascades

(b)

FIGURE 12.7 (*Continued*)
Estimated natural (preindustrial), baseline (2000–2004), and current (2006–2010) levels of haze (columns) and its composition (pie charts) on the 20% clearest, annual average, and 20% haziest visibility days for (a) MORA, (b) NOCA, and (c) OLYM. The baseline figures for NOCA use the RHR2_VS method of calculation since the years 2003 and 2004 were not valid years for this park. (From http://views.cira.colostate.edu/fed/Tools/RegionalHazeSummary.aspx, accessed October, 2012.) (*Continued*)

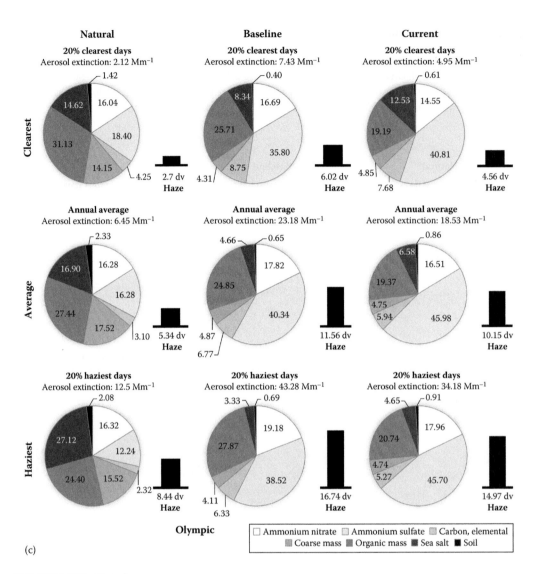

(c)

FIGURE 12.7 (*Continued*)
Estimated natural (preindustrial), baseline (2000–2004), and current (2006–2010) levels of haze (columns) and its composition (pie charts) on the 20% clearest, annual average, and 20% haziest visibility days for (a) MORA, (b) NOCA, and (c) OLYM. The baseline figures for NOCA use the RHR2_VS method of calculation since the years 2003 and 2004 were not valid years for this park. (From http://views.cira.colostate.edu/fed/Tools/RegionalHazeSummary.aspx, accessed October, 2012.)

visibility improvement on the haziest days, with the greatest improvements reported for MORA (−0.38 dv/yr), based on 18 years of monitoring (Figure 12.8). Improvements at OLYM and NOCA were less pronounced and included fewer years of monitoring data (Figure 12.8). The IMPROVE site that showed the largest improvements in visibility for the West Coast region over the two-decade period 1992–2011 on the 20% haziest days was MORA (−3.5%/yr; p ≤ 0.01; Hand et al. 2014).

12.6.4 State of Washington Regional Haze State Implementation Plan

The Washington Department of Ecology published their proposed Regional Haze State Implementation Plan in 2010 (http://www.ecy.wa.gov/programs/air/globalwarm_RegHaze/regional_haze.html). It summarized natural conditions (the 2064 goal of the Clean Air Act) and baseline conditions (2000–2004) for the clearest and haziest visibility days at MORA, NOCA, and OLYM (Table 12.4). The plan presented glideslopes to illustrate how visibility would be gradually improved over time from baseline conditions to

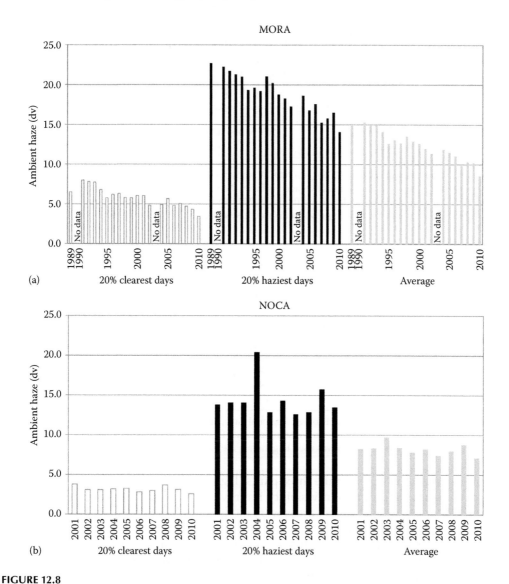

FIGURE 12.8
Trends in monitored visibility at (a) MORA, (b) NOCA, and (c) OLYM, based on IMPROVE measurements on the 20% clearest, 20% haziest, and annual average visibility days over the monitoring period of record. NOCA used the RHR2_VS method of calculation for the years 2003 and 2004. (From http://views.cira.colostate.edu/fed/Tools/FileExplorer.aspx?pathid=ImproveDataWarehouse&title=IMPROVE, accessed October, 2012.)

(Continued)

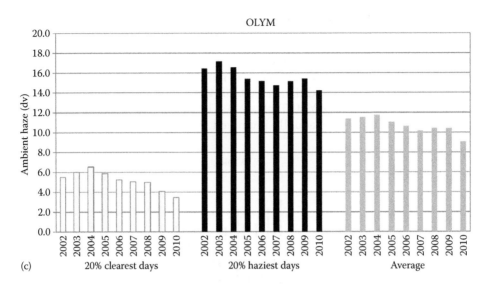

(c)   20% clearest days   20% haziest days   Average

FIGURE 12.8 (*Continued*)
Trends in monitored visibility at (a) MORA, (b) NOCA, and (c) OLYM, based on IMPROVE measurements on the 20% clearest, 20% haziest, and annual average visibility days over the monitoring period of record. NOCA used the RHR2_VS method of calculation for the years 2003 and 2004. (From http://views.cira.colostate.edu/fed/Tools/FileExplorer.aspx?pathid=ImproveDataWarehouse&title=IMPROVE, accessed October, 2012.)

TABLE 12.4

Summary of Visibility Baseline Conditions Averaged over the Period 2000–2004, Natural Background Conditions That Must Be Achieved by 2064 to Comply with the Regional Haze Rule, and Difference between Natural Conditions and the Baseline

National Park	20% Clearest Days (in Deciviews)			20% Haziest Days (in Deciviews)		
	2000–2004 Baseline	2064 Natural Conditions	Difference	2000–2004 Baseline	2064 Natural Conditions	Difference
MORA	5.47	2.56	2.91	18.24	8.54	9.70
NOCA[a]	3.37	1.93	1.44	16.01	8.39	7.62
OLYM	6.02	2.7	3.32	16.74	8.44	8.30

Source: Washington Department of Ecology, Regional haze, State implementation plan, Publication No. 10-02-041, Air Quality Program, Washington State Department of Ecology, Olympia, WA, 2010.
[a] RHR2_VS method of calculation was used to generate the baseline values.

estimated natural conditions in the three parks (Figure 12.9). The State is required to periodically demonstrate progress in visibility improvement. As part of Washington's plan, the contributions to haze at Class I areas were modeled, and results suggested that a wide range of sources were important.

Modeled impacts from major sulfur dioxide emission sources in the state suggested that reducing point source emissions would likely improve visibility at Washington's Class I areas. Emissions of sulfur dioxide in Washington are projected to decline almost 40% from the baseline period (2000–2004) used in the analyses. This projected emissions decline is expected to result from reduced point and mobile source emissions. Washington Department of Ecology (2010) concluded that improved progress toward the natural visibility goal in the future will require emissions reductions from Canadian and Pacific offshore, as well as Washington, sources.

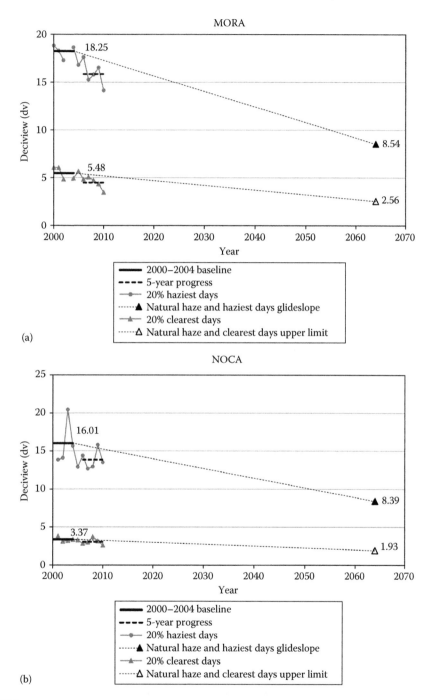

FIGURE 12.9
Glideslopes to achieving natural visibility conditions by 2064 for the 20% haziest (lower line) and the 20% clearest (upper line) days in (a) MORA, (b) NOCA, and (c) OLYM. In the Regional Haze Rule, the clearest days do not have a uniform rate of progress glideslope; the rule only requires that the clearest days do not get any worse than the baseline period. Also shown are measured values during the period 2000–2010. For NOCA, the RHR2_VS method of calculation was used to generate values for the baseline period of 2000–2004. (From http://vista.cira.colostate.edu/improve/Data/IMPROVE/summary_data.htm, accessed October, 2012.)   (*Continued*)

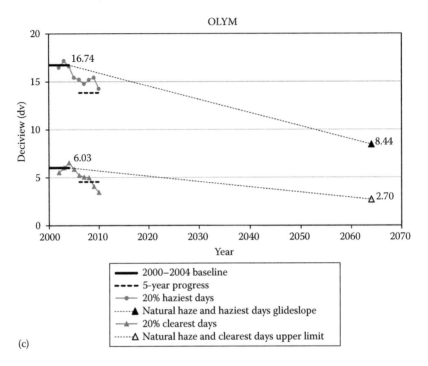

(c)

FIGURE 12.9 (Continued)
Glideslopes to achieving natural visibility conditions by 2064 for the 20% haziest (lower line) and the 20% clearest (upper line) days in (a) MORA, (b) NOCA, and (c) OLYM. In the Regional Haze Rule, the clearest days do not have a uniform rate of progress glideslope; the rule only requires that the clearest days do not get any worse than the baseline period. Also shown are measured values during the period 2000–2010. For NOCA, the RHR2_VS method of calculation was used to generate values for the baseline period of 2000–2004. (From http://vista.cira.colostate.edu/improve/Data/IMPROVE/summary_data.htm, accessed October, 2012.)

12.7 Toxic Airborne Contaminants

Environmental indicators were sampled and analyzed for Hg, pesticides, and other toxics at MORA, NOCA, and OLYM in the WACAP study (Landers et al. 2010). Only ambient air and vegetation were sampled at NOCA. Additional sampling, including fish, was conducted at MORA and OLYM. The WACAP study revealed that there were relatively high concentrations of semivolatile organic compounds in vegetation at NOCA, OLYM, and MORA compared to other western U.S. parks that were part of the WACAP study. Concentrations of all semivolatile organic compounds in NOCA vegetation were above the median for WACAP parks and were similar to those found in MORA and OLYM. The primary semivolatile organic compounds detected in the North Coast and Cascades Network were polycyclic aromatic hydrocarbons, followed by varying concentrations of endosulfans, hexachlorocyclohexane-α, hexachlorobenzenes, and dacthal. There were low concentrations of chlorpyrifos, trifluralin, and PCBs. The OLYM vegetation also had relatively high Hg concentrations. In vegetation at MORA, concentrations were especially high for Hg, polycyclic aromatic hydrocarbons, endosulfans, hexachlorocyclohexane-α, hexachlorobenzene, and dacthal. As forest productivity is relatively high in NOCA, OLYM, and MORA, significant transfer of contaminants to the soil in these parks likely arises from canopy throughfall and needle litterfall (Landers et al. 2008). Based on the analyses conducted in WACAP, current-use pesticide concentrations were generally lowest in surveyed parks in the Pacific Northwest (OLYM, MORA) and Alaska (Denali National Park and Preserve and Noatak National Preserve; Ackerman et al. 2008) compared to other WACAP parks.

Contaminant concentrations detected in snow in MORA were midrange among WACAP study sites. Snow samples collected near PJ Lake on the east side of OLYM had fairly high Hg concentrations compared to Hoh Lake on the west side of OLYM and to sites in other WACAP parks. Because there are few local Hg emission sources around OLYM, and none to the west of the park, most deposited Hg is likely transported to OLYM by wind from source areas to the east (Landers et al. 2008). Sources in Asia are also believed to contribute some Hg deposition to coastal areas in the Pacific Northwest (Landers et al. 2008).

Concentrations of pesticides and other semivolatile organic compounds in fish in MORA were generally midrange among WACAP sites, with the notable exception of polybrominated diphenyl ethers, which were higher in fish from Golden Lake in MORA than in fish from any other WACAP lake. One male fish from Golden Lake had elevated concentrations of vitellogenin, a blood protein that is a useful biomarker for environmental estrogen exposure. Dieldrin concentrations in fish from both MORA study lakes exceeded health thresholds for human subsistence fishers. The Hg levels in fish also exceeded thresholds for piscivorous birds (kingfishers) and mammals (otter and mink), and human health thresholds were exceeded by some fish from Lake LP19 at MORA. As compared to fish collected at other WACAP study lakes, fish in OLYM showed low concentrations of dieldrin, mirex, and chlordanes and average concentrations of other pesticides.

Moran et al. (2007) sampled nonnative cutthroat trout (*Oncorhynchus clarkii*) from lakes in OLYM, MORA, and NOCA for Hg and organochlorine contaminants. Of the 28 organochlorine compounds analyzed, only PCBs and the DDT breakdown product DDE were detected, and they were in low concentrations. Mercury, however, was found in fish tissue from all the lakes sampled. The highest fish tissue Hg concentration was measured in a trout collected from Green Lake in NOCA, well above human and wildlife health thresholds. Evidence from gene expression studies comparing fish from a lake with high contaminant loads to fish from a lake with low contaminant loads led the authors to speculate that Hg plays a role in disrupting reproduction and metabolic pathways in fish from the study lakes and perhaps other high-elevation lakes in western Washington (Moran et al. 2007).

Concentrations of Hg in lake sediments in MORA have increased since the 1900s, with a rapid spike in enrichment near the surface of the sediment, suggesting relatively high present-day Hg inputs. This may be a result of coal-fired power plant emissions, increasing global background levels associated with climate change and/or increasing emissions in Asia and trans-Pacific atmospheric transport of Hg (Landers et al. 2008).

In 1992 and 1994, Davis and Serdar (1996) collected fish from Lake Chelan in NOCA that exceeded human health screening values for some semivolatile organic compounds. Subsequently, the Washington Department of Ecology developed a Total Maximum Daily Load for Lake Chelan to reduce high concentrations of DDT and PCBs in fish (Anderson and Peterschmidt 2008). Lake Chelan has also been sampled for polybrominated diphenyl ether flame retardants in fish, but measured concentrations in fish tissue were low (Johnson et al. 2006).

Furl et al. (2010) documented high concentrations of Hg in largemouth bass (*Micropterus salmoides*) collected from two lakes in and adjacent to OLYM. The two study lakes were judged to be relatively unaffected by local and regional point sources of Hg. However, sediment core data suggested recent increases in sedimentation and Hg flux from the watersheds to the lakes. These results were attributed to erosion from logging, and perhaps road building, in the portions of the study watersheds that are outside park boundaries.

Ozette Lake in OLYM was placed on the Washington State List of Impaired Water Bodies because of high concentrations of Hg in northern pike minnow (Seiders and Deligeannis 2009) and largemouth bass (Furl and Meredith 2010). Fish consumption advisories have been issued for Hoh, Ozette, and PJ lakes in OLYM (Brenkman 2011).

A study by Frenzel et al. (1990) found higher concentrations of heavy metals in lichens at MORA as compared with lichens at OLYM, presumably due to the close proximity of a copper smelter to MORA. At that time, heavy metal concentrations were not in the range considered toxic to plants. However, the authors suggested the potential for greater contamination with increasing industrialization and development in the Puget Sound area (Frenzel et al. 1990).

Hageman et al. (2010) reported the results of pesticide analyses of snowpack at remote alpine, arctic, and subarctic sites in eight national parks, including OLYM, NOCA, and MORA. Various current-use pesticides (dacthal, chlorpyrifos, endosulfans, and γ-hexachlorocyclohexane) and historic-use pesticides (dieldrin, α-hexachlorocyclohexane, chlordanes, and hexachlorobenzene) were measured at OLYM in 2004 and NOCA in 2005. The pesticide concentration profiles were unique for individual parks, suggesting the importance of regional, rather than distant, sources.

The distributions of current-use pesticides among the parks were explained, using mass back trajectory analysis, based on the mass of individual current-use pesticides used in regions located one-day upwind of the parks. For most pesticides and parks, more than 75% of the snowpack pesticide burden was attributed to regional transport. The authors suggested that the majority of pesticide contamination in most U.S. national parks may be due to regional, rather than distant, pesticide applications.

Eagles-Smith et al. (2014) sampled fish in 21 national parks and analyzed them for Hg concentrations in tissue. Results varied substantially by park and by water body. The concentrations of Hg in fish collected from five lakes in OLYM were highly variable. Concentrations in fish from Hoh Lake (253 ng/g wet weight) were more than threefold higher than fish from Gladys Lake (71.5 ng/g wet weight). Model estimates suggested that Hoh Lake fish Hg levels likely approached or exceeded the EPA criterion for protecting human health and the avian benchmark for reproductive impairment in dietary fish longer than 180 mm. These results were similar to those reported by Landers et al. (2008), who also found high Hg concentrations in fish from Hoh Lake. Fish were sampled from 17 locations in MORA, many more than in most other parks in the study. Observed variation in Hg concentrations in fish collected in this park was very high, suggesting that local factors are important in determining the ecological risk of Hg bioaccumulation. Fish were sampled from three locations in NOCA. Concentrations of Hg were generally low in this park. Ecological risk appeared to be low.

Organochlorine chemicals can be estrogenic (Garcia-Reyero et al. 2007), contributing to the occurrence of intersex fish, and can accumulate in the aquatic food chains of remote mountain ecosystems (Blais et al. 1998). Biomarkers of exposure to estrogenic chemicals suggest the likelihood of reproductive dysfunction (Harries et al. 1997). Organochlorines are likely transported as atmospheric contaminants to high-elevation sites (Hageman et al. 2006). Work by Schreck and Kent (2013) followed up on some of the work performed in the WACAP Study. They expanded the range of coverage to additional western parks, including MORA and NOCA within North Coast and Cascades Network. No intersex fish were found in either of the study parks in the North Coast and Cascades Network.

12.8 Summary

The air quality in the North Coast and Cascades Network is generally very good compared to most other areas of the United States. The principal air masses that impact network parks are mainly derived from the atmosphere over the Pacific Ocean, and as a consequence, accumulated air pollutant concentrations are usually relatively low at most park locations. Nevertheless, local and regional sources of air pollutants impact sensitive resources in this network. Pollution sources mainly include motor vehicles; power plants; industrial facilities in the Portland, Seattle/Tacoma/Everett, and Victoria/Vancouver/Chilliwack metro areas, and other cities along the I-5 corridor; and marine shipping traffic. MORA, NOCA, and OLYM are located close to the Puget Sound urban zone.

Air pollutants of most concern in the North Coast and Cascades Network include N compounds and toxic airborne contaminants. Sulfur emissions, which contribute significantly to acidic deposition in the eastern United States, are relatively low upwind from network parks. Nitrogen emissions, which can contribute to both acidification and eutrophication effects, are elevated in some parts of the region. Contributions to total human-caused nitrogen oxides emissions in the network region are much higher from mobile sources than from area sources or point sources. Emissions of ammonia derive mainly from agricultural sources, including animal manure and fertilizer. Airborne contaminants, including pesticides and Hg from regional sources, also affect the parks. Measurements of atmospheric ozone, carbon monoxide, hydrocarbons, and aerosol chemistry in air arriving in the Pacific Northwest from Eurasia have shown that a variety of contaminants, including Hg, can be transported across the Pacific Ocean to the North Coast and Cascades Network region. Such long-range transported contaminants are typically found at relatively low concentrations in network parks (Jaffe et al. 2003).

Although S deposition levels are relatively low in the North Coast and Cascades Network, some aquatic and terrestrial ecosystems in this network are highly sensitive to acidification from the atmospheric deposition of S and N. A coarse screening risk assessment concluded that the overall level of concern for acidification effects on national park units within the North Coast and Cascades Network is very high, relative to other Inventory and Monitoring networks across the country (Sullivan et al. 2011a). Because S and N emissions and deposition in the network region are relatively low, acidification to date has been minimal (U.S. EPA 2009). High-elevation lakes are common within the three parks evaluated here. Each of these parks also contains an extensive network of potentially acid-sensitive high-elevation and low-order streams. Lakes and streams in MORA and NOCA, which are located in the Cascade Mountains, are especially sensitive to impacts from acidic deposition because of the predominance of base-poor bedrock and thin acidic soils, large amounts of precipitation, prevalence of coniferous vegetation, steep slopes, and the dilute nature of the surface waters (Eilers et al. 1994). Each of these characteristics limits the extent to which acidic inputs are neutralized by watershed soils. Substantial portions of the North Coast and Cascades Network contain areas of exposed bedrock, with little soil or vegetative cover to neutralize acidic inputs. As a consequence, surface waters in the Washington Cascade Mountains are among the most poorly buffered in the United States (Landers et al. 1987). Although surface water acidification has not been observed in MORA or NOCA, the high-elevation aquatic habitats in these parks and other areas of the Cascade Mountains provide conditions that may contribute to occasional episodic acidification with potentially lethal consequences to aquatic biota. In contrast to MORA and NOCA, all analyses of lakes and streams in OLYM have shown them to be well buffered and relatively insensitive to stress from acidic deposition, in part due to high rates of weathering and chemical composition of the local geology (Naiman et al. 1986). Comparing acid neutralizing capacity and pH levels between the Cascade and Olympic mountains of Washington illustrates the effect of bedrock type on surface waters. Bedrock in the Olympic Mountains is of marine sedimentary origin, while that in the Cascades is mostly igneous and more resistant to weathering. As a result, acid neutralizing capacity and pH levels in water draining from the Olympic Mountains are higher than in water draining from the Cascade Mountains.

Nitrogen pollutants can cause undesirable nutrient enrichment of natural ecosystems, leading to changes in plant species diversity. Remote oligotrophic waters are commonly N-limited, especially undisturbed northern temperate or boreal lakes that receive low levels of atmospheric N deposition. The overall level of concern for nutrient N enrichment effects on parks within the North Coast and Cascades Network is very high relative to other Inventory and Monitoring networks nationwide (Sullivan et al. 2011b). High-elevation lakes might be more prone than lakes at lower elevation to N limitation (Elser et al. 2009b) and therefore potentially more susceptible to eutrophication in response to atmospheric N input. Lakes in MORA are more likely to be N-limited than lakes in NOCA, due to lower nitrate concentrations in lake water. Eutrophication effects on freshwater ecosystems from the atmospheric deposition of N are of greatest concern in lakes and streams that have very low productivity and nutrient levels.

All three of the larger parks in the North Coast and Cascades Network contain extensive areas of alpine and subalpine vegetation, along with some meadow and wetland vegetation, which are among the predominant vegetation types found to be sensitive to nutrient N enrichment. Air pollution, specifically from N, has caused lichen communities in some areas of the Pacific Northwest to lose biodiversity, but this has not been documented in the national parks. In urban areas, intensive agricultural zones, and areas downwind of major urban and industrial centers in the North Coast and Cascades Network region, there are few air pollution–sensitive lichen species present. Replacement of sensitive lichens by nitrophilous species has undesirable ecological consequences. Lichens make important contributions to mineral cycling and soil fertility and comprise integral parts of the food web that supports mammals, insects, and birds (McCune and Geiser 1997, U.S. EPA 2005). Sensitive lichen species appear to be negatively affected by N inputs as low as 3–9 kg/ha/yr (Geiser et al. 2010). Available empirical data for ecoregions represented in North Coast and Cascades Network parks suggest that the critical load for protecting mycorrhizal fungi, herbaceous plants, and forests and for the prevention of nitrate leaching in various North Coast and Cascades Network parks may be exceeded by ambient N deposition.

Maximum atmospheric ozone concentrations are generally found downwind of urban centers in the Portland and Puget Sound areas and at higher elevations. As illustrated by the IMPROVE data summarized here, the highest weekly average ozone concentrations in western Washington were measured at the highest-elevation sites in MORA. Although ozone is of some concern with respect to air pollution effects on vegetation in national parks in the North Coast and Cascades Network, the overall risk to vegetation does not appear to be high at the present time. The ozone concentrations and overall ozone risk assessment rankings (Kohut 2007) were low across all parks within the network. Among the parklands in the North Coast and Cascades Network, MORA probably faces the highest risk from ozone.

Visibility at the monitored parks in North Coast and Cascades Network is impaired at times, with monitored haze on the 20% haziest days about 5 deciviews (NOCA), 7 deciviews (OLYM), and 8 deciviews (MORA) higher than estimated natural background conditions. Haze levels are generally trending downward and visibility is improving, especially at MORA, on the 20% haziest, 20% clearest, and average days. The majority of haze in the region is attributable to sulfate, largely from sulfur dioxide emissions and organics from fires, agriculture, and other sources.

Airborne toxics can harm humans, fish, and wildlife in the North Coast and Cascades Network. During the WACAP study (Landers et al. 2010), fish and/or other environmental indicators were sampled for Hg, pesticides, and other toxics at MORA, NOCA, and OLYM. Bioaccumulation of Hg was documented in some OLYM and MORA fish at levels that exceeded health thresholds for humans, piscivorous mammals (otter and mink), and piscivorous birds. High Hg concentrations have also been measured in fish sampled in NOCA by the Washington Department of Ecology and U.S. Geological Survey. Reasons for these high levels of Hg contamination in fish caught in the North Coast and Cascades Network parks are unclear but are likely a combination of elevated Hg deposition from both power plants and long-range transport (including from Asia) and also the presence of ecosystems, such as wetlands, that are conducive to Hg methylation, making this pollutant available for uptake by aquatic organisms.

References

Ackerman, L.K., A.R. Schwindt, S.L. Simonich, D.C. Koch, T.F. Blett, C.B. Schreck, M.L. Kent, and D.H. Landers. 2008. Atmospherically deposited PBDEs, pesticides, PCBs, and PAHs in western U.S. National Park fish: Concentrations and consumption guidelines. *Environ. Sci. Technol.* 42:2334–2341.

Anderson, R. and M. Peterschmidt. 2008. Lake Chelan DDT and PCB TMDL: Water quality implementation plan. Publication No. 08-10-048. Washington Department of Ecology, Olympia, WA.

Barna, M.G., B. Lamb, S. O'Neill, H. Westberg, C. Figueroa-Kaminsky, S. Otterson, C. Bowman, and J. DeMay. 2000. Modeling ozone formation and transport in the Cascadia Region of the Pacific Northwest. *J. Appl. Meteorol.* 39:349–366.

Basabe, F.A., R.L. Edmonds, W.L. Chang, and T.V. Larson. 1994. Fog and cloudwater chemistry in western Washington. *In* R.K. Olson and A.S. Lefohn (Eds.). *Effects of Air Pollution on Western Forests*. Air and Waste Management Association's TE-2 Ecological Effects Committee, Pittsburgh, PA, pp. 33–39.

Bergström, A., P. Blomqvist, and M. Jansson. 2005. Effects of atmospheric nitrogen deposition on nutrient limitation and phytoplankton biomass in unproductive Swedish lakes. *Limnol. Oceanogr.* 50(3):987–994.

Blais, J.M., D.W. Schindler, D.C.G. Muir, L.E. Kimpe, D.B. Donald, and B. Rosenberg. 1998. Accumulation of persistent organochlorine compounds in mountains of western Canada. *Nature* 395:585–588.

Brace, S. and D.L. Peterson. 1998. Spatial patterns of tropospheric ozone in the Mount Rainier region of the Cascade Mountains, U.S.A. *Atmos. Environ.* 32(21):3629–3637.

Brakke, D.F. 1984. Chemical surveys of North Cascade lakes. Report submitted to Washington State Department of Ecology. Western Washington University, Bellingham, WA.

Brakke, D.F. 1985. Chemical surveys of North Cascade lakes. Results of 1984 sampling. Report submitted to Washington State Department of Ecology. Western Washington University, Bellingham, WA.

Brenkman, S. 2011. Olympic National Park fish and shellfish regulations. National Park Service. Fisheries Management Division, Port Angeles, WA. Available at: https://www.nps.gov/olym/upload/Olympic-NP-Fish-Regs-2011-updated.pdf (accessed September, 2016).

Campbell, S.J., R. Wanek, and J.W. Coulston. 2007. Ozone injury in west coast forests: 6 years of monitoring. General Technical Report PNW-GTR-722. USDA Forest Service, Pacific Northwest Research Station, Portland, OR.

Clow, D.W. and D.H. Campbell. 2008. Atmospheric deposition and surface-water chemistry in Mount Rainier and North Cascades National Parks, U.S.A., water years 2000 and 2005–2006. Scientific Investigations Report 2008-5152. U.S. Geological Survey, Reston, VA.

Cobb, F.W., Jr. and R.W. Stark. 1970. Decline and mortality of smog injured ponderosa pine. *J. Forest.* 68:147–148.

Davis, D. and D. Serdar. 1996. Washington State Pesticide Monitoring Program: 1994 fish tissue and sediment sampling report. Publication No. 96-352. Washington Department of Ecology, Olympia, WA.

Drever, J.I. and D.R. Hurcomb. 1986. Neutralization of atmospheric acidity by chemical weathering in an alpine drainage basin in the North Cascade Mountains. *Geology* 14:221–224.

Eagles-Smith, C.A., J.J. Willacker, Jr., and C.M. Flanagan Pritz. 2014. Mercury in fishes from 21 national parks in the western United States—Inter- and intra-park variation in concentrations and ecological risk. U.S. Geological Survey Open-File Report 2014-1051, Reston, VA.

Edmonds, R.L. 1980. Litter decomposition and nutrient release in Douglas-fir, red alder, western hemlock, and Pacific silver-fir ecosystems in western Washington. *Can. J. For. Res.* 10:327–337.

Eilers, J.M., P. Kanciruck, R.A. McCord, W.S. Overton, L. Hook, D.J. Blick, D.F. Brakke et al. 1987. Characteristics of lakes in the western United States. Volume II: Data compendium for selected physical and chemical variables. EPA-600/3-86/054b. U.S. EPA, Washington, DC.

Eilers, J.M., C.L. Rose, and T.J. Sullivan. 1994. Status of air quality and effects of atmospheric pollutants on ecosystems in the Pacific Northwest Region of the National Park Service. National Park Service, Air Resources Division, Denver, CO.

Eilers, J.M. and K.B. Vaché. 1998. Lake response to atmospheric and watershed inputs in the Goat Rocks Wilderness, WA. Report prepared for Weyerhaeuser Paper Company, Longview, WA. E&S Environmental Chemistry, Inc., Corvallis, OR.

Ellis, R.A., D.J. Jacob, M.P. Sulprizio, L. Zhang, C.D. Holmes, B.A. Schichtel, T. Blett, E. Porter, L.H. Pardo, and J.A. Lynch. 2013. Present and future nitrogen deposition to national parks in the United States: Critical load exceedances. *Atmos. Chem. Phys.* 13(17):9083–9095.

Elser, J.J., T. Andersen, J.S. Baron, A.-K. Bergström, M. Jansson, M. Kyle, K.R. Nydick, L. Steger, and D.O. Hessen. 2009b. Shifts in lake N:P stoichiometry and nutrient limitation driven by atmospheric nitrogen deposition. *Science* 326:835–837.

Elser, J.J., M. Kyle, L. Steger, K.R. Nydick, and J.S. Baron. 2009a. Nutrient availability and phytoplankton nutrient limitation across a gradient of atmospheric nitrogen deposition. *Ecology* 90(11):3062–3073.

Fenn, M.E., J.S. Baron, E.B. Allen, H.M. Rueth, K.R. Nydick, L. Geiser, W.D. Bowman et al. 2003. Ecological effects of nitrogen deposition in the western United States. *BioScience* 53(4):404–420.

Fradkin, S.C., W. Baccus, R. Glesne, C. Welch, B. Samora, and R. Lofgren. 2012. Mountain lake study sites in the North Coast and Cascades Network. Version 1.1. Natural Resource Data Series NPS/NCCN/NRDS—2012/364.1. National Park Service, Fort Collins, CO.

Frenzel, R.W., G.W. Witmer, and E.E. Starkey. 1990. Heavy metal concentrations in a lichen of Mt. Rainier and Olympic National Parks, Washington USA. *Bull. Environ. Contam. Toxicol.* 44:158–164.

Furl, C. and C. Meredith. 2010. Perfluorinated compounds in Washington rivers and lakes. Publication No. 10-03-034. Washington Department of Ecology, Olympia, WA.

Furl, C.V., J.A. Colman, and M.H. Bothner. 2010. Mercury sources to Lake Ozette and Lake Dickey: Highly contaminated remote coastal lakes, Washington State, USA. *Water Air Soil Pollut.* 208:275–286.

Garcia-Reyero, N., J.O. Grimalt, I. Vives, P. Fernandez, and B. Piña. 2007. Estrogenic activity associated with organochlorine compounds in fish extracts from European mountain lakes. *Environ. Pollut.* 145(3):745–752.

Geiser, L.H., S.E. Jovan, D.A. Glavich, and M.K. Porter. 2010. Lichen-based critical loads for atmospheric nitrogen deposition in western Oregon and Washington forests, USA. *Environ. Pollut.* 158:2412–2421.

Geiser, L.H. and P.N. Neitlich. 2007. Air pollution and climate gradients in western Oregon and Washington indicated by epiphytic macrolichens. *Environ. Pollut.* 145:203–218.

Gibson, J.H., J.N. Galloway, C.L. Schofield, W. McFee, R. Johnson, S. McCorley, N. Dise, and D. Herzog. 1983. Rocky Mountain acidification study. FWS/OBS-80/40.17. U.S. Fish and Wildlife Service, Department of Biological Services Eastern Energy and Land Use Team, Washington, DC.

Gubala, C.P., C.T. Driscoll, R.M. Newton, and C.F. Schofield. 1991. The chemistry of a near-shore lake region during spring snowmelt. *Environ. Sci. Technol.* 25(12):2024–2030.

Hageman, K.J., W.D. Hafner, D.H. Campbell, D.A. Jaffe, D.H. Landers, and S.L.M. Simonich. 2010. Variability in pesticide deposition and source contributions to snowpack in western U.S. national parks. *Environ. Sci. Technol.* 44(12):4452–4458.

Hageman, K.J., S.L. Simonich, D.H. Campbell, G.R. Wilson, and D.H. Landers. 2006. Atmospheric deposition of current-use and historic-use pesticides in snow at national parks in the western United States. *Environ. Sci. Technol.* 40(10):3174–3180.

Hand, J.L., B.A. Schichtel, W.C. Malm, S. Copeland, J.V. Molenar, N. Frank, and M. Pitchford. 2014. Reductions in haze across the United States from the early 1990s through 2011. *Atmos. Environ.* 94:671–679.

Harries, J.E., D.A. Sheahan, S. Jobling, P. Matthiessen, P. Neall, J.P. Sumpter, T. Tylor, and N. Zaman. 1997. Estrogenic activity in five United Kingdom rivers detected by measurement of vitellogenesis in caged male trout. *Environ. Toxicol. Chem.* 16(3):534–542.

Hogsett, W.E., D.T. Tingey, C. Hendricks, and D. Rossi. 1989. Sensitivity of western conifers to SO_2 and season interaction of acid fog and ozone. *In* R.K. Olson and A.S. Lefohn (Eds.). *Transactions, Symposium on the Effects of Air Pollution on Western Forests*, Anaheim, CA, 1989 June. Air and Waste Management Association, Pittsburgh, PA, pp. 469–491.

Jaffe, D., I. McKendry, T. Anderson, and H. Price. 2003. Six 'new' episodes of trans-Pacific transport of air pollutants. *Atmos. Environ.* 37:391–404.

Johnson, A., K. Seiders, C. Deligeannis, K. Kinney, P. Sandvik, B. Era-Miller, and D. Alkire. 2006. PBDE flame retardants in Washington rivers and lakes: Concentrations in fish and water, 2005–06. Publication No. 06-03-027. Washington Department of Ecology, Olympia, WA.

Jovan, S. and B. McCune. 2005. Air quality bioindication in the greater Central Valley of California, with epiphytic macrolichen communities. *Ecol. Appl.* 15(5):1712–1726.

Kohut, R. 2007. Assessing the risk of foliar injury from ozone on vegetation in parks in the U.S. National Park Service's Vital Signs Network. *Environ. Pollut.* 149:348–357.

Landers, D.H., J.M. Eilers, D.F. Brakke, W.S. Overton, P.E. Kellar, W.E. Silverstein, R.D. Schonbrod et al. 1987. Characteristics of lakes in the western United States. Volume I: Population descriptions and physico-chemical relationships. EPA-600/3-86/054a. U.S. Environmental Protection Agency, Washington, D.C.

Landers, D.H., S.M. Simonich, D. Jaffe, L. Geiser, D.H. Campbell, A. Schwindt, C.B. Schreck et al. 2010. The Western Airborne Contaminant Project (WACAP): An interdisciplinary evaluation of the impacts of airborne contaminants in western U.S. National Parks. *Environ. Sci. Technol.* 44(3):855–859.

Landers, D.H., S.L. Simonich, D.A. Jaffe, L.H. Geiser, D.H. Campbell, A.R. Schwindt, C.B. Schreck et al. 2008. The fate, transport, and ecological impacts of airborne contaminants in western national parks (USA). EPA/600/R-07/138. U.S. Environmental Protection Agency, Office of Research and Development, NHEERL, Western Ecology Division, Corvallis, OR.

Larson, G.L., G. Lomnicky, R. Hoffman, W.J. Liss, and E. Deimling. 1999. Integrating physical and chemical characteristics of lakes into the glacially influenced landscape of the northern Cascade Mountains, Washington State, USA. *Environ. Manage.* 24(2):219–228.

Larson, G.L., A. Wones, C.D. McIntire, and B. Samora. 1994. Integrating limnological charac-
    teristics of high mountain lakes into the landscape of a natural area. *Environ. Manage.*
    18(6):871–888.

Liss, W.J., E.K. Deimling, R. Hoffman, G.L. Larson, G. Lomnicky, C.D. McIntire, and R. Truitt. 1991.
    Annual report 1990–1991. Ecological effects of stocked fish on naturally fishless high mountain
    lakes: North Cascade National Park Service Complex. Oregon State University, Corvallis, OR.

McCune, B. and L. Geiser. 1997. *Macrolichens of the Pacific Northwest*. Oregon State University Press,
    Corvallis, OR.

Moran, P.W., N. Aluru, R.W. Black, and M.M. Vijayan. 2007. Tissue contaminants and associated
    transcriptional response in trout liver from high elevation lakes of Washington. *Environ. Sci.
    Technol.* 41:6591–6597.

Naiman, R., R.E. Bilby, and S. Kantor. 1986. *River Ecology and Management: Lessons from the Pacific
    Coastal Region*. Springer-Verlag, New York.

National Park Service (NPS). 2010. Air quality in national parks: 2009 Annual performance and
    progress report. Natural Resource Report NPS/NRPC/ARD/NRR-2010/266. National Park
    Service, Air Resources Division, Denver, CO.

Nelson, P.O. 1991. Cascade mountains: Lake chemistry and sensitivity of acid deposition. *In* D.F.
    Charles (Ed.). *Acidic Deposition and Aquatic Ecosystems: Regional Case Studies*. Springer-Verlag,
    New York, pp. 531–563.

Panek, J.A. and A.H. Goldstein. 2001. Response of stomatal conductance to drought in ponderosa
    pine: Implications for carbon and ozone uptake. *Tree Physiol.* 21:337–344.

Pardo, L.H., M.J. Robin-Abbott, and C.T. Driscoll. 2011. Assessment of nitrogen deposition effects
    and empirical critical loads of nitrogen for ecoregions of the United States. General Technical
    Report NRS-80. U.S. Forest Service, Newtown Square, PA.

Peterson, D.L., D. Bowers, and S. Brace. 1999. Tropospheric ozone in the Nisqually River Drainage,
    Mount Rainier National Park. *Northwest Sci.* 73(4):241–254.

Pike, L.H. 1978. The importance of epiphytic lichens in mineral cycling. *Bryologist* 81:247–257.

Schoettle, A.W., K. Tonnessen, J. Turk, J. Vimont, and R. Amundson. 1999. An assessment of the
    effects of human-caused air pollution on resources within the interior Columbia River basin.
    PNW-GTR-447. USDA Forest Service, Pacific Northwest Research Station, Portland, OR.

Schreck, C.B. and M. Kent. 2013. Extent of endocrine disruption in fish of western and Alaskan
    national parks. Final report, NPS-OSU Task Agreement J8W07080024. Oregon State University,
    Corvallis, OR.

Schwede, D.B. and G.G. Lear. 2014. A novel hybrid approach for estimating total deposition in the
    United States. *Atmos. Environ.* 92:207–220.

Seiders, K. and C. Deligeannis. 2009. Washington state toxics monitoring program: Freshwater
    fish tissue component, 2007. Publication No. 09-03-003. Washington Department of Ecology,
    Olympia, WA.

Sheibley, R., M. Enache, P. Swarzenski, P. Moran, and J. Foreman. 2014. Nitrogen deposition effects
    on diatom communities in lakes from three national parks in Washington state. *Water Air Soil
    Pollut.* 225(2):1–23.

Sollins, P., C.C. Grier, F.M. McCorison, K. Cromack, Jr., R. Fogel, and R.L. Fredriksen. 1980. The inter-
    nal element cycles of an old-growth Douglas-fir ecosystem in western Oregon. *Ecol. Monogr.*
    50:261–285.

Suarez-Murias, T., J. Glass, E. Kim, L. Melgoza, and T. Najita. 2009. California regional haze plan.
    California Environmental Protection Agency, Air Resources Board, Sacramento, CA.

Sullivan, T.J. 2000. *Aquatic Effects of Acidic Deposition*. CRC Press, Boca Raton, FL.

Sullivan, T.J. and J.M. Eilers. 1994. Assessment of deposition levels of sulfur and nitrogen required to
    protect aquatic resources in selected sensitive regions of North America. Final report prepared
    for Technical Resources, Inc., Rockville, MD, under contract to U.S. Environmental Protection
    Agency, Environmental Research Laboratory-Corvallis. E&S Environmental Chemistry, Inc.,
    Corvallis, OR.

Sullivan, T.J., T.C. McDonnell, G.T. McPherson, S.D. Mackey, and D. Moore. 2011b. Evaluation of the sensitivity of inventory and monitoring national parks to nutrient enrichment effects from atmospheric nitrogen deposition. Natural Resource Report NPS/NRPC/ARD/NRR—2011/313. U.S. Department of the Interior, National Park Service, Denver, CO.

Sullivan, T.J., G.T. McPherson, T.C. McDonnell, S.D. Mackey, and D. Moore. 2011a. Evaluation of the sensitivity of inventory and monitoring national parks to acidification effects from atmospheric sulfur and nitrogen deposition. Natural Resource Report. NPS/NRPC/ARD/NRR—2011/349. U.S. Department of the Interior, National Park Service, Denver, CO.

Turner, R.S., R.B. Cook, H. van Miegroet, D.W. Johnson, J.W. Elwood, O.P. Bricker, S.E. Lindberg, and G.M. Hornberger. 1990. Watershed and lake processes affecting chronic surface water acid-base chemistry. State of the Science, SOS/T 10. National Acid Precipitation Assessment Program, Washington, DC.

U.S. Environmental Protection Agency. 2005. Review of the national ambient air quality standards for particulate matter: Policy assessment of scientific and technical information. OAQPS Staff Paper. EPA-452/R-05-005a. Office of Air Quality Planning and Standards, Research Triangle Park, NC.

U.S. Environmental Protection Agency. 2009. Risk and exposure assessment for review of the secondary national ambient air quality standards for oxides of nitrogen and oxides of sulfur: Final. EPA-452/R-09-008a. Office of Air Quality Planning and Standards, Health and Environmental Impacts Division, Research Triangle Park, NC.

van Dobben, H.F., T. Wolterbeek, G.W.W. Wamelink, and C.J.F. Ter Braak. 2001. Relationship between epiphytic lichens, trace elements and gaseous atmospheric pollutants. *Environ. Pollut.* 112:163–169.

Washington Department of Ecology. 2010. Regional haze. State implementation plan. Publication No. 10-02-041. Air Quality Program, Washington State Department of Ecology, Olympia, WA.

Wigington, P.J., Jr., T.D. Davies, M. Tranter, and K.N. Eshleman. 1990. Episodic acidification of surface waters due to acidic deposition. State of Science/Technology Report 12. National Acid Precipitation Assessment Program, Washington, DC.

13

Klamath Network

13.1 Background

There are six parks in the Klamath Network: Crater Lake National Park (CRLA), Lassen Volcanic National Park (LAVO), Lava Beds National Monument (LABE), Oregon Caves National Monument (ORCA), Redwood National Park (REDW), and Whiskeytown National Recreation Area (WHIS). Crater Lake NP, LABE, LAVO, and REDW are Class I air quality areas and receive special protection against air quality impacts under the Clean Air Act. These later four parks are the subjects of this analysis. The Klamath Network contains three parks larger than 100 mi^2 (259 km^2): CRLA, LAVO, and REDW. Larger parks generally have more available data with which to evaluate air pollution sensitivities and effects. In addition, the larger parks generally contain more extensive resources in need of protection against the adverse impacts of air pollution. Since air pollutants can be widespread, reduction of pollution emissions affecting large parks will often result in the protection of smaller parks located nearby as well.

Air quality in the Klamath Network is generally very good compared to most other areas of the United States (Eilers et al. 1994). The principal air masses that affect this region are largely derived from the atmosphere over the Pacific Ocean, and as a consequence, accumulated air pollutant levels tend to be low. Emissions of air pollutants within the Klamath Network region are also generally low (Sullivan et al. 2001).

There are no population centers in the Klamath Network region larger than 500,000 people and only two larger than 100,000. However, within a 300 mi (483 km) radius around the network boundary are the large cities of San Francisco, San Jose, and Sacramento to the south, and Portland and Seattle to the north. Figure 13.1 shows the Klamath Network region, indicating locations of each park and of population centers with more than 10,000 people.

Given REDW's location in remote coastal northern California, combined with the prevailing northwesterly winds, the park is generally upwind of most emissions sources. A portion of the monitored S deposition at REDW is derived from natural marine aerosols. Lassen Volcanic NP is located further inland than REDW in northern California, but is also relatively remote from urban areas. Nevertheless, emissions from the Sacramento Valley may affect air quality in LAVO (Sullivan et al. 2001). Lava Beds NM is also located in a remote region of northern California. Substantial transport of air pollutants to LABE from adjacent, more heavily developed areas in California is not expected (Sullivan et al. 2001). Based on a review of National Park Service Class I areas in California, Sullivan et al. (2001) concluded that there were no significant air pollution concerns in LABE or REDW at that time. Monitored and estimated pollutant exposures were uniformly low, and there were no known or suspected adverse effects on sensitive air quality-related values.

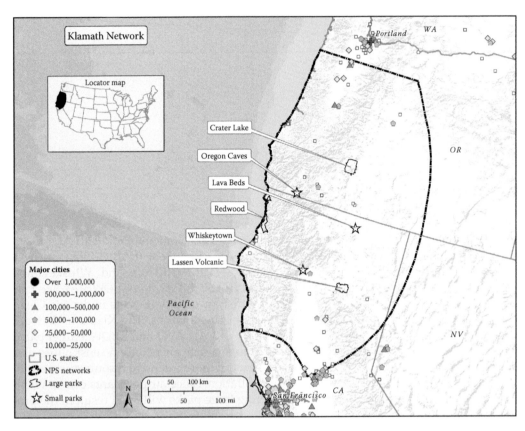

FIGURE 13.1
Network boundary and locations of parks and population centers greater than 10,000 people within the Klamath
Network region.

13.2 Atmospheric Emissions and Deposition

County-level S emissions within and near the Klamath Network region reported by
Sullivan et al. (2011a) were generally less than 1 ton of sulfur dioxide per square mile
per year, with only two counties having higher emissions, in the range 1–5 tons/mi^2/yr.
There were no individual sulfur dioxide point sources larger than 5000 tons/yr within the
Klamath Network region. The majority of the relatively small sulfur dioxide point sources
were located in the Central Valley of California in the southern half of the network region.

County-level N emissions within the region ranged from less than 1 ton/mi^2/yr to
between 5 and 20 tons/mi^2/yr. There were no individual large N point sources in the
region (Sullivan et al. 2011a). The major sources of N emissions in the western United
States, including the Klamath Network region, are transportation, agriculture, power
plants and industry. Trans-Pacific transport of N from southeast Asia is also a source, but
the amount is uncertain (Fenn et al. 2003b). According to data compiled by the state of
California and the EPA's National Emissions Inventory database, California was the only
western state that had a decrease in oxidized N emissions during the period 1985–1999
(Fenn et al. 2003b). Since 1999, oxidized N emissions have decreased further.

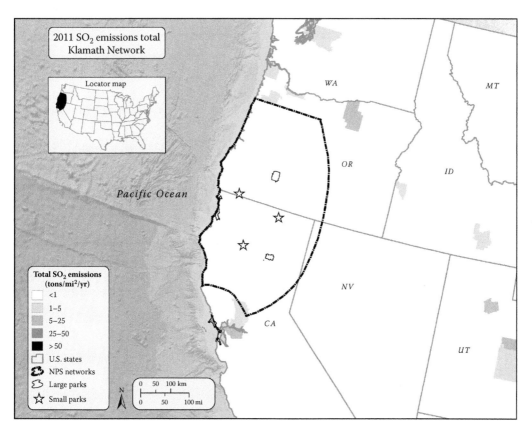

FIGURE 13.2
Total sulfur dioxide (SO$_2$) emissions, by county, near the Klamath Network for the year 2011. (Data from EPA's National Emissions Inventory, https://www.epa.gov/air-emissions-inventories, accessed January, 2014.)

County-level emissions near the Klamath Network, based on data from the EPA's National Emissions Inventory during a recent time period (2011), are depicted in Figures 13.2 through 13.4 for sulfur dioxide, oxidized N, and reduced N (ammonia), respectively. Throughout the network region, sulfur dioxide emissions were <1 ton/mi^2/yr. Spatial patterns in oxidized N emissions showed somewhat higher values, in many places between 1 and 5 tons/mi^2/yr. Emissions of ammonia near Klamath Network parks were between 0.5 and 2 tons/mi^2/yr at many locations.

Emissions of N and S in Asia are higher than in North America (Fenn et al. 2003b) and have increased dramatically in recent decades (Akimoto and Narita 1994). Emissions of N can be transported across the Pacific Ocean from Asia as peroxyacetyl nitrate, which has a lifetime in the atmosphere of days to weeks. Because of the comparatively short lifetime of oxidized N in the atmosphere (generally less than 24 hours), conversion of oxidized N to a longer-lived form (such as peroxyacetyl nitrate) is necessary for long-range transport of N. Once in the troposphere, peroxyacetyl nitrate can be transported long distances over a period of many days and then moved back into the boundary layer of the atmosphere, where it regenerates oxidized N and nitric acid (Kotchenruther et al. 2001). Long-range transport of peroxyacetyl nitrate from Asia to North America is well documented (Jaffe et al. 1999). Thus, it is likely that oxidized N emissions in Asia contribute some, albeit an unknown amount of, N deposition to the western United States, including the Klamath

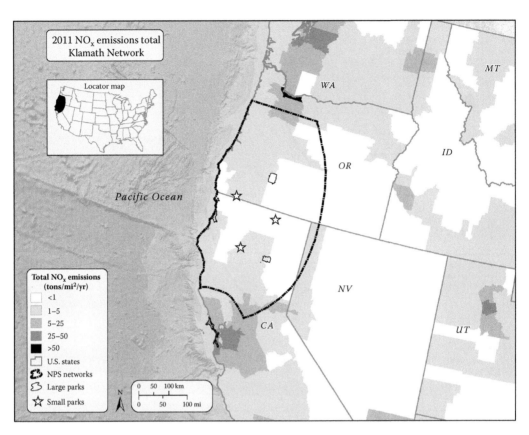

FIGURE 13.3
Total oxidized N (NO$_x$) emissions, by county, near the Klamath Network for the year 2011. (Data from EPA's National Emissions Inventory, https://www.epa.gov/air-emissions-inventories, accessed January, 2014.)

Network. If Asian oxidized N emissions continue to increase as predicted (cf., Streets and Waldhoff 2000, Klimont et al. 2001), trans-Pacific sources of N may become more important to west coast locations of the United States in the future.

Total modeled S deposition throughout most of the Klamath Network region was less than 2 kg S/ha/yr in 2002 (Sullivan et al. 2011a). Some portions of the Klamath Network region, mainly along the coastline, had S deposition values in the range of 2–5 kg S/ha/yr. Part of the coastal S deposition is undoubtedly of marine origin. Total modeled N deposition ranged from less than 2 kg N/ha/yr in the eastern portion of the Klamath Network region to as high as 5–10 kg N/ha/yr at scattered locations (Sullivan et al. 2011a). Atmospheric deposition in the Sierra Nevada and Cascade Mountains is largely influenced by local emissions (Sullivan et al. 2001).

Atmospheric S deposition levels have increased at all Klamath Network parks since 2001, based on Total Deposition project estimates (Table 13.1). Increases in total S deposition over the previous decade for the parks in this network were in several cases larger than 30% but still remained relatively low (<2 kg S/ha/yr). Estimated total N deposition over that same time period decreased at some parks and increased at others. Ammonium deposition increased at all parks in the network, in most cases by more than 30%. Oxidized and reduced N showed opposite patterns, with oxidized N decreasing in all parks, usually by more than 20%, since the monitoring period 2000–2002.

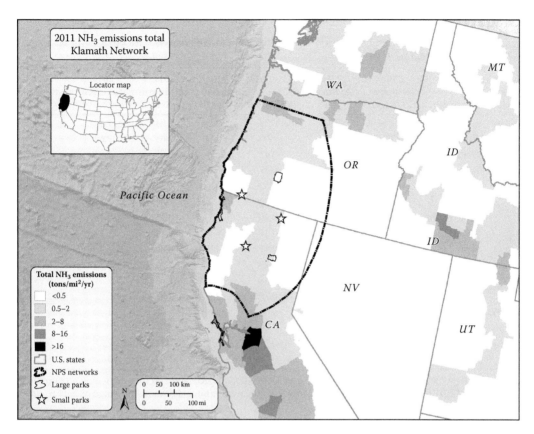

FIGURE 13.4

Total ammonia (NH$_3$) emissions, by county, near the Klamath Network for the year 2011. (Data from EPA's National Emissions Inventory, https://www.epa.gov/air-emissions-inventories, accessed January, 2014.)

Total S deposition for the period 2010–2012 was low (<2 kg S/ha/yr) at park locations within the Klamath Network. Some areas in the network region received slightly higher (2–5 kg N/ha/yr) deposition of total N. Oxidized inorganic N deposition for the period 2010–2012 was less than 2 kg N/ha/yr throughout most of the parklands within Klamath Network. Most areas also received less than 2 kg N/ha/yr of reduced inorganic N from atmospheric deposition during this same period; some areas to the south of LAVO received higher amounts. Total N deposition was less than 5 kg N/ha/yr at most park locations (Figure 13.5).

The WACAP (Landers et al. 2010) studied airborne contaminants in a number of national parks of the western United States from Alaska to southern California from 2002 to 2007, including two parks in the Klamath Network (CRLA and LAVO). Pollutant concentrations detected in the air at CRLA were moderate compared with other WACAP parks for most semivolatile organic compounds detected. These included dacthal, endosulfans, hexachlorocyclohexane-γ, hexachlorocyclohexane-α, and polycyclic aromatic hydrocarbons (PAHs). Concentrations of hexachlorobenzene were relatively high (Landers et al, 2010). Except for PAHs, all SOCs detected in air in LAVO (trifluralin, dacthal, hexachlorocyclohexane-γ, endosulfans, hexachlorobenzene, hexachlorocyclohexane-α, and chlordanes) were above median values for WACAP parks.

TABLE 13.1

Average Changes in S and N Deposition[a] between 2001 and 2011 across Park Grid Cells at Selected Klamath Network Parks

Park Code	Parameter	2001 Average (kg/ha/yr)	2011 Average (kg/ha/yr)	Absolute Change (kg/ha/yr)	Percent Change	2011 Minimum (kg/ha/yr)	2011 Maximum (kg/ha/yr)	2011 Range (kg/ha/yr)
CRLA	Total S	0.68	0.91	0.23	34.5	0.54	1.19	0.66
	Total N	2.88	2.94	0.06	1.6	2.04	3.68	1.64
	Oxidized N	1.59	1.17	−0.42	−26.3	0.87	1.44	0.57
	Reduced N	1.29	1.76	0.47	36.0	1.17	2.40	1.23
LABE	Total S	0.27	0.29	0.02	6.9	0.23	0.58	0.35
	Total N	1.77	1.56	−0.22	−10.3	1.34	2.33	0.99
	Oxidized N	1.07	0.68	−0.40	−37.2	0.55	1.11	0.56
	Reduced N	0.70	0.88	0.18	36.0	0.74	1.22	0.48
LAVO	Total S	1.16	1.27	0.11	14.0	0.76	1.73	0.97
	Total N	4.46	4.75	0.29	9.6	2.90	6.23	3.33
	Oxidized N	2.34	1.86	−0.48	−19.1	1.33	2.31	0.98
	Reduced N	2.11	2.89	0.78	43.1	1.57	3.92	2.35
REDW	Total S	0.86	1.42	0.56	65.5	1.15	1.70	0.55
	Total N	3.90	4.30	0.40	10.7	3.26	5.73	2.47
	Oxidized N	2.19	1.82	−0.37	−16.1	1.46	2.21	0.76
	Reduced N	1.71	2.48	0.76	45.2	1.80	3.76	1.95

[a] Deposition estimates were determined by the Total Deposition project, based on three-year averages centered on 2001 and 2011 for all ~4 km grid cells in each park. The minimum, maximum, and range of 2011 S and N deposition within each park are also shown.

13.3 Acidification

Sullivan et al. (2011a) conducted a coarse screening assessment of relative risk of acidification for all Inventory and Monitoring networks nationwide. The network rankings for acid pollutant exposure, ecosystem sensitivity to acidification, and park protection yielded an overall acidification summary risk ranking for the Klamath Network that was near the median among all networks. The overall level of concern for acidification effects on Klamath Network parks was judged to be moderate.

Acid pollutant exposure for Klamath Network parks ranked from low to very low (Table 13.2). Ecosystem sensitivity to acidification was more variable, with CRLA and LAVO ranking in the highest quintile (very high) and REDW in the second highest quintile (high), but LABE ranking in the lowest quintile (very low; Table 13.2). Differences among parks in ecosystem sensitivity to acidification were partly due to the presence of high-elevation lakes and streams in some of the parks (Sullivan et al. 2011a). Land slope also often influences the degree of acid neutralization provided by soils and bedrock within the watershed. Average slope in CRLA and LAVO ranges from 10° to 40°. REDW is predominately in the 30° to 40° range. In contrast, LABE has very low relief, with slopes less than 10° throughout the park. The occurrence of high-elevation lakes and streams in CRLA and LAVO contributes to acidification sensitivity in these parks. While rankings are an indication of risk, park-specific data, particularly data on ecosystem sensitivity, are needed to fully evaluate risk from acidification.

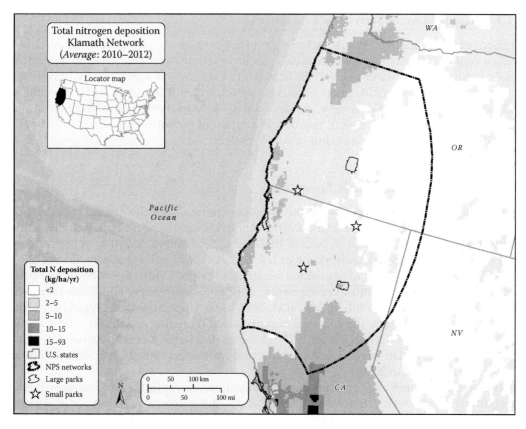

FIGURE 13.5

Total N deposition for the three-year period centered on 2011 in and around Klamath Network. (From Schwede, D.B. and Lear, G.G., *Atmos. Environ.*, 92, 207, 2014.)

TABLE 13.2

Estimated Inventory and Monitoring Park Rankings[a] according to Risk of Acidification Impacts on Sensitive Receptors

Park Name	Park Code	Estimated Acid Pollutant Exposure	Estimated Ecosystem Sensitivity to Acidification
Crater Lake	CRLA	Low	Very High
Lassen Volcanic	LAVO	Low	Very High
Lava Beds	LABE	Very Low	Very Low
Redwood	REDW	Low	High

Source: Sullivan, T.J. et al., Evaluation of the sensitivity of inventory and monitoring national parks to acidification effects from atmospheric sulfur and nitrogen deposition, U.S. Department of the Interior, National Park Service, Denver, CO, 2011a, available at: http://nature.nps.gov/air/Permits/ARIS/networks/acidification-eval.cfm.

[a] Relative park rankings are designated according to quintile ranking, among all Inventory and Monitoring Parks, from the lowest quintile (very low risk) to the highest quintile (very high risk).

13.3.1 Acidification of Terrestrial Ecosystems

Portions of the Klamath Network, including some of the high-elevation areas of LAVO and CRLA, contain areas of exposed bedrock, with little soil or vegetative cover to neutralize acidic inputs (Sullivan and Eilers 1994). It is likely that soils in some areas within these parks are sensitive to acidification. However, monitored and modeled acidic deposition to Klamath Network parks is very low and is not likely to increase substantially in the near future. Therefore, the overall risk of acidification of terrestrial resources in Klamath Network parks seems to be small but is poorly understood.

13.3.2 Acidification of Aquatic Ecosystems

The Sierra Nevada and Cascades are the mountain ranges with the greatest numbers of acid-sensitive aquatic resources in the western United States (Landers et al. 1987). Portions of the Klamath Network occur in both of these mountain ranges. Lakes (and presumably also streams) in mountainous portions of the Klamath Network are especially sensitive to potential impacts from acid deposition because of the predominance of base-poor bedrock, thin soils, large amounts of precipitation, coniferous vegetation, and dilute nature of the surface water (McColl 1981, Melack et al. 1985, Melack and Stoddard 1991). The hydrologic cycle is dominated by the annual accumulation and melting of a dilute, mildly acidic snowpack (Nelson 1991, Melack et al. 1997). In most lakes in the Klamath Network region, concentrations of sulfate are low, although natural watershed sources of S are substantial in some cases (Sullivan 2000).

High-elevation lakes are considered potentially more susceptible to atmospheric S and N input than are low-elevation lakes. Crater Lake is at high elevation and LAVO contains numerous high-elevation lakes. Both parks also contain a moderate length of first-through third-order streams at high elevation (Sullivan et al. 2011a). Low-order, high-elevation streams on steep terrain are often particularly sensitive to acidification impacts from both S and N deposition.

The available information on acid–base chemistry of surface waters in LAVO and CRLA is based mostly on synoptic data from the Western Lakes Survey (Landers et al. 1987) and some more localized studies. Data from intensive study sites elsewhere in the mountainous West suggested that episodic depression of stream pH may be more pronounced than that of lakes. There are no data with which to assess the sensitivity of streams to acid deposition in Klamath Network parks.

Crater Lake is well buffered from changes in acid–base chemistry, and acidification from atmospheric sources is not a concern (Eilers et al. 1994). Lassen Volcanic NP contains over 200 lakes and ponds and 15 perennial streams. Most of the lakes in the park are less than 1 ha in area, and many of those contain water only in spring and early summer. Lakes in the park are generally of glacial origin, although a few of the smaller ones are found in the craters of extinct volcanoes. Inlet and outlet streams are generally swift and intermittent, flowing primarily during the spring runoff period. Some lake and stream waters in LAVO are almost devoid of any buffering capacity (Sullivan et al. 2001). They are especially sensitive to acid deposition. Even moderate amounts of N or S deposition could lead to chronic, and especially episodic, acidification, with significant adverse effects on in-lake and in-stream biota (Sullivan et al. 2001). Current modeled and monitored acid deposition levels are very low, however, and are not likely to have caused adverse impacts to aquatic ecosystems. It appears that chronic acidification has not occurred to any significant degree.

It is possible, however, that limited episodic effects may occur under current deposition regimes (Sullivan et al. 2001).

The Western Lakes Survey sampled seven lakes in the 1980s within LAVO, plus an additional nine lakes within 25 km of the park (Landers et al. 1987). Five of the seven LAVO lakes had acid neutralizing capacity <50 µeq/L, and four of those had acid neutralizing capacity values between 18 and 27 µeq/L. These four lakes are highly sensitive to potential acidification. Fewer of the lakes sampled outside the park were highly sensitive, although there were two lakes with acid neutralizing capacity <50 µeq/L. The sensitive lakes were small (<8 ha) and had small watershed areas (between 31 and 114 ha). All had pH approximately equal to 6.5 and dissolved organic carbon ≤2.1 mg/L. Sulfate concentrations in the sensitive lakes ranged from 2 to 6 µeq/L, confirming the very low levels of S deposition that LAVO receives. Lake water nitrate concentrations were also very low, less than 1 µeq/L in all of the lakes sampled in and around the park. Concentrations of calcium and other base cations were very low (sum of base cation concentrations <35 µeq/L) in the most sensitive lakes.

Data from over 12,500 streams and lakes were used by the Critical Loads of Atmospheric Deposition Science Committee of the National Acid Deposition Program (http://nadp. sws.uiuc.edu/committees/clad/) to develop steady-state critical loads for the acidity of surface waters based on multiple approaches for estimating base cation weathering. Water quality data were obtained from a variety of sources including the EPA's Long Term Monitoring sites, lake surveys, Environmental Monitoring and Assessment Program assessments, and National Stream Surveys; U.S. Geological Surveys; National Park Service Vital Signs program; and U.S. Forest Service air program. The average water quality measurements from the most recent five years of data were used for sites with long-term water monitoring records. The Critical Loads of Atmospheric Deposition database included critical load estimates for seven sites in LAVO. The median estimate was 0 eq S/ha/yr, suggesting a high degree of acid sensitivity in the surface waters of this park.

A survey of water chemistry in 101 high-altitude lakes in seven national parks in the western United States was conducted by the U. S. Geological Survey during the fall of 1999 (Clow et al. 2000). Seventy-two of the lakes had previously been sampled as part of the Western Lakes Survey, including the seven lakes in LAVO. Water chemistry was generally similar to the results of the Western Lakes Survey. There was not a consistent pattern of change in acid neutralizing capacity between the two surveys. Nitrate concentration was 0 µeq/L in all of the lakes surveyed in LAVO by the U.S. Geological Survey.

On the whole, the lakes sampled by the Western Lakes Survey and the U.S. Geological Survey within LAVO reflected slightly greater acid sensitivity than did the lakes sampled further south. Sampled lakes in LAVO are among the most acid sensitive in the National Park system (Sullivan et al. 2001). In view of the extreme sensitivity of many of the lakes that have been studied, relatively little research has been conducted in LAVO on the response of aquatic ecosystems to atmospheric S and/or N deposition. More complete characterization of these resources is needed to better establish baseline conditions and determine if and when future impacts might occur.

Surface water resources are lacking at LABE, so air pollution impacts on water quality are not a concern at that park (Sullivan et al. 2001). In REDW, streams are moderately dilute, and some may be somewhat sensitive to acidification. Atmospheric inputs of N and S are very low, however, and there appears to be no real risk of future adverse aquatic effects unless acidic deposition increases substantially (Sullivan et al. 2001).

13.3.2.1 Episodic Acidification

Decreases in pH with increases in flow are common in drainage waters throughout the United States, especially in mountainous terrain such as is prevalent in the Klamath Network region (Wigington et al. 1990). Chemical changes during episodes are controlled in part by acidic deposition and in part by natural processes. Episodic acid pulses may last for hours to weeks and sometimes result in depletion of acid neutralizing capacity in acid-sensitive streams and lakes, such as those that occur in LAVO, to negative values with concomitant increases in inorganic aluminum in solution to toxic levels. Limited data collected during snowmelt suggest that spring concentrations in surface waters of sulfate, nitrate, and dissolved organic carbon throughout most of the Klamath Network are sometimes several times higher than concentrations in the fall (e.g., Reuss et al. 1995).

In the Klamath Network, episodic acidification has the potential to be an especially important issue for surface waters throughout high-elevation areas. A number of factors predispose watersheds in LAVO and, to a lesser extent, CRLA to potential episodic effects (Peterson et al. 1998, Sullivan 2000), including

1. The abundance of dilute to ultra-dilute lakes that exhibit very low concentrations of base cations, and therefore acid neutralizing capacity, throughout the year

2. Large seasonal snowpack accumulations at the high-elevation sites, thus causing substantial episodic acidification via the natural process of base cation dilution

3. Short hydraulic retention times for many of the high-elevation drainage lakes, thus enabling snowmelt to rapidly flush lake basins with highly dilute meltwater

The weight of evidence suggests that high-elevation lakes in the Klamath Network receive N deposition sufficiently high to cause some minimal nitrate leaching and likely some degree of associated acidification (Sullivan 2000). However, the effects, if they occur, are not thought to be large, and the data that would be needed for determining the extent and magnitude of episodic acidification have not been collected to a sufficient degree in acid-sensitive areas of the Klamath Network to support the regional assessment of episodic acidification (Sullivan 2000).

13.3.2.2 Effects on Aquatic Biota

If there are any effects of acidification on aquatic biota in the Klamath Network, they most likely occur episodically. Given the very low levels of S and N deposition, however, such biological effects are unlikely at the present time.

Serious declines and extirpation of frogs and toads have been documented in many areas of the United States. National parks in California have been affected, including LAVO. Possible causes for these declines include acid deposition, disease, habitat loss, increased ultraviolet light, or a combination of such factors (Sullivan et al. 2001). The decline and extinction of amphibians has also apparently been aggravated in many parks by the introduction of nonnative fish. Historically, it was possible for amphibians to use streams as dispersal corridors. However, the presence of introduced fish in many of the Sierran and Cascade lakes and streams has greatly reduced or eliminated this possibility (Bradford et al. 1993).

Redwood NP contains an abundant and diverse amphibian fauna, which could be sensitive to acidification. This park also contains an important anadromous fisheries resource, which is highly sensitive to acidification damage (Sullivan et al. 2001), but deposition levels are too low to cause substantial impact.

13.4 Nutrient Nitrogen Enrichment

The parks in the Klamath Network contain a wide diversity of habitats from wetlands and wet forests along the Pacific coast to desert conditions inland. Sullivan et al. (2011b) conducted a coarse screening assessment to evaluate the relative risk of nutrient enrichment from N deposition for all Inventory and Monitoring networks nationwide. The network rankings for nutrient N pollutant exposure, ecosystem sensitivity to nutrient N enrichment, and park protection yielded an overall network nutrient N summary risk ranking for the Klamath Network that was in the highest quintile among all networks. This suggests that parks in the Klamath Network may be highly susceptible to effects from nutrient enrichment caused by atmospheric N deposition.

13.4.1 Aquatic Ecosystem Nutrient Enrichment

Lassen Volcanic NP has 37 high-elevation lakes. Many may be N sensitive. Crater Lake is a high-elevation lake. The most sensitive aquatic indicator for Crater Lake is lake transparency, which is influenced to a large degree by primary productivity. An increase in N to the lake might increase algal growth and decrease transparency.

An examination of the EPA's Western Lakes Survey data (Eilers et al. 1987) found higher N concentrations in high-elevation lakes adjacent to and downwind from urban centers (Fenn et al. 2003a). A survey of 62 lakes that were the subject of N enrichment studies across a variety of regions reported a mean increase in phytoplankton biomass of 79% in response to N enrichment (average of 46.3 μeq/L N; Elser et al. 1990). Long-term measurements at Lake Tahoe, California, show that primary productivity doubled, while water clarity declined, partly as a result of atmospheric N deposition (Goldman 1988, Jassby et al. 1994). A meta-analysis of lakes from 42 regions of Europe and North America concluded that atmospheric N deposition was responsible at many locations for increased concentrations of N in lake water and elevated phytoplankton biomass (Bergström and Jansson 2006). Bergström et al. (2005) found a consistent pattern of N limitation at atmospheric deposition levels below approximately 2.5 kg N/ha/yr, colimitation of N and P at N deposition between ~2.5 and 5.0 kg N/ha/yr, and P limitation in areas with N deposition greater than 5.0 kg N/ha/yr. Throughout most of the Klamath Network, total modeled N deposition is less than 5 kg/ha/yr.

13.4.2 Terrestrial Ecosystem Nutrient Enrichment

Monitored wet N deposition in the western portion of the Pacific Northwest region, including parts of the Klamath Network, increased from generally less than 1 kg N/ha/yr in the mid-1980s to the range of 1–3 kg N/ha/yr in many locations in more recent years, mainly due to increased ammonium deposition (NADP/NTN 2006, Geiser and Neitlich 2007). Lichen communities in the region show evidence of changes in response to increased N pollution. The Pacific Northwest retains widespread populations of air pollution–sensitive lichens (Fenn et al. 2003a). However, in urban areas, intensive agricultural zones, and downwind of major urban and industrial centers, there are few air pollution-sensitive lichen species, and relatively high N concentrations have been measured in lichen tissue (Fenn et al. 2003a). Air pollution has been mainly associated with effects on community composition of lichens rather than species richness. With N enrichment, especially around urban and agricultural areas, there is a shift toward weedy, nitrophilous lichen species (Fenn et al. 2003a). Replacement of sensitive lichens by nitrophilous species has undesirable ecological consequences. In late-successional, naturally N-limited forests of the Oregon Coast Range and western Cascade Mountains, for example, epiphytic lichens make important contributions to mineral cycling and soil fertility (Pike 1978, Sollins et al. 1980). Together with other large, pollution-sensitive lichens, they constitute integral parts of the food web for mammals, insects, and birds (McCune et al. 1997, U.S. EPA 2005). Sensitive lichen species appear to be negatively affected by total N inputs as low as 3–8 kg/ha/yr (Fenn et al. 2003a), which occur in parts of the Klamath Network region.

Lichen taxa recommended as indicators of clean sites and polluted sites were used by Geiser and Neitlich (2007) to create six lichen zones of air quality within the Pacific Northwest region, from worst (all sensitive species absent) to best (all sensitive species present). The zone of worst air quality was associated with an absence of sensitive lichens; enhancement of nitrophilous lichens; mean wet ammonium deposition >0.06 mg N/L; lichen tissue N and S concentrations >0.6% and 0.07%, respectively, and sulfur dioxide levels harmful to sensitive lichens. The strongest relationship was with wet ammonium deposition, consistent with findings in California (Jovan and McCune 2005) and Europe (van Dobben et al. 2001).

Geiser et al. (2010) developed critical loads for nutrient N deposition to protect epiphytic lichen communities in maritime forests of western Washington and Oregon. They were based on observed relationships between epiphytic lichens and total N deposition estimates derived from the Community Multiscale Air Quality model. Estimated total N deposition values equal to 3–9 kg N/ha/yr were associated with 20%–40% declines in sensitive lichen species. A model was developed with which to predict N critical loads at any desired lichen community composition, expressed as the percent composition represented by oligotrophic (those adapted to low nutrient availability) versus eutrophic (nitrophilous) lichen species. The estimated critical load increased with increasing precipitation amount. This may have been because higher precipitation diluted pollutant levels. Oligotrophic lichens decreased in abundance and eutrophic lichens increased in abundance in response to higher N deposition (Jovan 2008, McCune and Geiser 2009). Geiser et al. (2010) concluded that no single lichen critical load was appropriate across land areas in this region that receive varying amounts of precipitation.

Despite the presumed high sensitivity of resources in LABE and LAVO to nutrient enrichment effects on both terrestrial (LABE and LAVO) and aquatic (LAVO) ecosystems, data are not available with which to fully assess nutrient enrichment effects on receptors other than lichens. Given the observed low levels of N emissions and deposition in the Klamath Network region, ecological effects are expected to be relatively minor at the present time.

Ellis et al. (2013) estimated the critical load for nutrient N deposition to protect the most sensitive ecosystem receptors in 45 national parks. They estimated the N critical load in the range of 2.5–7.1 kg N/ha/yr for the protection of lichens in CRLA and LAVO and 2.7–9.2 for the protection of lichens in REDW, based on Bowman et al. (2011) critical load estimates.

13.5 Ozone Injury to Vegetation

13.5.1 Ozone Exposure Indices and Levels

Ozone condition, as rated by the National Park Service, ranges from low to high in the Klamath Network parks (Table 13.3). Kohut's evaluation of risk to plants ranged from low to moderate. In LAVO, he assessed risk as low, while the National Park Service system rated exposure as high. Kohut found that during the period that he reviewed, very few hours had concentrations exceeding 100 ppb, and therefore he concluded that risk was low.

13.5.2 Ozone Formation

The ozone exposure in the Klamath Network region is partly due to trans-Pacific transport of pollutant emissions from Asia. The coastal ozone monitoring site at Trinidad Head in northern California is well suited to evaluate ozone and ozone precursor transport from Asia to the west coast of the United States (Oltmans et al. 2008). During the ozone seasonal maximum in the spring, measurements of ozone concentration at the Trinidad Head coastal location often exceeded 50 ppb in air that had not been influenced by emission sources in North America. During daytime hours, air that arrives at this headland site is nearly exclusively off the Pacific Ocean during the spring months. The average daytime ozone concentrations at Trinidad Head exceeded 40 ppb for the period 2002–2005 (Oltmans et al. 2008).

Jaffe et al. (2003) demonstrated that the concentration of ozone in air masses reaching the west coast of the United States from Asia increased by about 30% (10 ppb) from the mid-1980s to 2002. The study included sites in LAVO and REDW. The measured increase in ozone at these parks was correlated with increasing N emissions in Asia.

Forest fires emit considerable quantities of oxidized N and hydrocarbons and therefore can contribute to ozone formation in the Klamath Network region. The concentration of

TABLE 13.3

Ozone Assessment Results for Inventory and Monitoring Parks in the Klamath Network Based on Estimated Average 3-Month W126 and SUM06 Ozone Exposure Indices for the Period 2005–2009 and Kohut's (2007) Ozone Risk Ranking for the Period 1995–1999[a,b]

| Park Name | Park Code | W126 | | SUM06 | | Kohut Ozone Risk Ranking |
		Value (ppm-hr)	NPS Condition	Value (ppm-hr)	NPS Condition	
Crater Lake	CRLA	5.70	Low	7.41	Low	Moderate
Lassen Volcanic	LAVO	13.19	High	22.20	High	Low
Lava Beds	LABE	9.66	Moderate	14.64	Moderate	Moderate
Redwood	REDW	5.36	Low	7.32	Low	Moderate

[a] Parks are classified into one of three ranks (Low, Moderate, High) based on comparison with other Inventory and Monitoring parks.

[b] Degrees of concern for the W126 and SUM06 indices are based solely on levels of ozone exposure. Kohut's risk to vegetation is based on several factors that contribute to injury in plants, including ozone exposure and environmental variables, and considers the effects of soil moisture on the uptake of ozone.

ozone in the atmosphere in the nonurban western United States has increased since the late 1980s by about 5 ppb (Jaffe and Ray 2007). The summer burned area was significantly correlated with ozone concentration at atmospheric chemistry monitoring sites, including sites within LAVO and five other parks in the western United States (Jaffe et al. 2008). Increasing temperature in the future will likely further influence the effects of fire on ozone formation.

13.5.3 Ozone Exposure Effects

13.5.3.1 Field Survey Results

Forests of the central and eastern portions of the Klamath Network are primarily Rocky Mountain types where the dominant species include ponderosa pine at lower elevations and subalpine fir (*Abies amabilis*) at higher elevations. Ponderosa pine (var. *ponderosa*) is an excellent bioindicator of ozone exposure. Lower montane forests on the west side of the Klamath Network region, including at LAVO, are dominated by this species on xeric sites. At higher elevations, mixed conifer forest is prevalent and includes ponderosa pine along with several other coniferous tree species. Jeffrey pine (*Pinus jeffreyi*), also a good bioindicator, replaces ponderosa pine within the mixed conifer forest at higher elevations and on shallow soils. In addition to Jeffrey pine and ponderosa pine, there are other potential ozone bioindicators in LAVO and elsewhere within the Klamath Network, including quaking aspen. Foliar injury detected in these bioindicators in the Klamath Network was moderate (Arbaugh et al. 1998, Sullivan et al. 2001).

High ozone exposure along the western edge of the Sierra Nevada Mountains throughout late spring and summer provides sufficient stress in ponderosa pine and Jeffrey pine to induce a gradient in foliar injury from north to south; injury has been shown to be very severe at Sequoia and Kings Canyon National Parks, severe at Yosemite National Park, and moderate at LAVO (Sullivan et al. 2001). Sullivan et al. (2001) reported that LAVO experienced seasonal to chronic periods of poor air quality, especially relatively high ozone. Among the parks in the Klamath Network, LAVO had the highest monitored or interpolated W126 and SUM06 ozone exposure during the period 2005–2009 (Table 13.3). Nevertheless, ozone risk in LAVO was ranked low due to moisture constraints on stomatal opening.

There has been very limited research on, or monitoring of, the potential effects of air pollution on vegetation in LABE (Sullivan et al. 2001). This is despite the fact that the park has three good ozone bioindicators (Jeffrey pine, ponderosa pine [var. *ponderosa*], and quaking aspen) and that some ozone injury has been documented at LAVO, which is only 75 mi (120 km) to the south. Ozone weekly averages were measured at a single passive ozone monitoring site within LABE during the period 1995–1998. Ozone measurements from this sampler were among the lowest of all parks in California (Sullivan et al. 2001) and do not support the likelihood of appreciable risk to vegetation at LABE.

The Forest Ozone Response Study, conducted during the period 1991 to 1994, quantified the distribution of ozone injury to forest species throughout California (Rocchio et al. 1993, Cahill et al. 1996, Arbaugh et al. 1998), including at some national park locations. The data included results of ozone injury surveys at 11 sites (all with associated ozone monitoring data), ranging from LAVO in northern California to San Bernardino National Forest in southern California, with sites distributed throughout the western Sierra Nevada within the elevational range of mixed conifer forest. Visible ozone injury to ponderosa pine and

Jeffrey pine was detected at all sites, ranging from 26.8% of trees at the northernmost site in LAVO to 100% at the southernmost. The Ozone Injury Index (Schilling and Duriscoe 1996) ranged from 6.3 in the north to 65.1 in the south. There were significant correlations between the Ozone Injury Index and various ozone indices over the geographic range of the study. The Forest Ozone Response Study was one of the first field studies to demonstrate a quantitative relationship between tree injury and cumulative ozone exposure (expressed as SUM06).

The Forest Ozone Response Study corroborated results from a previous analysis of ozone injury (presence of chlorotic mottling and needle retention) to ponderosa pine in the Sierra Nevada (Peterson et al. 1991, Peterson and Arbaugh 1992), which found a gradient of injury, ranging from low injury at northern sites (Tahoe National Forest) to high injury at southern sites (Sequoia National Park and Sequoia National Forest). Strong correlations were found between injury level and ozone exposure. Confidence is high in the inference that ozone injury increases from north to south in the Sierra Nevada and that the injury is directly proportional to ambient ozone exposure (Sullivan et al. 2001). Nevertheless, foliar injury is only one manifestation of injury. Effects on growth and relative abundance of plant species are also possible.

13.5.3.2 Model Responses

Weinstein et al. (2005) simulated the effects on growth and competitive success of white fir (*Abies concolor*) and ponderosa pine at three sites in California, including LAVO, from ozone exposures of 0.5, 1.5, 1.75, and 2.0 times ambient. The response of individual trees was modeled using the TREGRO model (Weinstein et al. 1991), based on results obtained in controlled chamber studies. These results were then extrapolated to the overall forest using the ZELIG model (Urban 1990, Urban et al. 1991), which considers the effects of competition. Model results suggested little growth impact of ozone on white fir up to two times ambient exposure, but substantial growth reduction for ponderosa pine, mainly in southern California. Continued ozone exposure for 100 years at ambient levels at LAVO was not projected to decrease ponderosa pine abundance (Weinstein et al. 2005).

13.6 Visibility Degradation

13.6.1 Natural Background Visibility and Monitored Visibility Conditions

Natural background visibility has been estimated for Interagency Monitoring of Protected Visual Environments (IMPROVE) sites located in CRLA, LAVO, LABE, and REDW for the 20% clearest visibility days, the 20% haziest visibility days, and average days, in order to establish goals for visibility improvement. Estimated natural background and monitored visibility for Klamath Network parks are given in Table 13.4. REDW experienced much higher levels of natural haze than did the other parks, with natural haze values that were nearly twice as high as other parks in this network. Natural haze is not an important feature at the other monitored parks within the Klamath Network.

Class I national parks in Klamath Network have relatively good monitored visibility compared with Class I areas in other parts of California. These parks are far removed from

TABLE 13.4

Estimated Natural Haze and Measured Ambient Haze in Klamath Network Parks Averaged over the Period 2004–2008[a]

Park Name	Site ID	Estimated Natural Haze (in Deciviews)		
		20% Clearest Days	20% Haziest Days	Average Days
Crater Lake	CRLA1	0.10	7.62	3.26
Lassen Volcanic	LAVO1	1.00	7.31	3.77
Lava Beds	LABE1	1.29	7.85	4.13
Redwood	REDW1	3.46	13.91	8.11

Park Name	Site ID	Measured Ambient Haze (for Years 2004–2008)					
		20% Clearest Days		20% Haziest Days		Average Days	
		Deciview	Ranking	Deciview	Ranking	Deciview	Ranking
Crater Lake	CRLA1	1.58	Very Low	13.74	Low	6.72	Very Low
Lassen Volcanic	LAVO1	2.61	Very Low	15.32	Low	8.01	Low
Lava Beds	LABE1	2.95	Very Low	14.29	Low	8.06	Low
Redwood	REDW1	5.76	Moderate	19.08	Moderate	12.26	Moderate

[a] Parks are classified into one of five haze ranks (Very Low, Low, Moderate, High, or Very High) based on comparison with other monitored parks.

large urban population centers. CRLA is very remote from human air pollution sources and is not in close proximity to the ocean. It generally experiences the best visibility among all monitored Klamath Network parks (Table 13.4).

Although REDW had the highest ambient haze values among the monitored parks within Klamath Network, the difference between natural and ambient haze on the haziest 20% days was actually smaller for REDW than it was for any of the other monitored parks in Klamath Network. Thus, the human-caused contribution to haze on the poorest visibility days appears to be lower in REDW than in the other parks.

LAVO showed the greatest difference among monitored Klamath Network parks between natural background and monitored visibility (a difference of 8 dv on the 20% haziest days; Table 13.4). The 1988–1998 data showed the worst visibility at LAVO occurred during summer when the seasonal average visual range (i.e., how far a person can see) was only 73 mi (118 km). The best visibility occurred during winter when the seasonal average visual range was 139 mi (223 km).

Representative photos of selected vistas under three different visibility conditions are shown for REDW in Figure 13.6. Photos were selected to correspond with the clearest 20% of visibility conditions, the haziest 20% of visibility conditions, and annual average visibility conditions.

13.6.2 Composition of Haze

At IMPROVE sites throughout the interior Columbia River basin, from North Cascades NP and Glacier NP in the north to LAVO and Yellowstone NP in the south, carbon in various forms, including fine particulate organics and soot, contributes most to visibility impairment (Schoettle et al. 1999). The second most important contributor is sulfate. Savig and Morse (2001) reported that visibility impairment at LAVO in the 1990s was largely due to organics and sulfates. The clearest 20% days at LAVO probably approach natural

Redwood

20% clearest days
Taken = 3:00 PM
Haze = 6 deciviews
Extinction = 18 Mm^{-1}
Visual range = 220 km

20% haziest days
Taken = 3:00 PM
Haze = 21 deciviews
Extinction = 78 Mm^{-1}
Visual range = 50 km

Average days
Taken = 3:00 PM
Haze = 13 deciviews
Extinction = 36 Mm^{-1}
Visual range = 110 km

FIGURE 13.6
Three representative photos of the same view in REDW illustrating the 20% clearest visibility, the 20% haziest visibility, and the annual average visibility. Extinction is total particulate light extinction. (From http://vista. cira.colostate.edu/improve/Data/IMPROVE/Data_IMPRPhot.htm, accessed December, 2010.)

conditions (GCVTC 1996). Historically, visibility varied with patterns in weather, wind (and the effects of wind on coarse particles), and smoke from fires.

Lightning is the most important source of wildfire ignition in the Pacific Northwest (Rorig and Ferguson 1999, McKenzie et al. 2006). Managers of protected areas, such as national parks and wilderness, have adopted a policy whereby naturally ignited fires are allowed to burn unless they threaten structures or ambient air quality (Miller and Landres 2004, McKenzie et al. 2006). Prescribed fire is also widely used throughout the western United States, including in the Klamath Network region, for restoring and maintaining vegetation communities (Fulé et al. 1997, Stephenson 1999, Allen et al. 2002, Brockway et al. 2002, McKenzie et al. 2006).

A substantial component of regional haze in the Pacific Northwest is caused by prescribed and wildland fires. Smoke plumes can contribute to visibility impairment downwind from the fire locations (Malm 1999). In recent years, fires have become more severe than they were historically, largely because of more extreme droughts and buildup of fuel from fire suppression (Agee 1997, Flannigan et al. 1998, Covington 2000, Allen et al. 2002, McKenzie et al. 2006). This pattern may continue into the future in response to climate change. Both empirical (McKenzie et al. 2004) and process (Lenihan et al. 1998) models suggest that the area burned by wildfires is likely to increase in the western United States in response to a warming climate.

Data from IMPROVE monitors in the northern and central coastal locations in California, including at REDW, for the baseline period of 2000–2004, suggested that visibility-degrading pollutants on the 20% haziest days were primarily organics, sulfate, nitrate, and sea salt (Suarez-Murias et al. 2009). Sulfate and sea salt are the most important causes of haze on the haziest days at all of the coastal IMPROVE monitoring sites, except on the haziest winter days at Point Reyes NS to the south of the Klamath Network near San Francisco, when and where nitrate predominated (Suarez-Murias et al. 2009). The sulfate contribution to haze exhibited some seasonality at REDW, with highest contributions during summer. High contributions from sea salt occurred throughout the year at REDW. Emissions of sulfate from ocean vessels also constitute an important contributor to visibility impairment at coastal California Class I areas (Suarez-Murias et al. 2009).

Visibility modeling results reported by Suarez-Murias et al. (2009) for the IMPROVE sites in northern California suggested that a 48% improvement would be required on the haziest visibility days in order to meet Regional Haze Rule goals by 2064 at LABE and LAVO. A somewhat smaller improvement (25%) would be required for REDW (Table 13.5).

TABLE 13.5

IMPROVE Monitors and Haze (in Deciviews) at Class I Areas in Northern California

Improve Monitor	Class I Area(s)	Current Conditions (2000–2004 Baseline)		Future Natural Conditions (2064 Goals)		
		Haziest Days	Clearest Days (Maintain in Future Years)	Natural Haziest Days	Deciview Hurdle (Baseline to 2064)	Improvement Required from Current Visibility on Haziest Days (%)
LABE Lava Beds	Lava Beds National Monument	15.1	3.2	7.9	7.2	48
	South Warner Wilderness					
LAVO Lassen Volcanic	Lassen Volcanic National Park	14.1	2.7	7.3	6.8	48
	Caribou Wilderness					
	Thousand Lakes Wilderness					
REDW Redwood	Redwood National Park	18.5	6.1	13.9	4.6	25

Source: Suarez-Murias, T. et al., California regional haze plan, California Environmental Protection Agency, Air Resources Board, 2009.

Figure 13.7 shows estimated natural (preindustrial), baseline (2000–2004), and current (2006–2010) levels of haze and its composition for CRLA, LAVO, LABE, and REDW. On the 20% haziest days and average days, the majority of the impairment in all parks except REDW was attributable to organics, followed by sulfate (Figure 13.7). On the clearest 20% visibility days, the major contributions to current conditions were typically from sulfate and organics. The contributors to impairment at REDW differed from the other monitored Klamath Network parks (Figure 13.7). Sea salt played a much larger role, contributing the most to haze on the 20% haziest days, followed closely by sulfate. On the 20% clearest and average days, sulfate was the largest contributor, followed closely by sea salt. Consistent with the findings presented here, Savig and Morse (2001) found that organics (partly from

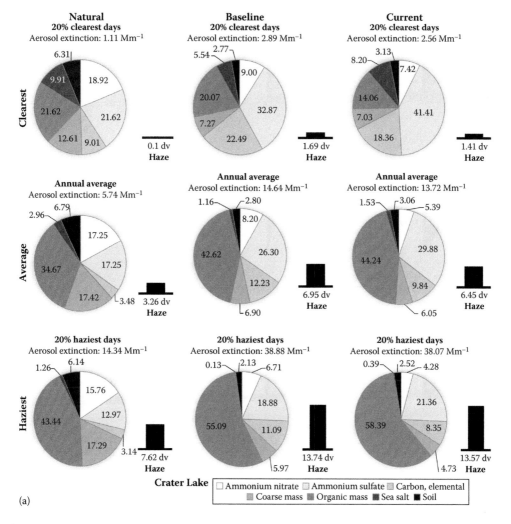

(a)

FIGURE 13.7
Estimated natural (preindustrial), baseline (2000–2004), and current (2006–2010) levels of haze (columns) and its composition (pie charts) on the 20% clearest, annual average, and 20% haziest visibility days for (a) CRLA, (b) LAVO, (c) LABE, and (d) REDW. LABE has no data for the year 2000, CRLA has no data for the years 2000 and 2001, and REDW has no measured data for the year 2010. (From http://views.cira.colostate.edu/fed/Tools/RegionalHazeSummary.aspx, accessed October, 2012.) *(Continued)*

wildfire) contributed the most to visibility impairment, on an annual basis, at parks in the Sierra-Humboldt area (including LAVO), whereas sulfate generally contributed the most to impairment in the Pacific Coastal Mountain parks (including REDW).

During the 2000–2004 baseline period, the 20% haziest days at the IMPROVE sites in LAVO and LABE were dominated by organic aerosols. These organic aerosols showed distinct seasonality, with peak values occurring during the summer, correlating with the prevalence of wildfire and increased natural biogenic emissions from vegetation (Suarez-Murias et al. 2009).

Dust storms commonly occur during the spring season in eastern Asia. With westerly air flows, dust particles can be transported to the western United States. Asian dust can

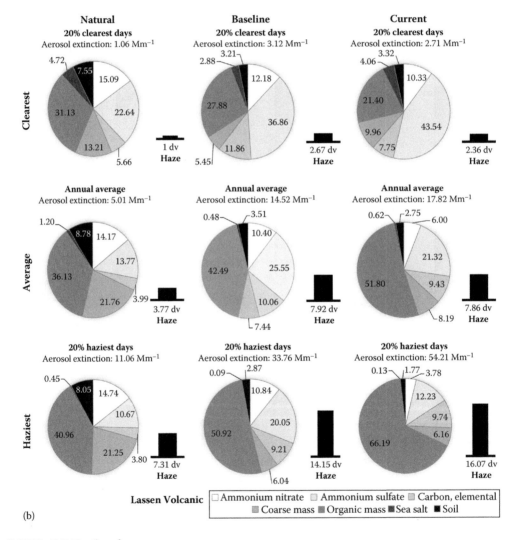

(b)

FIGURE 13.7 (*Continued*)
Estimated natural (preindustrial), baseline (2000–2004), and current (2006–2010) levels of haze (columns) and its composition (pie charts) on the 20% clearest, annual average, and 20% haziest visibility days for (a) CRLA, (b) LAVO, (c) LABE, and (d) REDW. LABE has no data for the year 2000, CRLA has no data for the years 2000 and 2001, and REDW has no measured data for the year 2010. (From http://views.cira.colostate.edu/fed/Tools/RegionalHazeSummary.aspx, accessed October, 2012.)   (*Continued*)

also be mixed with human-caused air pollutants emitted in industrial areas of China. Liu et al. (2003) determined that Asian dust contributed an estimated 16% and 11% of the fine particulate mass measured at the CRLA and LAVO IMPROVE sites, respectively.

Analyses conducted by the Western Regional Air Partnership indicated that organics from natural emissions sources, including wildfire and biogenic sources (vegetation), contribute to substantial visibility impairment throughout the western United States, including within the Klamath Network region. In addition, air pollution sources outside the Western Regional Air Partnership domain, including international offshore shipping and sources from Mexico, Canada, and Asia, can in some cases be substantial contributors to haze (Suarez-Murias et al. 2009).

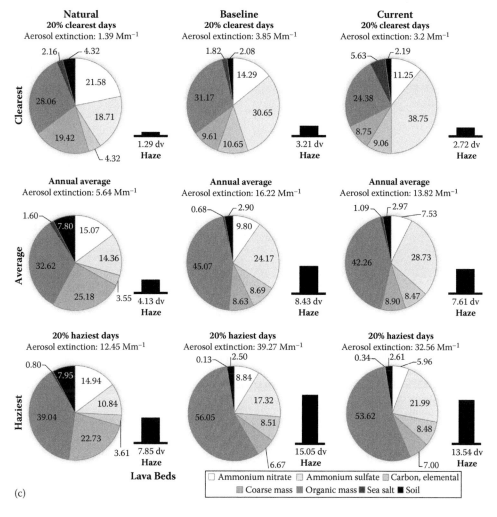

(c)

FIGURE 13.7 (*Continued*)
Estimated natural (preindustrial), baseline (2000–2004), and current (2006–2010) levels of haze (columns) and its composition (pie charts) on the 20% clearest, annual average, and 20% haziest visibility days for (a) CRLA, (b) LAVO, (c) LABE, and (d) REDW. LABE has no data for the year 2000, CRLA has no data for the years 2000 and 2001, and REDW has no measured data for the year 2010. (From http://views.cira.colostate.edu/fed/Tools/RegionalHazeSummary.aspx, accessed October, 2012.) (*Continued*)

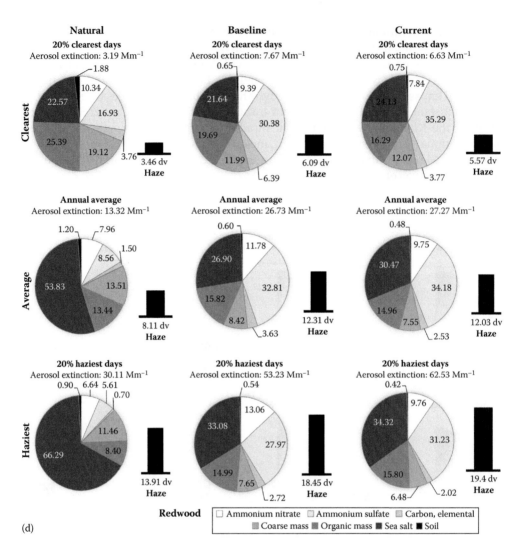

(d)

FIGURE 13.7 (Continued)
Estimated natural (preindustrial), baseline (2000–2004), and current (2006–2010) levels of haze (columns) and its composition (pie charts) on the 20% clearest, annual average, and 20% haziest visibility days for (a) CRLA, (b) LAVO, (c) LABE, and (d) REDW. LABE has no data for the year 2000, CRLA has no data for the years 2000 and 2001, and REDW has no measured data for the year 2010. (From http://views.cira.colostate.edu/fed/Tools/RegionalHazeSummary.aspx, accessed October, 2012.)

Source apportionment analyses were conducted for the California Regional Haze Plan (Suarez-Murias et al. 2009), using regional-scale, three-dimensional air quality models that simulated emissions, chemical transformations, and transport of criteria air pollutants and fine particulate matter. Effects on visibility in Class I areas in California were then evaluated for organic C, sulfate, and nitrate, which are the three major drivers of haze in California. The Particulate Matter Source Apportionment Technology algorithm was used for Western Regional Air Partnership model analysis. Directly emitted human-caused organic C was estimated to contribute half or less of the organic C in most areas of California. Human-caused sources of oxidized N contributed at least half of the nitrate in all Class I areas in California except REDW (where it only contributed 7%). The human-caused sulfate contribution ranged from 1% to 35% at various Class I areas, with highest contributions in southern California. Thus, important drivers of haze in California derive from natural sources (organic C), mobile sources (nitrate), and offshore and non-Western Regional Air Partnership regional sources of sulfate (Suarez-Murias et al. 2009).

13.6.3 Trends in Visibility

Over the full period of monitoring data for the parks in the Klamath Network region, there is only limited evidence of improvement in haze levels (Figure 13.8). Nevertheless, the most recent three years of monitoring data (2008–2010) do suggest some improvement in visibility on the 20% haziest days at each of the monitored parks.

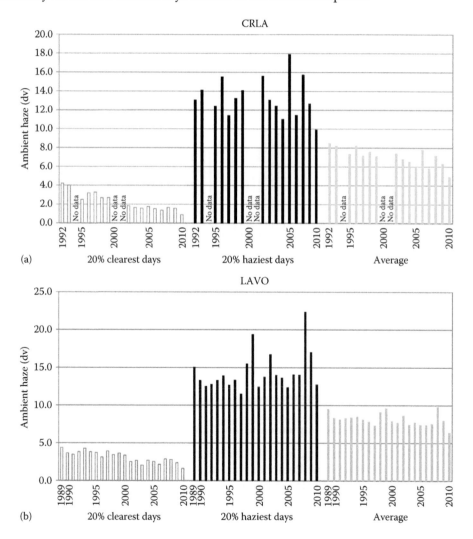

FIGURE 13.8
Trends in monitored visibility at (a) CRLA, (b) LAVO, (c) LABE, and (d) REDW, based on IMPROVE measurements on the 20% clearest, 20% haziest, and annual average visibility days over the monitoring period of record. (From http://vista.cira.colostate.edu/improve/Data/IMPROVE/summary_data.htm, accessed October, 2012.)
(Continued)

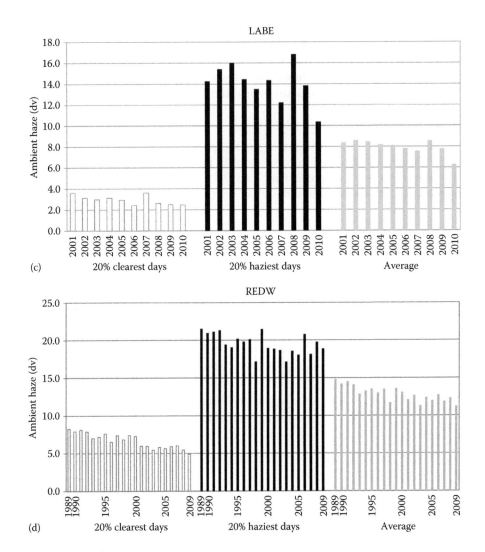

FIGURE 13.8 (*Continued*)
Trends in monitored visibility at (a) CRLA, (b) LAVO, (c) LABE, and (d) REDW, based on IMPROVE measurements on the 20% clearest, 20% haziest, and annual average visibility days over the monitoring period of record. (From http://vista.cira.colostate.edu/improve/Data/IMPROVE/summary_data.htm, accessed October, 2012.)

13.6.4 Development of State Implementation Plans

The Class I parks in California requiring the least amount of visibility improvement to meet the national goal specified by the Regional Haze Rule are REDW in the Klamath Network and Point Reyes NS to the south. These two parks are coastal and are relatively (compared with other parks in California) unaffected by land-based emissions sources in California. Sea salt provides an important natural cause of haze at these locations.

Progress to date in meeting the national visibility goal is illustrated in Figure 13.9 using a uniform rate of progress glideslope. Although some progress has been made since the baseline period (2000–2004), substantial additional visibility improvement is needed to eliminate human-caused haze by 2064. More data will be needed to evaluate compliance with requirements of the Regional Haze Rule.

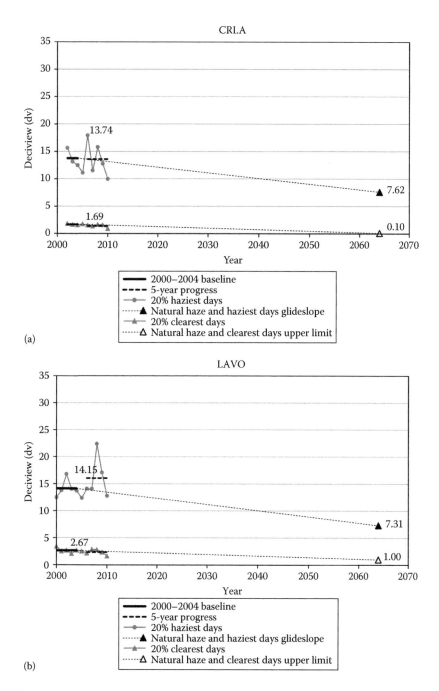

FIGURE 13.9
Glideslopes to achieving natural visibility conditions by 2064 for the 20% haziest (upper line) and the 20% clearest (lower line) days in (a) CRLA, (b) LAVO, (c) LABE, and (d) REDW. In the Regional Haze Rule, the clearest days do not have a uniform rate of progress glideslope; the rule only requires that the clearest days do not get any worse than the baseline period. Also shown are measured values during the period 2000–2010. LABE has no data for the year 2000; CRLA has no data for the years 2000 and 2001. (From http://vista.cira.colostate.edu/improve/Data/IMPROVE/summary_data.htm, accessed October, 2012.) *(Continued)*

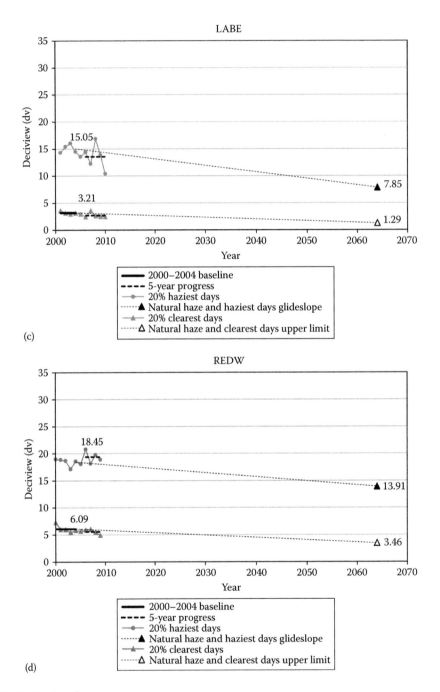

FIGURE 13.9 (*Continued*)
Glideslopes to achieving natural visibility conditions by 2064 for the 20% haziest (upper line) and the 20% clearest (lower line) days in (a) CRLA, (b) LAVO, (c) LABE, and (d) REDW. In the Regional Haze Rule, the clearest days do not have a uniform rate of progress glideslope; the rule only requires that the clearest days do not get any worse than the baseline period. Also shown are measured values during the period 2000–2010. LABE has no data for the year 2000; CRLA has no data for the years 2000 and 2001. (From http://vista.cira.colostate.edu/improve/Data/IMPROVE/summary_data.htm, accessed October, 2012.)

13.7 Toxic Airborne Contaminants

Data are limited on the effects of atmospherically deposited toxic substances on resources in Klamath Network parks. Concentrations of semivolatile organic compounds detected in CRLA vegetation were generally at or above medians for all WACAP parks studied in the western United States. In vegetation in CRLA, dominant semivolatile organic compounds detected were endosulfans and dacthal, followed by polycyclic aromatic hydrocarbons, hexachlorobenzene, hexachlorocyclohexane-α, chlordanes, and DDT. In LAVO, concentrations of semivolatile organic compounds detected in vegetation were relatively high and closely approached or exceeded medians for WACAP parks. Dominant semivolatile organic compounds detected in vegetation at LAVO were endosulfans and dacthal, followed by polycyclic aromatic hydrocarbons, hexachlorobenzene, hexachlorocyclohexane-α, chlordanes, and DDT. The extent to which these potentially toxic airborne contaminants actually impact biota in the parks within the Klamath Network is not known.

Krabbenhoft et al. (2002) sampled 90 lakes in seven national parks in the western United States, including LAVO, and analyzed their Hg and methyl Hg concentrations. Levels of methyl Hg were similar in most parks (~0.05 ng/L), including LAVO. The U.S. Geological Survey (last modified February 20, 2015) estimated methyl Hg concentrations in surface waters throughout the conterminous United States. Estimates for the Klamath Network region were low to moderate compared with other regions.

Eagles-Smith et al. (2014) sampled fish in 21 national parks and analyzed them for Hg concentrations in tissue. Results varied substantially by park and by water body. The concentrations of Hg in brook trout sampled in three lakes in LAVO were variable. Risk to fish and wildlife appeared to be limited to fish from Summit Lake. However, Hg concentrations in brook trout from Horseshoe Lake were estimated to reach concentrations of concern for tissue-based toxicity to fish and for the protection of human health.

Kokanee (*Oncorhynchus nerka*) and rainbow trout were sampled from two locations in CRLA. These fish had among the lowest Hg concentrations recorded in the survey. There was no evidence that piscivorous predators were at risk from Hg contamination in this park.

Organochlorine chemicals can be estrogenic (Garcia-Reyero et al. 2007), contributing to the occurrence of intersex fish, and can accumulate in the aquatic food webs of remote mountain ecosystems (Blais et al. 1998). Biomarkers of exposure to estrogenic chemicals suggest the likelihood of reproductive dysfunction (Harries et al. 1997). Organochlorines are likely transported as atmospheric contaminants from lower agricultural areas to high-elevation sites (Hageman et al. 2006). Work by Schreck and Kent (2013) followed up on some of the work performed in the WACAP study. They expanded the range of coverage to additional western parks, including CRLA and LAVO within the Klamath Network. No intersex fish were found in CRLA; in LAVO, 1 out of 45 sampled male fish was intersex. The low observed frequency of intersex fish in these parks and water bodies may be a natural phenomenon (Schwindt et al. 2009). The extent to which human-caused contaminants contribute to an increased frequency is difficult to determine (Schreck and Kent 2013).

13.8 Summary

Air quality in the portions of California and Oregon within the Klamath Network region is among the best in the United States, in particular in remote sections of the Cascade Mountain Range, northern portions of the Sierra Nevada, and at some near-coastal locations (Sullivan et al. 2001). This is because the principal air masses that influence these areas are largely derived from the atmosphere over the Pacific Ocean. Air quality generally deteriorates further to the south in California, especially in the southern Sierra Nevada and near Los Angeles.

Sulfur and N air pollutants can cause acidification of streams, lakes, and soils. Parklands in the Klamath Network vary in their sensitivity to acidification. In a coarse screening assessment that evaluated the relative risk of acidification in all Inventory and Monitoring parks nationwide, CRLA and LAVO ranked as very sensitive (Sullivan et al. 2011a) while REDW ranked as sensitive. Differences in ecosystem sensitivity to acidification within Klamath Network parks were largely due to differing slopes and to the presence of high-elevation lakes and streams in some of the parks. Acid-sensitive waters are often located at relatively high elevation, on steep slopes having shallow, base-poor (having low amounts of adsorbed base cations) soils. All of the streams in LAVO and CRLA are low order, which increases their potential sensitivity to acidification. A very high percentage of the lakes in LAVO are highly sensitive. Such lakes tend to be small, are situated in rather small watersheds, and are almost devoid of buffering capacity because of a combination of factors, including steep slopes, the predominance of base-poor bedrock that is resistant to weathering, thin soils, large amounts of precipitation, deep seasonal snowpacks, coniferous vegetation, and extremely dilute waters. Current modeled and monitored acidic deposition levels in the Klamath Network are very low, however, and are not likely to have caused acidification impacts to aquatic ecosystems.

A coarse screening assessment that evaluated the relative risk of N enrichment effects in all Inventory and Monitoring parks nationwide (Sullivan et al. 2011b) found that ecosystem sensitivity to nutrient N enrichment was variable throughout the Klamath Network, with LABE ranked as very highly sensitive and LAVO as highly sensitive. Differences in ecosystem sensitivity among parks were due to differing coverage of vegetation types thought to be most sensitive to N deposition, and also to the presence of high-elevation lakes, which are often N limited and therefore sensitive to increased N. The most sensitive indicator for aquatic ecosystems at CRLA is the transparency of Crater Lake. Transparency is influenced to a large degree by primary productivity. An increase in N contribution to the lake might further increase algal growth and decrease transparency.

Ozone-induced foliar injury was documented at LAVO as part of the Forest Ozone Response Study in the early 1990s (Arbaugh et al. 1998), but the amount of plant injury was relatively low compared to national parks in the southern Sierra Nevada (Sullivan et al. 2001). Nevertheless, this injury represented a significant impact of air pollution at a location quite remote from urban air pollution sources in California. In 2012, LAVO experienced several exceedances of the ozone National Ambient Air Quality Standard. However, because ozone concentrations did not exceed the standard during the two previous years, LAVO was not in violation and was not designated as nonattainment.

Particulate pollution can cause haze, reducing visibility. Monitoring indicates that haze impairs visibility to some extent at all Klamath Network parks. Impacts are relatively modest.

Atmospherically deposited toxic substances were evaluated in the Klamath Network as part of the WACAP research effort (Landers et al. 2010). Concentrations of semivolatile organic compounds were generally comparable to those in other surveyed parks in the western United States. Potential effects on park biota are not known.

References

Agee, J.K. 1997. The severe weather wildfire: Too hot to handle? *Northwest Sci.* 71:153–157.

Akimoto, H. and H. Narita. 1994. Distribution of SO_2, NO_x, and CO_2 emissions from fuel combustion and industrial activities in Asia with $1° \times 1°$ resolution. *Atmos. Environ.* 28:213–225.

Allen, C.D., M. Savage, D.A. Falk, K.F. Suckling, T.W. Swetnam, P. Schulke, P.B. Stacey, P. Morgan, M. Hoffman, and J.T. Klingel. 2002. Ecological restoration of southwestern ponderosa pine ecosystems: A broad perspective. *Ecol. Appl.* 12:1418–1433.

Arbaugh, M.J., P.R. Miller, J.J. Carroll, B.J. Takemoto, and T. Procter. 1998. Relationships of ozone exposure to pine injury in the Sierra Nevada and San Bernardino Mountains of California, USA. *Environ. Pollut.* 101:291–301.

Bergström, A., P. Blomqvist, and M. Jansson. 2005. Effects of atmospheric nitrogen deposition on nutrient limitation and phytoplankton biomass in unproductive Swedish lakes. *Limnol. Oceanogr.* 50(3):987–994.

Bergström, A. and M. Jansson. 2006. Atmospheric nitrogen deposition has caused nitrogen enrichment and eutrophication of lakes in the northern hemisphere. *Glob. Change Biol.* 12:635–643.

Blais, J.M., D.W. Schindler, D.C.G. Muir, L.E. Kimpe, D.B. Donald, and B. Rosenberg. 1998. Accumulation of persistent organochlorine compounds in mountains of western Canada. *Nature* 395:585–588.

Bowman, W.D., J.S. Baron, L.H. Geiser, M.E. Fenn, and E.A. Lilleskov. 2011. Northwestern forested mountains. *In* L.H. Pardo, M.J. Robin-Abbott, and C.T. Driscoll (Eds.). *Assessment of Nitrogen Deposition Effects and Empirical Critical Loads of Nitrogen for Ecoregions of the United States.* General Technical Report NRS-80. U.S. Forest Service, Newtown Square, PA, pp. 75–88.

Bradford, D.F., F. Tabatabai, and D.M. Graber. 1993. Isolation of remaining populations of the native frog, *Rana muscosa*, by introduced fishes in Sequoia and Kings Canyon National Parks, California. *Conserv. Biol.* 7(4):882–888.

Brockway, D.G., R.G. Gatewood, and R.B. Paris. 2002. Restoring fire as an ecological process in shortgrass prairie ecosystems: Initial effects of prescribed burning during the dormant and growing seasons. *J. Environ. Manage.* 65:135–152.

Cahill, T.A., J.J. Carroll, D. Campbell, and T.E. Gill. 1996. Air quality. *In Sierra Nevada Ecosystem Project: Final Report to Congress.* Volume II: Assessments and Scientific Basis for Management Options. Centers for Water and Wildland Resources, University of California, Davis, CA, pp. 1227–1261.

Clow, D.W., R.G. Striegl, D.H. Campbell, and M.A. Mast. 2000. Survey of high-altitude lake chemistry in national parks in the western United States. *Proceedings of the International Symposium on High Mountain Lakes and Streams*, Innsbruck, Austria.

Covington, W.W. 2000. Helping western forests heal: The prognosis is poor for United States forest ecosystems. *Nature* 408:135–136.

Eagles-Smith, C.A., J.J. Willacker, Jr., and C.M. Flanagan Pritz. 2014. Mercury in fishes from 21 national parks in the western United States—Inter- and intra-park variation in concentrations and ecological risk. U.S. Geological Survey Open-File Report 2014-1051, Reston, VA.

Eilers, J.M., P. Kanciruck, R.A. McCord, W.S. Overton, L. Hook, D.J. Blick, D.F. Brakke et al. 1987. *Characteristics of Lakes in the Western United States. Volume II: Data Compendium for Selected Physical and Chemical Variables.* EPA-600/3-86/054b. U.S. EPA, Washington, DC.

Eilers, J.M., C.L. Rose, and T.J. Sullivan. 1994. Status of air quality and effects of atmospheric pollutants on ecosystems in the Pacific Northwest Region of the National Park Service. National Park Service, Air Resources Division, Denver, CO.

Ellis, R.A., D.J. Jacob, M.P. Sulprizio, L. Zhang, C.D. Holmes, B.A. Schichtel, T. Blett, E. Porter, L.H. Pardo, and J.A. Lynch. 2013. Present and future nitrogen deposition to national parks in the United States: Critical load exceedances. *Atmos. Chem. Phys.* 13(17):9083–9095.

Elser, J.J., E.R. Marzolf, and C.R. Goldman. 1990. Phosphorus and nitrogen limitation of phytoplankton growth in the freshwaters of North America: A review and critique of experimental enrichments. *Can. J. Fish. Aquat. Sci.* 47:1468–1477.

Fenn, M.E., J.S. Baron, E.B. Allen, H.M. Rueth, K.R. Nydick, L. Geiser, W.D. Bowman et al. 2003a. Ecological effects of nitrogen deposition in the western United States. *BioScience* 53(4):404–420.

Fenn, M.E., R. Haeuber, G.S. Tonnesen, J.S. Baron, S. Grossman-Clark, D. Hope, D.A. Jaffe et al. 2003b. Nitrogen emissions, deposition, and monitoring in the western United States. *BioScience* 53(4):391–403.

Flannigan, M.D., Y. Bergeron, O. Engelmark, and B.M. Wotton. 1998. Future wildfire in circumboreal forests in relation to global warming. *J. Veg. Sci.* 9:469–476.

Fulé, P.Z., W.W. Covington, and M.M. Moore. 1997. Determining reference conditions for ecosystem management of southwestern ponderosa pine forests. *Ecol. Appl.* 7:895–908.

Garcia-Reyero, N., J.O. Grimalt, I. Vives, P. Fernandez, and B. Piña. 2007. Estrogenic activity associated with organochlorine compounds in fish extracts from European mountain lakes. *Environ. Pollut.* 145(3):745–752.

Geiser, L.H., S.E. Jovan, D.A. Glavich, and M.K. Porter. 2010. Lichen-based critical loads for atmospheric nitrogen deposition in western Oregon and Washington forests, USA. *Environ. Pollut.* 158:2412–2421.

Geiser, L.H. and P.N. Neitlich. 2007. Air pollution and climate gradients in western Oregon and Washington indicated by epiphytic macrolichens. *Environ. Pollut.* 145:203–218.

Goldman, C.R. 1988. Primary productivity, nutrients, and transparency during the early onset of eutrophication in ultra-oligotrophic Lake Tahoe, California-Nevada. *Limnol. Oceanogr.* 33:1321–1333.

Grand Canyon Visibility Transport Commission (GCVTC). 1996. Recommendations for improving western vistas. Report to the U.S. Environmental Protection Agency. Available at: http://www.westgov.org/images/dmdocuments/GCVTCFinal.PDF (accessed January, 2014).

Hageman, K.J., S.L. Simonich, D.H. Campbell, G.R. Wilson, and D.H. Landers. 2006. Atmospheric deposition of current-use and historic-use pesticides in snow at national parks in the western United States. *Environ. Sci. Technol.* 40(10):3174–3180.

Harries, J.E., D.A. Sheahan, S. Jobling, P. Matthiessen, P. Neall, J.P. Sumpter, T. Tylor, and N. Zaman. 1997. Estrogenic activity in five United Kingdom rivers detected by measurement of vitellogenesis in caged male trout. *Environ. Toxicol. Chem.* 16(3):534–542.

Jaffe, D., T. Anderson, D.S. Covert, R. Kotchenruther, B. Trost, J. Danielson, W. Simpson et al. 1999. Transport of Asian air pollution to North America. *Geophys. Res. Lett.* 26:711–714.

Jaffe, D., D. Chand, W. Hafner, A. Westerling, and D. Spracklen. 2008. Influence of fires on O_3 concentrations in the western U.S. *Environ. Sci. Technol.* 42:5885–5891.

Jaffe, D., H. Price, D. Parrish, A. Goldstein, and J. Harris. 2003. Increasing background ozone during spring on the west coast of North America. *Geophys. Res. Lett.* 30(12):4.

Jaffe, D.A. and J. Ray. 2007. Increase in ozone at rural sites in the western U.S. *Atmos. Environ.* 41(26):5452–5463.

Jassby, A.D., J.E. Reuter, R.P. Axler, C.R. Goldman, and S.H. Hackley. 1994. Atmospheric deposition of N and phosphorus in the annual nutrient load of Lake Tahoe (California-Nevada). *Water Resour. Res.* 30:2207–2216.

Jovan, S. 2008. Lichen bioindication of biodiversity, air quality, and climate: Baseline results from monitoring in Washington, Oregon, and California. General Technical Report PNW-GTR-737. U.S. Department of Agriculture, Forest Service, Pacific Northwest Research Station, Portland, OR.

Jovan, S. and B. McCune. 2005. Air quality bioindication in the greater Central Valley of California, with epiphytic macrolichen communities. *Ecol. Appl.* 15(5):1712–1726.

Klimont, Z., J. Cofala, W. Schöpp, M. Amann, D.G. Streets, Y. Ichikawa, and S. Fujita. 2001. Projections of SO_2, NO_x, NH_3, and VOC emissions in East Asia up to 2030. *Water Air Soil Pollut.* 130:193–198.

Kohut, R. 2007. Assessing the risk of foliar injury from ozone on vegetation in parks in the U.S. National Park Service's Vital Signs Network. *Environ. Pollut.* 149:348–357.

Kotchenruther, R.A., D.A. Jaffe, and L. Jaegle. 2001. Ozone photochemistry and the role of peroxy-acetyl nitrate in the springtime northeastern Pacific troposphere: Results from the PHOEBEA campaign. *J. Geophys. Res.* 106:28731–28741.

Krabbenhoft, D.P., M.L. Olson, J.F. DeWild, D.W. Clow, R.G. Striegl, M.M. Dornblaser, and P. VanMetre. 2002. Mercury loading and methylmercury production and cycling in high-altitude lakes from the western United States. *Water Air Soil Pollut.* 2:233–249.

Landers, D.H., J.M. Eilers, D.F. Brakke, W.S. Overton, P.E. Kellar, W.E. Silverstein, R.D. Schonbrod et al. 1987. *Characteristics of Lakes in the Western United States. Volume I: Population Descriptions and Physico-Chemical Relationships.* EPA-600/3-86/054a. U.S. Environmental Protection Agency, Washington, DC.

Landers, D.H., S.M. Simonich, D. Jaffe, L. Geiser, D.H. Campbell, A. Schwindt, C.B. Schreck et al. 2010. The Western Airborne Contaminant Project (WACAP): An interdisciplinary evaluation of the impacts of airborne contaminants in Western U.S. National Parks. *Environ. Sci. Technol.* 44(3):855–859.

Lenihan, J.M., C. Daly, D. Bachelet, and R.P. Neilson. 1998. Simulating broad-scale fire severity in a dynamic global vegetation model. *Northwest Sci.* 72:91–103.

Liu, W., P.K. Hopke, and R.A. VanCuren. 2003. Origins of fine aerosol mass in the western United States using positive matrix factorization. *J. Geophys. Res.* 108(D23, 4716):18.

Malm, W.C. 1999. *Introduction to Visibility.* Cooperative Institute for Research in the Atmosphere (CIRA), Fort Collins, CO.

McColl, J.G. 1981. Effects of acid rain on plants and soils in California. Final report. Contract A7-169-30. California Air Resources Board, Sacramento, CA.

McCune, B., J. Dey, J. Peck, K. Heiman, and S. Will-Wolf. 1997. Regional gradients in lichen communities of the southeast United States. *Bryologist* 100(2):145–158.

McCune, B. and L. Geiser. 2009. *Macrolichens of the Pacific Northwest.* Oregon State University Press, Corvallis, OR.

McKenzie, D., Z.M. Gedalof, D.L. Peterson, and P. Mote. 2004. Climatic change, wildfire, and conservation. *Conserv. Biol.* 18:890–902.

McKenzie, D., S.M. O'Neill, N.K. Larkin, and R.A. Norheim. 2006. Integrating models to predict regional haze from wildland fire. *Ecol. Model.* 199:278–288.

Melack, J.M., J.O. Sickman, F. Setaro, and D. Dawson. 1997. Monitoring of wet deposition in alpine areas in the Sierra Nevada. Final report. California Air Resources Board Contract A932-081. Marine Sciences Institute, University of California, Santa Barbara, CA.

Melack, J.M. and J.L. Stoddard. 1991. Sierra Nevada: Unacidified, very dilute waters and mildly acidic atmospheric deposition. *In* D.F. Charles (Ed.). *Acidic Deposition and Aquatic Ecosystems: Regional Case Studies.* Springer-Verlag, New York, pp. 503–530.

Melack, J.M., J.L. Stoddard, and C.A. Ochs. 1985. Major ion chemistry and sensitivity to acid precipitation of Sierra Nevada lakes. *Water Resour. Res.* 21:27–32.

Miller, C. and P. Landres. 2004. Exploring information needs for wildland fire and fuels management. USDA Forest Service General Technical Report RMRS-GTR-127. U.S. Forest Service Rocky Mountain Research Station, Fort Collins, CO.

National Atmospheric Deposition Program/National Trends Network. 2006. Isopleth maps. Program Office, Illinois State Water Survey, Champaign, IL.

Nelson, P.O. 1991. Cascade mountains: Lake chemistry and sensitivity of acid deposition. *In* D.F. Charles (Ed.). *Acidic Deposition and Aquatic Ecosystems: Regional Case Studies.* Springer-Verlag, New York, pp. 531–563.

Oltmans, S.J., A.S. Lefohn, J.M. Harris, and D.S. Shadwick. 2008. Background ozone levels of air entering the west coast of the US and assessment of longer-term changes. *Atmos. Environ.* 42:6020–6038.

Peterson, D.L. and M.J. Arbaugh. 1992. Coniferous forests of the Colorado front range. Part B: Ponderosa pine second-growth stands. *In* M.J. Mitchell and S.E. Lindberg (Eds.). *Atmospheric Deposition and Forest Nutrient Cycling: A Synthesis of the Integrated Forest Study.* Springer-Verlag, New York, pp. 433–460.

Peterson, D.L., M.J. Arbaugh, and L.J. Robinson. 1991. Regional growth changes in ozone-stressed ponderosa pine (*Pinus ponderosa*) in the Sierra Nevada, California, USA. *Holocene* 1:50–61.

Peterson, D.L., T.J. Sullivan, J.M. Eilers, S. Brace, D. Horner, K. Savig, and D. Morse. 1998. Assessment of air quality and air pollutant impacts in national parks of the Rocky Mountains and northern Great Plains. NPS D-657. U.S. Department of the Interior, National Park Service, Air Resources Division, Denver, CO.

Pike, L.H. 1978. The importance of epiphytic lichens in mineral cycling. *Bryologist* 81:247–257.

Reuss, J.O., F.A. Vertucci, R.C. Musselman, and R.A. Sommerfeld. 1995. Chemical fluxes and sensitivity to acidification of two high-elevation catchments in southern Wyoming. *J. Hydrol.* 173:165–189.

Rocchio, J.E., D.M. Ewell, C.T. Procter, and B.K. Takemoto. 1993. Project FOREST: The forest ozone response study. *In* W.E. Brown and S.D. Veirs (Eds.). *Partners in Stewardship.* George Wright Society, Hancock, MI, pp. 112–119.

Rorig, M.L. and S.A. Ferguson. 1999. Characteristics of lightning and wildland fire ignition in the Pacific Northwest. *J. Appl. Meteorol.* 38:1565–1575.

Savig, K. and D. Morse. 2001. Visibility sections. *In* T.J. Sullivan, D.L. Peterson, C.L. Blanchard, S.J. Tanenbaum, K. Savig and D. Morse (Eds.). *Assessment of Air Quality and Air Pollutant Impacts in Class I National Parks of California.* U.S. Department of the Interior, National Park Service, Air Resources Division, Denver, CO.

Schilling, S. and D. Duriscoe. 1996. Data management and analysis of ozone injury to pines. *In* P.R. Miller, K.W. Stolte, D.M. Duriscoe and J. Pronos (Technical Advisors.). *Evaluating Ozone Air Pollution Effects on Pines in the Western United States.* USDA Forest Service, Albany, CA. pp. 63–69.

Schoettle, A.W., K. Tonnessen, J. Turk, J. Vimont, and R. Amundson. 1999. An assessment of the effects of human-caused air pollution on resources within the interior Columbia River basin. PNW-GTR-447. USDA Forest Service, Pacific Northwest Research Station, Portland, OR.

Schreck, C.B. and M. Kent. 2013. Extent of endocrine disruption in fish of western and Alaskan national parks. Final report. NPS-OSU Task Agreement J8W07080024, Oregon State University, Corvallis, OR.

Schwede, D.B. and G.G. Lear. 2014. A novel hybrid approach for estimating total deposition in the United States. *Atmos. Environ.* 92:207–220.

Schwindt, A.R., M.L. Kent, L.K. Ackerman, S.L.M. Simonich, D.H. Landers, T. Blett, and C.B. Schreck. 2009. Reproductive abnormalities in trout from western U.S. national parks. *Trans. Am. Fish. Soc.* 138(3):522–531.

Sollins, P., C.C. Grier, F.M. McCorison, K. Cromack, Jr., R. Fogel, and R.L. Fredriksen. 1980. The internal element cycles of an old-growth Douglas-fir ecosystem in western Oregon. *Ecol. Monogr.* 50:261–285.

Stephenson, N. 1999. Reference conditions for giant sequoia forest restoration. *Ecol. Appl.* 9:1253–1265.

Streets, D.G. and S.T. Waldhoff. 2000. Present and future emissions of air pollutants in China: SO_2, NO_x, and CO. *Atmos. Environ.* 34:363–374.

Suarez-Murias, T., J. Glass, E. Kim, L. Melgoza, and T. Najita. 2009. California regional haze plan. California Environmental Protection Agency, Air Resources Board, Sacramento, CA.

Sullivan, T.J. 2000. *Aquatic Effects of Acidic Deposition.* CRC Press, Boca Raton, FL.

Sullivan, T.J. and J.M. Eilers. 1994. Assessment of deposition levels of sulfur and nitrogen required to protect aquatic resources in selected sensitive regions of North America. Final report prepared for Technical Resources, Inc., Rockville, MD, under contract to U.S. Environmental Protection Agency, Environmental Research Laboratory-Corvallis. E&S Environmental Chemistry, Inc., Corvallis, OR.

Sullivan, T.J., T.C. McDonnell, G.T. McPherson, S.D. Mackey, and D. Moore. 2011b. Evaluation of the sensitivity of inventory and monitoring national parks to nutrient enrichment effects from atmospheric nitrogen deposition. Natural Resource Report NPS/NRPC/ARD/NRR—2011/313. U.S. Department of the Interior, National Park Service, Denver, CO.

Sullivan, T.J., G.T. McPherson, T.C. McDonnell, S.D. Mackey, and D. Moore. 2011a. Evaluation of the sensitivity of inventory and monitoring national parks to acidification effects from atmospheric sulfur and nitrogen deposition. Natural Resource Report. NPS/NRPC/ARD/NRR—2011/349. U.S. Department of the Interior, National Park Service, Denver, CO.

Sullivan, T.J., D.L. Peterson, C.L. Blanchard, K. Savig, and D. Morse. 2001. Assessment of air quality and air pollutant impacts in Class I national parks of California. NPS D-1454. U.S. Department of the Interior, National Park Service, Air Resources Division, Denver, CO.

U.S. Environmental Protection Agency. 2005. Review of the national ambient air quality standards for particulate matter: Policy assessment of scientific and technical information. OAQPS Staff Paper. EPA-452/R-05-005a. Office of Air Quality Planning and Standards, Research Triangle Park, NC.

U.S. Geological Survey (USGS). Last modified February 20, 2015. Predicted surface water methylmercury concentrations in National Park Service Inventory and Monitoring Program Parks. U.S. Geological Survey. Wisconsin Water Science Center, Middleton, WI. Available at: http://wi.water.usgs.gov/mercury/NPSHgMap.html (accessed February 26, 2015).

Urban, D.L. 1990. A versatile model to simulate forest pattern: A user's guide to Zelig version 1.0. Environmental Sciences Department, University of Virginia, Charlottesville, VA.

Urban, D.L., G.B. Bonan, T.M. Smith, and H.H. Shugart. 1991. Spatial applications of gap models. *For. Ecol. Manage.* 42:95–110.

van Dobben, H.F., T. Wolterbeek, G.W.W. Wamelink, and C.J.F. Ter Braak. 2001. Relationship between epiphytic lichens, trace elements and gaseous atmospheric pollutants. *Environ. Pollut.* 112:163–169.

Weinstein, D.A., R.M. Beloin, and R.D. Yanai. 1991. Modeling changes in red spruce carbon balance and allocation in response to interacting ozone and nutrient stress. *Tree Physiol.* 9:127–146.

Weinstein, D.A., J.A. Laurence, W.A. Retzlaff, J.S. Kern, E.H. Lee, W.E. Hogsett, and J. Weber. 2005. Predicting the effects of tropospheric ozone on regional productivity of ponderosa pine and white fir. *For. Ecol. Manage.* 205:73–89.

Wigington, P.J., Jr., T.D. Davies, M. Tranter, and K.N. Eshleman. 1990. Episodic acidification of surface waters due to acidic deposition. State of Science/Technology Report 12. National Acid Precipitation Assessment Program, Washington, DC.

14

Hawaii

14.1 Background

The Pacific Island Network includes islands at diverse locations in the Pacific Ocean. It includes the Hawaiian Islands, Northern Mariana Islands, Guam, and American Samoa. This analysis focuses on the parks in Hawaii. Figure 14.1 shows the Pacific Island Network boundaries across the Hawaiian Islands, along with locations of each park and population centers with more than 10,000 people around the network. There are nine parks in the Pacific Island Network. This chapter focuses on Hawaii Volcanoes National Park (HAVO) and Haleakala National Park (HALE). Only one, HAVO, is larger than 100 mi^2 (259 km^2). It is located on the island of Hawaii. HALE and HAVO are Class I air quality areas, which receive the highest level of protection under the Clean Air Act from effects attributable to poor air quality; the rest of the Pacific Island Network parks are Class II areas. The NPS air quality efforts focus largely on the Class I areas, but all national parks are protected. The only population center in the Pacific Island Network region of any magnitude (more than 100,000 people) is Honolulu on Oahu.

Some of the largest areas of protected humid tropical forest and humid subtropical forest in the United States are located within HAVO and HALE. These forests contain a large number of endemic species. In the United States, the tropical and subtropical humid forest ecoregion occurs only in southern Florida, Puerto Rico, and Hawaii. Biodiversity is very high in this ecoregion and includes many species of epiphytes, which have been found to be highly sensitive to N inputs in other biomes.

14.2 Atmospheric Emissions and Deposition

In a screening analysis by Sullivan et al. (2011), human-caused county-level sulfur dioxide emissions within the Hawaii portion of the Pacific Island Network ranged from less than 1 ton per square mile per year (ton/mi^2/yr) of sulfur dioxide on Kauai to the range of 25–50 tons/mi^2/yr on Oahu. Other islands in the Hawaiian archipelago generally had human-caused sulfur dioxide emissions in the range of 1–5 tons/mi^2/yr. There were very few individual human-caused point sources of sulfur dioxide on any of the Hawaiian Islands and none of any magnitude. All emitted less than 5000 tons of sulfur dioxide per year. Natural S emissions sources include the Kilauea Volcano, whose emissions ranged

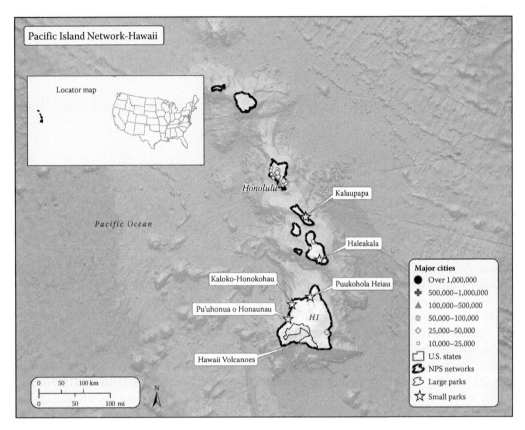

FIGURE 14.1
Network boundary and locations of parks and population centers with more than 10,000 people within the Hawaiian Islands in the Pacific Island Network region.

from about 600,000 to 1.4 million tons per year of sulfur dioxide from 2008 to 2010 (Elias and Sutton 2012). High levels of atmospheric sulfur dioxide concentrations, caused by the Kilauea volcano, have led to closures of parts of HAVO because of human health concerns. The HAVO alert site provides real-time air quality data for sulfur dioxide along with wind direction and speed (http://www.hawaiiso2network.com/).

Human-caused county-level N oxide emissions within the Pacific Island Network region ranged from less than 1 ton/mi^2/yr on the big island of Hawaii to more than 20 tons/mi^2/yr on Oahu. Important sources of N oxide emissions include motor vehicles and industry. Annual emissions of N from the other Hawaiian Islands were intermediate, between 1 and 5 tons/mi^2/yr. There are relatively few N point sources in the Hawaiian Islands. Most are located on Oahu.

County-level emissions near the Pacific Island Network, based on data from the EPA's National Emissions Inventory during a recent time period (2011), are depicted in Figures 14.2 through 14.4 for sulfur dioxide, oxidized N, and ammonia, respectively. Counties in the vicinity of Hawaiian parks had relatively low sulfur dioxide emissions (1–5 tons/mi^2/yr; Figure 14.2). Emissions of oxidized N were lowest on the island of Hawaii (Figure 14.3). Emissions of ammonia near the Hawaiian parks were below 8 tons/mi^2/yr (Figure 14.4).

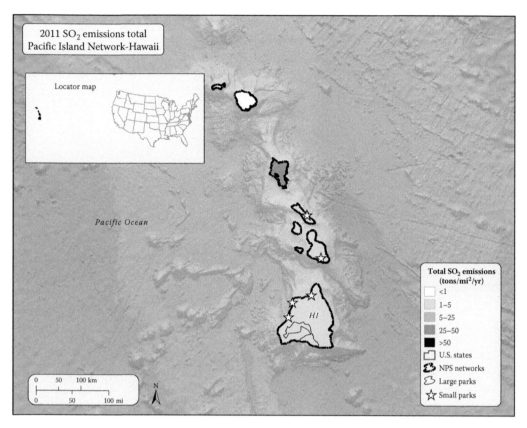

FIGURE 14.2

Total sulfur dioxide (SO$_2$) emissions, by county, near the Pacific Island Network for the year 2011. (Data from EPA's National Emissions Inventory, https://www.epa.gov/air-emissions-inventories, accessed January, 2014.)

Atmospheric deposition data are not available for recent years in the Pacific Island Network region, but deposition of N and S are likely relatively high at some locations on some of the Hawaiian Islands, especially Oahu and Hawaii, and lower elsewhere. There was a National Atmospheric Deposition Program wet deposition monitoring station at HAVO that reported deposition data as recently as 2003 and 2004. Wet N deposition was low, less than 1 kg N/ha/yr, during both years. Wet S deposition was considerably higher, about 6 and 11 kg S/ha/yr during those years. There was also a deposition monitoring site in operation at Mauna Loa in HAVO through 1993. Total N and S deposition values at that site were below 1 kg/ha/yr for both N and S during the two most recent years of data collection.

14.3 Acidification

Many species of moss occur in Pacific Island Network parks. Moss is known to be highly sensitive to acidic deposition in other biomes. Little is known, however, about the acid

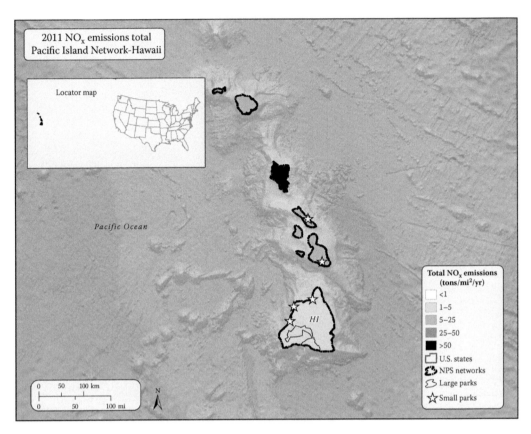

FIGURE 14.3
Total oxidized N (NO_x) emissions, by county, near the Pacific Island Network for the year 2011. (Data from EPA's National Emissions Inventory, https://www.epa.gov/air-emissions-inventories, accessed January, 2014.)

sensitivity of moss species in the Pacific Island Network region. Waite (2007) compiled a list of moss species in HALE, which identified 110 species in and near the park. An estimated 43% of all moss species that occur in the Hawaiian Islands are represented at this park. Thus, potential effects of S and N (and heavy metal) deposition on the rich moss flora of this and perhaps other parks in the Pacific Island Network may be important ecological concerns.

Information is not available regarding the possible effects of human-caused acidic deposition on aquatic or terrestrial resources in the Pacific Island Network. However, more data are available regarding the effects of natural volcanic emissions of S and other constituents in, and downwind of, HALE (Nelson and Sewake 2008). The Kilauea volcano has been erupting continually since 1983, with increased emissions since late 2007 and 2008. The caldera area and land to the southwest are largely devoid of vegetation, covered mainly by recent lava flows. There are also localized areas affected by high emissions and deposition of S and other volcanic constituents, including heavy metals. Trade winds carry gases from the volcano in a southwesterly direction. Plant damage has been reported for a variety of crop and other plant species (Nelson and Sewake 2008).

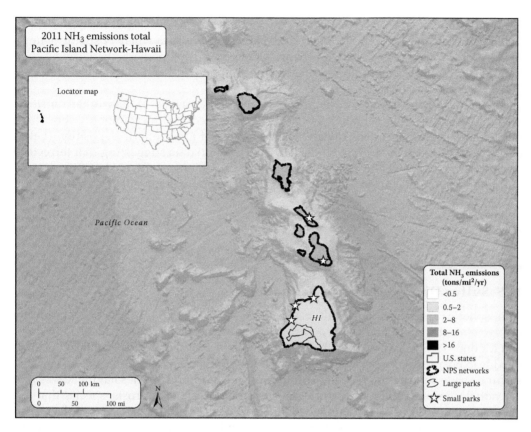

FIGURE 14.4

Total ammonia (NH_3) emissions, by county, near the Pacific Island Network for the year 2011. (Data from EPA's National Emissions Inventory, https://www.epa.gov/air-emissions-inventories, accessed January, 2014.)

14.4 Nutrient Nitrogen Enrichment

Some limited experimental work has been conducted in the Pacific Island Network region on nutrient N enrichment. Fertilization with ammonium nitrate in a P-limited forest on Kaua'i caused increased emissions from soils of N oxides subsequent to addition of 15 kg N/ha/yr, but not subsequent to addition of 5 kg N/ha/yr (Hall and Matson 1999, Carillo et al. 2002). Ambient wet N deposition at this site was <1 kg N/ha/yr. Experimental addition of a large amount (>100 kg/ha/yr) of N did not stimulate plant growth or affect species composition in the forest (Harrington et al. 2001, Ostertag and Verville 2002). Similarly, high N addition did not affect epiphytic (aboveground) lichen abundance or diversity in P-limited or N + P-limited lower montane rain forests (Benner et al. 2007, Benner and Vitousek 2007).

On tropical or subtropical sites where N is not limiting, atmospheric N deposition would not be expected to alter productivity or plant species composition. However, loss of N to the atmosphere as gases produced by denitrification or to drainage water as nitrate may be stimulated by increased N deposition (Herbert and Fownes 1995, Hall and Matson 1999, Lohse and Matson 2005, Templer et al. 2008, Hall 2011). On sites where N supply is limiting, added N would be expected to increase plant growth and perhaps change plant community composition, eventually leading to N saturation (Hall and Matson 1999, Erickson et al. 2001, Feller et al. 2007).

Tropical and subtropical forests are often relatively rich in N, and other nutrients are more often limiting (Chadwick et al. 1999). Such ecosystems commonly show high rates of N leaching and denitrification, irrespective of atmospheric deposition (Lewis et al. 1999, Davidson et al. 2007, Hall 2011). There are no data suggesting that terrestrial or aquatic resources in the Pacific Island Network region are affected at this time by nutrient enrichment caused by atmospheric N deposition at current atmospheric loading rates.

14.5 Sulfur Dioxide Injury to Vegetation

Atmospheric concentrations of sulfur dioxide in the range of 0.1–1 parts per million can change stomatal conductance for a variety of plant species (Farrar et al. 1977). Winner and Mooney (1980) showed that changes in stomatal conductance in response to sulfur dioxide exposure are associated with foliar injury. Hawaiian plants exposed to volcanic sulfur dioxide showed interspecific differences in leaf injury. Species having leaves that did not close stomata upon exposure developed chlorosis or necrosis; species that did close stomata did not show damage symptoms. Tanner et al. (2007) found decreased stomatal index (number of stomata relative to number of epidermal cells) of swordfern (*Nephrolepis exaltata*) in the plumes of outgassing vents on the Kilauea volcano, where atmospheric concentrations of both sulfur dioxide and carbon dioxide were much higher than background and also at a site where sulfur dioxide (but not carbon dioxide) was elevated.

14.6 Visibility Degradation

14.6.1 Natural Background Visibility and Monitored Visibility Conditions

Visibility is monitored in HALE (HALE1) and HAVO (HAVO1). In 2007, a second Interagency Monitoring of Protected Visual Environments (IMPROVE) monitor was installed in HALE near the crater; the new site is designated HACR1 (Haleakala Crater). The two monitors were run simultaneously for several years for comparison. In 2012, the HALE1 monitor was decommissioned, and the HACR1 monitor now serves to characterize visibility at HALE. Data from the HALE1 monitor have been used in this analysis to represent visibility conditions for HALE. Haze from volcanic smog (called vog locally) is caused by gaseous and particulate emissions from the Kilauea volcano (Nelson and Sewake 2008). Such haze can be substantial in western portions of Hawaii Island.

TABLE 14.1

Estimated Natural Haze and Measured Ambient Haze in Selected Pacific Island Network Parks Averaged over the Period 2004–2008[a]

Park Name	Park Code	Site ID	Estimated Natural Haze (in Deciviews)		
			20% Clearest Days	20% Haziest Days	Average Days
Haleakala	HALE	HALE1	2.66	7.43	4.95
Hawaii Volcanoes	HAVO	HAVO1	2.20	7.17	4.54
Park Name	Park Code	Site ID	Measured Ambient Haze (for Years 2004–2008) (in Deciviews)		
			20% Clearest Days	20% Haziest Days	Average Days
Haleakala	HALE	HALE1	4.39	14.24	8.88
Hawaii Volcanoes	HAVO	HAVO1	3.76	22.59	10.37

[a] Visibility conditions have deteriorated significantly since that time due to increased volcanic activity.

Data from 2004 to 2008 were used to document haze conditions at parks in the Pacific Island Network in order to provide information on spatial differences in visibility and air pollution (Table 14.1). Haze levels ranged from very low haze (very good visibility) to very high haze (very poor visibility). Haze is reported on the 20% clearest, 20% haziest, and average days for the monitored parks in this network. Since the period 2004–2008 was used for this ranking, it does not reflect changes in haze since the late 2007 enhanced eruption of Kilauea. A more recent analysis showed that haze increased significantly on the haziest days at HAVO and HALE from 2000 to 2009 (NPS 2013).

Representative photos of a selected vista in HAVO under three different visibility conditions are shown in Figure 14.5. Photos were selected to correspond with the clearest 20% visibility conditions, the haziest 20% visibility conditions, and annual average visibility conditions at that location. The impact of haze on the 20% haziest days is pronounced.

IMPROVE data allow estimation of visual range, which reflects how far a person can see. Data from HALE indicate that haze has reduced the average visual range from about 150 to 95 mi (242 to 152 km). On the haziest days, visual range has been reduced from 110 to 65 mi (177 to 105 km). Severe haze episodes occasionally reduce visibility to 21 mi (34 km) or less. At HAVO, haze has reduced the average visual range from 150 to 75 mi (242 to 120 km). On the haziest days, visual range has been reduced from 120 to 30 mi (193 to 48 km). Severe haze episodes occasionally reduce visibility to 4 mi (6 km) or less.

14.6.2 Composition of Haze

Figure 14.6 shows estimated natural (preindustrial), baseline (2000–2004), and current (2006–2010) levels of haze and its composition for HALE and HAVO. The majority of the haze in both monitored parks for all groups (20% clearest days, annual average, and 20% haziest days) was attributable to sulfate, sea salt, and coarse mass (Figure 14.6). By far, the largest percentage was attributable to sulfate. For the 20% haziest days and annual average, sulfate was responsible for more than 50% of the haze in HALE and more than 85% in HAVO. Results were similar for the HACR1 site in HALE (http://www.wrapair2.org/documents/6.0%20STATE%20AND%20CLASS%20I%20AREA%20SUMMARIES/6.05%20

FIGURE 14.5
Three representative photos of the same view in HAVO illustrating the 20% clearest visibility, the 20% haziest visibility, and the annual average visibility. Extinction is total particulate light extinction. (From http://vista. cira.colostate.edu/improve/Data/IMPROVE/Data_IMPRPhot.htm, accessed December, 2010.)

Hawaii/WRAP_RHRPR_Appendix_E_Hawaii.pdf). Sea salt played a larger role on the 20% clearest days, contributing 27.5% in HALE and 33.3% in HAVO. Nitrates and organics were also responsible for appreciable portions of the haze at these parks.

14.6.3 Trends in Visibility

Visibility monitoring data are shown in Figure 14.7 for the period of record at HALE and HAVO. Conditions from 2001 to 2009 degraded significantly on the haziest days at HAVO (NPS 2013). Other trends were more modest.

14.6.4 Development of State Implementation Plans

Variability in volcanic emissions make it difficult to estimate natural conditions, but the State of Hawaii can reduce emissions from power plants and other sources to try to achieve the Regional Haze Rule goals. States must evaluate progress by 2018 (and every 10 years thereafter) based on a baseline period of 2000–2004 (Air Resource Specialists 2007). In 2012, the National Park Service Air Resources Division recommended that federally enforceable sulfur dioxide limits be established for several large power plants on Maui, in addition to reductions required at other power plants in the State (National Park Service 2012).

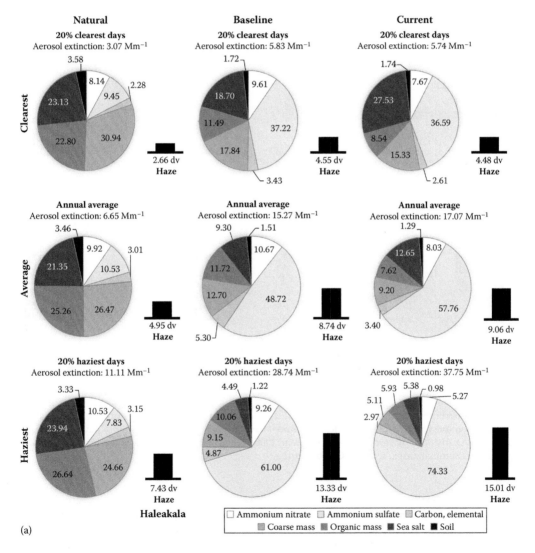

(a)

FIGURE 14.6

Estimated natural (preindustrial), baseline (2000–2004), and current (2006–2010) levels of haze (columns) and its composition (pie charts) on the 20% clearest, annual average, and 20% haziest visibility days for (a) HALE and (b) HAVO (the sites have no data for the year 2000). (From http://views.cira.colostate.edu/fed/Tools/RegionalHazeSummary.aspx, accessed October, 2012.)   (Continued)

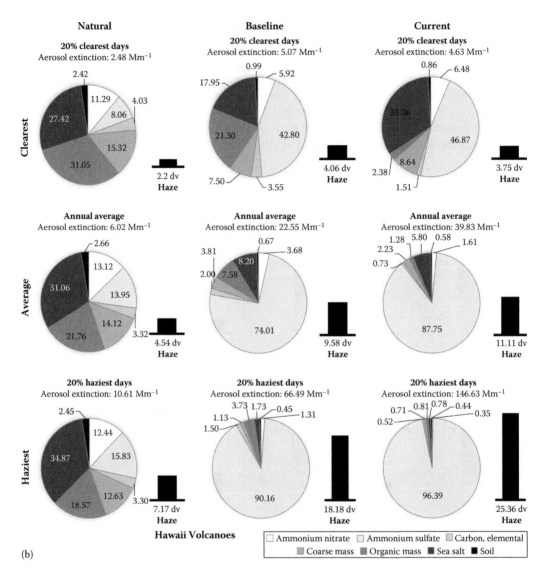

(b)

FIGURE 14.6 (*Continued*)
Estimated natural (preindustrial), baseline (2000–2004), and current (2006–2010) levels of haze (columns) and its composition (pie charts) on the 20% clearest, annual average, and 20% haziest visibility days for (a) HALE and (b) HAVO (the sites have no data for the year 2000). (From http://views.cira.colostate.edu/fed/Tools/RegionalHazeSummary.aspx, accessed October, 2012.)

Progress to date in meeting the national visibility goal is illustrated in Figure 14.8 using a uniform rate of progress glideslope. Although haze on the 20% clearest days shows essentially no change relative to the 2000–2004 baseline, haze on the 20% haziest days at HALE and HAVO shows increasing departure over time between the baseline and the glideslope of required progress toward natural background visibility by 2064. Although increases in haze are expected because of increased volcanic activity, it may be possible to further reduce emissions from human-caused sources.

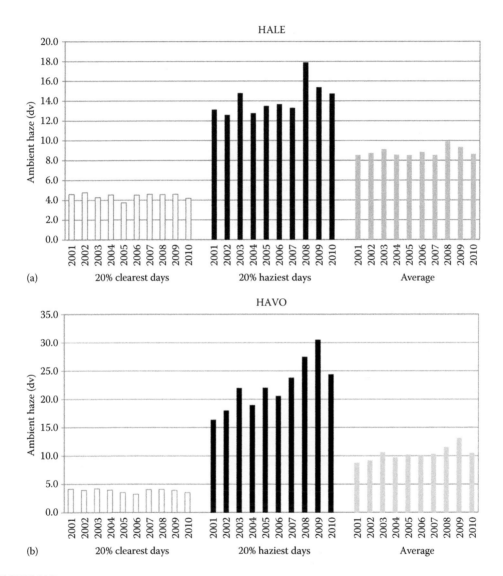

FIGURE 14.7

Trends in monitored visibility for (a) HALE and (b) HAVO, based on IMPROVE measurements on the 20% clearest, 20% haziest, and annual average visibility days over the monitoring period of record. (From http://vista. cira.colostate.edu/improve/Data/IMPROVE/summary_data.htm, accessed October, 2012.)

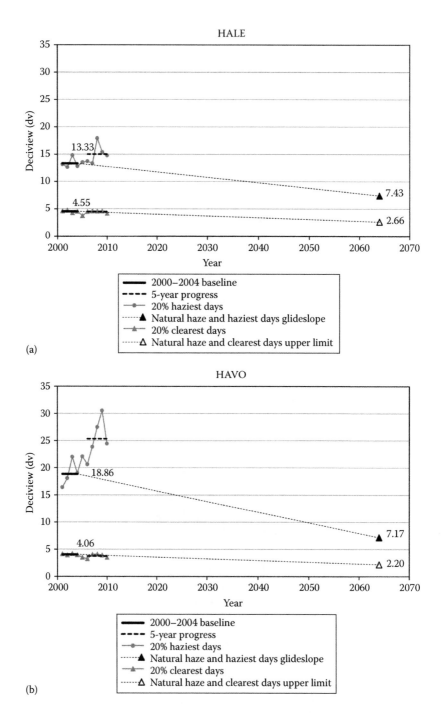

FIGURE 14.8
Glideslopes to achieving natural visibility conditions by 2064 for the 20% haziest (upper line) and the 20% clearest (lower line) days for (a) HALE and (b) HAVO (the sites have no data for the year 2000). In the Regional Haze Rule, the clearest days do not have a uniform rate of progress glideslope; the rule only requires that the clearest days do not get any worse than the baseline period. Also shown are measured values during the period 2000–2010. (From http://vista.cira.colostate.edu/improve/Data/IMPROVE/summary_data.htm, accessed October, 2012.)

14.7 Summary

Air pollution levels are relatively low in most of the Pacific Island Network parks. The exception is HAVO, where sulfur dioxide emissions from volcanic activity frequently contribute to high levels of air pollution. The Pacific Island Network parks are spread across a large portion of the Pacific Ocean. The only human population center in the Pacific Island Network region of any magnitude (more than 100,000 people) is Honolulu on the island of Oahu.

Emissions sources include vehicles, power plants, and industry. These human-caused emissions have local island effects but are now set against the background of large sulfur dioxide emissions from the Kilauea Volcano, which generally range from 1600 to 4000 tons per day since the most recent eruption in 2008 (Elias and Sutton 2012). Volcanic emissions affect primarily the Big Island, Hawaii, but are occasionally transported to the other islands, particularly Maui, Molokai, and Oahu.

Current atmospheric N and S deposition data are not available for Pacific Island Network parks, but it is expected that deposition is relatively high at some locations, especially on Oahu, because of nitrogen oxide emissions from vehicles and fuel combustion for power production, and on Hawaii, because of sulfur dioxide emissions from the volcano. Deposition measurements from 2000 to 2005 confirmed that S deposition resulting from sulfur dioxide emissions was high at HAVO, not surprisingly. Deposition is likely much higher now (at the time of this writing in 2015) downwind of the volcano, as emissions have increased significantly since the 2008 opening of a new vent in the volcano.

Sulfur and N pollutants can cause acidification of streams, lakes, and soils. Nitrogen pollutants can also cause undesirable nutrient enrichment of natural ecosystems, leading to changes in plant species diversity and soil nutrient cycling. The extent of such changes in Pacific Island Network parks to date is not known.

Particulate pollution can cause haze, reducing visibility. Monitoring has found high levels of haze at times in HAVO and HALE. Sulfate from volcanic emissions causes the majority of haze at both parks, but sulfate from fuel oil combustion (associated with power production) also contributes to haze in these parks.

References

Air Resource Specialists. 2007. VISTAS conceptual description support document. Report prepared for Visibility Improvement State and Tribal Association of the Southeast. Fort Collins, CO.

Benner, J.W., S. Conroy, C.K. Lunch, N. Toyoda, and P.M. Vitousek. 2007. Phosphorus fertilization increases the abundance and nitrogenase activity of the cyanolichen *Pseudocyphellaria crocata* in Hawaiian montane forests. *Biotropica* 39(3):400–405.

Benner, J.W. and P.M. Vitousek. 2007. Development of a diverse epiphyte community in response to phosphorus fertilization. *Ecol. Lett.* 10(7):628–636.

Carillo, J., M.G. Hastings, S.M. Sigman, and B.J. Huebert. 2002. Atmospheric deposition of inorganic and organic nitrogen and base cations in Hawaii. *Glob. Biogeochem. Cycles* 16(4):1076. doi:1010.1029/2002GB001892.

Chadwick, O.A., L.A. Derry, P.M. Vitousek, B.J. Huebert, and L.O. Hedin. 1999. Changing sources of nutrients during four million years of ecosystem development. *Nature* 397(6719):491–497.

Davidson, E.A., C.J.R. de Carvalho, A.M. Figueira, F.Y. Ishida, J. Ometto, G.B. Nardoto, R.T. Saba et al. 2007. Recuperation of nitrogen cycling in Amazonian forests following agricultural abandonment. *Nature* 447(7147):995–998.

Elias, T. and A.J. Sutton. 2012. Sulfur dioxide emission rates from Kïlauea Volcano, Hawai'i, 2007–2010. Open-File Report 2012–1107. U.S. Department of the Interior, U.S. Geological Survey, Reston, VA.

Erickson, H., M. Keller, and E.A. Davidson. 2001. Nitrogen oxide fluxes and nitrogen cycling during postagricultural succession and forest fertilization in the humid tropics. *Ecosystems* 4(1):67–84.

Farrar, J.F., J. Relton, and A.J. Rutter. 1977. Sulphur dioxide and the growth of *Pinus sylvestris*. *J. Appl. Ecol.* 14(3):861–875.

Feller, I.C., C.E. Lovelock, and K.L. McKee. 2007. Nutrient addition differentially affects ecological processes of *Avicennia germinans* in nitrogen versus phosphorus limited mangrove ecosystems. *Ecosystems* 10(3):347–359.

Hall, S.J. 2011. Tropical and subtropical humid forests. *In* L.H. Pardo, M.J. Robin-Abbott, and C.T. Driscoll (Eds.). *Assessment of Nitrogen Deposition Effects and Empirical Critical Loads of Nitrogen for Ecoregions of the United States.* General Technical Report NRS-80. U.S. Forest Service, Newtown Square, PA, pp. 181–192.

Hall, S.J. and P.A. Matson. 1999. Nitrogen oxide emissions after nitrogen additions in tropical forests. *Nature* 400(6740):152–155.

Harrington, R.A., J.H. Fownes, and P.M. Vitousek. 2001. Production and resource use efficiencies in N and P-limited tropical forests: A comparison of responses to long-term fertilization. *Ecosystems* 4(7):646–657.

Herbert, D.A. and J.H. Fownes. 1995. Phosphorus limitation of forest leaf area and net primary production on a highly weathered soil. *Biogeochemistry* 29:223–235.

Lewis, W.M., J.M. Melack, W.H. McDowell, M. McClain, and J.E. Richey. 1999. Nitrogen yields from undisturbed watersheds in the Americas. *Biogeochemistry* 46(1–3):149–162.

Lohse, K.A. and P.A. Matson. 2005. Consequences of nitrogen additions for soil losses from wet tropical forests. *Ecol. Appl.* 15(5):1629–1648.

National Park Service. 2012. Letter from S. Johnson to G. Nudd, U.S. EPA, dated July 2, 2012. Denver, CO.

National Park Service–Air Resources Division (NPS–ARD). 2013. Air quality in national parks: Annual trends (2000–2009) and conditions (2005–2009) report. Natural Resource Report NPS/NRSS/ARD/NRR—2013/683. National Park Service, Denver, CO.

Nelson, S. and K. Sewake. 2008. Volcanic emissions injury to plant foliage. Plant Disease PD-47. Cooperative Extension Service, College of Tropical Agriculture and Human Resources, University of Hawaii, Manoa, HI.

Ostertag, R. and J.H. Verville. 2002. Fertilization with nitrogen and phosphorus increases abundance of non-native species in Hawaiian montane forests. *Plant Ecol.* 162(1):77–90.

Sullivan, T.J., G.T. McPherson, T.C. McDonnell, S.D. Mackey, and D. Moore. 2011. Evaluation of the sensitivity of inventory and monitoring national parks to acidification effects from atmospheric sulfur and nitrogen deposition. U.S. Department of the Interior, National Park Service, Denver, CO.

Tanner, L.H., D.L. Smith, and A. Allan. 2007. Stomatal response of swordfern to volcanogenic CO_2 and SO_2 from Kilauea volcano. *Geophys. Res. Lett.* 34:L15807.

Templer, P.H., W.L. Silver, J. Pett-Ridge, K.M. DeAngelis, and M.K. Firestone. 2008. Plant and microbial controls on nitrogen retention and loss in a humid tropical forest. *Ecology* 89(11):3030–3040.

Waite, M. 2007. Mosses of Hawaii Volcanoes National Park. Technical Report 153. Pacific Cooperative Studies Unit, University of Hawaii, Manoa, HI.

Winner, W.E. and H.A. Mooney. 1980. Responses of Hawaiian plants to volcanic sulfur dioxide: Stomatal behavior and foliar injury. *Science* 210(4471):789–791.

15

Mojave Desert

15.1 Background

There are five parks in the Mojave Desert Network considered in this book that are larger than 100 mi^2 (259 km^2): Death Valley National Park (DEVA), Great Basin National Park (GRBA), Joshua Tree National Park (JOTR), Lake Mead National Recreation Area (LAKE), and Mojave National Preserve (MOJA; Figure 15.1). Larger parks generally have more available data with which to evaluate air pollution sensitivities and effects. Most available data have been collected for JOTR. Since air pollutants can be widespread, reduction of pollution emissions affecting large parks will often result in the protection of smaller parks located nearby as well.

The western Mojave Desert is impacted by air pollution that originates mainly in urban areas and then moves inland with the prevailing westerly winds (Edinger et al. 1972, Fenn et al. 2003, Allen et al. 2006). Inputs of atmospheric N to JOTR and other parklands in the Mojave Desert Network may affect plant production and distribution differentially. Nonnative species, especially some invasive grasses, often have higher N uptake rates and growth than many native species (Allen et al. 1998, Yoshida and Allen 2001, Brooks 2003, Yoshida and Allen 2004, Rao et al. 2010).

Figure 15.1 shows the network boundary along with locations of each park considered in this analysis and population centers with more than 10,000 people. There are several population centers larger than 100,000 people near Mojave Desert Network parks, including Los Angeles, San Diego, and Las Vegas. Air pollutants generated in the Los Angeles Basin have substantial impact on resources in the southern portion of the network, especially in JOTR.

15.2 Atmospheric Emissions and Deposition

County-level emissions near the Mojave Desert Network, based on data from the EPA's National Emissions Inventory during a recent time period (2011), are depicted in Figures 15.2 and 15.3 for oxidized nitrogen and reduced N (ammonia), respectively. Most counties in the vicinity of Mojave Desert Network parks had sulfur dioxide emissions lower than 1 ton/mi^2/yr. Emissions of oxidized N were higher, with values at some locations

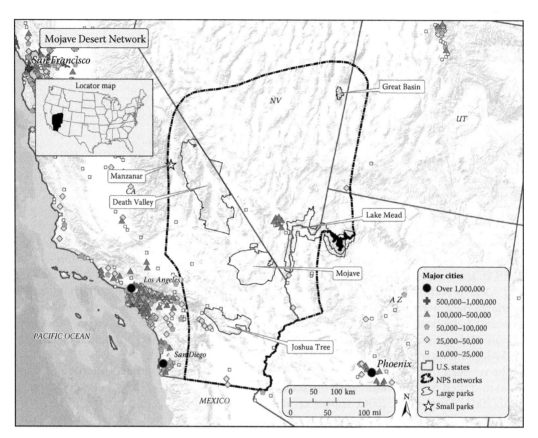

FIGURE 15.1
Network boundary and locations of parks and population centers larger than 10,000 people near the Mojave
Desert Network.

higher than 5 tons/mi^2/yr (Figure 15.2). Emissions of ammonia near Mojave Desert
Network parks were <2 tons/mi^2/yr at most locations (Figure 15.3). Emissions of S are
low and are not shown.

Atmospheric S deposition levels have increased slightly at most Mojave Desert
Network parks since 2001, based on Total Deposition project estimates (Table 15.1).
The largest increase (23.8%) was at GRBA. Estimated total N deposition over that same
time period increased at some parks and decreased at others. Ammonia deposition
increased at all network parks in all cases by considerable amounts (>40%). Oxidized N
deposition declined in all parks except GRBA. Total modeled S deposition throughout
the Mojave Desert Network was generally less than 2 kg S/ha/yr. Such low levels of S
deposition would not contribute to any appreciable soil or water acidification in this
network.

Oxidized inorganic N deposition for the period 2010–2012 was less than 5 kg N/ha/yr
throughout the parklands within the Mojave Desert Network. Most areas received less
than 2 kg N/ha/yr of reduced inorganic N from atmospheric deposition during this same

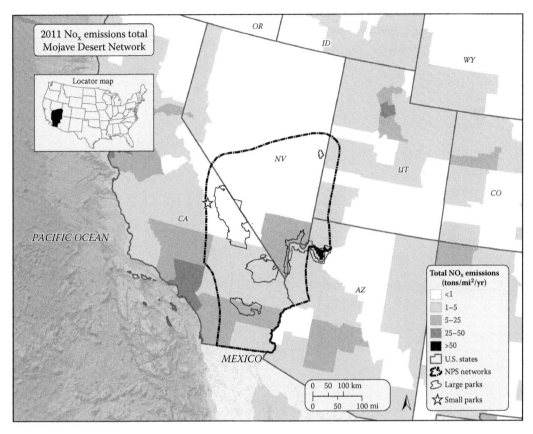

FIGURE 15.2

Total oxidized N (NOₓ) emissions, by county, near the Mojave Desert Network for the year 2011. (Data from EPA's National Emissions Inventory, https://www.epa.gov/air-emissions-inventories, accessed January, 2014.)

period; a few areas received higher amounts. Total modeled N deposition in 2011 was less than about 4 kg N/ha/yr in LAKE but was as high as 6.4 kg N/ha/yr in parts of JOTR and DEVA (Table 15.1; Figure 15.4).

Dry deposition predominates in arid ecosystems, such as are prevalent in the Mojave Desert Network. Dry deposition is difficult to quantify, however, in large part because the deposition is influenced by a multitude of factors, including the mix of air pollutants present, surface characteristics of soil and vegetation, and meteorological conditions (Weathers et al. 2006). Because the vegetation is sparse in arid ecosystems, soil surfaces are potentially exposed to substantial direct dry deposition of atmospheric pollutants. Padgett et al. (1999) found that inorganic N can accumulate on the soil surface of coastal sage scrub vegetation communities in southern California during dry smoggy periods. Upon onset of the rainy season, the concentration of nitrate in the surface soil decreased and was leached down the soil profile (Padgett et al. 1999, Padgett and Bytnerowicz 2001).

Allen et al. (2006) measured throughfall (water from precipitation that has dripped through the vegetation canopy prior to reaching the ground surface) N at two sites

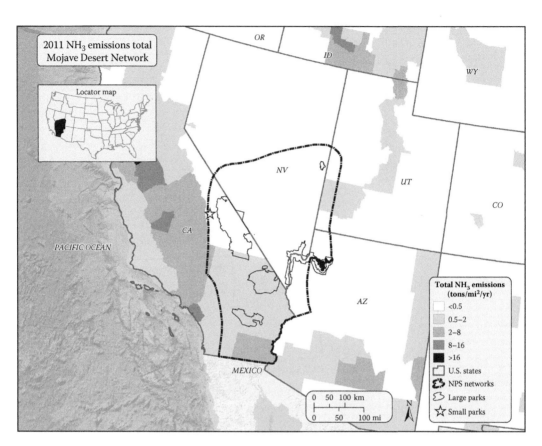

FIGURE 15.3
Total ammonia (NH_3) emissions, by county, near the Mojave Desert Network for the year 2011. (Data from EPA's National Emissions Inventory, https://www.epa.gov/air-emissions-inventories, accessed January, 2014.)

(Covington Flat and Pine City) in the western portion of JOTR. Under pinyon pine (*Pinus edulis*), throughfall N deposition measurements were 36.2 and 23.1 kg N/ha/yr. Assuming 27% and 20% tree cover, respectively, at the two sites, estimated stand-level N deposition averages were 12.4 and 6.2 kg N/ha/yr.

Emerging atmospheric deposition issues at GRBA concern deposition of dust and other contaminants to the snowpack during winter and early spring. Dust deposited on the snowpack can alter local hydrology by increasing melting and spring runoff and evaporation/sublimation from the snowpack. Dust deposition can not only alter hydrology but also affect the delivery of contaminants and nutrients to plants and to surface waters.

The Los Angeles Basin is not the only source of air pollutants that travel from outside the network boundaries to parklands within the Mojave Desert Network. Based on Western Regional Air Partnership analyses (Suarez-Murias et al. 2009), emissions of sulfate from ocean vessels constitute an important contributor to visibility impairment at coastal California Class I areas and may influence air quality at parks in the Mojave Desert Network.

TABLE 15.1

Average Changes in S and N Deposition[a] between 2001 and 2011 across Park Grid Cells at the Larger Mojave Desert Network Parks

Park Code	Parameter	2001 Average (kg/ha/yr)	2011 Average (kg/ha/yr)	Absolute Change (kg/ha/yr)	Percent Change	2011 Minimum (kg/ha/yr)	2011 Maximum (kg/ha/yr)	2011 Range (kg/ha/yr)
DEVA	Total S	0.45	0.49	0.05	11.5	0.26	1.06	0.79
	Total N	3.19	3.51	0.31	9.6	1.94	6.35	4.42
	Oxidized N	2.40	2.37	−0.03	−1.3	1.34	4.24	2.90
	Reduced N	0.79	1.13	0.34	47.4	0.60	2.11	1.52
GRBA	Total S	1.07	1.30	0.23	23.8	0.71	1.64	0.92
	Total N	3.29	3.92	0.63	20.4	2.47	4.72	2.25
	Oxidized N	1.97	1.99	0.02	1.7	1.18	2.30	1.12
	Reduced N	1.32	1.93	0.61	49.0	1.29	2.42	1.13
JOTR	Total S	0.42	0.46	0.04	9.9	0.29	0.63	0.34
	Total N	5.62	4.87	−0.75	−13.4	2.86	6.44	3.58
	Oxidized N	4.83	3.41	−1.42	−29.2	2.02	4.49	2.47
	Reduced N	0.80	1.47	0.67	85.6	0.84	2.10	1.25
LAKE	Total S	0.56	0.52	−0.04	−6.8	0.39	0.99	0.59
	Total N	3.22	2.70	−0.52	16.0	1.77	4.07	2.30
	Oxidized N	2.68	1.83	−0.85	−31.7	0.79	2.66	1.87
	Reduced N	0.53	0.87	0.33	63.7	0.69	1.46	0.78
MOJA	Total S	0.51	0.53	0.02	5.7	0.26	0.70	0.44
	Total N	4.53	4.60	0.07	1.6	3.02	5.66	2.64
	Oxidized N	3.77	2.99	−0.78	−20.6	1.86	3.77	1.90
	Reduced N	0.75	1.61	0.85	114.2	1.06	2.00	0.93

[a] Deposition estimates were determined by the Total Deposition project, based on three-year averages centered on 2001 and 2011 for all ~4 km grid cells in each park. The minimum, maximum, and range of 2011 S and N deposition within each park are also shown.

15.3 Acidification

Although surface water resources are generally scarce in the parks in the Mojave Desert Network, GRBA has several high-elevation lakes that are fairly low in acid neutralizing capacity. These lakes were surveyed in the 1980s as part of the EPA's Western Lakes Survey. All of the sampled lakes in the park were considered acid sensitive (acid neutralizing capacity less than 200 µeq/L), according to the EPA's classification criteria at that time (Landers et al. 1987). The most sensitive lake included in the study was Baker Lake at 3,238 m (10,620 ft elevation), with an acid neutralizing capacity of 73 µeq/L. Conductivity measurements and correlation analysis suggest that the acid neutralizing capacity of Baker Lake is, at times, less than 50 µeq/L, a level considered to be very acid sensitive. Nevertheless, the overall risk of acidification of both water and soil is low in view of the very low levels of S deposition to GRBA. Depletion of acid neutralizing capacity increases the risk to lakes and streams in the park from episodic or chronic acidification. Baker Lake has a listed population of Lahontan cutthroat trout (*Onchorhynchus clarki henshawi*), as well as other fish and invertebrates, that could be negatively affected by water acidification.

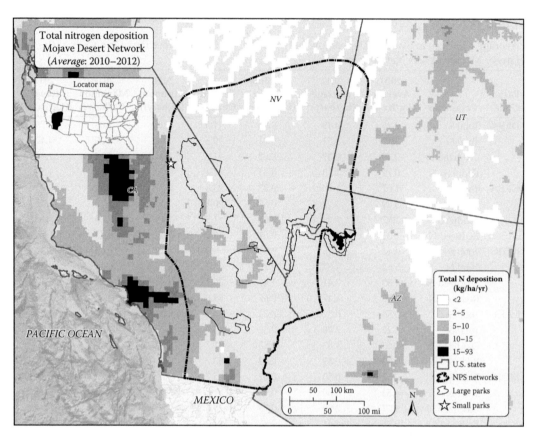

FIGURE 15.4
Total N deposition for the three-year period centered on 2011 in and around the Mojave Desert Network. (From Schwede, D.B. and Lear, G.G., *Atmos. Environ.*, 92, 207, 2014.)

15.4 Nutrient Nitrogen Enrichment

15.4.1 Risk Ranking

The network rankings developed in a coarse screening analysis by Sullivan et al. (2011) for nutrient N pollutant exposure, ecosystem sensitivity to nutrient N enrichment, and park protection yielded an overall network nutrient N enrichment summary risk ranking for the Mojave Desert Network that was in the top quintile among all networks. The overall level of concern for nutrient N enrichment effects on the parks within this network was judged by Sullivan et al. (2011) to be very high. Although rankings provide an indication of risk, park-specific data, particularly regarding nutrient enrichment sensitivity, are needed to fully evaluate risk from nutrient N addition.

Ecosystem sensitivity to nutrient N enrichment for most of the parks in Mojave Desert Network was ranked in the highest quintile among all national parks evaluated due to the preponderance of desert vegetation (Figure 15.5), which is presumed to be highly sensitive to nutrient N enrichment. GRBA has cheatgrass (a nonnative invasive annual grass) at low elevations. This species is very sensitive to N enrichment. If N levels

FIGURE 15.5
Desert vegetation, such as is found throughout JOTR, is sensitive to impacts on plant biodiversity as a consequence of N inputs. (Photo courtesy of National Park Service.)

increase in the future, we might expect to see higher cheatgrass densities and expansion into currently unoccupied areas, perhaps contributing to a cascade of ecological changes.

15.4.2 Field Studies

Studies in desert environments have indicated possible plant community changes resulting from elevated nutrient N input. A fertilization experiment in the Chihuahuan desert, with very high inputs of 100 kg N/ha/yr over about a decade, resulted in a 30% increase in cover of warm season grasses and a 52% reduction in cover of legumes (Báez et al. 2007). Somewhat similar responses might be expected in the Mojave Desert Network. From 1989 to 2004, Baez et al. (2007) observed a 43% increase in ambient N deposition, from 1.71 to 2.45 kg N/ha/yr, resulting in an additional 5.88 kg N/ha deposition over that time period. These increases in N deposition may result in significant plant community changes, as indicated by fertilization studies of blue grama and black grama (*Bouteloua eriopoda*). With additions of 20 kg N/ha in one season, blue grama was favored over black grama, the previously dominant species (Báez et al. 2007).

Plant productivity on arid land typically increases with both increasing precipitation (Romney et al. 1978, Bowers 2005) and N availability (Salo et al. 2005, Allen et al. 2009, Rao et al. 2010). In desert ecosystems, water is generally more limiting than N. Brooks (2003) found that plant responses were influenced by specific rainfall events rather than by average annual rainfall. The annual plants thrived in a year when high rainfall events triggered germination. In the Mojave Desert, the shrub creosote bush showed no increased growth response to experimental N additions (at 10 and 40 kg N/ha/yr as calcium nitrate) but did respond to increased water alone (Barker et al. 2006). Invasive annuals showed a greater response to elevated N than did native plant species.

Fertilization experiments by Brooks (2003) in the Mojave Desert showed that increased levels of N deposition could favor the establishment of nonnative species where nonnatives were already prevalent. At very high N application rates of 32 kg N/ha/yr over two years, both density and biomass of nonnative plants increased. Nonnative biomass increased by 54%, while native species biomass declined by about 39%.

Fire risk in desert vegetation communities is largely controlled by interactions among water and N availability, soil texture, and the presence of invasive grasses. Grasses can create a continuous fire fuel bed in the interspaces between shrubs. Exotic grass litter breaks down slowly, creating a highly flammable spatially continuous fire fuel load during the dry season (Brooks and Minnich 2006). Because of the historical rarity of fire in arid ecosystems, arid land shrubs are typically not fire adapted and experience high plant mortality and slow reestablishment subsequent to fire (Brown and Minnich 1986). Slow recovery of shrubs and fast recovery of grasses after fire contribute to increased fire frequency and a shift from shrub-dominated to exotic grass-dominated vegetation (D'Antonio and Vitousek 1992, Brooks et al. 2004, Steers 2008, Rao et al. 2010). Soil texture also affects fire frequency by modifying soil water holding capacity, infiltration, and hydraulic conductivity (Austin et al. 2004, Schwinning et al. 2004, Rao et al. 2010).

Allen et al. (2009) measured reactive N in the air and soils along an N deposition gradient at JOTR to determine the effects on invasive and native plant species. Because invasive plant species such as Mediterranean split grass (*Schismus barbatus*), red brome (*Bromus rubens*), and stork's bill (*Erodium cicutarium*) had become widespread over the preceding two decades, it was postulated that increasing N deposition might influence invasive plant range expansion. Nitrogen fertilization was added at levels of 5 and 30 kg N/ha/yr for each of two years, in addition to about 3–12 kg N/ha/yr of ambient atmospheric deposition (Tonnesen et al. 2007, Allen and Geiser 2011). Nonnative grass biomass increased significantly upon addition of 30 kg N/ha/yr at three of four treatment sites but not at the lower application level. Native forb species richness declined at a site with high nonnative grass cover, but richness and cover of native forbs increased in response to N addition at a site having low nonnative grass cover (Allen et al. 2009).

The Allen et al. (2009) study also suggested that the response of desert vegetation to added N varied by wetness condition. During 2003, which was a dry year, vegetation biomass remained unchanged. During 2004, a moderate precipitation year, two of four experimental plots showed increased biomass of nonnative invasive grasses at addition of 30 kg N/ha/yr but not at 5 kg N/ha/yr. During the wettest year (2005), invasive grass biomass increased at both N treatment levels (Allen et al. 2009). The productivity of native forbs decreased in response to N addition at these sites but increased at another treatment site that had only sparse invasive grass cover.

Some native species are able to use added N efficiently. For example, greasewood (*Sarcobatus vermiculatus*), a native desert shrub found in the Great Basin Desert, demonstrated a 2- to 3-fold increase in stem growth, a 2.5- to 4-fold increase in viable seed production, and a 17%–35% increase in leaf N with N additions at sites around Mono Lake, CA (Drenovsky and Richards 2005).

The studies by Allen et al. (2009) at JOTR suggested that the critical load of N to protect desert vegetation in this park against invasive grass biomass increase due to N enrichment during wet years may be as low as about 8 kg N/ha/yr (Allen and Geiser 2011). Modeling results using the Daily Century biogeochemical model (DayCent; Brooks and Matchett 2006, Allen and Geiser 2011) were similar, suggesting a critical load less than

8.2 kg N/ha/yr for low-elevation desert dominated by invasive Mediterranean grass and a critical load less than 5.7 kg N/ha/yr at higher-elevation sites dominated by invasive red brome (Rao et al. 2010). The model results further suggested that, at higher N input levels above these critical loads, fire risk is controlled more by precipitation than by grass productivity as influenced by N input (Allen and Geiser 2011).

15.4.3 Modeling Studies

The DayCent model was applied to plant communities in the deserts of southern California by Rao et al. (2009, 2010). Invasive nonnative grass production was simulated from 2003 to 2008 under varying levels of N input and precipitation, resulting in predictions of changing fire frequency. Simulated fire risk increased at levels of N deposition above 3 kg N/ha/yr. The simulated risk leveled off at N deposition 5.7 kg N/ha/yr in pinyon–juniper and at 8.2 kg N/ha/yr in creosote bush scrub (Rao et al. 2010).

Rao et al. (2010) applied DayCent to estimate the critical loads of N deposition in JOTR to control effects on fire risk in two different arid land vegetation types, creosote bush and pinyon–juniper. Fire risk was expressed as the probability that annual biomass production exceeded the general fire threshold of 1000 kg/ha. Critical loads were calculated as the amount of N deposition at the point along the deposition continuum where modeled fire risk began to increase exponentially. The mean estimated critical loads for all soil types, under annual precipitation less than 21 cm/yr, were 3.2 and 3.9 kg N/ha/yr for creosote bush and pinyon–juniper plant communities, respectively. Critical loads decreased (more nutrient sensitive) with decreasing soil clay content and increasing precipitation. The wettest areas with clay content of 6%–14% had estimated critical loads as low as 1.5 kg N/ha/yr (Rao et al. 2010). These values fall at the lower end of the critical load range identified by Pardo et al. (2011) for herbaceous vegetation in North American deserts. Fire risks in the two vegetation types were highest under N deposition of 9.3 and 8.7 kg N/ha/yr, respectively; above these deposition levels, fire risk was driven by precipitation amount.

Ellis et al. (2013) estimated the critical load for nutrient N deposition to protect the most sensitive ecosystem receptors in 45 national parks, based on the data of Pardo et al. (2011). The lowest terrestrial critical load of N is generally estimated for the protection of lichens (Geiser et al. 2010). Changes to lichen communities may signal the beginning of other changes to the ecosystem that might affect structure and function (Pardo et al. 2011). Ellis et al. (2013) estimated that the N critical loads for DEVA, GRBA, and JOTR were in the range of 3–8.4 kg N/ha/yr for the protection of herbaceous plants.

15.5 Ozone Injury to Vegetation

Experimental studies using controlled exposures have identified desert plant species that are potentially sensitive to ozone (Thompson et al. 1980, 1984a,b, Temple 1989). Bytnerowicz et al. (1988) developed an open-air system for measuring ozone exposure in the desert. They then exposed 16 Mojave Desert winter annual plant species to a range of ozone conditions. Of the species studied, *Camissonia claviformis*, *C. hirtella*, and *Erodium cicutarium* were the most sensitive to developing foliar symptoms due to ozone exposures. In a major air pollutant screening effort, Thompson et al. (1984a,b) used controlled exposures to determine the sensitivity of 49 Mojave Desert species to ozone and sulfur dioxide. They found considerable variation in sensitivity. Evening primrose (*Oenothera* spp.) and catseye (*Cryptantha* spp.) had some of the highest symptoms in response to exposure to ozone and sulfur dioxide.

Temple (1989) tested the sensitivity of the woody perennials catclaw acacia (*Acacia greggii*), desert willow (*Chilopsis linearis*), skunkbush sumac (*Rhus trilobata*), and Goodding's willow (*Salix gooddingii*) to controlled exposures of ozone ranging up to 200 parts per billion (ppb) over 4-hour periods. This study found that skunkbush sumac exhibited foliar injury at concentrations as low as 100 ppb (still well above the standard of 70 ppb), while Goodding's willow exhibited foliar injury at 200 ppb; other species had no injury symptoms. Surprisingly, skunkbush sumac exposed to elevated levels of ozone also grew faster, but the fact that it had such clear foliar symptoms indicates a relatively high sensitivity and possibly good potential as a bioindicator. There appears to be high potential for vegetation injury in this network caused by ozone, although injury symptoms are known for relatively few desert plant species and can be difficult to diagnose (Sullivan et al. 2001). Nevertheless, some plant species that occur in Mojave Desert Network parks are known to be sensitive to ozone and/or serve as bioindicators of ozone symptoms.

The W126 and SUM06 exposure indices calculated by National Park Service staff are given in Table 15.2, along with Kohut's (2007a) ozone risk ranking. Kohut (2007b) assessed the risk of foliar injury from ozone exposure in Mojave Desert Network parks, with three parks (DEVA, GRBA, and JOTR) assessed based on in-park monitoring data and the others based on kriging of data from surrounding monitoring stations. In addition, data for 2000–2004 were analyzed for DEVA and GRBA. The SUM06 index generally exceeded the threshold for foliar injury in DEVA and GRBA during the initial monitoring period (1995–1999) and more definitively exceeded the threshold in parks (including DEVA and GRBA) for the monitoring period 2000–2004. The threshold for the W126 index was exceeded in JOTR and MOJA, but generally not in the other parks. There was an apparent relationship between ozone concentrations and soil moisture, whereby

TABLE 15.2

Ozone Assessment Results for Parks in the Mojave Desert Network Based on Estimated Average 3-Month W126 and SUM06 Ozone Exposure Indices for the Period 2005–2009 and Kohut's (2007a) Ozone Risk Ranking for the Period 1995–1999[a,b]

| Park Name | Park Code | W126 | | SUM06 | | Kohut Ozone Risk Ranking |
		Value (ppm-h)	NPS Condition	Value (ppm-h)	NPS Condition	
Death Valley	DEVA	28.96	High	40.96	High	Low
Great Basin	GRBA	15.55	High	21.20	High	Low
Joshua Tree	JOTR	29.53	High	39.95	High	High
Lake Mead	LAKE	19.58	High	28.46	High	Low
Mojave	MOJA	25.92	High	36.49	High	High

[a] Parks are classified into one of three ranks (Low, Moderate, High), based on comparison with other Inventory and Monitoring parks.

[b] Degrees of concern for the W126 and SUM06 indices are based solely on levels of ozone exposure. Kohut's risk to vegetation is based on several factors that contribute to injury in plants, including ozone exposure and environmental variables, and considers the effects of soil moisture on the uptake of ozone.

when ozone exposure was high, soil moisture was low. This would be expected to reduce ozone uptake and therefore the likelihood of foliar injury. This observed relationship was most pronounced for JOTR and MOJA. Kohut (2007b) classified JOTR and MOJA as having high risk of foliar injury and the remaining three parks as having low risk. For JOTR and MOJA, there were many hours each year during the monitoring period having ozone exposure greater than 100 ppb. During some years, drought would significantly restrict ozone uptake and therefore foliar injury. During other years (e.g., 1995, 1998), however, ozone exposure was high and soil moisture was normal. The hours of exposure to ozone concentrations at DEVA that were greater than 60 ppb increased over the full period of record (1995–2004), and exposure to concentrations greater than 80 ppb increased slightly (Kohut 2007b). Although the W126 index was substantially higher than the threshold in DEVA, there were only 2 hours of exposure greater than 100 ppb over the nine-year period of record. Therefore, the W126 criteria were not met. Frequent drought in this park further constrained ozone uptake and reduced the likelihood of foliar injury.

During more recent years (2000–2009), no significant trends in ozone concentration were detected in monitored parks in this network (DEVA, GRBA, JOTR, LAKE; NPS–ARD 2013). The EPA has set national standards for ozone to protect both human and plant health. The 2008 standard was set at 75 ppb for protecting both humans and plants, based on an 8-hour average. It was decreased to 70 ppb in 2015. JOTR and MOJA are both located in areas that have been designated nonattainment by the EPA because ozone concentrations violated the standard, and air quality has been considered unhealthy at times. The EPA has recognized that the 8-hour form for the standard is probably not adequate to protect plants, which respond to longer-term ozone exposures. In 2010, the EPA proposed a new secondary ozone standard to protect plant health (Federal Register Vol. 75, No. 11, 40 CFR Parts 50 and 58, National Ambient Air Quality Standards for Ozone, Proposed Rules, January 19, 2010, p. 2938). It was based on an index of the total plant ozone exposure, the W126. For the W126 index, hourly values are weighted according to magnitude and then summed for daylight hours over three months, approximately a growing season. The EPA proposed to set the level of the new standard in the range of 7–15 ppm-hr. There were 17 parks that exceeded the upper end of this range in 2008. The highest values were reported for Sequoia-Kings Canyon National Parks (57 ppm-hr), JOTR (53 ppm-hr), Yosemite National Park (34 ppm-hr), and DEVA (29 ppm-hr), all in California. The standard was not finalized, but the EPA may consider the W126 standard again in subsequent reviews of the secondary ozone standard.

15.6 Visibility Degradation

15.6.1 Estimated Natural Background and Monitored Visibility Conditions

Although JOTR is the only Class I area in the Mojave Desert Network (the rest are classified as Class II by the Clean Air Act), visibility is also monitored in DEVA, GRBA, and MOHA. Improvements required by the Regional Haze Rule are anticipated to benefit both Class I and Class II parks.

The JOTR Interagency Monitoring of Protected Visual Environments (IMPROVE) monitor (JOSH1) is considered by the EPA to be representative of MOJA because it is located

TABLE 15.3

Estimated Natural Haze and Measured Ambient Haze in Parks in the Mojave Desert Network Averaged over the Period 2004–2008[a]

Park Name	Park Code	Site ID	Estimated Natural Haze (in Deciviews)		
			20% Clearest Days	20% Haziest Days	Average Days
Death Valley	DEVA	DEVA1	2.22	7.90	4.68
Great Basin	GRBA	GRBA1	0.85	6.24	3.18
Joshua Tree	JOTR	JOSH1	1.68	7.19	4.17
Mojave[b]	MOJA	JOSH1	1.68	7.19	4.17

Park Name	Park Code	Site ID	Measured Ambient Haze (for Years 2004–2008)					
			20% Clearest Days		20% Haziest Days		Average Days	
			Deciview	Ranking	Deciview	Ranking	Deciview	Ranking
Death Valley	DEVA	DEVA1	4.69	Low	15.41	Low	9.63	Low
Great Basin	GRBA	GRBA1	2.06	Very Low	10.71	Very Low	5.94	Very Low
Joshua Tree	JOTR	JOSH1	5.51	Moderate	18.29	Moderate	11.87	Moderate
Mojave[b]	MOJA	JOSH1	5.51	Moderate	18.29	Moderate	11.87	Moderate

[a] Parks are classified into one of five haze ranks (Very Low, Low, Moderate, High, or Very High) based on comparison with other monitored parks.

[b] Data from the JOTR IMPROVE monitor are used to represent conditions at MOJA. A monitoring site is considered by IMPROVE to be representative of an area if it is within 60 mi (100 km) and 425 ft (130 m) in elevation of that area.

within 60 mi (100 km) and 425 ft (130 m) elevation of MOJA. Estimated natural haze was relatively low in all four parks (Table 15.3).

Natural background visibility assumes no human-caused pollution but varies with natural processes such as windblown dust, fire, volcanic activity, and natural biogenic emissions from vegetation. Monitored visibility reflects recent pollution levels and was used to rank conditions at parks. Given the substantially greater amount of human-caused emissions in the eastern United States and environmental factors such as generally higher humidity in the East, visibility tends to be much better at western parks than at eastern parks. Relative rankings presented in Table 15.3 can range from very low haze (very good visibility) to very high haze (very poor visibility). Only parks with on-site or representative IMPROVE monitors were used to generate the visibility ranking. Table 15.3 gives the relative park visibility rankings on the 20% clearest, 20% haziest, and average days. Measured visibility for the period 2004–2008 was somewhat worse at each of the monitored parks than the estimated natural condition. Measured visibility impairment in DEVA and GRBA was considered low and very low, respectively, for all groups (20% clearest days, 20% haziest days, and average days). JOTR and MOJA were ranked moderate for all groups and had the worst visibility among the monitored parks in the Mojave Desert Network (Table 15.3).

IMPROVE data allow estimation of visual range, that is, how far a person can see. Data indicate that at the DEVA monitoring site (DEVA1), air pollution has reduced average visual range from 150 to 80 mi (241 to 129 km). On the haziest days, visual range has been reduced from 110 to 50 to mi (177 to 80 km). Severe haze episodes occasionally reduce visibility to 19 mi (30 km). At the GRBA monitoring site (GRBA1), air pollution has reduced average visual range from 170 to 120 mi (274 to 193 km). On the haziest days, visual range has been reduced from 130 to 85 mi (209 to 137 km). Severe haze episodes occasionally reduce visibility to 16 mi (26 km). At the JOTR monitoring site (JOSH1), also representative of MOJA, air pollution has reduced average visual range from 160 to 90 mi (257 to 145 km). On the haziest days, visual range has been reduced from 120 to 35 mi (193 to 56 km). Severe haze episodes occasionally reduce visibility to 13 mi (21 km).

15.6.2 Composition of Haze

Figure 15.6 shows estimated natural (preindustrial), baseline (2000–2004), and current (2006–2010) levels of haze and its composition for DEVA, GRBA, JOTR, and MOJA (values for MOJA are based on JOTR data). The figure illustrates that sulfate is the primary component of current haze at all four parks on the 20% clearest days, when human-caused emissions are relatively low. Some of that sulfate is likely derived from marine sources. On the 20% haziest days, organics contribute the most to haze in DEVA and GRBA. In JOTR and MOJA, the largest contribution is from nitrate. For the average days, sulfate is the largest contributor to haze in DEVA, JOTR, and MOJA; organics is the largest contributor in GRBA. Park staff at JOTR have noted that visibility impairment frequently occurs at the park, particularly during summer (Sullivan et al. 2001).

15.6.3 Trends in Visibility

The National Park Service (2010) reported long-term trends in visibility on the clearest and haziest 20% of days at monitoring sites in 29 national parks. The average difference between measured visibility and estimated natural condition was 8.3 deciviews, but several western parks had measured haze on the haziest days well above (more than 8 deciviews) estimated natural conditions. Such large differences between ambient and estimated natural haze are reflected in the 2004–2008 monitoring results shown in Table 15.4 for JOTR and MOJA. All monitored parks in the Mojave Desert Network showed some evidence of improvement in visibility in recent years on the 20% clearest days, 20% haziest days, and annual average condition (Figure 15.7). Such improvements have been most pronounced on the haziest days for JOTR and MOJA.

15.6.4 Development of State Implementation Plans

Source apportionment analyses were conducted for the California Regional Haze Plan (Suarez-Murias et al. 2009), using regional-scale, three-dimensional air quality models that simulated emissions, chemical transformations, and transport of criteria air pollutants and fine particulate matter. Effects on visibility in Class I areas in California were then evaluated for organic C, sulfate, and nitrate, which are the three major drivers of haze in California. The Particulate Matter Source Appointment Technology algorithm was used for Western Regional Air Partnership model analysis.

Directly emitted anthropogenic organic C was estimated to contribute half or less of the organic C in most areas of California except Point Reyes National Seashore (67%) and Pinnacles National Park (73%), both well to the northwest of the Mojave Desert Network. Anthropogenic sources of oxidized N contributed at least half of the nitrate in all Class I areas in California except Redwood National Park (where it contributed only 7%). The anthropogenic sulfate contribution ranged from 1% to 35% at various Class I areas, with highest contributions in southern California. The primary drivers of haze in California

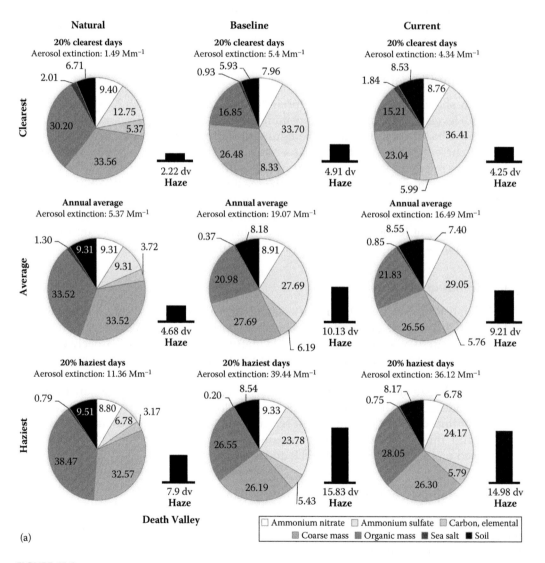

(a)

FIGURE 15.6
Estimated natural (preindustrial), baseline (2000–2004), and current (2006–2010) levels of haze (columns) and its composition (pie charts) on the 20% clearest, annual average, and the 20% haziest days for (a) DEVA, (b) GRBA, and (c) JOTR and MOJA. Data for MOJA were taken from a representative site (JOSH1). DEVA has no data for the year 2000 or 2003; JOTR and MOJA have no data for the year 2000. (From National Park Service–Air Resources Division, http://views.cira.colostate.edu/fed/Tools/RegionalHazeSummary.aspx, accessed October, 2012.)   *(Continued)*

derive from a variety of sources including organic C, mobile sources of nitrate, and off-shore and regional sources of sulfate, both human and natural (Suarez-Murias et al. 2009). Modeling results generated by Suarez-Murias et al. (2009) for JOTR and the U.S. Forest Service Agua Tibia Wilderness suggested that a 63% improvement would be needed in haze-forming atmospheric pollutants in order to comply with the national visibility goal of no human-caused visibility impairment by 2064 (Table 15.4). The California Regional Haze Plan (Suarez-Murias et al. 2009) also concluded that urban emissions were key sources of

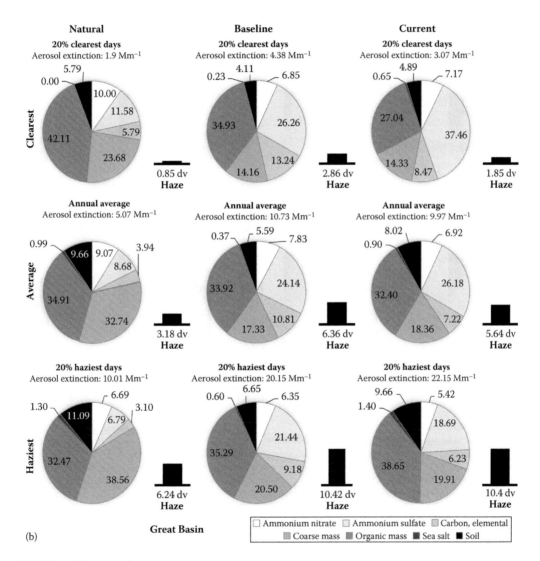

FIGURE 15.6 (*Continued*)
Estimated natural (preindustrial), baseline (2000–2004), and current (2006–2010) levels of haze (columns) and its composition (pie charts) on the 20% clearest, annual average, and the 20% haziest days for (a) DEVA, (b) GRBA, and (c) JOTR and MOJA. Data for MOJA were taken from a representative site (JOSH1). DEVA has no data for the year 2000 or 2003; JOTR and MOJA have no data for the year 2000. (From National Park Service–Air Resources Division, http://views.cira.colostate.edu/fed/Tools/RegionalHazeSummary.aspx, accessed October, 2012.) (*Continued*)

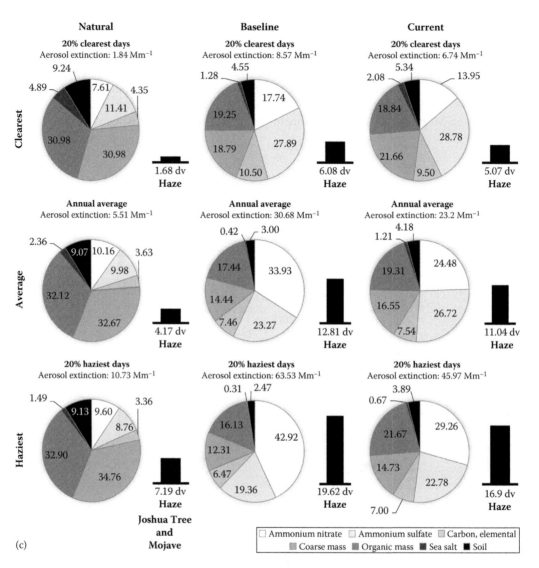

FIGURE 15.6 (*Continued*)
Estimated natural (preindustrial), baseline (2000–2004), and current (2006–2010) levels of haze (columns) and its composition (pie charts) on the 20% clearest, annual average, and the 20% haziest days for (a) DEVA, (b) GRBA, and (c) JOTR and MOJA. Data for MOJA were taken from a representative site (JOSH1). DEVA has no data for the year 2000 or 2003; JOTR and MOJA have no data for the year 2000. (From National Park Service–Air Resources Division, http://views.cira.colostate.edu/fed/Tools/RegionalHazeSummary.aspx, accessed October, 2012.)

haze throughout the South Coast Air Basin and the central and western portions of the Mojave Desert Air Basin, which includes JOTR. Emissions from offshore shipping and international transport were also judged to be important. Glideslope analyses shown in Figure 15.8 suggest that improvements on the haziest days at the monitored parks in the Mojave Desert Network are progressing along (or perhaps even better than) the trajectories needed for attaining natural visibility by 2064.

TABLE 15.4

IMPROVE Monitors and Visibility at Class 1 Areas in the Mojave Desert Network Region

Improve Monitor (Name and Elevation in Meters)		Class I Area(s)	Current Conditions (2000–2004 Baseline)		Future Natural Conditions (2064 Goals)		
			Worst Days (Deciviews)	Best Days (Deciviews; Maintain in Future Years)	Natural Worst Days (Deciviews)	Deciview Hurdle (Baseline to 2064)	Improvement Needed from Current Visibility on Worst Days (%)
AGTI (508 m)	Aqua Tibia	Agua Tibia	23.5	9.6	7.6	15.9	69
JOSH (1235 m)	Joshua Tree	Joshua Tree National Park	19.6	6.1	7.2	12.4	63

Source: Suarez-Murias, T. et al., California regional haze plan, California Environmental Protection Agency, Air Resources Board, 2009.

15.7 Toxic Airborne Contaminants

Little information is available regarding the possible effects of air toxics deposition on sensitive park resources in the Mojave Desert Network. However, elevated levels of both banned and current-use pesticides and industrial contaminants were found in vegetation, fish, snow, and sediments in Sequoia-Kings Canyon National Parks to the west of the Mojave Desert Network during the WACAP study (Landers et al. 2008). Analyses and modeling results suggested that intensive regional agriculture was at least partly responsible for these contaminants, and it is possible that Mojave Desert Network parks have also received airborne toxics deposition.

Eagles-Smith et al. (2014) sampled fish in 21 national parks and analyzed them for Hg concentrations in tissue. Results varied substantially by park and by water body. Concentrations of Hg in brook trout sampled from two streams and one lake in GRBA were very low compared with other parks included in the study.

15.8 Summary

Air pollutant emissions vary throughout the Mojave Desert Network region. Some of the parks, notably GRBA in Nevada, are quite distant from air pollutant sources and enjoy relatively good air quality. Other Mojave Desert Network parks, including JOTR in southeastern California, experience much higher levels of air pollution.

California is the most populous state in the nation, with its largest population center located in the Los Angeles Basin. Because the prevailing winds are from the west and northwest, many of California's national parks, including those located in the southeastern California deserts within the Mojave Desert Network, are often downwind of the most populated portion of the state. Pollutant transport to the more remote regions of California has been studied extensively, and urban area and agricultural emissions are known to affect air quality in southeastern California's national parks. Prevailing winds carry pollutants along known transport routes into the mountains and deserts to the south and east of heavily populated areas, including the Los Angeles Basin. LAKE, located on the Arizona/Nevada border, also receives air pollutants from Las Vegas, Nevada.

Ozone, N, and particulate matter are problematic air pollutants in the Mojave Desert Network and are of concern for their effects on public health, visibility, and vegetation (Sullivan et al. 2001). Atmospheric S and N pollutants can cause acidification of streams, lakes, and soils. DEVA and GRBA are potentially sensitive to acidification because of their

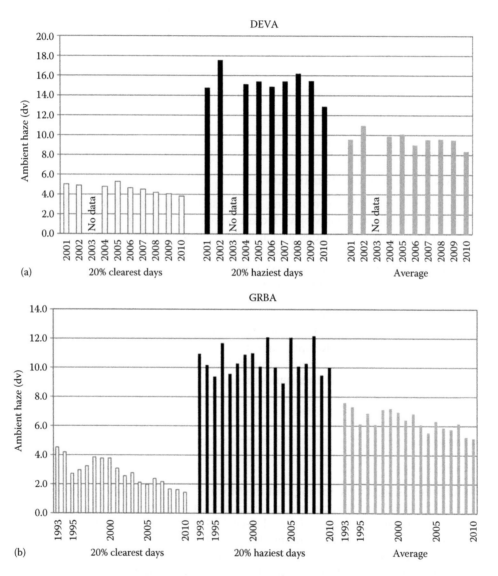

FIGURE 15.7
Trends in monitored visibility at (a) DEVA, (b) GRBA, and (c) JOTR and MOJA, based on IMPROVE measurements on the 20% clearest, 20% haziest, and annual average visibility days over the monitoring period of record. Data for MOJA were taken from a representative site (JOSH1). (From http://vista.cira.colostate.edu/improve/Data/IMPROVE/summary_data.htm, accessed October, 2012.) *(Continued)*

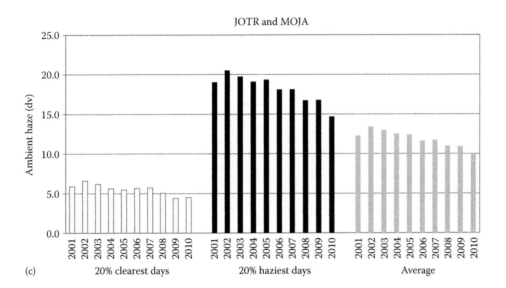

(c)   20% clearest days   20% haziest days   Average

FIGURE 15.7 (Continued)
Trends in monitored visibility at (a) DEVA, (b) GRBA, and (c) JOTR and MOJA, based on IMPROVE measurements on the 20% clearest, 20% haziest, and annual average visibility days over the monitoring period of record. Data for MOJA were taken from a representative site (JOSH1). (From http://vista.cira.colostate.edu/improve/Data/IMPROVE/summary_data.htm, accessed October, 2012.)

steep slopes, which allow limited opportunity for incoming acidic deposition to be buffered by base cations in rocks and soils. Lakes in GRBA are considered to be somewhat acid-sensitive, with Baker Lake thought to be the most sensitive lake in the park (T. Cummings, National Park Service, personal communication, July 2014).

Nitrogen deposition can also cause undesirable nutrient enrichment of natural ecosystems, leading to changes in plant species composition and soil nutrient cycling. Ecosystems in Mojave Desert Network parks are considered to be highly sensitive to nutrient N enrichment because of the preponderance of desert vegetation in the parks, which is very responsive to N inputs. Enhanced N has been found to facilitate the recently observed invasion of some exotic and invasive plant species within parts of the Mojave and Sonoran deserts (Allen and Geiser 2011).

Water availability affects plant community response to N input, especially in desert environments. Ecosystem modeling and empirical evidence suggest that N deposition of 3–9 kg/ha/yr was sufficient to increase the biomass of invasive annual grasses, thereby significantly increasing fire risk, in creosote bush and pinyon–juniper communities in JOTR during average precipitation years (<21 cm/yr). During wetter years, N deposition as low as 1.5 kg/ha/yr induced the same response (Rao et al. 2010). Modeled N deposition ranges from <2 kg/ha/yr to values in the range of 5–10 kg/ha/yr in the network parks, suggesting that some areas may be at risk for increased buildup of invasive grass biomass and consequent wildfire.

Ozone pollution can harm human health, reduce plant growth, and cause visible symptoms on plant foliage. JOTR and MOJA are located in areas that have been designated nonattainment by the EPA because ozone concentrations have violated the national ozone standard to protect human health, and air quality is unhealthy at times. Risk to plants is assessed using metrics that reflect exposure over three or five months of the growing season. Risk to plants in Mojave Desert Network parks varied from low to high in an assessment conducted for all parks nationwide (Kohut 2007a,b).

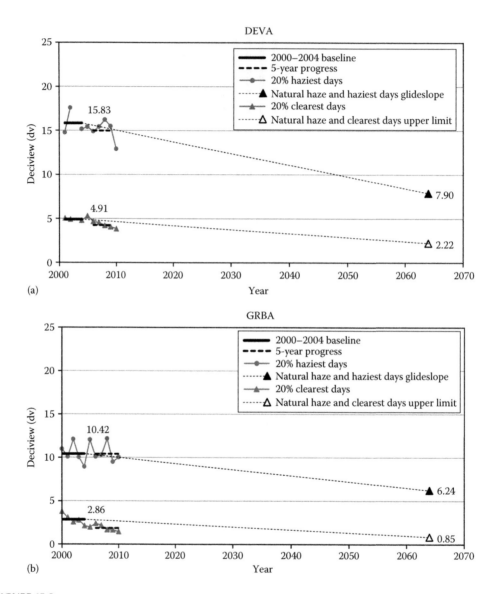

FIGURE 15.8
Glideslopes to achieving natural visibility conditions by 2064 for the 20% haziest (upper line) and the 20% clearest (lower line) days in (a) DEVA, (b) GRBA, and (c) JOTR and MOJA. In the Regional Haze Rule, the clearest days do not have a uniform rate of progress glideslope; the rule only requires that the clearest days do not get any worse than the baseline period. Also shown are measured values during the period 2000–2010. Data for MOJA were taken from a nearby site (JOSH1). DEVA has no data for the year 2000 or 2003; JOTR and MOJA have no data for the year 2000. (From http://vista.cira.colostate.edu/improve/Data/IMPROVE/summary_data.htm, accessed October, 2012.)   (*Continued*)

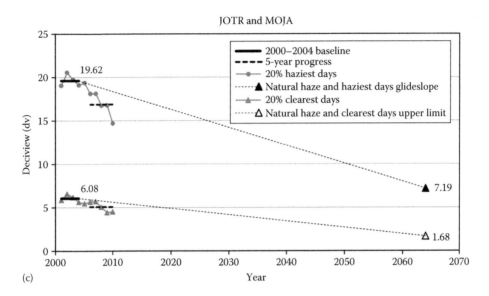

(c)

FIGURE 15.8 (*Continued*)
Glideslopes to achieving natural visibility conditions by 2064 for the 20% haziest (upper line) and the 20% clearest (lower line) days in (a) DEVA, (b) GRBA, and (c) JOTR and MOJA. In the Regional Haze Rule, the clearest days do not have a uniform rate of progress glideslope; the rule only requires that the clearest days do not get any worse than the baseline period. Also shown are measured values during the period 2000–2010. Data for MOJA were taken from a nearby site (JOSH1). DEVA has no data for the year 2000 or 2003; JOTR and MOJA have no data for the year 2000. (From http://vista.cira.colostate.edu/improve/Data/IMPROVE/summary_data.htm, accessed October, 2012.)

Particulate and gaseous air pollution can cause haze, reducing visibility. JOTR has the highest levels of haze among the four monitored parks (DEVA, GRBA, JOTR, and MOJA) in the Mojave Desert Network. Haze in network parks is primarily caused by sulfate, nitrate, organics, and coarse mass.

Airborne toxics, including Hg and other heavy metals, can accumulate in food webs, reaching toxic levels in top predators. Effects have been documented in some areas, including parts of California, in piscivorous fish and wildlife (Landers et al. 2010). Data on bioaccumulation of toxic substances in Mojave Desert Network parks and associated effects on sensitive receptors are scarce.

References

Allen, E.B., A. Bytnerowicz, M.E. Fenn, R.A. Minnich, and M.F. Allen. 2006. Impacts of anthropogenic N deposition on weed invasion, biodiversity and the fire cycle at Joshua Tree National Park. *In* R.H. Webb, L.F. Fenstermaker, J.S. Heaton, D.L. Hughson, E.V. McDonald, and D.M. Miller (Eds.). *The Mojave Desert: Ecosystem Processes and Sustainability.* University of Nevada Press, Las Vegas, NV.

Allen, E.B. and L.H. Geiser. 2011. North American deserts. *In* L.H. Pardo, M.J. Robin-Abbott, and C.T. Driscoll (Eds.). *Assessment of Nitrogen Deposition Effects and Empirical Critical Loads of Nitrogen for Ecoregions of the United States.* General Technical Report NRS-80. U.S. Forest Service, Newtown Square, PA, pp. 133–142.

Allen, E.B., P.E. Padgett, A. Bytnerowicz, and R.A. Minnich. 1998. Nitrogen deposition effects on coastal sage vegetation of southern California. *Proceedings of the International Symposium on Air Pollution and Climate Change Effects on Forest Ecosystems*, USDA Forest Service, Pacific Southwest Research Station, Riverside, CA, pp. 131–140.

Allen, E.B., L.E. Rao, and R.J. Steers. 2009. Impacts of atmospheric nitrogen deposition on vegetation and soils at Joshua Tree National Park. *In* R.H. Webb, J.S. Fenstermaker, J.S. Heaton, D.L. Hughson, E.V. McDonald, and D.M. Miller (Eds.). *The Mojave Desert: Ecosystem Processes and Sustainability.* University of Nevada Press, Las Vegas, NV, pp. 78–100.

Austin, A.T., L. Yahdjian, J.M. Stark, J. Belnap, A. Porporato, U. Norton, D.A. Ravetta, and S.M. Schaeffer. 2004. Water pulses and biochemical cycles in arid and semiarid ecosystems. *Oecologia* 141:221–245.

Báez, S., J. Fargione, D.I. Moore, S.L. Collins, and J.R. Gosz. 2007. Atmospheric nitrogen deposition in the northern Chihuahuan Desert: Temporal trends and potential consequences. *J. Arid Environ.* 68:640–651.

Barker, D.H., C. Vanier, E. Naumburg, T.N. Charlet, K.M. Nielsen, B.A. Newingham, and S.D. Smith. 2006. Enhanced monsoon precipitation and nitrogen deposition affect leaf traits and photosynthesis differently in spring and summer in the desert shrub *Larrea tridentata. New Phytol.* 169:799–808.

Bowers, J.E. 2005. El Niño and displays of spring-flowering annuals in the Mojave and Sonoran Deserts. *J. Torrey Bot. Soc.* 132:38–49.

Brooks, M.L. 2003. Effects of increased soil nitrogen on the dominance of alien annual plants in the Mojave Desert. *J. Appl. Ecol.* 40:344–353.

Brooks, M.L., C.M. D'Antonio, D.M. Richardson, J.B. Grace, J.E. Keeley, J.M. DiTomaso, R.J. Hobbs, M. Pellant, and D. Pyke. 2004. Effects of invasive alien plants on fire regimes. *BioScience* 54:677–688.

Brooks, M.L. and J.R. Matchett. 2006. Spatial and temporal patterns of wildfires in the Mojave Desert, 1980–2004. *J. Arid Environ.* 67:148–164.

Brooks, M.L. and R.A. Minnich. 2006. Southeastern deserts bioregion. *In* N.G. Sugihara, J.W.V. Wagtendonk, K.E. Shaffer, J. Fites-Kaufman, and A.E. Thode (Eds.). *Fire in California's Ecosystems.* University of California Press, Berkeley, CA, pp. 391–414.

Brown, D.E. and R.A. Minnich. 1986. Fire and changes in creosote bush scrub of the Western Sonoran Desert, California. *Am. Midl. Nat.* 116:411–422.

Bytnerowicz, A., D.M. Olszyk, C.A. Fox, P.J. Dawson, G. Kats, C.L. Morrison, and J. Wolf. 1988. Responses of desert annual plants to ozone and water stress in an *in situ* experiment. *J. Air Pollut. Control Assoc.* 38:1145–1151.

D'Antonio, C.M. and P.M. Vitousek. 1992. Biological invasions by exotic grasses: The grass-fire cycle and global change. *Ann. Rev. Ecol. Syst.* 23:63–87.

Drenovsky, R.H. and J.H. Richards. 2005. Nitrogen addition increases fecundity in the desert shrub *Sarcobatus vermiculatus. Oecologia* 143:349–356.

Eagles-Smith, C.A., J.J. Willacker, Jr., and C.M. Flanagan Pritz. 2014. Mercury in fishes from 21 national parks in the western United States: Inter- and intra-park variation in concentrations and ecological risk. U.S. Geological Survey Open-File Report 2014–1051, Reston, VA.

Edinger, J.G., M.H. McCutchan, P.R. Miller, B.C. Ryan, M. Schroeder, and J.V. Behar. 1972. Penetration and duration of oxidant air pollution in the South Coast Air Basin of California. *J. Air Pollut. Control Assoc.* 22:881–886.

Ellis, R.A., D.J. Jacob, M.P. Sulprizio, L. Zhang, C.D. Holmes, B.A. Schichtel, T. Blett, E. Porter, L.H. Pardo, and J.A. Lynch. 2013. Present and future nitrogen deposition to national parks in the United States: Critical load exceedances. *Atmos. Chem. Phys.* 13(17):9083–9095.

Fenn, M.E., R. Haeuber, G.S. Tonnesen, J.S. Baron, S. Grossman-Clark, D. Hope, D.A. Jaffe et al. 2003. Nitrogen emissions, deposition, and monitoring in the western United States. *BioScience* 53(4):391–403.

Geiser, L.H., S.E. Jovan, D.A. Glavich, and M.K. Porter. 2010. Lichen-based critical loads for atmospheric nitrogen deposition in western Oregon and Washington forests, USA. *Environ. Pollut.* 158:2412–2421.

Kohut, R. 2007a. Assessing the risk of foliar injury from ozone on vegetation in parks in the U.S. National Park Service's Vital Signs Network. *Environ. Pollut.* 149:348–357.

Kohut, R.J. 2007b. Ozone risk assessment for vital signs monitoring networks, Appalachian National Scenic Trail, and Natchez Trace National Scenic Trail. Natural Resource Technical Report NPS/ NRPC/ARD/NRTR—2007/001. National Park Service, Natural Resource Program Center, Fort Collins, CO.

Landers, D.H., J.M. Eilers, D.F. Brakke, W.S. Overton, P.E. Kellar, W.E. Silverstein, R.D. Schonbrod et al. 1987. Characteristics of lakes in the western United States. Volume I: Population descriptions and physico-chemical relationships. EPA-600/3-86/054a. U.S. Environmental Protection Agency, Washington, DC.

Landers, D.H., S.L. Simonich, D.A. Jaffe, L.H. Geiser, D.H. Campbell, A.R. Schwindt, C.B. Schreck et al. 2008. The fate, transport, and ecological impacts of airborne contaminants in western national parks (USA). EPA/600/R-07/138. U.S. Environmental Protection Agency, Office of Research and Development, NHEERL, Western Ecology Division, Corvallis, OR.

Landers, D.H., S.M. Simonich, D. Jaffe, L. Geiser, D.H. Campbell, A. Schwindt, C.B. Schreck et al. 2010. The Western Airborne Contaminant Project (WACAP): An interdisciplinary evaluation of the impacts of airborne contaminants in western U.S. national parks. *Environ. Sci. Technol.* 44(3):855–859.

National Park Service (NPS). 2010. Air quality in national parks: 2009 annual performance and progress report. Natural Resource Report NPS/NRPC/ARD/NRR-2010/266. National Park Service, Air Resources Division, Denver, CO.

National Park Service–Air Resources Division (NPS–ARD). 2013. Air quality in national parks: Annual trends (2000–2009) and conditions (2005–2009) report. Natural Resource Report NPS/ NRSS/ARD/NRR—2013/683. National Park Service, Denver, CO.

Padgett, P.E., E.B. Allen, A. Bytnerowicz, and R.A. Minnich. 1999. Changes in soil inorganic nitrogen as related to atmospheric nitrogenous pollutants in southern California. *Atmos. Environ.* 33:769–781.

Padgett, P.E. and A. Bytnerowicz. 2001. Deposition and adsorption of the air pollutant HNO_3 vapor to soil surfaces. *Atmos. Environ.* 35:2405–2415.

Pardo, L.H., M.E. Fenn, C.L. Goodale, L.H. Geiser, C.T. Driscoll, E.B. Allen, J.S. Baron et al. 2011. Effects of nitrogen deposition and empirical nitrogen critical loads for ecoregions of the United States. *Ecol. Appl.* 21(8):3049–3082.

Rao, L.E., E.B. Allen, and T. Meixner. 2010. Risk-based determination of critical nitrogen deposition loads for fire spread in southern California deserts. *Ecol. Appl.* 20(5):1320–1335.

Rao, L.E., D.R. Parker, A. Bytnerowicz, and E.B. Allen. 2009. Nitrogen mineralization across an atmospheric nitrogen deposition gradient in southern California deserts. *J. Acid Environ.* 73:920–930.

Romney, E.M., A. Wallace, and R.B. Hunter. 1978. Plant response to nitrogen fertilization in the northern Mohave Desert and its relationship to water manipulation. *In* N.E. West and J.J. Skujins (Eds.). *Nitrogen in Desert Ecosystems.* Dowden, Hutchinson and Ross, Stroudsburg, PA, pp. 232–243.

Salo, L.F., G.R. McPherson, and D.G. Williams. 2005. Sonoran Desert winter annuals affected by density of red brome and soil nitrogen. *Am. Midl. Nat.* 153:95–109.

Schwede, D.B. and G.G. Lear. 2014. A novel hybrid approach for estimating total deposition in the United States. *Atmos. Environ.* 92:207–220.

Schwinning, S., O.E. Sala, M.E. Loik, and J.R. Ehleringer. 2004. Thresholds, memory, and seasonality: Understanding pulse dynamics in arid/semi-arid ecosystems. *Oecologia* 141:191–193.

Steers, R.J. 2008. Invasive plants, fire succession, and restoration of creosote bush scrub in southern California. PhD, University of California, Riverside, CA.

Suarez-Murias, T., J. Glass, E. Kim, L. Melgoza, and T. Najita. 2009. California regional haze plan. California Environmental Protection Agency, Air Resources Board, Sacramento, CA.

Sullivan, T.J., T.C. McDonnell, G.T. McPherson, S.D. Mackey, and D. Moore. 2011. Evaluation of the sensitivity of inventory and monitoring national parks to nutrient enrichment effects from atmospheric nitrogen deposition. Natural Resource Report NPS/NRPC/ARD/NRR—2011/313. U.S. Department of the Interior, National Park Service, Denver, CO.

Sullivan, T.J., D.L. Peterson, C.L. Blanchard, K. Savig, and D. Morse. 2001. Assessment of air quality and air pollutant impacts in class I national parks of California. NPS D-1454. U.S. Department of the Interior, National Park Service, Air Resources Division, Denver, CO.

Temple, P.J. 1989. Oxidant air pollution effects on plants of Joshua Tree National Monument. *Environ. Pollut.* 57:35–47.

Thompson, C.R., D.M. Olszyk, G. Kats, A. Bytnerowicz, P.J. Dawson, and D.C. Wolf. 1984a. Effects of ozone and sulfur dioxide on annual plants of the Mojave Desert. *J. Air Pollut. Control Assoc.* 34:1017–1022.

Thompson, C.R., D.M. Olszyk, G. Kats, A. Bytnerowicz, P.J. Dawson, D.C. Wolf, and C.A. Fox. 1984b. Air pollutant impacts on plants of the Mojave Desert. Rosemead, CA.

Thompson, M.E., M.C. Elder, A.R. Davis, W. Whitlow, D. Drablos, and E. Tollan. 1980. Evidence of acidification of rivers of eastern Canada. Ecological impact of acid precipitation. *Proceedings from an International Conference in Sandefjord, Norway*, Oslo, Norway, pp. 244–245.

Tonnesen, G., Z. Wang, M. Omary, and C.J. Chien. 2007. Assessment of nitrogen deposition: Modeling and habitat assessment. CEC-500-2005-032. California Energy Commission, PIER Energy-Related Environmental Research, Sacramento, CA.

Weathers, K.C., S.M. Simkin, G.M. Lovett, and S.E. Lindberg. 2006. Empirical modeling of atmospheric deposition in mountainous landscapes. *Ecol. Appl.* 16(4):1590–1607.

Yoshida, L.C. and E.B. Allen. 2001. Response to ammonium and nitrate by a mycorrhizal annual invasive grass and native shrub in southern California. *Am. J. Bot.* 88:1430–1436.

Yoshida, L.C. and E.B. Allen. 2004. N-15 uptake by mycorrhizal native and invasive plants from a N-eutrophied shrubland: A greenhouse experiment. *Biol. Fertil. Soils* 39:243–248.

16

Chihuahuan Desert

16.1 Background

Two national parks in the Chihuahuan Desert Network are discussed here: Big Bend National Park (BIBE) and Guadalupe Mountains National Park (GUMO). Both are designated as Class I, giving them a heightened level of protection against harm caused by poor air quality under the Clean Air Act. The only large population center within the Chihuahuan Desert Network is El Paso. There are, however, several large urban centers near the network boundary, including Dallas, Fort Worth, Austin, San Antonio, Houston, Phoenix, Tucson, and Albuquerque, plus human population centers in Mexico. Figure 16.1 shows the network boundary along with locations of each park and population centers with more than 10,000 people.

16.2 Atmospheric Emissions and Deposition

In general, county-level emissions of S and N in the Chihuahuan Desert Network have been low. Most counties in the network region had estimated annual emissions of both sulfur dioxide and N that were less than 5 tons/mi^2/yr in 2002 (Sullivan et al. 2011). Point sources of both sulfur dioxide and N were widely scattered throughout the network, but none were large compared with those in the surrounding networks.

County-level emissions near the Chihuahuan Desert Network, based on data from the EPA's National Emissions Inventory during a recent time period (2011), are depicted in Figures 16.2 through 16.4 for sulfur dioxide, oxidized N, and ammonia, respectively. Most counties near Chihuahuan Desert Network parks had low sulfur dioxide emissions (<1 ton/mi^2/yr; Figure 16.2). Emissions of oxidized N were slightly higher, with most counties having <5 tons/mi^2/yr (Figure 16.3). Emissions of ammonia near Chihuahuan Desert Network parks were somewhat lower, with all counties showing emissions levels below 2 tons/mi^2/yr (Figure 16.4).

Estimated total S and N deposition levels in and around the network, including both wet and dry forms of deposition and both the oxidized N and ammonium species, have been generally low to moderate. Dry deposition is proportionately more important than wet deposition in these desert environments. Total S deposition in and around the Chihuahuan Desert Network for the period 2010–2012 was generally <2.5 kg S/ha/yr at

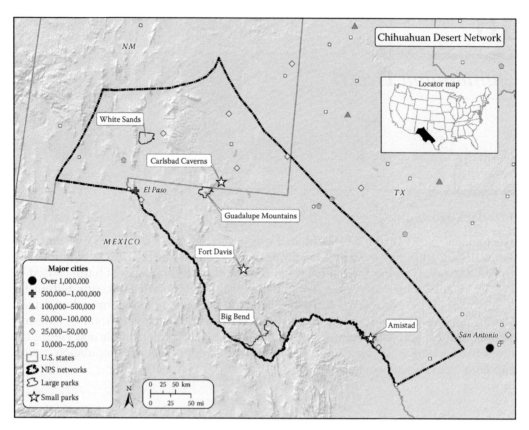

FIGURE 16.1
Network boundary and locations of parks and population centers with more than 10,000 people near the Chihuahuan Desert Network.

the location of parklands. Oxidized inorganic N deposition for the period 2010–2012 was less than 2.5 kg N/ha/yr at BIBE and GUMO. These parks received less than 1.5 kg N/ha/yr of reduced inorganic N (ammonium) from atmospheric deposition during this same period (Table 16.1); a few areas received higher amounts. Total N deposition was less than 3.5 kg N/ha/yr at most park locations.

In the southwestern United States, atmospheric N deposition increased during the last two decades of the twentieth century (Fenn et al. 2003). This increase in N deposition was attributed to urban and agricultural development (Báez et al. 2007). When precipitation effects were removed statistically, N deposition in central New Mexico increased at an annual rate of 0.049 kg N/ha/yr between 1989 and 2004. Nitrogen deposition has increased along the border with Mexico (Lehmann et al. 2015). Emissions from recent oil and gas development likely contribute to haze and N deposition in parks in the Chihuahuan Desert Network.

Atmospheric S deposition levels have declined at BIBE but increased at GUMO, since 2001, based on Total Deposition project estimates (Table 16.1). Estimated total N deposition decreased at both parks over that same time period. Oxidized N decreased more than reduced N. Modeled deposition values for this recent period were less than 2.50 kg S/ha/yr and less than 3.50 kg N/ha/yr at both BIBE and GUMO on average.

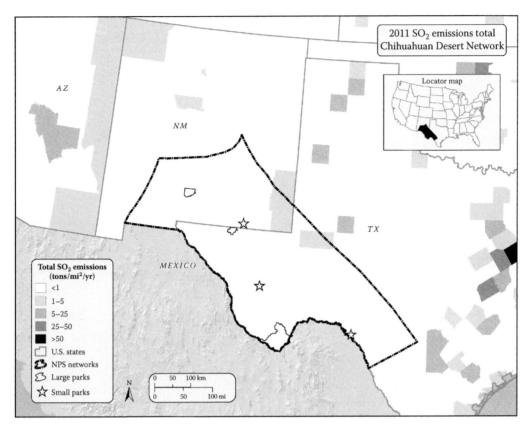

FIGURE 16.2
Total sulfur dioxide (SO_2) emissions, by county, near the Chihuahuan Desert Network for the year 2011. (Data from EPA's National Emissions Inventory, https://www.epa.gov/air-emissions-inventories, accessed January, 2014.)

16.3 Acidification

No information is available indicating the possible effects of acidic deposition on aquatic or terrestrial resources in the Chihuahuan Desert Network. In view of the low precipitation and levels of S and N air pollutant exposure for the parks in this network, such effects are considered unlikely.

16.4 Nutrient Nitrogen Enrichment

There are few published studies documenting nutrient enrichment impacts on either aquatic or terrestrial resources in the Chihuahuan Desert Network. Given the low levels of N pollutant exposure of parklands in this network, such effects are not likely to be substantial at the present time. However, as is the case for other desert ecosystems, the

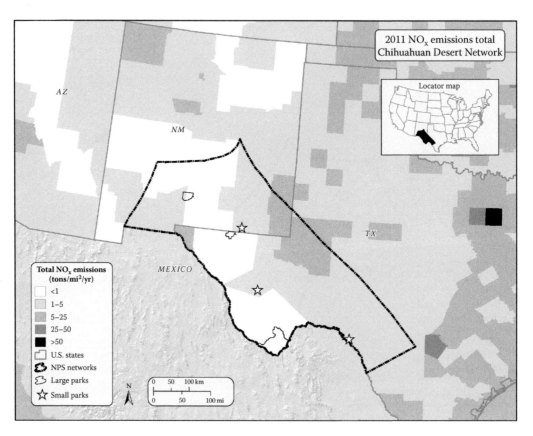

FIGURE 16.3
Total oxidized N (NO$_x$) emissions, by county, near the Chihuahuan Desert Network for the year 2011. (Data from EPA's National Emissions Inventory, https://www.epa.gov/air-emissions-inventories, accessed January, 2014.)

ecosystems in these parks appear to be highly sensitive to nutrient N enrichment and may be threatened by eutrophication if atmospheric N loads, such as those associated with local oil and gas development, increase substantially in the future.

Despite the fact that water is typically the key limiting driver of ecosystem processes in the Chihuahuan Desert and other arid regions (Figure 16.5), N can limit plant production where and when moisture is available, especially after rainstorms (Austin et al. 2004). Báez et al. (2007) quantified long-term trends in N deposition and its impacts on desert plant community structure. Based on two independent fertilization studies, Báez et al. (2007) concluded that continued inputs of N are likely to increase grass cover and decrease legume abundance. Even though N deposition levels are relatively low, increased N deposition in the future might cause changes in plant community structure (Báez et al. 2007).

Nitrogen addition experiments at BIBE found that N affected the functioning of soil fungi (Grizzle and Zak 2006). Results suggested that precipitation and N interact to affect plant productivity differently depending on dry versus wet years and vegetation type (especially shrub versus grass; Robertson et al. 2009).

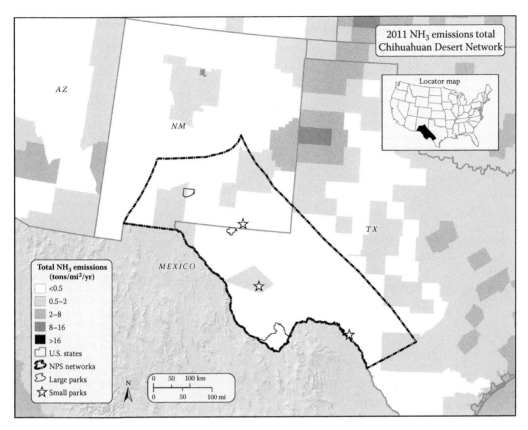

FIGURE 16.4

Total ammonia (NH_3) emissions, by county, near the Chihuahuan Desert Network for the year 2011. (Data from EPA's National Emissions Inventory, https://www.epa.gov/air-emissions-inventories, accessed January, 2014.)

TABLE 16.1

Average Changes in S and N Deposition[a] between 2001 and 2011 across Park Grid Cells at BIBE and GUMO

Park Code	Parameter	2001 Average (kg/ha/yr)	2011 Average (kg/ha/yr)	Absolute Change (kg/ha/yr)	Percent Change	2011 Minimum (kg/ha/yr)	2011 Maximum (kg/ha/yr)	2011 Range (kg/ha/yr)
BIBE	Total S	1.44	1.30	−0.15	−10.4	0.96	2.03	1.07
	Total N	2.79	2.39	−0.39	−14.2	1.87	3.59	1.72
	Oxidized N	1.89	1.53	−0.36	−19.1	1.19	2.22	1.03
	Reduced N	0.90	0.87	−0.03	−3.7	0.68	1.37	0.70
GUMO	Total S	1.74	2.32	0.57	31.6	1.38	2.93	1.56
	Total N	3.92	3.46	−0.46	−12.1	2.34	4.42	2.08
	Oxidized N	2.73	2.34	−0.39	−14.6	1.66	2.97	1.31
	Reduced N	1.19	1.12	−0.07	−6.5	0.68	1.45	0.77

[a] Deposition estimates were determined by the Total Deposition project, based on three-year averages centered on 2001 and 2011 for all ~4 km grid cells in each park. The minimum, maximum, and range of 2011 S and N deposition within each park are also shown.

FIGURE 16.5
Arid land plant community in GUMO. (Photo courtesy of National Park Service.)

Pardo et al. (2011) compiled data on the empirical critical load for protecting sensitive resources in Level I ecoregions across the conterminous United States against nutrient enrichment effects caused by atmospheric N deposition. Available data on empirical critical loads of nutrient N in the Chihuahuan Desert Network suggested that the critical load for resource protection was consistently greater than or equal to 3 kg N/ha/yr for the protection of lichens and herbaceous plants. Ellis et al. (2013) estimated the critical load for nutrient N deposition to protect the most sensitive ecosystem receptors in 45 national parks, based on data from Pardo et al. (2011). They estimated the N critical load for BIBE and GUMO to be in the range of 3–8.4 kg N/ha/yr for the protection of herbaceous plants.

16.5 Ozone Injury to Vegetation

Ozone exposures were rated as moderate by the National Park Service in BIBE and GUMO. However, Kohut (2007) concluded that, because of generally dry soil conditions during the summer, the ozone risk to vegetation is low in these parks (Table 16.2).

TABLE 16.2

Ozone Assessment Results for BIBE and GUMO Based on Estimated Average 3-Month W126 and SUM06 Ozone Exposure Indices for the Period 2005–2009 and Kohut's (2007) Ozone Risk Ranking for the Period 1995–1999[a,b]

Park Name	Park Code	W126 Value (ppm-h)	W126 NPS Condition	SUM06 Value (ppm-h)	SUM06 NPS Condition	Kohut Ozone Risk Ranking
Big Bend	BIBE	9.64	Moderate	10.05	Moderate	Low
Guadalupe Mountains	GUMO	11.64	Moderate	13.87	Moderate	Low

[a] Parks are classified into one of three ranks (Low, Moderate, High), based on comparison with other Inventory and Monitoring parks.

[b] Degrees of concern for the W126 and SUM06 indices are based solely on levels of ozone exposure. Kohut's risk to vegetation is based on several factors that contribute to injury in plants, including ozone exposure and environmental variables, and considers the effects of soil moisture on the uptake of ozone.

16.6 Visibility Degradation

Visibility is an especially important air quality–related value in BIBE, with views of the canyons cut by the Rio Grande River and the Chisos and Sierra Del Carmen Mountains arising from the Chihuahuan Desert (Figure 16.5). It has been estimated that visibility at BIBE decreased about 20% during the 1990s in response to increased sulfur dioxide emissions in Texas and neighboring states and construction of the Carbón coal-fired electric generating stations to the south near Piedras Negras, Coahuila in Mexico. During this time period, particulate sulfate concentrations on the 20% of days having the highest sulfate concentrations increased about 30% (Schichtel et al. 2005).

16.6.1 Natural Background and Ambient Visibility Conditions

Natural haze has been estimated by Interagency Monitoring of Protected Visual Environments (IMPROVE) for BIBE and GUMO. Both of these parks experience natural haze levels that are relatively low compared with other monitored parks for the 20% clearest days, the 20% haziest days, and the average of the results for all days (Table 16.3). Natural haze does not appear to be a large component of visibility for these parks.

TABLE 16.3

Estimated Natural Haze and Measured Ambient Haze in BIBE and GUMO Averaged over the Period 2004–2008

Park Name	Park Code	Site ID	Estimated Natural Haze (in Deciviews) 20% Clearest Days	20% Haziest Days	Average Days
Big Bend	BIBE	BIBE1	1.62	7.16	4.01
Guadalupe Mountains	GUMO	GUMO1	0.99	6.65	3.41

Park Name	Park Code	Site ID	Measured Ambient Haze (for Years 2004–2008) (in Deciviews) 20% Clearest Days	20% Haziest Days	Average Days
Big Bend	BIBE	BIBE1	5.69	17.00	11.19
Guadalupe Mountains	GUMO	GUMO1	5.50	16.16	10.46

The highest 90th percentile summer concentrations of atmospheric sulfate in the western United States during the period 1995–1999 were at BIBE (Malm et al. 2002). This would be expected to contribute to visibility impairment. Measured ambient haze values in BIBE and GUMO for the period 2004–2008 were considerably higher than the estimated natural haze conditions (Table 16.3).

Representative photos of selected vistas in BIBE under three differing visibility conditions are shown in Figure 16.6. Photos were selected to correspond with the clearest 20% of visibility conditions, the haziest 20% of visibility conditions, and annual average visibility conditions. This series of photos illustrates the consequences of haze on visibility at a representative vista in this park.

Big Bend

20% clearest days
Taken = 9:00 AM
Haze = 5 deciviews
Extinction = 16 Mm^{-1}
Visual range = 240 km

20% haziest days
Taken = 9:00 AM
Haze = 17 deciviews
Extinction = 56 Mm^{-1}
Visual range = 70 km

Average days
Taken = 9:00 AM
Haze = 10 deciviews
Extinction = 28 Mm^{-1}
Visual range = 140 km

FIGURE 16.6
Three representative photos of the same view in BIBE, illustrating the 20% clearest visibility, the 20% haziest visibility, and the annual average visibility. Extinction is total particulate light extinction. (From http://vista. cira.colostate.edu/improve/Data/IMPROVE/Data_IMPRPhot.htm, accessed December, 2010.)

16.6.2 Composition of Haze

Extinction charts are shown for BIBE and GUMO in Figure 16.7. Under the estimated natural conditions at these parks, sulfate contributed less than 14% of the haze on clear, average, and hazy days. On average, sulfate, derived primarily from coal-burning power plants, contributed 53.9% of ambient annual total particulate light extinction in BIBE. Other substantial contributors in this park were coarse mass (15.7%) and organics (15.0%). On the current 20% haziest days, the contribution of sulfate to light extinction increased to 61.7%, coarse mass contributed 11.7%, and organics contributed 14.1%. On the clearest 20%

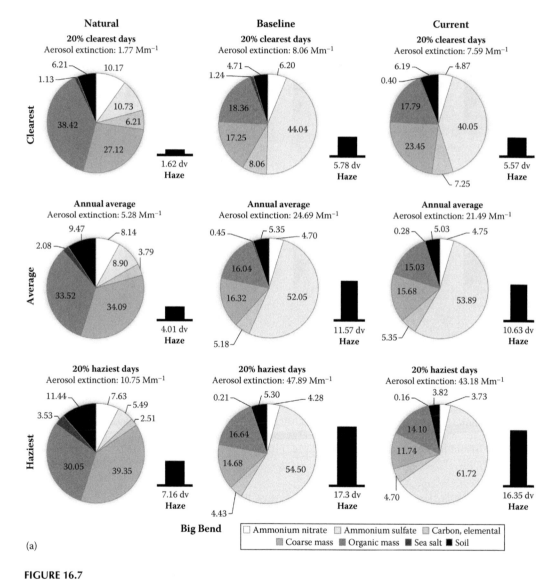

(a)

FIGURE 16.7
Estimated natural (preindustrial), baseline (2000–2004), and current (2006–2010) levels of haze (columns) and its composition (pie charts) on the 20% clearest, annual average, and the 20% haziest visibility days for (a) BIBE and (b) GUMO. (From http://views.cira.colostate.edu/fed/Tools/RegionalHazeSummary.aspx, accessed October, 2012.) *(Continued)*

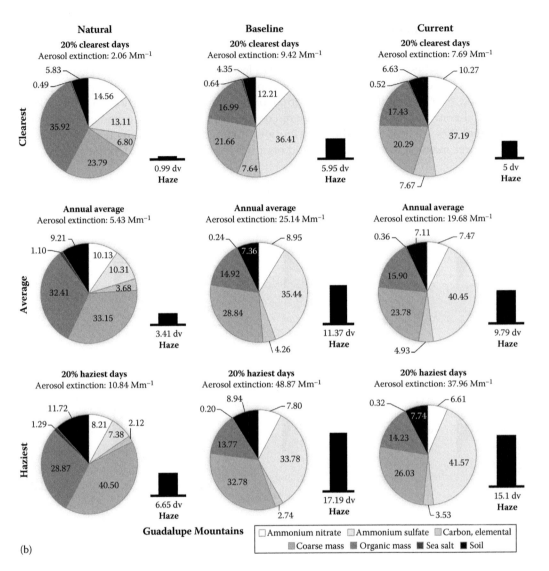

FIGURE 16.7 (Continued)
Estimated natural (preindustrial), baseline (2000–2004), and current (2006–2010) levels of haze (columns) and its composition (pie charts) on the 20% clearest, annual average, and the 20% haziest visibility days for (a) BIBE and (b) GUMO. (From http://views.cira.colostate.edu/fed/Tools/RegionalHazeSummary.aspx, accessed October, 2012.)

visibility days, sulfate contributed 40.1%, organics 17.8%, and coarse mass 23.5% of the light extinction in BIBE.

Much of the annual average light extinction in GUMO was also contributed by sulfate. On average, 40.5% of annual light extinction was due to sulfate, 23.8% to coarse mass, and 15.9% to organics (Figure 16.7). On the 20% haziest days, sulfate contributed 41.6% of the light extinction, coarse mass 26.0%, and organics 14.2%. On the clearest 20% visibility days, the contribution of sulfate decreased to 37.2% of light extinction, organics increased to 17.4%, and coarse mass decreased to 20.3%. Analyses conducted by the Western Regional Air Partnership indicated that organics from natural emissions sources, including wildfire and biogenic sources (vegetation), contribute to substantial visibility impairment throughout the western United States, including within the Chihuahuan Desert Network.

The EPA and the National Park Service initiated the Big Bend Regional Aerosol and Visibility Observational (BRAVO; Pitchford et al. 2004) study to monitor possible changes in aerosol concentrations and visibility at BIBE over time and to determine the identity and origins of particulates that were causing the observed haze. Atmospheric haze at BIBE had been increasing since the late 1980s (http://www.nps.gov/bibe/nature-science/upload/Bravo_Fact_Sheet.pdf). The BRAVO study was intended to determine the major sources and source regions contributing to haze at BIBE (Schichtel et al. 2005). The major focus was on fine particulate sulfate for a four-month period (July–October 1999). Several approaches were developed using air quality models, receptor models, and various sources of meteorological data (Schichtel et al. 2005, Barna et al. 2006, Gebhart et al. 2006).

In agreement with the data for baseline and current conditions shown in Figure 16.7, the results of the BRAVO study suggested that haze in BIBE was caused mainly by sulfate, organic C, and coarse mass, followed by lesser contributions by fine particles of light-absorbing C, fine soils, and nitrates. The composition and air concentrations of these pollutants varied seasonally; high haze levels were most common in the spring, when carbonaceous particles were most prevalent, and in the late summer to mid-fall, when sulfate compounds dominated (Gebhart et al. 2001). This seasonal cycle was attributed to different wind patterns throughout the year, leading to variations in atmospheric transport of particulate matter from emissions sources to BIBE. Spring haze was predominantly caused by smoke from fires in Mexico and Central America. Summer haze was influenced by easterly winds carrying dust from Africa across the Atlantic Ocean and Gulf of Mexico into the Chihuahuan Desert Network. Nevertheless, the dominant source of haze-forming particulate air pollution in BIBE during the late summer and early fall was sulfur dioxide emissions from sources within the United States (Pitchford et al. 2005). During intense episodes of relatively high particulate sulfate concentrations in the late summer and fall, sulfur dioxide emissions were derived mainly from sources in Texas and states to the east of Texas. In the spring and early summer, however, sources in northeastern Mexico were responsible for the majority of particulate sulfate in BIBE (Pitchford et al. 2005).

Chemical transport and receptor models were used in BRAVO to apportion observed sulfate concentrations at BIBE to various source regions and to the Carbón power plants (Schichtel et al. 2005). Opposing biases in the various modeling approaches were identified, and a hybrid receptor model was developed. The best estimates from the model/data reconciliation process suggested that 55% of the sulfur dioxide emissions impacting visibility at BIBE originated in the United States and 38% originated in Mexico. Among U.S. source regions, the distribution was 16% Texas, 30% eastern United States, and 9% western United States. Contributions from Texas and the eastern United States were generally episodic, with their greatest contribution during high sulfate episodes. Sources in Mexico were more chronic in nature.

On the 20% of the days having the highest fine particulate sulfate concentrations in the atmosphere at BIBE, sources in the United States contributed an estimated 71% of the sulfate. In contrast, on the 20% of the days having lowest sulfate concentrations, sources in the United States accounted for only 40% of the ambient sulfate.

Schichtel et al. (2005) concluded that Mexican sources of sulfur dioxide contributed more frequently than sources in the United States. Nevertheless, the highest sulfate concentrations and highest haze events at BIBE resulted from emissions derived from Texas and the eastern United States. To substantially reduce haze during the intensive haze episodes that occur at BIBE during late summer and fall would require sulfate emissions reductions in these areas.

The work of Pitchford et al. (2005) illustrated the considerable variability in the nature of haze levels and atmospheric transport pathways from diverse source regions to BIBE on daily, weekly, seasonal, and interannual time scales. The relative importance of the various pollutant source areas changed dramatically depending on the time scale and period considered. Therefore, short-term results of source attribution studies might provide unrepresentative results. Fine particulate sulfate concentrations at BIBE are seasonal; concentrations are highest during summer and fall when air commonly travels to the park from the southeast after passing over Mexico.

The Trajectory Mass Balance model was applied by Gebhart et al. (2006) to estimate the source apportionment of particulate sulfate in BIBE during the period July–October 1999. Mean sulfate source apportionment results suggested a 39%–50% contribution from Mexico, 12%–45% from Texas, 7%–26% from the eastern United States, and 3%–25% from the western United States.

Episodically high concentrations of fine mass and high light extinction also occur periodically at BIBE due to fine particles of organic C attributed to seasonal agricultural burning in Mexico, or blowing soil dust, likely from the Sahara Desert in Africa (Gebhart et al. 2001). Nevertheless, light scattering by sulfate accounts for about half of the non-Raleigh light extinction at BIBE.

16.6.3 Trends in Visibility

In much of the western United States, atmospheric sulfate concentrations decreased during the 1990s. In west Texas, however, concentrations increased. There were only two national park areas in the United States that showed a statistically significant increase in atmospheric sulfate concentrations during the period 1988–1999. One was BIBE (Malm et al. 2002).

Available haze-monitoring data suggest that the highest haze values were reached at BIBE and GUMO around the late 1990s. There is evidence that haze levels have decreased since then (Figure 16.8). The National Park Service (2010) reported long-term trends in annual haze on the clearest and haziest 20% of days at monitoring sites in 29 national parks. All 27 parks that showed statistically significant ($p \leq 0.05$) trends on the clearest days for the 11–20-year monitoring periods through 2008 exhibited decreases in haze over time. None of the sites showed increasing trends on the clearest days. One of the parks (GUMO) showed a statistically significant increase in haze on the haziest days. From 2000 to 2009, visibility improved on the clearest days at GUMO but was unchanged on the haziest days (NPS-ARD 2013).

16.6.4 Development of State Implementation Plans

Monitoring data suggest that recent measured haze values at BIBE and GUMO thus far are in compliance with the Regional Haze Rule and roughly follow the glideslopes needed to reach natural haze levels of about 7.7 deciviews on the 20% haziest days by 2064 (Figure 16.9). Substantial additional reductions in haze will be required to meet the goal.

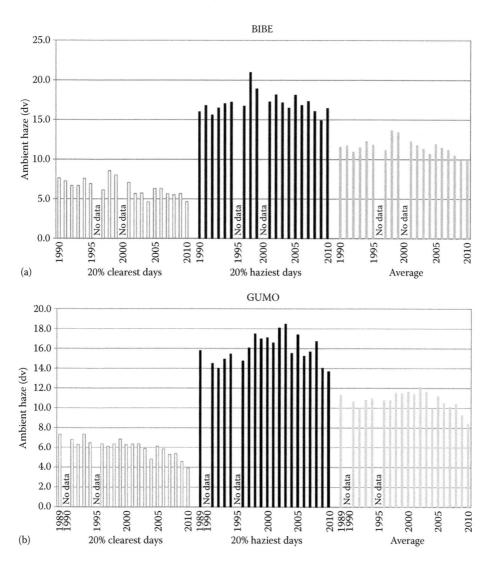

FIGURE 16.8

Trends in ambient haze levels at (a) BIBE and (b) GUMO, based on IMPROVE measurements on the 20% clearest, 20% haziest, and annual average visibility days over the monitoring period of record. (From http://vista.cira. colostate.edu/improve/Data/IMPROVE/summary_data.htm, accessed October, 2012.)

16.7 Toxic Airborne Contaminants

Limited data are available regarding the potential influence of airborne toxic substances on natural resources in the Chihuahuan Desert Network. Concentrations of current-use pesticides in vegetation in BIBE were at or slightly above the median for WACAP parks (Landers et al. 2008); these current-use pesticides included chlorpyrifos, dacthal, endosulfans, and hexachlorocyclohexane-γ. Concentrations of historic-use pesticides, PCBs, and polycyclic aromatic hydrocarbons were at or slightly below the median among WACAP

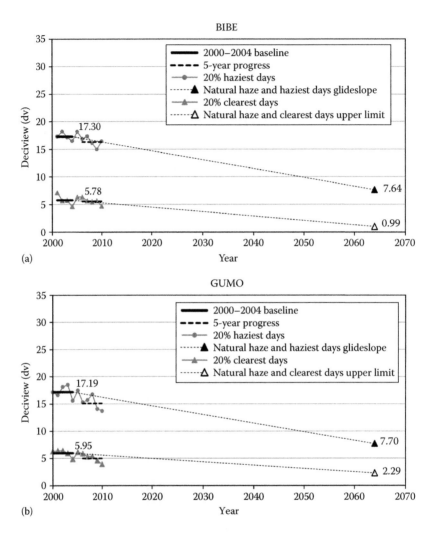

FIGURE 16.9
Glideslopes to achieving natural visibility conditions in 2064 for the 20% haziest (upper line) and the 20% clearest (lower line) days in (a) BIBE and (b) GUMO. In the Regional Haze Rule, the clearest days do not have a uniform rate of progress glideslope; the rule only requires that the clearest days do not get any worse than the baseline period. Also shown are measured values during the period 2000–2010. (From http://vista.cira.colostate.edu/improve/Data/IMPROVE/summary_data.htm, accessed October, 2012.)

study sites. Data are not available for other parks in the Chihuahuan Desert Network or with which to evaluate the effects of atmospherically deposited toxic substances on sensitive receptors in this network.

16.8 Summary

One of the most important air quality–related values in the Chihuahuan Desert Network is visibility. Parks in this network are affected by haze that is primarily due to sulfate and to a lesser extent coarse mass and organic particles. Relative contributions of those constituents to light extinction vary with season and reporting period. The highest haze values observed within the network occurred at BIBE. Prior to 1990, BIBE had the poorest visibility among the monitored Class I national parks in the western United States (Malm et al. 1990). At times, visibility is still significantly degraded in BIBE but is now worse at some other western parks, notably in parts of California and Arizona (Hand et al. 2011).

The parks in the Chihuahuan Desert Network also appear to be highly sensitive to potential terrestrial eutrophication in response to nutrient N enrichment. Nevertheless, nutrient N pollutant exposure is relatively low at the parks in this network, and effects from N deposition will likely be confounded by future changes in precipitation timing and amount (Weltzin et al. 2003). Resource sensitivity to acidification in response to atmospheric contributions of S and N appears to be relatively high in BIBE and GUMO, although acid pollutant exposure in these parks is low and, as a consequence, acidification impacts are unlikely at this time.

Ozone exposure and risk rankings for the parks in the Chihuahuan Desert Network are low to moderate; the generally dry soil conditions that predominate in this network are likely to limit ozone uptake into foliage. Nevertheless, ozone-sensitive species such as Goodding's willow (*Salix gooddingii*) occupy riparian sites where soil moisture is higher and effects of ozone exposure may be more pronounced.

Concentrations of airborne toxic contaminants in vegetation measured in BIBE in the WACAP effort were generally typical of other inventoried parks in the western United States with respect to current-use pesticides, historic-use pesticides, polycyclic aromatic hydrocarbons, and PCBs. Data regarding emissions of semivolatile organic compounds near Chihuahuan Desert Network parks are not available. However, compared with other WACAP study parks, BIBE had elevated concentrations in the air of endosulfans and the DDT breakdown product DDE. Also detected were midrange concentrations of dacthal, hexachlorocyclohexane-α, hexachlorocyclohexane-γ, and chlordanes and low concentrations of trifluralin. It is not known if these atmospheric contaminants cause adverse impacts on park resources.

References

Austin, A.T., L. Yahdjian, J.M. Stark, J. Belnap, A. Porporato, U. Norton, D.A. Ravetta, and S.M. Schaeffer. 2004. Water pulses and biochemical cycles in arid and semiarid ecosystems. *Oecologia* 141:221–245.

Báez, S., J. Fargione, D.I. Moore, S.L. Collins, and J.R. Gosz. 2007. Atmospheric nitrogen deposition in the northern Chihuahuan Desert: Temporal trends and potential consequences. *J. Arid Environ.* 68:640–651.

Barna, M.G., B.A. Schichtel, K.A. Gebhart, and W.C. Malm. 2006. Modeling regional sulfate during the BRAVO study: Part 2. Emission sensitivity simulations and source apportionment. *Atmos. Environ.* 2006:2423–2435.

Ellis, R.A., D.J. Jacob, M.P. Sulprizio, L. Zhang, C.D. Holmes, B.A. Schichtel, T. Blett, E. Porter, L.H. Pardo, and J.A. Lynch. 2013. Present and future nitrogen deposition to national parks in the United States: Critical load exceedances. *Atmos. Chem. Phys.* 13(17):9083–9095.

Fenn, M.E., R. Haeuber, G.S. Tonnesen, J.S. Baron, S. Grossman-Clark, D. Hope, D.A. Jaffe et al. 2003. Nitrogen emissions, deposition, and monitoring in the western United States. *BioScience* 53(4):391–403.

Gebhart, K., S. Kreidenweis, and W. Malm. 2001. Back-trajectory analyses of fine particulate matter measured at Big Bend National Park in the historical database and the 1996 scoping study. *Sci. Total Environ.* 276:185–204.

Gebhart, K.A., B.A. Schichtel, M.G. Barna, and W.C. Malm. 2006. Quantitative back-trajectory apportionment of sources of particulate sulfate at Big Bend National Park, TX. *Atmos. Environ.* 40:2823–2834.

Grizzle, H.W. and J.C. Zak. 2006. A microtiter plate procedure for evaluating fungal functional diversity on nitrogen substrates. *Mycologia* 98(2):353–363.

Hand, J.L., S.A. Copeland, D.E. Day, A.M. Dillner, H. Idresand, W.C. Malm, C.E. McDade et al. 2011. IMPROVE (Interagency Monitoring of Protected Visual Environments): Spatial and seasonal patterns and temporal variability of haze and its constituents in the United States. Report V. Cooperative Institute for Research in the Atmosphere, Fort Collins, CO. Available at: http://vista.cira.colostate.edu/Improve/wp-content/uploads/2016/04/Cover_TOC.pdf (accessed September, 2016).

Kohut, R. 2007. Assessing the risk of foliar injury from ozone on vegetation in parks in the U.S. National Park Service's Vital Signs Network. *Environ. Pollut.* 149:348–357.

Landers, D.H., S.L. Simonich, D.A. Jaffe, L.H. Geiser, D.H. Campbell, A.R. Schwindt, C.B. Schreck et al. 2008. The fate, transport, and ecological impacts of airborne contaminants in western national parks (USA). EPA/600/R-07/138. U.S. Environmental Protection Agency, Office of Research and Development, NHEERL, Western Ecology Division, Corvallis, OR.

Lehmann, C.M.B., B.M. Kerschner, and D.A. Gay. 2015. Impact of sulfur dioxide (SO_2) and nitrogen oxides (NO_x) emissions reductions on acidic deposition in the United States. *EM Magazine*, July 2015, pp. 6–11.

Malm, W., K. Gebhart, and R. Henry. 1990. An investigation of the dominant source regions of fine sulfur in the western United States and their areas of influence. *Atmos. Environ.* 1990(24A):3047–3060.

Malm, W.C., B.A. Schichtel, R.B. Ames, and K.A. Gebhart. 2002. A 10-year spatial and temporal trend of sulfate across the United States. *J. Geophys. Res.* 107(D22, 4627).

National Park Service (NPS). 2010. Air quality in national parks: 2009 annual performance and progress report. Natural Resource Report NPS/NRPC/ARD/NRR-2010/266. National Park Service, Air Resources Division, Denver, CO.

National Park Service–Air Resources Division (NPS–ARD). 2013. Air quality in national parks: Annual trends (2000–2009) and conditions (2005–2009) report. Natural Resource Report NPS/NRSS/ARD/NRR—2013/683. National Park Service, Denver, CO.

Pardo, L.H., M.J. Robin-Abbott, and C.T. Driscoll. 2011. Assessment of nitrogen deposition effects and empirical critical loads of nitrogen for ecoregions of the United States. General Technical Report NRS-80. U.S. Forest Service, Newtown Square, PA.

Pitchford, M.L., B.A. Schichtel, K.A. Gebhart, M.G. Barna, W.C. Malm, I.H. Tombach, and E.M. Knipping. 2005. Reconciliation and interpretation of the Big Bend National Park light extinction source apportionment: Results from the Big Bend Regional Aerosol and Visibility Observational Study—Part II. *J. Air Waste Manage. Assoc.* 55:1726–1732.

Pitchford, M.L., I.H. Tombach, M.G. Barna, K.A. Gebhart, M.C. Green, E.M. Knipping, N. Kumar et al. 2004. Big Bend Regional Aerosol and Visibility Observational Study (BRAVO) Study. Final Report. Official Report of the U.S. Environmental Protection Agency, National Park Service, Texas Commission on Environmental Quality, Electrical Power Research Institute, and National Oceanic and Atmospheric Administration. Available at https://www.dri.edu/images/stories/editors/eafeditor/Pitchfordetal2004BRAVOReport.pdf (accessed September, 2016).

Robertson, T.R., C.W. Bell, J.C. Zak, and D.T. Tissue. 2009. Precipitation timing and magnitude differentially affect aboveground annual net primary productivity in three perennial species in a Chihuahuan Desert grassland. *New Phytol.* 181:230–242.

Schichtel, B.A., K.A. Gebhart, W.C. Malm, M.G. Barna, M.L. Pitchford, E.M. Knipping, and I.H. Tombach. 2005. Reconciliation and interpretation of Big Bend National Park particulate sulfur source apportionment: Results from the Big Bend Regional Aerosol and Visibility Observational Study—Part I. *J. Air Waste Manage. Assoc.* 55:1709–1725.

Sullivan, T.J., G.T. McPherson, T.C. McDonnell, S.D. Mackey, and D. Moore. 2011. Evaluation of the sensitivity of inventory and monitoring national parks to acidification effects from atmospheric sulfur and nitrogen deposition. Natural Resource Report. NPS/NRPC/ARD/NRR—2011/349. U.S. Department of the Interior, National Park Service, Denver, CO.

Weltzin, J.F., M.E. Loik, S. Schwinning, D.G. Williams, P.A. Fay, B.M. Haddad, J. Harte et al. 2003. Assessing the response of terrestrial ecosystems to potential changes in precipitation. *BioScience* 53(10):941–952.

17

Sierra Nevada Network

17.1 Background

There are four national parks in the Sierra Nevada Network: Devils Postpile National Monument (DEPO), Yosemite National Park (YOSE), and Sequoia (SEQU) and Kings Canyon (KICA) national parks, which are jointly managed and referred to as SEKI; they are combined for most analyses in this chapter. The SEKI parks and YOSE are each larger than 100 mi^2 (259 km^2). They are the focus of this chapter. Since air pollutants can be widespread, the reduction of pollution emissions affecting large parks will often result in the protection of smaller parks located nearby as well.

There are several human population centers larger than 100,000 people within the Sierra Nevada Network region. There are also several large population centers on the perimeter of the network boundary, including San Francisco, San Jose, and Sacramento. California is the most populous state in the nation, with major population centers located along the Pacific coast and in the Sacramento and San Joaquin valleys. The Central Valley of California is also one of the most intensively cultivated agricultural regions of the United States. Air pollution is an important concern in portions of the parklands in the Sierra Nevada Network that are in proximity to major emissions sources. Figure 17.1 shows the network boundary along with locations of each park and population centers with more than 10,000 people. The two areas of California that show the most severe air quality problems have been the South Coast and the San Joaquin Valley (Suarez-Murias et al. 2009).

The Sierra Nevada region has had elevated concentrations of ozone and N air pollutants in some areas (Bytnerowicz and Fenn 1996, Bytnerowicz et al. 1999). Serious concerns have been raised regarding the effects on natural resources in SEKI and YOSE due to air pollutants emitted in the San Francisco Bay area and the San Joaquin Valley (Bytnerowicz et al. 2002), coupled with generally high resource sensitivity.

Bytnerowicz et al. (2002) characterized N, S, and ozone air pollutants along an elevational gradient in Sequoia National Park. They found significant declines in atmospheric concentrations of ozone, nitric acid, ammonia, and sulfur dioxide with increasing elevation and distance from pollution source areas. In contrast, the highest concentrations of particulate nitrate and ammonium were observed at mid-elevation, likely due to formation of ammonium nitrate during atmospheric transport from pollution source areas.

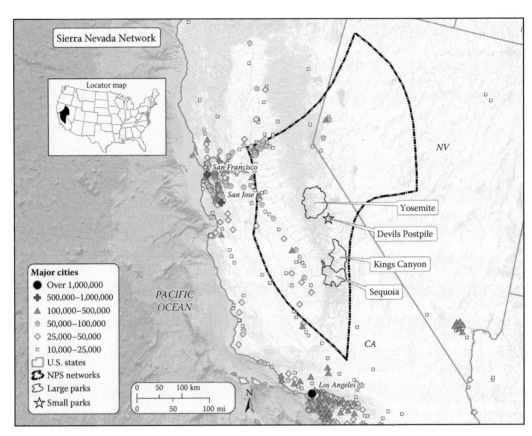

FIGURE 17.1
Network boundary and locations of parks and population centers larger than 10,000 people in the Sierra Nevada Network region.

17.2 Atmospheric Emissions and Deposition

Because the prevailing winds are from the west and northwest, many of California's national parks and Class I areas, including those located in the Sierra Nevada Mountains, are downwind of some of the major urban and agricultural areas of the state. Sulfur dioxide emissions have been low in California for several decades and do not currently pose a significant concern for parks in the Sierra Nevada Network. County-level sulfur dioxide emissions within the region were mostly less than 1 ton/mi²/yr in 2002, with somewhat higher emissions in the westernmost portion of the region and in the San Francisco Bay area. Most of the emissions from individual sulfur dioxide point source emissions identified by Sullivan et al. (2011a) near the Sierra Nevada Network parks were relatively small (less than 5000 tons/yr) and located to the west of the parks, within about 100 mi (161 km) of the coast.

County-level emissions near the Sierra Nevada Network, based on data from the EPA's National Emissions Inventory during a recent time period (2011), are depicted in Figures 17.2 through 17.4 for sulfur dioxide, oxidized N, and ammonia, respectively. Counties in the vicinity of Sierra Nevada Network parks had low sulfur dioxide emissions (<1 ton/mi²/yr;

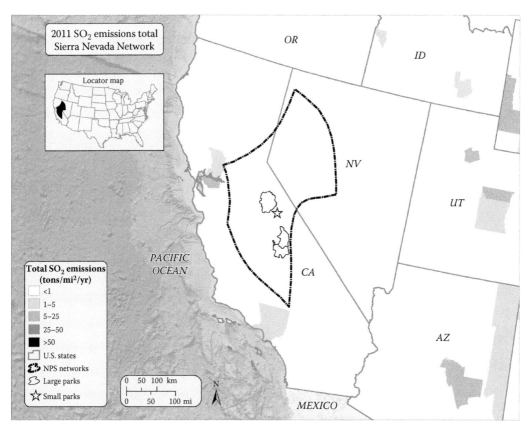

FIGURE 17.2
Total sulfur dioxide (SO_2) emissions, by county, near the Sierra Nevada Network for the year 2011. (Data from EPA's National Emissions Inventory, https://www.epa.gov/air-emissions-inventories, accessed January, 2014.)

Figure 17.2). Oxidized N emissions were higher, mostly in the range of 1–5 tons/mi^2/yr (Figure 17.3). Emissions of ammonia near Sierra Nevada Network parks were mostly less than 8 tons/mi^2/yr (Figure 17.4).

Nitrogen emissions and deposition are a more substantial concern than S for Sierra Nevada Network parks (Figure 17.5). County-level N emissions near the network generally ranged from less than 1 ton/mi^2/yr around YOSE to 20 tons/mi^2/yr near SEKI. Most sources are mobile or areal. There were very few individual point sources of N of any magnitude in the Sierra Nevada Network region. Most emissions of ozone precursors (i.e., oxidized N and volatile organic compounds) come from source areas in the South Coast, San Francisco Bay Area, and San Joaquin Valley (http://www.epa.gov/ttnchie1/trends/). These emissions are mainly from motor vehicles, with other significant sources of N from fossil-fueled power plants and of reactive organic gases from cleaning and surface coatings and solvent evaporation. Additionally, the land in the western portion of the Sierra Nevada Network region is a mix of row crops, pasture/hay, and developed areas, which are potential sources of air pollutants (especially ammonia, ozone precursors, and pesticides) to the parks in the Sierra Nevada Network. Southern airsheds on the west side of the Sierra Nevada (including those upwind of SEKI and YOSE) experience elevated N emissions and deposition, especially during summer (http://epa.gov/castnet/javaweb/index.html).

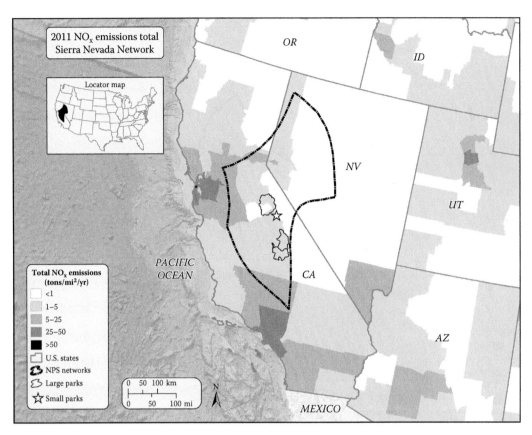

FIGURE 17.3
Total oxidized N (NO_x) emissions, by county, near the Sierra Nevada Network for the year 2011. (Data from EPA's National Emissions Inventory, https://www.epa.gov/air-emissions-inventories, accessed January, 2014.)

Total modeled S deposition in 2002, using the Community Multiscale Air Quality Model, was less than 2 kg/ha/yr throughout virtually the entire Sierra Nevada Network region (Sullivan et al. 2011a). Estimated total N deposition in 2002 ranged from less than 2 kg N/ha/yr to higher than 10 kg N/ha/yr.

Atmospheric S and N deposition levels have increased at all Sierra Nevada Network parks since 2002, based on Total Deposition project estimates (Table 17.1). Increases in total S deposition over the previous decade for the parks in this network were typically about 20%. Estimated total N deposition also increased over that same time period, and the increases were larger than those for S. Oxidized N deposition was relatively stable, but ammonium (which is primarily from agricultural sources) increased, in most cases approximately doubling since the monitoring period 2000–2002.

Total S deposition in and around the Sierra Nevada Network estimated by the Total Deposition project for the period 2010–2012 was low (<2 kg S/ha/yr) at park locations within the network area. Oxidized inorganic N deposition for the period 2010–2012 was in the range of 2–5 kg N/ha/yr throughout the parklands. Most areas received less than 6 kg N/ha/yr of ammonium from atmospheric deposition during this same period. Total N deposition was in the range of 5–10 kg N/ha/yr at most park locations (Figure 17.6); parts of SEKI were estimated to have received higher amounts.

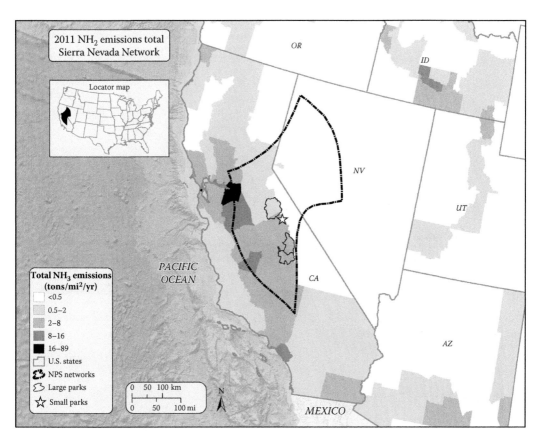

FIGURE 17.4

Total ammonia (NH_3) emissions, by county, near the Sierra Nevada Network for the year 2011. (Data from EPA's National Emissions Inventory, https://www.epa.gov/air-emissions-inventories, accessed January, 2014.)

Upwind emissions of pesticides can also constitute an important concern for parks in the Sierra Nevada Network. Pollutants can be emitted from agricultural source areas into the atmosphere as spray drift during pesticide application, via volatilization, and associated with windblown dust. Pesticides applied in California's Central Valley can be transported to sensitive high-elevation ecosystems in the nearby Sierra Nevada Mountains by prevailing west-to-east summer winds (Zabik and Seiber 1993, Aston and Seiber 1997).

Snow accounts for most of the precipitation that falls at high elevation in the Sierra Nevada and therefore is an important factor influencing the wet deposition of atmospheric contaminants. For example, at Emerald Lake (2800 meters [m]) near treeline in SEKI, snow accounted for 88% of annual precipitation from 1983 to 2000 (Sickman et al. 2001, Fenn et al. 2009). Concentrations of N in snow in the Sierra Nevada can be an order of magnitude less than concentrations in rain (Fenn et al. 2009). National Atmospheric Deposition Program wet deposition data illustrate increased nitrate and ammonium deposition from 1981 to 2001, although the increase was statistically significant only for ammonium (0.037 kg N/ha/yr; $p \leq 0.01$; Fenn et al. 2003c).

In the southern Sierra Nevada, winter storms are derived from relatively clean air masses that move onshore from Pacific frontal systems. Nonwinter precipitation is more

FIGURE 17.5
Terrestrial and aquatic ecosystems at high elevation in SEKI are sensitive to N inputs. (Photo courtesy of National Park Service by Rose Goodchild.)

commonly associated with localized thunderstorms that form from relatively N-enriched air from the Central Valley. As a consequence, N deposition exhibits seasonal variation, with lowest levels during winter (Fenn et al. 2003c).

McConnell et al. (1998) measured the atmospheric inputs of current-use pesticides to two locations (533 and 1920 m elevation) in Sequoia National Park by analyzing rain and snow samples. The highest pesticide concentration at both sites was for chlorothalonil; other quantitatively important organic chemicals in the wet deposition included the pesticides malathion, diazinon, and chlorpyrifos.

Both sulfur dioxide emissions and atmospheric sulfate concentrations at Interagency Monitoring of Protected Visual Environments (IMPROVE) visibility monitoring sites decreased during the 1990s throughout much of the western United States (Malm et al. 2002). Decreases in S emissions were largest in California (−37%). This may have contributed to visibility improvement.

TABLE 17.1

Average Changes in S and N Deposition[a] between 2001 and 2011 across Park Grid Cells at Three Sierra Nevada Network Parks

Park Code	Parameter	2001 Average (kg/ha/yr)	2011 Average (kg/ha/yr)	Absolute Change (kg/ha/yr)	Percent Change	2011 Minimum (kg/ha/yr)	2011 Maximum (kg/ha/yr)	2011 Range (kg/ha/yr)
KICA	Total S	1.31	1.55	0.24	19.3	1.20	1.90	0.71
	Total N	5.75	7.21	1.46	25.2	5.01	12.15	7.14
	Oxidized N	3.49	3.37	−0.13	−3.3	2.45	4.39	1.94
	Ammonium	2.26	3.84	1.58	69.4	2.34	7.80	5.46
SEQU	Total S	1.31	1.54	0.23	18.4	1.06	1.90	0.84
	Total N	6.31	9.16	2.84	44.3	5.91	13.27	7.37
	Oxidized N	3.71	3.72	0.01	0.4	2.84	4.96	2.12
	Ammonium	2.60	5.43	2.83	106.7	2.70	8.32	5.61
YOSE	Total S	1.11	1.30	0.18	16.5	0.97	1.61	0.63
	Total N	4.40	5.76	1.36	31.0	4.24	7.13	2.89
	Oxidized N	2.74	2.45	−0.29	−10.4	2.03	2.93	0.90
	Ammonium	1.66	3.31	1.65	99.6	2.20	4.20	1.99

[a] Deposition estimates were determined by the Total Deposition project based on three-year averages centered on 2001 and 2011 for all ~4 km grid cells in each park. The minimum, maximum, and range of 2011 S and N deposition within each park are also shown.

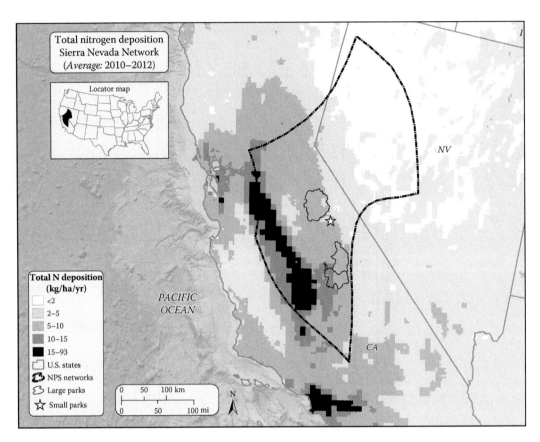

FIGURE 17.6
Total N deposition estimated by the Total Deposition project for the three-year period centered on 2011 in and around the Sierra Nevada Network. (From Schwede, D.B. and Lear, G.G., *Atmos. Environ.*, 92, 207, 2014.)

17.3 Acidification

Sullivan et al. (2011a) evaluated the relative risk of acidification for all National Park Service Inventory and Monitoring networks nationwide. The network rankings developed by Sullivan et al. (2011a) for acid Pollutant Exposure, Ecosystem Sensitivity to acidification, and Park Protection yielded an overall network acidification Summary Risk ranking for the Sierra Nevada Network that was among the highest of all 32 Inventory and Monitoring networks. The overall level of concern for acidification effects on Sierra Nevada Network parks was judged by Sullivan et al. (2011a) to be very high.

17.3.1 Acidification of Terrestrial Ecosystems

Portions of the mountainous West contain large areas of exposed bedrock, with little soil or vegetative cover to neutralize acidic inputs. This is particularly true of alpine regions of the Sierra Nevada (Charles 1991). It is likely that alpine terrestrial ecosystems in the Sierra Nevada Network are highly sensitive to acidification. In view of the relatively low levels of sulfur deposition, however, effects to date are expected to have been minimal.

Data on the possible effects of S and N deposition on the acid–base characteristics of coniferous forest ecosystems in the Sierra Nevada Network region are spotty and inconclusive. What data are available suggest that California mixed coniferous forests may be less susceptible to N deposition stress than coniferous forests in the eastern United States. For example, a study by Bytnerowicz (2002) demonstrated increased bole growth in ponderosa pine and black oak with additions of 50 kg N/ha/yr. A study by Temple (1993) found that ponderosa pine seedlings exposed to acidic precipitation (pH 5.3, 4.4, and 3.5) showed no significant changes in growth. Fenn et al. (2003b) reported that deposition levels of 20–35 kg N/ha/yr contributed to increased nitrate leaching and soil acidity and decreased base saturation in southern California ecosystems, but they did not report quantitative measures of plant growth. Most coniferous forests in this network receive much lower levels of atmospheric N deposition and are unlikely to be measurably impacted by acidification.

17.3.2 Acidification of Aquatic Ecosystems

The Cascade and Sierra Nevada Mountains (including within the Sierra Nevada Network region) constitute the mountain ranges with the greatest number of acid-sensitive aquatic resources in the country (Landers et al. 1987). Surface waters in these regions are among the most poorly buffered in the United States (Landers et al. 1987, Melack and Stoddard 1991). Lakes in the Sierra Nevada are especially sensitive to impacts from acidic deposition because of the predominance of granitic bedrock, thin naturally acidic soils, large amounts of precipitation, sparse coniferous vegetation, and dilute nature of the lakes (McColl 1981, Melack et al. 1985, Melack and Stoddard 1991). The hydrologic cycle is dominated by the annual accumulation and melting of a dilute, mildly acidic (pH 5.5) snowpack (Melack et al. 1997). The acid–base chemistry of lake (and presumably also stream) waters in the Sierra Nevada Network region appears to be primarily a function of the interactions among several key parameters and associated processes: atmospheric deposition; bedrock geology; the depth and composition of surficial deposits and associated hydrologic flowpaths; and the occurrence of soils, tundra, and forest vegetation (Sullivan 2000).

The weight of evidence suggests that many high-elevation lakes in Sierra Nevada Network parks receive N deposition sufficiently high to cause limited chronic nitrate leaching. It does not appear that widespread chronic lake or stream acidification has occurred to any significant degree, although some episodic acidification has likely occurred (Sullivan 2000). It is important to note that even low to moderate concentrations of nitrate in surface waters within Sierra Nevada Network parks might be significant in view of (1) the low base cation concentrations in many surface waters, (2) the potential for continuing N deposition to eventually exhaust natural assimilative capabilities, and (3) the fact that these observations are typically based on summer or fall data. Acid anion concentrations in most western mountain lakes are relatively low during summer and fall but can be higher during spring snowmelt (Melack et al. 1989).

A large number of water quality studies have been conducted within the parks in the Sierra Nevada Network, and several were summarized by Sullivan et al. (2001). The National Acid Precipitation Assessment Program State of Science and Technology Reports and the Integrated Assessment (NAPAP 1991) provided only a cursory treatment of aquatic effects issues in the Sierra Nevada Network region, largely because it was well known that atmospheric depositions of S and N were generally low compared to areas in the East (Sullivan 2000). The available information on acid–base chemistry of surface waters in the Sierra Nevada Network region is based on synoptic data from the EPA's 1985 Western Lakes Survey (Landers et al. 1987) and some more localized studies (cf., Shaw et al. 2014). The Western Lakes Survey sampled 11 lakes in Kings Canyon National Park and 9 lakes in Sequoia National Park, plus an additional 14 lakes within 16 mi (25 km) of SEKI (Sullivan et al. 2001). Nitrate concentrations were virtually undetectable in most western lakes sampled by the Western Lakes Survey during the fall season (Landers et al. 1987). However, in a few cases, fall nitrate concentrations were surprisingly high. In the Sierra Nevada, about 10% of the Western Lakes Survey lakes had nitrate concentrations above 5 µeq/L (Sullivan and Eilers 1994).

Data from over 12,500 streams and lakes were used by the Critical Loads of Atmospheric Deposition Science Committee of the National Atmospheric Deposition Program (http:// nadp.sws.uiuc.edu/committees/clad/) to develop steady-state critical loads for acidity of surface waters based on multiple approaches for estimating base cation weathering. Water quality data were obtained from a variety of sources, including the EPA Long Term Monitoring sites, lake surveys, Environmental Monitoring and Assessment Program studies, and National Stream Surveys; U.S. Geological Survey; National Park Service Vital Signs program; and U.S. Forest Service air program. The average water quality measurements from the most recent five years of data were used for sites with long-term water monitoring records. The Critical Loads of Atmospheric Deposition Science Committee database included 11 sites in Kings Canyon National Park and 12 sites in each of Sequoia National Park and YOSE. All three parks had some lakes with modeled estimates of critical load = 0 kg S/ha/yr to attain lake acid neutralizing capacity of 20 µeq/L, suggesting substantial acid sensitivity of surface waters in these parks.

Shaw et al. (2014) used the Steady State Water Chemistry model to estimate the steady-state critical load of acidity for 208 lakes in the Sierra Nevada. Study lakes were generally dilute (mean specific conductance 8 µS/cm), with low acid neutralizing capacity (mean 57 µeq/L). Most were located in Forest Service wilderness areas, some in proximity to Sierra Nevada Network parks. Using a critical acid neutralizing capacity limit of 10 µeq/L to protect aquatic biota from the effects of water acidification, the Steady State Water Chemistry model predicted a critical load of 149 eq/ha (14.9 meq/m^2) of acidity for watersheds situated on granitic bedrock. More than one-third of the lakes received acidic deposition in exceedance of the critical load. The median study lake showed a critical load exceedance of about 80 eq/ha (8 meq/m^2). Based on these calculations, Shaw et al. (2014) concluded that high-elevation lakes in the Sierra Nevada have not fully recovered from the effects of acidic deposition despite large improvements in air quality over the last several decades.

17.3.2.1 Sequoia and Kings Canyon National Parks

The potential for chronic, and especially episodic, acidification of lakes and streams in SEKI from N and S deposition is an important water quality concern (Figure 17.5). This concern is based largely on the moderate levels of N deposition that occur in the parks, extreme sensitivity to acidification of many high-elevation surface waters, potential short-term effects of acidic rain-on-snow events during the snowmelt period, and documented importance of episodic hydrological processes in alpine and subalpine watersheds within the parks. Significant effort has been expended to determine the extent to which acidic deposition may have contributed to chronic and/or episodic acidification in these parks and the extent of acidification that would be required to elicit adverse biological impacts. The results of these studies paint a rather complete picture of the acid–base status of freshwater resources in SEKI. These results are also relevant to YOSE.

About one-third of the lakes sampled by the Western Lakes Survey in and near SEKI had acid neutralizing capacity <50 µeq/L, and two of the lakes in the park had acid neutralizing capacity <20 µeq/L (Figure 17.7). These lakes, especially those having acid neutralizing capacity <20 µeq/L, are sensitive to potential acidification from acidic deposition. All of the most sensitive lakes had pH between 6.1 and 7.0, and most had sum of base cation concentrations between about 25 and 50 µeq/L. Most were small (<about 7 ha), occurred in watersheds less than about 100 ha in area, and were situated at elevations ranging from about 2800 to 3700 m.

A chemical survey of 101 high-altitude lakes in 7 national parks in the western United States was conducted by the U.S. Geological Survey during the fall of 1999 (Clow and Sueker 2000); 72 of the lakes had previously been sampled during the fall of 1985 as part of the Western Lakes Survey. This study sampled 24 lakes in SEKI, including all of the previously sampled Western Lakes Survey lakes. Five lakes had acid neutralizing capacity <30 µeq/L, with the lowest being 7 µeq/L. That same lake (4A1-056) was the only one having relatively high nitrate concentration (9 µeq/L). Eleven lakes had acid neutralizing capacity <50 µeq/L. Of the 10 lakes sampled in YOSE, 9 of which had been included in the Western Lakes Survey, all had acid neutralizing capacity ≤75 µeq/L and 4 had acid neutralizing capacity ≤30 µeq/L. Only one lake in YOSE exhibited nitrate concentration higher than 0.1 µeq/L (Clow and Sueker 2000).

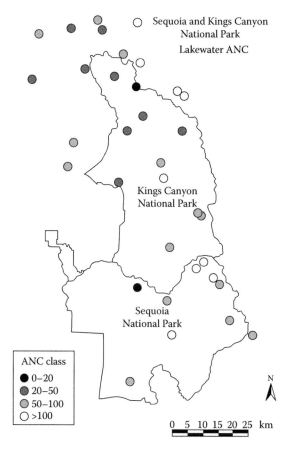

FIGURE 17.7
Lake water acid neutralizing capacity (ANC) in and around SEKI, based on the results of the Western Lakes Survey.

The concentrations of sulfate in the most acid-sensitive lakes sampled by Clow and Sueker (2000) tended to be relatively low; lake water sulfate concentration ranged between about 3 and 10 µeq/L in most cases. Such concentrations are approximately what would be expected, assuming average sulfate concentrations in precipitation of about 3–5 µeq/L, negligible dry deposition, and less than 50% evapotranspiration (Sullivan et al. 2001). Although concentrations of sulfate in lakes in the Sierra Nevada Network region are generally low, at some locations geologic sources contribute substantial amounts of sulfate to lake waters (Sullivan et al. 2001). Clow et al. (1996) presented a comparison of median major ion chemistry from four surface water surveys in the Sierra Nevada. Water chemistry measured in the various surveys was similar. Granitic watersheds were more acid sensitive than nongranitic watersheds and were less likely to exhibit high sulfate concentrations suggestive of watershed sources of S. Many of the Western Lakes Survey lakes in SEKI, including two of those having low acid neutralizing capacity (<50 µeq/L), had relatively high concentrations of sulfate (>10 µeq/L), which were likely the result of watershed sources of S. Nitrate concentrations were variable in the acid-sensitive lakes, ranging from near 0 to 10 µeq/L.

Paleolimnological reconstructions of lake water pH and acid neutralizing capacity were calculated by Holmes et al. (1989) at 24 depth intervals at Emerald Lake in Sequoia National Park for the period 1825 to the present. Significant trends were not found for either pH or acid neutralizing capacity, and the researchers concluded that Emerald Lake had not acidified in response to acidic deposition. Whiting et al. (1989) completed paleolimnological analyses of three additional lakes in the Sierra Nevada. Eastern Brook Lake (pH = 7.06) showed evidence of both long-term alkalization (~0.3 pH units over the past 200 years) and pH fluctuations since 1970. Lake 45 (pH = 5.16) may have acidified slightly (~0.2 pH units) over the previous 60 years. Lake Harriet (pH = 6.52) showed no significant change.

Williams and Melack (1989) examined variation in the onset of snowmelt in portions of the Emerald Lake watershed and effects on the magnitude and timing of ionic pulses in the various subbasins. Source areas of snowmelt runoff varied spatially and temporally in response to slope and aspect. Concentrations of nitrate and sulfate in streams were correlated with the amount of snowmelt in each subbasin. Inflows to Emerald Lake had elevated concentrations of nitrate (~18 µeq/L) and sulfate (~9 µeq/L) at initiation of snowmelt, which then decreased as snowmelt progressed. The onset of snowmelt, and accompanying ionic pulses in stream water, shifted from subbasins with southwesterly aspect to those with northerly aspect (Williams and Melack 1989).

Thirty lakes were selected at random for study by Jenkins et al. (1994) from the subset of Sierra Nevada lakes above approximately 2400 m elevation. The lakes were selected using the EPA's Environmental Monitoring and Assessment Program area-based sampling technique that allowed extrapolations to be performed to a population of 1404 lakes that are larger than 1 ha in surface area. The lakes were sampled for their water chemistry and to assess the populations of fish and macroinvertebrates that were present in the lakes and their associated streams. The results of water chemistry analyses were generally similar to those found in the Western Lakes Survey, although lakes in the survey of Jenkins et al. (1994) were higher in elevation and smaller in surface area overall as compared with the Western Lakes Survey lakes. Median pH and acid neutralizing capacity values were slightly lower than the medians reported in the Western Lakes Survey, and nitrate concentrations were higher. Several lakes at the southeastern border of Kings Canyon National Park were acidic from watershed sources of S. These lakes had high sulfate and Al concentrations. In the lakes represented by the study, 8% had pH < 6.

17.3.2.2 Yosemite National Park

High-elevation lakes and streams in YOSE are extremely sensitive to acidification from even moderate levels of S or N deposition. There are about 268 lakes and 92 rivers and streams (1415 km in length) within YOSE. Geological factors, including jointing of the bedrock and the distribution of glacial till, appear to exert strong controls on water chemistry (Clow et al. 1996).

Water chemistry data from three surface water surveys in the upper Merced River watershed, conducted in August 1981, June 1988, and August 1991, were analyzed and compared by Clow et al. (1996) with mapped geological, hydrological, and topographic features to try to identify solute sources and major processes that control water chemistry during base flow periods. A total of 23 sites were sampled during the three surveys. The water at most of the sample sites was dilute, with acid neutralizing capacity values ranging from 26 to 77 μeq/L at most sites. The acid neutralizing capacity was considerably higher in two of the sampled subwatersheds, however, ranging from 51 to 302 μeq/L. Concentrations of weathering products in surface water were correlated with the fraction of each subwatershed that was underlain by surficial material, primarily glacial till.

The Western Lakes Survey sampled 9 lakes within YOSE and an additional 14 lakes within 25 km of YOSE. Two-thirds of the surveyed lakes in YOSE, and nearly half of those near the park, had acid neutralizing capacity <50 μeq/L. Three of the sampled lakes in and near the park had acid neutralizing capacity <20 μeq/L. Many of the sensitive low acid neutralizing capacity lakes were small (<10 ha), as is typically the case for acid-sensitive Sierra Nevada lakes, although several were considerably larger (16–30 ha). The pH values in the sensitive lakes were generally between 6.2 and 7.0, although one lake outside the park had pH < 6. Lake water sulfate concentrations were generally between about 3 and 12 μeq/L in the sensitive lakes, which is roughly consistent with the level of S deposition to the watersheds. There were also three lakes within YOSE with slightly higher sulfate concentration (14 and 18 μeq/L), which could be partly attributable to watershed sources of S, and one lake outside the park with very high sulfate (92 μeq/L), which is almost certainly attributable to geologic S. The three lakes having higher-than-expected sulfate concentration were not among the most acid sensitive of the sampled lakes (acid neutralizing capacity = 71, 57, 38 μeq/L, respectively). Although most of the sampled lakes had very low nitrate concentration (<2 μeq/L), there was one lake in YOSE with relatively high nitrate (8 μeq/L) and two lakes outside the park with similarly high nitrate (6 and 10 μeq/L). All of the lakes having high nitrate were low in acid neutralizing capacity (27–38 μeq/L) and were over 3100 m elevation The most acid-sensitive lakes in and near YOSE, that is, those having acid neutralizing capacity less than about 30 μeq/L, had very low concentrations of base cations (about 20–35 μeq/L), about half of which was Ca, and low dissolved organic carbon (<2 mg/L). All are presumably highly sensitive to chronic, and especially episodic, acidification if S or N deposition increases substantially in the future.

Surface waters in YOSE generally consist of chains of relatively small lakes connected by high-gradient streams (Clow et al. 2002, 2003). Clow et al. (2010) sampled the water chemistry of 21 streams and the outlets of 31 small (typically <10 ha) lakes in YOSE during September 2003. Samples were collected during baseflow conditions at the end of the growing season and therefore generally represented nitrate concentration annual minima and acid neutralizing capacity maxima. Surface water nitrate concentrations ranged from <1 to 14 μeq/L and were highest in small, high-elevation watersheds on the eastern side of the park. Surface water acid neutralizing capacity values were low at sites located at high elevation and increased in a downstream direction. The minimum measured acid neutralizing capacity was 11 µeq/L.

Clow et al. (2010) used multiple linear regression to estimate fall nitrate concentration and acid neutralizing capacity in surface waters across YOSE. The multiple linear regression models explained 84% of the variation in nitrate and 70% of the variation in acid neutralizing capacity. Surface water nitrate concentration was positively associated with elevation, abundance of neoglacial till and talus deposits, and amount of unvegetated terrain and negatively correlated with amount of alluvium and riparian areas. The positive association with elevation reflects cold temperature and short growing season, which limit N uptake by vegetation and aquatic biota at high elevation. Unglaciated areas and areas of neoglacial till and talus would be expected to lack well-developed soil and exhibit limited N assimilation. Nitrate concentrations were also negatively associated with the percent of mixed conifer forest in the watershed in response to high N uptake in such forests (Sickman et al. 2002). Winter N deposition was not correlated with surface water nitrate concentration. Thus, watershed features seemed to exert greater control on surface water nitrate concentration than N input rate. Surface water acid neutralizing capacity was positively correlated with watershed area and abundance of metamorphic rocks and negatively correlated with the amount of unvegetated terrain, water, and N deposition in the watershed. The positive relationship between surface water acid neutralizing capacity and watershed area reflected the influence of water residence time and water/rock interactions. Similarly, areas of alluvium and Pleistocene till would be expected to have deeper, better-developed soils allowing greater interaction with drainage water (Clow et al. 2010). Acid neutralizing capacity was positively correlated with forest cover due to weathering in forest soil and negatively correlated with winter N and S deposition and snow duration, reflecting the influence of deposition and snow dynamics.

In YOSE, talus and unvegetated terrain are most common at high elevation. Clow et al. (2010) grouped the major geological materials into four classes, ranging from high to low sensitivity to acidic deposition: (1) neoglacial till and talus; (2) felsic intrusive rocks such as granite, granodiorite, and diorite; (3) mafic intrusive rocks; and (4) metamorphic rocks, alluvium and Pleistocene glacial till. On the basis of this geologic classification, most surface waters in YOSE are highly sensitive to acidic deposition, consistent with their known low ionic strength (Clow et al. 2002). There are also scattered areas of low sensitivity throughout the park. Many are associated with alluvium and Pleistocene glacial till. There are metamorphic rock outcrops near the Sierran Crest north of Tioga Pass that contain carbonate and sulfide minerals. In this area, acid neutralizing capacity tends to be high, along with Ca and sulfate. Many waters in this area have sulfate concentration >30 µeq/L, much of which is derived from mineral weathering rather than atmospheric deposition.

17.3.2.3 Episodic Acidification

High-elevation watersheds with steep topography, extensive areas of exposed bedrock, deep snowpack accumulation, and shallow, base-poor soils tend to be most sensitive to episodic acidification. Such features are common in parklands within the Sierra Nevada Network.

Temporal variability in surface water and soil solution chemistry, and patterns in nutrient uptake by terrestrial and aquatic biota, influence acidification processes and pathways. Thus, conditions are constantly changing in response to episodic, seasonal, and interannual cycles and processes. In particular, climatic fluctuations that govern the amount and timing of precipitation inputs, snowmelt, vegetative growth, depth to groundwater tables, and evapoconcentration of solutes influence soil and surface water chemistry and the interactions between pollution stress and sensitive aquatic and terrestrial biological receptors. Because of this variability, many years of data are required to establish the existence of trends in surface water chemistry. Assignment of causality to changes that are found to occur is a challenge. The data that would be needed for determining the extent and magnitude of episodic acidification have not been collected to a sufficient degree in acid-sensitive areas of the Sierra Nevada Network parks to support the regional assessment of episodic acidification.

The hydrology of alpine and subalpine ecosystems in the Sierra Nevada is dominated by snowfall and snowmelt, with over 90% of the annual precipitation falling as snow. The relatively small loads of acidic deposition can supply relatively high concentrations of sulfate and nitrate to lakes and streams during the early phase of snowmelt (Stoddard 1995) through the process of preferential elution (Johannessen and Henriksen 1978). Acidic deposition can contribute to episodic acidification of surface water both by supplying N that can produce pulses of nitrate during high flow periods, contributing hydrologically mobile sulfate, and by lowering baseline pH and acid neutralizing capacity so that episodes that do occur are sufficient to produce biologically harmful conditions (Stoddard et al. 2003).

Lake water pH and acid neutralizing capacity in the Sierra Nevada generally decrease with increasing runoff, reaching minima near peak snowmelt discharge (Melack et al. 1998). Most other solutes exhibit temporal patterns that indicate dilution or a pulse of increased concentration followed by either dilution or biological uptake (Melack et al. 1998). Melack et al. (1998) found that nitrate peaks of 515 μeq/L were common, although they were usually less than 2 μeq/L in the N-limited study lakes (Chrystal and Lost Lakes). Increases in sulfate concentration were smaller in magnitude. Except in watersheds thought to have bedrock sources of S (Spuller and Ruby Lakes), the differences between sulfate maxima and minima were generally within 2 μeq/L. Nitrate and sulfate then declined at peak runoff. Outflow acid neutralizing capacity declined by 24%–80% during the spring, with an average decline of 50%. Lowest acid neutralizing capacity was generally between about 15 μeq/L (Lost and Pear Lakes) and 30 μeq/L (Ruby and Crystal Lakes). Seasonal acid neutralizing capacity depressions were greatest during years with deep snowpacks and high snowmelt runoff (Melack et al. 1998).

Melack et al. (1998) found that the relationship between minimum acid neutralizing capacity during snowmelt and fall overturn acid neutralizing capacity in the Sierra Nevada was linear:

$$\text{Minimum acid neutralizing capacity} = (0.88 \pm 0.03) \times (\text{Overturn acid}$$

$$\text{neutralizing capacity}) - (8.6 \pm 2.8)$$

$$(r^2 = 0.98, \quad se = 8.0 \; \mu eq/L) \tag{17.1}$$

Application of this equation to the lakes in the Sierra Nevada that were included in the Western Lakes Survey statistical frame suggested that none of the lakes represented by the Western Lakes Survey were acidified to acid neutralizing capacity below 0 μeq/L by snowmelt under ambient levels of acidic deposition (Melack et al. 1998). However, the confidence limit of this empirical relationship allowed the possibility that up to 1.8% of lakes in the Sierra Nevada were acidified to acid neutralizing capacity ≤0 μeq/L during snowmelt.

Williams and Melack (1991) and Williams et al. (1995) documented ionic pulses (2–10 days in duration) in meltwater concentrations in the Emerald Lake watershed 2- to 12-fold greater than the snowpack average. Sulfate and nitrate concentrations in meltwater decreased to below the initial bulk concentrations after about 30% of the snowpack had melted. The initial meltwater draining from the snowpack had concentrations of nitrate and ammonium as high as 28 µeq/L, compared to bulk snowpack concentrations <5 µeq/L (Williams et al. 1995). Stream water nitrate concentrations peaked during the early snowmelt period, with maximum stream water concentrations of 18 µeq/L. During summer, stream water nitrate concentrations were always near or below the detection limit.

Leydecker et al. (1999) investigated episodic lake acidification from 1990 to 1994 in seven high-elevation lake watersheds in the Sierra Nevada Mountains, including three lakes in Sequoia National Park (Emerald, Pear, Topaz). Base cation dilution accounted for 75%–97% of the episodic acid neutralizing capacity depression during snowmelt. Where increasing anion concentration was important, sulfate and nitrate were equal contributors during early snowmelt, and sulfate predominated during late snowmelt. A model based on the observed relationship between snowmelt minimum and fall season acid neutralizing capacity estimated that none of the lakes sampled in the Sierra Nevada during the Western Lakes Survey had been subject to episodic acidification to acid neutralizing capacity ≤0 µeq/L (Leydecker et al. 1999).

17.3.2.4 Effects on Aquatic Biota

Episodic acidification is often the limiting condition for aquatic organisms in streams that can be suitable for aquatic life under baseflow conditions. Potential biological effects of acidic deposition on surface waters in the Sierra Nevada are primarily associated with possible acidification from high nitrate concentrations. In general, such effects tend to be episodic rather than chronic.

Emerald Lake in Sequoia National Park has been the focus of a considerable amount of work on potential biological effects of acidification due to S and N deposition. *In situ* enclosure studies were conducted for 35 days at Emerald Lake by Barmuta et al. (1990). Contrary to previous studies (e.g., Melack et al. 1987), the lake sediments were included within the experimental enclosures. This allowed the investigators to document the response of zoobenthos as well as zooplankton. Treatments included a control (pH 6.3) and acid addition to reach pH levels of 5.8, 5.5, 5.3, 5.0, and 4.7. Results indicated that zooplankton were sensitive to acidification, but zoobenthos were unaffected by the experimental treatment. *Daphnia rosea* and *Diaptomus signicauda* decreased in abundance below the range of pH 5.5–5.8 and were eliminated below about pH 5.0. *Bosnia longirostris* and *Keratella taurocephala* generally became more abundant with decreasing pH. Barmuta et al. (1990) concluded that even slight acidification of high-elevation lakes in the Sierra Nevada might alter the structure of the zooplankton community.

Melack et al. (1989) concluded that Emerald Lake was not showing serious chemical or biological effects of acidification. Many species of aquatic biota known to be acid sensitive were found in Emerald Lake and associated streams, and the brook trout population did not show signs of acid-induced stress. However, Melack et al. (1989) also concluded that Emerald Lake and associated streams are extraordinarily sensitive to acidification because of their extremely dilute ionic chemistry. If acidic deposition was to increase in the future, stream biota would likely be impacted. Seasonal patterns in the chemistry of Emerald Lake are controlled largely by thermal stratification, flushing during snowmelt, and nutrient uptake by phytoplankton during the ice-free season (Melack et al. 1989). Interannual variation in the chemistry of Emerald Lake and the streams in its watershed were largely explained by variation in the annual quantity of snowfall, which in turn determines the degree of flushing of the lake during snowmelt and also the duration of the subsequent ice-free season.

Responses of aquatic macroinvertebrates to acidification were evaluated by Kratz et al. (1994) in 12 streamside channels in Sequoia National Park. Replicated treatments included a control (pH 6.5–6.7) and experimental exposure at pH levels of 5.1–5.2 and 4.4–4.6. Invertebrate drift was monitored continuously, and benthic macroinvertebrate densities were determined before and after acidification. Single 8-hour acid pulses increased the drift of sensitive taxa, and their densities were reduced. Mayflies (*Baetis* sp.) showed reduced density post treatment to less than 25% of control densities in both pH reduction treatments (5.2, 4.6) and two different experimental exposures. Densities of *Paraleptophlebia* sp. appeared to be reduced by the acidification, but most treatment effects were not statistically significant. Kratz et al. (1994) suggested that the effects of acid inputs on benthic species densities depended on microhabitat preferences. *Baetis* sp. nymphs are epibenthic and active. They are often found on the upper surfaces of rocks where they are directly exposed to acidified water. This may have been responsible for their greater response to acidification.

In acid-sensitive lakes in the western United States, the focus of concern regarding the biological effects of acidification is often mainly on native salmonid fish. It is important to note, however, that many high-elevation lakes and streams in the Sierra Nevada Network region were historically fishless. The top predators in such aquatic ecosystems were often amphibians or crustaceans. Thus, even though fish might be considered native to high-elevation waters in the Sierra Nevada Network region, they were not necessarily native to a particular lake or stream. At least 30 nonnative fish species have been introduced to Sierran waters, 10 of which have become widespread (Sullivan et al. 2001). Fish introductions have been especially pronounced at the higher elevations (>1800 m) that lacked fish prior to the trout introductions that began during the nineteenth century.

Serious declines and extirpation of frogs and toads have been documented in many areas of the United States. National parks in California have been affected, including YOSE and SEKI. Perhaps the greatest justification for focusing attention on amphibians in the high Sierra is that dramatic population declines appear to have already occurred or are in progress for at least two of the five amphibian species that still occur in the high Sierra (Pierce 1985, Bradford and Gordon 1992). These are mountain yellow-legged frog and Yosemite toad (*Anaxyrus canorus* or *Bufo canorus*), both of which are restricted to high elevation. Many of the population declines of these species have occurred in seemingly pristine environments, including areas within SEKI, YOSE, and numerous wilderness areas in the Sierra Nevada Network region. Possible causes for these declines include acidic deposition, disease, habitat loss, increased ultraviolet light, climatic changes, toxics, or a combination of such factors (Boiano et al. 2005). The decline and extinction of amphibians has also apparently been aggravated at many locations by the introduction of nonnative fish. Historically, it was possible for amphibians to use streams as dispersal corridors. However, the presence of introduced fish in many of the Sierran lakes and streams greatly reduced or eliminated this possibility (Bradford et al. 1993). Predatory trout have dramatically altered lake and stream ecosystems and have been implicated in the decline of a number of amphibian species, in particular the pronounced decline of the mountain yellow-legged frog, whose historic distribution corresponded very closely with location of the zone of historically fishless lakes and streams (Jennings 1996). Prior to the introduction of game fish, the mountain yellow-legged frog was probably the most prevalent aquatic vertebrate in mid- and high-elevation lakes in the Sierra Nevada (Bradford 1989). Surveys have indicated, however, that the species disappeared from much of its former range (Phillips 1990, Bradford and Gordon 1992). Surveys for Yosemite toad indicated that it disappeared from approximately half of its former range (cf., Bradford and Gordon 1992).

The sensitivity of aquatic amphibians to air pollution impacts is noteworthy for a number of reasons (Bradford and Gordon 1992). Many of the physiological and ecological characteristics of amphibians render them especially sensitive to environmental change or degradation (Blaustein and Wake 1990). Additionally, aquatic amphibians at high elevation in the Sierra Nevada breed during or shortly after snowmelt and are exposed to the acidifying influence of the Sierran snowmelt during early life stages that are most sensitive to acidification (Freda 1990). Laboratory dose-response studies for the mountain yellow-legged frog, Yosemite toad, Pacific tree frog, and long-toed salamander showed high survival for all four species at higher pH levels, but survival declined dramatically as a function of pH in the mid to low 4s.

Bradford and Gordon (1992) studied several species of amphibian as potential indicators of adverse ecological effects of acidic deposition in the Sierra Nevada (see also Bradford et al. 1993, 1994). They conducted laboratory dose-response studies to determine the sensitivity of four amphibian species to low pH and elevated Al concentration in early life stages, and they also conducted field surveys to characterize the abundance and associated water chemistry of amphibian populations at high elevation. The results suggested that amphibians were at little risk from low pH in waters that might be acidified to pH 5.0. The authors concluded that the possibility exists that observed sublethal effects in response to pH as high as 5.25 might represent a threat to amphibian populations due to reduced growth rate and early hatching. Potential amphibian breeding sites were surveyed for two declining and one non-declining species at high elevation within 30 randomly selected survey areas. No significant water chemistry differences were found between sites that contained the species versus sites that lacked the species. It appeared that the water chemistry was not different among the sites inhabited by the three species. These findings implied that acidic deposition was unlikely to have been a major cause of amphibian population decline in the Sierra Nevada (Bradford and Gordon 1992, Bradford et al. 1993, 1994). If acidification of high-elevation surface water does indeed play a role in the observed amphibian decline, such a role is likely highly episodic in nature.

17.4 Nutrient Nitrogen Enrichment

17.4.1 Aquatic Ecosystem Enrichment

Sullivan et al. (2011b) evaluated the relative risk of nutrient enrichment from N deposition for all Inventory and Monitoring networks nationwide. The coarse network rankings developed by Sullivan et al. (2011b) for nutrient N pollutant exposure, ecosystem sensitivity to nutrient N enrichment, and park protection yielded an overall network nutrient N enrichment summary risk ranking for Sierra Nevada Network parks that was highest among the 32 inventory and Monitoring networks. The overall level of concern for nutrient N enrichment effects on Inventory and Monitoring parks within the Sierra Nevada Network was judged by Sullivan et al. (2011b) to be very high.

An examination of EPA Western Lakes Survey data compiled by Eilers et al. (1987) found enhanced N concentrations in high-elevation lakes adjacent to and downwind of urban centers (Fenn et al. 2003b). An additional survey of 62 lakes that were the subject of N enrichment studies reported a mean increase in phytoplankton biomass of 79% in response to N enrichment (average of 46.3 µeq/L N; Elser et al. 1990). Long-term measurements at Lake Tahoe, California, to the north of the Sierra Nevada Network region, showed that primary productivity doubled, while water clarity declined, mostly as a result of atmospheric N deposition (Goldman 1988, Jassby et al. 1994).

High-elevation lakes in the Sierra Nevada are primarily oligotrophic. Even small changes in the nutrient supply can impact algal productivity (Sickman et al. 2003) and cause important changes in other environmental parameters. Chamise Creek in SEKI was reported to have throughfall N deposition of 10 kg N/ha/yr and high nitrate leaching, suggesting relatively high atmospheric N loading and N saturation (Fenn et al. 2003a,b). The U.S. Forest Service has suggested policy thresholds for stream nitrate concentration in N-limited ecosystems in the western United States. A concentration less than 2 µeq/L might suggest concern for possible over-enrichment of N (Fenn et al. 2011b).

Data from 28 Sierra Nevada lakes sampled in 1985 and again in 1999 suggested that nitrate concentrations decreased during that period and total P concentrations increased in more than 70% of the lakes sampled. Sickman et al. (2003) concluded that lakes throughout the Sierra Nevada appear to be experiencing measureable eutrophication in response to atmospheric deposition of nutrients, but N deposition is only part of the process.

At Emerald Lake, the concentration of nitrate during spring snowmelt runoff and during the growing season declined from 1983 to 1995 (Sickman et al. 2003). This was caused in part by changes in the snow regime; during the 1987–1992 drought, years having shallow and early-melting snowpack had lower nitrate concentration in runoff due to decreased labile N production in soils and greater N uptake during longer plant growing seasons. Continued nitrate declines in lake water subsequent to the drought during the period 1993–2000 were likely caused by increased atmospheric P loading to the Emerald Lake watershed and the release of lake phytoplankton from previous P limitation (Sickman et al. 2003). The researchers also showed that these trends measured at Emerald Lake were reflected in the set of 28 Sierra Nevada lakes sampled in Sequoia National Park and YOSE during synoptic lake surveys in 1985 and 1999. The most sensitive lakes are those located at high elevation (e.g., >2400 m) and that are highly oligotrophic. Atmospheric deposition of both N and P represents a large fraction of the nutrient inputs to the watersheds of these lakes. Small changes in nutrient supply can therefore have substantial impacts on lake productivity.

Based on the results of bioassay experiments and application of coarse nutrient limitation indices, Sickman et al. (2003) suggested that phytoplankton growth in Emerald Lake during the early 1980s was limited by P supply, but that there subsequently developed a pattern of colimitation by N + P and N limitation as P concentrations increased. The total P concentration in the lake doubled from 1983 to 1999. Particulate C concentration in the lake in the autumn of 1999 was fourfold higher than the average during the period 1983–1998. Together with the observed increase in total P, this shows that measureable eutrophication has occurred (Sickman et al. 2003). The researchers further proposed that P loading has increased to most Sierra Nevada lakes, causing decreased nitrate concentration and greater incidence of N limitation. The increased P loading may be due to increased emissions and deposition of P and/or accelerated internal P cycling due to changes in the timing of snowmelt and runoff patterns (Dettinger and Cayan 1995, Johnson 1998).

Sickman et al. (2001) calculated that the annual yield of N from the Emerald Lake watershed varied by a factor of 8 (0.4–3.2 kg N/ha/yr) and was a linear function of runoff, with runoff explaining 89% and 74% of the variation in dissolved inorganic and organic N, respectively. Ecosystem processes increased the N content of runoff during years having high runoff. Nitrate pulses were larger during years having deep, late melting snowpacks. Thus, it is expected that a warming climate, with associated earlier snowmelt, might increase N retention in high-elevation watersheds in the Sierra Nevada (Sickman et al. 2001).

Because there was evidence of increased P loading throughout the Sierra Nevada, Sickman et al. (2003) concluded that site-specific P sources were unlikely to be the cause of observed trends. They proposed that atmospheric deposition and accelerated internal cycling of P in response to changes in climatic factors were the most likely sources of increased P loading to the Sierra Nevada lakes. It is not known why atmospheric deposition of P to these lakes has increased over time. Possibilities include the use of organophosphate pesticides (Kegley et al. 2000) and aeolian transport of soils and dust that are high in P from the San Joaquin Valley to the Sierra Nevada (Bergametti et al. 1992, Lesack and Melack 1996, Sickman et al. 2003). Research conducted at 1900 m elevation in Giant Forest, Sequoia National Park showed that Fe/Al and Fe/Ca ratios suggested a mixture of mineral dust from regional agricultural activities and long-range transport of dust from Asia. Asian sources made up an estimated 40%–90% in midsummer and then declined during the late summer and early fall (Vicars and Sickman 2011).

Increased atmospheric deposition of N has the potential of altering the algal biomass of lakes in the Sierra Nevada to the extent that N is the growth-limiting nutrient in those lakes. To examine this issue, Sickman and Melack (1989) conducted experimental manipulations of the plankton in Emerald Lake *in situ* using artificial enclosures. Experiments were conducted using five microcosms (volume 10 L) and four mesocosms (volume about 3500 L) during the summers of 1983 and 1984. These experiments were designed to determine the limiting nutrients for phytoplankton growth in Emerald Lake and separate out the effects of increased acidity from those of increased nutrient concentration on the lake's phytoplankton. Results suggested that phytoplankton growth in Emerald Lake was strongly limited by P at that time. Five out of seven experiments indicated that chlorophyll levels of the seston (algae and particulate matter in the water column) increased with P additions. Nitrogen was never found to be the sole limiting element and was colimiting in only one experiment. The extent to which these results are applicable to other lakes throughout the Sierra Nevada is unclear.

Baron et al. (2011) synthesized the critical loads of N deposition for protecting against nutrient enrichment of high-elevation lakes, including within the Sierra Nevada Network region. For sites in the Rocky Mountains of Colorado and Wyoming and the Sierra Nevada, relationships between both mean and peak annual nitrate concentration and N deposition suggested a critical load near 2 kg N/ha/yr to prevent nitrate leaching.

17.4.2 Terrestrial Ecosystem Enrichment

Herbaceous plants in alpine communities, such as are found in SEKI and YOSE, are considered very sensitive to changes in N deposition. A combination of short growing season, strong seasonal variation in moisture and temperature, shallow and poorly developed soils, steep terrain, sparse vegetation, and low rates of primary productivity generally limit the N uptake and retention capacity of herbaceous plant species in alpine ecosystems (Fisk et al. 1998, Burns 2004). Alpine plant communities have often developed under conditions of low nutrient supply, in part because soil-forming processes are poorly developed, and this also contributes to their N sensitivity. Alpine herbaceous plants are generally considered N-limited and changes in alpine plant productivity and species composition have been noted in response to increased N inputs (Vitousek et al. 1997, Bowman et al. 2006).

Nitrogen cycling in alpine environments is strongly tied to variations in moisture regime (Bowman et al. 1993, Bowman 1994, Fisk et al. 1998). Blowing snow is transported across alpine landscapes by wind and tends to accumulate in depression areas. Nutrient enrichment is also an important concern in forested ecosystems in Sierra Nevada Network parks. Future increases in the two most limiting resources in Sierran mixed conifer forests (water and N) can potentially change the understory plant community and increase the risk of wildfire in response to understory fuels buildup (Fenn et al. 1998, Witty et al. 2003, Hurteau and North 2009). It is important to land managers to determine how such changes are likely to affect the distribution and abundance of rare species in the national parks. Hurteau and North (2009) assessed potential effects on mock leopardbane (*Arnica dealbata*), a listed sensitive species in YOSE. Increasing N deposition by experimental application of 12 kg N/ha/yr over a three-year period negatively affected mock leopardbane coverage.

The most notable effect of N deposition on vegetation in the Sierra Nevada is alteration of the lichen community (Fenn et al. 2008, Bowman et al. 2011). Lichen communities in chaparral and oak woodlands in parts of California have been substantially changed by atmospheric N deposition. These affected plant communities occur along the western edge of SEKI. Impacts on lichen distribution in California have been especially evident near the Tahoe Basin and Modoc Plateau and within SEKI and YOSE (Jovan and McCune 2006, Bowman et al. 2011). However, the response of lichens to N has been confounded by simultaneous exposure to ozone (Fenn et al. 2011a).

The critical load to protect lichens in the western Sierra Nevada has been estimated to lie in the range of 3.1–5.2 kg N/ha/yr (Fenn et al. 2008, Bowman et al. 2011). Fenn et al. (2008) concluded that community composition of epiphytic macrolichens begins shifting from oligotrophic to more mesotrophic species at around 3.1 kg N/ha/yr and shifts further to domination by eutrophic species above about 5.2 kg N/ha/yr. Oligotrophic lichens may be completely eliminated at N deposition near 10 kg N/ha/yr (Bowman et al. 2011).

Eutrophic lichen richness and abundance have been shown to decrease with ammonia emissions and N deposition in the Central Valley (Jovan and McCune 2005, Jovan 2008). Fenn et al. (2011a) reported that, of the 53 lichen survey sites in California that had relatively clean air scores, only three occurred in locations where the modeled deposition was more than 5.5 kg N/ha/yr.

Jovan and McCune (2006) developed a model of ammonia exposure in forests of the Sierra Nevada. Nonmetric multidimensional scaling was used to extract gradients in lichen community composition from surveys at 115 sites. The observed lichen communities suggested relatively high ammonium deposition to forests of the southern Sierra Nevada, including portions of YOSE and Sequoia National Park.

Fenn et al. (2008) examined eutrophication responses attributable to N deposition to ponderosa pine forests in California. Epiphytic lichens were used as bioindicators of N effects (Blett et al. 2003, Bobbink et al. 2003, Fenn et al. 2003b). Fifteen of the 24 study sites were located in the Sierra Nevada and 9 in the San Bernardino Mountains. Lichen species were classified into three groups based on their responses to atmospheric deposition of N (Jovan and McCune 2006, Geiser and Neitlich 2007, Jovan 2008). Acidophytes are sensitive to small increases in N deposition (Van Herk et al. 2003) and can be eliminated from the lichen community at moderate levels of N deposition. This group includes beard lichens (e.g., *Bryoria fremontii*) and other species used by wildlife for forage and nesting material in the Sierra Nevada (McCune et al. 2007). Neutrophytes are more tolerant of N input. Nitrophytes are fast growing, respond dramatically to increased ammonium deposition, and favor high pH substrates, including the bark of hardwood trees.

Ellis et al. (2013) estimated the critical load for nutrient N deposition to protect the most sensitive ecosystem receptors in 45 national parks, based on the data of Pardo et al. (2011). The lowest terrestrial critical load of N was generally estimated for the protection of lichens (Geiser et al. 2010). Changes to lichen communities may signal the beginning of other changes to the ecosystem that might affect structure and function (Pardo et al. 2011).

Ellis et al. (2013) estimated the N critical load for SEKI and YOSE in the range of 2.5–7.1 kg N/ha/yr for the protection of lichens.

17.5 Ozone Injury to Vegetation

The ozone-sensitive plant species that are known or thought to occur within SEKI and YOSE are listed in Table 17.2. Those considered to be bioindicators, because they exhibit distinctive symptoms when injured by ozone (e.g., dark stipple), are designated by an asterisk. Each park contained at least 13 ozone-sensitive and/or bioindicator species.

Air pollution injury in California, especially that associated with exposure of sensitive plants to ozone, varies from very little to virtually none in National Park Service networks in the north and northwest to significant within the Sierra Nevada Network, especially in SEKI (Sullivan et al. 2001), and substantial in the Los Angeles Basin. Campbell et al. (2007) documented ozone-induced foliar injury in 25%–37% of Forest Inventory Analysis monitoring biosites in forested ecosystems in California during the period 2000–2005.

There have been more vegetation monitoring and research activities related to air pollution at the SEKI parks than at any other western national park (Sullivan et al. 2001). The most comprehensive ozone research has been conducted in the mixed conifer forest (Peterson et al. 1987, 1991, 1992a,b, Miller et al. 1989, Horner and Peterson 1993), in part because of the known high sensitivities to ozone harm of ponderosa pine, Jeffrey pine, and quaking aspen. Fewer data are available to relate ozone dose or exposure to herbaceous plant species in this network.

TABLE 17.2

Ozone-Sensitive and Bioindicator (*) Plant Species Known or Thought to Occur in SEKI and YOSE[a]

		Park	
Species	Common Name	SEKI	YOSE
Amelanchier alnifolia var. *pumila*	Saskatoon serviceberry		×
*Apocynum androsaemifolium**	Spreading dogbane	×	×
Apocynum cannabinum	Dogbane, Indian hemp	×	×
*Artemisia douglasiana**	Mugwort	×	×
*Physocarpus capitatus**	Ninebark	×	×
*Pinus jeffreyi**	Jeffrey pine	×	×
*Pinus ponderosa**	Ponderosa pine	×	×
*Populus tremuloides**	Quaking aspen	×	×
Quercus kelloggii	California black oak	×	×
*Rhus trilobata**	Skunkbush	×	×
Rubus parviflorus	Thimbleberry	×	×
Salix gooddingii	Goodding's willow	×	
*Salix scouleriana**	Scouler's willow	×	×
*Sambucus mexicana**	Blue elderberry	×	×
*Vitis vinifera**	European wine grape	×	

[a] Lists are available and periodically updated at https://irma.nps.gov/NPSpecies/Report

Giant sequoia (*Sequoiadendron giganteum*) seedlings in Sequoia National Park are exposed to relatively high atmospheric ozone levels and have been found to exhibit visible ozone damage. Grulke and Miller (1994) investigated the effects of ozone exposure on the gas exchange characteristics of giant sequoia trees of different ages using open top and branch exposure chambers. Seedlings were sensitive to ozone exposure until they reached the age of about five years, after which low conductance, high water-use efficiency, and compact mesophyll all contributed to natural ozone tolerance (Grulke and Miller 1994).

17.5.1 Ozone Exposure Indices and Levels

The W126 and SUM06 exposure indices calculated by National Park Service staff are given in Table 17.3, along with Kohut's (2007) ozone risk ranking. Ozone levels have been high for many years at all of the monitoring sites in SEKI. The annual daily 1-hour maximum ozone concentration exceeded the state standard (90 ppb) in every year that data were available for the analysis reported by Sullivan et al. (2001) and regularly exceeded the federal standard (https://www.epa.gov/ozone-pollution/table-historical-ozone-national-ambient-air-quality-standards-naaqs). Passive ozone samplers showed that the highest ozone levels occurred in the western portions of Sequoia National Park, adjacent to the San Joaquin Valley.

At 13 sites in YOSE where ozone was monitored, measurements rarely exceeded the federal 1-hour ozone standard (120 ppb) between 1988 and 2013, but the California ozone standard of 90 ppb was exceeded at all sites (cf., Sullivan et al. 2001). The National Park Service (2010) evaluated the fourth highest 8-hour ozone concentrations for national parks that have consistently been at or above the standard of 75 ppb, based on 11–20 years of monitoring data running through 2008. There were six parks that routinely had such high concentrations, including SEKI and YOSE. In 2015, the standard was reduced to 70 ppb.

A high seasonal average ozone concentration was measured by Bytnerowicz et al. (2002) at Three Pole Corner (71.5 ppb) at mid-elevation in the lower Kaweah/Marble Fork drainage in Sequoia National Park. At higher elevation and increasing distance from the San Joaquin Valley, ozone concentrations decreased. Back trajectory analyses conducted by Burley and Ray (2007) suggested that air masses having high ozone concentrations at YOSE often originate in the San Francisco Bay area and pass through the Central Valley prior to reaching YOSE.

TABLE 17.3

Ozone Assessment Results for Selected Parks in the Sierra Nevada Network Based on Estimated Average 3-Month W126 and SUM06 Ozone Exposure Indices for the Period 2005–2009 and Kohut's (2007) Ozone Risk Ranking for the Period 1995–1999[a,b]

Park Name	Park Code	W126		SUM06		Kohut O_3 Risk Ranking
		Value (ppm-hr)	NPS Condition	Value (ppm-hr)	NPS Condition	
Kings Canyon	KICA	40.39	High	51.20	High	High
Sequoia	SEQU	42.62	High	54.42	High	High
Yosemite	YOSE	27.77	High	29.80	High	High

[a] Parks are classified into one of three ranks (Low, Moderate, High) based on comparison with other Inventory and Monitoring parks.

[b] Degrees of concern for the W126 and SUM06 indices are based solely on levels of ozone exposure. Kohut's risk to vegetation is based on several factors that contribute to injury in plants, including ozone exposure and environmental variables, and considers the effects of soil moisture on the uptake of ozone.

Burley and Ray (2007) sampled atmospheric ozone concentrations using portable ozone monitors at multiple locations in YOSE during the summers of 2003 and 2005. The goal was to provide data that would help interpolate measurements from fixed monitoring stations at Turtleback Dome and Merced River to remote park locations in complex terrain. Most study sites exhibited similar ozone exposure during well-mixed daytime periods. However, concentrations were more variable at night. The ozone exposures tended to be highest in the western and southern portions of the park. As discussed in the following text, sampled locations generally fell into one of two groups based on the magnitude of the observed diurnal variation in ozone levels.

Some sites experienced frequent middle-of-the-night ozone spikes, which suggested that they were well exposed to the free troposphere during the evening. Because the nocturnal ozone levels tended to be high at such sites, their average exposures over a 24-hour period were generally 50 ppb or higher. Other sites typically had predawn ozone levels less than 50 ppb. They had few nocturnal ozone spikes and were not well exposed to the troposphere during evening hours.

The EPA measurements of atmospheric ozone concentrations at 568 monitoring sites nationwide showed generally stable levels throughout the 1990s, with declines between 2002 and 2007, attributed largely to oxidized N emissions reductions (U.S. EPA 2008). Even though there have been widespread improvements in atmospheric ozone levels, there were still 51 sites classified as nonattainment areas in 2007 because of failure to meet the ozone National Ambient Air Quality Standard. Some of the highest ozone concentrations in the United States in 2007 were in California (U.S. EPA 2008).

The EPA proposed a new ozone standard in January 2010 (Federal Register Vol. 75, No. 11, 40 CFR Parts 50 and 58, National Ambient Air Quality Standards for Ozone, Proposed Rules, January 19, 2010, p. 2938). It was based on an index of the total plant ozone exposure during the daytime, whereby hourly values were weighted according to magnitude and then summed, as the three-month period having the highest cumulative exposure. The proposed standard was based on a three-year average of the annual W126 metric. The EPA proposed to set the level of the new standard in the range of 7–15 parts per million per hour (ppm-hr). Although this proposal was not adopted, there were 17 parks that exceeded the upper end of this range in 2008. The highest values were reported for SEKI (57 ppm-hr), Joshua Tree National Park (53 ppm-hr), YOSE (34 ppm-hr), and Death Valley National Park (29 ppm-hr). All of these parks are located in California.

17.5.2 Ozone Exposure Effects

High ozone exposures in the Sierra Nevada Network region have affected trees in the parks. Lower montane forests in the region are dominated by ponderosa pine on xeric sites. Ponderosa pine is highly sensitive to ozone exposure. At mid-elevation, mixed conifer forest is prevalent, including some ponderosa pine, along with white fir, incense cedar (*Calocedrus decurrens*), sugar pine, and California black oak (*Quercus kelloggii*). Jeffrey pine replaces ponderosa pine within this forest type at higher locations and on shallow soils. The change from black oak woodland to ponderosa pine forest is abrupt on relatively dry sites at mid-elevation. Lower montane forests are widespread on the west side of Sequoia National Park and are dominated by ponderosa pine. Groves of giant sequoia are found at scattered locations mixed with various combinations of the other lower montane coniferous tree species.

Upper montane coniferous forests vary considerably in species and stand structure. The upper montane zone is found between about 2000 and 2600 m elevation and occupies 23%

of YOSE. Dry and rocky areas, especially on south-facing slopes, are occupied by Jeffrey pine. Grulke et al. (2003) tested whether this species had different transpirational patterns (with associated differing ozone uptake potential) in mesic versus xeric microsites in Sequoia National Park. Based on leaf gas exchange measurements, trees on mesic sites had an estimated 46% decrease in ozone uptake from June to August; trees on xeric sites had a 72% decrease over the same time period. Trees on mesic sites showed greater ozone injury, as indicated by lower foliar N content and reduced needle retention in mid-canopy (Grulke et al. 2003). The only deciduous tree that is common in the upper montane zone is quaking aspen, which is found in dense stands near streams and other areas that have relatively high soil moisture.

Chaparral vegetation, which occurs at lower elevations along the western edge of SEKI, is relatively ozone tolerant (Stolte 1982, Temple 1999). In addition, chaparral vegetation is quiescent during summer, when ozone levels are relatively high (Fenn et al. 2011a).

Just as ozone exposure varies in magnitude spatially within the Sierra Nevada Network region, it has varying effects on different plant species. Field data suggest a shift in community composition with elevated ozone concentration that favors grass species over legumes (U.S. EPA 2006). A study by Weinstein et al. (2005) showed that white fir was relatively unaffected by ozone as compared to ponderosa pine, which showed significantly reduced growth with increased ozone.

The mixed conifer forest at SEKI is part of a broad region of the Sierra Nevada that has significant impacts to vegetation from chronic exposure to ozone. Spatial patterns of ozone injury and spatial and temporal patterns of tree growth in SEKI have been quantified in several studies. In addition, intensive physiological investigations have provided evidence to support clear cause-effect relationships.

One of the first studies to quantify ozone injury to ponderosa pine and Jeffrey pine in SEKI was completed in the early 1980s (Warner et al. 1983). This study documented widespread foliar injury as part of a broader regional pattern encompassing adjacent national forests (Pronos et al. 1978, Pronos and Vogler 1981, Allison 1982). It motivated periodic surveys of oxidant injury to these bioindicator species at SEKI and YOSE (Duriscoe and Stolte 1989, Duriscoe 1990, Stolte et al. 1992, Ewell and Gay 1993). An extensive survey of sites along the western edge of SEKI (where ozone exposure is highest) indicated that approximately 90% of pines had some evidence of foliar injury (including high injury on first-year needles), with symptomatic trees having lower needle retention than asymptomatic trees (Peterson and Arbaugh 1988, 1992, Peterson et al. 1991). Injury at SEKI was higher than any other location in the Sierra Nevada. The Forest-Ozone Response Study (Arbaugh et al. 1998) corroborated this pattern, determining that 93% of pines at SEKI had visible injury, again higher than at any other location in the Sierra Nevada. Physiological studies in and near SEKI in the field (Patterson and Rundel 1989, Bytnerowicz and Grulke 1992, Grulke 1999) and in controlled exposure studies of seedlings (Temple et al. 1993) confirmed that the observed injury was related to physiological effects in pines.

The Forest-Ozone Response Study was conducted during 1991–1994. It quantified the distribution of ozone injury to forest species throughout California (Rocchio et al. 1993, Cahill et al. 1996, Arbaugh et al. 1998). This study was linked to the Sierra Cooperative Ozone Impact Study, another regional study that focused primarily on the monitoring of ambient ozone and associated weather variables in the Sierra Nevada (Van Ooy and Carroll 1995). The Forest-Ozone Response Study added to the spatial resolution of long-term research and monitoring of ozone effects by Miller (1992) and Miller and McBride (1999) and to extensive surveys of injury and growth patterns in the Sierra Nevada by Peterson et al. (1987, 1991) and Peterson and Arbaugh (1988, 1992). Other assessments of ozone injury have been conducted for individual national forests and national parks (e.g., Pronos and Vogler 1981, Vogler 1982, Warner et al. 1983), but the Forest-Ozone Response Study results are more spatially comprehensive.

The Forest-Ozone Response Study included ozone injury surveys at 11 sites (all with associated monitoring data), ranging from Lassen Volcanic National Park in northern California to San Bernardino National Forest in southern California, with sites distributed throughout the westside Sierra Nevada within the elevational range of mixed conifer forest. The Forest-Ozone Response Study included two sites in YOSE and two sites in SEKI.

Visible ozone injury to ponderosa pine and Jeffrey pine was detected at all Forest-Ozone Response Study sites, ranging from 26.8% of trees at the northernmost site to 100% at the southernmost. Mean whorl retention ranged from 6.5 to 2.7 for these same sites (although Jeffrey pine, which is more dominant in the north, has more needles than ponderosa pine). The computed Ozone Injury Index (Schilling and Duriscoe 1996) ranged from 6.3 in the north to 65.1 in the south. It was determined that the Ozone Injury Index was strongly correlated with the Forest Pest Management injury index used by the Forest Service (Pronos et al. 1978), which indicated that data collected by personnel from different organizations could be combined for robust analyses. Most importantly, there were significant correlations between Ozone Injury Index and various ozone indices over the geographic range of the study. This suggested that (1) ozone-monitoring data could be used to predict ozone injury in sensitive pine species and (2) injury data for pines (much less expensive to collect than ambient exposure data) provided a reasonable reflection of the level of ozone exposure at a particular geographic location. Thus, ponderosa pine and Jeffrey pine, which are present in several Class I national parks in California and elsewhere, can be used as true bioindicator species for ozone sensitivity and exposure. The Forest-Ozone Response Study was one of the first field studies to demonstrate a quantitative relationship between tree injury and cumulative ozone exposure.

Even more compelling evidence of the effects of ozone on mixed conifer forest was provided by data that indicated that growth has been reduced in injured trees. Ewell et al. (1989) and Duriscoe and Stolte (1992) found that the level of ozone injury in ponderosa pine foliage was inversely correlated with foliar biomass. Reduced stem growth was documented for both ponderosa pine and Jeffrey pine attributable to ozone injury in SEKI. Open-grown Jeffrey pine with injury symptoms was found to have 11% lower radial growth since 1950, as compared with asymptomatic trees from nearby locations (Peterson et al. 1987). As part of a study that examined tree growth at 56 sites ranging from the Tahoe National Forest in the northern Sierra Nevada to the Sequoia National Forest in the south, 8 sites (4 symptomatic, 4 asymptomatic) were examined at SEKI (Peterson et al. 1991, Peterson and Arbaugh 1992). Older trees and trees with lower needle retention tended to have the lowest growth. Prolonged exposure to elevated ozone may have reduced root biomass in mature ponderosa pine (Grulke et al. 1998, Grulke 1999), which may have predisposed the trees to increased stress during periods of low soil moisture. Given that summer drought is a normal part of the Mediterranean climate of California, this could be a significant long-term stress.

SEKI consistently has been observed to have a high level of injury, with the most substantial effects occurring within the mixed conifer zone (Sullivan et al. 2001). Documentation of severe ozone injury in dominant trees, reduced tree growth, and associated physiological data for mature trees in the western portion of SEKI presented a clear picture of widespread impacts from prolonged exposure to ozone and other air pollutants.

YOSE has also experienced severe ozone-induced foliar injury to ponderosa pine. Ozone injury was widespread in the western portion of the park, and foliar biomass was lower in symptomatic trees. YOSE is located adjacent to the San Joaquin Valley, which experiences high levels of ozone during summer months. Concentrations of gases and particulate air pollutants are transported to the park from the valley by prevailing westerly winds. Ozone injury was first documented in YOSE by Duriscoe and Stolte (1989) and was followed up by another survey that determined that about 30% of the pines at four symptomatic sites had ozone injury in needles that were two years old or less (Peterson et al. 1991, Peterson and Arbaugh 1992). The study of Arbaugh et al. (1998) found that about 40% of pines at two sites at YOSE had ozone injury.

An analysis of tree growth at YOSE (Peterson et al. 1991, Peterson and Arbaugh 1992) found that the growth of symptomatic ponderosa pine was significantly lower since 1950, including trees with some of the largest growth reductions measured in the Sierra Nevada. However, two sites included in the YOSE study had growth decreases throughout the twentieth century, with the most probable cause being annosus root rot, which is common throughout Yosemite Valley (Parmeter et al. 1978). The differential effects of ozone and fungal pathogens on tree growth at YOSE are difficult to discern.

17.6 Visibility Degradation

17.6.1 Natural Background Visibility and Monitored Visibility Conditions

IMPROVE monitors are located in Sequoia National Park and YOSE. The Sequoia National Park monitor (SEQU1) is also used to characterize conditions at Kings Canyon National Park. Natural background visibility has been estimated by IMPROVE for Sequoia National Park and YOSE (Table 17.4). Monitored visibility on the haziest days at 47 park locations reported by the National Park Service (2010) during the period 2006–2008 ranged from 1.5 to 21 deciviews higher than the estimated natural background condition at each site. This differential was at the upper end of the distribution (16.2 deciviews) in Sequoia National Park.

Impaired visibility frequently occurs at SEKI, largely due to organics, sulfates, and nitrates. Visibility conditions vary by season. Smoke occasionally causes reduced visibility, especially during the summer months. Reconstructed extinction generated from aerosol data (Savig and Morse 2001) showed that the worst visibility occurred during summer, when the seasonal average visual range (i.e., how far a person can see) was 37 mi

TABLE 17.4

Estimated Natural Background Visibility and Monitored Visibility in SEKI and YOSE Averaged over the Period 2004–2008

| Park Name | Park Code | Site ID | Estimated Natural Background Visibility (in Deciviews) | | |
			20% Clearest Days	20% Haziest Days	Average Days
Sequoia and Kings Canyon	SEKI	SEQU1	2.29	7.70	4.92
Yosemite	YOSE	YOSE1	0.99	7.64	4.08
Park Name	Park Code	Site ID	Monitored Visibility (for Years 2004–2008) (in Deciviews)		
			20% Clearest Days	20% Haziest Days	Average Days
Sequoia and Kings Canyon	SEKI	SEQU1	8.44	23.91	16.58
Yosemite	YOSE	YOSE1	3.10	16.99	9.73

(60 km). Visibility impairment at YOSE was largely due to organics and sulfates. Smoke from frequent fires typically reduces visibility during the summer months. Reconstructed extinction generated from the aerosol data at YOSE showed seasonal variation in visibility conditions. The best seasonal average visual range (131 mi [211 km]) occurred during the winter months, and the worst seasonal average visual range (67 mi [107 km]) occurred during the summer months. Emissions of organic aerosols from wildfire and biogenic emissions from vegetation in the Sierra Nevada Network region peaked during summer months in the baseline period 2000–2004 (Suarez-Murias et al. 2009).

IMPROVE data indicate that air pollution at SEKI has reduced average natural visual range from 150 to 35 mi (241 to 56 km). On the haziest days, visual range has been reduced from 110 to 20 mi (177 to 32 km). Severe haze episodes occasionally reduce visibility to 5 mi (8 km). At YOSE, pollution has reduced average natural visual range from 160 to 75 mi (258 to 121 km). On the haziest days, visual range has been reduced from 110 to 35 mi (177 to 56 km). Severe haze episodes occasionally reduce visibility to 7 mi (11 km).

Impairment on the 20% haziest days during the baseline period 2000–2004 was more than twice as high at Sequoia National Park as compared with YOSE and with most other Class I areas in the Sierra Nevada (Suarez-Murias et al. 2009). The largest difference was for nitrate, which was responsible for about four times as much extinction in Sequoia National Park, as compared with YOSE, on the 20% haziest days.

The baseline days having worst air quality were dominated by organic aerosols associated with biogenic emissions and wildfire. The monitor at Sequoia National Park is situated at relatively low elevation (519 m) and is therefore in proximity to urban, agricultural, and transportation corridor emissions from the San Joaquin Valley that is located just west of the monitoring site.

17.6.2 Composition of Haze

Figure 17.8 shows estimated natural (preindustrial), baseline (2000–2004), and current (2006–2010) levels of haze and its composition for Sequoia National Park and YOSE. The majority of the haze in Sequoia National Park for the 20% haziest and average days was attributable to nitrate and organics (Figure 17.8), followed by sulfate. On the clearest 20% visibility days, the contribution of sulfate increased, and nitrate and organics continued to contribute substantially. The biggest difference between the 20% clearest and 20% haziest visibility days at Sequoia National Park was the increased role played by nitrate on the 20% haziest days (Figure 17.8). The majority of visibility impairment in YOSE on the 20% haziest and average days was attributable to organics (Figure 17.8). On the 20% clearest days, sulfate was the largest contributor. The biggest difference between the 20% clearest and 20% haziest visibility days at YOSE was the increased role played by organics on the haziest days (Figure 17.8).

Analyses conducted by the Western Regional Air Partnership indicated that organics from natural emissions sources, including wildfire and biogenic sources (vegetation), contribute to substantial visibility impairment throughout the western United States, including within the Sierra Nevada Network region. In addition, air pollution sources outside the Western Regional Air Partnership domain, including international offshore shipping and sources from Mexico, Canada, and Asia, can in some cases be substantial contributors to haze (Suarez-Murias et al. 2009).

Some visibility impairment occurs at all of the National Park Service Class I monitoring sites in California (Sullivan et al. 2001). Monitoring data showed that coarse mass concentrations, particles between the size of 2.5 and 10 µm, were highest along the Pacific

Coastal Mountains and lowest in the Sierra-Humbolt region. Fine aerosol concentrations were highest in the Southern California region. On an annual basis, organics contributed the most to light extinction at parks in the Sierra-Humbolt area (Lassen Volcanic National Park) and Sierra Nevada area (YOSE and SEKI), whereas sulfates contributed the most to extinction in the Pacific Coastal Mountain parks (Pinnacles National Monument, Point Reyes National Seashore, Redwood National Park), and nitrates were the largest contributors to extinction in southern California (Sullivan et al. 2001).

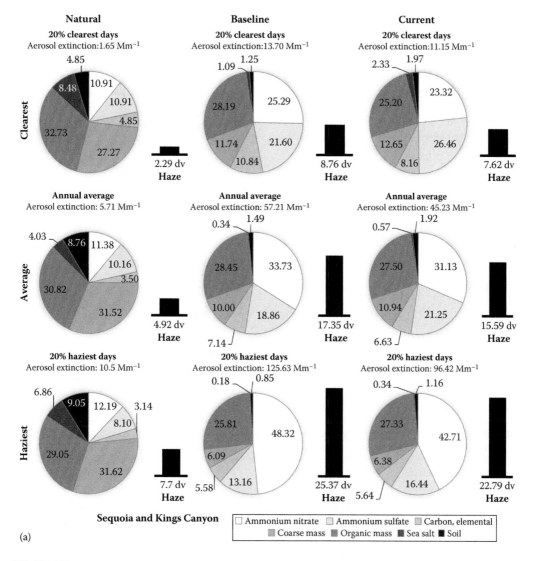

(a)

FIGURE 17.8
Estimated natural (preindustrial), baseline (2000–2004), and current (2006–2010) levels of haze (columns) and its composition (pie charts) on the 20% clearest, annual average, and 20% haziest visibility days for (a) SEKI. and (b) YOSE. Sequoia National Park has substituted data for the years 2002 and 2003. Data for Sequoia National Park are used to represent Kings Canyon National Park. (From http://views.cira.colostate.edu/fed/Tools/RegionalHazeSummary.aspx, accessed October, 2012.) (*Continued*)

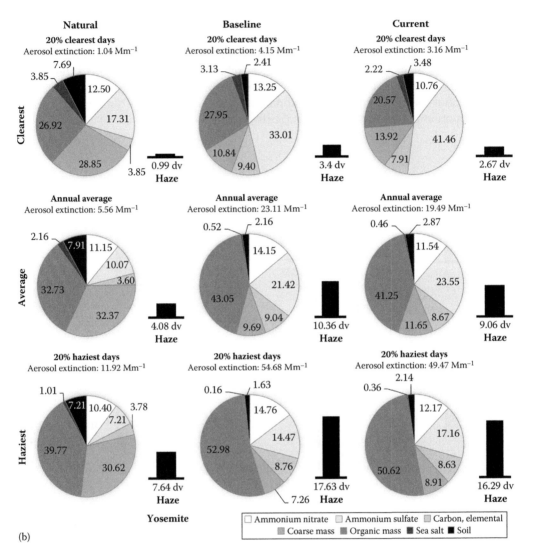

(b)

FIGURE 17.8 (*Continued*)
Estimated natural (preindustrial), baseline (2000–2004), and current (2006–2010) levels of haze (columns) and its composition (pie charts) on the 20% clearest, annual average, and 20% haziest visibility days for (a) SEKI. and (b) YOSE. Sequoia National Park has substituted data for the years 2002 and 2003. Data for Sequoia National Park are used to represent Kings Canyon National Park. (From http://views.cira.colostate.edu/fed/Tools/RegionalHazeSummary.aspx, accessed October, 2012.)

Emissions of carbonaceous particles (organic C and elemental C) from wildfires can have strong and sustained regional impacts on aerosol concentrations, air quality, and visibility (McMeeking et al. 2006). This issue was investigated in YOSE as part of the Yosemite Aerosol Characterization Study, which was conducted during the summer of 2002 (McMeeking et al. 2005). Air masses arriving at YOSE were sampled to characterize visibility and to define aerosol sources contributing haze to the park. Results indicated that organic C was the dominant aerosol species, especially during periods identified as smoke impacted. Meteorological conditions suggested that fires burning in Oregon had the largest impact on aerosol concentrations in YOSE during the study. Thus, it appears that emissions from regional wildfires can have substantial impacts on visibility in YOSE and other western parklands (McMeeking et al. 2006).

During the Yosemite Aerosol Characterization Study, there were several local wildfires and prescribed burns, in addition to two regional haze episodes that were strongly influenced by smoke from biomass burning that was transported long distances (Engling et al. 2006). Several types of biomass burning smoke tracers were used to quantify contributions of primary biomass burning smoke to fine particulate matter. The fine aerosols in YOSE during the summer of 2002 were dominated by natural sources, mainly by smoke from wildfires and by secondary organic aerosols of biogenic origin (Engling et al. 2006). Low concentrations of anthropogenic organic compounds suggested that motor vehicles only contributed about 10%, on average, of the organic C during that study. There remains uncertainty regarding the quantitative role of organic C from smoke episodes and its influence on visibility. The primary objective of the Yosemite Aerosol Characterization Study was to characterize the physical, chemical, and optical properties of C-dominated aerosols. Results suggested that the EPA's recommended approach for evaluating smoke-related light scattering, for tracking progress toward achieving the national visibility goal, may underestimate smoke-related light scattering by about a factor of 2, if YOSE results are universally applicable (Engling et al. 2006).

Lightning is an important source of wildfire ignition in the Sierra Nevada Network region. Prescribed fire is also widely used for restoring and maintaining vegetation communities (Fulé et al. 1997, Stephenson 1999, Allen et al. 2002, Brockway et al. 2002, McKenzie et al. 2006). Managers of protected areas, such as national parks and wilderness, have adopted a policy whereby naturally ignited fires are managed for resource benefit when they do not threaten structures or ambient air quality (Miller and Landres 2004, McKenzie et al. 2006). Smoke plumes can contribute to visibility impairment downwind from fire locations (Malm 1999). Fires have become more severe than they were historically, because of more extreme weather and buildup of fuel from previous fire suppression (Agee 1997, Flannigan et al. 1998, Covington 2000, Allen et al. 2002, McKenzie et al. 2006). This pattern may continue into the future in response to climate change. Both empirical (McKenzie et al. 2004) and process (Lenihan et al. 1998) models suggest that the area burned by wildfire is likely to increase in the western United States in response to a warming climate.

17.6.3 Trends in Visibility

Most of the investigated national park sites in the western United States showed decreasing 20th percentile atmospheric sulfate concentrations from 1989 to 1999 (Malm et al. 2002). The trends were statistically significant at only 6 of 30 sites, including YOSE, where the reduction was about 40%.

No trend in haze was observed on the clearest or haziest 20% days for the visibility monitoring site in Sequoia National Park during the period 1994–2008 (NPS 2010). However, visibility conditions in both Sequoia National Park and YOSE show evidence of decreasing haze in recent years on the 20% clearest, average, and 20% haziest days (Figure 17.9). These apparent improvements in visibility have been most pronounced on the 20% haziest days since 2002.

17.6.4 Development of State Implementation Plans

Glideslope analyses shown in Figure 17.10 suggest that haze levels on the 20% clearest days at both Sequoia National Park and YOSE appear to be moving toward the attainment of the national goal of natural background visibility by 2064 (~7.7 deciviews in each park).

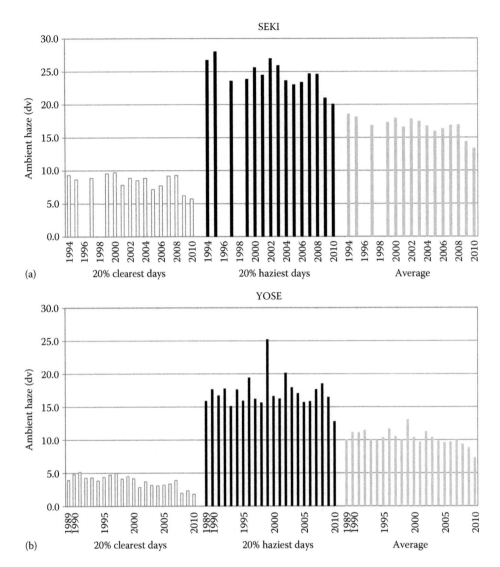

FIGURE 17.9
Trends in monitored visibility at (a) SEKI and (b) YOSE, based on IMPROVE measurements on the 20% clearest, 20% haziest, and annual average visibility days over the monitoring period of record. Note that SEKI has substituted data for the years 2002 and 2003. (From http://vista.cira.colostate.edu/improve/Data/IMPROVE/summary_data.htm, accessed October, 2012.)

Nevertheless, large additional improvements will be needed for full compliance with the Regional Haze Rule. Additional monitoring will be needed.

17.7 Toxic Airborne Contaminants

In addition to criteria pollutants (ozone, particulate matter, nitrogen dioxide, sulfur dioxide, and carbon monoxide), a large number of chemical compounds can be found in ambient

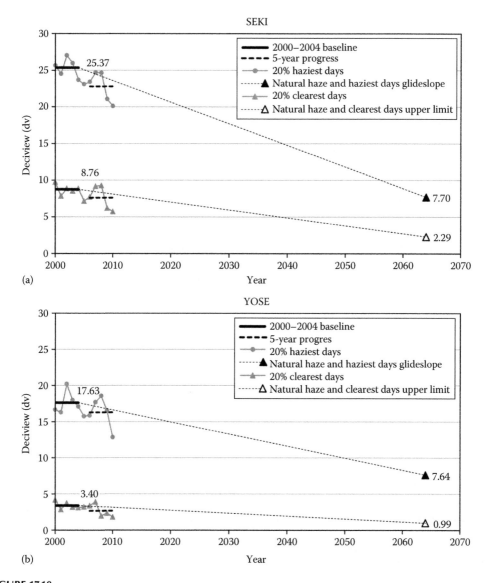

FIGURE 17.10
Glideslopes to achieving natural visibility conditions by 2064 for the 20% haziest (upper line) and the 20% clearest (lower line) days in (a) SEKI and (b) YOSE. In the Regional Haze Rule, the clearest days do not have a uniform rate of progress glideslope; the rule only requires that the clearest days do not get any worse than the baseline period. Also shown are measured values during the period 2000–2010. SEKI has substituted data for the years 2002 and 2003. (From http://vista.cira.colostate.edu/improve/Data/IMPROVE/summary_data.htm, accessed October, 2012.)

air monitoring samples collected in the Sierra Nevada Network at concentrations that are of concern from either a human health effects or an environmental perspective. Among these compounds are a variety of pesticides that are used extensively in California's heavily farmed San Joaquin and Sacramento valleys. Pesticides vary in their levels of volatility. Potential impacts are also affected by the degree to which a particular pesticide breaks down in the environment and the levels of toxicity of the breakdown products. The release and transport of pesticides from their point of application to downwind areas involve numerous other important physical and chemical processes. Although California only contains 3% of the agricultural land in the United States, it accounts for about 15% of all pesticide use (Census of Agriculture 2002), mostly in the Central Valley, which is generally upwind of the Sierra Nevada Mountains.

The snowpack is efficient at scavenging both particulate and gas-phase pesticides from the atmosphere. However, until recently, little was known about the occurrence, distribution, or sources of pesticides in high-elevation ecosystems. The analysis of pesticides in snow-pack samples from seven national parks in the western United States by Hageman et al. (2006) illustrated the deposition and fate of 47 pesticides and their degradation products. Correlation analysis with latitude, temperature, elevation, particulate matter, and two indicators of regional pesticide use suggested that regional patterns in historic and current agricultural practices are largely responsible for the distribution of pesticides in the western national parks in the conterminous United States. Pesticide deposition to parks in Alaska was attributed to long-range atmospheric transport rather than regional sources (Hageman et al. 2006).

A known air pollution transport route into the Sierra Nevada is from the San Joaquin Valley. Pesticides applied to crops can volatilize under warm valley temperatures and subsequently be transported via upslope air movement to be deposited in the cooler, high-elevation regions of the Sierra Nevada Network. Zabik (1993) demonstrated the atmospheric transport of organophosphate pesticides from the Central Valley to two sites within Sequoia National Park. The first site was located at Ash Mountain at an elevation of 533 m. Organophosphate and other current-use pesticides and PCBs have been found in surface waters, fish, and amphibians at this location (Datta et al. 1998). The second site was located at Lower Kaweah, in Giant Forest, at an elevation of 1920 m. A third site outside the park was located at a lower elevation, at the base of the Sierra Nevada foothills. Samples were collected of air, and estimates were made of wet and dry deposition of pesticides during the winters of 1990–1991 and 1991–1992 and also during the spring of 1992. Zabik (1993) concluded that organophosphate compounds were detected in dry deposition, but it appeared that wet deposition accounted for the majority of organophosphate insecticide deposition at all three of the sites during the winter months. The observed concentrations decreased substantially between Ash Mountain and Lower Kaweah (Zabik and Seiber 1993).

Air and dry deposition samples were collected during the period May to September, 1996 at three elevations in Sequoia National Park and analyzed for eight pesticides by LeNoir et al. (1999). Air samples showed a decrease in pesticide concentration with elevation at the lower elevations. Pesticide residues found at highest concentration in Sequoia National Park were generally those most heavily used in the Central Valley during summer: chlorpyrifos and endosulfan. Concentrations in surface waters decreased significantly above 2040 m elevation. Pesticide concentrations were below acute toxicity values (LeNoir et al. 1999).

The WACAP study (Landers et al. 2008) measured the inputs of airborne contaminants in a number of national parks in the western United States, including SEKI and YOSE, from 2002 to 2007. Along with Glacier and Rocky Mountain National Parks, SEKI had some of the highest measured pesticide concentrations in many environmental compartments. All of these parks have relatively high regional agricultural development, which is the likely primary source of the observed contaminants (Landers et al. 2008). The semivolatile organic compounds detected in the air at SEKI by the WACAP study were mainly the current-use pesticides endosulfan and dacthal. Except for polycyclic aromatic hydrocarbons and dacthal, all semivolatile organic compounds in air (hexachlorocyclohexane-γ, endosulfans, hexachlorocyclohexane-α, and chlordanes) at YOSE were above median values for WACAP parks.

Pesticide concentrations in SEKI and YOSE vegetation were among the highest measured by the WACAP study, along with vegetation in Glacier and Great Sand Dunes National Parks (Landers et al. 2008). Dominant semivolatile organic compounds in SEKI vegetation included polycyclic aromatic hydrocarbons, endosulfans, dacthal, DDT, chlorpyrifos, hexachlorobenzene, hexachlorocyclohexane-γ, dieldrin, hexachlorocyclohexane-α, and PCBs. In YOSE, dominant semivolatile organic compounds in vegetation were the current-use pesticides endosulfan and dacthal, followed by polycyclic aromatic hydrocarbons, chlorpyrifos, DDT, and hexachlorobenzene. PCBs and selected pesticides (including chlorpyrifos and chlorothalonil) had previously been reported in brook trout and Pacific tree frogs (*Hyla regilla*) in the Kaweah River basin in Sequoia National Park by Datta et al. (1998).

The presence of pesticides in the bodies of Sierra Nevadan amphibians is well documented (Cory et al. 1970, Datta et al. 1998, Sparling et al. 2001, Fellers et al. 2004). The disappearance of the mountain yellow-legged and other frog species from sites in California that they had occupied historically has been documented (Davidson et al. 2002, Davidson 2004, Davidson and Knapp 2007). The latter study quantified habitat characteristics and presence/absence of mountain yellow-legged frogs and fish at 6831 sites in California. Upwind pesticide use was estimated from application records. After accounting for habitat effects, the probability of mountain yellow-legged frog presence was significantly reduced by both fish and pesticides, with the effect of pesticides much stronger than the effect of fish presence. These results suggest that wind-borne pesticides might contribute to amphibian decline in the Sierra Nevada (Davidson and Knapp 2007, Bradford et al. 2011).

Based on analyses conducted in WACAP, current-use pesticides concentrations were generally highest in fish collected from lakes in Sequoia National Park, followed by Rocky Mountain and Glacier National Parks, and lowest in surveyed parks in the Pacific Northwest (Olympic and Mount Rainier National Parks) and Alaska (Denali National Park and Noatak National Preserve; Ackerman et al. 2008). DDT concentrations in several fish from Emerald and Pear Lakes in Sequoia National Park exceeded health thresholds for kingfishers. These results suggest that organic contaminants deposited from the atmosphere onto remote national park watersheds in the western United States can accumulate in fish to levels that are of concern regarding wildlife health for some, but not all, wildlife species. Contaminant health thresholds for mink and river otter, which are higher than those for kingfisher, were not exceeded by fish contaminant levels in any of the park lakes studied (Ackerman et al. 2008).

Cholinesterase activity is a bioindicator of exposure to organophosphate pesticides (Ludke et al. 1975). Organophosphorus insecticides, such as malathion, chlorpyrifos, and diazinon, bind with cholinesterase and interfere with neural function. Sparling et al. (2001) measured cholinesterase activity and body residues in adult and tadpole life stages of Pacific tree frogs in California. Cholinesterase activity in tadpoles was lower in mountainous areas east of the Central Valley compared with habitats to the north or near the coast. Cholinesterase was also lower where ranid amphibian population status was poor or moderate. The frequency of detection and concentration of organophosphate pesticides in Pacific tree frog tissue followed north–south and west–east patterns that were consistent with agricultural sources of pesticides upwind of the areas having most severe amphibian decline. Reduced cholinesterase in tadpoles has been shown to be associated with reduced activity, decreased growth, and increased mortality (Rosenbaum et al. 1988, Cowman and Mazanti 2000).

Fellers et al. (2004) measured pesticide concentrations in mountain yellow-legged frogs collected in the Sixty Lakes Basin in Kings Canyon National Park and in the Tablelands area in Sequoia National Park in 1997. The population in Kings Canyon National Park appeared to be large and healthy; the Tablelands area once had a large population of this species, but it had been eliminated by the 1980s. Tablelands is exposed directly to the prevailing winds from agricultural areas located to the west. An experimental reintroduction of the frogs in 1994–1995 was not successful. The remaining 20 reintroduced frogs were captured and analyzed for tissue pesticide concentrations. Both γ-chlordane and trans-nonachlor were present at significantly higher concentration in the frogs collected from Tablelands as compared with Sixty Lakes Basin. The highest concentration of organochlorine pesticides was for the DDT breakdown product dichlorodiphenyldichloroethylene, which was 27 times higher, on average, in Tablelands (Fellers et al. 2004).

Fish were found in lakes in Sequoia National Park with DDT concentrations that exceeded contaminant health thresholds for one or more piscivorous wildlife species. These concentrations exceeded levels detected in many fish collections around the world, even fish in Lake Malawi in East Africa, where DDT is sprayed as a mosquito control (Landers et al. 2010). Fish from similar ecosystems in Canada had lower dieldrin concentrations as compared with fish from Sequoia National Park. Both WACAP lakes in Sequoia National Park had average dieldrin concentrations that exceeded human thresholds for subsistence fishers, and one fish exceeded the threshold for recreational human fishers (Landers et al. 2008).

A study was conducted of pesticide residues in brook trout and Pacific tree frogs in the Kaweah River watershed by Datta et al. (1998). Samples were collected along Sycamore Creek, at 610 m elevation, and compared with results for brook trout from Lake Tahoe and tree frog egg masses from Upper Meadows in Lassen Volcanic National Park to the north of the Sierra Nevada Network. Concentrations of PCBs in brook trout taken from Sycamore Creek (4.8–8.1 ppb, wet weight) were similar to those found in lake trout in Lake Tahoe and rainbow trout in Marlette Lake. However, tree frog tadpoles and egg masses showed enriched levels of tri- and tetrachlorinated biphenyl congeners (PCBs). Analyses for current-use pesticides upwind from SEKI revealed the presence of chlorpyrifos, an organophosphate pesticide, and chlorothalanil, a chloronitrile fungicide. This may have implications for the observed decline in populations of some amphibian species in and around the Sierra Nevada Network national parks.

Ohyama et al. (2004) chose salmonid fish, mainly rainbow trout (*Oncorhyncus mykiss*), as an indicator to evaluate the transport and bioaccumulation of organochloride compounds in the northern and central Sierra Nevada. They found that elevation was an important factor affecting the residual levels of PCBs in fish muscle tissue. On this basis, Ohyama et al. (2004) concluded that PCB residues in rainbow trout, a widely distributed salmonid species, provided a good monitoring tool for studying the effects of mountainous topography on the long-range transport and distribution of persistent organic pollutants.

Organochlorine chemicals can be estrogenic (Garcia-Reyero et al. 2007), contributing to the occurrence of intersex fish (male and female reproductive structures occurring in the same fish), and can accumulate in the aquatic food chains of remote mountain ecosystems (Blais et al. 1998). Biomarkers of exposure to estrogenic chemicals suggest the likelihood of reproductive dysfunction (Harries et al. 1997). Organochlorines are likely transported as atmospheric contaminants to high-elevation sites (Hageman et al. 2006). Work by Schreck and Kent (2013) followed up on some of the work performed in the WACAP study. They expanded the range of coverage to additional western parks. The extent of endocrine disruption in fish was assessed in YOSE and SEKI. Of fish sampled in YOSE, one male brook trout was intersex. None of the fish sampled in SEKI (n = 50 male fish) were intersex. The low observed frequency of intersex fish in study parks and water bodies in the Sierra Nevada Network may be a natural phenomenon (Schwindt et al. 2009). The extent to which human-caused contaminants contribute to an increased frequency is difficult to determine (Schreck and Kent 2013).

Hageman et al. (2010) reported the results of pesticide analyses of snowpack at remote alpine, arctic, and subarctic sites in eight national parks, including SEKI. Various current-use pesticides (dacthal, chlorpyrifos, endosulfans, and γ-hexachlorocyclohexane) and historic-use pesticides (dieldrin, el hexachlorocyclohexane, chlordanes, and hexachlorobenzene) were commonly measured at all sites and years (2003–2005). The pesticide concentration profiles were unique for individual parks, suggesting the importance of regional sources. The distributions of current-use pesticides among the parks were explained, using mass back trajectory analysis, based on the mass of individual current-use pesticides used in regions located one-day upwind of the parks. For most pesticides and parks, more than 75% of the snowpack pesticide burden was attributed to regional transport. The researchers concluded that the majority of pesticide contamination in U.S. national parks is due to regional pesticide applications.

WACAP measurements of Hg deposition fluxes to the snowpack were relatively low, consistent with the presence of few local or regional sources of Hg near SEKI. However, Hg concentrations in vegetation were comparatively high as compared with other WACAP parks, and fish were found that exceeded Hg health thresholds for wildlife and humans (Landers et al. 2008). Krabbenhoft et al. (2002) sampled 90 lakes in seven of the WACAP national parks, including SEKI and YOSE, and analyzed their Hg and methyl Hg concentrations. Levels of methyl Hg were lowest in Glacier National Park (0.02 ng/L) and similar among all of the other parks (~0.05 ng/L).

Wright et al. (2014) used surrogate artificial deposition surfaces, passive samplers, and temporal changes in Hg deposition and atmospheric concentrations to improve understanding of sources of Hg to remote western United States locations. Lowest Hg deposition (0.2–0.4 ng/m^2/hr) occurred at low-elevation coastal sites. Highest deposition values (0.5–2.4 ng/m^2/hr) were recorded in Great Basin National Park in the adjacent Mojave Desert Network and at a high-elevation (1279 m) coastal site. Intermediate deposition values were recorded at YOSE and SEKI (0.9–1.2 ng/m^2/hr).

Eagles-Smith et al. (2014) sampled fish in 21 national parks and analyzed them for Hg concentrations in tissue. Results varied substantially by park and by water body. Fish were sampled from four high-elevation lakes in SEKI. Concentrations of Hg in fish were generally low compared to other parks included in the study. Fish were analyzed for Hg in only two lakes in YOSE. At Spillway Lake, Hg concentrations were very low (mean 39 nanograms per gram wet weight [ng/g ww]). In contrast, fish from Mildred Lake had much higher concentrations of Hg (mean 174 ng/g ww), and the levels in individual fish varied 15-fold. One fish had more than 1100 ng/g ww Hg. The cause(s) of this extreme variation is/are not known.

17.8 Summary

Urban and agricultural emissions affect air quality in California's national parks, particularly parks in the Sierra Nevada. The San Joaquin Valley, upwind of the Sierra Nevada Network parks, is one of the most polluted areas in California and is a primary source of air pollutants transported to these parks (Lin and Jao 1995). The Central Valley, with a population of 6.5 million, has been the fastest-growing region of California. Wildland fires, common occurrences in this region, are also large contributors to air pollution in the parks (McMeeking et al. 2006).

Sulfur and N pollutants can cause acidification of streams, lakes, and soils. The levels of N deposition received in high-elevation portions of SEKI and YOSE have contributed to some episodic acidification of surface waters during snowmelt and rain storms (Sickman and Melack 1989). Slight chronic acidification may also have occurred in the most acid-sensitive aquatic systems (Sickman and Melack 1989). Such effects on water chemistry, though small in magnitude, may have had subtle long-term effects on aquatic biota.

Nitrogen pollutants can also cause undesirable enrichment of natural ecosystems, leading to changes in plant species diversity and soil nutrient cycling. Certain lichen species now found in YOSE are suggestive of N pollution, and modeled N deposition in all Sierra Nevada Network parks exceeds levels known to harm natural diversity in lichen communities and herbaceous plant and shrub communities (Fenn et al. 2011b). In high-elevation lakes in the eastern Sierra Nevada, increased N from atmospheric deposition has induced changes in diatom communities from species adapted to low N conditions to species typical of more nutrient-rich disturbed lakes. To date, these changes have not been observed in Sierra Nevada Network park lakes (cf., Sickman and Bennett 2011).

Ozone pollution can harm human health, reduce plant growth, injure foliage, and render sensitive plant species more vulnerable to drought and other stressors (Grulke et al. 1998, Grulke and Balduman 1999). All parks in the Sierra Nevada Network are located in counties that have been designated as nonattainment for the ozone National Ambient Air Quality Standard. Sequoia and Kings Canyon national parks have had ozone health advisory programs that alerted visitors when the air was expected to be unhealthy because of high ozone levels. Parts of SEKI experienced some of the worst air quality in the National Park system. Vegetation at SEKI has exhibited among the most severe effects of ozone in California outside the Los Angeles Basin, with more documented impacts than at any other western national park (Bytnerowicz et al. 2008). Ozone-induced foliar injury in ponderosa pine and Jeffrey pine has been widespread and pronounced in the western regions of SEKI and was present to a lesser extent in YOSE. Foliar injury has also been documented in giant sequoia seedlings. Reduced foliar biomass and tree growth at some locations suggest that ecosystem productivity may have been affected by chronic, long-term exposure to elevated levels of ozone in the mixed conifer zone.

Particulate pollution can cause haze, reducing visibility even at levels well below the National Ambient Air Quality Standard for fine particulate matter (Pitchford and Malm 1994). Visibility in Sierra Nevada Network parks is often reduced substantially from estimated natural background conditions.

Pesticides are used heavily in California's agricultural valleys, especially the San Joaquin Valley. Some are highly volatile and are transported through the atmosphere to sites in the Sierra Nevada, including within SEKI and YOSE. Other toxic substances, including Hg, are also deposited in national park ecosystems in the Sierra Nevada Network. Current-use pesticides were found to be especially prevalent in Sequoia National Park, as compared with other parklands in the western United States (Landers et al. 2008). Biological impacts are not well understood.

References

Ackerman, L.K., A.R. Schwindt, S.L. Simonich, D.C. Koch, T.F. Blett, C.B. Schreck, M.L. Kent, and D.H. Landers. 2008. Atmospherically deposited PBDEs, pesticides, PCBs, and PAHs in western U.S. national park fish: Concentrations and consumption guidelines. *Environ. Sci. Technol.* 42:2334–2341.

Agee, J.K. 1997. The severe weather wildfire: Too hot to handle? *Northwest Sci.* 71:153–157.

Allen, C.D., M. Savage, D.A. Falk, K.F. Suckling, T.W. Swetnam, P. Schulke, P.B. Stacey, P. Morgan, M. Hoffman, and J.T. Klingel. 2002. Ecological restoration of southwestern ponderosa pine ecosystems: A broad perspective. *Ecol. Appl.* 12:1418–1433.

Allison, J.R. 1982. Evaluation of ozone injury on the Stanislaus National Forest. USDA Forest Service, Pacific Southwest Region, Forest Pest Management Report 82-7, San Francisco, CA.

Arbaugh, M.J., P.R. Miller, J.J. Carroll, B.J. Takemoto, and T. Procter. 1998. Relationships of ozone exposure to pine injury in the Sierra Nevada and San Bernardino Mountains of California, USA. *Environ. Pollut.* 101:291–301.

Aston, L. and J.N. Seiber. 1997. The fate of summertime airborne organophosphate pesticide residues in the Sierra Nevada Mountains. *J. Environ. Qual.* 26:1483–1492.

Barmuta, L.A., S.D. Cooper, S.K. Hamilton, K.W. Kratz, and J.M. Melack. 1990. Responses of zooplankton and zoobenthos to experimental acidification in a high-elevation lake (Sierra Nevada, California, U.S.A.). *Freshw. Biol.* 23:571–586.

Baron, J.S., C.T. Driscoll, and J.L. Stoddard. 2011. Inland surface water. *In* L.H. Pardo, M.J. Robin-Abbott, and C.T. Driscoll (Eds.). *Assessment of Nitrogen Deposition Effects and Empirical Critical Loads of Nitrogen for Ecoregions of the United States.* General Technical Report NRS-80. U.S. Forest Service, Newtown Square, PA, pp. 209–228.

Bergametti, G., E. Remoudaki, R. Losno, E. Steiner, and B. Chatenet. 1992. Source, transport and deposition of atmospheric phosphorus over the northwestern Mediterranean. *J. Atmos. Chem.* 14:501–513.

Blais, J.M., D.W. Schindler, D.C.G. Muir, L.E. Kimpe, D.B. Donald, and B. Rosenberg. 1998. Accumulation of persistent organochlorine compounds in mountains of western Canada. *Nature* 395:585–588.

Blaustein, A.R. and D.B. Wake. 1990. Declining amphibian populations: A global phenomenon. *Trends Ecol. Evol.* 5:203–204.

Blett, T., L. Geiser, and E. Porter. 2003. Air pollution-related lichen monitoring in national parks, forests, and refuges: Guidelines for studies intended for regulatory and management purposes. Technical Report NPS-D2292. U.S. Department of the Interior, Denver, CO. USDA Forest Service, Corvallis, OR.

Bobbink, R., M. Ashmore, S. Braun, W. Flückiger, and I.J.J. van den Wyngaert. 2003. Empirical nitrogen critical loads for natural and semi-natural ecosystems: 2002 update. *In* B. Achermann and R. Bobbink (Eds.). *Empirical Critical Loads for Nitrogen.* Swiss Agency for Environment, Forest and Landscape SAEFL, Berne, Switzerland, pp. 43–170.

Boiano, D.M., D.P. Weeks, and T. Hemry. 2005. Sequoia and Kings Canyon National Parks, California: Water resources information and issues overview report. Technical Report NPS/NRWRD/NRTR-2005/333. National Park Service, Water Resources Division, Fort Collins, CO.

Bowman, W.D. 1994. Accumulation and use of nitrogen and phosphorus following fertilization in two alpine tundra communities. *Oikos* 70:261–270.

Bowman, W.D., J.S. Baron, L.H. Geiser, M.E. Fenn, and E.A. Lilleskov. 2011. Northwestern forested mountains. *In* L.H. Pardo, M.J. Robin-Abbott, and C.T. Driscoll (Eds.). *Assessment of Nitrogen Deposition Effects and Empirical Critical Loads of Nitrogen for Ecoregions of the United States.* General Technical Report NRS-80. U.S. Forest Service, Newtown Square, PA, pp. 75–88.

Bowman, W.D., J.R. Gartner, K. Holland, and M. Wiedermann. 2006. Nitrogen critical loads for alpine vegetation and terrestrial ecosystem response: Are we there yet? *Ecol. Appl.* 16(3):1183–1193.

Bowman, W.D., T.A. Theodose, J.C. Schardt, and R.T. Conant. 1993. Constraints of nutrient availability on primary production in two alpine tundra communities. *Ecology* 74:2085–2097.

Bradford, D.F. 1989. Allotropic distribution of native frogs and introduced fishes in high Sierra Nevada lakes of California: Implication of the negative effect of fish introductions. *Copeia* 1989:775–778.

Bradford, D.F. and M.S. Gordon. 1992. Aquatic amphibians in the Sierra Nevada: Current status and potential effects of acidic deposition on populations. California Air Resources Board, Sacramento, CA.

Bradford, D.F., M.S. Gordon, D.F. Johnson, R.D. Andrews, and W.B. Jennings. 1994. Acidic deposition as an unlikely cause for amphibian population declines in the Sierra Nevada, California. *Biol. Conserv.* 69(2):155–161.

Bradford, D.F., K. Stanley, L.L. McConnell, N.G. Tallent-Halsel, M.S. Nash, and S.M. Simonich. 2011. Spatial patterns of atmospherically deposited organic contaminants at high elevation in the southern Sierra Nevada Mountains, California, USA. *Environ. Toxicol. Chem.* 29(5):1056–1066.

Bradford, D.F., F. Tabatabai, and D.M. Graber. 1993. Isolation of remaining populations of the native frog, *Rana muscosa*, by introduced fishes in Sequoia and Kings Canyon National Parks, California. *Conserv. Biol.* 7(4):882–888.

Brockway, D.G., R.G. Gatewood, and R.B. Paris. 2002. Restoring fire as an ecological process in short-grass prairie ecosystems: Initial effects of prescribed burning during the dormant and growing seasons. *J. Environ. Manage.* 65:135–152.

Burley, J.D. and J.D. Ray. 2007. Surface ozone in Yosemite National Park. *Atmos. Environ.* 41:6048–6062.

Burns, D.A. 2004. The effects of atmospheric nitrogen deposition in the Rocky Mountains of Colorado and southern Wyoming, USA—A critical review. *Environ. Pollut.* 127:257–269.

Bytnerowicz, A. 2002. Physiological/ecological interactions between ozone and nitrogen deposition in forest ecosystems. *Phyton* 42(3):13–28.

Bytnerowicz, A., M. Arbaugh, S. Schilling, W. Frączek, and D. Alexander. 2008. Ozone distribution and phytotoxic potential in mixed conifer forests of the San Bernardino Mountains, southern California. *Environ. Pollut.* 155(3):398–408.

Bytnerowicz, A., M.E. Fenn, P.R. Miller, and M.J. Arbaugh. 1999. Wet and dry pollutant deposition to the mixed conifer forest. *In* P.R. Miller and J.R. McBride (Eds.). *Oxidant Air Pollution Impacts in the Montane Forests of Southern California*. Springer-Verlag, New York, pp. 235–269.

Bytnerowicz, A. and N.E. Grulke. 1992. Physiological effects of air pollutants on western trees. *In* R.K. Olson, D. Binkley, and M. Böhm (Eds.). *Response of Western Forests to Air Pollution*. Springer-Verlag, New York, pp. 183–233.

Bytnerowicz, A., M. Tausz, R. Alonso, D. Jones, R. Johnson, and N. Grulke. 2002. Summer-time distribution of air pollutants in Sequoia National Park, California. *Environ. Pollut.* 118:187–203.

Bytnerowicz, A.B. and M.E. Fenn. 1996. Nitrogen deposition in California forests: A review. *Environ. Pollut.* 92(2):127–146.

Cahill, T.A., J.J. Carroll, D. Campbell, and T.E. Gill. 1996. Air quality. Sierra Nevada Ecosystem Project: Final report to Congress. Volume II: Assessments and Scientific Basis for Management Options. Centers for Water and Wildland Resources, University of California, Davis, CA, pp. 1227–1261.

Campbell, S.J., R. Wanek, and J.W. Coulston. 2007. Ozone injury in west coast forests: 6 years of monitoring. General Technical Report PNW-GTR-722. USDA Forest Service, Pacific Northwest Research Station, Portland, OR.

Census of Agriculture. 2002. Geographic area series, Part 51. National Agricultural Statistics Service, U.S. Department of Agriculture, Washington, DC.

Charles, D.F. 1991. *Acidic Deposition and Aquatic Ecosystems: Regional Case Studies*. Springer-Verlag, New York.

Clow, D.W., A. Mast, and D.H. Campbell. 1996. Controls on surface water chemistry in the Upper Merced River Basin, Yosemite National Park, California. *Hydrol. Process.* 10:727–746.

Clow, D.W., L. Nanus, and B. Huggett. 2010. Use of regression-based models to map sensitivity of aquatic resources to atmospheric deposition in Yosemite National Park, USA. *Water Resour. Res.* 46(W09529):14. doi:10.1029/2009WR008316.

Clow, D.W., J.O. Sickman, R.G. Striegl, D.P. Krabbenhoft, J.G. Elliott, M.M. Dornblaser, D.A. Roth, and D.H. Campbell. 2003. Changes in the chemistry of lakes and precipitation in high-elevation national parks in the western United States. *Water Resour. Res.* 39:1985–1999.

Clow, D.W., R. Striegl, L. Nanus, M.A. Mast, D.H. Campbell, and D.P. Krabbenhoft. 2002. Chemistry of selected high-elevation lakes in seven national parks in the western United States. *Water Air Soil Pollut.* 2(2):139–164.

Clow, D.W. and J.K. Sueker. 2000. Relations between basin characteristics and stream water chemistry in alpine/subalpine basins in Rocky Mountain National Park, Colorado. *Water Resour. Res.* 36(1):49–61.

Cory, L., P. Fjerd, and W. Serat. 1970. Distribution patterns of DDT residues in the Sierra Nevada Mountains. *Pestic. Monit. J.* 3:204–211.

Covington, W.W. 2000. Helping western forests heal: The prognosis is poor for United States forest ecosystems. *Nature* 408:135–136.

Cowman, D.F. and L.E. Mazanti. 2000. Ecotoxicology of new generation pesticides to amphibians. *In* D.W. Sparling, G. Linder and C.A. Bishop (Eds.). *Ecotoxicology of Amphibians and Reptiles. Society of Environmental Toxicology and Chemistry*, Pensacola, FL, pp 233–268.

Datta, S., L. McConnell, J.L. Baker, J.S. LeNoir, and J.N. Seiber. 1998. Pesticides and PCB contaminants in fish and tadpoles from Kaweah River Basin, California. *Bull. Environ. Contam. Toxicol.* 60:829–836.

Davidson, C. 2004. Declining downwind: Amphibian population declines in California and historic pesticide use. *Ecol. Appl.* 14:1894–1902.

Davidson, C. and R.A. Knapp. 2007. Multiple stressors and amphibian declines: Dual impacts of pesticides and fish on yellow-legged frogs. *Ecol. Appl.* 17(2):587–597.

Davidson, C., H.B. Shaffer, and M. Jennings. 2002. Spatial tests of the pesticide drift, habitat destruction, UV-B and climate change hypotheses for California amphibian declines. *Conserv. Biol.* 16:1588–1601.

Dettinger, M.D. and D.R. Cayan. 1995. Large-scale atmospheric forcing of recent trends toward early snowmelt runoff in California. *J. Clim.* 8:606–623.

Duriscoe, D.M. 1990. Cruise Survey of oxidant air pollution injury to *Pinus ponderosa* and *Pinus jeffreyi* in Saguaro National Monument, Yosemite National Park, and Sequoia and Kings Canyon National Parks. Report NPS/AQD-90/003. National Park Service, Air Quality Division, Denver, CO.

Duriscoe, D.M. and K.W. Stolte. 1989. Photochemical oxidant injury to ponderosa (*Pinus ponderosa* Laws.) and Jeffrey pine (*Pinus jeffreyi* Grey, and Balf.) in the national parks of the Sierra Nevada. *In* R.K. Olson and A.S. Lefohn (Eds.). *Effects of Air Pollution on Western Forests.* Air and Waste Management Association, Anaheim, CA, pp. 261–278.

Duriscoe, D.M. and K.W. Stolte. 1992. Decreased foliage production and longevity observed in ozone-injured Jeffrey and ponderosa pines in Sequoia National Park, California. Air and Waste Management Association, Pittsburgh, PA.

Eagles-Smith, C.A., J.J. Willacker, Jr., and C.M. Flanagan Pritz. 2014. Mercury in fishes from 21 national parks in the western United States—Inter- and intra-park variation in concentrations and ecological risk. U.S. Geological Survey Open-File Report 2014-1051, Reston, VA.

Eilers, J.M., P. Kanciruck, R.A. McCord, W.S. Overton, L. Hook, D.J. Blick, D.F. Brakke et al. 1987. Characteristics of lakes in the western United States. Volume II: Data compendium for selected physical and chemical variables. EPA-600/3-86/054b. U.S. EPA, Washington, DC.

Ellis, R.A., D.J. Jacob, M.P. Sulprizio, L. Zhang, C.D. Holmes, B.A. Schichtel, T. Blett, E. Porter, L.H. Pardo, and J.A. Lynch. 2013. Present and future nitrogen deposition to national parks in the United States: Critical load exceedances. *Atmos. Chem. Phys.* 13(17):9083–9095.

Elser, J.J., E.R. Marzolf, and C.R. Goldman. 1990. Phosphorus and nitrogen limitation of phytoplankton growth in the freshwaters of North America: A review and critique of experimental enrichments. *Can. J. Fish. Aquat. Sci.* 47:1468–1477.

Engling, G., P. Herckes, S.M. Kreidenweis, W.C. Malm, and J.L. Collett Jr. 2006. Composition of the fine organic aerosol in Yosemite National Park during the 2002 Yosemite Aerosol Characterization Study. *Atmos. Environ.* 40:2959–2972.

Ewell, D.M. and D.T. Gay. 1993. Long-term monitoring of ozone injury to Jeffrey and ponderosa pines in Sequoia and Kings Canyon National Parks. *Proceedings of the 86th Annual Meeting of the Air and Waste Management Association.* Reprint 93-TA-42.04, Pittsburgh, PA.

Ewell, D.M., L.C. Mazzu, and D.M. Duriscoe. 1989. Specific leaf weight and other characteristics of ponderosa pine as related to visible ozone injury. *In* R.K. Olson and A.S. Lefohn (Eds.). *Effects of Air Pollution on Western Forests.* Transactions Series No.16. Air and Waste Management Association, Pittsburgh, PA, pp. 411–418.

Fellers, G.M., L.L. McConnell, D. Pratt, and S. Datta. 2004. Pesticides in mountain yellow-legged frog (*Rana muscosa*) from the Sierra Nevada Mountains of California, USA. *Environ. Toxicol. Chem.* 23(9):2170–2177.

Fenn, M.E., E.B. Allen, and L.H. Geiser. 2011a. Mediterranean California. *In* L.H. Pardo, M.J. Robin-Abbott, and C.T. Driscoll (Eds.). *Assessment of Nitrogen Deposition Effects and Empirical Critical Loads of Nitrogen for Ecoregions of the United States*. General Technical Report NRS-80. U.S. Department of Agriculture, Forest Service, Northern Research Station, Newtown Square, PA, pp. 143–170.

Fenn, M.E., J.S. Baron, E.B. Allen, H.M. Rueth, K.R. Nydick, L. Geiser, W.D. Bowman et al. 2003b. Ecological effects of nitrogen deposition in the western United States. *BioScience* 53(4):404–420.

Fenn, M.E., R. Haeuber, G.S. Tonnesen, J.S. Baron, S. Grossman-Clark, D. Hope, D.A. Jaffe et al. 2003c. Nitrogen emissions, deposition, and monitoring in the western United States. *BioScience* 53(4):391–403.

Fenn, M.E., S. Jovan, F. Yuan, L. Geiser, T. Meixner, and B.S. Gimeno. 2008. Empirical and simulated critical loads for nitrogen deposition in California mixed conifer forests. *Environ. Pollut.* 155:492–511.

Fenn, M.E., K.F. Lambert, T. Blett, D.A. Burns, L.H. Pardo, G.M. Lovett, R.A. Haeuber, D.C. Evers, C.T. Driscoll, and D.S. Jefferies. 2011b. Setting limits: Using air pollution thresholds to protect and restore U.S. ecosystems. Issues in Ecology, Report No. 14. Ecological Society of America, Washington, DC.

Fenn, M.E., M.A. Poth, J.D. Aber, J.S. Baron, B.T. Bormann, D.W. Johnson, A.D. Lemly, S.G. McNulty, D.F. Ryan, and R. Stottlemyer. 1998. Nitrogen excess in North American ecosystems: Predisposing factors, ecosystem responses, and management strategies. *Ecol. Appl.* 8:706–733.

Fenn, M.E., M.A. Poth, A. Bytnerowicz, J.O. Sickman, and B. Takemoto. 2003a. Effects of ozone, nitrogen deposition, and other stressors on montane ecosystems in the Sierra Nevada. *In* A. Bytnerowicz, M.J. Arbaugh, and R. Alonso (Eds.). *Ozone Air Pollution in the Sierra Nevada: Distribution and Effects on Forests*. Volume 2: Developments in Environmental Sciences. Elsevier, Amsterdam, the Netherlands, pp. 111–155.

Fenn, M.E., J.O. Sickman, A. Bytnerowicz, D.W. Clow, N.P. Molotch, J.E. Pleim, G.S. Tonnesen, K.C. Weathers, P.E. Padgett, and D.H. Campbell. 2009. Methods for measuring atmospheric nitrogen deposition inputs in arid and montane ecosystems of western North America. *In* A.H. Legge (Ed.). *Developments in Environmental Science*. Elsevier, Amsterdam, the Netherlands, pp. 179–228.

Fisk, M.C., S.K. Schmidt, and T.R. Seastedt. 1998. Topographic patterns of above-and belowground production and nitrogen cycling in alpine tundra. *Ecology* 79(7):2253–2266.

Flannigan, M.D., Y. Bergeron, O. Engelmark, and B.M. Wotton. 1998. Future wildfire in circumboreal forests in relation to global warming. *J. Veg. Sci.* 9:469–476.

Freda, J. 1990. Effects of acidification on amphibians. *In* J.P. Baker, D.P. Bernard, S.W. Christensen, M.J. Sale, J. Freda, K. Heltcher, D. Marmorek et al. (Eds.). *Biological Effects of Changes in Surface in Surface Water Acid-Base Chemistry*. NAPAP Report 13. National Acid Precipitation Assessment Program, Acid Deposition. State-of-Science and Technology, Washington, DC.

Fulé, P.Z., W.W. Covington, and M.M. Moore. 1997. Determining reference conditions for ecosystem management of southwestern ponderosa pine forests. *Ecol. Appl.* 7:895–908.

Garcia-Reyero, N., J.O. Grimalt, I. Vives, P. Fernandez, and B. Piña. 2007. Estrogenic activity associated with organochlorine compounds in fish extracts from European mountain lakes. *Environ. Pollut.* 145(3):745–752.

Geiser, L.H., S.E. Jovan, D.A. Glavich, and M.K. Porter. 2010. Lichen-based critical loads for atmospheric nitrogen deposition in western Oregon and Washington forests, USA. *Environ. Pollut.* 158:2412–2421.

Geiser, L.H. and P.N. Neitlich. 2007. Air pollution and climate gradients in western Oregon and Washington indicated by epiphytic macrolichens. *Environ. Pollut.* 145:203–218.

Goldman, C.R. 1988. Primary productivity, nutrients, and transparency during the early onset of eutrophication in ultra-oligotrophic Lake Tahoe, California-Nevada. *Limnol. Oceanogr.* 33:1321–1333.

Grulke, N.E. 1999. Physiological responses of ponderosa pine to gradients of environmental stress-ors. *In* P.R. Miller and J.R. McBride (Eds.). *Oxidant Air Pollution Impacts in the Montane Forests of Southern California: A Case Study of the San Bernardino Mountains.* Springer, New York, pp. 126–163.

Grulke, N.E., C.P. Anderson, M.E. Fenn, and P.R. Miller. 1998. Ozone exposure and nitrogen depo-sition lowers root biomass of ponderosa pine in the San Bernardino Mountains, California. *Environ. Pollut.* 103:63–73.

Grulke, N.E. and L. Balduman. 1999. Deciduous conifers: High N deposition and O_3 exposure effects on growth and biomass allocation in ponderosa pine. *Water Air Soil Pollut.* 116:235–248.

Grulke, N.E., R. Johnson, A. Esperanza, D. Jones, T. Nguyen, S. Posch, and M. Tausz. 2003. Canopy transpiration of Jeffrey pine in mesic and zeric microsites: O_3 uptake and injury response. *Trees* 17:292–298.

Grulke, N.E. and P.R. Miller. 1994. Changes in gas exchange characteristics during the life span of giant sequoia: Implications for response to current and future concentrations of atmospheric ozone. *Tree Physiol.* 14:659–668.

Hageman, K.J., W.D. Hafner, D.H. Campbell, D.A. Jaffe, D.H. Landers, and S.L.M. Simonich. 2010. Variability in pesticide deposition and source contributions to snowpack in western U.S. national parks. *Environ. Sci. Technol.* 44(12):4452–4458.

Hageman, K.J., S.L. Simonich, D.H. Campbell, G.R. Wilson, and D.H. Landers. 2006. Atmospheric deposition of current-use and historic-use pesticides in snow at national parks in the western United States. *Environ. Sci. Technol.* 40(10):3174–3180.

Harries, J.E., D.A. Sheahan, S. Jobling, P. Matthiessen, P. Neall, J.P. Sumpter, T. Tylor, and N. Zaman. 1997. Estrogenic activity in five United Kingdom rivers detected by measurement of vitellogen-esis in caged male trout. *Environ. Toxicol. Chem.* 16(3):534–542.

Holmes, R.S., M.L. Whiting, and J.L. Stoddard. 1989. Changes in diatom-inferred pH and acid neutraliz-ing capacity in a dilute, high elevation, Sierra Nevada lake since A.D. 1825. *Freshw. Biol.* 21:295–310.

Horner, D. and D.L. Peterson. 1993. Goat rocks wilderness monitoring plan. Report to Gifford National Forest. USDA Forest Service, Packwood, WA.

Hurteau, M. and M. North. 2009. Response of *Arnica dealbata* to climate change, nitrogen deposition, and fire. *Plant Ecol.* 202:191–194.

Jassby, A.D., J.E. Reuter, R.P. Axler, C.R. Goldman, and S.H. Hackley. 1994. Atmospheric deposition of N and phosphorus in the annual nutrient load of Lake Tahoe (California-Nevada). *Water Resour. Res.* 30:2207–2216.

Jenkins, T.M., R.A. Knapp, K.W. Kratz, and S.D. Cooper. 1994. Aquatic biota in the Sierra Nevada: Current status and potential effects of acid deposition on populations. Final Report to the California Air Resources Board, Contract A932-138. Marine Science Institute, University of California, Santa Barbara, CA.

Jennings, M.R. 1996. Status of amphibians. Status of the Sierra Nevada. Volume II: Assessments and Scientific Basis for Management Options. Sierra Nevada Ecosystem Project, Final Report to Congress. Center for Water and Wildland Resources, University of California, Davis, CA, pp. 921–944.

Johannessen, M. and A. Henriksen. 1978. Chemistry of snow meltwater: Changes in concentrations during melting. *Water Resour. Res.* 14:615–619.

Johnson, T.R. 1998. Climate change and Sierra Nevada snowpack. Masters thesis, University of California, Santa Barbara, CA.

Jovan, S. 2008. Lichen bioindication of biodiversity, air quality, and climate: Baseline results from monitoring in Washington, Oregon, and California. General Technical Report PNW-GTR-737. U.S. Department of Agriculture, Forest Service, Pacific Northwest Research Station, Portland, OR.

Jovan, S. and B. McCune. 2005. Air quality bioindication in the greater Central Valley of California, with epiphytic macrolichen communities. *Ecol. Appl.* 15(5):1712–1726.

Jovan, S. and B. McCune. 2006. Using epiphytic macrolichen communities for biomonitoring ammo-nia in forests of the greater Sierra Nevada, California. *Water Air Soil Pollut.* 170:69–93.

Kegley, S., S. Orme, and L. Neumeister. 2000. Hooked on poison: Pesticide use in California, 1991–1998. Californians for Pesticide Reform, Pesticide Action Network, San Francisco, CA.

Kohut, R. 2007. Assessing the risk of foliar injury from ozone on vegetation in parks in the U.S. National Park Service's Vital Signs Network. *Environ. Pollut.* 149:348–357.

Krabbenhoft, D.P., M.L. Olson, J.F. DeWild, D.W. Clow, R.G. Striegl, M.M. Dornblaser, and P. VanMetre. 2002. Mercury loading and methylmercury production and cycling in high-altitude lakes from the western United States. *Water Air Soil Pollut.* 2:233–249.

Kratz, K.W., S.D. Cooper, and J.M. Melack. 1994. Effects of single and repeated experimental acid pulses on invertebrates in a high altitude Sierra Nevada stream. *Freshw. Biol.* 32:161–183.

Landers, D.H., J.M. Eilers, D.F. Brakke, W.S. Overton, P.E. Kellar, W.E. Silverstein, R.D. Schonbrod et al. 1987. *Characteristics of Lakes in the Western United States. Volume I: Population Descriptions and Physico-Chemical Relationships.* EPA-600/3-86/054a. U.S. Environmental Protection Agency, Washington, DC.

Landers, D.H., S.L. Simonich, D.A. Jaffe, L.H. Geiser, D.H. Campbell, A.R. Schwindt, C.B. Schreck et al. 2008. The fate, transport, and ecological impacts of airborne contaminants in western national parks (USA). EPA/600/R-07/138. U.S. Environmental Protection Agency, Office of Research and Development, NHEERL, Western Ecology Division, Corvallis, OR.

Landers, D.H., S.M. Simonich, D. Jaffe, L. Geiser, D.H. Campbell, A. Schwindt, C.B. Schreck et al. 2010. The Western Airborne Contaminant Project (WACAP): An interdisciplinary evaluation of the impacts of airborne contaminants in western U.S. national parks. *Environ. Sci. Technol.* 44(3):855–859.

Lenihan, J.M., C. Daly, D. Bachelet, and R.P. Neilson. 1998. Simulating broad-scale fire severity in a dynamic global vegetation model. *Northwest Sci.* 72:91–103.

LeNoir, J.S., L.L. McConnell, G.M. Fellers, T.M. Cahill, and J.N. Seiber. 1999. Summertime transport of current-use pesticides from California's Central Valley to the Sierra Nevada mountain range, USA. *Environ. Toxicol. Chem.* 18(12):2715–2722.

Lesack, L.F.W. and J.M. Melack. 1996. Mass balance of major solutes in a rainforest catchment in the central Amazon: Implications for nutrient budgets in tropical rainforests. *Biogeochemistry* 32:115–142.

Leydecker, A., J.O. Sickman, and J.M. Melack. 1999. Episodic lake acidification in the Sierra Nevada, California. *Water Resour. Res.* 35(9):2793–2804.

Lin, Y.L. and I.C. Jao. 1995. A numerical study of flow circulations in the Central Valley of California and formation mechanisms of the Fresno eddy. *Mon. Weather Rev.* 123:3227–3239.

Ludke, J.L., E.F. Hill, and M.P. Dieter. 1975. Cholinesterase (ChE) response and related mortality among birds fed ChE inhibitors. *Arch. Environ. Contam. Toxicol.* 3:1–21.

Malm, W.C. 1999. *Introduction to Visibility.* Cooperative Institute for Research in the Atmosphere (CIRA), Fort Collins, CO.

Malm, W.C., B.A. Schichtel, R.B. Ames, and K.A. Gebhart. 2002. A 10-year spatial and temporal trend of sulfate across the United States. *J. Geophys. Res.* 107(D22, 4627).

McColl, J.G. 1981. Effects of acid rain on plants and soils in California. Final report. Contract A7-169-30. California Air Resources Board, Sacramento, CA.

McConnell, L.L., J.S. LeNoir, S. Datta, and J.N. Seiber. 1998. Wet deposition of current-use pesticides in the Sierra Nevada mountain range, California, USA. *Environ. Toxicol. Chem.* 17(10):1908–1916.

McCune, B., J. Grenon, L.S. Mutch, and E.P. Martin. 2007. Lichens in relation to management issues in the Sierra Nevada national parks. *Pacific Northwest Fungi* 2(3):1–39.

McKenzie, D., Z.M. Gedalof, D.L. Peterson, and P. Mote. 2004. Climatic change, wildfire, and conservation. *Conserv. Biol.* 18:890–902.

McKenzie, D., S.M. O'Neill, N.K. Larkin, and R.A. Norheim. 2006. Integrating models to predict regional haze from wildland fire. *Ecol. Model.* 199:278–288.

McMeeking, G.R., S.M. Kreidenweis, C.M. Carrico, T. Lee, J.L. Collett, Jr., and W.C. Malm. 2005. Observations of smoke-influenced aerosol during the Yosemite Aerosol Characterization Study: Size distributions and chemical composition. *J. Geophys. Res.* 110:11.

McMeeking, G.R., S.M. Kreidenweis, M. Lunden, J. Carrillo, C.M. Carrico, T. Lee, P. Herckes et al. 2006. Smoke-impacted regional haze in California during the summer of 2002. *Agric. For. Meteorol.* 137:25–42.

Melack, J.M., S.C. Cooper, R.S. Holmes, J.O. Sickman, K. Kratz, P. Hopkins, H. Hardenbergh, M. Thieme, and L. Meeker. 1987. Chemical and biological survey of lakes and streams located in the Emerald Lake watershed, Sequoia National Park. Final Report, Contract A3–096–32. California Air Resources Board. Sacramento, CA.

Melack, J.M., S.C. Cooper, T.M. Jenkins, L. Barmuta, Jr., S. Hamilton, K. Kratz, J. Sickman, and C. Soiseth. 1989. Chemical and biological characteristics of Emerald Lake and the streams in its watershed, and the response of the lake and streams to acidic deposition. Final report to the California Air Resource Board, Contract A6-184-32, Sacramento, CA.

Melack, J.M. and J.L. Stoddard. 1991. Sierra Nevada: Unacidified, very dilute waters and mildly acidic atmospheric deposition. *In* D.F. Charles (Ed.). *Acidic Deposition and Aquatic Ecosystems: Regional Case Studies.* Springer-Verlag, New York, pp. 503–530.

Melack, J.M., J.L. Stoddard, and C.A. Ochs. 1985. Major ion chemistry and sensitivity to acid precipitation of Sierra Nevada lakes. *Water Resour. Res.* 21:27–32.

Melack, J.M., J.O. Sickman, and A. Leydecker. 1998. Comparative analyses of high-altitude lakes and catchments in the Sierra Nevada: Susceptibility to acidification. Final report prepared for the California Air Resources Board, Contract A032-188. Marine Science Institute and Institute for Computational Earth System Science, University of California, Santa Barbara, CA.

Melack, J.M., J.O. Sickman, F. Setaro, and D. Dawson. 1997. Monitoring of wet deposition in alpine areas in the Sierra Nevada. Final report. California Air Resources Board Contract A932-081. Marine Sciences Institute, University of California, Santa Barbara, CA.

Miller, C. and P. Landres. 2004. Exploring information needs for wildland fire and fuels management. USDA Forest Service General Technical Report RMRS-GTR-127. U.S. Forest Service Rocky Mountain Research Station, Fort Collins, CO.

Miller, P.R. 1992. Mixed conifer forests of the San Bernardino Mountains, California. *In* R. Olson, D. Binkley, and M. Böhm (Eds.). *The Response of Western Forests to Air Pollution.* Springer-Verlag, New York.

Miller, P.R. and J.R. McBride. 1999. *Oxidant Air Pollution Impacts in the Montane Forests of Southern California: A Case Study of the San Bernardino Mountains.* Springer-Verlag, New York.

Miller, P.R., J.R. McBride, S.L. Schilling, and A.P. Gomez. 1989. Trend of ozone damage to conifer forests between 1974 and 1988 in the San Bernardino Mountains of southern California. *In* R.K. Olson and A.S. Lefohn (Eds.). *Effects of Air Pollution on Western Forests.* Transaction Series. Air and Waste Management Association, Pittsburgh, PA, pp. 309–323.

National Acid Precipitation Assessment Program (NAPAP). 1991. Integrated assessment report. National Acid Precipitation Assessment Program, Washington, DC.

National Park Service (NPS). 2010. Air quality in national parks: 2009 annual performance and progress report. Natural Resource Report NPS/NRPC/ARD/NRR-2010/266. National Park Service, Air Resources Division, Denver, CO.

Ohyama, K., J. Angermann, D.Y. Dunlap, and F. Matsumura. 2004. Distribution of polychlorinated biphenyls and chlorinated pesticide residues in trout in the Sierra Nevada. *J. Environ. Qual.* 33(5):1752–1764.

Pardo, L.H., M.E. Fenn, C.L. Goodale, L.H. Geiser, C.T. Driscoll, E.B. Allen, J.S. Baron et al. 2011. Effects of nitrogen deposition and empirical nitrogen critical loads for ecoregions of the United States. *Ecol. Appl.* 21(8):3049–3082.

Parmeter, J.R., N.J. McGregor, and R.S. Smith. 1978. An evaluation of *Fomes annosus* in Yosemite National Park. Forest Insect and Disease Management Report 78-2. USDA Forest Service, Southwest Region.

Patterson, M.T. and P.W. Rundel. 1989. Seasonal physiological responses of ozone stressed Jeffrey pine in Sequoia National Park, California. *In* R.K. Olson and A.S. Lefohn (Eds.). *Effects of Air Pollution on Western Forests.* Transactions Series 16. Air and Waste Management Association, Pittsburgh, PA, pp. 419–428.

Peterson, D.L. and M.J. Arbaugh. 1988. Growth patterns of ozone-injured ponderosa pine (*Pinus ponderosa*) in the southern Sierra Nevada. *J. Air Pollut. Control Assoc.* 38:921–927.

Peterson, D.L. and M.J. Arbaugh. 1992. Coniferous forests of the Colorado Front Range. Part B: Ponderosa Pine second-growth stands. *In* M.J. Mitchell and S.E. Lindberg (Eds.). *Atmospheric Deposition and Forest Nutrient Cycling: A Synthesis of the Integrated Forest Study.* Springer-Verlag, New York, pp. 433–460.

Peterson, D.L., M.J. Arbaugh, and L.J. Robinson. 1991. Regional growth changes in ozone-stressed ponderosa pine (*Pinus ponderosa*) in the Sierra Nevada, California, USA. *Holocene* 1:50–61.

Peterson, D.L., M.J. Arbaugh, V.A. Wakefield, and P.R. Miller. 1987. Evidence of growth reduction in ozone injured Jeffrey pine in Sequoia and Kings Canyon National Parks. *J. Air Pollut. Control Assoc.* 37:908–912.

Peterson, D.L., D.L. Schmoldt, J.M. Eilers, R.W. Fisher, and R.D. Doty. 1992a. Guidelines for evaluating air pollution impacts on class I wilderness areas in California. General Technical Report PSW-GTR-136. USDA Forest Service, Pacific Southwest Research Station, Albany, CA.

Peterson, J., D.L. Schmoldt, D. Peterson, J.M. Eilers, R. Fisher, and R. Bachman. 1992b. Guidelines for evaluating air pollution impacts on class I wilderness areas in the Pacific Northwest. PNW-GTR-299. USDA Forest Service, Portland, OR.

Phillips, K. 1990. Where have all the frogs and toads gone? *BioScience* 40:422–424.

Pierce, B.A. 1985. Acid tolerance in amphibians. *BioScience* 35:239–243.

Pitchford, M.L. and W.C. Malm. 1994. Development and applications of a standard visual index. *Atmos. Environ.* 28:1049–1055.

Pronos, J. and D.R. Vogler. 1981. of ozone injury to pines in the southern Sierra Nevada, 1979/1980. USDA Forest Service Pacific Southwest Region, San Francisco, CA.

Pronos, J., D.R. Vogler, and R.S. Smith. 1978. An evaluation of ozone injury to pines in the southern Sierra Nevada. Forest Pest Management Report 78-1. USDA Forest Service Pacific Southwest Region, Albany, CA.

Rocchio, J.E., D.M. Ewell, C.T. Procter, and B.K. Takemoto. 1993. Project FOREST: The forest ozone response study. *In* W.E. Brown and S.D. Veirs (Eds.). *Partners in Stewardship.* George Wright Society, Hancock, MI, pp. 112–119.

Rosenbaum, E.A., A. Caballero de Castro, L. Gauna, and A.M. Pechen de D'angelo. 1988. Early biochemical changes produced by malathion on toad embryos. *Arch. Environ. Contam. Toxicol.* 17:831–835.

Savig, K. and D. Morse. 2001. Visibility sections. *In* T.J. Sullivan, D.L. Peterson, C.L. Blanchard, S.J. Tanenbaum, K. Savig and D. Morse (Eds.). *Assessment of Air Quality and Air Pollutant Impacts in Class I National Parks of California.* U.S. Department of the Interior, National Park Service, Air Resources Division, Denver, CO. Available at: http://www.nature.nps.gov/air/Pubs/pdf/reviews/ca/CAreport.pdf (accessed September, 2016).

Schilling, S. and D. Duriscoe. 1996. Data management and analysis of ozone injury to pines. *In* Miller, P.R., K.W. Stolte, D.M. Duriscoe, and J. Pronos, technical coordinators. *Evaluating Ozone Air Pollution Effects on Pines in the Western United States.* Gen. Tech. Rep. PSW–GTR–155. USDA Forest Service Pacific Southwest Research Station, Albany, CA. pp. 57–62.

Schreck, C.B. and M. Kent. 2013. Extent of endocrine disruption in fish of western and Alaskan national parks. Final report. NPS-OSU Task Agreement J8W07080024, Oregon State University, Corvallis, OR.

Schwede, D.B. and G.G. Lear. 2014. A novel hybrid approach for estimating total deposition in the United States. *Atmos. Environ.* 92:207–220.

Schwindt, A.R., M.L. Kent, L.K. Ackerman, S.L.M. Simonich, D.H. Landers, T. Blett, and C.B. Schreck. 2009. Reproductive abnormalities in trout from western U.S. national parks. *Trans. Am. Fish. Soc.* 138(3):522–531.

Shaw, G., R. Cisneros, D. Schweizer, J. Sickman, and M. Fenn. 2014. Critical loads of acid deposition for wilderness lakes in the Sierra Nevada (California) estimated by the steady-state water chemistry model. *Water Air Soil Pollut.* 225(1):1–15.

Sickman, J.O. and D.M. Bennett. 2011. Development of critical loads for atmospheric nitrogen deposition in high elevation lakes in the Sierra Nevada (PMIS 119189). Final report. National Park Service, Riverside, CA.

Sickman, J.O., A. Leydecker, and J.M. Melack. 2001. Nitrogen mass balances and abiotic controls on N retention and yield in high-elevation catchments of the Sierra Nevada, California, United States. *Water Resour. Res.* 37(5):1445–1461.

Sickman, J.O. and J.M. Melack. 1989. Characterization of year-round sensitivity of California's montane lakes to acidic deposition. Report prepared for California Air Resources Board under Contract A5-203-32. University of California, Santa Barbara, CA.

Sickman, J.O., J.M. Melack, and D.W. Clow. 2003. Evidence for nutrient enrichment of high-elevation lakes in the Sierra Nevada, California. *Limnol. Oceanogr.* 48(5):1885–1892.

Sickman, J.O., J.M. Melack, and J.L. Stoddard. 2002. Regional analysis of inorganic nitrogen yield and retention in high-elevation ecosystems of the Sierra Nevada and Rocky Mountains. *Biogeochemistry* 57/58:341–374.

Sparling, D.W., G.M. Fellers, and L.L. McConnell. 2001. Pesticides and amphibian population declines in California, USA. *Environ. Toxicol. Chem.* 20(7):1591–1595.

Stephenson, N. 1999. Reference conditions for giant sequoia forest restoration. *Ecol. Appl.* 9:1253–1265.

Stoddard, J., J.S. Kahl, F.A. Deviney, D.R. DeWalle, C.T. Driscoll, A.T. Herlihy, J.H. Kellogg, P.S. Murdoch, J.R. Webb, and K.E. Webster. 2003. Response of surface water chemistry to the Clean Air Act amendments of 1990. EPA 620/R-03/001. U.S. Environmental Protection Agency, Office of Research and Development, National Health and Environmental Effects Research Laboratory, Research Triangle Park, NC.

Stoddard, J.L. 1995. Episodic acidification during snowmelt of high elevation lakes in the Sierra Nevada Mountains of California. *Water Air Soil Pollut.* 85:353–358.

Stolte, K.W. 1982. The effects of ozone on chaparral plants in the California South Coast Air Basin. MS University of California, Riverside, CA.

Stolte, K.W., M.I. Flores, D.R. Mangis, and D.B. Joseph. 1992. Decreased foliage production and longevity observed in ozone-injured Jeffrey and ponderosa pines in Sequoia National Park, California. *Tropospheric Ozone and the Environment: Effects Modeling and Control.* Air and Waste Management Association, Pittsburgh, PA, pp. 637–662.

Suarez-Murias, T., J. Glass, E. Kim, L. Melgoza, and T. Najita. 2009. California regional haze plan. California Environmental Protection Agency, Air Resources Board, Sacramento, CA.

Sullivan, T.J. 2000. *Aquatic Effects of Acidic Deposition.* CRC Press, Boca Raton, FL.

Sullivan, T.J. and J.M. Eilers. 1994. Assessment of deposition levels of sulfur and nitrogen required to protect aquatic resources in selected sensitive regions of North America. Final Report prepared for Technical Resources, Inc., Rockville, MD, under contract to U.S. Environmental Protection Agency, Environmental Research Laboratory-Corvallis. E&S Environmental Chemistry, Inc., Corvallis, OR.

Sullivan, T.J., T.C. McDonnell, G.T. McPherson, S.D. Mackey, and D. Moore. 2011b. Evaluation of the sensitivity of inventory and monitoring national parks to nutrient enrichment effects from atmospheric nitrogen deposition. Natural Resource Report NPS/NRPC/ARD/NRR—2011/313. U.S. Department of the Interior, National Park Service, Denver, CO.

Sullivan, T.J., G.T. McPherson, T.C. McDonnell, S.D. Mackey, and D. Moore. 2011a. Evaluation of the sensitivity of inventory and monitoring national parks to acidification effects from atmospheric sulfur and nitrogen deposition. Natural Resource Report. NPS/NRPC/ARD/NRR—2011/349. U.S. Department of the Interior, National Park Service, Denver, CO.

Sullivan, T.J., D.L. Peterson, C.L. Blanchard, K. Savig, and D. Morse. 2001. Assessment of air quality and air pollutant impacts in class I national parks of California. U.S. Department of the Interior, National Park Service, Air Resources Division, Denver, CO. Available at: http://www.nature.nps.gov/air/Pubs/pdf/reviews/ca/CAreport.pdf (accessed September, 2016).

Temple, P.J. 1999. Effects of ozone on understory vegetation in the mixed conifer forest. *In* P.R. Miller and J.R. McBride (Eds.). *Oxidant Air Pollution Impacts in the Montane Forests of Southern California.* Springer-Verlag, New York, pp. 208–222.

Temple, P.J., G.H. Reichers, P.R. Miller, and R.W. Lennox. 1993. Growth responses of ponderosa pine to long-term exposure to ozone, wet and dry acidic deposition, and drought. *Can. J. For. Res.* 23:59–66.

U.S. Environmental Protection Agency. 2006. Air quality criteria for ozone and related photochemical oxidants. Volumes I–III. EPA 600/R-05/004 a-cF. U.S. Environmental Protection Agency, Research Triangle Park, NC.

U.S. Environmental Protection Agency. 2008. Integrated science assessment for oxides of nitrogen and sulfur—Ecological criteria. EPA/600/R-08/082F. National Center for Environmental Assessment, Office of Research and Development, Research Triangle Park, NC.

Van Herk, C.M., E.A.M. Mathijssen-Spiekman, and D. de Zwart. 2003. Long distance nitrogen air pollution effects on lichens in Europe. *Lichenologist* 35:347–359.

Van Ooy, D.J. and J.J. Carroll. 1995. Spatial variation of ozone climatology on the western slope of the Sierra Nevada. *Atmos. Environ.* 29:1319–1330.

Vicars, W.C. and J.O. Sickman. 2011. Mineral dust transport to the Sierra Nevada, California. Loading rates and potential source areas. *J. Geophys. Res.* 116:G01018.

Vitousek, P.M., J.D. Aber, R.W. Howarth, G.E. Likens, P.A. Matson, D.W. Schindler, W.H. Schlesinger, and D.G. Tilman. 1997. Human alteration of the global nitrogen cycle: Sources and consequences. *Ecol. Appl.* 7(3):737–750.

Vogler, D.R. 1982. Ozone injury and height growth of planted ponderosa pines on the Sequoia National Forest. Forest Pest Management Report 82-18. USDA Forest Service Pacific Southwest Region, Albany, CA.

Warner, T.E., D.W. Wallner, and D.R. Vogler. 1983. Ozone injury to ponderosa and Jeffrey pines in Sequoia-Kings Canyon National Parks. *In* C. Van Riper, L. D. Whittig and M. L. Murphy, Davis, CA (Eds.). *Proceedings of the First Biennial Conference of Research in California's National Parks*, pp. 1–7.

Weinstein, D.A., J.A. Laurence, W.A. Retzlaff, J.S. Kern, E.H. Lee, W.E. Hogsett, and J. Weber. 2005. Predicting the effects of tropospheric ozone on regional productivity of ponderosa pine and white fir. *For. Ecol. Manage.* 205:73–89.

Whiting, M.C., D.R. Whitehead, R.W. Holmes, and S.A. Norton. 1989. Paleolimnological reconstruction of recent acidity changes in four Sierra Nevada lakes. *J. Paleolimnol.* 2:285–304.

Williams, M.R. and J.M. Melack. 1989. Effects of spatial and temporal variation in snow melt on nitrate ion and sulfate ion pulses in melt waters within an alpine basin. *Ann. Glaciol.* 13:285–288.

Williams, M.W. and J.M. Melack. 1991. Solute chemistry of snowmelt and runoff in an alpine basin, Sierra Nevada. *Water Resour. Res.* 27(7):1575–1588.

Williams, M.W., R.C. Bales, A.D. Brown, and J.M. Melack. 1995. Fluxes and transformations of nitrogen in a high-elevation catchment, Sierra Nevada. *Biogeochemistry* 28:1–31.

Witty, J.H., R.C. Graham, K.R. Hubbert, J.A. Doolittle, and J.A. Wald. 2003. Contributions of water supply from the weathered bedrock zone to forest soil quality. *Geoderma* 114:389–400.

Wright, G., M.S. Gustin, P. Weiss-Penzias, and M.B. Miller. 2014. Investigation of mercury deposition and potential sources at six sites from the Pacific Coast to the Great Basin, USA. *Sci. Total Environ.* 470–471:1099–1113.

Zabik, J.M. 1993. Atmospheric transport and deposition of pesticides from California's Central Valley into the Sierra Nevada Mountains. Masters, University of California, Davis, CA.

Zabik, J.M. and J.N. Seiber. 1993. Atmospheric transport of organophosphate pesticides from California's Central Valley to the Sierra Nevada Mountains. *J. Environ. Qual.* 22:80–90.

18

Rocky Mountain and Glacier National Parks

18.1 Background

Atmospheric deposition of air pollutants represents an important potential threat to aquatic and terrestrial resources in Rocky Mountain (ROMO) and Glacier (GLAC) national parks, especially at higher elevations. Terrestrial resources in ROMO and GLAC are varied and include alpine and subalpine plant communities, boreal forests, wetlands, and meadows. Aquatic resources in these two parks include a wealth of lakes and streams of exceptional quality. The natural lakes and stream valleys were formed by glaciation. The majority of the surface waters in these parks are found in alpine and subalpine settings, most of which are accessible only on foot or horseback. Many high-elevation surface waters are fed by small glaciers. Because of the remoteness of so many surface waters in ROMO and GLAC, some of the human impacts on water quality are minimized. Human population centers around Rocky Mountain Network parks are sparse, except along the eastern edge of the Colorado Front Range (Figure 18.1). ROMO is located within about 25 mi (40 km) of the Colorado Front Range urban corridor, which has the highest human population density in the Rocky Mountain region (Dennehy et al. 1993). This corridor experienced an increase in human population of about 53% from 1980 to 2000 (Porter and Johnson 2007). ROMO has been especially well studied with respect to impacts from air pollutants. With the exception of anthropogenic atmospheric contributions of pollutants and climate change, direct human impacts on most lakes, streams, forests, and alpine areas in these parks are limited. Especially at remote locations, potential impacts from causes other than air pollution and climate change are restricted mainly to a few dams and irrigation channels, as well as the impacts of hiking, camping, and horseback-riding activities.

Atmospheric inputs of S and N are higher at ROMO than at GLAC. More than one-fourth of ROMO is above treeline, which is located at an elevation of about 11,483 ft (3,500 m) in this park. Precipitation increases with elevation, and the higher peaks receive most of their precipitation as snow (Doesken et al. 2003).

A great deal of research has been conducted on the interactions between atmospheric pollutants and water quality in ROMO at an integrated watershed study site at The Loch, a small subalpine lake at the base of the Loch Vale watershed. Biogeochemical and hydrological processes have been studied intensively at Loch Vale since 1983 (e.g., Denning et al. 1991, Baron 1992, Campbell et al. 1995b, Baron and Campbell 1997, Baron et al. 2009). The Loch Vale watershed is a 2.7 mi^2 (7 km^2) basin situated along the Continental Divide in the southeastern portion of ROMO, located 50 mi (80 km) northwest of Denver. The watershed ranges in elevation from about 10,171 to 13,124 ft (3,100 to 4,000 m).

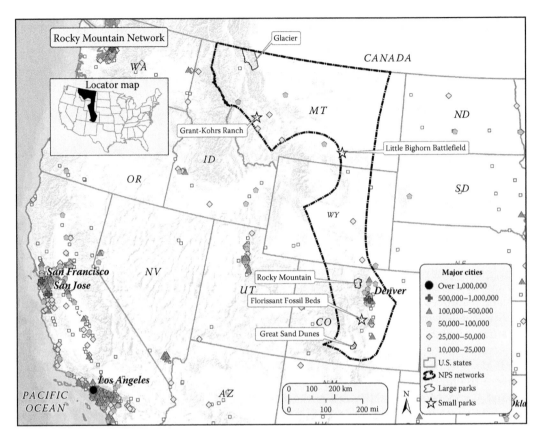

FIGURE 18.1
Network boundary and locations of parks and population centers greater than 10,000 people near the Rocky Mountain Network.

18.2 Atmospheric Emissions and Deposition

18.2.1 Emissions

In general, air quality in the Rocky Mountains is considerably better than in most other areas of the conterminous United States. National parks in the Rocky Mountains receive generally low levels of most atmospheric pollutants (Sisterson et al. 1990, Smith 1990). Nevertheless, sensitive aquatic and terrestrial ecosystems, especially those at high elevations, are degraded by existing pollution. Elevated inputs of nitrogen oxides and elevated concentrations of ozone have been measured (Sisterson et al. 1990, Benedict et al. 2013a,b). ROMO is within the boundaries of Denver's nonattainment area for ozone (https://www.colorado.gov/pacific/sites/default/files/AP_PO_ozone-nonattainment-area-map.pdf). Agricultural activities, including feedlots, are significant sources of reduced nitrogen (ammonia) in some areas. The parks also have impaired visibility at times (Savig and Morse 1998). GLAC and ROMO had two of the three highest average snowpack pesticide concentrations of all western parks included in an EPA survey of air toxics (Landers et al. 2008).

Emissions of N oxides from the western states that lie between ROMO and the west coast decreased about 13% between 1990 and 2001. Emissions in Colorado decreased 8.7%

during that time period. Further decreases are ongoing but may be counteracted to some extent by the relatively recent and pronounced oil and gas development in the region. Federal mobile source standards should reduce mobile source N oxide emissions in the Denver metropolitan area by an additional 71% by the year 2022. Although stationary and area source emissions are projected to increase, total N oxide emissions in the Denver metropolitan area are predicted to decrease between 2001 and 2022.

County-level emissions near the Rocky Mountain Network, based on data from the EPA's National Emissions Inventory during a recent time period (2011), are depicted in Figures 18.2 through 18.4 for sulfur dioxide, N oxide, and ammonia, respectively. Most counties near the network parks had sulfur dioxide emissions in the range of 1–5 tons/mi^2/yr (Figure 18.2). Patterns in N oxide emissions were generally similar, with highest values in the range of 5–25 tons/mi^2/yr (Figure 18.3). Emissions of ammonia near parks were somewhat lower, with most counties showing emissions levels below 2 tons/mi^2/yr (Figure 18.4).

The predominant direction of air mass movement over the Front Range is from west to east (Barry 1973), with periodic upslope movement from the east and southeast (Kelley and Stedman 1980). Wind rose data from ROMO during the period 1989–1995 showed a distinct pattern of predominant air movement from the northwest. However, a second frequent

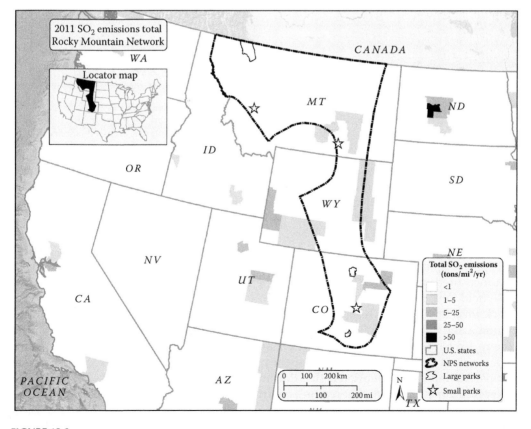

FIGURE 18.2
Total sulfur dioxide (SO$_2$) emissions, by county, near the Rocky Mountain Network for the year 2011. (Data from EPA's National Emissions Inventory, https://www.epa.gov/air-emissions-inventories, accessed January, 2014.)

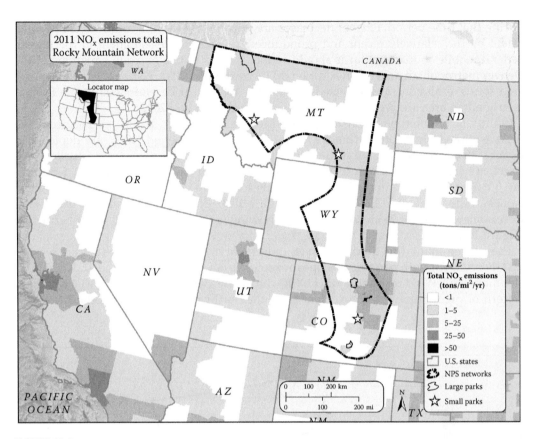

FIGURE 18.3
Total oxidized N (NO$_x$) emissions, by county, near the Rocky Mountain Network for the year 2011. (Data from EPA's National Emissions Inventory, https://www.epa.gov/air-emissions-inventories, accessed January, 2014.)

wind direction was from the south and southeast, from the general direction of the Denver metropolitan area (Peterson et al. 1998). This is important because air masses that move directly from the Denver area to ROMO have the potential to transport high levels of N, S, and ozone-forming compounds to the park. The easterly upslope storm track also carries air across agricultural (livestock and fertilized cropland) and industrial and metropolitan areas of Colorado before reaching the vicinity of ROMO (Bowman 1992). Higher atmospheric concentrations of ammonia, N oxide gases, and nitric acid particulates have been measured near the park during upslope events (Parrish et al. 1986, Langford and Fehsenfeld 1992, Benedict et al. 2013a,b).

Back trajectory analysis has been used to identify major source regions of atmospheric ammonia and sulfur dioxide aerosol emissions that impact Class I national parks in the western United States (Xu et al. 2006, Gebhart et al. 2011, 2014). This approach estimates the amount of time that an air mass spent over each of a group of prespecified source regions. An implicit assumption is that the amount of time that an air mass spends over a region is linearly related to that region's contribution to the pollutant concentration in the air measured at the receptor site location. Results of this modeling study, conducted using 2000–2002 data, suggested that ocean shipping and other port emissions along the Pacific coast contributed substantially to atmospheric pollutant aerosol concentrations over large portions of the western United States, including the Rocky Mountain region. The largest

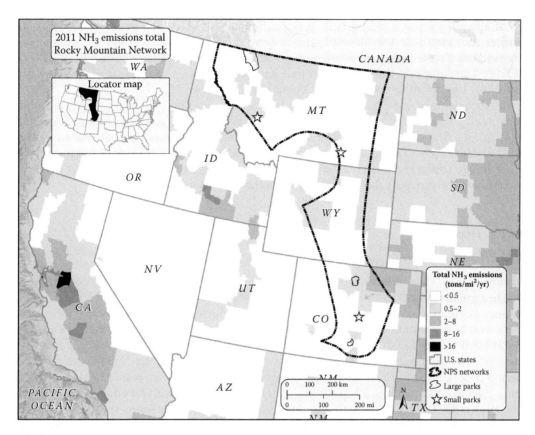

FIGURE 18.4
Total ammonia (NH₃) emissions, by county, near the Rocky Mountain Network for the year 2011. (Data from EPA's National Emissions Inventory, https://www.epa.gov/air-emissions-inventories, accessed January, 2014.)

source areas for atmospheric pollutants in the Rocky Mountain region, however, appeared to be the northwestern United States, northwestern Colorado, southeastern Colorado, and Arizona (Xu et al. 2006, Gebhart et al. 2011, 2014).

Production of oil and gas across the western United States contributes substantial N and volatile organic compound emissions. Oil and gas production accelerated in recent years up to about 2015. The Western Regional Air Partnership has compiled emissions data from these sources, which are individually minor, but collectively substantial. Western Regional Air Partnership inventories identified over 100,000 tons per year of N oxide emissions in the western region, which had not previously been quantified (Gribovicz 2013).

ROMO is located just to the west of large areas of livestock, crop, and pasture lands. Pesticide use in this region is more than 1.3 million kg/yr, mainly applied to corn and hay (Kimbrough and Litke 1998). Dieldrin was produced at facilities along the Colorado Front Range until the 1970s (T. Blett, personal communication, April 2016). The heaviest agricultural pesticide use around GLAC is to the west of the park in central Washington and southern Idaho and also to the northeast in Alberta, Canada (Mast et al. 2006). Pesticides applied to agricultural lands in proximity to these national parks have the potential to volatize, be transported in the atmosphere, and then deposit on sensitive ecosystems.

The highest atmospheric concentrations of ammonia in the vicinity of ROMO are found on the Colorado eastern plains. The highest concentrations of atmospheric oxidized N occur along the Front Range urban corridor (Benedict et al. 2013b). Benedict et al. (2013b) further indicated that the atmospheric concentrations of reactive N species east of the Continental Divide were, on average, more than 50% higher than those found west of ROMO. This supports the conclusion that emissions sources to the east of the park are especially important to the pollutant deposition at ROMO.

Measurements of gas phase and particulate N (reduced and oxidized forms) and other constituents were made at ground level at nine sites located across Colorado (Benedict et al. 2013b). The highest concentrations of ammonia were found in the eastern plains, attributable largely to agricultural sources. Oxidized N concentrations were highest in the Front Range urban corridor. Both of these source areas lie to the east of ROMO. Upslope easterly winds transport these pollutants into the park (Benedict et al. 2013b).

18.2.2 Deposition

18.2.2.1 Sulfur and Nitrogen

Atmospheric N deposition in the Colorado Front Range has increased in recent decades due to a greater prevalence of large industrial animal feedlots and increased urbanization, human population, and distance driven by motor vehicles (Fenn et al. 2003b). Total inorganic wet N deposition at the high elevation National Atmospheric Deposition Program site at Loch Vale in ROMO was estimated to be about 3.1 kg N/ha/yr, based on a five-year average centered on 2001 (Blett and Morris 2004) and about 2.9 kg N/ha/yr during the period 2008–2012 (Morris et al. 2014). Wet inorganic N deposition at Loch Vale increased from 1985 to 2000 by about 2% per year, largely due to increased wet ammonium deposition (Burns 2003, Clow et al. 2003). Dry inorganic N deposition is less certain, with an estimate based on the Clean Air Status and Trends Network equal to 0.8 kg N/ha/yr (Blett and Morris 2004). This estimate is probably biased low (Porter and Johnson 2007) because the Clean Air Status and Trends Network monitoring site is located at lower elevation than the National Atmospheric Deposition Program/National Trends Network site that measures wet deposition and because Clean Air Status and Trends Network does not include the measurement of all N species.

Total estimated S deposition in 2002 was generally less than 2 kg S/ha/yr throughout the Rocky Mountain region, with scattered small areas estimated to receive up to 5 kg S/ha/yr (Sullivan et al. 2011). Wet S and N deposition hot spots in northern Colorado were probably due in large part to emissions of sulfur dioxide and N oxide from coal-fired power plants in northwestern Colorado and southwestern Wyoming (Turk and Campbell 1997, Mast et al. 2001, Nanus et al. 2003). Total N deposition within the network region ranged from less than 2 kg N/ha/yr to as high as 5–10 kg N/ha/yr, especially along the Front Range and at higher elevations. Throughout most of the network region, estimated total N deposition was in the range of 2–5 kg N/ha/yr (Sullivan et al. 2011).

Atmospheric S deposition levels have decreased at ROMO and GLAC since 2001 while N deposition levels have increased, based on Total Deposition project estimates (Table 18.1). Decreases in total S deposition over the previous decade were less than 20% for these parks. Oxidized and reduced N showed opposite patterns, with oxidized N decreasing and reduced N increasing since the monitoring period 2000–2002.

Total S deposition in and around the Rocky Mountain Network for the period 2010–2012 was generally less than 2 kg S/ha/yr at park locations within the network area. Oxidized inorganic N deposition for the period 2010–2012 was less than 4 kg N/ha/yr throughout the parklands within the network. Most areas received less than 4 kg N/ha/yr of reduced

TABLE 18.1

Average Changes in S and N Deposition[a] between 2001 and 2011 across Park Grid Cells at ROMO and GLAC

Park Code	Parameter	2001 Average (kg/ha/yr)	2011 Average (kg/ha/yr)	Absolute Change (kg/ha/yr)	Percent Change	2011 Minimum (kg/ha/yr)	2011 Maximum (kg/ha/yr)	2011 Range (kg/ha/yr)
GLAC	Total S	1.75	1.42	−0.33	−18.4	0.69	2.24	1.54
	Total N	3.59	3.70	0.11	3.8	1.84	5.47	3.63
	Oxidized N	2.28	2.03	−0.26	−10.9	1.04	2.92	1.88
	Reduced N	1.31	1.68	0.37	30.7	0.80	2.59	1.79
ROMO	Total S	1.79	1.62	−0.17	−7.3	1.08	2.20	1.12
	Total N	5.53	7.27	1.74	34.5	4.19	11.98	7.79
	Oxidized N	3.37	2.78	−0.60	−16.8	1.91	3.54	1.63
	Reduced N	2.16	4.49	2.33	115.0	2.21	8.97	6.76

[a] Deposition estimates were determined by the Total Deposition project, based on three-year averages centered on 2001 and 2011 for all ~4 km grid cells in each park. The minimum, maximum, and range of 2011 S and N deposition within each park are also shown.

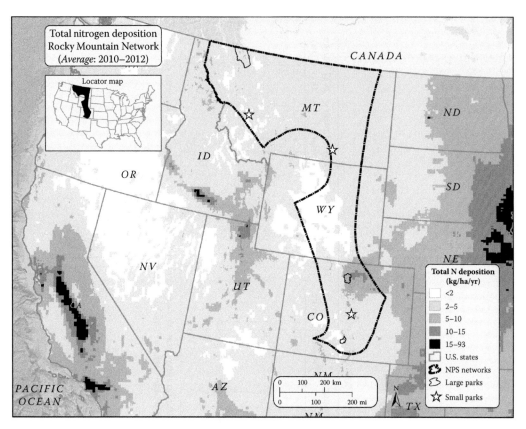

FIGURE 18.5
Total N deposition for the three-year period centered on 2011 in and around the Rocky Mountain Network. (From Schwede, D.B. and Lear, G.G., *Atmos. Environ.*, 92, 207, 2014.)

inorganic N from atmospheric deposition during this same period. Estimated total N deposition was higher than 5 kg N/ha/yr at ROMO but generally lower elsewhere within the Network (Figure 18.5; Table 18.1).

In the Rocky Mountains, deposition chemistry is influenced by a complex collection of emissions sources. Some parks are subject to deposition of pollutants from urban areas, some are more heavily affected by long-distance transport of pollutants from industrial facilities and electric utilities, and some are affected by local sources, such as feedlots and oil and gas development. Therefore, the quantity of emissions received and the potential threat to natural resources differ within the Rocky Mountain region (Peterson et al. 1998). For example, in the Mt. Zirkel Wilderness of northwestern Colorado, sulfate and nitrate in the snow appeared to originate mainly from large power plant emissions sources in the Yampa Valley, about 75 km to the west (Turk et al. 1992), whereas ROMO is more affected by emissions from the Front Range to the southeast of the park. A comparison of winter snowpack and National Atmospheric Deposition Program/National Trends Network precipitation chemistry in Colorado (Heuer et al. 2000) suggested that high-elevation watersheds in ROMO are influenced by air pollution sources located on both sides of the Continental Divide. Concentrations of nitrate and sulfate in the snowpack were higher on the eastern slope during winter, but there were no significant differences between east and west in snowpack ammonium concentrations. During summer, National Atmospheric

Deposition Program/National Trends Network precipitation chemistry showed that concentrations of nitrate and ammonium were higher on the eastern slope, with no significant difference for sulfate. For all three solutes, summer concentrations in precipitation were about twice as high as winter concentrations. Therefore, annual patterns in wet deposition are more strongly driven by summer conditions (Heuer et al. 2000). Winter precipitation in Colorado originates mainly from the northwest (Hansen et al. 1978, Parrish et al. 1990). During spring and summer, there are often upslope movements of air from the Denver-Boulder-Fort Collins urban corridor to the vicinity of ROMO (Parrish et al. 1990).

The Rocky Mountain Atmospheric Nitrogen and Sulfur Study was conducted to improve the understanding of S and N deposition to ROMO and their sources. The study involved two five-week sampling periods in 2006, during spring and summer. These seasons typically had relatively high levels of S and N deposition at ROMO. During spring, large upslope events transported pollutants to the park from the east. During summer, frequent (almost daily) afternoon precipitation due to convection activity brought increased wet deposition.

A weight of evidence approach was used in the Rocky Mountain Atmospheric Nitrogen and Sulfur Study to apportion deposited S and N to source regions. It included both simple qualitative spatial data analysis and quantitative hybrid receptor modeling. Results indicated that the pollutant concentrations generally increased in ROMO when air transport was from the east. The final inorganic N deposition apportionment budget indicated that the source areas responsible for most N deposition in ROMO during spring were northeastern Colorado (~40%) and the Denver/Front Range region (~25%). In contrast, during summer, N source regions were more varied, with about 25% from southwesterly transport, 20% from local source areas, 13% from western Colorado, and 8% from the greater Denver area (Malm et al. 2009).

A detailed N deposition budget was calculated by Benedict et al. (2013a) at the Loch Vale watershed in ROMO over a one-year period from November 2008 to November 2009. The estimated total reactive N deposition was 3.65 kg N/ha/yr, with highest deposition in July and lowest deposition in January. The two largest contributors were wet deposition of nitrate and ammonium, which together accounted for 54% of the total. The next two largest contributors were wet deposition of organic N and dry deposition of ammonium (combined 37% of total). These latter two deposition pathways are rarely included in deposition budgets, including those used to assess critical loads and their exceedances (Benedict et al. 2013a).

Snow accounts for about 50%–80% of total precipitation at high-elevation snowpack telemetry sites in Colorado. However, nitrate and ammonium concentrations in snow at Loch Vale in ROMO are only about half of the concentrations of nitrate and ammonium in rain (Fenn et al. 2009). Thus, both snow and rainfall are important contributors to wet nitrate deposition in ROMO.

Based on assumed preindustrial background atmospheric N deposition of 0.5 kg N/ha/yr in 1900 and two decades of measured wet N deposition at the Loch Vale deposition monitoring site, Baron (2006) reconstructed the N deposition history of ROMO, based on correlation with N oxide emissions data for Colorado and 10 other western states. The estimated mean annual wet N deposition for the period 1950–1964 was about 1.5 kg N/ha/yr. This time period corresponded with the alteration of lake diatom assemblages in ROMO that were attributed to N deposition (Baron 2006).

Retention of S in alpine watersheds of ROMO, such as Loch Vale, is limited (Michel et al. 2000). Total S losses in drainage water from the Loch Vale basin were considerably higher than atmospheric wet S inputs and ranged from 3.3 to 4.2 kg S/ha/yr (Baron et al. 1995). This information, coupled with the discovery of small pyrite deposits within the basin, suggests a significant mineral source of S in the Loch Vale basin (Mast et al. 1990). Interpretation of potential ecosystem effects of decreased S emissions and deposition since 1984 is obscured by these internal watershed sources of S (Baron et al. 1995).

Grenon and Story (2009) reported increasing trends in wet deposition of ammonium within at least one season of the year since the 1980s for all monitoring sites in the northern Rocky Mountains region. This result agreed with the analyses reported by Ingersoll et al. (2008) and Lehmann et al. (2005). This trend of increasing ammonium wet deposition has occurred across much of the western United States, likely due in part to increased agricultural emissions of ammonia. The concentrations of nitrate in precipitation at monitoring sites in the Rocky Mountains have also generally increased over time, especially near urban areas (Fenn et al. 2003b, Lehmann et al. 2005, Morris et al. 2014). In contrast, wet S deposition decreased between 1985 and 2002 at National Atmospheric Deposition Program/National Trends Network sites throughout the western United States (Lehmann et al. 2005), likely due to reduced power plant emissions.

18.2.2.2 Toxics

Various studies have identified organochlorine compounds in remote arctic or alpine ecosystems (cf., Heit et al. 1984, Simonich and Hites 1995, Wania and Mackay 1996, Blais 2005). Compounds detected commonly include DDT, hexachlorocyclohexane, hexachlorobenzene, and PCBs. Such compounds tend to concentrate in aquatic biota in response to biomagnification processes and have been associated with disruption of endocrine systems in aquatic species. Based on the results of earlier studies in remote mountainous areas of Europe (Grimalt et al. 2001) and Canada (Blais et al. 1998), Mast et al. (2006) suggested that some remote areas of the Rocky, Cascade, and Sierra Nevada mountains might be susceptible to organochlorine compound accumulation due to low temperatures, high precipitation, and proximity to agricultural and urban source areas.

Because relatively little was known about atmospheric deposition of organochlorines to these sensitive ecosystems, the U.S. Geological Survey, in cooperation with the National Park Service, conducted a baseline study in ROMO and GLAC (Mast et al. 2006). They measured organochlorine compounds and current-use pesticides in snow and lake sediment samples. Measurements of snow chemistry can provide important information regarding atmospheric deposition. The pesticides most frequently found in snow samples were endosulfan, dacthal, and chlorothalonil. In sediment samples, low concentrations of DDT breakdown products were commonly found. DDT has been banned for use in the United States since 1972. Sediment core measurements for Mills Lake in ROMO indicated that atmospheric deposition of DDT to high-elevation ecosystems in ROMO has been decreasing since the 1970s. Dacthal and endosulfan were detected in nearly all snow samples analyzed, however, confirming that these current-use pesticides are deposited in high-elevation ecosystems and accumulate in lake sediments (Mast et al. 2006).

The WACAP studied airborne contaminants in a number of national parks in the western United States, including ROMO and GLAC from 2002 to 2007 (Landers et al. 2008). In ROMO, the Mills Lake watershed on the east side of the Continental Divide had higher semivolatile organic compounds and Hg fluxes in snow and lake sediment, as compared with the Lone Pine Lake watershed on the west side. Air monitors on either side of the Continental Divide did not detect obvious differences in semivolatile organic compound concentrations between the east and the west. Dominant semivolatile organic compounds in Rocky Mountain parks were polycyclic aromatic hydrocarbons, dacthal, endosulfans, hexachlorobenzene, hexachlorocyclohexane-α, and hexachlorocyclohexane-γ.

Atmospheric input of toxic materials to high-elevation ecosystems has been evaluated through the study of snowpack chemistry. Hageman et al. (2010) analyzed snowpack samples collected from 56 remote alpine and arctic locations in eight western national parks, including ROMO and GLAC, during the period 2003–2005. Four current-use pesticides were commonly measured at all sites and years: dacthal, chlorpyriphos, endosulfans, and hexachlorocyclohexane-γ. The relative pesticide concentration profiles were consistent from year to year, but unique for individual parks, indicating the importance of regional sources. The historic-use pesticides were strongly correlated with regional cropland presence. The amount of current-use pesticides used in the regions located one day upwind, based on mass back trajectory analysis, helped to explain the distribution of current-use pesticides among the study parks.

18.3 Acidification

There are many high-elevation lakes and first-order streams in GLAC and ROMO. Nearly all streams in these parks are first through third order and occur on steep terrain. The vast majority are first-order streams, with relatively small drainage areas. Such streams tend to be more likely to be sensitive to acidification than the larger, higher-order streams found at lower elevation.

Within the Rocky Mountain Network, episodic acidification attributable to atmospheric S and N deposition is of concern primarily in ROMO. Moderate levels of N deposition, coupled with high watershed sensitivity to acidification, may have contributed to some episodic acidification of surface waters in this park at existing levels of N deposition. However, neither episodic nor chronic acidification has been documented in the park (Sullivan et al. 2005).

The National Acid Precipitation Assessment Program State of Science and Technology Reports and the Integrated Assessment (NAPAP 1991) concluded that high-elevation areas of the West contained many of the watersheds most sensitive to the potential effects of acidic deposition in the United States. These highly sensitive surface waters are especially numerous in ROMO and elsewhere within the Colorado Front Range. The acid–base chemistry of lake and stream waters in these areas appears to be primarily a function of the interactions among several key parameters and associated processes: atmospheric deposition; bedrock geology; the depth and composition of surficial deposits and associated hydrologic flowpaths; and the occurrence of soils, tundra, and forest vegetation (Sullivan 2000). High-elevation areas in the Rocky Mountains can contain extensive areas of exposed bedrock and alpine meadows, with little soil or vegetative cover to neutralize acidic inputs. Sensitivity to adverse effects from acidic deposition in the Rocky Mountains varies widely. The individual ranges that comprise the Rocky Mountains are discontinuous, with highly variable geological composition. Lakes having lowest acid neutralizing capacity occur in clusters, such as in the Bitterroot Mountains in Montana and the Wind River Range in Wyoming (Schoettle et al. 1999). For that reason, assessments of the sensitivity of Rocky Mountain aquatic resources to acidification should be specific to individual ranges (Turk and Spahr 1991, Sullivan and Eilers 1994). For example, high concentrations of base cations, acid neutralizing capacity, and silica occur in the upper Colorado River basin, a portion of ROMO underlain by highly weatherable ash-flow tuff and andesite. In contrast, the acid neutralizing capacity and base cation concentrations are much lower in Glacier Creek, a watershed underlain by Silver Plume granite (Gibson et al. 1983).

Nanus et al. (2009) described the sensitivity of high mountain lakes to acidification in GLAC, ROMO, and Great Sand Dunes National Park within the Rocky Mountains and of Yellowstone and Grand Teton National Parks within the Greater Yellowstone Network. They developed statistical relations between lake acid neutralizing capacity and basin characteristics for 151 surveyed lakes. Results were confirmed through lake sampling in 2004 at 58 lakes. Lakes having acid neutralizing capacity below 100 μeq/L were generally located in watersheds above 9843 ft (3000 m) elevation with <30% of the watershed area having northeast aspect and with >80% of the watershed area having bedrock with low buffering capacity. The most acid-sensitive lakes were generally located in ROMO and Grand Teton national parks (Nanus et al. 2009).

Data from over 12,500 streams and lakes were used by the Critical Loads of Atmospheric Deposition Science Committee of the National Atmospheric Deposition Program (http://nadp.sws.uiuc.edu/committees/clad/) to develop steady-state critical loads for acidity of surface waters based on multiple approaches for estimating base cation weathering. Water quality data were obtained from a variety of sources. The Critical Loads of Atmospheric Deposition Science Committee database included 9 sites in GLAC and 26 sites in ROMO. The critical load estimates were as low as 0 eq S/ha/yr at ROMO but were above 377 eq S/ha/yr at all modeled sites in GLAC.

18.3.1 Studies Conducted in ROMO

Most lakes sampled in the Rocky Mountain region as part of the EPA's Western Lakes Survey had nitrate concentrations near 0 μeq/L during fall (Landers et al. 1987). However, fall nitrate concentrations were high in some areas in the Rocky Mountains. For example, nearly one fourth of the lakes in northwestern Wyoming had nitrate >5 μeq/L. In the Colorado Rocky Mountain subregion, about 10% of the lakes had fall nitrate concentrations above 5 μeq/L.

Analyses of fall samples from 22 lakes in ROMO and 14 lakes in adjacent wilderness areas (Eilers et al. 1988) showed that the median acid neutralizing capacity for these lakes was 80 μeq/L, with 20% of the lakes having acid neutralizing capacity <41 μeq/L. The minimum acid neutralizing capacity value measured in this subpopulation was 19 μeq/L. Minimum pH and base cation values were 6.48 and 47 μeq/L, respectively. Sulfate concentrations ranged from 10 to 113 μeq/L. The wide range in sulfate concentrations suggested that some lakes in ROMO may receive S inputs from geologic sources rather than purely atmospheric inputs (e.g., Turk and Spahr 1991). Nitrate concentrations ranged from 0 to 16 μeq/L in the Front Range lakes, with a population-weighted mean of 4 μeq/L.

Musselman et al. (1996) reported the results of a synoptic survey of surface water chemistry in the mountainous areas along the eastern edge of the Continental Divide in Colorado and southeastern Wyoming that are exposed to relatively high (by Rocky Mountain standards) deposition of N. A total of 267 high-elevation lakes situated in watersheds having high percentage of exposed bedrock or glaciated landscape were selected for sampling. None of the lakes were chronically acidic (acid neutralizing capacity <0 μeq/L), although several had acid neutralizing capacity <10 μeq/L, and more than 10% of the lakes had acid neutralizing capacity <50 μeq/L.

In 1999, Clow et al. (2003) resampled 69 of the lakes located within western national parks that had been surveyed earlier in the Western Lakes Survey. In general, lake sulfate concentrations decreased between 1985 and 1999, in response to regional decreases in sulfur dioxide emissions and S deposition. In addition, lake nitrate concentrations were slightly lower in 1999, probably at least in part because rain prior to the 1985 survey may have caused elevated nitrate concentration in some lakes (Clow et al. 2003). This finding reinforced the idea that changes in precipitation should be included in an assessment of water chemistry trends (cf., Bayley et al. 1992, Webster and Brezonik 1995). Lake acid neutralizing capacity was relatively consistent between the two surveys in most parks. However, acid neutralizing capacity increased in lakes sampled in ROMO because concentrations of strong acid anions (sulfate and nitrate) decreased while base cations remained stable.

In order to understand the response of watersheds in ROMO and elsewhere in the Front Range to atmospheric N deposition, it is important to consider a variety of hydrologic and biogeochemical processes. Campbell et al. (1995b) studied the water chemistry of the two major tributaries to The Loch: Andrews Creek and Icy Brook. The catchments for the two streams are entirely alpine, consisting of rock outcrops, talus slopes, and some tundra. Only 5%–15% of the catchments are covered by well-developed soil. Total storage of soil water was estimated to be less than 5% of the total outflow at The Loch (Baron and Denning 1992). Volume-weighted mean annual concentrations of nitrate in the streams were 21 and 23 µeq/L, respectively. Total N export was approximately equivalent to atmospheric inputs, assuming about 25% evapotranspiration. Nitrate concentrations in individual samples of stream water ranged from 12 µeq/L in late summer to 39 µeq/L during snowmelt. Given the steep slopes, porosity of soil, and high elevation, it is not surprising that the Loch Vale watershed leaches relatively high amounts of nitrate under only moderate levels of N deposition.

Baron et al. (2009) explored the relationships between recent temperature warming and stream nitrate concentrations at Loch Vale. The mean annual nitrate concentration since 2000 was about 50% higher than during the period 1991–1999. Precipitation was below average between 2000 and 2006, and both summer and fall air temperature increased steadily since 1991. Baron et al. (2009) concluded that recent increases in stream nitrate concentrations at Loch Vale were caused largely by warmer summer and fall temperatures that are melting glaciers in the watershed.

The Loch Vale watershed can be considered N-saturated (e.g., Aber et al. 1989, Stoddard 1994). It is not clear to what extent the terrestrial and aquatic systems are receiving N inputs in excess of the assimilative capacities of watershed biota, however. The apparent N saturation may be hydrologically mediated. Hydrologic flowpaths and brief soil water residence times may limit the opportunity for the biological uptake of N. Ecosystems may be N-limited but still be unable to utilize the atmospheric inputs of N (Baron et al. 1994, Campbell et al. 1995b). The implications of this apparent N saturation are important for the estimation of critical loads of N deposition (Williams et al. 1996b).

A number of factors predispose watersheds in ROMO, including Loch Vale, and elsewhere throughout the Front Range to potential acidification in response to N deposition. These include

- Steep watershed gradient
- Short hydrologic residence time of lake waters
- Large input of N to lakes and streams during the early phases of snowmelt
- High percentage of watershed covered by exposed bedrock and talus; small percentage of watershed covered by forest
- Phosphorus limitation of aquatic ecosystem primary production in some surface waters

Baron et al. (1986) investigated metal and diatom stratigraphy and inferred pH profiles of four subalpine lakes in ROMO. They found no evidence of historical influence on pH attributable to atmospheric deposition at that time. Other paleolimnological studies of Rocky Mountain lakes reported similar results: stratigraphy of metals (primarily lead) exhibited temporal dynamics related to the increase and decline of precious metal mining in the region, but these were asynchronous with other metal or biological indicators of acidification (Wolfe et al. 2003). Both the study by Wolfe et al. (2003) and a study by Saros et al. (2003) showed no paleolimnological evidence of chronic acidification of high-elevation Rocky Mountain lake waters over time but did show evidence of eutrophication from atmospheric N deposition.

The DayCent-Chem model was used to project a timeline to acidification for an alpine watershed in ROMO (Hartman et al. 2005). At ambient levels of N deposition of 4–6 kg N/ha/yr, acidification was not projected to occur over 48 years of simulation. Higher assumed N deposition contributed to simulated episodic acidification over time at a future deposition level of 7.0–7.5 kg N/ha/yr (Hartman et al. 2005). Simulation results from the Model of Acidification of Groundwater in Catchments suggested that a sustained N deposition load of 12.2 kg N/ha/yr would be required over a period of 50 years to cause chronic acidification of the Andrews Creek watershed in ROMO to acid neutralizing capacity = 0 μeq/L (Sullivan et al. 2005).

18.3.2 Studies Conducted in GLAC

The central portion of GLAC is dominated by two mountain ranges that run northwest to southeast and contain many small glaciers and snowfields. Extensive portions of both ranges lie above timberline (about 6,560 ft [2,000 m]), and many of the peaks extend above 9,186 ft (2,800 m). Elevation ranges from about 3,115 ft (950 m) along the western boundary to 10,465 ft (3,190 m) in the central mountains and back down to about 1,370 m along the eastern boundary. Glaciers provide substantial amounts of base cations to many drainage waters in this park. This may change in the future if glaciers in the park continue to melt in response to climate warming.

Many of the lakes and streams within GLAC have characteristics generally associated with acid sensitivity. They tend to be high in elevation, with little or no soil development in their watersheds, have steep slopes and flashy hydrology, and are hydrologically dominated by spring snowmelt. The majority of these surface waters, however, are actually not at all sensitive to acidification from acidic deposition, due in part to the preponderance of glaciers within their watersheds. Glaciers contribute buffering in the form of base cations (Ca, Mg, Na, K) to drainage waters in GLAC in sufficient quantity to neutralize any amount of sulfate and nitrate that might be reasonably expected to be deposited from acidic deposition (Peterson et al. 1998). There are some waters in the park that receive only modest contributions of base cations, however. These do not receive glacial meltwater contribution to any significant extent and are situated in watersheds with relatively inert bedrock. These sensitive waters are relatively rare within GLAC. However, analyses have not been conducted to date to determine the potential effects of glacial melting on the acid-base chemistry of surface waters in GLAC.

The EPA's Western Lakes Survey sampled 5 lakes in GLAC and 10 other lakes in surrounding areas in the fall of 1985 (Landers et al. 1987). Measured values of selected important physical and chemical variables are listed in Table 18.2 for these 15 lakes and their watersheds. The lowest pH value measured in the park was 7.1, although three of the lakes in surrounding areas had pH between 6.5 and 7.0. One of the lakes having lowest pH (6.6) contained significant natural organic acidity (dissolved organic carbon = 10 mg/L); the others were low in pH as a consequence of their dilute chemistry. Sulfate concentrations

TABLE 18.2

Results of Lake Water Chemistry Analyses by the Western Lakes Survey for Selected Variables in GLAC and Adjacent Areas

Lake Name	Lake ID	Lake Area (ha)	Watershed Area (ha)	Elevation (m)	pH	Acid Neutralizing Capacity (µeq/L)	Sulfate (µeq/L)	Nitrate (µeq/L)	Calcium (µeq/L)	Sum of Base Cations (µeq/L)	Dissolved Organic Carbon (mg/L)
Lakes within GLAC											
No Name	4C3-004	2.8	88	1930	8.1	1210	28.6	11.6	929	1212	0.4
Feather Woman L.	4C3-010	3.7	44	2298	7.3	79	5.7	4.9	49	81	0.3
Harrison L.	4C3-011	162.0	5475	1126	8.0	543	32.1	4.5	375	613	0.5
Cobalt L.	4C3-013	4.4	62	2003	7.1	83	10.1	0.5	50	93	0.7
Glenns L.	4C3-062	104.8	4302	1482	8.1	1142	48.7	3.0	774	1210	0.7
Lakes Outside GLAC											
	4C3-053	3.7	20	1979	6.6	21	8.0	0.1	29	57	1.5
	4C3016	5.4	88	1932	8.1	1388	20.1	0.0	1096	1393	1.0
	4C3-017	6.2	173	1828	8.0	1209	32.3	0.7	832	1218	0.8
	4C3-021	1.8	108	2104	7.9	492	27.4	1.5	768	1092	0.3
	4C3-022	8.7	51	2050	7.5	360	10.0	2.1	295	386	0.8
	4C3-026	167.9	2188	1229	8.3	1409	52.4	0.1	965	1435	4.5
	4C3-055	2.3	7	1921	7.6	426	8.0	0.7	360	453	4.3
	4C3-060	12.7	77	1228	6.6	152	3.0	0.0	81	213	10.0
	4C3-031	6.4	54	2006	6.9	72	8.2	0.1	40	78	1.2
	4C3-059	1.8	31	2226	7.4	292	10.5	4.3	170	319	2.7

Source: Peterson, D.L. et al., Assessment of air quality and air pollutant impacts in national parks of the Rocky Mountains and Northern Great Plains, NPS D-657, U.S. Department of the Interior, National Park Service, Air Resources Division, Lakewood, CO, 1998.

in lake water were very low in lakes having low base cation concentrations. For example, the four lakes with total base cation concentrations less than 100 µeq/L all had sulfate concentrations in the range of 5.7–10.1 µeq/L. Such concentrations of sulfate are approximately what would be expected, based on sulfate concentration in the precipitation at the time of lake sampling (~6 to 8 µeq/L), negligible dry deposition of S, and 30%–50% evapotranspiration. These lakes are moderately to highly acid sensitive, with acid neutralizing capacity values of 21–84 µeq/L, although the two most sensitive were located outside the park boundaries. Many other lakes inside and outside the park had moderately elevated concentrations of sulfate in the range of 20–52 µeq/L. These relatively high concentrations of sulfate are not attributable to atmospheric S deposition. They reflect geological sources of S in drainage waters, as also evidenced by the much higher concentrations (>500 µeq/L) of base cations in all of the lakes that had sulfate concentration greater than 20 µeq/L. Based on these data, it appears that GLAC and surrounding areas contain lakes that exhibit a mixture of acid sensitivities. Some lakes that have low concentrations of sulfate (less than about 10 µeq/L) that can be reasonably attributed to atmospheric deposition inputs also have low concentrations of base cations. These lakes tend to be relatively acid sensitive, and many have acid neutralizing capacity values below 100 µeq/L. The lowest measured acid neutralizing capacity in the park was 79 µeq/L. Other lakes are characterized by higher concentrations of base cations and sulfate of geologic origin. These lakes are not acid sensitive and have acid neutralizing capacity values greater than 500 µeq/L. Nitrate concentrations were generally below 5 µeq/L in Western Lakes Survey lakes sampled in GLAC. One lake exhibited relatively high nitrate concentration (11.6 µeq/L), but this lake was not acid sensitive (Peterson et al. 1998, Sullivan 2000).

Ellis et al. (1992) monitored the water quality of eight small backcountry lakes and five large front country lakes in GLAC. The objective was to establish a water quality baseline for the park. Data were collected during the period 1984–1990. The backcountry lakes were located in remote alpine headwater areas across the various regions and geologies of the park. Three of the lakes (Cobalt, Snyder, and Upper Dutch) had acid neutralizing capacity less than about 200 µeq/L. Cobalt Lake had the lowest acid neutralizing capacity (~100 µeq/L on average) and specific conductance (~10 µS/cm) of the study lakes and would be expected to be sensitive to episodic effects of acidic deposition if S or N deposition increased substantially in the future. Cobalt Lake is situated in the southeastern portion of GLAC at an elevation of 6560 ft (2000 m). It lies immediately below a very steep alpine ridge and therefore receives runoff that has limited contact with geological materials prior to entering the lake. Based on the analysis of nine samples, mean sulfate concentration was 9.6 µeq/L and mean sum of base cation concentration was 85.2 µeq/L. These values suggest that watershed sources of S were not significant and that there were insufficient base cations to neutralize substantial amounts of acidic deposition (Peterson et al. 1998, Sullivan 2000).

18.3.3 Studies of Potential Episodic Acidification of Aquatic Ecosystems

During hydrologic episodes, which are driven by rainstorms and/or snowmelt events, both discharge and water chemistry change, sometimes substantially. These processes have been well studied in ROMO.

Available data from intensive study sites in the Rocky Mountain region (e.g., Loch Vale in ROMO and the Glacier Lakes Watershed, Wyoming) suggested that episodic depression of stream pH may be more pronounced than for lakes. However, there are no available systematic regional stream chemistry data with which to assess regional sensitivity of streams to acidic deposition in the Rocky Mountain Network. Spatial variability in water chemistry can be considerable in lakes, and this complicates efforts to quantify the magnitude of episodic effects (Gubala et al. 1991). Moreover, synoptic lake surveys are typically conducted during the summer or autumn "index period," during which time lake water chemistry exhibits relatively low temporal variability. Acid anion concentrations in most Rocky Mountain lakes are low during fall but can be higher during snowmelt (Williams and Melack 1991). Although summer or autumn is an ideal time for surveying lake water chemistry in terms of minimizing variability, samples collected under low-flow conditions provide little relevant data on episodic processes and, in particular, on the dynamics or importance of N as a cause of acidification. Nitrate concentrations in lake water are elevated during the autumn season only in lakes having watersheds that exhibit fairly advanced symptoms of N saturation (Stoddard 1994, Sullivan 2000).

In ROMO, episodic acidification is a concern for surface waters throughout high-elevation areas. In the Green Lakes Valley, located below Niwot Ridge near ROMO, atmospheric N deposition has been shown to contribute to episodic stream water acidification (Williams and Tonnessen 2000). Episodic effects probably occur to a limited extent in ROMO.

Based on measurements of microbial biomass, carbon dioxide flux through the snowpack, and soil N pools, Williams et al. (1996a) concluded that N cycling under the snowpack in Colorado during the winter and spring was sufficient to supply the nitrate measured in stream waters. Brooks et al. (1996) investigated soil N dynamics throughout the snow-covered season on Niwot Ridge. Sites with consistent snow cover had a 3–8 cm layer of thawed soil under the snowpack for several months before snowmelt began. Nitrogen mineralization in this thawed layer contributed N to reactive N pools that were significantly larger than the pool of N stored in the snowpack. As snowmelt began, soil inorganic N pools decreased sharply, concurrent with a large increase in microbial biomass N. As snowmelt continued, both microbial N and soil inorganic N decreased, presumably due to increased demand by growing vegetation (Brooks et al. 1996).

Baron and Bricker (1987) documented episodic pH and acid neutralizing capacity depressions during snowmelt in Loch Vale during three successive years, but surface waters did not become acidic. Similarly, Denning et al. (1991) showed a dramatic decline in the acid neutralizing capacity of The Loch between mid-April and mid- to late May in 1987 and 1988 to acid neutralizing capacity values as low as 28 µeq/L (and pH around 6.2). This was in spite of the fact that meltwater acid neutralizing capacity dropped to between 0 and −10 µeq/L for extended periods during snowmelt (Denning et al. 1991). If a large component of the snowmelt had been transported directly to surface waters, the latter would have become acidic during snowmelt. Because surface water does not become acidic or exhibit the low pH of meltwater (often 4.8–5.0), direct pathways from the snowpack to the streams are not dominant (Denning et al. 1991) or are offset by more alkaline drainage from watershed soils.

Williams et al. (1996a) sampled 53 ephemeral streams in the Green Lakes Valley of Colorado during snowmelt in 1994. They also sampled an additional 76 sites from the central Colorado Rocky Mountains to the Wyoming border in 1995. Nitrate concentrations in stream water during snowmelt ranged to 44 µeq/L in the Green Lakes Valley and during the growing season ranged to 23 µeq/L in the regional sampling conducted in 1995. Tundra areas had significantly lower nitrate concentrations than talus and bedrock areas, suggesting that tundra ecosystems were still N-limited (Sullivan 2000) and that nitrification combined with limited plant uptake accounted for the high concentrations of nitrate observed in waters draining talus and bedrock areas (Williams et al. 1996a).

Brooks et al. (1995) reported the inputs to the soil inorganic N pool at Niwot Ridge due to mineralization and nitrification (17–20 kg N/ha) under deep snowpack that were an order of magnitude higher than inputs directly from snowmelt (~1.5 kg N/ha). Mineralization varied with severity of the freeze and the length of time the soils were insulated by snowpack. Mineralization was often higher under deeper, earlier-accumulating snowpacks. Under shallower, late-accumulating snowpacks, mineralization was lower and more variable (5–15 kg N/ha, Brooks et al. 1995). The severity with which the soils freeze may be an important determining factor of the amount of N mineralization. Highest mineralization inputs were found under a shallow snowpack that experienced a severe freeze, followed by an extended period of snow cover.

Time-intensive discharge and chemical data for two alpine streams in Loch Vale watershed identified strong seasonal control on stream water nitrate concentrations (Campbell et al. 1995b). In spite of the paucity of soil cover, the chemical composition of streams was regulated much as in typical forested watersheds. Soils and other shallow groundwater matrices such as boulder fields appeared more important in controlling surface water chemistry than their spatial coverage would indicate (Campbell et al. 1995a). Spring stream water nitrate concentrations were as high as 40 µeq/L, compared with summer values near 10 µeq/L. Elution of acidic waters from snowpack along with dilution of base cations originating in shallow groundwater caused episodes of decreased acid neutralizing capacity in the alpine streams (Campbell et al. 1995b).

It does not appear that any appreciable amount of chronic acidification has occurred in ROMO or GLAC, although episodic acidification has been reported for lakes throughout the Colorado Front Range (Williams and Tonnessen 2000). The data that would be needed for determining the extent and magnitude of episodic acidification have not been collected to a sufficient degree in acid-sensitive areas of ROMO to support the regional assessment of episodic acidification (Sullivan 2000).

The critical load for lake acidification in the Green Lakes Valley in Colorado was estimated empirically to be about 4 kg N/ha/yr (Williams and Tonnessen 2000). Baron et al. (2011b) also estimated an empirical critical load of 4.0 kg N/ha/yr for western mountain lakes to protect against episodic N pulses in lakes having low acid neutralizing capacity.

18.3.4 Effects of Acidification on Biota

In acid-sensitive lakes in the western United States, one focus of studies on aquatic effects of acidification has been on native cutthroat trout. Native trout are sensitive to short-term increases in acidity (Woodward et al. 1989). It is important to note, however, that many high-elevation western lakes and streams were historically fishless. The top predators in such aquatic ecosystems were often amphibians or crustaceans. Thus, even though cutthroat trout might be considered native to the region, this species is not necessarily native to a particular lake or stream.

Episodic acidification can be the limiting condition for aquatic organisms in streams or lakes that exhibit chronic chemistry that is suitable for aquatic biota, but nevertheless experience occasional episodic acidification (cf., Wigington et al. 1993). The EPA's Episodic Response Project suggested that the chemical response to acidification that has the greatest effect on biota is usually increased inorganic aluminum (Al_i) concentration. There is no evidence at present to indicate that increased Al_i concentration in response to episodic acidification has harmed fish or other aquatic biota in the ROMO region, but such effects are possible.

It is unlikely that aquatic biota in GLAC have experienced adverse impacts from acidic deposition. This is because N and S deposition are low, the available lake water chemistry data do not suggest chronic acidification, and most sampled lakes appear to be relatively insensitive to both chronic and episodic acidification at any foreseeable future levels of acidic deposition.

18.4 Nutrient Nitrogen Enrichment

18.4.1 Aquatic Nitrogen Enrichment

Some freshwater aquatic ecosystems in the United States are sensitive to nutrient enrichment effects from atmospheric N deposition. In order to be sensitive to such effects, the lake or stream must be N-limited. Conventional wisdom previously held that most lakes and streams in the United States are P-, rather than N-limited. More recent research suggests that this may not always be the case. Fertilization experiments and lake studies found that oligotrophic waters are commonly N-limited, especially undisturbed northern temperate or boreal lakes that receive low levels of atmospheric N deposition (Bergström et al. 2005, Elser et al. 2009a,b). There is increasing evidence from surveys and paleolimnological research suggesting that N limitation is common or was common in many lakes prior to human settlement. Bergström et al. (2005) found a consistent pattern showing N limitation for watersheds that receive N deposition below approximately 2.5 kg N/ha/yr, colimitation of N and P for deposition between ~2.5 and 5.0 kg N/ha/yr, and P limitation in areas with N deposition greater than 5.0 kg N/ha/yr. An examination of Western Lakes Survey data (Eilers et al. 1987) found enhanced N concentrations in high-elevation lakes adjacent to and downwind of urban centers (Fenn et al. 2003a). Therefore, eutrophication effects on freshwater ecosystems from atmospheric deposition of N are of greatest concern in lakes and streams that have very low productivity and nutrient levels and that are located in remote areas. Baron et al. (2011a) estimated that 45% of 6666 lakes represented in the Rocky Mountain region of the Western Lakes Survey were likely N-limited, based on having dissolved inorganic N:total P ratio less than 4.

Research on ecosystem responses to N fertilization within the Rocky Mountain Network has mainly been conducted in the Front Range of Colorado, including in ROMO. Lakes and streams in ROMO tend to be clearwater, low ionic strength, oligotrophic aquatic systems. Concentrations of virtually all dissolved constituents except oxygen (e.g., nutrients, organic material, major ions, weathering products) tend to be very low (Sullivan 2000). ROMO surface waters are clear, cold, dilute systems that are highly sensitive to degradation by human activities.

Lakes in the Colorado Front Range tend to have higher nitrate concentrations than lakes elsewhere in Colorado (Baron et al. 2000, Musselman et al. 2004, Elser et al. 2009a). In addition to the presence of N emissions sources along the Front Range in Colorado, it has been hypothesized that N retention in soils of the Front Range may be constrained by soil freezing, as compared with the Sierra Nevada, for example (Sickman et al. 2002).

Nitrate concentrations in lake water in ROMO are higher on the east side than the west side of the park (Baron et al. 2000). Data from a survey of 44 lakes east and west of the Continental Divide in Colorado indicated that lakes on the western side of the Continental Divide averaged 6.6 µeq/L of nitrate, whereas lakes on the eastern side of the Continental Divide, adjacent to the Front Range urban corridor, averaged 10.5 µeq/L of nitrate concentration. Nitrogen deposition appears to have stimulated productivity and altered algal species assemblages at low deposition rates of 1.5–2.2 kg N/ha/yr (Baron 2006). In the Colorado Front Range, nitrate concentrations in lakes above 15 µeq/L have commonly been measured, suggesting some degree of N saturation (Baron 1992).

In situ N addition field experiments showed that some low-nutrient lakes in ROMO are N-limited (cf., Lafrancois et al. 2004). Collectively, these studies suggest that atmospheric N deposition is changing the structure and function of sensitive high-elevation aquatic and terrestrial ecosystems on the east side of the park (Porter and Johnson 2007).

Additions of N can stimulate algal growth in N-limited lakes. Studies have shown an increase in lake phytoplankton biomass with increasing N deposition in the Snowy Range in Wyoming (Lafrancois et al. 2003), the Sierra Nevada Mountains in California (Sickman et al. 2003), Sweden (Bergström et al. 2005), and across Europe (Bergström and Jansson 2006). However, not all species of diatoms or other algae and bacteria are equally responsive to N supply. Differences in resource requirements allow some species to gain a competitive edge over others upon nutrient addition, and as a consequence, shifts in algal assemblages have been observed (Wolfe et al. 2001, 2003, Lafrancois et al. 2004, Saros et al. 2005). Interlandi and Kilham (2001) demonstrated that species diversity declines with increasing availability of N; maximum species diversity was maintained when nitrate levels were extremely low (<3 µM) in lakes in the Yellowstone National Park (Wyoming, Montana) region. Sediment cores from lakes in the Colorado Front Range showed increasing representation of mesotrophic diatom taxa in recent times, as compared with predevelopment conditions (Wolfe et al. 2001). Community shifts in phytoplankton were observed in the Snowy Range, with chrysophytes favored in lakes having lower N and cyanophytes and chlorophytes favored in lakes having higher N (Lafrancois et al. 2003). These results corroborate earlier work on resource requirements for these algal species (Tilman 1981).

Chlorophytes generally prefer high N concentrations and are able to rapidly dominate the flora when N concentrations increase (Findlay et al. 1999). This occurs in both circumneutral and acidified waters (Wilcox and Decosta 1982, Findlay et al. 1999). Two species of diatom, *Asterionella formosa* and *Fragilaria crotonensis*, now dominate the flora of some alpine and montane Rocky Mountain lakes (Interlandi and Kilham 1998, Baron et al. 2000, Wolfe et al. 2001, 2003, Saros et al. 2003, 2005). *A. formosa* and *F. crotonensis* have extremely low resource requirements for P and moderate requirements for N, allowing for rapid response to increased N availability (Michel et al. 2006). They were among the first diatoms to increase in abundance following watershed settlement and agricultural development in European lake watersheds in the twelfth and thirteenth centuries (Anderson et al. 1995, Lotter 1998) and North American settlements in the eighteenth and nineteenth centuries (Christie and Smol 1993, Hall et al. 1999). In these studies, as well as in a Swedish lake influenced by acidic deposition, these two diatom species expanded following initial disturbance and were later replaced by other species more tolerant of either acidification or eutrophication (Renberg et al. 1993, Hall et al. 1999). Moreover, the growth of *A. formosa* has been stimulated with N amendments during *in situ* incubations, using bioassays and mesocosms (6.4–1616 µM N/L, McKnight et al. 1990, 76 µM N/L, Lafrancois et al. 2004, 18 µM N/L, Saros et al. 2005). Mesocosm enrichments in Wyoming lakes found positive responses of *A. formosa* and *F. crotonensis* to N, but not to P or Si enrichment (Saros et al. 2005). Furthermore, studies of diatom remains in lake sediments (Wolfe et al. 2003) have shown declines in the oligotrophic diatom species *Aulacoseria perglabra*, *Cycotella steligera*, and *Achnanthes* spp. coincident with increases in abundance of *A. formosa* and *F. crotonensis*.

Increased N deposition can cause changes in the species composition of algal communities in sensitive oligotrophic lakes. For example, Baron (2006) found that diatom reconstructions from lake sediment cores in ROMO during the period 1850–1964 suggested changes in algal abundance associated with wet N deposition of only about 1.5 kg N/ha/yr. Similar results were found by Saros et al. (2003) in the Beartooth Mountains in Wyoming. The freshwater algae thought to be most sensitive to effects of increased N deposition included *A. formosa* and *F. crotonensis*. These are considered to be opportunistic algae that can respond rapidly to slight nutrient enrichment.

Lake sediment records, including diatom stratigraphies, suggest that changes attributable to atmospheric N inputs began in Rocky Mountain lakes during the period 1950–1960 (Wolfe et al. 2001, 2003, Das et al. 2005, Enders et al. 2008). Changes in diatom abundance included decreases in oligotrophic species such as *Aulacoseria perglabra*, *Cyclotella steligera*, and *Achnanthes* spp. and increases in the more mesotrophic species *A. formosa* and *F. crotonensis* (Wolfe et al. 2001, 2003).

Documented and potential future impacts from atmospheric N deposition on sensitive aquatic and terrestrial resources in ROMO led to the formation of the Rocky Mountain National Park Initiative, a collaborative process involving the National Park Service, the EPA, and the Colorado Department of Public Health and Environment. Through this initiative process, a critical load target of N deposition has been developed in order to protect aquatic resources in the park against eutrophication (Porter and Johnson 2007). The National Park Service established a wet N deposition critical load of 1.5 kg N/ha/yr for protecting high-elevation lakes in ROMO against biological effects associated with nutrient enrichment. The National Park Service also entered into a Memorandum of Understanding with the EPA and Colorado Department of Public Health and Environment to address harmful impacts of N deposition in this park (Cheatham 2011). The Memorandum of Understanding was intended to facilitate interagency cooperation in reversing N critical load exceedance in ROMO. The Nitrogen Deposition Reduction Plan was endorsed by the three participating agencies and the Colorado Air Quality Control Commission (http://www.colorado.gov/cdphe/rmnpinitiative). The National Park Service adopted, and the Memorandum of Understanding agencies endorsed, a goal of wet N deposition of 1.5 kg N/ha/yr at the Loch Vale deposition monitoring site. In order to achieve this resource management goal, a glidepath approach was selected, as described by Morris et al. (2014). The baseline wet N deposition for the period 2002–2006 was 3.1 kg N/ha/yr. Although some progress has been made, the target five-year rolling average value for 2012 of 2.7 kg N/ha/yr was not met; the measured five-year rolling average in 2012 was slightly higher (2.9 kg N/ha/yr (Morris et al. 2014).

Nanus et al. (2012) calculated spatially explicit estimates of nutrient N deposition for the Rocky Mountains using a geostatistical approach. To do this, they established the response of sensitive diatoms to variation in surface water nitrate concentration at many site locations and identified a threshold nitrate concentration above which ecological effects were observed. Response of the diatom *A. formosa* to changes in nitrate concentration during nutrient addition experiments was determined. Nanus et al. (2012) found that surface water nitrate concentration was positively correlated with north-facing aspect, elevation, slope, and N deposition. Growth experiments using *A. formosa* showed maximum algal growth at 0.5 μM nitrate, and this level was specified as the threshold for algal growth, which was used to determine nutrient critical loads. The lowest critical load levels (<1.5 kg N/ha/yr) occurred at high-elevation locations having steep slope, sparse vegetation, abundant exposed rock, and talus. Such areas were commonly in exceedance of the critical load by more than 1.5 kg N/ha/yr. Atmospheric N deposition exceeded the critical load in 21% ± 8% of the Rocky Mountain study area; this estimate was sensitive to selection of the nitrate threshold of ecological effects (Nanus et al. 2012).

Changes to aquatic food webs have not been as thoroughly explored as changes to algal assemblages. However, a few studies have suggested declines in zooplankton biomass (Paul et al. 1995, Lafrancois et al. 2004) in response to N-related shifts in phytoplankton biomass toward less palatable taxa with higher C:P ratios (Elser et al. 2001). Thus, nutrient N input can potentially disrupt food webs in ways that scientists are only beginning to understand.

18.4.2 Terrestrial Nitrogen Enrichment

The potential impacts of nutrient N deposition on terrestrial resources within ROMO are an important concern. The major issues appear to be (1) "terrestrial eutrophication," whereby excess fertilization leads to increased ecosystem productivity, increased spread of invasive grass species, and decreased native plant species diversity (cf., Huston 1994); (2) N saturation, whereby N supply exceeds the vegetative uptake capacity and nitrate leaches out of the soil in high concentrations; and (3) effects of N enrichment on lichens, which are highly sensitive to N deposition.

Factors that govern the sensitivity of high-elevation plant communities to N deposition include low rates of primary production, short growing season, low temperature, and wide variation in moisture availability (Bowman and Fisk 2001; Figure 18.6). Alpine plant communities in particular have developed under conditions of low nutrient supply, in part because soil-forming processes are poorly developed in the alpine zone, and this also contributes to their N sensitivity. Alpine vegetation in the southern Rocky Mountains responds to increased N supply by increasing plant productivity for some species. This increase in productivity is accompanied by changes in species composition and abundance (Bowman et al. 1993, 2012). Many of the dominant plant species do not respond to additional N supply with increased production. Rather, many subdominant species, primarily grasses and some forbs, tend to increase in abundance when the N supply is increased (Fenn et al. 2003a).

FIGURE 18.6
Alpine and subalpine vegetation in ROMO is sensitive to modest levels of N deposition. (Photo courtesy of Rocky Mountain National Park.)

Nitrogen cycling in Rocky Mountain alpine environments is strongly tied to variations in moisture regime (Bowman et al. 1993, Bowman 1994, Fisk et al. 1998). Blowing snow is transported across alpine landscapes by wind and tends to accumulate in depression areas. These areas receive much higher levels of moisture and winter season N deposition than other more wind-exposed portions of the alpine environment (Bowman 1992). Fenn et al. (2003a) suggested that as much as 10 kg N/ha/yr may leach through the snow during the initial phases of snowmelt in some of the alpine areas in Colorado that accumulate substantial snowpack. It appears that moist meadow areas are highly affected by N deposition, and they are also likely to show changes in plant species composition and impacts on N cycling (Bowman and Steltzer 1998).

Baron et al. (2000) showed that small differences in N deposition between the east (3–5 kg N/ha/yr) and the west (1–2 kg N/ha/yr) side of the Continental Divide in Colorado were associated with substantial declines in foliar Mg levels and increased foliar N:Mg and N:Ca ratios in old-growth stands of Engelmann spruce (*Picea engelmanii*). It is not known whether such differences in foliar nutrient ratios affected the health or growth of these forests. Nevertheless, analyses by Baron et al. (2000) suggested that the eastern slope of the Colorado Front Range may be at the beginning of a trajectory of N saturation change. Baron et al. (1994) estimated N uptake by plants and soils in Colorado using the CENTURY model. Simulated N export increased in alpine ecosystems at relatively low levels of N deposition (3.4 kg N/ha/yr). The adjacent subalpine forests had more substantial forest N retention.

Most major ecosystem types, including temperate forest, grassland, and tropical forest, tend to be dominated by a single physiognomic type of vegetation. In contrast, arctic and alpine tundra tend to be dominated by multiple types of plant communities (Chapin et al. 1980). Across a relatively small area of tundra, there may be a wide variety of plant community types in which graminoids, forbs, mosses, lichens, deciduous shrubs, or evergreen shrubs dominate (Bliss et al. 1973). There can be important differences among these plant growth forms in their use of, and response to, addition of nutrients (Thomas and Grigal 1976, Schlesinger and Chabot 1977).

Alpine ecosystems are adapted to cold temperature, short growing season, high soil moisture, and periodically low soil oxygen level. Plants respond to reduced N availability by changing the allocation of biomass to favor root growth (Bloom et al. 1985) or changing the efficiency with which N is used or stored (Chapin 1980). Increased abundance of nitrophilous (high N demand) plant species has been demonstrated in alpine plant communities at the Niwot Ridge Long Term Ecological Research site, about 10 km south of ROMO (Korb and Ranker 2001). The Niwot Ridge experimental site receives slightly higher atmospheric N deposition than does Loch Vale (Burns 2003). Results of fertilization experiments suggested that the lowest amount of atmospheric N deposition expected to alter alpine plant communities at Niwot Ridge and Loch Vale is about 3–4 kg N/ha/yr (Bowman et al. 2006, 2012). This deposition level is similar to ambient deposition measured at Loch Vale.

Changes in alpine plant species composition precede detectable changes in soil chemistry in response to increased N deposition (Sverdrup et al. 2012). Changes in species composition of dry meadows (among the most sensitive alpine plant community types) are probably ongoing along the Front Range (Bowman et al. 2006, 2012), in response to ambient N deposition, which varies from about 3 to 6 kg N/ha/yr. Results of a modeling study by Baron et al. (1994) suggested that subalpine forest soils in the Front Range would exhibit increased N leaching at N deposition above about 4 kg N/ha/yr.

Rueth and Baron (2002) compared N dynamics of Engelmann spruce (*Picea engelmannii*) forest stands east and west of the Continental Divide in Colorado. Nitrogen deposition, arising mainly from agricultural and urban areas of the South Platte River Basin, was moderate (3–5 kg N/ha/yr) on the east slope but only 1–2 kg N/ha/yr on the west slope. East slope sites showed lower soil organic horizon C:N, lower foliar C:N, higher potential net mineralization, and higher percent N and N:Mg and N:P ratios in foliage. These results suggested that even moderate levels of N deposition input can cause measurable changes in spruce forest biogeochemistry. It is unclear, however, to what extent such biogeochemical changes affect forest species composition, growth, or health.

Research on experimental N enrichment effects on alpine and subalpine ecosystems has been conducted at Loch Vale and Niwot Ridge, both located east of the Continental Divide (see review by Burns 2004). Addition of 25 kg N/ha during summer caused a community shift toward greater dominance of hairgrass (*Deschampsia* sp.) in wet alpine meadows, but the increase in plant biomass (+67%) and plant N content (+107%) following N fertilization was higher in graminoid-dominated dry meadows than in forb-dominated wet meadows (+53% plant biomass, +64% standing N crop; Bowman et al. 1995, Burns 2004).

Additions of 20, 40, and 60 kg N/ha/yr at Niwot Ridge (on top of ambient N deposition near 5 kg N/ha/yr) over an eight-year period to a dry alpine meadow led to a change in plant species composition, an increase in species diversity and plant biomass, and an increase in tissue N concentration at all treatment levels within three years of application. Much of the response was due to increased cover and total biomass of sedges (*Carex* spp.). There was a significant decrease in *Kobresia* sp. biomass with increasing N input. Vegetation composition appeared to respond at lower N input levels than those that caused measurable changes in soil inorganic N content.

The effects of N deposition on terrestrial alpine ecosystems in the Rocky Mountain Network region are thought to include community-level changes in plants, lichens, and mycorrhizae. Subtle effects have been shown to occur at what would be considered relatively low levels of N deposition in the eastern United States (about 4 kg N/ha/year; Bowman et al. 2006). Bowman et al. (2006, 2012) concluded that alpine plants may be more sensitive indicators of the effects of increased N inputs than soils. Changes in plant species composition occurred at all treatment levels within three years. Changes in an individual species (*Carex rupestris*) were estimated to occur at deposition levels near 4 kg N/ha/yr. Changes in the plant community, based on the first axis of a detrended correspondence analysis, were estimated to occur at deposition levels near 10 kg N/ha/yr. In contrast, increases in nitrate leaching, soil solution nitrate concentration, and net nitrification occurred at levels above 20 kg N/ha/yr. The researchers concluded that changes in vegetation composition preceded detectable changes in soil indicators of ecosystem response to N deposition.

Nitrogen deposition to the alpine tundra of Niwot Ridge altered N cycling and provided the potential for the replacement of some plant species by more competitive, faster-growing species (Bowman and Steltzer 1998, Baron et al. 2000, Bowman 2000). Many plants that grow in alpine tundra, as is true of plants growing in other low resource environments, tend to have some similar characteristics, including slow growth rate, low photosynthetic rate, low capacity for nutrient uptake, and low soil microbial activity (Bowman and Steltzer 1998, Bowman 2000). Such plants generally continue to grow slowly when provided with an optimal supply and balance of resources (Pearcy et al. 1987, Chapin 1991). In addition, plants adapted to cold, moist environments grow more leaves than roots as the relative availability of N increases. These patterns of vegetative development and their response to added N affect plant capacity to respond to variation in available resources and to environmental stresses such as frost, high winds, and drought.

Changes in alpine plant species composition on Niwot Ridge have included increased cover of the plant species that tend to be most responsive to N fertilization in some of the long-term monitoring plots (Korb and Ranker 2001, Fenn et al. 2003a). These changes have probably developed in response to changes in N deposition. However, the influences of climatic change, particularly changes in precipitation (Williams et al. 1996b), and pocket gopher disturbance (Sherrod and Seastedt 2001) could not be ruled out as contributors to vegetation change (Fenn et al. 2003a). Other environmental factors also affect the species makeup of alpine ecosystems, but long-term experimental fertilization plots demonstrated a clear response of alpine flora to N, including shifts toward graminoid plants that shade smaller flowering species, and accompanying changes in soil N cycling (Bowman et al. 2006).

Changes in plant species in response to N deposition to the alpine zone can result in increased leaching of nitrate from the soils because the plant species favored by higher N supply are often associated with greater rates of N mineralization and nitrification than the preexisting species (Bowman et al. 1993, 2006, Steltzer and Bowman 1998, Suding et al. 2006). Total organic N pools in the soils of dry alpine meadows are large compared to pools of ammonium and nitrate (Fisk and Schmidt 1996). However, positive response to inorganic N fertilization has been demonstrated, and thus, some plant species appear to be restricted in their ability to take up organic N from the soil and are growth limited by the availability of inorganic N (Bowman et al. 1993, 1995, Theodose and Bowman 1997).

Bowman et al. (2012) added ammonium nitrate at rates of 5, 10, and 30 kg N/ha/yr to an alpine dry meadow plant community in ROMO. Three years after fertilization, they measured aboveground biomass, plant N concentration, and soil and soil solution chemistry. Plant species composition was measured annually. Plant species richness and diversity did not change in response to the N addition, but one species of sedge, *Carex rupestris*, increased in cover by 34%–125%. Based on the rate of change in *C. rupestris* in the N treatment plots and the ambient control plots (receiving 4 kg N/ha/yr), and assuming that the change in sedge cover was attributable solely to N addition, Bowman et al. (2012) estimated the N critical load of nutrient N loading to protect plant community composition up to about 3 kg N/ha/yr. Inorganic N in soil solution increased above ambient levels at input rates (experimental addition plus ambient loading) between 9 (resin bag measurements) and 14 (lysimeter measurements) kg N/ha/yr, suggesting N saturation at these levels of N loading. There was no indication of change after three years in soil pH or extractable base cations, indicating that soil acidification had not occurred.

McDonnell et al. (2014) evaluated potential long-term impacts of N deposition and climate change on a subalpine plant community at Loch Vale, using the ForSAFE-VEG model. ForSAFE-VEG is a coupled biogeochemical, vegetation niche and plant competition model. The model had earlier been applied in a generalized fashion to the Rocky Mountain region (Porter et al. 2012, Sverdrup et al. 2012). Simulated changes in N deposition, temperature, and precipitation over the previous century caused pronounced changes in model projections of plant species cover. The model estimate of the critical load of N deposition required to protect against a change in plant cover of 10% was between 1.9 and 3.5 kg N/ha/yr. Ambient N deposition is slightly higher than that, suggesting that the critical load for protection against nutrient N enrichment has been exceeded at Loch Vale.

Pardo et al. (2011a) compiled data on empirical critical load for protecting sensitive resources in Level I ecoregions across the conterminous United States against nutrient enrichment effects caused by atmospheric N deposition. Data compiled by Pardo et al. (2011a) suggest that ambient N deposition may exceed the lower limit of the expected critical load to protect against nutrient enrichment effects in GLAC and ROMO. These potential exceedances were reported for the protection of mycorrhizal fungi (GLAC only), lichens, herbaceous plants, and forest vegetation and to limit nitrate leaching in drainage waters.

Ellis et al. (2013) estimated the critical load for nutrient N deposition to protect the most sensitive ecosystem receptors in 45 national parks, based on the data of Pardo et al. (2011a).

The lowest terrestrial critical load of N is generally estimated for the protection of lichens (Geiser et al. 2010). Changes to lichen communities may signal the beginning of other changes to the ecosystem that might affect structure and function (Pardo et al. 2011b). Ellis et al. (2013) estimated the N critical load for GLAC and ROMO in the range of 2.5–7.1 kg N/ha/yr for the protection of lichens.

18.5 Ozone Injury to Vegetation

18.5.1 Ozone Exposure Indices and Levels

GLAC is known to contain eight ozone-sensitive plant species, four of which are recognized as bioindicators. ROMO contains nine sensitive species, three of which are recognized as bioindicators.

The W126 (a measure of cumulative ozone exposure that preferentially weights higher concentrations) and SUM06 (a measure of cumulative exposure that includes only hourly concentrations over 60 ppb ozone) exposure indices calculated by National Park Service staff are given in Table 18.3, along with Kohut's (2007) ozone risk ranking. Using these criteria, ozone levels are rated uniformly low at GLAC and low (Kohut ranking) to high (W126 and SUM06) at ROMO (Table 18.3).

18.5.2 Ozone Formation

Vegetation in ROMO is likely at higher risk for ozone symptoms than vegetation in GLAC, given the relatively high values of W126 and SUM06 in ROMO. Ozone levels in GLAC are much lower (Table 18.3). The ROMO area has been designated as a nonattainment area for ozone. Ozone advisories are posted when thresholds for human health are exceeded.

The National Park Service (2010) reported long-term trends in the annual fourth highest 8-hour daily maximum ozone concentration for 31 monitoring sites in 27 national parks having more than 10 years of data through 2008. Statistically significant increases were reported for only four parks, including ROMO. This park is the recipient of relatively

TABLE 18.3

Ozone Assessment Results for GLAC and ROMO, Based on Estimated Average 3-Month W126 and SUM06 Ozone Exposure Indices for the Period 2005–2009 and Kohut's (2007) Ozone Risk Ranking for the Period 1995–1999[a,b]

		W126		SUM06		
Park Name	Park Code	Value (ppm-h)	NPS Condition	Value (ppm-h)	NPS Condition	Kohut Ozone Risk Ranking
Glacier	GLAC	2.30	Low	1.67	Low	Low
Rocky Mountain	ROMO	17.35	High	23.75	High	Low

[a] Parks are classified into one of three ranks (Low, Moderate, High) based on comparison with other Inventory and Monitoring parks.
[b] Degrees of concern for the W126 and SUM06 indices are based solely on levels of ozone exposure. Kohut's risk to vegetation is based on several factors that contribute to injury in plants, including ozone exposure and environmental variables, and considers the effects of soil moisture on the uptake of ozone.

ozone-rich air masses originating from the Denver-to-Fort Collins area, especially during the summer (Peterson et al. 1998). One of the greatest threats to vegetation in ROMO is ozone pollution from urban areas southeast of the park and from valley and foothill areas where ozone is synthesized in transit from local sources of N oxides and volatile organic compounds.

Exposure of plants to ozone is relatively high in ROMO based on both the W126 and SUM06 statistics (Table 18.3). Sources of N oxide that contribute to these high exposures include motor vehicles, power plants, and other human sources. At ROMO, the largest number of exceedances of the 8-hour 0.075 ppm ozone standard that was established in 2008 occurred in 2002 and 2003 (six and seven days in exceedance, respectively). These were also two of the top three years in terms of greatest summer burned area by forest fire in the surrounding region. Jaffe et al. (2008) concluded that fire plays an important role in elevating the ozone level and increasing the likelihood of an exceedance of the 8-hour ozone standard.

Forest fires emit considerable quantities of N oxide and hydrocarbons and therefore can contribute to ozone formation throughout the Rocky Mountain Network. The concentration of ozone in the atmosphere in the nonurban western United States has increased since the late 1980s by about 5 ppb (Jaffe and Ray 2007). Jaffe et al. (2008) investigated the role of forest fire in this trend. The summer burned area was significantly correlated with ozone concentration at Clean Air Status and Trends Network and National Park Service atmospheric chemistry monitoring sites within six national parks, including ROMO and GLAC. For mean and maximum fire years, the concentration of ozone in the atmosphere appeared to be increased by fire by an average of 3.5 and 8.8 ppb, respectively. The estimated amount of biomass consumed was a slightly better predictor of ozone concentration than burned area. These relationships between atmospheric ozone concentration and burned area or biomass consumed are especially important because the frequency of extreme fire years in the western United States appears to be increasing (Jaffe et al. 2004) due to increased spring and summer temperature, earlier snowmelt, past fire suppression, and generally dryer forest conditions (Cook et al. 2004, Westerling et al. 2006). Jaffe et al. (2008) concluded that the increase in number and size of fires in recent years in the western United States has largely been responsible for the observed increase in summer ozone concentrations reported by Jaffe and Ray (2007), above the levels caused by vehicular, power plant, and industrial emissions. Increasing temperature in the future will likely further influence the effects of fire on ozone formation.

18.5.3 Ozone Exposure Effects

In 1980, the Forest Service conducted a survey of ponderosa pine in the Front Range west of Denver in order to determine if any trees had evidence of ozone injury. No symptoms were found (James and Staley 1980). In 1987, the National Park Service conducted an extensive survey of ponderosa pine pathological condition in ROMO, with data collected at plots throughout the range of the species in the park (Stolte 1987). No symptoms of ozone injury were noted in any trees in that survey. Similarly, a study of ponderosa pine at 30 stands throughout the Front Range (20 stands east side, 10 west of the Rampart Range with presumed lower ozone) determined that there were no visible symptoms of ozone injury at any locations and that long-term tree growth was unaffected by elevated ozone levels (Graybill et al. 1993, Peterson et al. 1993). The Rocky Mountain variety of ponderosa pine (var. *scopulosum*) is known to be somewhat more tolerant to ozone and has a higher threshold for symptoms of injury under experimental exposures than var. *ponderosa* (Aitken et al. 1984), which is found further to the west, including in California.

Quaking aspen, an ozone-sensitive hardwood species, grows at various locations in riparian ecosystems and in fire- or avalanche-disturbed areas in ROMO. Numerous studies have documented the sensitivity of this species to ozone under field and experimental conditions (Wang et al. 1986, Karnosky et al. 1992, Coleman et al. 1996), although there is considerable variability in sensitivity among different genotypes (Berrang et al. 1986). To date, however, there are no data indicating ozone injury to this species in ROMO. Of the tree species present at GLAC, quaking aspen is probably the most sensitive to ozone.

Recent research by Kohut et al. (2012) discovered foliar ozone symptoms on cutleaf coneflower, but not on spreading dogbane or quaking aspen, in ROMO. This was the first documentation of ozone symptoms on vegetation in the Rocky Mountain region. They reported SUM06 and W126-3-month levels in ROMO that were higher than the common thresholds for observing foliar symptoms. The increase in oil and gas drilling in Wyoming in recent years may have caused or contributed to relatively high concentrations of ozone in some remote areas that previously had relatively low ozone levels (Wyoming Department of Environmental Quality 2010, Kohut et al. 2012).

18.6 Visibility Degradation

18.6.1 Natural Background and Ambient Visibility Conditions

GLAC and ROMO are classified as Class I under the Clean Air Act (CAA). Thus, these parks are subject to requirements of the Regional Haze Rule. Some of the best visibility in the conterminous United States occurs in the Rocky Mountains (Savig and Morse 1998). Nevertheless, some visibility impairment does occur.

Haze is monitored by the Interagency Monitoring of Protected Visual Environments (IMPROVE) program for GLAC and ROMO. Table 18.4 gives the relative park haze rankings on the 20% clearest, 20% haziest, and average days in GLAC and ROMO. Haze measurements for the period 2004–2008 were higher than the estimated natural condition for both parks. Measured ambient haze in GLAC was considered moderate for all groups (20% clearest, average, and 20% haziest days); haze in ROMO was ranked very low for all groups.

TABLE 18.4

Estimated Natural Haze and Measured Ambient Haze in GLAC and ROMO Averaged over the Period 2004–2008[a]

		Estimated Natural Haze (in Deciviews)		
Park Name	Site ID	20% Clearest Days	20% Haziest Days	Average Days
Glacier	GLAC1	2.42	9.18	5.08
Rocky Mountain	ROMO1	0.28	7.15	3.28

		Measured Ambient Haze (for Years 2004–2008)					
		20% Clearest Days		20% Haziest Days		Average Days	
Park Name	Site ID	Deciviews	Ranking	Deciviews	Ranking	Deciviews	Ranking
Glacier	GLAC1	7.02	Moderate	18.90	Moderate	12.47	Moderate
Rocky Mountain	ROMO1	2.06	Very Low	12.66	Very Low	7.26	Very Low

[a] Parks are classified into one of five ranks (Very Low, Low, Moderate, High, Very High) based on a relative comparison of visibility conditions at all monitored parks.

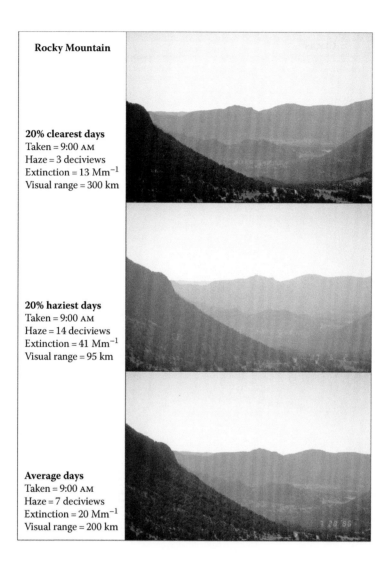

FIGURE 18.7
Three representative photos of the same view in ROMO, illustrating the 20% clearest visibility, the 20% haziest visibility, and the annual average visibility. Extinction is total particulate light extinction. (From http://vista. cira.colostate.edu/improve/Data/IMPROVE/Data_IMPRPhot.htm, accessed December, 2010.)

Representative photos of selected vistas under three different visibility conditions are shown in Figures 18.7 and 18.8 for GLAC and ROMO. Photos were selected to correspond with the clearest 20% of visibility conditions, haziest 20% of visibility conditions, and annual average visibility conditions at each location. This series of photos provides a graphic illustration of the visual effect of these differences in haze level on a representative vista in each of these parks.

In GLAC, air pollution has reduced average visual range (compared with background levels) from 140 to 45 mi (225 to 72 km). On the haziest days, visual range has been reduced from 95 to 20 mi (153 to 32 km). Severe haze episodes occasionally reduce visibility to 6 mi (10 km). At ROMO, pollution has reduced average visual range from 170 to 100 mi (274 to 161 km). On the haziest days, visual range has been reduced from 120 to 60 mi (193 to 97 km). Severe haze episodes occasionally reduce visibility to 18 mi (29 km).

FIGURE 18.8
Three representative photos of the same view in GLAC, illustrating the 20% clearest visibility, the 20% haziest visibility, and the annual average visibility. Extinction is total particulate light extinction. (From http://vista. cira.colostate.edu/improve/Data/IMPROVE/Data_IMPRPhot.htm, accessed December, 2010.)

18.6.2 Composition of Haze

At IMPROVE sites throughout the interior Columbia River basin, from North Cascades National Park and GLAC in the north to Lassen Volcanic and Yellowstone national parks in the south, carbon in various forms, including fine particulate organics and soot, dominates the light extinction budget (Schoettle et al. 1999). The second most important contributor is typically sulfate.

Figure 18.9 shows estimated natural (preindustrial), baseline (2000–2004), and current (2006–2010) levels of haze and its composition for GLAC and ROMO. GLAC has no valid measured data for the years 2002, 2003, and 2009; substituted data were used for

2002 and 2003. For more information on this haze monitoring network, see http://vista.cira.colostate.edu/improve/Data/IMPROVE/summary_data.htm.

In GLAC, on the 20% haziest days and average days, organics were the largest contributors to the light extinction coefficient; on the 20% clearest days, the largest contributor was sulfate. Organics account for about one-third (20% clearest days) to one half (20% haziest days) of the light extinction in GLAC. Elemental C, coarse mass, and nitrate were also significant contributors to haze in this park.

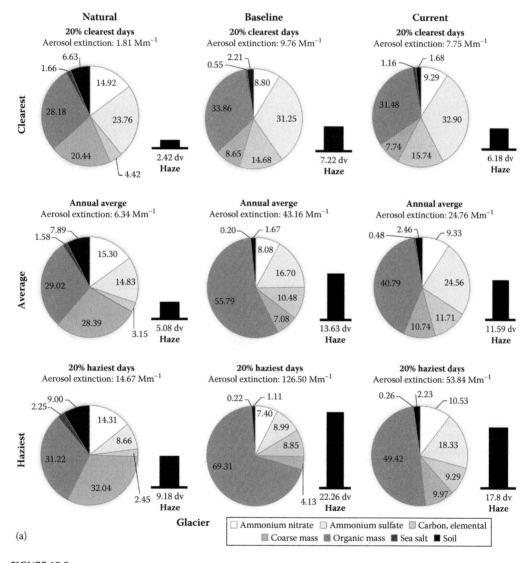

FIGURE 18.9

Estimated natural (preindustrial), baseline (2000–2004), and current (2006–2010) levels of haze (columns) and its composition (pie charts) on the 20% clearest, annual average, and 20% haziest visibility days for (a) GLAC and (b) ROMO. GLAC has no valid, measured data for the years 2002, 2003, and 2009, but has substituted data for the years 2002 and 2003. (From http://views.cira.colostate.edu/fed/Tools/RegionalHazeSummary.aspx, accessed October, 2012.) *(Continued)*

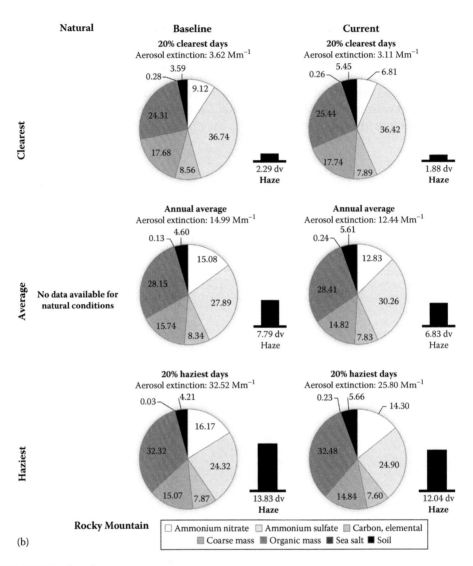

FIGURE 18.9 (*Continued*)
Estimated natural (preindustrial), baseline (2000–2004), and current (2006–2010) levels of haze (columns) and its composition (pie charts) on the 20% clearest, annual average, and 20% haziest visibility days for (a) GLAC and (b) ROMO. There were no data available for ROMO for natural haze levels. GLAC has no valid, measured data for the years 2002, 2003, and 2009, but has substituted data for the years 2002 and 2003. (From http://views.cira.colostate.edu/fed/Tools/RegionalHazeSummary.aspx, accessed October, 2012.)

Non-Rayleigh (mainly human-caused) atmospheric light extinction at ROMO, unlike many rural western areas, can have a large nitrate component during the winter and spring when the poorest visibility occurs. However, at other times, like in most areas, atmospheric light extinction in the Rocky Mountain Network is typically associated with sulfate, organics, and soil (Savig and Morse 1998). Nitrate accounts for a greater proportion of light extinction coefficient at ROMO than at other monitored parks in the Rocky Mountain Network, perhaps due to the high density of motor vehicles in the greater Denver/Ft. Collins area.

A substantial component of the regional haze observed in GLAC is caused by prescribed and wildland fires. Smoke plumes can contribute to visibility impairment downwind from the fire location (Malm 1999). Fires have become more severe than they were historically, and this has been attributed to more extreme weather and buildup of fuel from fire suppression (Agee 1997, Flannigan et al. 1998, Covington 2000, Allen et al. 2002, McKenzie et al. 2006). This pattern may continue into the future in response to climate change. Both empirical (McKenzie et al. 2004) and process (Lenihan et al. 1998) models suggest that the area burned by wildfires is likely to increase in the western United States in response to a warming climate.

McKenzie et al. (2006) modeled the occurrence of wildland fires and consequent haze. They based the likelihood of fire and its severity on environmental conditions conducive to wildfire. They included three modules in their model system:

1. Climate–fire–vegetation module to estimate the effects of climate on fire regimes and vegetation succession
2. Emissions module to calculate particulate and aerosol emissions
3. Smoke dispersion module to track spatial patterns of impact

The highest light extinction coefficients caused by smoke for Class I areas were in the Bob Marshall Wilderness and GLAC.

Analyses conducted by the Western Regional Air Partnership indicated that organics from natural emissions sources, including wildfire and biogenic sources (vegetation), contribute to substantial visibility impairment throughout the western United States, including within the Rocky Mountain Network. In addition, air pollution sources outside the Western Regional Air Partnership domain, including international offshore shipping and sources from Mexico, Canada, and Asia, can in some cases be substantial contributors to haze (Suarez-Murias et al. 2009).

18.6.3 Trends in Visibility

The National Park Service (2010) reported long-term trends in annual haze on the clearest and haziest 20% of days at monitoring sites in 29 national parks (http://www.nature.nps.gov/air/data/products/index.cfm). All 27 parks that showed statistically significant ($p \leq 0.05$) deciview trends on the clearest days for the 11–20-year monitoring periods through 2008, including GLAC and ROMO, exhibited decreases in deciviews over time. None of the sites showed increasing trends on the clearest days. Available haze monitoring data are shown in Figure 18.10 for the period of record at each park. In general, haze levels appear to be decreasing at both parks over the last decade or more.

18.6.4 Development of State Implementation Plans

Progress to date in meeting the national visibility goal for GLAC and ROMO is illustrated in Figure 18.11 using a uniform rate of progress glideslope. Haze on the 20% haziest days at the monitored parks appears to be decreasing sufficiently to comply with the glideslope required by the Regional Haze Rule. Additional data will be needed to ensure continued progress.

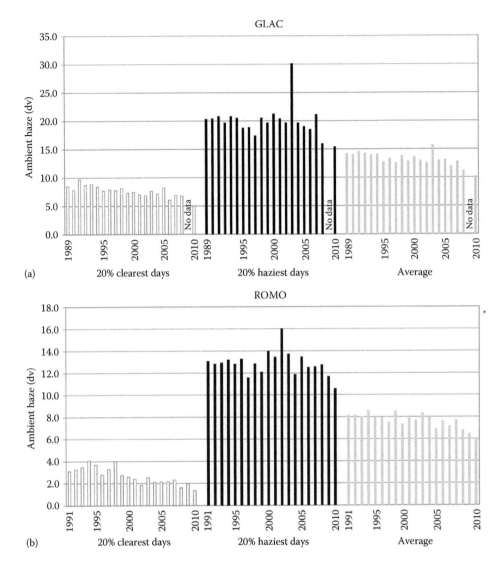

FIGURE 18.10
Trends in ambient haze levels at (a) GLAC and (b) ROMO, based on IMPROVE measurements on the 20% clearest, 20% haziest, and annual average visibility days over the monitoring period of record. GLAC has no valid, measured data for the years 2002, 2003, and 2009, but has substituted data for the years 2002 and 2003. (From http://vista.cira.colostate.edu/improve/Data/IMPROVE/summary_data.htm, accessed October, 2012.)

18.7 Toxic Airborne Contaminants

18.7.1 Semivolatile Organic Compounds

The WACAP study (Landers et al. 2008) assessed airborne toxic contaminants in snow, air, water, sediment, vegetation, and fish in eight western parks (http://www.nature.nps.gov/air/Studies/air_toxics/wacap.cfm) and found relatively high concentrations of pesticides in GLAC. Among all WACAP parks, GLAC had some of the highest current-use pesticide concentrations in snow (along with ROMO) and vegetation (Landers et al. 2008, 2010).

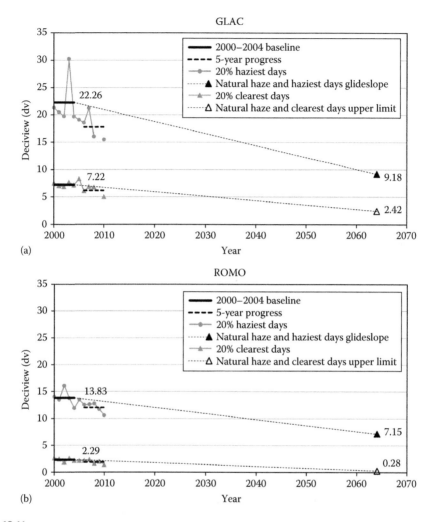

FIGURE 18.11
Glideslopes to achieving natural visibility conditions in 2064 for the 20% haziest (upper line) and the 20% clearest (lower line) days in (a) GLAC and (b) ROMO. In the Regional Haze Rule, the clearest days do not have a uniform rate of progress glideslope; the rule only requires that the clearest days do not get any worse than the baseline period. Also shown are measured values during the period 2000–2010. GLAC has no valid, measured data for the years 2002, 2003, and 2009, but has substituted data for the years 2002 and 2003. (From http://vista.cira.colostate.edu/improve/Data/IMPROVE/summary_data.htm, accessed October, 2012.)

The dominant current-use pesticides detected were endosulfans and dacthal. Polycyclic aromatic hydrocarbon concentrations in snow, sediments, and vegetation were 10–100 times higher in GLAC than at any other WACAP park. Notably, polycyclic aromatic hydrocarbon concentrations were higher in the Snyder Lake watershed on the west side of GLAC than in the Oldman Lake watershed on the east side, likely the result of emissions from an Al smelter (no longer operating) in nearby Columbia Falls, Montana (Usenko et al. 2010). Concentrations of dacthal, endosulfans, hexachlorobenzene, hexachlorocyclohexane-α, chlorpyrifos, and PCBs in GLAC ranged from mid to high as compared with other WACAP parks; along with polycyclic aromatic hydrocarbon, these pollutants were highest on the west side of the park. On the east side, where there is more agricultural development, the pesticides chlorpyrifos and hexachlorocyclohexane-γ were higher than on the west side of the park.

Fish in Snyder Lake had lower concentrations of pesticides than fish in Oldman Lake (Landers et al. 2008). Concentrations of dieldrin and p,p'-DDT breakdown products in fish in Oldman Lake exceeded contaminant health thresholds for subsistence fishers; dieldrin in one fish exceeded the threshold for recreational fishers. These concentrations were significantly higher than those in fish from similar ecosystems in Canada (Landers et al. 2008). In Oldman Lake, the contaminant health thresholds for piscivorous birds were exceeded for DDT and chlordane (in one fish). The DDT concentrations exceeded levels detected in many fish collections around the world, even in regions of Africa that spray DDT as a mosquito control (Landers et al. 2010).

Deposition fluxes for endosulfans and dacthal in snow were comparatively high at ROMO; accordingly, concentrations of these pollutants were also high in fish (Landers et al. 2008). Dieldrin concentrations in fish in ROMO were the highest of all fish analyzed in the WACAP study. This probably represents, in large part, a legacy of dieldrin production in the Denver area prior to the 1970s. All sampled fish exceeded contaminant health thresholds for human subsistence fishers. Some fish also exceeded thresholds for recreational fishers. However, thresholds for wildlife were not exceeded. Fish from similar ecosystems in Canada had significantly lower dieldrin concentrations than fish in ROMO (Landers et al. 2008).

Various studies have identified organochlorine compounds in remote tundra ecosystems (cf., Simonich and Hites 1995, Wania and Mackay 1996, Blais 2005). Compounds detected commonly include DDT, hexachlorocyclohexane, hexachlorobenzene, and PCBs. Such compounds tend to concentrate in aquatic biota in response to biomagnification processes and have been associated with disruption of endocrine systems in aquatic species.

Exposure to semivolatile organic compounds can potentially disrupt natural hormonal systems of fish (Kidd et al. 2007). Biological endpoints that can be used to document reproduction abnormalities include the presence of elevated levels of plasma vitellogenin (a protein that indicates estrogen exposure) in males and intersexuality (displaying both male and female reproductive structures; van der Oost et al. 2003). The WACAP study found evidence for endocrine disruption in fish in GLAC and ROMO. Samples from Oldman Lake in GLAC contained one intersex trout. Additionally, two male fish, one from each sampled lake in GLAC, had high concentrations of vitellogenin. In the fish from Oldman Lake, this high vitellogenin coincided with the only measured concentration of the endocrine disrupting compound o,p'-DDT and the highest concentration found in any fish sampled in the WACAP study of the DDT breakdown product DDE. Lakes in ROMO also contained intersex trout, male trout with poorly developed reproductive organs, and male fish with elevated concentrations of vitellogenin. The WACAP data suggested that the occurrence of intersex and increased vitellogenin in fish may be influenced by contaminants present in the lakes (Landers et al. 2008). However, they also acknowledged possible alternative explanations. Enhanced vitellogenin levels may be influenced by the species of fish sampled; elevated vitellogenin was found in all lakes containing fish of the genus *Oncorhynchus* (Mills Lake in ROMO and both sampled lakes in GLAC). Additionally, all intersex fish found in the WACAP study were confined to the Rocky Mountains at ROMO and GLAC.

Based on the analyses conducted in the WACAP study, current-use pesticide concentrations were generally highest in fish collected from lakes in Sequoia National Park, followed by ROMO and GLAC, and lowest in fish from surveyed parks in the Pacific Northwest (Olympic, Mount Rainier) and Alaska (Denali and Noatak; Ackerman et al. 2008). Lake average DDT (plus metabolic breakdown products) concentrations in fish exceeded wildlife contaminant health thresholds for belted kingfisher in Oldman Lake in GLAC. In addition, chlordane concentrations in several fish from Oldman lake exceeded health thresholds for kingfishers. These results suggest that organic contaminants deposited from the atmosphere to remote national park watersheds in the western United States, including GLAC and ROMO, can accumulate in fish to levels that are of concern regarding piscivorous wildlife health. However, health thresholds for mink and river otter were not exceeded by fish contaminant levels in any of the park lakes studied (Ackerman et al. 2008).

As part of the WACAP study, Schwindt et al. (2009) observed intersex male cutthroat trout and brook trout at frequencies of 9%–33% in a majority of subalpine lakes sampled in ROMO and GLAC. In addition, male cutthroat trout, brook trout, and rainbow trout produced elevated levels of vitellogenin. In contrast, reproductive abnormalities were not found in fish in national parks of the Sierra Nevada, Cascade, Olympic, Brooks, or Alaska mountain ranges. Schwindt et al. (2009) also sampled various fish species of the family Salmonidae collected prior to the era of organic pollutants (pre-1930s). In these museum specimens, they found intersex male greenback cutthroat trout (*O. clarkia stomias*) collected in the late 1800s from Twin Lakes, Colorado. This latter finding suggests that although semivolatile organic compounds may be associated with some reproductive abnormalities, organic pollutants may not be the only factors involved in reproduction disruption in fish in high-elevation Rocky Mountain lakes (Schwindt et al. 2009).

Low rates of the intersex condition are probably normal (Schwindt et al. 2009). It is difficult to determine the levels of intersex occurrence rate that would signify that the fish had experienced an endocrine disrupting exposure. However, Schreck and Kent (2013) judged that, because the sampled water bodies in ROMO show such a high intersex frequency (~50%) and because the condition occurs in multiple species, there appears to be a sound basis for concluding that the rate of intersex occurrence in ROMO is not the natural condition.

A follow-on study to WACAP was conducted during the summer of 2009 in high-elevation lakes and streams in several western national parks including ROMO and GLAC (Keteles 2011). Keteles (2011) analyzed surface water samples from high-elevation areas of ROMO and GLAC and analyzed them for pesticides. Increased contaminant deposition can occur at high elevation in the Rocky Mountains due to high precipitation, upslope winds, and the process of cold condensation, whereby chemicals volatilize in warm locations (i.e., low-elevation agricultural areas) and subsequently condense when they reach colder areas in the mountains (Blais et al. 1998). Concentrations of pesticides in water were compared with EPA Office of Pesticide Programs Aquatic life benchmarks for freshwater species and to toxicity values determined in other programs (Kegley et al. 2011). The measured concentrations in this screening study suggested low pesticide concentration values in these national parks, far below the EPA aquatic life benchmarks (Keteles 2011).

Hageman et al. (2010) reported the results of pesticide analyses of snowpack during the WACAP study at remote alpine, arctic, and subarctic sites in eight national parks, including ROMO and GLAC. Various current-use pesticides (dacthal, chlorpyrifos, endosulfans, and hexachlorocyclohexane-γ) and historic-use pesticides (dieldrin, hexachlorocyclohexane-α, chlordanes, and hexachlorobenzene) were commonly measured at all sites and years (2003–2005). The pesticide concentration profiles were unique for individual parks, suggesting the importance of regional sources. The distribution of current-use pesticides among the parks was explained, using mass back trajectory analysis, based on the mass of individual current-use pesticides used in regions located within one-day upwind of the parks. For most pesticides and parks, more than 75% of the snowpack pesticide burden was attributed to regional transport. The researchers concluded that the majority of pesticide contamination in most U.S. national parks (outside Alaska) is due to regional pesticide applications.

18.7.2 Mercury

In the ROMO snowpack, Hg concentration was relatively high among parks that were included in the WACAP study; however, Hg burdens in fish were relatively low, suggesting low rates of Hg methylation and bioaccumulation. Nevertheless, Hg concentrations in some fish at both ROMO study lakes exceeded the contaminant health threshold for piscivorous mammals and birds (Landers et al. 2008). In ROMO, lake sediment analysis showed that there have been decreases in Hg deposition fluxes to the lakes since about 1990. The fluxes of Hg to most other WACAP lakes have either remained constant or increased since 1990.

Krabbenhoft et al. (2002) sampled 90 lakes in seven national parks in the western United States and analyzed their Hg and methyl Hg concentrations. The parks included GLAC and ROMO. Levels of methyl Hg were lowest in GLAC (0.02 ng/L) and similar among the other parks (~0.05 ng/L).

Watras et al. (1995) measured concentrations of Hg and methyl Hg in 12 lakes within GLAC. The dissolved organic C concentration explained much of the variability in Hg levels among lakes in GLAC. Other factors, including sulfate concentration, may also partially explain the differences observed.

Elevated levels of Hg have been measured in fish caught from several lakes in GLAC, resulting in fish consumption advisories (Downs and Stafford 2009). Relationships between Hg concentrations and fish size varied by lake and species. Especially strong relationships were found for lake trout (*Salvelinus namaycush*) in Harrison, McDonald, and St. Mary lakes (Stafford et al. 2004, Brinkmann 2007). Downs and Stafford (2009) sampled fish in four lakes in GLAC and analyzed them for Hg content. Study lakes included Lake McDonald, Bowman Lake, and Harrison Lake on the west side and St. Mary Lake on the east side of the park. Atmospheric Hg deposition is an important source of the Hg found in park fish. In McDonald and St. Mary lakes, five and seven fish species were sampled, respectively. The concentrations of Hg in fish tissue varied by species and lake. Lake trout were captured in all four lakes. In general, the top predators, such as lake trout and burbot (*Lota lota*), had the highest Hg concentrations (both absolute and normalized by size).

Eagles-Smith et al. (2014) sampled fish in 21 national parks and analyzed them for Hg concentrations in tissue. Results varied substantially by park and by water body. Fish from 19 lakes in ROMO were sampled. The mean Hg concentration in fish collected in ROMO was slightly lower than the mean across all parks studied. However, variability was high, with concentrations of Hg in fish ranging from about 20 ng/g wet weight (Lake Haiyaha) to 121 ng/g wet weight (Mirror Lake). Relatively large (>250 mm) brook trout from Mirror Lake and suckers from the Colorado and Fall rivers exceeded the EPA human health criterion and the avian reproductive impairment benchmark.

18.8 Summary

This chapter describes the air quality–related values of two national parks in the National Park Service's Rocky Mountain Network: ROMO and GLAC. Both are designated as Class I, giving them a heightened level of protection against harm caused by poor air quality under the Clean Air Act.

Although ROMO and GLAC are situated in relatively remote areas, they have experienced air pollutants from regional and local sources, including large power plants, urban areas, agriculture, industry, oil and gas development, and fires. ROMO is located in close proximity to several relatively large population centers along the eastern edge of the Colorado Front Range, including Denver and Fort Collins. GLAC is more remote, although it is adjacent to a former Al smelter and to pesticide application areas in Canada.

Levels of S deposition are relatively low in the Rocky Mountain Network region; N deposition is somewhat higher, especially along the Colorado Front Range. High-elevation lakes, soils, and streams are common in these parks. They are vulnerable to acidification as they often have limited buffering capacity, mainly in ROMO. However, because S and N deposition levels are not high, chronic acidification has not been documented in the network parks. Episodic acidification is likely more prevalent, again mainly in ROMO.

In addition to contributing to the potential for acidification, N is a nutrient and therefore N deposition can also cause undesirable enrichment of natural ecosystems, leading to changes in plant and algal species diversity and nutrient cycling. ROMO receives moderately high levels of N deposition, and ecosystem impacts from N inputs have been documented in this park, including

- Increased concentration of nitrate in surface water
- Change in the species composition and abundance of diatoms in alpine lakes
- Changes in the chemistry of soil and tree foliage, consistent with the onset of N saturation
- Changes in the species composition and abundance of alpine plants

Ozone pollution can harm human health, reduce plant growth, and cause visible injury to foliage. Both short-term concentrations, important to visitor and park staff health, and long-term cumulative exposures, important to plant health, are elevated at times in ROMO. Portions of ROMO are located in counties that have previously been designated as nonattainment for the national ozone standard and periodically experienced unhealthy air.

Particulate pollution can cause haze, reducing visibility. This is an important concern in the Rocky Mountains because the parks are known for their magnificent vistas. Haze has reduced the clarity of these views. Sulfates from large power plant emissions; N emissions from vehicles, power plants, industry, and agriculture; and organics from fires and other sources all contribute to haze in these parks.

Airborne contaminants, including Hg and other heavy metals, can accumulate in food webs, reaching toxic levels in top predators. Contaminants deposited from the atmosphere onto remote national park watersheds in the western United States can accumulate in fish to levels that are of concern regarding human and wildlife health. The WACAP study analyzed fish, vegetation, lake sediments, and other ecosystem compartments in GLAC and ROMO and found relatively high concentrations of toxics in both parks. Relevant findings include the following:

- Polycyclic aromatic hydrocarbon concentrations in snow, sediments, and vegetation were significantly elevated in GLAC and attributed partly to emissions from a local aluminum smelter (Usenko et al. 2010). Fire is also an important source of polycyclic aromatic hydrocarbon in ROMO and GLAC.
- Concentrations of banned pesticides were high in fish from both parks, with some fish exceeding health thresholds for humans and wildlife.
- Evidence of endocrine disruption was found in some fish from both parks.

References

Aber, J.D., K.J. Nadelhoffer, P. Steudler, and J.M. Mellilo. 1989. Nitrogen saturation in northern forest ecosystems. *BioScience* 39(6):378–386.

Ackerman, L.K., A.R. Schwindt, S.L. Simonich, D.C. Koch, T.F. Blett, C.B. Schreck, M.L. Kent, and D.H. Landers. 2008. Atmospherically deposited PBDEs, pesticides, PCBs, and PAHs in western U.S. national park fish: Concentrations and consumption guidelines. *Environ. Sci. Technol.* 42:2334–2341.

Agee, J.K. 1997. The severe weather wildfire: Too hot to handle? *Northwest Sci.* 71:153–157.

Aitken, W.M., W.R. Jacobi, and J.M. Staley. 1984. Ozone effects on seedlings of Rocky Mountain ponderosa pine. *Plant Dis.* 68:398–401.

Allen, C.D., M. Savage, D.A. Falk, K.F. Suckling, T.W. Swetnam, P. Schulke, P.B. Stacey, P. Morgan, M. Hoffman, and J.T. Klingel. 2002. Ecological restoration of southwestern ponderosa pine ecosystems: A broad perspective. *Ecol. Appl.* 12:1418–1433.

Anderson, N.J., I. Renberg, and U. Segerstrom. 1995. Diatom production responses to the development of early agriculture in a boreal forest lake-catchment (Kassjon, Northern Sweden). *J. Ecol.* 83:809–822.

Baron, J. 1992. Surface waters. *In* J. Baron (Ed.). *Biogeochemistry of a Subalpine Ecosystem: Loch Vale Watershed*. Springer-Verlag, New York, pp. 142–186.

Baron, J. and O.P. Bricker. 1987. Hydrologic and chemical flux in Loch Vale watershed, Rocky Mountain National Park. *In* R.C. Averett and D.M. McKnight (Eds.). *Chemical Quality of Water and the Hydrologic Cycle*. Lewis Publishers, Chelsea, MI, pp. 141–156.

Baron, J. and A.S. Denning. 1992. Hydrologic budget estimates. *In* J. Baron (Ed.). *Biogeochemistry of a Subalpine Ecosystem: Loch Vale Watershed*. Springer-Verlag, New York, pp. 28–47.

Baron, J., S.A. Norton, D.R. Beeson, and R. Hermann. 1986. Sediment diatom and metal stratigraphy from Rocky Mountain lakes with special reference to atmospheric deposition. *Can. J. Fish. Aquat. Sci.* 43:1350–1362.

Baron, J.S. 2006. Hindcasting nitrogen deposition to determine ecological critical load. *Ecol. Appl.* 16(2):433–439.

Baron, J.S., J. Allstott, and B.K. Newkirk. 1995. Analysis of long term sulfate and nitrate budgets in a Rocky Mountain basin. *In* K.A. Tonnesson, M.W. Williams, and M. Tranter (Eds.). *Biogeochemistry of Seasonally Snow-Covered Catchments*. IASH Publication No. 228. International Association of Hydrologic Sciences Press, Institute of Hydrology, Oxfordshire, U.K., pp. 255–261.

Baron, J.S. and D.H. Campbell. 1997. Nitrogen fluxes in a high elevation Colorado Rocky Mountain basin. *Hydrol. Process.* 11:783–799.

Baron, J.S., C.T. Driscoll, and J.L. Stoddard. 2011a. Inland surface water. *In* L.H. Pardo, M.J. Robin-Abbott, and C.T. Driscoll (Eds.). *Assessment of Nitrogen Deposition Effects and Empirical Critical Loads of Nitrogen for Ecoregions of the United States*. General Technical Report NRS-80. U.S. Forest Service, Newtown Square, PA, pp. 209–228.

Baron, J.S., C.T. Driscoll, J.L. Stoddard, and E.E. Richer. 2011b. Empirical critical loads of atmospheric nitrogen deposition for nutrient enrichment and acidification of sensitive US lakes. *BioScience* 61(8):602–613.

Baron, J.S., D.S. Ojima, E.A. Holland, and W.J. Parton. 1994. Analysis of nitrogen saturation potential in Rocky Mountain tundra and forest: Implications for aquatic systems. *Biogeochemistry* 27:61–82.

Baron, J.S., H.M. Rueth, A.M. Wolfe, K.R. Nydick, E.J. Allstott, J.T. Minear, and B. Moraska. 2000. Ecosystem responses to nitrogen deposition in the Colorado Front Range. *Ecosystems* 3:352–368.

Baron, J.S., T.M. Schmidt, and M.D. Hartman. 2009. Climate-induced changes in high elevation stream nitrate dynamics. *Glob. Change Biol.* 15(7):1777–1789.

Barry, R.G. 1973. A climatological transect of the east slope of the Front Range, Colorado. *Arct. Alp. Res.* 5:89–110.

Bayley, S.E., D.W. Schindler, B.R. Parker, M.P. Stainton, and K.G. Beaty. 1992. Effects of forest fire and drought on acidity of a base-poor boreal forest stream: Similarities between climatic warming and acidic precipitation. *Biogeochemistry* 17:191–204.

Benedict, K.B., C.M. Carrico, S.M. Kreidenweis, B. Schichtel, W.C. Malm, and J.L. Collett. 2013a. A seasonal nitrogen deposition budget for Rocky Mountain National Park. *Ecol. Appl.* 23(5):1156–1169.

Benedict, K.B., D. Day, F.M. Schwandner, S.M. Kreidenweis, B. Schichtel, W.C. Malm, and J.L. Collett Jr. 2013b. Observations of atmospheric reactive nitrogen species in Rocky Mountain National Park and across northern Colorado. *Atmos. Environ.* 64(0):66–76.

Bergström, A., P. Blomqvist, and M. Jansson. 2005. Effects of atmospheric nitrogen deposition on nutrient limitation and phytoplankton biomass in unproductive Swedish lakes. *Limnol. Oceanogr.* 50(3):987–994.

Bergström, A. and M. Jansson. 2006. Atmospheric nitrogen deposition has caused nitrogen enrichment and eutrophication of lakes in the northern hemisphere. *Glob. Change Biol.* 12:635–643.

Berrang, P., D.F. Karnosky, R.A. Mickler, and J.P. Bennett. 1986. Natural selection for ozone tolerance in *Populus tremuloides*. *Can. J. For. Res.* 16:1214–1216.

Blais, J.M. 2005. Biogeochemistry of persistent bioaccumulative toxicants—Processes affecting the transport of contaminants to remote areas. *Can. J. Fish. Aquat. Sci.* 62(1):236–243.

Blais, J.M., D.W. Schindler, D.C.G. Muir, L.E. Kimpe, D.B. Donald, and B. Rosenberg. 1998. Accumulation of persistent organochlorine compounds in mountains of western Canada. *Nature* 395:585–588.

Blett, T. and K. Morris. 2004. Nitrogen deposition: Issues and effects in Rocky Mountain National Park technical background document. Rocky Mountain Park Initiative, Denver, CO.

Bliss, L.C., G.M. Courtin, D.L. Pattie, R.R. Riewe, D.W.A. Whitfield, and P. Widden. 1973. Arctic tundra ecosystems. *Ann. Rev. Ecol. Syst.* 4:359–399.

Bloom, A.J., F.S.I. Chapin, and H.A. Mooney. 1985. Resource limitation in plants—An economic analogy. *Ann. Rev. Ecol. Syst.* 16:363–392.

Bowman, W.D. 1992. Inputs and storage of nitrogen in winter snowpack in an alpine ecosystem. *Arct. Alp. Res.* 24:211–215.

Bowman, W.D. 1994. Accumulation and use of nitrogen and phosphorus following fertilization in two alpine tundra communities. *Oikos* 70:261–270.

Bowman, W.D. 2000. Biotic controls over ecosystem response to environmental change in alpine tundra of the Rocky Mountains. *Ambio* 29(7):396–400.

Bowman, W.D. and M.C. Fisk. 2001. Primary production. *In* W.D. Bowman and T.R. Seastedt (Eds.). *Structure and Function of an Alpine Ecosystem: Niwot Ridge, Colorado.* Oxford University Press, Oxford, U.K., pp. 177–197.

Bowman, W.D., J.R. Gartner, K. Holland, and M. Wiedermann. 2006. Nitrogen critical loads for alpine vegetation and terrestrial ecosystem response: Are we there yet? *Ecol. Appl.* 16(3):1183–1193.

Bowman, W.D., J. Murgel, T. Blett, and E. Porter. 2012. Nitrogen critical loads for alpine vegetation and soils in Rocky Mountain National Park. *J. Environ. Manage.* 103:165–171.

Bowman, W.D. and H. Steltzer. 1998. Positive feedbacks to anthropogenic nitrogen deposition in Rocky Mountain alpine tundra. *Ambio* 27(7):514–517.

Bowman, W.D., T.A. Theodose, and M.C. Fisk. 1995. Physiological and production responses of plant growth forms to increases in limiting resources in alpine tundra: Implications for differential community response to environmental change. *Oecologia* 101:217–227.

Bowman, W.D., T.A. Theodose, J.C. Schardt, and R.T. Conant. 1993. Constraints of nutrient availability on primary production in two alpine tundra communities. *Ecology* 74:2085–2097.

Brinkmann, L. 2007. Mercury biomagnification in the upper south Saskatchewan River Basin. Masters, University of Lethbridge, Lethbridge, Alberta, Canada.

Brooks, P.D., M.W. Williams, and S.K. Schmidt. 1995. Snowpack controls on soil nitrogen dynamics in the Colorado alpine. *In* K.A. Tonnessen, M.W. Williams, and M. Tranter (Eds.). *Biogeochemistry of Seasonally Snow Covered Catchments. Proceedings of a Boulder Symposium July 1995.* IAHS Publication No. 228. International Association of Hydrological Sciences, Oxfordshire, U.K.

Brooks, P.D., M.W. Williams, and S.K. Schmidt. 1996. Microbial activity under alpine snowpacks, Niwot Ridge, Colorado. *Biogeochemistry* 32:93–113.

Burns, D.A. 2003. Atmospheric nitrogen deposition in the Rocky Mountains of Colorado and southern Wyoming—A review and new analysis of past study results. *Atmos. Environ.* 37:921–932.

Burns, D.A. 2004. The effects of atmospheric nitrogen deposition in the Rocky Mountains of Colorado and southern Wyoming, USA—A critical review. *Environ. Pollut.* 127:257–269.

Campbell, D.H., D.W. Clow, G.P. Ingersoll, M.A. Mast, N.E. Spahr, and J.T. Turk. 1995a. Nitrogen deposition and release in alpine watersheds, Loch Vale, Colorado, USA. *In* K.A. Tonnessen, M.W. Williams, and M. Tranter (Eds.). *Biogeochemistry of Seasonally Snow Covered Catchments. Proceedings of a Boulder Symposium July 1995.* IAHS Publication No. 228. International Association of Hydrological Sciences, Oxfordshire, U.K., pp. 243–253.

Campbell, D.H., D.W. Clow, G.P. Ingersoll, M.A. Mast, N.E. Spahr, and J.T. Turk. 1995b. Processes controlling the chemistry of two snowmelt-dominated streams in the Rocky Mountains. *Water Resour. Res.* 31(11):2811–2821.

Chapin, F.S. 1991. Integrated responses of plants to stress: A centralized system of physiological responses. *BioScience* 41:29–36.

Chapin, F.S., III, D.A. Johnson, and J.D. McKendrick. 1980. Seasonal movement of nutrients in plants of differing growth form in an Alaskan tundra ecosystem: Implications for herbivory. *J. Ecol.* 68:189–209.

Chapin, F.S.I. 1980. The mineral nutrition of wild plants. *Ann. Rev. Ecol. Syst.* 11:233–260.

Cheatham, J. 2011. Partnership to reduce the ecological effects of atmospheric nitrogen deposition in Rocky Mountain National Park. Abstract. *George Wright Society Conference on Parks*, Protected Areas, & Cultural Sites, New Orleans, LA.

Christie, C.E. and J.P. Smol. 1993. Diatom assemblages as indicators of lake trophic status in southeastern Ontario lakes. *J. Phycol.* 29:575–586.

Clow, D.W., J.O. Sickman, R.G. Striegl, D.P. Krabbenhoft, J.G. Elliott, M.M. Dornblaser, D.A. Roth, and D.H. Campbell. 2003. Changes in the chemistry of lakes and precipitation in high-elevation national parks in the western United States. *Water Resour. Res.* 39:1985–1999.

Coleman, M.D., R.E. Dickson, J.G. Isebrands, and D.F. Karnosky. 1996. Root growth and physiology of potted and field-grown trembling aspen exposed to tropospheric ozone. *Tree Physiol.* 16:145–152.

Cook, E.R., C.A. Woodhouse, C.M. Eakin, D.M. Meko, and D.W. Stahle. 2004. Long-term aridity changes in the western United States. *Science* 306(5698):1015–1018.

Covington, W.W. 2000. Helping western forests heal: The prognosis is poor for United States forest ecosystems. *Nature* 408:135–136.

Das, B., R.D. Vinebrooke, A. Sanchez-Azofeifa, B. Rivard, and A.P. Wolfe. 2005. Inferring sedimentary chlorophyll concentrations with reflectance spectroscopy: A novel approach to reconstructing historical changes in the trophic status of mountain lakes. *Can. J. Fish. Aquat. Sci.* 62:1067–1078.

Dennehy, K.F., D.W. Litke, C.M. Tate, and J.S. Heiny. 1993. South Platte River Basin—Colorado, Nebraska, and Wyoming. *Water Resour. Bull.* 29(4):647–683.

Denning, A.S., J. Baron, M.A. Mast, and M. Arthur. 1991. Hydrologic pathways and chemical composition of runoff during snowmelt in Loch Vale watershed, Rocky Mountain National Park, Colorado, USA. *Water Air Soil Pollut.* 59:107–123.

Doesken, N.J., R.A. Pielke, and A.P. Bliss. 2003. Climate of Colorado. Climatography of the United States No. 60 (updated January 2003). Colorado Climate Center, Colorado State University, Fort Collins, CO.

Downs, C.C. and C. Stafford. 2009. Glacier National Park fisheries inventory and monitoring annual report, 2008. National Park Service, Glacier National Park, West Glacier, MT.

Eagles-Smith, C.A., J.J. Willacker, Jr., and C.M. Flanagan Pritz. 2014. Mercury in fishes from 21 national parks in the western United States—Inter- and intra-park variation in concentrations and ecological risk. Open-File Report 2014-1051. U.S. Geological Survey, Reston, VA.

Eilers, J.M., D.F. Brakke, D.H. Landers, and P.E. Keller. 1988. Characteristics of lakes in mountainous areas of the western United States. *Int. Ver. Theor. Angew. Limnol. Verh.* 23:144–151.

Eilers, J.M., P. Kanciruck, R.A. McCord, W.S. Overton, L. Hook, D.J. Blick, D.F. Brakke et al. 1987. Characteristics of lakes in the western United States. Volume II: Data compendium for selected physical and chemical variables. EPA-600/3-86/054b. U.S. EPA, Washington, DC.

Ellis, B.K., J.A. Stanford, J.A. Craft, D.W. Chess, G.R. Gregory, and L.F. Marnell. 1992. Monitoring water quality of selected lakes in Glacier National Park, Montana: Analysis of data collected 1984–1990. Open file report 129-92. University of Montana, Flathead Lake Biol. Station, Polson, MT.

Ellis, R.A., D.J. Jacob, M.P. Sulprizio, L. Zhang, C.D. Holmes, B.A. Schichtel, T. Blett, E. Porter, L.H. Pardo, and J.A. Lynch. 2013. Present and future nitrogen deposition to national parks in the United States: Critical load exceedances. *Atmos. Chem. Phys.* 13(17):9083–9095.

Elser, J.J., T. Andersen, J.S. Baron, A.-K. Bergström, M. Jansson, M. Kyle, K.R. Nydick, L. Steger, and D.O. Hessen. 2009b. Shifts in lake N:P stoichiometry and nutrient limitation driven by atmospheric nitrogen deposition. *Science* 326:835–837.

Elser, J.J., K. Hayakawa, and J. Urabe. 2001. Nutrient limitation reduces food quality for zooplankton: Daphnia response to seston phosphorus enrichment. *Ecology* 82:898–903.

Elser, J.J., M. Kyle, L. Steger, K.R. Nydick, and J.S. Baron. 2009a. Nutrient availability and phytoplankton nutrient limitation across a gradient of atmospheric nitrogen deposition. *Ecology* 90(11):3062–3073.

Enders, S.K., M. Pagani, S. Pantoja, J.S. Baron, A.P. Wolfe, N. Pedentchouk, and L. Nuñez. 2008. Compound-specific stable isotopes of organic compounds from lake sediments track recent environmental changes in an alpine ecosystem, Rocky Mountain National Park (United States of America). *Limnol. Oceanogr.* 53(4):1468–1478.

Fenn, M.E., J.S. Baron, E.B. Allen, H.M. Rueth, K.R. Nydick, L. Geiser, W.D. Bowman et al. 2003a. Ecological effects of nitrogen deposition in the western United States. *BioScience* 53(4):404–420.

Fenn, M.E., R. Haeuber, G.S. Tonnesen, J.S. Baron, S. Grossman-Clark, D. Hope, D.A. Jaffe et al. 2003b. Nitrogen emissions, deposition, and monitoring in the western United States. *BioScience* 53(4):391–403.

Fenn, M.E., J.O. Sickman, A. Bytnerowicz, D.W. Clow, N.P. Molotch, J.E. Pleim, G.S. Tonnesen, K.C. Weathers, P.E. Padgett, and D.H. Campbell. 2009. Methods for measuring atmospheric nitrogen deposition inputs in arid and montane ecosystems of western North America. *In* A.H. Legge (Ed.). *Developments in Environmental Science.* Elsevier, Amsterdam, the Netherlands, pp. 179–228.

Findlay, D.L., R.E. Hecky, S.E.M. Kasian, M.P. Stainton, L.L. Hendzel, and E.U. Schindler. 1999. Effects on phytoplankton of nutrients added in conjunction with acidification. *Freshw. Biol.* 41:131–145.

Fisk, M.C. and S.K. Schmidt. 1996. Microbial responses to nitrogen additions in alpine tundra soil. *Soil Biol. Biogeochem.* 28:751–755.

Fisk, M.C., S.K. Schmidt, and T.R. Seastedt. 1998. Topographic patterns of above-and belowground production and nitrogen cycling in alpine tundra. *Ecology* 79(7):2253–2266.

Flannigan, M.D., Y. Bergeron, O. Engelmark, and B.M. Wotton. 1998. Future wildfire in circumboreal forests in relation to global warming. *J. Veg. Sci.* 9:469–476.

Gebhart, K.A., W.C. Malm, M.A. Rodriguez, M.G. Barna, B.A. Schichtel, K.B. Benedict, J.L. Collett, and C.M. Carrico. 2014. Meteorological and back trajectory modeling for the Rocky Mountain Atmospheric Nitrogen and Sulfur Study II. *Adv. Meteorol.* 2014:19.

Gebhart, K.A., B.A. Schichtel, W.C. Malm, M.G. Barna, M.A. Rodriguez, and J.L. Collett, Jr. 2011. Back-trajectory-based source apportionment of airborne sulfur and nitrogen concentrations at Rocky Mountain National Park, Colorado, USA. *Atmos. Environ.* 45(3):621–633.

Geiser, L.H., S.E. Jovan, D.A. Glavich, and M.K. Porter. 2010. Lichen-based critical loads for atmospheric nitrogen deposition in western Oregon and Washington forests, USA. *Environ. Pollut.* 158:2412–2421.

Gibson, J.H., J.N. Galloway, C.L. Schofield, W. McFee, R. Johnson, S. McCorley, N. Dise, and D. Herzog. 1983. Rocky Mountain Acidification Study. FWS/OBS-80/40.17. U.S. Fish and Wildlife Service, Department of Biological Services Eastern Energy and Land Use Team, Washington, DC.

Graybill, D.A., D.L. Peterson, and M.J. Arbaugh. 1993. Coniferous forests of the Colorado Front Range. *In* R.K. Olson, D. Binkley, and M. Böhm (Eds.). *Response of Western Forests to Air Pollution.* Springer-Verlag, New York, pp. 365–401.

Grenon, J. and M. Story. 2009. U.S. Forest Service region 1. Lake chemistry, NADP, and IMPROVE air quality data analysis. General Technical Report RMRS-GTR-230WWW. USDA Forest Service, Rocky Mountain Research Station, Fort Collins, CO.

Gribovicz, L. 2013. *Analysis of States' and EPA Oil & Gas Air Emissions Control Requirements for Selected Basins in the Western United States (2013 Update).* Western Region Air Partnership. Available at: http://www.wrapair2.org/pdf/2013-11x_O&G%20Analysis%20(master%20w%20State%20 Changes%2011–08).pdf (accessed October, 2016).

Grimalt, J.O., P. Fernandez, L. Berdie, R.M. Vilanova, J. Catalan, R. Psenner, R. Hofer et al. 2001. Selective trapping of organochlorine compounds in mountain lakes of temperate areas. *Environ. Sci. Technol.* 35(13):2690–2697.

Gubala, C.P., C.T. Driscoll, R.M. Newton, and C.F. Schofield. 1991. The chemistry of a near-shore lake region during spring snowmelt. *Environ. Sci. Technol.* 25(12):2024–2030.

Hageman, K.J., W.D. Hafner, D.H. Campbell, D.A. Jaffe, D.H. Landers, and S.L.M. Simonich. 2010. Variability in pesticide deposition and source contributions to snowpack in western U.S. national parks. *Environ. Sci. Technol.* 44(12):4452–4458.

Hall, R.I., P.R. Leavitt, R. Quinlan, A.S. Dixit, and J.P. Smol. 1999. Effects of agriculture, urbanization, and climate on water quality in the northern Great Plains. *Limnol. Oceanogr.* 44:739–756.

Hansen, W.R., J. Chronic, and J. Matelock. 1978. Climatography of the front range urban corridor and vicinity, Colorado. Geological Survey Professional Paper No. 1019. U.S. Government Printing Office, Washington, DC.

Hartman, M.D., J.S. Baron, D.S. Ojima, and W.J. Parton. 2005. Modeling the timeline for surface water acidification from excess nitrogen deposition for Rocky Mountain National Park. Abstracts of the *George Wright Society Biennial Conference*, Philadelphia, PA, p. 21.

Heit, M., C. Klusek, and J. Baron. 1984. Evidence of deposition of anthropogenic pollutants in remote rocky mountain lakes. *Water Air Soil Pollut.* 22(4):403–416.

Heuer, K., K.A. Tonnessen, and G.P. Ingersoll. 2000. Comparison of precipitation chemistry in the central Rocky Mountains, Colorado, USA. *Atmos. Environ.* 34:1713–1722.

Huston, M.A. 1994. *Biological Diversity: The Coexistence of Species on Changing Landscapes.* Cambridge University Press, New York, 681pp.

Ingersoll, G.P., M.A. Mast, D.H. Campbell, D.W. Clow, L. Nanus, and J.T. Turk. 2008. Trends in snowpack chemistry and comparison to national atmospheric deposition program results for the Rocky Mountains, US 1993–2004. *Atmos. Environ.* 42:6098–6113.

Interlandi, S.J. and S.S. Kilham. 1998. Assessing the effects of nitrogen deposition on mountain waters: A study of phytoplankton community dynamics. *Water Sci. Technol.* 38:139–146.

Interlandi, S.J. and S.S. Kilham. 2001. Limiting resources and the regulation of diversity in phytoplankton communities. *Ecology* 82:1270–1282.

Jaffe, D., I. Bertschi, L. Jaegle, P. Novelli, J.S. Reid, H. Tanimoto, R. Vingarzan, and D.L. Westphal. 2004. Long-range transport of Siberian biomass burning emissions and impact on surface ozone in eastern North America. *Geophys. Res. Lett.* 31:L161606. doi: 10.1029/2004GL020093.

Jaffe, D., D. Chand, W. Hafner, A. Westerling, and D. Spracklen. 2008. Influence of fires on O_3 concentrations in the western U.S. *Environ. Sci. Technol.* 42:5885–5891.

Jaffe, D.A. and J. Ray. 2007. Increase in ozone at rural sites in the western U.S. *Atmos. Environ.* 41(26):5452–5463.

James, R.L. and J.M. Staley. 1980. Photochemical air pollution damage survey of ponderosa pine within and adjacent to Denver, Colorado: A preliminary report. USDA Forest Service Forest Insect and Disease Management Biological Evaluation R2-80-6. USDA Forest Service Rocky Mountain Region, Lakewood, CO.

Karnosky, D.F., Z.E. Gagnon, D.D. Reed, and J.A. Witter. 1992. Threshold levels for foliar injury to *Populus tremuloides* by sulfur dioxide and ozone. *Can. J. For. Res.* 6:166–169.

Kegley, S.E., B.R. Hill, O. S., and A.H. Choi. 2011. PAN pesticide database. Pesticide Action Network, North America, San Francisco, CA.

Kelley, T.J. and D.H. Stedman. 1980. Effects of urban sources on acid precipitation in the western United States. *Science* 210:1043.

Keteles, K. 2011. Screening for pesticides in high elevation lakes in federal lands. U.S. EPA Region 8, Denver, CO.

Kidd, K.A., P.J. Blanchfield, K.H. Mills, V.P. Palace, R.E. Evans, J.M. Lazorchak, and R.W. Flick. 2007. Collapse of a fish population after exposure to a synthetic estrogen. *Proc. Natl. Acad. Sci.* 104:8897–8901.

Kimbrough, R.A. and D.W. Litke. 1998. Pesticides in surface water in agricultural and urban areas of the South Platte River Basin, from Denver, Colorado, to North Platte, Nebraska, 1993–94. Water-Resources Investigations Report 97-4230. U.S. Geological Survey, Denver, CO.

Kohut, R. 2007. Assessing the risk of foliar injury from ozone on vegetation in parks in the U.S. National Park Service's Vital Signs Network. *Environ. Pollut.* 149:348–357.

Kohut, R., C. Flanagan, J. Cheatham, and E. Porter. 2012. Foliar ozone injury on cutleaf coneflower at Rocky Mountain National Park, Colorado. *West. N. Am. Nat.* 72(1):32–42.

Korb, J.E. and T.A. Ranker. 2001. Changes in stand composition and structure between 1981 and 1996 in four Front Range plant communities in Colorado. *Plant Ecol.* 157:1–11.

Krabbenhoft, D.P., M.L. Olson, J.F. DeWild, D.W. Clow, R.G. Striegl, M.M. Dornblaser, and P. VanMetre. 2002. Mercury loading and methylmercury production and cycling in high-altitude lakes from the western United States. *Water Air Soil Pollut.* 2:233–249.

Lafrancois, B.M., K.R. Nydick, and B. Caruso. 2003. Influence of nitrogen on phytoplankton biomass and community composition in fifteen Snowy Range lakes (Wyoming, U.S.A.). *Arct. Anarct. Alp. Res.* 35(4):499–508.

Lafrancois, B.M., K.R. Nydick, B.M. Johnson, and J.S. Baron. 2004. Cumulative effects of nutrients and pH on the plankton of two mountain lakes. *Can. J. Fish. Aquat. Sci.* 61:1153–1165.

Landers, D.H., J.M. Eilers, D.F. Brakke, W.S. Overton, P.E. Kellar, W.E. Silverstein, R.D. Schonbrod et al. 1987. Characteristics of lakes in the western United States. Volume I: Population descriptions and physico-chemical relationships. EPA-600/3-86/054a. U.S. Environmental Protection Agency, Washington, DC.

Landers, D.H., S.L. Simonich, D.A. Jaffe, L.H. Geiser, D.H. Campbell, A.R. Schwindt, C.B. Schreck et al. 2008. The fate, transport, and ecological impacts of airborne contaminants in western national parks (USA). EPA/600/R-07/138. U.S. Environmental Protection Agency, Office of Research and Development, NHEERL, Western Ecology Division, Corvallis, OR.

Landers, D.H., S.M. Simonich, D. Jaffe, L. Geiser, D.H. Campbell, A. Schwindt, C.B. Schreck et al. 2010. The Western Airborne Contaminant Project (WACAP): An interdisciplinary evaluation of the impacts of airborne contaminants in western U.S. national parks. *Environ. Sci. Technol.* 44(3):855–859.

Langford, A.O. and F.C. Fehsenfeld. 1992. Natural vegetation as a source or sink for atmospheric ammonia: A case study. *Science* 255:581–583.

Lehmann, C.M.B., V.C. Bowersox, and S.M. Larson. 2005. Spatial and temporal trends of precipitation chemistry in the United States, 1985–2002. *Environ. Pollut.* 135:347–361.

Lenihan, J.M., C. Daly, D. Bachelet, and R.P. Neilson. 1998. Simulating broad-scale fire severity in a dynamic global vegetation model. *Northwest Sci.* 72:91–103.

Lotter, A.F. 1998. The recent eutrophication of Baldeggersee (Switzerland) as assessed by fossil diatom assemblages. *Holocene* 8:395–405.

Malm, W.C. 1999. *Introduction to Visibility.* Cooperative Institute for Research in the Atmosphere (CIRA), Fort Collins, CO.

Malm, W.C., J.L. Collett, Jr., M.G. Barna, K. Beem, C.M. Carrico, K.A. Gebhart, J.L. Hand et al. 2009. RoMANS: Rocky Mountain Atmospheric Nitrogen and Sulfur Study. Final report. Colorado State University, Fort Collins, CO.

Mast, M.A., J.I. Drever, and J. Baron. 1990. Chemical weathering in the Loch Vale watershed, Rocky Mountain National Park, Colorado. *Water Resour. Res.* 26:2971–2978.

Mast, M.A., W.T. Foreman, and S.V. Skaates. 2006. Organochlorine compounds and current-use pesticides in snow and lake sediment in Rocky Mountain National Park, Colorado, and Glacier National Park, Montana, 2002–03. U.S. Geological Survey Scientific Investigations Report 2006-5119. U.S. Department of the Interior, Reston, VA.

Mast, M.A., J.T. Turk, G.P. Ingersoll, D.W. Clow, and C.L. Kester. 2001. Use of stable sulfur isotopes to identify sources of sulfate in Rocky Mountain snowpacks. *Atmos. Environ.* 35:3303–3313.

McDonnell, T.C., S. Belyazid, T.J. Sullivan, H. Sverdrup, W.D. Bowman, and E.M. Porter. 2014. Modeled subalpine plant community response to climate change and atmospheric nitrogen deposition in Rocky Mountain National Park, USA. *Environ. Pollut.* 187:55–64.

McKenzie, D., Z.M. Gedalof, D.L. Peterson, and P. Mote. 2004. Climatic change, wildfire, and conservation. *Conserv. Biol.* 18:890–902.

McKenzie, D., S.M. O'Neill, N.K. Larkin, and R.A. Norheim. 2006. Integrating models to predict regional haze from wildland fire. *Ecol. Model.* 199:278–288.

McKnight, D.M., R.L. Smith, J.P. Bradbury, J.S. Baron, and S. Spaulding. 1990. Phytoplankton dynamics in 3 Rocky Mountain lakes, Colorado USA. *Arct. Alp. Res.* 22(3):264–274.

Michel, R.L., D. Campbell, D. Clow, and J.T. Turk. 2000. Timescales for migration of atmospherically derived sulphate through an alpine/subalpine watershed, Loch Vale, Colorado. *Water Resour. Res.* 36(1):27–36.

Michel, T.J., J.E. Saros, S.J. Interlandi, and A.P. Wolfe. 2006. Resource requirements of four freshwater diatom taxa determined by *in situ* growth bioassays using natural populations from alpine lakes. *Hydrobiologia* 568:235–243.

Morris, K., A. Mast, D. Clow, G. Wetherbee, J. Baron, C. Taipale, T. Blett, D. Gay, and J. Heath. 2014. 2012 Monitoring and tracking wet nitrogen deposition at Rocky Mountain National Park. NPS/NRSS/ARD/NRR-2014/757. National Park Service, Denver, CO.

Musselman, R.C., L. Hudnell, M.W. Williams, and R.A. Sommerfeld. 1996. Water chemistry of Rocky Mountain Front Range aquatic ecosystems. Research Paper RM-RP-325. USDA Rocky Mountain Forest and Range Experiment Station, Fort Collins, CO.

Musselman, R.C., L. Hudnell, M.W. Williams, and R.A. Sommerfeld. 2004. Water chemistry of Rocky Mountain Front Range aquatic ecosystems. RM-RP-325. U.S. Department of Agriculture, Forest Service, Fort Collins, CO.

Nanus, L., D.H. Campbell, G.P. Ingersoll, D.W. Clow, and M.A. Mast. 2003. Atmospheric deposition maps for the Rocky Mountains. *Atmos. Environ.* 37:4881–4892.

Nanus, L., D.W. Clow, J.E. Saros, V.C. Stephens, and D.H. Campbell. 2012. Mapping critical loads of nitrogen deposition for aquatic ecosystems in the Rocky Mountains, USA. *Environ. Pollut.* 166:125–135.

Nanus, L., M.W. Williams, D.H. Campbell, K.A. Tonnessen, T. Blett, and D.W. Clow. 2009. Assessment of lake sensitivity to acidic deposition in national parks of the Rocky Mountains. *Ecol. Appl.* 19(4):961–973.

National Acid Precipitation Assessment Program (NAPAP). 1991. Integrated assessment report. National Acid Precipitation Assessment Program, Washington, DC.

National Park Service (NPS). 2010. Air quality in national parks: 2009 annual performance and progress report. Natural Resource Report NPS/NRPC/ARD/NRR-2010/266. National Park Service, Air Resources Division, Denver, CO.

Pardo, L.H., M.E. Fenn, C.L. Goodale, L.H. Geiser, C.T. Driscoll, E.B. Allen, J.S. Baron et al. 2011b. Effects of nitrogen deposition and empirical nitrogen critical loads for ecoregions of the United States. *Ecol. Appl.* 21(8):3049–3082.

Pardo, L.H., M.J. Robin-Abbott, and C.T. Driscoll. 2011a. Assessment of nitrogen deposition effects and empirical critical loads of nitrogen for ecoregions of the United States. General Technical Report NRS-80. U.S. Forest Service, Newtown Square, PA.

Parrish, D.D., C.H. Hahn, D.W. Fahey, E.J. Williams, M.J. Bollinger, G. Hubler, M.P. Buhr et al. 1990. Systematic variations in the concentration of NO_x at Niwot Ridge, Colorado. *J. Geophys. Res.* 95(D2):1817–1836.

Parrish, D.D., R.B. Norton, M.J. Bollinger, S.C. Liu, P.C. Murphy, D.L. Albritton, F.C. Fehsenfeld, and B.J. Huebert. 1986. Measurements of HNO_3 and NO_3^- particulates at a rural site in the Colorado Mountains. *J. Geophys. Res.* 91:5379–5393.

Paul, A.J., P.R. Leavitt, D.W. Schindler, and A.K. Hardie. 1995. Direct and indirect effects of predation by a calanoid copepod (subgenus: *Hesperodiaptomus*) and of nutrients in a fishless alpine lake. *Can. J. Fish. Aquat. Sci.* 52:2628–2638.

Pearcy, R.W., O. Björkman, M.M. Caldwell, J.E. Keeley, R.K. Monson, and B.R. Strain. 1987. Carbon gain by plants in natural environments. *BioScience* 37:21–29.

Peterson, D.L., M.J. Arbaugh, and L.J. Robinson. 1993. Effects of ozone and climate on ponderosa pine (*Pinus ponderosa*) growth in the Colorado Rocky Mountains. *Can. J. For. Res.* 23:1750–1759.

Peterson, D.L., T.J. Sullivan, J.M. Eilers, S. Brace, D. Horner, K. Savig, and D. Morse. 1998. Assessment of air quality and air pollutant impacts in national parks of the Rocky Mountains and northern Great Plains. NPS D-657. U.S. Department of the Interior, National Park Service, Air Resources Division, Lakewood, CO.

Porter, E. and S. Johnson. 2007. Translating science in policy: Using ecosystem thresholds to protect resources in Rocky Mountain National Park. *Environ. Pollut.* 149:268–280.

Porter, E., H. Sverdrup, and T.J. Sullivan. 2012. Estimating and mitigating the impacts of climate change and air pollution on alpine plant communities in national parks. *Park Sci.* 28(2):58–64.

Renberg, I., T. Korsman, and N.J. Anderson. 1993. A temporal perspective of lake acidification in Sweden. *Ambio* 22:264–271.

Rueth, H.M. and J.S. Baron. 2002. Differences in Englemann spruce forest biogeochemistry east and west of the continental divide in Colorado, USA. *Ecosystems* 5:45–57.

Saros, J.E., S.J. Interlandi, A.P. Wolfe, and D.R. Engstrom. 2003. Recent changes in the diatom community structure of lakes in the Beartooth Mountain Range, U.S.A. *Arct. Anarct. Alp. Res.* 35(1):18–23.

Saros, J.E., T.J. Michel, S.J. Interlandi, and A.P. Wolfe. 2005. Resource requirements of *Asterionella formosa* and *Fragilaria crotonensis* in oligotrophic alpine lakes: Implications for recent phytoplankton community reorganizations. *Can. J. Fish. Aquat. Sci.* 62:1681–1689.

Savig, K. and D. Morse. 1998. Visibility sections. *In* D.L. Peterson, T.J. Sullivan, J.M. Eilers, S. Brace and D. Horner (Eds.). *Assessment of Air Quality and Air Pollutant Impacts in National Parks of the Rocky Mountains and Northern Great Plains.* U.S. Department of the Interior, National Park Service, Air Resources Division, Denver, CO.

Schlesinger, W.H. and B.F. Chabot. 1977. The use of water and minerals by evergreen and deciduous shrubs in Okefenokee Swamp. *Bot. Gaz.* 138:490–497.

Schoettle, A.W., K. Tonnessen, J. Turk, J. Vimont, and R. Amundson. 1999. An assessment of the effects of human-caused air pollution on resources within the interior Columbia River Basin. PNW-GTR-447. USDA Forest Service, Pacific Northwest Research Station, Portland, OR.

Schreck, C.B. and M. Kent. 2013. Extent of endocrine disruption in fish of western and Alaskan national parks. Final report. NPS-OSU Task Agreement J8W07080024, Oregon State University, Corvallis, OR.

Schwede, D.B. and G.G. Lear. 2014. A novel hybrid approach for estimating total deposition in the United States. *Atmos. Environ.* 92:207–220.

Schwindt, A.R., M.L. Kent, L.K. Ackerman, S.L.M. Simonich, D.H. Landers, T. Blett, and C.B. Schreck. 2009. Reproductive abnormalities in trout from western U.S. national parks. *Trans. Am. Fish. Soc.* 138:522–531.

Sherrod, S.K. and T.R. Seastedt. 2001. Effects of the northern pocket gopher (*Thomomys talpoides*) on alpine soil characteristics, Niwot Ridge, CO. *Biogeochemistry* 55:195–218.

Sickman, J.O., J.M. Melack, and D.W. Clow. 2003. Evidence for nutrient enrichment of high-elevation lakes in the Sierra Nevada, California. *Limnol. Oceanogr.* 48(5):1885–1892.

Sickman, J.O., J.M. Melack, and J.L. Stoddard. 2002. Regional analysis of inorganic nitrogen yield and retention in high-elevation ecosystems of the Sierra Nevada and Rocky Mountains. *Biogeochemistry* 57/58:341–374.

Silverstein, M. and C. Taipale. 2006. Rocky Mountain national park initiative: Response to questions and benefits of planned programs. Presentation to the *Rocky Mountain National Park Subcommittee of the Colorado Air Quality Control Commission*, February 15, 2006.

Simonich, S.L. and R.A. Hites. 1995. Global distribution of persistent organochlorine compounds. *Science* 269:1851–1854.

Sisterson, D.L., V.C. Bowersox, T.P. Meyers, A.R. Olsen, and R.J. Vong. 1990. Deposition monitoring: Methods and results. State of Science and Technology Report 6. National Acid Precipitation Assessment Program, Washington, DC.

Smith, W.H. 1990. *Air Pollution and Forests: Interactions between Air Contaminants and Forest Ecosystems*. Springer-Verlag, New York, 618pp.

Stafford, C.P., B. Hansen, and J.A. Stanford. 2004. Mercury in fishes and their diet items from Flathead Lake, Montana. *Trans. Am. Fish. Soc.* 133(2):349–357.

Steltzer, H. and W.D. Bowman. 1998. Differential influence of plant species on soil N transformations within the alpine tundra. *Ecosystems* 1:464–474.

Stoddard, J.L. 1994. Long-term changes in watershed retention of nitrogen: Its causes and aquatic consequences. *In* L.A. Baker (Ed.). *Environmental Chemistry of Lakes and Reservoirs*. American Chemical Society, Washington, DC, pp. 223–284.

Stolte, K. 1987. Summary of ponderosa pine data for Rocky Mountain National Park. National Park Service, Air Quality Division, Lakewood, CO.

Suarez-Murias, T., J. Glass, E. Kim, L. Melgoza, and T. Najita. 2009. California regional haze plan. California Environmental Protection Agency, Air Resources Board, Sacramento, CA.

Suding, K.N., A.E. Miller, H. Bechtold, and W.D. Bowman. 2006. The consequence of species loss on ecosystem nitrogen cycling depends on community compensation. *Oecologia* 149:141–149.

Sullivan, T.J. 2000. *Aquatic Effects of Acidic Deposition*. CRC Press, Boca Raton, FL.

Sullivan, T.J., B.J. Cosby, K.A. Tonnessen, and D.W. Clow. 2005. Surface water acidification responses and critical loads of sulfur and nitrogen deposition in Loch Vale watershed, Colorado. *Water Resour. Res.* 41:W01021. doi:10.1029/2004 WR 003414.

Sullivan, T.J. and J.M. Eilers. 1994. Assessment of deposition levels of sulfur and nitrogen required to protect aquatic resources in selected sensitive regions of North America. Final report prepared for Technical Resources, Inc., Rockville, MD, under contract to U.S. Environmental Protection Agency, Environmental Research Laboratory-Corvallis. E&S Environmental Chemistry, Inc., Corvallis, OR.

Sullivan, T.J., G.T. McPherson, T.C. McDonnell, S.D. Mackey, and D. Moore. 2011. Evaluation of the sensitivity of inventory and monitoring national parks to acidification effects from atmospheric sulfur and nitrogen deposition. Natural Resource Report. NPS/NRPC/ARD/NRR—2011/349. U.S. Department of the Interior, National Park Service, Denver, CO.

Sverdrup, H., T.C. McDonnell, T.J. Sullivan, B. Nihlgard, S. Belyazid, B. Rihm, E. Porter, W.D. Bowman, and L. Geiser. 2012. Testing the feasibility of using the ForSAFE-VEG model to map the critical load of nitrogen to protect plant biodiversity in the Rocky Mountains region, USA. *Water Air Soil Pollut.* 23:371–387.

Theodose, T.A. and W.D. Bowman. 1997. Nutrient availability, plant abundance, and species diversity in two alpine tundra communities. *Ecology* 78:1861–1872.

Thomas, B. and D.F. Grigal. 1976. Phosphorus conservation by evergreenness of mountain laurel. *Oikos* 27:19–26.

Tilman, D. 1981. Tests of resource competition theory using 4 species of Lake Michigan algae. *Ecology* 62:802–815.

Turk, J.T. and D.H. Campbell. 1997. Are aquatic resources of the Mt. Zirkel wilderness area affected by acid deposition and what will emissions reductions at the local power plants do? USGS Fact Sheet 043-97. U.S. Department of the Interior, U.S. Geological Survey, Denver, CO.

Turk, J.T., D.H. Campbell, and N.E. Spahr. 1992. Initial findings of synoptic snowpack sampling in Colorado Rocky Mountains. Open-file report 92-645. U.S. Geological Survey, Denver, CO.

Turk, J.T. and N.E. Spahr. 1991. Rocky Mountains. *In* D.F. Charles (Ed.). *Acidic Deposition and Aquatic Ecosystems: Regional Case Studies*. Springer-Verlag, New York, pp. 471–501.

Usenko, S., S.L. Simonich, K.J. Hageman, J.E. Schrlau, L. Geiser, D.H. Campbell, P.G. Appleby, and D.H. Landers. 2010. Sources and deposition of polycyclic aromatic hydrocarbons to western U.S. national parks. *Environ. Sci. Technol.* 44(12):4512–4518.

van der Oost, R., J. Beyer, and N.P.E. Vermeulen. 2003. Fish bioaccumulation and biomarkers in environmental risk assessment: A review. *Environ. Toxicol. Pharmacol.* 13:57–149.

Wang, D., D.F. Karnosky, and F.H. Bormann. 1986. Effects of ambient ozone on the productivity of *Populus tremuloides* Michx. grown under field conditions. *Can. J. For. Res.* 16:47–55.

Wania, F. and D. Mackay. 1996. Tracking the distribution of persistent organic pollutants. *Environ. Sci. Technol.* 30:390A–396A.

Watras, C.J., K.A. Morrison, and N.S. Bloom. 1995. Mercury in remote Rocky Mountain lakes of Glacier National Park, Montana, in comparison with other temperate North American regions. *Can. J. Fish. Aquat. Sci.* 52:1220–1228.

Webster, K.E. and L. Brezonik. 1995. Climate confounds detection of chemical trends related to acid deposition in upper midwest lakes in the USA. *Water Air Soil Pollut.* 85:1575–1580.

Westerling, A.L., H.G. Hidalgo, D.R. Cayan, and T.W. Swetnam. 2006. Warming and earlier spring increase western U.S. forest wildfire activity. *Science* 313(5789):940–943.

Wigington, P.J., J.P. Baker, D.R. DeWalle, W.A. Kretser, P.S. Murdoch, H.A. Simonin, J. Van Sickle, M.K. McDowell, D.V. Peck, and W.R. Barchet. 1993. Episodic acidification of streams in the northeastern United States: Chemical and biological results of the Episodic Response Project. EPA/600/R-93/190. U.S. Environmental Protection Agency, Washington, DC.

Wilcox, G. and J. Decosta. 1982. The effect of phosphorus and nitrogen addition on the algal biomass and species composition of an acidic lake. *Arch. Hydrobiol.* 94:393–424.

Williams, M.W., J.S. Baron, N. Caine, R. Sommerfeld, and J.R. Sanford. 1996b. Nitrogen saturation in the Rocky Mountains. *Environ. Sci. Technol.* 30:640–646.

Williams, M.W. and J.M. Melack. 1991. Precipitation chemistry in and ionic loading to an alpine basin, Sierra Nevada. *Water Resour. Res.* 27:1563–1574.

Williams, M.W., T. Platts-Mills, and N. Caine. 1996a. Landscape controls on surface water nitrate concentrations at catchment and regional scales in the Colorado Rocky Mountains. *Chapman Conference: Nitrogen Cycling in Forested Catchments*, Sun River, OR.

Williams, M.W. and K.A. Tonnessen. 2000. Critical loads for inorganic nitrogen deposition in the Colorado Front Range, USA. *Ecol. Appl.* 10:1648–1665.

Wolfe, A.P., J.S. Baron, and R.J. Cornett. 2001. Anthropogenic nitrogen deposition induces rapid ecological changes in alpine lakes of the Colorado Front Range (USA). *J. Paleolimnol.* 25:1–7.

Wolfe, A.P., A.C. Van Gorpe, and J.S. Baron. 2003. Recent ecological and biogeochemical changes in alpine lakes of Rocky Mountain National Park (Colorado, USA): A response to anthropogenic nitrogen deposition. *Geobiology* 1(2):153–168.

Woodward, D.F., A.M. Farag, M.E. Mueller, E.E. Little, and F.A. Vertucci. 1989. Sensitivity of endemic Snake River cutthroat trout to acidity and elevated aluminum. *Trans. Am. Fish. Soc.* 118:630–641.

Wyoming Department of Environmental Quality. 2010. Validated data. Ozone data products for Jonah Field. Cheyenne, WY.

Xu, J., D. DuBois, M. Pitchford, M. Green, and V. Etyemezian. 2006. Attribution of sulfate aerosols in federal class I areas of the western United States based on trajectory regression analysis. *Atmos. Environ.* 40:3433–3447.

19

Greater Yellowstone Network

19.1 Background

This chapter addresses air quality issues in two parks in the Greater Yellowstone area: Yellowstone (YELL) and Grand Teton (GRTE). Figure 19.1 shows the Greater Yellowstone Network boundary, along with locations of each park and the population centers with more than 10,000 people around the network. Although there are no human population centers of any magnitude in this area, Salt Lake City, Boise, and Denver are all within 300 mi (483 km) of the network boundary.

19.2 Atmospheric Emissions and Deposition

In general, air quality in the Greater Yellowstone Network is considerably better than in most other areas of the United States. This is partly due to the absence of high levels of fossil fuel combustion associated with power plants and metropolitan areas. Although current air quality is good in most areas, resource sensitivity is generally high, and some pollutants may have potential impacts on national park resources in both the short and long term.

In 2006, the Greater Yellowstone Area Clean Air Partnership updated its earlier (1999) assessment of air quality in the greater Yellowstone area. The focus was on new information regarding (1) urban and industrial emissions, (2) oil and gas development in southwestern Wyoming, (3) smoke from prescribed fire and wildfire, and (4) snowmobile emissions. Urban and industrial emissions were identified as important concerns. They consisted largely of industrial, petroleum refinery, gas transmission, agricultural processing, wood processing, mining, power generation, sand and gravel, and mining emissions sources. The oil and gas development, in particular, has been expanded in southwestern Wyoming, including in the Upper Green River Basin.

Wildfire smoke was identified as having the most dramatic air quality impact. Wildfire in this region commonly has both natural and human causes. Emissions of C and particulate matter from prescribed fire and wildfire contribute to periodic visibility impairment. The increased popularity of snowmobiles in the Greater Yellowstone Area during the late 1980s and 1990s raised concerns about air pollution and other impacts on natural resources and the visitor experience. By 2000, visitors were making about 75,000 snowmobile trips and 1,300 snowcoach trips into YELL during the 90-day winter season. The majority entered the park through the West Entrance from West Yellowstone.

Sullivan et al. (2011a) reported that, in general, annual sulfur dioxide and N emissions across the network region in 2002 were both less than 1 ton/mi^2, although emissions were

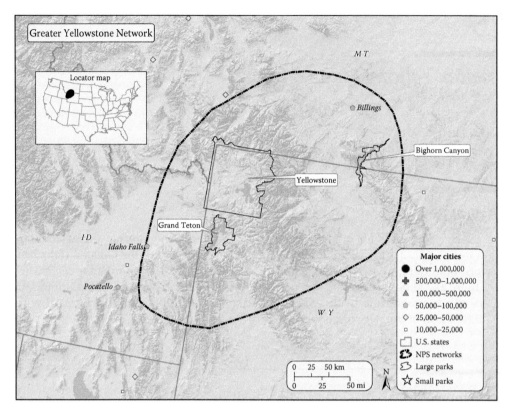

FIGURE 19.1
The Greater Yellowstone Network boundary and locations of parks and population centers larger than 10,000 people near the network.

higher close to GRTE and YELL as compared with other portions of the Greater Yellowstone Network region. There were no large individual sulfur dioxide point sources in close proximity to Greater Yellowstone Network parks, although there were two point sources located to the south of YELL and GRTE that emitted S in the range of 5,000–20,000 tons/yr.

In the central part of the United States from Texas to North Dakota, sulfur dioxide emissions increased during the 1990s, with statistically significant increases in several states, including Wyoming (31%; Malm et al. 2002). Nevertheless, all parks within the network had S deposition values less than 2 kg S/ha/yr in 2002. Total N deposition within the network region ranged from below 2 kg N/ha/yr in the southeast to over 5 kg N/ha/yr in and around portions of GRTE and the southwestern portion of YELL (Sullivan et al. 2011a). There were few N emissions point sources of any magnitude near this network. The few relatively large point sources that did exist were sources of oxidized N.

County-level emissions near the Greater Yellowstone Network, based on data from the EPA's National Emissions Inventory during a recent time period (2011), were generally low. Most counties near network parks had relatively low sulfur dioxide emissions (<1 ton/mi²/yr), although higher emissions occurred in some areas, especially south of YELL and GRTE. Oxidized N emissions were generally higher, especially to the southwest of the parks. Emissions of ammonia near the parks were somewhat lower, with most counties showing emissions levels below 2 tons/mi²/yr.

Most S and N deposition in this network is due to long-distance transport from emissions sources to the west. Industrial and electric utility facilities in Idaho produce oxidized N and volatile organic compounds that can contribute to ozone formation in or upwind from the Greater Yellowstone Network. The Salt Lake City, Utah, region to the southwest may also contribute ozone precursors. Most emissions of oxidized N and sulfur dioxide in Wyoming are from industrial and power-generation facilities located to the east of YELL and GRTE; only relatively uncommon easterly winds would transport these pollutants into the parks.

Grenon and Story (2009) reported increasing trends in the wet deposition of ammonium within at least one season of the year since the 1980s for all monitoring sites in the northern Rocky Mountains region. This result agreed with the analyses reported by Ingersoll et al. (2008) and Lehmann et al. (2005). This trend of increasing ammonium wet deposition has likely been due in part to increased agricultural emissions of reduced N (ammonia). The concentrations of nitrate in precipitation at National Atmospheric Deposition Program monitoring sites in the Rocky Mountains generally increased over time until about the year 2000, especially near urban areas (Fenn et al. 2003, Lehmann et al. 2005). In contrast, wet S deposition decreased between 1985 and 2002 at National Atmospheric Deposition Program sites throughout the western United States (Lehmann et al. 2005), likely due in large part to reductions in power plant emissions required by the Clean Air Act.

The Grand Teton Reactive Nitrogen Deposition study investigated atmospheric concentrations and deposition of various reactive N constituents in and around GRTE. Wet and dry deposition of ammonium were the main reactive N species, together accounting for about 60% of the N deposition in GRTE (Benedict et al. 2013). The park is closer to N emissions source regions than YELL. These include agricultural activities in the Snake River Valley and northern Utah, oil and gas extraction and power plants in Wyoming, and both wildfire and prescribed fire. Benedict et al. (2013) estimated total N deposition at the National Oceanographic and Atmospheric Administration Climate Center in 2010–2011 of 2.5 kg N/ha/yr.

The WACAP (Landers et al. 2008) studied airborne contaminants in a number of national parks of the western United States, including GRTE, from 2002 to 2007. Concentrations of dacthal, endosulfans, hexachlorocyclohexane-γ, hexachlorobenzene, hexachlorocyclohexane-α, and chlordanes in the air sampled at GRTE were relatively high among the western parks and exceeded the median for the 20 parks investigated in that study.

Snowfall that accumulates in the seasonal snowpack generally accounts for about 50%–70% of the annual precipitation in headwater areas of the Greater Yellowstone Network (Western Regional Climate Center 2004 [http://www.wrcc.dri.edu, accessed December, 2004], Ingersoll et al. 2005). Atmospherically deposited pollutants accumulate in these snowpacks and are released to surface waters during the snowmelt. The U.S. Geological Survey monitors the chemistry of the snowpack at multiple mountain locations from New Mexico to Montana. Concentrations of ammonium in snow are highest in the vicinity of YELL and GRTE. Nitrate and sulfate concentrations in snow tend to be highest to the south of the Greater Yellowstone Network (Ingersoll et al. 2005, Ingersoll et al. 2011). These contaminants can contribute episodic pulses of pollutants to surface waters during snowmelt.

Snowmobile use in YELL has led to concern about possible effects of snowmobile emissions on snowpack chemistry and air quality. Ingersoll (1999) sampled the snowpack at approximately the time of maximum snow accumulation at a variety of locations in and near YELL to determine the association between snowmobile use and snowpack chemistry. Concentrations of ammonium, nitrate, sulfate, benzene, and toluene were positively correlated with snowmobile use. Concentrations decreased rapidly, however, with distance from roadways. These pollutants are generally dispersed into surrounding areas at concentrations below levels likely to threaten ecosystem health, although local episodic effects might be possible (Ingersoll 1999). The two most common air pollutants emitted by snowmobiles are carbon monoxide and fine particulate matter. Winter-time air quality monitoring has found that, over time, concentrations of carbon monoxide and fine particulate matter have decreased at the West Entrance, where most snowmobiles enter the park, and at Old Faithful, a popular destination for the snowmobiles. The decreases are likely due to the implementation of Best Available Technology regulation for snowmobiles and a decrease in the number of snowmobiles entering the park. The Best Available Technology regulation was introduced in winter 2002–2003 and has significantly reduced snowmobile emissions, with many machines now having cleaner four-stroke engines as compared to the more polluting two-stroke engines. Air quality in YELL meets the EPA's health standards for carbon monoxide and fine particulate matter (Ray 2012).

Schwede and Lear (2014) recently reported total S and N deposition estimates developed by the National Atmospheric Deposition Program Total Deposition Science Committee. This approach combined monitoring and modeling data. Total S deposition in and around the Greater Yellowstone Network for the period 2010–2012 was generally <3.5 kg S/ha/yr. Oxidized inorganic N deposition for the period 2010–2012 was also less than 3.5 kg N/ha/yr throughout the parklands. Most areas received 2–5 kg N/ha/yr of ammonium deposition during this same period. Total N deposition was generally about 2–3 kg N/ha/yr, although slightly higher values were estimated for portions of southwestern YELL and western GRTE (Figure 19.2; Table 19.1).

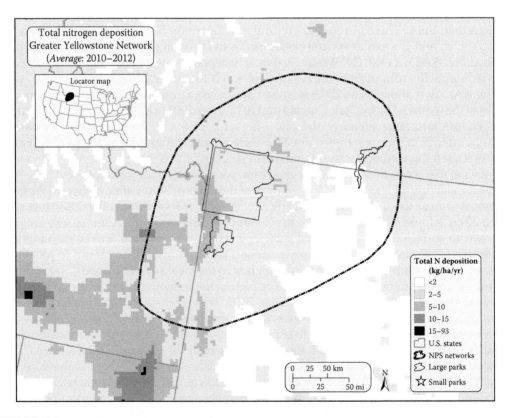

FIGURE 19.2
Total N deposition for the three-year period centered on 2011 in and around the Greater Yellowstone Network. (From Schwede, D.B. and Lear, G.G., *Atmos. Environ.*, 92, 207, 2014.)

TABLE 19.1

Average Changes in S and N Deposition[a] between 2001 and 2011 across Park Grid Cells at Selected Greater Yellowstone Network Parks

Park Code	Parameter	2001 Average (kg/ha/yr)	2011 Average (kg/ha/yr)	Absolute Change (kg/ha/yr)	Percent Change	2011 Minimum (kg/ha/yr)	2011 Maximum (kg/ha/yr)	2011 Range (kg/ha/yr)
GRTE	Total S	1.74	2.01	0.27	16.1	0.94	3.24	2.30
	Total N	4.07	4.70	0.63	15.3	2.58	7.79	5.22
	Oxidized N	2.28	1.94	−0.33	−15.6	1.17	3.33	2.16
	Reduced N	1.79	2.76	0.97	56.2	1.41	4.47	3.06
YELL	Total S	1.78	1.94	0.16	8.9	1.09	2.76	1.67
	Total N	3.71	4.08	0.37	9.6	1.89	6.78	4.89
	Oxidized N	2.03	1.73	−0.30	−15.2	0.86	2.65	1.79
	Reduced N	1.68	2.36	0.68	40.1	1.03	4.13	3.10

[a] Deposition estimates were determined by the Total Deposition project, based on three-year averages centered on 2001 and 2011 for all ~4 km grid cells in each park. The minimum, maximum, and range of 2011 S and N deposition within each park are also shown.

Atmospheric S deposition levels have increased at YELL and GRTE since 2001, based on Total Deposition project estimates (Table 19.1). Total N deposition also increased at YELL (9.6%) and GRTE (15.3%). Oxidized N decreased and ammonium deposition increased (both substantially) at both parks over that time period.

19.3 Acidification

This network is known to contain acid-sensitive surface waters and geology, and there are numerous high-elevation lakes and streams and low-order streams, especially in GRTE, that might be sensitive to acidification. The overall level of concern for acidification effects on Inventory and Monitoring parks within this network was judged by Sullivan et al. (2011a) to be high. While these rankings provide an indication of risk, park-specific data, particularly data on ecosystem sensitivity, are needed to fully evaluate the risk from acidification.

Many high-elevation, low-order streams are found above 2500 m elevation in the Teton Range, which runs north to south along the western border of GRTE. YELL also contains numerous high-elevation, low-order streams, especially along the eastern border of the park and in the northwest. These high-elevation water bodies might be more prone than lakes and streams at lower elevation to acidification effects and therefore are potentially more susceptible to acidification in response to current and future atmospheric S and N inputs.

There is no information available on the possibility of terrestrial acidification attributable to acidic deposition in the Greater Yellowstone Network. Sensitivity to soil acidification may be relatively high. However, given the generally low levels of S and N deposition to this network, such effects are unlikely under current deposition levels.

The available information on acid–base chemistry of surface waters in the Greater Yellowstone Network is based in part on synoptic data from the EPA's Western Lakes Survey collected in the 1980s (Landers et al. 1987) and some more localized studies. Acid anion concentrations tend to be low during fall, the season of Western Lakes Survey sampling, but can be higher during snowmelt. Concentrations of sulfate in lake waters are generally low, although watershed sources of S in this network are substantial in some cases. Nitrate concentrations in most Western Lakes Survey lakes in the Greater Yellowstone Network were near 0 µeq/L during fall sampling (Landers et al. 1987).

The Western Lakes Survey sampled six lakes in YELL and nine lakes in surrounding areas (Landers et al. 1987). One of the lakes in the park was acidic (acid neutralizing capacity = −24 µeq/L). This acidity was attributable to geothermal inputs as evidenced by the extremely high concentrations of sulfate (818 µeq/L) and sum of base cations (1330 µeq/L). One lake had an acid neutralizing capacity of 139 µeq/L; its pH was also relatively low (6.6), mainly as a consequence of high concentration of naturally occurring dissolved organic carbon (11 mg/L). Several of the lakes in surrounding areas surveyed by the Western Lakes Survey were more acid sensitive than those in the park; four of the nine surrounding lakes had acid neutralizing capacity less than 100 µeq/L and one was below 50 µeq/L. Most of these had low base cation concentrations and low sulfate concentrations (~8–10 µeq/L) that could reasonably be attributed to atmospheric inputs.

High-elevation lakes constitute an important sensitive aquatic resource of concern in the Greater Yellowstone Network, especially within GRTE. There are about 90 alpine and subalpine lakes and ponds in GRTE located above about 2700 m elevation. Many lakes in the park were formed behind the terminal moraines of glaciers. The majority are in remote areas that are difficult to access (Gulley and Parker 1985). Most are less than 10 ha in surface area. Larger lakes are found at lower elevation and are not expected to be acid sensitive. Surface water acid neutralizing capacity values tend to be high throughout most of the low-elevation areas of GRTE. Lakes and streams with acid neutralizing capacity less than 400 μeq/L are generally restricted to the high mountain areas near the western border of the park (Peterson et al. 1998). Alpine lakes in GRTE exhibit a range of characteristics that contribute to their sensitivity to potential acidic deposition impacts (cf., Marcus et al. 1983): bedrock resistant to weathering, shallow soil, steep slope, low watershed to lake surface area ratio, high lake flushing rate, high precipitation, high snow accumulation, and short growing season.

Corbin and Woods (2004) sampled nine high-elevation lakes in GRTE plus three in the adjacent Targhee National Forest in 2002. The focus of the study was on status and trends in lake water quality and development of a predictive model to identify lakes likely to be sensitive to acidification. Three of the sampled lakes (Amphitheater, Surprise, and Delta Lakes) had acid neutralizing capacity <50 μeq/L. Presence of limestone (positive association) and granite (negative association) bedrock were the strongest predictors of lake acid neutralizing capacity and explained 86.5% of the variance in the acid neutralizing capacity predictions (Corbin and Woods 2004).

Peterson et al. (1998) summarized data collected by M. Williams and K. Tonnessen in August 1996 for 17 high-elevation headwater lakes throughout and adjacent to GRTE. Although none were acidic, about half had acid neutralizing capacity values near 50 μeq/L, and almost all had acid neutralizing capacity below 200 μeq/L. About one-third had pH in the range of 5.8–6.0. These data do not suggest that chronic lake water acidification has occurred to any significant degree, but do suggest sensitivity, especially to episodic acidification. Of particular importance were the observed concentrations of nitrate, which were relatively high in many of the lakes sampled. Six had nitrate concentrations in the range of 5–10 μeq/L, and three had nitrate concentrations higher than 10 μeq/L (to a maximum of 13 μeq/L). Given that the lakes were sampled at the end of the growing season, these data suggested that the capacity for the biological uptake of N may have been reached in response to ambient N inputs. It is likely that future increase in N deposition to these watersheds would contribute to higher lake water nitrate concentrations and possible chronic acidification.

Peterson et al. (1998) analyzed an NPS database of water quality and flow data for YELL, based on retrievals from five of the EPA's national databases, mostly from STORET. Measured values of pH were available in this database on one or more sampling occasion(s) from 201 sites, 5 of which had pH < 5.5. However, none of the lakes or streams having pH < 5.5 exhibited other data that reflected sensitivity to acidification from acidic deposition. For example, sulfate concentrations ranged from 18 to 191 mg/L in the low-pH waters, much higher than would be attributable to atmospheric deposition caused by air pollution. Chloride (Cl⁻) concentrations were also high in most of these low-pH surface waters, ranging from 1 to 121 mg/L. Four of the five lakes with pH < 5.5 had Cl⁻ concentration >57 mg/L. It is likely that all of these low-pH surface waters were impacted by geothermal discharge that caused the water to be low in pH.

Gibson et al. (1983) reported the chemistry of 106 lakes in YELL. Data were evaluated by Peterson et al. (1998); lakes were included from every region of the park and from every major geological formation. A spatial sensitivity map was created for the park based on measurements of acid neutralizing capacity. Excluding lakes located in the midst of major geothermal areas, the lowest acid neutralizing capacity measured was 40 μeq/L. Most of the lakes with acid neutralizing capacity less than 200 μeq/L occurred within a large rhyolite flow, which rose from the southwest and spread along the west central and central portions of the park. The greatest proportion of high pH and high acid neutralizing capacity lakes were found in the northern portions of the park (Gibson et al. 1983).

In 1999, the USGS resampled many of the lakes that had been surveyed as part of the EPA's Western Lakes Survey, including six lakes in YELL and one lake in GRTE. Most had high acid neutralizing capacity, although one lake sampled in YELL was acidic as a consequence of geothermal influence (Clow et al. 2002).

Nanus et al. (2005) assessed the sensitivity of 400 lakes in GRTE and YELL to acidification from atmospheric deposition of S and N using measured lake chemistry at 52 lakes in GRTE and 23 lakes in YELL. Lakes were classified into three general acidification classes (acid neutralizing capacity of 0–50, 0–100, and 0–200 µeq/L). They considered a range of potential explanatory predictor variables, including atmospheric deposition (inorganic N, S, H), basin characteristics (elevation, slope, aspect, area), bedrock type, soil type, vegetation type, precipitation, and lake area. The data were fit to logistic regression models and applied to lake watersheds larger than 1 ha (106 in GRTE and 294 in YELL). Lake acid sensitivity was higher in GRTE than in YELL. In GRTE, 36% of the lakes were estimated to have acid neutralizing capacity ≤100 µeq/L, compared with only 13% in YELL. In GRTE, 7% of the lakes had greater than a 60% probability of having acid neutralizing capacity less than 50 µeq/L, compared with 0% in YELL. For both parks, elevation was the principal predictor variable; for GRTE, the watershed area facing to the northeast was also a statistically significant predictor variable. Lakes that exceeded the 60% probability of having acid neutralizing capacity ≤100 µeq/L were located above 2790 m in GRTE and above 2590 m in YELL (Nanus et al. 2005). Lakes predicted to have acid neutralizing capacity ≤50 µeq/L (n = 7 lakes) with a probability greater than 60% were located at high elevation in the southern half of GRTE.

19.3.1 Episodic Acidification

Episodic surface water acidification in the mountainous West is most commonly associated with N deposition, and effects tend to be most pronounced during snowmelt. Snowmelt can flush into surface waters N that was deposited from the atmosphere to the snowpack as well as N that was mineralized within the soil under the snowpack during winter (Campbell et al. 1995). A substantial component of the nitrate flux may have been derived from the mineralization of organic N (Ley et al. 2004). Much of the N released from the snowpack during the melting period is retained in underlying soils, and only a portion of that is flushed to surface waters. Where soils are sparse, as in alpine regions of the Greater Yellowstone Network, much of the snowpack N is likely flushed to surface waters. Snowpack N has been reported to cause temporary acidification of alpine streams (Williams and Tonnessen 2000, Campbell et al. 2002).

The weight of evidence suggests that many high-elevation lakes in the Greater Yellowstone Network receive N deposition sufficiently high to cause some degree of episodic water acidification. The mechanisms that produce acidic episodes can also include dilution of base cations and flushing of nitrate, sulfate, and/or organic acids from soils or snowpack to drainage water (Kahl et al. 1992, Wigington et al. 1996, Wigington 1999, Lawrence 2002). However, there has not been sufficient data collected for the Greater Yellowstone Network to support a regional assessment of episodic acidification.

It is important to note that even low to moderate concentrations of nitrate in many Greater Yellowstone Network lakes might be significant in view of the low base cation concentrations in many lakes, the potential for continuing N deposition to eventually exhaust natural assimilative capabilities, and the fact that these distributions are often based on fall data and are therefore insufficient to make a conclusive determination (Sullivan 2000).

Future increases in N deposition, especially if they are substantial, might result in increased surface water acidification in these parks.

19.3.2 Effects on Aquatic Biota

There have been no documented changes in the aquatic biological communities in GRTE or YELL due to surface water acidification. It is unlikely that aquatic biota in the Greater Yellowstone Network has experienced adverse impacts to date from acidification. This is because depositions of S and N are low, and the available lake water chemistry data are not indicative of chronic acidification. However, in view of the high sensitivity of lake water chemistry to future increases in acidic deposition, aquatic biota in the Greater Yellowstone Network constitute important air quality–related values, especially within GRTE.

19.4 Nutrient Enrichment

Based on a coarse screening assessment, the overall level of concern for nutrient N enrichment effects on Inventory and Monitoring parks within this network was considered by Sullivan et al. (2011b) to be very high. In general, the predominant vegetation types thought to be most responsive to nutrient N enrichment effects include wetland, grassland and meadow, alpine, and arid and semiarid vegetation types. There are extensive wetland areas in GRTE and YELL. In GRTE, alpine vegetation is abundant at the higher elevations of the Grand Teton Mountains. Arid and semiarid land is found at the lower elevations to the southeast. Meadow vegetation types are scattered throughout the park. A large proportion of GRTE is covered by these vegetation types thought to be highly sensitive to nutrient N enrichment effects (Sullivan et al. 2011b). A relatively high proportion of YELL is also covered with sensitive vegetation types, with a broad mix of meadow, arid and semiarid, alpine, and wetland vegetation types. Although the parks in the Greater Yellowstone Network contain extensive coverage of vegetation community types expected to be highly sensitive to eutrophication caused by atmospheric N inputs, the extent to which such effects have actually occurred under ambient N deposition loading rates is not known. It is also likely that future changes in atmospheric N deposition will interact with changes in temperature and precipitation caused by climate change in Greater Yellowstone Network parks (cf., McDonnell et al. 2014). Alpine plant community monitoring has begun in YELL and GRTE through the Global Observation Research Initiative in Alpine Environments (GLORIA; http://www.gloria.ac.at/?a=2). This monitoring effort has been initiated by the Colorado National Heritage Program and the National Park Service.

Algal communities in high-elevation lakes within the Greater Yellowstone Network are sensitive to potential eutrophication. *In situ* N enrichment experiments in lakes within YELL stimulated the growth of the diatom *Fragilaria crotonensis*. Maximum algal species diversity in lakes in the YELL region occurred at experimental nitrate concentrations <3 µeq/L (Interlandi and Kilham 1998). More recently, N addition experiments found that *Asterionella formosa* reached a maximum growth rate at a nitrate concentration of 0.5 µeq/L (Nanus et al. 2012). This finding suggests the possibility that lakes in this network having nitrate concentration higher than 0.5 µeq/L may be experiencing changes in algal diversity due to nutrient N enrichment.

Nanus et al. (2012) used a geostatistical approach to map surface water nitrate concentration in lakes throughout the Rocky Mountain region, using basin characteristics, climate,

N deposition, and other factors. They mapped regional N critical load and exceedance. The lowest critical load estimates (<1.5 kg N/ha/yr) were for high-elevation watersheds having steep slopes, sparse vegetation, and abundant exposed rock and talus, such as predominate in GRTE. Many such areas were estimated to be in exceedance of the critical load to protect diatoms, widely believed to be among the most sensitive high-elevation aquatic life-forms.

Saros et al. (2011) used fossil diatom remains in lake sediment cores to determine changes in diatom species abundance over the last century. Observed changes were suggestive of N enrichment in the Greater Yellowstone Ecosystem, starting in about 1980. They modeled wet N deposition for the period during which diatom species shifts first occurred. From these analyses, a critical load of N deposition to prevent nutrient-mediated changes in diatom species assemblages was estimated to be about 1.4 kg N/ha/yr.

Ellis et al. (2013) estimated the critical load for N deposition to protect the most sensitive ecosystem receptors in 45 national parks, based on the data of Pardo et al. (2011a). The lowest terrestrial critical load of N is generally estimated for the protection of lichens (Geiser et al. 2010). Changes to lichen communities may signal the beginning of other changes to the ecosystem that might affect structure and function (Pardo et al. 2011b). Ellis et al. (2013) estimated the N critical load for GRTE and YELL in the range of 2.5–7.1 kg N/ha/yr for the protection of lichens.

19.5 Ozone Injury to Vegetation

The ozone-sensitive plant species that are known or thought to occur within YELL and GRTE are listed in Table 19.2. Those considered to be bioindicators, because they exhibit distinctive symptoms when injured by ozone (e.g., dark stipple), are designated by an asterisk. Each park contains at least seven ozone-sensitive and/or bioindicator species.

TABLE 19.2

Ozone-Sensitive and Bioindicator (*) Plant Species Known or Thought to Occur in YELL and GRTE

| | | Park | |
| | | GRTE | YELL |
Species	**Common Name**		
Amelanchier alnifolia	Saskatoon serviceberry	×	
*Apocynum androsaemifolium**	Spreading dogbane	×	×
Apocynum cannabinum	Dogbane, Indian hemp	×	
*Artemisia ludoviciana**	Silver wormwood	×	
*Physocarpus malvaceus**	Pacific ninebark	×	×
*Populus tremuloides**	Quaking aspen	×	×
Rhus aromatica var. *trilobata**	Skunkbush		×
Rubus parviflorus	Thimbleberry	×	
Rubus parviflorus var. *parviflorus*	Thimbleberry		×
*Salix scouleriana**	Scouler's willow	×	×
*Sambucus racemosa**	Red elderberry	×	
*Symphoricarpos albus**	Common snowberry	×	
*Vaccinium membranaceum**	Huckleberry	×	×

Note: Lists are periodically updated at https://irma.nps.gov/NPSpecies/Report.

The W126 and SUM06 ozone exposure indices calculated by National Park Service staff are given in Table 19.3, along with Kohut's (2007a) ozone risk ranking. Using these National Park Service criteria, ozone exposure risks at GTYE and YELL are rated moderate. However, the levels of ozone exposure in this network were considered by Kohut (2007b) to be unlikely to injure vegetation. In general, soil moisture was low when ozone exposure was high in YELL, but the pattern was not consistent. In YELL, low soil moisture constrains ozone uptake into foliage during higher exposure years. In GRTE, soil moisture conditions were generally favorable for ozone uptake, but the ozone exposure was low enough that there was not a substantial risk of ozone injury. Kohut (2007b) classified both of these parks as having low risk of foliar injury in response to ozone exposure.

In YELL, the levels of ozone exposure reported by Kohut (2007b) generally increased over the 10-year monitoring period (1995–2004). However, there were more months of drought stress during the second half of the period of record, thereby constraining the uptake of ozone into foliage and reducing the likelihood of injury (Kohut 2007b).

Natural background ozone concentrations can be above 40 ppb at YELL under appropriate atmospheric conditions. For example, a high-elevation site at YELL exhibited the largest number of coincidences (more than 19 days in one month) of average ozone concentration in excess of 50 ppb and the occurrence of stratospheric intrusions (Lefohn et al. 2011). This finding suggests that naturally occurring stratospheric intrusion can be an important source of elevated ozone levels at YELL. The maximum daily 8-hour average concentrations mostly ranged from about 50 to 60 ppb at this site (Lefohn et al. 2011).

Forest fires can also be an important source of ozone precursors in the Greater Yellowstone Network (Jaffe et al. 2008). The recent increase in fire frequency in the western United States has been identified as an important cause of increased ozone concentrations. This trend may continue in response to global warming.

The National Park Service (2010) evaluated the fourth highest 8-hour ozone concentrations for national parks that have consistently been at or above the 2008 standard of 75 ppb, based on 11–20 years of monitoring data running through 2008. Six parks, including YELL, had ozone levels that were typically below the 2008 standard of 75 ppb but were within the EPA's proposed range for a possible new standard (60–70 ppb; NPS 2010). The standard was changed to 70 ppb in 2015.

TABLE 19.3

Ozone Assessment Results for GRTE and YELL Based on Estimated Average 3-Month W126 and SUM06 Ozone Exposure Indices for the Period 2005–2009 and Kohut's (2007a) Ozone Risk Ranking for the Period 1995–1999[a,b]

| Park Name | Park Code | W126 | | SUM06 | | Kohut Ozone Risk Ranking |
		Value (ppm-h)	NPS Condition	Value (ppm-h)	NPS Condition	
Grand Teton	GRTE	10.78	Moderate	12.97	Moderate	Low
Yellowstone	YELL	9.40	Moderate	10.19	Moderate	Low

[a] Parks are classified into one of three ranks (Low, Moderate, High) based on comparison with other Inventory and Monitoring parks.

[b] Degrees of concern for the W126 and SUM06 indices are based solely on levels of ozone exposure. Kohut's risk to vegetation is based on several factors that contribute to injury in plants, including ozone exposure and environmental variables, and considers the effects of soil moisture on the uptake of ozone.

19.6 Visibility Degradation

19.6.1 Natural Background and Ambient Visibility Conditions

Haze is monitored at YELL. Data from the BRID1 (Bridger Wilderness, Sublette County) site are considered representative of GRTE. Natural haze levels are estimated to be very low in Greater Yellowstone Network parks (Table 19.4). Ambient visibility estimates reflect current pollution levels and were used to rank conditions at parks in order to provide information on spatial differences in visibility and air pollution. Rankings range from very low haze (very good visibility) to very high haze (very poor visibility).

Measured ambient haze values for the period 2004–2008 were only slightly higher than the estimated natural condition. Measured ambient haze rankings were ranked very low for both parks (Table 19.4). However, because of the importance of scenic views to the visitor experience in Greater Yellowstone Network parks, visibility remains an important air quality–related value that is of concern to the National Park Service.

IMPROVE data allow estimation of visual range. Data indicate that at the BRID1 site representative of GRTE, both natural and human-caused air pollution have reduced average visual range from 180 to 120 mi (290 to 193 km). On the haziest days, visual range has been reduced from 120 to 75 mi (193 to 121 km). Severe haze episodes occasionally reduce visibility to 14 mi (23 km). At the YELL2 site in YELL, air pollution has reduced average visual range from 180 to 120 mi (290 to 193 km). On the haziest days, visual range has been reduced from 120 to 70 mi (193 to 113 km). Severe haze episodes occasionally reduce visibility to 13 mi (21 km).

TABLE 19.4

Estimated Natural Haze and Measured Ambient Haze in Inventory and Monitoring Parks in the Greater Yellowstone Network Averaged over the Period 2004–2008[a]

Park Name	Park Code	Site ID	Estimated Natural Haze (in Deciviews)		
			20% Clearest Days	20% Haziest Days	Average Days
Grand Teton[b]	GRTE	BRID1	0.28	6.45	2.89
Yellowstone	YELL	YELL2	0.43	6.44	2.93

Park Name	Park Code	Site ID	Measured Ambient Haze (for Years 2004–2008)					
			20% Clearest Days		20% Haziest Days		Average Days	
			Deciviews	Ranking	Deciviews	Ranking	Deciviews	Ranking
Grand Teton[b]	GRTE	BRID1	1.62	Very Low	10.96	Very Low	5.79	Very Low
Yellowstone	YELL	YELL2	2.18	Very Low	11.76	Very Low	6.50	Very Low

[a] Parks are classified into one of five ranks (Very Low, Low, Moderate, High, Very High) based on comparison with other monitored parks.

[b] Data are borrowed from a nearby site. A monitoring site is considered by Interagency Monitoring of Protected Visual Environments (IMPROVE) to be representative of an area if it is within 60 mi (100 km) and 425 ft (130 m) in elevation of that area.

19.6.2 Composition of Haze

At IMPROVE sites throughout the interior Columbia River basin, from North Cascades National Park and Glacier National Park in the north to Lassen Volcanic National Park and YELL in the south, carbon in various forms, including fine particulate organics and soot, dominates the light extinction budget (Schoettle et al. 1999). The second most important contributor to extinction is sulfate. Extinction charts are shown for GRTE and YELL in Figure 19.3. The largest share of the annual average total particulate light extinction

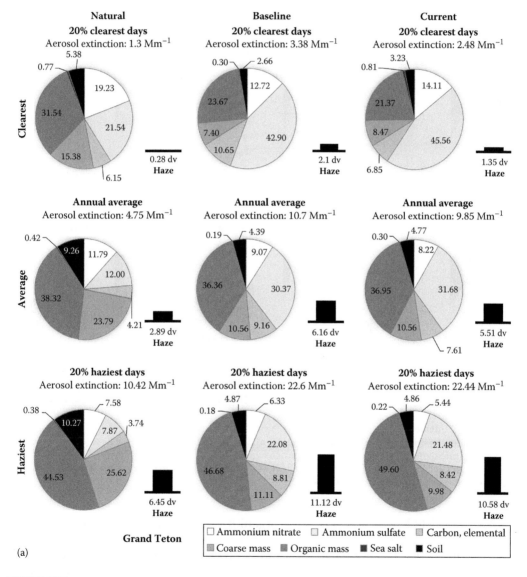

FIGURE 19.3

Estimated natural (preindustrial), baseline (2000–2004), and current (2006–2010) levels of haze (columns) and its composition (pie charts) on the 20% clearest, annual average, and 20% haziest visibility days for (a) GRTE and (b) YELL. Data for GRTE were taken from a nearby site (BRID1). (From http://views.cira.colostate.edu/fed/Tools/RegionalHazeSummary.aspx, accessed October, 2012.) *(Continued)*

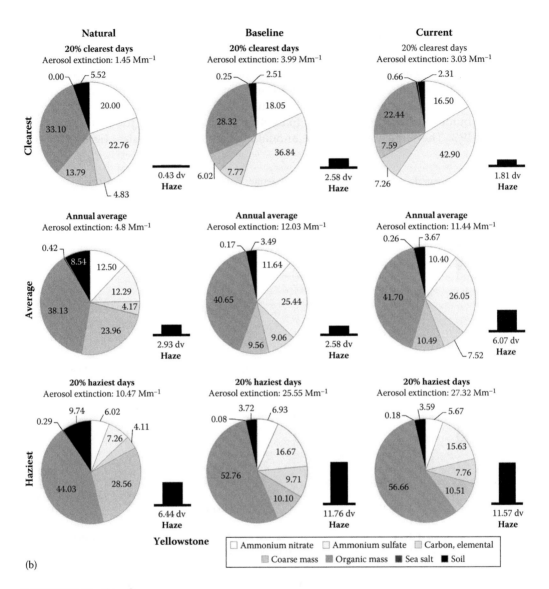

FIGURE 19.3 (*Continued*)
Estimated natural (preindustrial), baseline (2000–2004), and current (2006–2010) levels of haze (columns) and its composition (pie charts) on the 20% clearest, annual average, and 20% haziest visibility days for (a) GRTE and (b) YELL. Data for GRTE were taken from a nearby site (BRID1). (From http://views.cira.colostate.edu/fed/Tools/RegionalHazeSummary.aspx, accessed October, 2012.)

in GRTE and YELL was attributable to organics (32.5% and 41.7%, respectively), followed closely by sulfate (Figure 19.3). On the 20% haziest days, organics contributed over 40%–light extinction in both parks. In contrast, on the clearest 20% visibility days, sulfate accounted for more than 40% of light extinction in these parks. Analyses conducted by the Western Regional Air Partnership indicated that organics from natural emissions sources, including wildfire and biogenic sources (vegetation), contribute to substantial visibility impairment throughout the western United States, including within the Greater Yellowstone Network.

19.6.3 Trends in Visibility

There are not clear trends in haze at YELL or GRTE on the 20% clearest, 20% haziest, or average days over the full period of available monitoring data. However, in each of the parks, the haze level shows some evidence of decreasing over the last several years (Figure 19.4).

The glideslope analysis depicted in Figure 19.5 shows some evidence of decreasing haze. Measured values are generally close to following the glideslope required for conformance with the Regional Haze Rule goal of zero human-caused visibility impairment in 2064. More data will be needed for determining the extent of needed improvement.

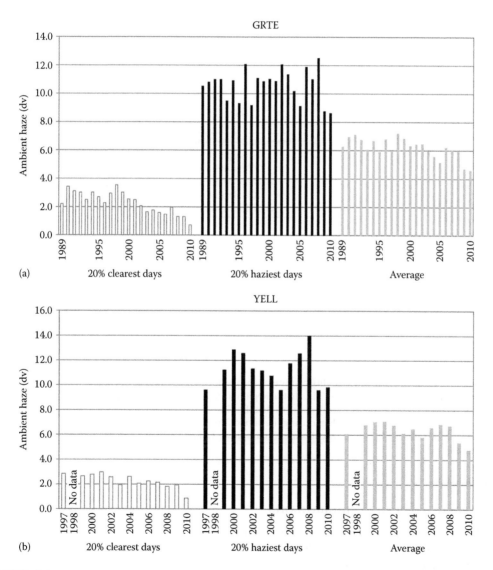

FIGURE 19.4
Trends in ambient haze levels at (a) GRTE and (b) YELL, based on IMPROVE measurements on the 20% clearest, 20% haziest, and annual average visibility days over the monitoring period of record. Data for GRTE were taken from a nearby monitoring site (BRID1). (From http://vista.cira.colostate.edu/improve/Data/IMPROVE/summary_data.htm, accessed October, 2012.)

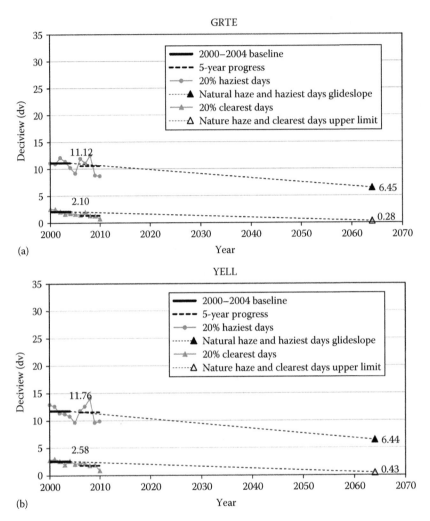

FIGURE 19.5
Glideslopes to achieving natural visibility conditions in 2064 for the 20% haziest (top line) and the 20% clearest (bottom line) days in (a) GRTE and (b) YELL. In the Regional Haze Rule, the clearest days do not have a uniform rate of progress glideslope; the rule only requires that the clearest days do not get any worse than the baseline period. Also shown are measured values during the period 2000–2010. Data for GRTE were taken from a nearby site (BRID1). (From http://vista.cira.colostate.edu/improve/Data/IMPROVE/summary_data.htm, accessed October, 2012.)

19.7 Toxic Airborne Contaminants

Pesticide and other semivolatile organic compound levels in vegetation at GRTE, as measured by WACAP (Landers et al. 2008), were dominated by polycyclic aromatic hydrocarbons and the current-use pesticides endosulfan and dacthal. Likely sources of polycyclic aromatic hydrocarbons to these parks include forest fires and perhaps woodburning stoves. Lichen data reported by WACAP parks in GRTE showed relatively high semivolatile organic compounds concentrations. Some work has been done on the presence of PCBs and organochlorine insecticides and their metabolites in and near YELL as part of the

U.S. Geological Survey National Water Quality Assessment (Peterson and Boughton 2000), although the connection to atmospheric deposition is unclear. They found DDT, and its metabolites DDD and DDE, in fish collected in YELL. This probably reflects residual effects from historical DDT spraying for spruce budworm in and near the park. Fish appeared to be more sensitive indicators of contamination than river bed sediment in YELL, likely due to the lipophilic nature of these compounds, chemical partitioning, and bioaccumulation (Peterson and Boughton 2000).

Krabbenhoft et al. (2002) sampled 90 lakes in seven national parks in the western United States and analyzed their Hg and methyl Hg concentrations. YELL and GRTE were included in the study. Levels of methyl Hg were similar among YELL, GRTE, and most of the other surveyed parks (~0.05 ng/L). Data are available regarding Hg in lake trout in Yellowstone Lake, but much of that Hg may be attributed to geothermal, rather than atmospheric, sources.

Eagles-Smith et al. (2014) sampled fish in 21 national parks and analyzed them for Hg concentrations in fish tissue. Results varied substantially by park and by water body. Mean fish Hg concentrations in YELL (101 ng/g wet weight) were higher than the mean across all study parks. The size-adjusted values in YELL were among the highest measured in this study. However, natural geothermal concentrations of Hg might play a role in this observation. Concentrations of Hg in fish collected from two lakes and one stream in GRTE were very low, compared with other study parks.

Keteles (2011) collected surface water samples from high-elevation areas of GRTE and three other parks during the summer of 2009 and analyzed them for pesticides. Concentrations in water were compared with the EPA Office of Pesticide Programs Aquatic life benchmarks for freshwater species and to toxicity values determined in other programs (Kegley et al. 2011). The measured concentrations in this screening study suggested low pesticide concentration values in GRTE, far below the aquatic life benchmarks (Keteles 2011).

19.8 Summary

In general, air quality in GRTE and YELL is excellent, largely because the network area is not heavily developed, and has limited emissions sources and good air dispersion (Story et al. 2006). Nevertheless, regional emissions from power plants, agricultural sources, oil and gas development, and on-road and off-road vehicles affect air quality in these parks. In addition, geothermal sources in YELL are natural sources of S, Hg, and other pollutants. The greatest human-caused air quality concerns in the Greater Yellowstone Area include gas development in southwestern Wyoming, wildfire, and emissions from energy-related industries (Story et al. 2006).

The region receives relatively low levels of S and N deposition. However, many lakes within GRTE are highly sensitive to potential acidic deposition effects. Although it does not appear that chronic surface water acidification has occurred to any significant degree, episodic acidification has been reported for lakes in the Colorado Front Range and probably occurs in the most acid-sensitive lakes and streams in GRTE as well. Episodic acidification can be an especially important issue for surface waters throughout high-elevation areas such as occur widely within GRTE.

Nitrogen deposition can also cause undesirable nutrient enrichment of natural ecosystems, leading to changes in plant species composition and diversity and soil nutrient cycling. Parks in the Greater Yellowstone Network receive generally low levels of N deposition, but sensitive aquatic and terrestrial ecosystems, including those at high elevation and those containing plant communities that are very sensitive to nutrient N input, may be affected by current or future levels of N deposition.

Elevated levels of ozone have been measured in southern Wyoming. However, ozone levels in the Greater Yellowstone Area are usually relatively low and unlikely to induce plant injury.

Particulate pollution can cause haze, reducing visibility. For many visitors, scenic vistas are an important reason for their visit to the Greater Yellowstone Area parks. Some of the regional and local air pollution sources that likely contribute to visibility impairment include on-road and off-road vehicles, coal- and oil-fired power plants, agricultural activities, oil and gas development, and particulates from unpaved roads. Visibility may also be affected by prescribed fires and wildfires, with the specific effects of fire on visibility varying interannually depending on climatic conditions and the frequency and intensity of fires in and adjacent to a specific park. Seasonal increases in carbon monoxide and particulates associated with woodburning stoves in Jackson, WY (40 km south of GRTE) and high snowmobile use in YELL and GRTE are also recognized as important air pollution concerns.

Airborne toxics, including Hg and other heavy metals, can accumulate in food webs, reaching toxic levels in top predators. Sampling and analyses conducted in the WACAP documented the deposition and bioaccumulation of airborne toxics in several western national parks, including GRTE. Semivolatile organic compounds in GRTE were dominated by polycyclic aromatic hydrocarbons and current-use pesticides. The former derive from combustion, including forest fire. The latter derive from agriculture.

References

Benedict, K.B., C.M. Carrico, S.M. Kreidenweis, B. Schichtel, W.C. Malm, and J.L. Collett. 2013. A seasonal nitrogen deposition budget for Rocky Mountain National Park. *Ecol. Appl.* 23(5):1156–1169.

Campbell, D.H., D.W. Clow, G.P. Ingersoll, M.A. Mast, N.E. Spahr, and J.T. Turk. 1995. Nitrogen deposition and release in alpine watersheds, Loch Vale, Colorado, USA. In K.A. Tonnessen, M.W. Williams, and M. Tranter (Eds.). *Biogeochemistry of Seasonally Snow Covered Catchments. Proceedings of a Boulder Symposium July 1995.* IAHS Pub. No. 228. International Association of Hydrological Sciences, Oxfordshire, U.K., pp. 243–253.

Campbell, D.H., C. Kendall, C.C.Y. Chang, S.R. Silva, and K.A. Tonnessen. 2002. Pathways for nitrate release from an alpine watershed: Determination using $\delta^{15}N$ and $\delta^{18}O$. *Water Resour. Res.* 31:2811–2821.

Clow, D.W., R. Striegl, L. Nanus, M.A. Mast, D.H. Campbell, and D.P. Krabbenhoft. 2002. Chemistry of selected high-elevation lakes in seven national parks in the western United States. *Water Air Soil Pollut.* 2(2):139–164.

Corbin, J.A. and S.W. Woods. 2004. Effects of atmospheric deposition on water quality in high alpine lakes of Grand Teton National Park, Wyoming. University of Montana, Missoula, MT.

Eagles-Smith, C.A., J.J. Willacker, Jr., and C.M. Flanagan Pritz. 2014. Mercury in fishes from 21 national parks in the western United States—Inter- and intra-park variation in concentrations and ecological risk. Open-File Report 2014-1051. U.S. Geological Survey, Reston, VA.

Ellis, R.A., D.J. Jacob, M.P. Sulprizio, L. Zhang, C.D. Holmes, B.A. Schichtel, T. Blett, E. Porter, L.H. Pardo, and J.A. Lynch. 2013. Present and future nitrogen deposition to national parks in the United States: Critical load exceedances. *Atmos. Chem. Phys.* 13(17):9083–9095.

Fenn, M.E., R. Haeuber, G.S. Tonnesen, J.S. Baron, S. Grossman-Clark, D. Hope, D.A. Jaffe et al. 2003. Nitrogen emissions, deposition, and monitoring in the western United States. *BioScience* 53(4):391–403.

Geiser, L.H., S.E. Jovan, D.A. Glavich, and M.K. Porter. 2010. Lichen-based critical loads for atmospheric nitrogen deposition in western Oregon and Washington forests, USA. *Environ. Pollut.* 158:2412–2421.

Gibson, J.H., J.N. Galloway, C.L. Schofield, W. McFee, R. Johnson, S. McCorley, N. Dise, and D. Herzog. 1983. Rocky Mountain acidification study. FWS/OBS-80/40.17. U.S. Fish and Wildlife Service, Department of Biological Services Eastern Energy and Land Use Team, Washington, DC.

Grenon, J. and M. Story. 2009. U.S. Forest Service region 1. Lake chemistry, NADP, and IMPROVE air quality data analysis. General Technical Report RMRS-GTR-230WWW. USDA Forest Service, Rocky Mountain Research Station, Fort Collins, CO.

Gulley, D.D. and M. Parker. 1985. A limnological survey of 70 small lakes and ponds in Grand Teton National Park. Department of Zoology and Physiology, University of Wyoming, Laramie, WY.

Ingersoll, G.P. 1999. Effects of snowmobile use on snowpack chemistry in Yellowstone National Park, 1998. Water-Resources Investigations Report 99-4148. U.S. Geological Survey, in cooperation with the National Park Service, Denver, CO.

Ingersoll, G.P., M.A. Mast, D.H. Campbell, D.W. Clow, L. Nanus, and J.T. Turk. 2008. Trends in snowpack chemistry and comparison to national atmospheric deposition program results for the Rocky Mountains, US 1993–2004. *Atmos. Environ.* 42:6098–6113.

Ingersoll, G.P., M.A. Mast, L. Nanus, D.J. Manthorne, H.H. Handran, D.M. Hulstrand, and J. Winterringer. 2005. Rocky Mountain snowpack chemistry at selected sites, 2003. Open-File Report 2005-1332. U.S. Department of the Interior, Reston, VA.

Ingersoll, G.P., M.A. Mast, J.M. Swank, and C.D. Campbell. 2011. Rocky Mountain snowpack physical and chemical data for selected sites, 2010. Data Series 570. U.S. Geological Survey, Reston, VA.

Interlandi, S.J. and S.S. Kilham. 1998. Assessing the effects of nitrogen deposition on mountain waters: A study of phytoplankton community dynamics. *Water Sci. Technol.* 38:139–146.

Jaffe, D., D. Chand, W. Hafner, A. Westerling, and D. Spracklen. 2008. Influence of fires on O_3 concentrations in the western U.S. *Environ. Sci. Technol.* 42:5885–5891.

Kahl, J.S., S.A. Norton, T.A. Haines, E.A. Rochette, R.C. Heath, and S.C. Nodvin. 1992. Mechanisms of episodic acidification in low-order streams in Maine, USA. *Environ. Pollut.* 78:37–44.

Kegley, S.E., B.R. Hill, O. S., and A.H. Choi. 2011. PAN pesticide database. Pesticide Action Network, North America, San Francisco, CA.

Keteles, K. 2011. Screening for pesticides in high elevation lakes in federal lands. U.S. EPA Region 8, Denver, CO.

Kohut, R. 2007a. Assessing the risk of foliar injury from ozone on vegetation in parks in the U.S. National Park Service's Vital Signs Network. *Environ. Pollut.* 149:348–357.

Kohut, R.J. 2007b. Ozone risk assessment for vital signs monitoring networks, Appalachian National Scenic Trail, and Natchez Trace National Scenic Trail. Natural Resource Technical Report NPS/NRPC/ARD/NRTR—2007/001. National Park Service, Natural Resource Program Center, Fort Collins, CO.

Krabbenhoft, D.P., M.L. Olson, J.F. DeWild, D.W. Clow, R.G. Striegl, M.M. Dornblaser, and P. VanMetre. 2002. Mercury loading and methylmercury production and cycling in high-altitude lakes from the western United States. *Water Air Soil Pollut.* 2:233–249.

Landers, D.H., J.M. Eilers, D.F. Brakke, W.S. Overton, P.E. Kellar, W.E. Silverstein, R.D. Schonbrod et al. 1987. *Characteristics of Lakes in the Western United States. Volume I: Population Descriptions and Physico-Chemical Relationships.* EPA-600/3-86/054a. U.S. Environmental Protection Agency, Washington, DC.

Landers, D.H., S.L. Simonich, D.A. Jaffe, L.H. Geiser, D.H. Campbell, A.R. Schwindt, C.B. Schreck et al. 2008. The fate, transport, and ecological impacts of airborne contaminants in western national parks (USA). EPA/600/R-07/138. U.S. Environmental Protection Agency, Office of Research and Development, NHEERL, Western Ecology Division, Corvallis, OR.

Lawrence, G.B. 2002. Persistent episodic acidification of streams linked to acid rain effects on soil. *Atmos. Environ.* 36:1589–1598.

Lefohn, A.S., H. Wernli, D. Shadwick, S. Limbach, S.J. Oltmans, and M. Shapiro. 2011. The importance of stratospheric-tropospheric transport in affecting surface ozone concentrations in the western and northern tier of the United States. *Atmos. Environ.* 45:4845–4857.

Lehmann, C.M.B., V.C. Bowersox, and S.M. Larson. 2005. Spatial and temporal trends of precipitation chemistry in the United States, 1985–2002. *Environ. Pollut.* 135:347–361.

Ley, R., M.W. Williams, and S.K. Schmidt. 2004. Microbial population dynamics in an extreme environment: Controlling factors in talus soils at 3750 m in the Colorado Rocky Mountains. *Biogeochemistry* 68(3):297–311.

Malm, W.C., B.A. Schichtel, R.B. Ames, and K.A. Gebhart. 2002. A 10-year spatial and temporal trend of sulfate across the United States. *J. Geophys. Res.* 107(D22, 4627).

Marcus, M.D., B.R. Parkhurst, R.W. Brocksen, and F.E. Payne. 1983. An assessment of the relationship among acidifying deposition, surface water acidification, and fish populations in North America. Volume 1. Electric Power Research Institute, Palo Alto, CA.

McDonnell, T.C., S. Belyazid, T.J. Sullivan, H. Sverdrup, W.D. Bowman, and E.M. Porter. 2014. Modeled subalpine plant community response to climate change and atmospheric nitrogen deposition in Rocky Mountain National Park, USA. *Environ. Pollut.* 187:55–64.

Nanus, L., D.H. Campbell, and M.W. Williams. 2005. Sensitivity of alpine and subalpine lakes to acidification from atmospheric deposition in Grand Teton National Park and Yellowstone National Park, Wyoming. Scientific Investigations Report 2005–5023. U.S. Department of Interior, U.S. Geological Survey, Reston, VA.

Nanus, L., D.W. Clow, J.E. Saros, V.C. Stephens, and D.H. Campbell. 2012. Mapping critical loads of nitrogen deposition for aquatic ecosystems in the Rocky Mountains, USA. *Environ. Pollut.* 166:125–135.

National Park Service (NPS). 2010. Air quality in national parks: 2009 annual performance and progress report. Natural Resource Report NPS/NRPC/ARD/NRR-2010/266. National Park Service, Air Resources Division, Denver, CO.

Pardo, L.H., M.E. Fenn, C.L. Goodale, L.H. Geiser, C.T. Driscoll, E.B. Allen, J.S. Baron et al. 2011b. Effects of nitrogen deposition and empirical nitrogen critical loads for ecoregions of the United States. *Ecol. Appl.* 21(8):3049–3082.

Pardo, L.H., M.J. Robin-Abbott, and C.T. Driscoll. 2011a. Assessment of nitrogen deposition effects and empirical critical loads of nitrogen for ecoregions of the United States. General Technical Report NRS-80. U.S. Forest Service, Newtown Square, PA.

Peterson, D.A. and G.K. Boughton. 2000. Organic compounds and trace elements in fish tissue and bed sediment from streams in the Yellowstone River Basin, Montana and Wyoming, 1998. U.S. Department of the Interior, U.S. Geological Survey, Cheyenne, WY.

Peterson, D.L., T.J. Sullivan, J.M. Eilers, S. Brace, D. Horner, K. Savig, and D. Morse. 1998. Assessment of air quality and air pollutant impacts in national parks of the Rocky Mountains and northern Great Plains. NPS D-657. U.S. Department of the Interior, National Park Service, Air Resources Division, Denver, CO.

Ray, J.D. 2012. Winter air quality in Yellowstone National Park: 2009–2011. Natural Resource Technical Report NPS/NRSS/ARD/NRTR—2012/551. National Park Service, Denver, CO.

Saros, J.E., D.W. Clow, T. Blett, and A.P. Wolfe. 2011. Critical nitrogen deposition loads in high-elevation lakes of the western US inferred from paleolimnological records. *Water Air Soil Pollut.* 216:193–202.

Schoettle, A.W., K. Tonnessen, J. Turk, J. Vimont, and R. Amundson. 1999. An assessment of the effects of human-caused air pollution on resources within the interior Columbia River Basin. PNW-GTR-447. USDA Forest Service, Pacific Northwest Research Station, Portland, OR.

Schwede, D.B. and G.G. Lear. 2014. A novel hybrid approach for estimating total deposition in the United States. *Atmos. Environ.* 92:207–220.

Story, M.T., J. Shea, T. Svalberg, M. Hektner, and G. Ingersoll. 2006. Greater Yellowstone area air quality assessment update. *In* A.W. Biel (Ed.). *Greater Yellowstone Public Lands: A Century of Discovery, Hard Lessons, and Bright Prospects: Proceedings of the Eighth Biennial Scientific Conference on the Greater Yellowstone Ecosystem,* Mammoth Hot Springs Hotel, Yellowstone National Park, WY, October 17–19, 2005. Yellowstone Center for Resources, Yellowstone National Park, WY, pp. 159–169.

Sullivan, T.J. 2000. *Aquatic Effects of Acidic Deposition.* CRC Press, Boca Raton, FL.

Sullivan, T.J., T.C. McDonnell, G.T. McPherson, S.D. Mackey, and D. Moore. 2011b. Evaluation of the sensitivity of inventory and monitoring national parks to nutrient enrichment effects from atmospheric nitrogen deposition. Natural Resource Report NPS/NRPC/ARD/NRR—2011/313. U.S. Department of the Interior, National Park Service, Denver, CO.

Sullivan, T.J., G.T. McPherson, T.C. McDonnell, S.D. Mackey, and D. Moore. 2011a. Evaluation of the sensitivity of inventory and monitoring national parks to acidification effects from atmospheric sulfur and nitrogen deposition. Natural Resource Report. NPS/NRPC/ARD/NRR—2011/349. U.S. Department of the Interior, National Park Service, Denver, CO.

Wigington, P.J. 1999. Episodic acidification: Causes, occurrence and significance to aquatic resources. *In* J.R. Drohan (Ed.). *The Effects of Acidic Deposition on Aquatic Ecosystems in Pennsylvania. 1998 PA Acidic Deposition Conference.* Environmental Resources Research Institute, University Park, PA, pp. 1–5.

Wigington, P.J., Jr., D.R. DeWalle, P.S. Murdoch, W.A. Kretser, H.A. Simonin, J. Van Sickle, and J.P. Baker. 1996. Episodic acidification of small streams in the northeastern United States: Ionic controls of episodes. *Ecol. Appl.* 6(2):389–407.

Williams, M.W. and K.A. Tonnessen. 2000. Critical loads for inorganic nitrogen deposition in the Colorado Front Range, USA. *Ecol. Appl.* 10:1648–1665.

20

Denali National Park and Central Alaska Network

20.1 Background

The Central Alaska Network contains three park units: Denali National Park (DENA), Wrangell-St. Elias National Park and Preserve (WRST), and Yukon-Charley Rivers National Preserve (YUCH). All are larger than 100 mi² (259 km²). This analysis focuses mostly on DENA. Only two urban centers, Anchorage and Fairbanks, are located near the Central Alaska Network. Figure 20.1 shows the network boundary along with locations of each park and population centers with more than 10,000 people. Emissions from urban centers in Alaska are not expected to be particularly important to the parks in the Central Alaska Network, although wildfires and international pollution from urban, industrial, and agricultural sources can affect air quality–related values, including visibility, in these parks.

20.2 Atmospheric Emissions and Deposition

Emissions of S and N near Central Alaska Network, based on data from the EPA's National Emissions Inventory during a recent time period (2011), were relatively low. All boroughs near Central Alaska Network parks had very low sulfur dioxide emissions (<1 ton/mi²/yr). Emissions of nitrogen oxides and ammonia were also low in most boroughs near network parks.

Regionally estimated deposition data are not available for Alaska because of the scarcity of monitors. Total S and N deposition within the Central Alaska Network are both expected to be low because there are few point sources and urban areas and because estimated emissions levels from the various boroughs that comprise the network region are low (Sullivan et al. 2011).

There are five active National Atmospheric Deposition Program/National Trends Network wet deposition monitoring sites in Alaska: Poker Creek, Juneau, DENA, Gates of the Arctic National Park, and Katmai National Park. Data have been collected since 1980 at DENA and since 1993 at Poker Creek. The other three monitoring sites have been added within the last decade. There have also been Clean Air Status and Trends Network dry deposition measurements at DENA and Poker Flat. The latter operated through January 2004. At all monitored sites in Alaska, wet N deposition has consistently been less than 1 kg N/ha/yr, and it has been less than 0.5 kg N/ha/yr at all monitored sites except Juneau. Wet S deposition has been slightly higher than 1 kg S/ha/yr at Juneau but less than that at the other monitoring sites. The Clean Air Status and Trends Network dry deposition

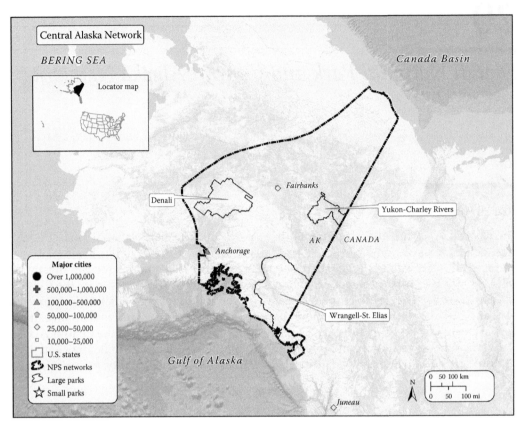

FIGURE 20.1
Network boundary and locations of parks and population centers greater than 10,000 people near the Central Alaska Network.

measurements have also been low, below about 0.25 kg N/ha/yr for each site and year measured. Thus, the sparse available atmospheric deposition data for DENA and Alaska in general are consistent with the general understanding that atmospheric deposition of both N and S tends to be very low. It can be assumed that both S and N deposition across the Central Alaska Network would be lower than about 1 kg/ha/yr, on average.

The WACAP (Landers et al. 2008) studied airborne contaminants in a number of national parks of the western United States, including two parks in the Central Alaska Network (DENA and WRST) from 2002 to 2007. The primary semivolatile organic compounds contaminants detected in the air at DENA were hexachlorobenzene and hexachlorocyclohexane-α, both historic-use pesticides. This finding is consistent with contaminant composition found in the air by WACAP elsewhere in Alaska, at Noatak and Gates of the Arctic national parks. Research conducted as part of WACAP measured snowpack semivolatile organic compounds concentrations at DENA that were among the lowest of all WACAP parks. Air sampled in WRST contained the fewest identified semivolatile organic compound contaminants of any WACAP park. The only semivolatile organic compounds detected were low levels of polycyclic aromatic hydrocarbons and hexachlorocyclohexane-γ. Lichen data collected by WACAP showed no evidence of elevated N deposition in this park.

Smoke from wildfires is common in Central Alaska Network parks, and the fire frequency and area burned is increasing in response to climate change. Smoke can contribute to visibility degradation. International air pollution sources also contribute air contaminants to Central Alaska Network parks. Direct pollutant transport across the Pacific Ocean and development of arctic haze contribute to decreased air quality.

20.3 Acidification

Low weathering rates attributable partly to cold temperatures, the presence of some relatively impervious rock types, and thin poorly developed soils having high organic content contribute to acid sensitivity in this network. Despite the presumed acid sensitivity of parks in the Central Alaska Network, however, aquatic and terrestrial acidification from acidic deposition have not been documented in this network. Such effects are considered unlikely in view of the very low levels of S and N deposition and long distances between the parks and emissions sources. There are no surface water chemistry data that suggest that there are low acid neutralizing capacity lakes or streams in Central Alaska Network parks.

20.4 Nutrient Enrichment

The availability of N can have a major influence on plant communities in tundra ecosystems. Plant species community composition, plant growth, and a variety of ecosystem processes are partially controlled by N supply (Shaver et al. 1992, Mack et al. 2004). Bryophytes and lichens have a major influence on water and energy balances in arctic tundra vegetation communities. High inputs of N can reduce their growth, while increasing the growth of competing grass and sometimes also shrub species (Berendse et al. 2001, van der Wal et al. 2005, Bubier et al. 2007).

Competitive interactions among plants can be just as important as species-specific growth stimulation effects in response to N addition. For example, competition by erect vascular plants can decrease moss cover despite the N limitation of the mosses (Klanderud 2008). Such competitive effects are more likely to occur in low and mid-arctic regions than in high arctic regions where growth is more strongly limited by harsh climatic conditions (Cornelissen et al. 2001, van Wijk et al. 2004, Madan et al. 2007).

Much of interior Alaska lies within the Taiga ecoregion, which is underlain by sandstone, limestone, and shale, creating a generally flat rolling plain. There are abundant organic deposits and substantial permafrost in the north. Lowlands are primarily covered with peatlands. Wetlands interspersed with shrublands grade into tundra at the northern extent of the Taiga (Geiser and Nadelhoffer 2011). Most plants in this ecoregion are adapted to low N supply. Increased N deposition to Taiga plant communities, provided the increase is sufficiently large, is expected to increase plant growth in the short term, followed by the likelihood of a longer-term change in species composition and richness (Gough et al. 2000, Geiser and Nadelhoffer 2011).

Any change from the natural state in species composition or richness in a Class I area may be viewed as adverse. In peatland bogs, species richness commonly decreases in response to N addition (Allen 2004), perhaps in part because the low pH and high water table inhibit nitrophyte invasion (Geiser and Nadelhoffer 2011). In contrast, forested areas may exhibit increases in nitrophilic (nitrogen-loving) plant species and decreases in species that evolved under low N supply, with little change in species richness (Bobbink 2004, Geiser and Nadelhoffer 2011). Increased N supply, especially in combination with a warming climate, might cause an advance of the location of treeline to the north or to higher elevation (Chapin and Körner 1996, Sverdrup et al. 2012).

Tundra lichens that occur near the southern extent of arctic tundra vegetation and in alpine areas may be very susceptible to the effects of climate warming because higher temperatures may increase the availability of N and P and this favors vascular plants more than bryophytes and lichens (Cornelissen et al. 2001). Wet tundra sites are more likely to be P-limited, and dry tundra sites tend to be N-limited (Shaver et al. 1998). Lichens were completely eliminated subsequent to more than 10 years of very high experimental N addition at a rate of 100 kg N/ha/yr at the Toolik Lake Tundra Long Term Ecological Research site (Weiss et al. 2005).

Van Wijk et al. (2004) conducted a meta-analysis in mid-arctic regions of Alaska and Sweden to examine ecosystem response to experimental N addition. They found that the growth of vascular plants, primary productivity, and total biomass all increased substantially subsequent to addition of N alone and also addition of N plus P and other nutrients. Growth of shrubs and graminoids increased, while growth of forbs, bryophytes, and lichens decreased. Shrub response appeared to vary with soil condition. At locations having base-rich soils, N addition did not necessarily contribute to increasing shrub dominance (Gough and Hobbie 2003).

Many plant species in arctic tundra are N-limited (Shaver and Chapin 1980). Soluble N in tundra soil solution is often dominated by organic N, including free amino acids, rather than ammonium or nitrate (Kielland 1995). Tundra plants appear to exhibit a range of interspecific differences that allow coexistence under conditions that reflect a single limiting element. Species differ in rooting depth, phenology, and uptake preferences for organic and inorganic forms of N (Shaver and Billings 1975, Chapin et al. 1993, Kielland 1994, McKane et al. 2002). McKane et al. (2002) demonstrated, based on ^{15}N isotope field experiments, that arctic tundra plant species were differentiated in timing, depth, and chemical form of N utilization. Furthermore, the species that exhibited greatest productivity were those that efficiently used the most locally abundant N forms.

Arctic ecosystems appear to respond to increased N supply at levels less than 5 kg N/ha/yr by changing the level of primary productivity and the structure of the plant community (Arens et al. 2008). There has been a paucity of studies on Taiga responses to N addition in North America. However, changes in ground vegetation occurred in Swedish boreal forests at relatively low levels of atmospheric N deposition. Whortleberry (*Vaccinium myrtillus*) cover decreased at N deposition ≥6 kg N/ha/yr (Strengbom et al. 2003, Nordin et al. 2005); the growth of wavy hairgrass (*Deschampsia flexuosa*) increased at ≥5 kg N/ha/yr (Kellner and Redbo-Torstensson 1995, Nordin et al. 2005).

Geiser and Nadelhoffer (2011) recommended a critical load of atmospheric N deposition to protect lichen and mosses of the Taiga ecoregion in North America less than 1 to 3 kg N/ha/yr. To protect shrub and grass community composition, the critical load would likely be higher, in the range of about 6 kg N/ha/yr (Strengbom et al. 2003, Nordin et al. 2005, Geiser and Nadelhoffer 2011).

Classification of arctic plants into functional types or growth forms provides a useful framework for evaluating the effects of added N and climate change on plant community composition, functioning, and ecosystem processes (Chapin et al. 1996). Functional plant types such as deciduous and evergreen shrubs, forbs, graminoids, mosses, and lichens have been used extensively in arctic environments to describe plant responses to changes in nutrient supply (Walker et al. 1989, Chapin et al. 1996). Typically, woody plants can achieve greater canopy height than nonwoody plants. Large stature, provided by woodiness, allows individual plants to capture a disproportionate amount of available light (Tilman 1988). Nonvascular plants tend to dominate where conditions limit plant growth: too wet, too dry, too nutrient poor. Vascular plants dominate under more favorable growth conditions (Chapin et al. 1996). Deciduous woody plants have a shorter season of photosynthetic activity and require proportionately more nutrient resources than do evergreen vascular plants. As a consequence, deciduous shrubs tend to dominate fertile nutrient-rich upland sites; evergreen shrubs are more common on dry and infertile sites (Shaver and Chapin 1991). Grasses dominate steep south-facing slopes (Walker et al. 1991); forbs generally prefer moist, nutrient-rich sites (Chapin et al. 1996). Ecological sorting results in species of greater stature and higher rates of resource acquisition dominating nutrient-rich sites. Species having low rates of resource capture and relatively low nutrient turnover dominate infertile sites and sites subjected to frequent disturbance (Chapin and Körner 1996).

In the short term, most arctic and alpine plant communities increase productivity in response to increased N supply. In the long term, however, added N is expected to change the species composition, typically resulting in a community that is less resistant to stress (Shaver and Chapin 1986, Körner 1989, Chapin and Körner 1996). Plant species diversity provides insurance against dramatic changes in ecosystem processes or nutrient cycles. If there are multiple species present within a given functional group, severe disturbance is less likely to have serious ecosystem consequences (Chapin and Körner 1996). Thus, species richness, on its own, can be important for maintaining ecosystem structure and function, mainly in the case of a dramatic change in a key ecosystem forcing function such as N supply.

Most major ecosystem types, including temperate forest, grassland, and tropical forest, tend to be dominated by a single physiognomic type of vegetation. In contrast, arctic and alpine tundra tend to be dominated by multiple plant communities (Chapin et al. 1980). Across a relatively small area of tundra, there may be a wide variety of plant community types in which graminoids, forbs, mosses, lichens, deciduous shrubs, or evergreen shrubs dominate (Bliss et al. 1973). There can be important differences among these plant growth forms in their use of, and response to, added nutrients (Thomas and Grigal 1976, Schlesinger and Chabot 1977).

Future climate warming could have important effects on N cycling in arctic tundra ecosystems in the Central Alaska Network. In the past, organic materials have accumulated in tundra soils, largely because decomposition has been slower than plant growth. Climate warming may increase the decomposition of soil organic matter, thereby increasing the availability of stored N (Weintraub and Schimel 2005). The distributions of woody plant species are also increasing in response to warming, with likely feedbacks on C and N cycling. For example, a dominant shrub birch species (*Betula nana*) in the arctic tundra in Alaska is expanding its distribution in tussock vegetation communities (Weintraub and Schimel 2005).

Arctic tundra is characterized by the presence of continuous permafrost, which limits root penetration into deeper soil layers, and maintains high water content in the overlying soil during summer. This modifies the response of the tundra environment to deposition of contaminants, especially N. If the arctic climate continues to warm, widespread melting of permafrost may contribute N to surface waters. This conversion of stored N to a more highly available form may augment atmospherically deposited N, leading to greater eutrophication effects in the future under a warming climate. Loss of permafrost in response to climate warming may have dramatic effects on the distribution of wetland vegetation in Alaska, especially along the southern edge of the zone of continuous permafrost. For example, in Kobuk Valley National Park in the Arctic Network, preliminary analysis of monitoring data indicated that lake surface area has decreased by about 14% in the Ahnewetut wetlands and by 20% in the Nigeruk Plain over the past two decades. Lake drainage in these areas is caused by permafrost melting, terrain subsidence, and subsequent stream channel incision (Larsen 2011). Such changes may alter the distribution and abundance of plant communities that are sensitive to air quality.

Hobbs et al. (2010) synthesized diatom records from lakes in western North America and west Greenland. They found that both temperature and N deposition drove species composition changes in diatoms, with temperature driving changes in the Arctic and N deposition driving changes in mid-latitude alpine lakes. The researchers concluded that remote lakes will continue to experience diatom species shifts, particularly in regions where increasing temperatures intersect with increasing N deposition.

Relatively little information is available regarding the effects of atmospheric N inputs on freshwaters in arctic settings. However, Levine and Whalen (2001) reported the results of nutrient enrichment bioassays in 39 lakes and ponds in the Arctic Foothills region of Alaska. Significant ^{14}C uptake responses were observed subsequent to experimental addition of N + P, N alone, and P alone in 83%, 35%, and 22% of the bioassays, respectively. Overall, the data suggested that N was somewhat more important than P in regulating phytoplankton production in lakes and ponds in this Arctic region. As the summer progressed, the strength of the response to nutrient addition decreased, but this decrease was not related to irradiance or water temperature. This suggested secondary limitation by a micronutrient such as iron during the latter parts of the growing season (Levine and Whalen 2001).

Arctic lake diatom flora and nutrient balances have been shown to have been altered by N deposition in the Canadian Arctic (Wolfe et al. 2006). Anadromous salmonid fish constitute an important source of N to nutrient-poor streams and lakes in the Pacific Northwest and major parts of Alaska. These fish accumulate nutrients, including N, during adult life in the ocean and then return to spawn and die in freshwaters that can be considerable distances inland. Thus, anadromous salmonids transport to inland freshwaters the N and other nutrients that they accumulate while in the ocean, thereby potentially altering responses to N input and nutrient cycling.

20.5 Ozone Injury to Vegetation

An assessment to evaluate risk to vegetation from ozone exposure found that atmospheric ozone concentrations at DENA, expressed as cumulative exposures using the SUM06 (a measure of cumulative exposure that includes only hourly concentrations over 60 ppb ozone) and W126 (a measure of cumulative ozone exposure that preferentially weights higher concentrations) metrics, were below levels known to induce foliar injury. Both DENA and WRST contain several ozone-sensitive species, including quaking aspen, Saskatoon serviceberry (*Amelanchier alnifolia*), and Scouler's willow (*Salix scouleriana*). Soil moisture data were not available for the evaluation of potential drought constraints on foliar ozone uptake (Kohut 2007).

Studies have documented the influence of wildfire in Eurasia on ozone levels in western North America during spring and summer (Bertschi et al. 2004, Bertschi and Jaffe 2005). During April 2008, a severe ozone episode occurred at western park locations in response to particularly active and early wildfires in Russia, coupled with favorable transport meteorology (Fuelberg et al. 2010, Oltmans et al. 2010). Unprecedented high ozone concentrations were recorded from northern Alaska to northern California. The highest ozone concentrations recorded in 37 years of monitoring were recorded for Barrow, Alaska, in the northernmost portion of the United States (hourly average ozone concentration >55 ppbv). At DENA, an hourly average of 79 ppbv was recorded during the 8-hour period in which the average was more than 75 ppbv, exceeding the ambient air quality standards for 2008 and 2015. Based on extensive trajectory calculations, wildfire smoke from Russia was identified as the likely source of the observed high ozone concentrations (Oltmans et al. 2010).

Although ozone levels in DENA are generally low other than a few extreme events, there is evidence that they may be increasing. The National Park Service (2010) reported long-term trends in the annual fourth highest 8-hour daily maximum ozone concentration for 31 monitoring sites in 27 national parks having more than 10 years of data through 2008. Statistically significant increases were reported for only four parks, including DENA. From 1990 to 2008, ozone increased about 0.33 ppb/yr, or about 6 ppb over 19 years. More recently, ozone concentrations showed no trend from 2000 to 2009.

20.6 Visibility Degradation

20.6.1 Natural Background and Ambient Visibility Conditions

DENA is the only park in the Central Alaska Network that has an Interagency Monitoring of Protected Visual Environments (IMPROVE) monitoring site. DENA experienced relatively low natural haze levels for the 20% clearest days, the 20% haziest days, and for the average of all natural haze conditions (Table 20.1). Measured ambient haze values for the period 2004–2008 were only slightly higher than the estimated natural haze condition (Table 20.1). Measured ambient haze within DENA was classified as Very Low for each category compared with other Inventory and Monitoring parks.

IMPROVE data allow estimation of visual range. Data from DENA indicate that air pollution has reduced average visual range in the park from 160 to 130 miles (257 to 209 km). On the haziest days, visual range has been reduced from 110 to 80 miles (177 to 129 km). Severe haze episodes occasionally reduce visibility to 10 miles or less.

20.6.2 Composition of Haze

Extinction charts are shown for DENA in Figure 20.2. The majority of total particulate light extinction in DENA was contributed by sulfate and organics, followed by coarse mass and light-absorbing C. On an annual average basis, 42.3% of light extinction was attributable to sulfate and 32.5% to organics. On the 20% haziest days, sulfate contributed 36.4% of light extinction and organics contributed 43.6%. On the clearest 20% visibility days, sulfate contributed 49.2% of light extinction, organics 18.4%, and coarse mass 15.7%.

TABLE 20.1

Estimated Natural Haze and Measured Ambient Haze in DENA Averaged over the Period 2004–2008[a]

	Estimated Natural Haze (in Deciviews)		
Site ID	20% Clearest Days	20% Haziest Days	Annual Average
DENA1	1.77	7.31	3.79

	Measured Ambient Haze (for Years 2004–2008)					
	20% Clearest Days		20% Haziest Days		Annual Average	
Site ID	Deciview	Ranking	Deciview	Ranking	Deciview	Ranking
DENA1	2.41	Very Low	9.55	Very Low	5.34	Very Low

[a] Parks are classified into one of five ranks (Very Low, Low, Moderate, High, Very High) based on comparison with other monitored parks.

20.6.3 Trends in Visibility

Available monitoring data from DENA suggest improving trends in visibility over the period of record, although haze was relatively high on the 20% haziest and average days during one recent year (2009; Figure 20.3). There is indication that haze levels may be decreasing slightly on the 20% clearest, average, and 20% haziest days. Other than the one relatively high haze year in 2009, ambient measured values are generally following the glideslope to attain zero human-caused haze (7.31 dv) by the year 2064 (Figure 20.4).

20.7 Toxic Airborne Contaminants

Within the Central Alaska Network, the effects of atmospheric deposition of toxic materials have been investigated in DENA as part of the WACAP study (Landers et al. 2008). Vegetation concentrations of semivolatile organic compounds, nutrients, and toxic metals measured at DENA were lower than at any other WACAP park except the other two WACAP sites in other Alaskan networks (Noatak and Gates of the Arctic national parks). Contaminant concentrations in vegetation were also low in WRST, compared with other WACAP parks, and mostly included polycyclic aromatic hydrocarbons, hexachlorobenzene, and hexachlorocyclohexane-γ. Howe et al. (2004) found that concentrations of polycyclic aromatic hydrocarbons and hexachlorobenzene in spruce (*Picea* spp.) needles at 36 sites in eastern Alaska varied by an order of magnitude. Samples collected near the city of Fairbanks generally had higher concentrations than samples collected from rural areas.

Fish in both WACAP study lakes in DENA had dieldrin concentrations that exceeded health thresholds for mammals (river otter and mink), and both lakes exceeded Hg thresholds for piscivorous birds. In Wonder Lake, concentrations of Hg and Pb in sediments showed increasing trends since the 1920s, likely a result of changes in global emissions (Landers et al. 2008).

Flanagan Pritz et al. (2014) analyzed semivolatile organic compound data for fish collected from 14 national parks in Alaska and the western conterminous United States.

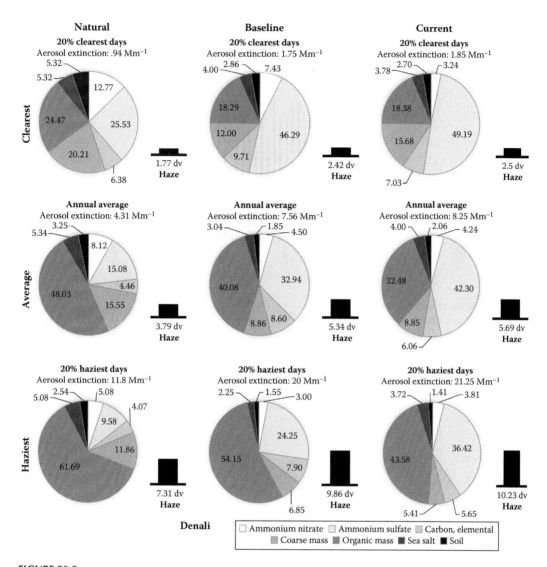

FIGURE 20.2

Estimated natural (preindustrial), baseline (2000–2004), and current (2006–2010) levels of haze (columns) and its composition (pie charts) on the 20% clearest, annual average, and 20% haziest visibility days for DENA. (From http://views.cira.colostate.edu/fed/Tools/RegionalHazeSummary.aspx, accessed October, 2012.)

The study included data for fish sampled in WRST and DENA in the Central Alaska Network. Contaminant loading was highest in fish from Alaskan and Sierra Nevada parks. Historic-use compounds were highest in Alaskan parks, whereas current-use pesticides were higher in Rocky Mountain and Sierra Nevada parks. Concentrations of the historic-use pesticides dieldrin, the DDT breakdown product p,p′-DDE, and/or chlordanes in fish exceeded EPA guidelines for wildlife (kingfisher) and for human subsistence fish consumer health thresholds at 13 of 14 parks. Additional research is needed to determine how best to manage the risk of toxic substances in these parks and to judge the role played by anadromous fish in defining that risk (Flanagan Pritz et al. 2014).

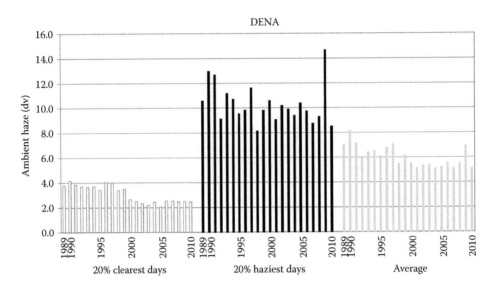

FIGURE 20.3
Trends in ambient haze levels at DENA, based on IMPROVE measurements on the 20% clearest, 20% haziest, and annual average visibility days over the monitoring period of record. (From http://vista.cira.colostate.edu/improve/Data/IMPROVE/summary_data.htm, accessed October, 2012.)

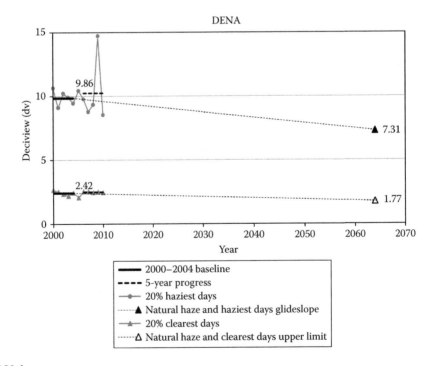

FIGURE 20.4
Glideslopes to achieving natural visibility conditions in 2064 for the 20% haziest (upper line) and the 20% clearest (lower line) days in DENA. In the Regional Haze Rule, the clearest days do not have a uniform rate of progress glideslope; the rule only requires that the clearest days do not get any worse than the baseline period. (From http://vista.cira.colostate.edu/improve/Data/IMPROVE/summary_data.htm, accessed October, 2012.)

Hageman et al. (2010) reported the results of pesticide analyses of snowpack at remote alpine, arctic, and subarctic sites in eight national parks, including DENA. Various current-use pesticides (dacthal, chlorpyrifos, endosulfans, and hexachlorocyclohexane-γ) and historic-use pesticides (dieldrin, hexachlorocyclohexane-α, chlordanes, and hexachlorobenzene) were commonly measured at all sites and years (2003–2005). The pesticide concentration profiles were unique for individual parks, suggesting the importance at some parks of regional sources.

Eagles-Smith et al. (2014) sampled fish in 21 national parks and analyzed them for Hg concentrations in tissue. Results varied substantially by park and by water body. Lake-specific average fish Hg content from fish collected from three lakes in WRST varied nearly eightfold, from rainbow trout collected in Summit Lake (53 ng/g wet weight) to lake trout collected in Tanada Lake (417 ng/g wet weight). The latter fish had the highest Hg concentration of any fish included in the study. The average size-adjusted Hg concentration in 400 mm fish from Copper Lake (242 ng/g wet weight) and Tanada Lake (321 ng/g wet weight) approached or exceeded the EPA fish tissue criterion, the avian reproductive impairment benchmark, and the tissue-based criterion for fish toxicity (Eagles-Smith et al. 2014). The cause(s) of these high concentrations of Hg in fish from WRST is/are not known. Northern pike were sampled from one lake in DENA. The concentrations of Hg in fish from that lake were among the lowest recorded in the study.

Organochlorine chemicals can be estrogenic (Garcia-Reyero et al. 2007), contributing to the occurrence of intersex fish, and can accumulate in the aquatic food chains of remote ecosystems (Blais et al. 1998). Biomarkers of exposure to estrogenic chemicals suggest the likelihood of reproductive dysfunction (Harries et al. 1997). Organochlorines are likely transported as atmospheric contaminants to high-elevation sites (Hageman et al. 2006). Work by Schreck and Kent (2013) followed up on some of the work performed in the WACAP study. They expanded the range of coverage to additional western parks, including WRST within Central Alaska Network. Two of 20 male kokanee salmon sampled in WRST were intersex. The low observed frequency of intersex fish in water bodies in study parks may be a natural phenomenon (Schwindt et al. 2009). The extent to which human-caused contaminants contribute to an increased frequency is difficult to determine (Schreck and Kent 2013).

20.8 Summary

Parks in the Central Alaska Network are relatively isolated from human activities and air pollutant emissions, in comparison to parks in the contiguous 48 states. However, DENA is located 6.1 km from the Healy Power Plant, perhaps the closest proximity of a coal-fired power plant to a Class I national park. Nevertheless, total S and N deposition near this network are both expected to be relatively low because there are few other point sources or urban areas. The sparse available atmospheric deposition data are consistent with the general understanding that atmospheric deposition of both N and S tends to be very low at national parklands within Alaska and more specifically within the Central Alaska Network. In agreement with measured values at DENA, it can be assumed that S and N deposition across this network would each be lower than about 1 kilogram per hectare per year (kg/ha/yr), on average.

Sulfur and N pollutants can cause acidification of streams, lakes, and soils. Despite the presumed acid sensitivity of some resources in parks in the Central Alaska Network, aquatic and terrestrial acidification have not been documented in this network and are considered unlikely in view of the very low levels of S and N deposition. Acid-sensitive resources constitute potentially important air quality related values, but are at little risk of damage under any reasonable future deposition scenario.

Nitrogen pollutants can also cause undesirable enrichment of natural ecosystems, leading to changes in plant species diversity and soil nutrient cycling. Vegetation communities in high-latitude ecosystems, such as are commonly found within the parks in the Central Alaska Network, may be highly sensitive to relatively low levels of N addition. Both arctic and alpine plant communities dominated by lichens, graminoids, and herbaceous plants are likely to be especially sensitive. Less information is available for evaluating the relative sensitivity of woody plants at high-latitude locations. Lichens and mosses in barren areas may be highly sensitive to N addition, responding at relatively low levels of N deposition with changes in species composition. Lichens vary from those adapted to very low N levels to lichens more tolerant of pollution and disturbance. Geiser and Nadelhoffer (2011) summarized the effects on lichens and bryophytes in ecosystems of the Taiga ecoregion. They suggested that N deposition levels of 1–3 kg/ha/yr were sufficient to cause changes in lichen species composition and cover.

In the short term, most arctic and alpine plant communities are expected to increase productivity in response to increased N supply. In the long term, however, added N is expected to change the species composition, typically resulting in a community that is less resistant to stress. If the arctic climate continues to warm, widespread melting of permafrost may contribute additional N to soil waters and surface waters. This conversion of stored N to a more highly available form may augment atmospherically deposited N, leading to greater eutrophication effects in the future under a warming climate.

Measured ozone levels at DENA, which are probably typical of the other Central Alaska Network parks, are relatively low. An assessment to evaluate the risk to vegetation from ozone concluded that the risk was low at DENA. There is no information to suggest that ozone levels at the other Central Alaska Network parks are sufficient to harm plants. Ozone exposure is not considered to be an important threat to vegetation in the Central Alaska Network.

Visibility is monitored at DENA. Measured ambient haze values for the period 2004–2008 were slightly higher than the estimated natural haze condition that would exist in the absence of human-caused pollution. Measured ambient haze within DENA was classified as very low on clear, hazy, and average days compared with other Inventory and Monitoring parks elsewhere in the United States.

Concentrations of semivolatile organic compounds (e.g., pesticides, industrial contaminants), nutrients, and toxic metals in vegetation were measured at two parks in the Central Alaska Network as part of the WACAP study. Concentrations of these substances at DENA and WRST were mostly lower than in parks in the lower 48 states. Nevertheless, fish in both WACAP study lakes in DENA had dieldrin pesticide concentrations that exceeded EPA contaminant health thresholds for people who consume at least 19 servings of fish per month. Wonder Lake had Hg concentrations that exceeded health thresholds for mammals (river otter and mink), and both study lakes exceeded Hg thresholds for piscivorous birds. In Wonder Lake, concentrations of Hg and lead in sediments showed increasing trends since the 1920s, likely a result of global emissions.

References

Allen, E. 2004. Effects of nitrogen deposition on forests and peatlands: A literature review and discussion of the potential impacts of nitrogen deposition in the Alberta Oil Sands region. Report to the Wood Buffalo Association, Fort McMurray, Alberta, Canada.

Arens, S.J.T., P.F. Sullivan, and J.M. Welker. 2008. Nonlinear responses to nitrogen and strong interactions with nitrogen and phosphorus additions drastically alter the structure and function of a high arctic ecosystem. *J. Geophys. Res.* 113:G03509.

Berendse, F., N. van Breemen, H. Rydin, A. Buttler, M. Heijmans, M.R. Hoosbeek, J.A. Lee et al. 2001. Raised atmospheric CO_2 levels and increased N deposition cause shifts in plant species composition and production in Sphagnum bogs. *Glob. Change Biol.* 7(5):591–598.

Bertschi, I.T. and D.A. Jaffe. 2005. Long-range transport of ozone, carbon monoxide, and aerosols to the NE Pacific troposphere during the summer of 2003: Observations of smoke plumes from Asian boreal fires. *J. Geophys. Res.* 110(D05303).

Bertschi, I.T., D.A. Jaffe, L. Jaegle, H.U. Price, and J.B. Dennison. 2004. PHOBEA/ITCT 2002 airborne observations of transpacific transport of ozone, CO, volatile organic compounds, and aerosols to the northeast Pacific: Impacts of Asian anthropogenic and Siberian boreal fire emissions. *J. Geophys. Res.* 109(D23S12).

Blais, J.M., D.W. Schindler, D.C.G. Muir, L.E. Kimpe, D.B. Donald, and B. Rosenberg. 1998. Accumulation of persistent organochlorine compounds in mountains of western Canada. *Nature* 395:585–588.

Bliss, L.C., G.M. Courtin, D.L. Pattie, R.R. Riewe, D.W.A. Whitfield, and P. Widden. 1973. Arctic tundra ecosystems. *Ann. Rev. Ecol. Syst.* 4:359–399.

Bobbink, R. 2004. Plant species richness and the exceedance of empirical nitrogen critical loads: An inventory. Landscape Ecology, Utrecht University, Utrecht, the Netherlands.

Bubier, J.L., T.R. Moore, and L.A. Bldezki. 2007. Effects of nutrient addition on vegetation and carbon cycling in an ombrotrophic bog. *Glob. Change Biol.* 13:1168–1186.

Chapin, F.S., III, M.S. Bret-Harte, S.E. Hobbie, and H. Zhong. 1996. Plant functional types as predictors of transient responses of arctic vegetation to global change. *J. Veg. Sci.* 7:347–358.

Chapin, F.S., III, D.A. Johnson, and J.D. McKendrick. 1980. Seasonal movement of nutrients in plants of differing growth form in an Alaskan tundra ecosystem: Implications for herbivory. *J. Ecol.* 68:189–209.

Chapin, F.S., III and C. Körner. 1996. Arctic and alpine biodiversity: Its patterns, causes and ecosystem consequences. *In* H.A. Mooney, J.H. Cushman, E. Medina, O.E. Sala, and E.-D. Schulze (Eds.). *Functional Roles of Biodiversity: A Global Perspective.* John Wiley & Sons Ltd., Chichester, U.K., pp. 7–32.

Chapin, F.S.I., L. Moilanen, and K. Kielland. 1993. Preferential use of organic nitrogen for growth by a non-mycorrhizal arctic sedge. *Nature* 361:150–153.

Cornelissen, J.H.C., T.V. Callaghan, J.M. Alatalo, A. Michelsen, E. Graglia, A.E. Hartley, D.S. Hik et al. 2001. Global change and arctic ecosystems: Is lichen decline a function of increases in vascular plant biomass? *J. Ecol.* 89:984–994.

Eagles-Smith, C.A., J.J. Willacker, Jr., and C.M. Flanagan Pritz. 2014. Mercury in fishes from 21 national parks in the western United States—Inter- and intra-park variation in concentrations and ecological risk. Open-File Report 2014-1051. U.S. Geological Survey, Reston, VA.

Flanagan Pritz, C.M., J.E. Schrlau, S.L. Massey Simonich, and T.F. Blett. 2014. Contaminants of emerging concern in fish from western U.S. and Alaskan national parks—Spatial distribution and health thresholds. *J. Am. Water Resour. Assoc.* 50(2):309–323.

Fuelberg, H.E., D.L. Harrigan, and W. Sessions. 2010. A meteorological overview of the ARCTAS 2008 mission. *Atmos. Chem. Phys.* 10:817–842.

Garcia-Reyero, N., J.O. Grimalt, I. Vives, P. Fernandez, and B. Piña. 2007. Estrogenic activity associated with organochlorine compounds in fish extracts from European mountain lakes. *Environ. Pollut.* 145(3):745–752.

Geiser, L.H. and K. Nadelhoffer. 2011. Taiga. *In* L.H. Pardo, M.J. Robin-Abbott, and C.T. Driscoll (Eds.). *Assessment of Nitrogen Deposition Effects and Empirical Critical Loads of Nitrogen for Ecoregions of the United States.* General Technical Report NRS-80. U.S. Forest Service, Newtown Square, PA, pp. 49–60.

Gough, L. and S.E. Hobbie. 2003. Responses of moist non-acidic arctic tundra to altered environment: Productivity, biomass, and species richness. *Oikos* 103:204–216.

Gough, L., C.W. Osenberg, K.L. Gross, and S.L. Collins. 2000. Fertilization effects on species density and primary productivity in herbaceous plant communities. *Oikos* 89:428–439.

Hageman, K.J., W.D. Hafner, D.H. Campbell, D.A. Jaffe, D.H. Landers, and S.L.M. Simonich. 2010. Variability in pesticide deposition and source contributions to snowpack in western U.S. national parks. *Environ. Sci. Technol.* 44(12):4452–4458.

Hageman, K.J., S.L. Simonich, D.H. Campbell, G.R. Wilson, and D.H. Landers. 2006. Atmospheric deposition of current-use and historic-use pesticides in snow at national parks in the western United States. *Environ. Sci. Technol.* 40(10):3174–3180.

Harries, J.E., D.A. Sheahan, S. Jobling, P. Matthiessen, P. Neall, J.P. Sumpter, T. Tylor, and N. Zaman. 1997. Estrogenic activity in five United Kingdom rivers detected by measurement of vitellogenesis in caged male trout. *Environ. Toxicol. Chem.* 16(3):534–542.

Hobbs, W.O., R.J. Telford, H.J.B. Birks, J.E. Saros, R.R.O. Hazewinke, B.B. Perren, É. Saulnier-Talbo, and A.P. Wolfe. 2010. Quantifying recent ecological changes in remote lakes of North America and Greenland using sediment diatom assemblages. *PLOS ONE* 5(4):e10026.

Howe, T.S., S. Billings, and R.J. Stolzberg. 2004. Sources of polycyclic aromatic hydrocarbons and hexachlorobenzene in spruce needles of eastern Alaska. *Environ. Sci. Technol.* 38(12):3294–3298.

Kellner, P.S. and P. Redbo-Torstensson. 1995. Effects of elevated nitrogen deposition on the field-layer vegetation in coniferous forests. *Ecol. Bull.* 44:227–237.

Kielland, K. 1994. Amino acid absorption by arctic plants: Implications for plant nutrition and nitrogen cycling. *Ecology* 75:2373–2383.

Kielland, K. 1995. Landscape patterns of free amino acids in arctic tundra soils. *Biogeochemistry* 31:85–98.

Klanderud, K. 2008. Species-specific responses of an alpine plant community under simulated environmental change. *J. Veg. Sci.* 19:363–372.

Kohut, R.J. 2007. Ozone risk assessment for vital signs monitoring networks, Appalachian National Scenic Trail, and Natchez Trace National Scenic Trail. Natural Resource Technical Report NPS/NRPC/ARD/NRTR—2007/001. National Park Service, Natural Resource Program Center, Fort Collins, CO.

Körner, C. 1989. The nutritional status of plants from high altitudes. A worldwide comparison. *Oecologia* 81:379–391.

Landers, D.H., S.L. Simonich, D.A. Jaffe, L.H. Geiser, D.H. Campbell, A.R. Schwindt, C.B. Schreck et al. 2008. The fate, transport, and ecological impacts of airborne contaminants in western national parks (USA). EPA/600/R-07/138. U.S. Environmental Protection Agency, Office of Research and Development, NHEERL, Western Ecology Division, Corvallis, OR.

Larsen, A. 2011. Impacts of permafrost degradation on shallow lakes and wetlands in Kobuk Valley National Park. Abstract. *George Wright Society Conference on Parks*, Protected Areas, & Cultural Sites, New Orleans, LA.

Levine, M.A. and S.C. Whalen. 2001. Nutrient limitation of phytoplankton production in Alaskan Arctic foothill lakes. *Hydrobiologia* 455:189–201.

Mack, M.C., E.A.G. Schuur, M.S. Bret-Harte, G.R. Shaver, and F.S. Chapin. 2004. Ecosystem carbon storage in arctic tundra reduced by long-term nutrient fertilization. *Nature* 431:440–443.

Madan, N.J., L.J. Deacon, and C.H. Robinson. 2007. Greater nitrogen and/or phosphorus availability increase plant species' cover and diversity at a high Arctic polar semidesert. *Polar Biol.* 30:559–570.

McKane, R.B., L.C. Johnson, G.R. Shaver, K.J. Nadelhoffer, E.B. Rastetter, B. Fry, A.E. Giblin et al. 2002. Resource-based niches provide a basis for plant species diversity and dominance in arctic tundra. *Nature* 415:68–72.

National Park Service (NPS). 2010. Air quality in national parks: 2009 annual performance and progress report. Natural Resource Report NPS/NRPC/ARD/NRR-2010/266. National Park Service, Air Resources Division, Denver, CO.

Nordin, A., J. Strengbom, J. Witzell, T. Nasholm, and L. Ericson. 2005. Nitrogen deposition and the biodiversity of boreal forests: Implication for nitrogen critical loads. *Ambio* 34:20–24.

Oltmans, S.J., A.S. Lefohn, J.M. Harris, D.W. Tarasick, A.M. Thompson, H. Wernli, B.J. Johnson et al. 2010. Enhanced ozone over western North America from biomass burning in Eurasia during April 2008 as seen in surface and profile observations. *Atmos. Environ.* 44:4497–4509.

Schlesinger, W.H. and B.F. Chabot. 1977. The use of water and minerals by evergreen and deciduous shrubs in Okefenokee Swamp. *Bot. Gaz.* 138:490–497.

Schreck, C.B. and M. Kent. 2013. Extent of endocrine disruption in fish of western and Alaskan national parks. Final report, NPS-OSU Task Agreement J8W07080024. Oregon State University, Corvallis, OR.

Schwindt, A.R., M.L. Kent, L.K. Ackerman, S.L.M. Simonich, D.H. Landers, T. Blett, and C.B. Schreck. 2009. Reproductive abnormalities in trout from western U.S. national parks. *Trans. Am. Fish. Soc.* 138(3):522–531.

Shaver, G.R. and W.D. Billings. 1975. Root production and root turnover in a wet tundra ecosystem, Barrow, Alaska. *Ecology* 56:401–410.

Shaver, G.R., W.D. Billings, F.S. Chapin, A.E. Giblin, K.J. Nadelhoffer, W.C. Oechel, and E.B. Rastetter. 1992. Global change and the carbon balance of arctic ecosystems. *BioScience* 42:433–441.

Shaver, G.R. and F.S. Chapin. 1980. Response to fertilization by various plant growth forms in an Alaskan tundra: Nutrient accumulation and growth. *Ecology* 61:662–675.

Shaver, G.R. and F.S. Chapin, III. 1986. Effect of fertilizer on production and biomass of tussock tundra, Alaska, U.S.A. *Arct. Alp. Res.* 18:261–268.

Shaver, G.R. and F.S. Chapin, III. 1991. Production: Biomass relationships and element cycling in contrasting arctic vegetation types. *Ecol. Monogr.* 61:1–31.

Shaver, G.R., L.C. Johnson, D.H. Cades, G. Murray, J.A. Laundre, E.B. Rastetter, K.J. Nadelhoffer, and A.E. Giblin. 1998. Biomass and CO_2 flux in wet sedge tundras: Responses to nutrients, temperature, and light. *Ecol. Monogr.* 68:75–97.

Strengbom, J., M. Walheim, T. Näsholm, and L. Ericson. 2003. Regional differences in the occurrence of understory species reflect nitrogen deposition in Swedish forests. *Ambio* 32:91–97.

Sullivan, T.J., G.T. McPherson, T.C. McDonnell, S.D. Mackey, and D. Moore. 2011. Evaluation of the sensitivity of inventory and monitoring national parks to acidification effects from atmospheric sulfur and nitrogen deposition. Natural Resource Report. NPS/NRPC/ARD/NRR—2011/349. U.S. Department of the Interior, National Park Service, Denver, CO.

Sverdrup, H., T.C. McDonnell, T.J. Sullivan, B. Nihlgård, S. Belyazid, B. Rihm, E. Porter, W.D. Bowman, and L. Geiser. 2012. Testing the feasibility of using the ForSAFE-VEG model to map the critical load of nitrogen to protect plant biodiversity in the Rocky Mountains region, USA. *Water Air Soil Pollut.* 23:371–387.

Thomas, B. and D.F. Grigal. 1976. Phosphorus conservation by evergreenness of mountain laurel. *Oikos* 27:19–26.

Tilman, D. 1988. *Plant Strategies and the Dynamics and Function of Plant Communities.* Princeton University Press, Princeton, NJ.

van der Wal, R., I.S.K. Pearce, and R.W. Brooker. 2005. Mosses and the struggle for light in a nitrogen-polluted world. *Ecophysiology* 142:159–168.

van Wijk, M.T., K.E. Clemmensen, G.R. Shaver, M. Williams, T.V. Callaghan, F.S. Chapin, J.H.C. Cornelissen et al. 2004. Long-term ecosystem level experiments at Toolik Lake, Alaska, and at Abisko, northern Sweden: Generalizations and differences in ecosystem and plant type responses to global change. *Glob. Change Biol.* 10:105–123.

Walker, D.A., E. Binnian, B.M. Evans, N.D. Lederer, E. Nordstrand, and P.J. Webber. 1989. Terrain, vegetation, and landscape evolution of the R4D research site, Brooks Range foothills, Alaska. *Holarct. Ecol.* 12:238–261.

Walker, M.D., D.A. Walker, K.R. Everett, and S.K. Short. 1991. Steppe vegetation on south-facing slopes of pingos, central Arctic Coastal Plain, Alaska, U.S.A. *Arct. Alp. Res.* 23:170–188.

Weintraub, M.N. and J.P. Schimel. 2005. Nitrogen cycling and the spread of shrubs control changes in the carbon balance of the arctic tundra ecosystems. *BioScience* 55(5):408–415.

Weiss, M., S.E. Hobbie, and G.M. Gettel. 2005. Contrasting responses of nitrogen-fixation in arctic lichens to experimental and ambient nitrogen and phosphorus availability. *Arct. Anarct. Alp. Res.* 37:396–401.

Wolfe, A.P., C.A. Cooke, and W.O. Hobbs. 2006. Are current rates of atmospheric nitrogen deposition influencing lakes in the eastern Canadian Arctic? *Arct. Anarct. Alp. Res.* 38(3):465–476.

21

Arctic Network Parks

21.1 Background

There are five parks in the Arctic Network, ranging in size from about 900 to 12,000 mi^2 (2,330 to 31,080 km^2). Two of them are addressed here: Gates of the Arctic National Park and Preserve (GAAR) and Noatak National Preserve (NOAT). Figure 21.1 shows the network boundary along with locations of each park and population centers having more than 10,000 people around the network. There are no population centers larger than 25,000 people within the region.

Much of the land area of the Arctic Network parks lies within the tundra and taiga ecoregions in interior Alaska. Almost all of GAAR is mountainous, as are large portions of NOAT. There is an abundance of organic material and substantial permafrost in the northern portions of the network. Lowlands are primarily covered with peatlands. Wetlands interspersed with shrublands grade into tundra at the northern extent of the taiga. Most plants in this ecoregion are adapted to low N supply (Geiser and Nadelhoffer 2011).

21.2 Atmospheric Emissions and Deposition

Sulfur dioxide and total N emissions in recent years were both estimated to be less than 1 ton/mi^2/yr throughout the entire network region. Nevertheless, Red Dog Mine, near NOAT, is a major point source of sulfur dioxide, N oxides, and heavy metal–laden dusts. There are few other S point sources around the network. Most individual point sources in the region emitted sulfur dioxide in the range of 1000–5000 tons per year or less (Sullivan et al. 2011). Many of the point sources of N oxides were on the coast of the Beaufort Sea, in the easternmost portion of the network region. There were no substantial point sources of reduced N (ammonia; Sullivan et al. 2011).

All boroughs near Arctic Network parks had very low sulfur dioxide and N oxide emissions in 2011 (<1 ton/mi^2/yr). Emissions of ammonia near Arctic Network parks were also low, mostly less than 0.5 tons/mi^2/yr; one area south of the parks showed slightly higher ammonia emissions, between 0.5 and 2 tons/mi^2/yr.

Data regarding regional emissions of Hg and semivolatile organic compounds near the Arctic Network are generally not available. The lifetime in the atmosphere of gaseous elemental Hg, which constitutes about 95% of atmospheric Hg, is generally about one year (Lin and Pehkonen 1999). However, during spring (March through June), the lifetime of

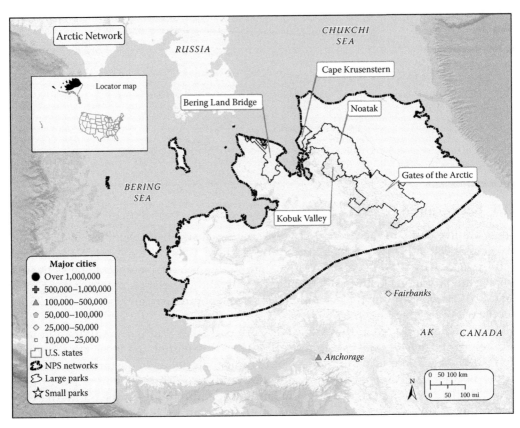

FIGURE 21.1
Network boundary and locations of parks and population centers greater than 10,000 people near the Arctic
Network region.

gaseous elemental Hg in the Arctic is much shorter, and atmospheric gaseous elemental
Hg can be depleted in less than one day during what is called an atmospheric Hg depletion episode (Lindberg et al. 2002, Skov et al. 2004). During an atmospheric Hg depletion
episode, gaseous elemental Hg is rapidly photooxidized to reactive gaseous Hg that can be
efficiently deposited to the ground surface (Skov et al. 2004).

The WACAP (Landers et al. 2008) studied airborne contaminants in a number of national
parks of the western United States from Alaska to southern California from 2002 to 2007,
including NOAT and GAAR. At these two parks, atmospheric concentrations of semivolatile organic compounds were generally low compared with other WACAP parks.
Because there are no regional air pollutant sources near the WACAP sampling sites in
NOAT and GAAR, the presence of pesticides and metals in environmental compartments within the parks suggested the atmospheric transport of pollutants from global
anthropogenic sources. The primary semivolatile organic compound contaminants
detected in the air at these parks were the historic-use pesticides hexachlorobenzene and
hexachlorocyclohexane-α. There were highly variable semivolatile organic compound
concentrations in snow across sites and time, but NOAT and GAAR had mid to high levels of endosulfans and hexachlorocyclohexane-α compared to other WACAP study sites.
Many of the semivolatile organic compound levels were below detection limits in lake

sediments. Spheroidal carbonaceous particles were not present, consistent with the scarcity of nearby coal combustion emissions sources.

The snowpack is efficient at scavenging both particulate and gas-phase pesticides from the atmosphere. However, little was known prior to the WACAP study about the occurrence, distribution, or sources of semivolatile organic compounds in alpine, subarctic, and arctic ecosystems. The analysis of pesticides in snowpack samples from seven national parks in the western United States by Hageman et al. (2006) illustrated the deposition and fate of 47 pesticides and their degradation products. Pesticide deposition to parks in Alaska was attributed to long-range atmospheric transport rather than local or regional sources (Hageman et al. 2006).

It is believed that persistent organic pollutants can be atmospherically transported across the globe because of their volatility and response to changes in temperature. This "global distillation theory" (Wania and Mackay 1993, Holmqvist et al. 2006) predicts that persistent organic pollutants in the northern hemisphere are generally transported toward the Arctic, and in the southern hemisphere they are transported toward the Antarctic.

Perfluorinated acids can exhibit substantial long-range transport even though they have very low volatility. Some of the highest perfluorinated acid concentrations in wildlife were reported for polar bears (Martin et al. 2004), and the contamination levels have been increasing since the 1990s (Dietz et al. 2008, Clair et al. 2011). Some of the perfluorinated acids in the Arctic may have derived from the degradation of chemical precursors that were transported atmospherically in the vapor phase (Ellis et al. 2004, Martin et al. 2006).

Organochlorine pesticide residues found in arctic regions are apparently attributable to long-range atmospheric transport (Gregor and Gummar 1989, Cotham and Bidleman 1991, Fellin et al. 1996, Rice and Chernyak 1997). Chlorinated persistent organic pollutants can be transported through the atmosphere as particles from industrial and agricultural sources and deposited in remote regions. They have been detected at all levels of the arctic food web (Oehme et al. 1995).

A wet deposition sampler for S and N has been operating at Bettles AK, in GAAR since fall 2008, as part of the National Acid Deposition Program/National Trends Network. Limited data are available, but they indicate that S and N deposition are both very low in the area around GAAR and likely throughout most of the Arctic Network. Both N and S wet deposition were less than 0.5 kg/ha/yr for the two years with valid data, 2010–2011. Other active wet deposition monitoring sites in Alaska include Poker Creek, Juneau, Denali National Park, and Katmai National Park. Data have been collected since 1980 at Denali and since 1993 at Poker Creek. The other three monitoring sites have been added within the last decade. There are also Clean Air Status and Trends Network dry deposition measurements at Denali and Poker Flat. At all monitored sites in Alaska, wet N deposition has consistently been less than 1 kg N/ha/yr, and it has been less than 0.5 kg N/ha/yr at all monitored sites except Juneau. Wet S deposition has been slightly higher than 1 kg S/ha/yr at Juneau but less than that at the other monitoring sites. The Clean Air Status and Trends Network dry deposition estimates have also been low, below about 0.25 kg N/ha/yr for each site and year measured. Thus, the sparse available atmospheric deposition data for Alaska are consistent with the general understanding that total (wet plus dry) atmospheric deposition of both N and S tends to be very low at national parklands within Alaska. It can be assumed that S and N deposition across the Arctic Network parks would each be lower than about 1 kg/ha/yr, on average.

21.3 Acidification

Data on potential effects of acidic deposition on aquatic and terrestrial resources in the Arctic Network are generally lacking. The WACAP study found that Burial Lake in NOAT and Matcharak Lake in GAAR were both very well buffered, with acid neutralizing capacity of about 280 and 2000 μeq/L, respectively. In view of the very low levels of emissions and deposition, acidification effects are expected to be minimal or nonexistent, even in lakes with lower acid neutralizing capacity.

Forsius et al. (2010) reported the critical loads of acidity (S + N) for terrestrial ecosystems north of 60° latitude in Europe and North America using the simple Mass Balance (SMB) model and the critical chemical criteria base cation to Al ratio = 1 and acid neutralizing capacity = 0 μeq/L. Critical loads were exceeded in large areas of northern Europe and the Norilsk region of arctic western Siberia during the 1990s, largely due to S emissions from metal smelters in arctic northern Russia. There were no acidity exceedances in northern North America because S deposition was estimated to be uniformly low. The minimum calculated critical load of acidity for North America was three to four times higher than the maximum modeled S + N deposition (Forsius et al. 2010).

21.4 Nutrient Nitrogen Enrichment

21.4.1 Ecosystem Response to Nitrogen

Areas of arctic and alpine tundra tend to be dominated by multiple types of plant communities (Chapin et al. 1980). Across a relatively small area of arctic tundra, there may be plant community types in which graminoids (grasses and sedges), forbs, mosses, lichens, deciduous shrubs, or evergreen shrubs dominate (Bliss et al. 1973; Figure 21.2). There can be important differences among these plant growth forms in their use of, and response to, addition of N (Thomas and Grigal 1976, Schlesinger and Chabot 1977).

Plant species distributions in arctic tundra are often N-limited (Shaver and Chapin 1980). Soluble N in tundra soil solution is typically dominated by organic N, including free amino acids, rather than ammonium or nitrate (Kielland 1995). Tundra plants appear to exhibit a range of interspecific differences that allow coexistence under conditions that reflect a single limiting element. Species differ in rooting depth, phenology, and uptake preferences for organic and inorganic forms of N (Shaver and Billings 1975, Chapin et al. 1993, Kielland 1994, McKane et al. 2002). McKane et al. (2002) demonstrated, based on ^{15}N field experiments, that arctic tundra plant species were differentiated in timing, depth, and chemical form of N utilization. Furthermore, the species that exhibited the greatest productivity were those that efficiently used the most abundant N forms at the site.

The distribution of arctic plant communities is largely controlled by the presence of permafrost and moisture gradients. Where they are present, soils tend to be acidic and vegetation is largely restricted to tussock-shrub tundras and wetlands. Ecological sorting results in species of greater stature and higher rates of resource acquisition dominating nutrient-rich sites. Species having low rates of resource capture and relatively low nutrient turnover dominate infertile sites and sites subjected to frequent

FIGURE 21.2
Plant communities in GAAR are highly sensitive to N inputs and change in climate. (Photo courtesy of National Park Service.)

disturbance (Chapin and Körner 1996). These differences modify responses to atmospheric N input.

In the short term, most arctic and alpine plant communities would be expected to increase productivity in response to increased N supply. In the long term, however, added N might be expected to change the species composition, perhaps resulting in a community that is less resistant to stress (Shaver and Chapin 1986, Körner 1989, Chapin and Körner 1996).

Increased N could contribute to increased deciduous biomass at the expense of the diverse lichen flora. Lichens are the primary winter food source for the western Arctic caribou herd, the largest herd in North America, with numbers ranging from 250,000 to 500,000 animals. Herds like this may not be possible in Alaskan areas with less continuous lichen cover.

Plant species diversity provides insurance against dramatic changes in ecosystem processes or nutrient cycles. If there are multiple species present within a given functional group, severe disturbance is less likely to have serious ecosystem consequences (Chapin and Körner 1996). Thus, species richness, on its own, can be important for maintaining ecosystem structure and function, mainly in the case of a dramatic change in a key ecosystem forcing function. Most commonly, however, changes in the abundance of one species result in compensatory changes in other species, likely with little impact on overall ecosystem functioning (Chapin and Körner 1996).

Bryophytes and lichens have major influence on water and energy balances in arctic tundra vegetation communities. High inputs of N can reduce their growth, while increasing the growth of competing grass and sometimes also shrub species (Berendse et al. 2001, van der Wal et al. 2005, Bubier et al. 2007). Wet tundra sites are more likely to be P-limited, and dry tundra sites tend to be N-limited (Shaver et al. 1998). Therefore, changes in N deposition are more likely to affect dry tundra sites.

Shrub and forest vegetation communities in high-latitude locations may be highly sensitive to relatively low levels of N addition. Unfortunately, experimental data are generally lacking. Arctic and alpine plant communities dominated by graminoids and herbaceous plants are likely to be especially sensitive, but there is not an adequate basis for evaluating the relative sensitivity of woody plants at high-latitude locations. In addition, much of the land coverage in some of these parks is snow and ice or barren land, with a scarcity of vascular plants. Lichens and mosses in barren areas are probably highly sensitive to N addition. Holt and Neitlich (2010) inventoried lichens in Arctic Network parks. They identified more than 400 taxa. The strongest community gradient was associated with site moisture and elevation (alpine versus lowlands). The next strongest gradient was pH, which was partly driven by organic acidity associated with *Sphagnum* spp. and alkaline bedrock chemistry. There is concern that future increases in S, and especially N oxide, emissions could affect the distribution and abundance of nonvascular plants in the Arctic Network.

Some experimental work has been done in tundra ecosystems. In dry heath tundra in the northern Brooks range near Toolik Lake, Alaska, primary productivity increased in response to very high (100 kg N/ha/yr) N fertilization (Gough et al. 2002). Dwarf birch became dominant, while species of grass, other shrubs, forbs, bryophytes, and lichens decreased in coverage. Van Wijk et al. (2004) conducted a meta-analysis in mid-Arctic regions of Alaska and Sweden to examine ecosystem response to experimental N addition. They found that the growth of vascular plants, primary productivity, and total biomass all increased substantially subsequent to addition of N alone and also addition of N plus P and other nutrients. The growth of shrubs and graminoids increased, while the growth of forbs, bryophytes, and lichens decreased. Shrub response appeared to vary with soil condition. At locations having base-rich soils, N addition did not necessarily contribute to shrub dominance (Gough and Hobbie 2003). Lichens were completely eliminated subsequent to over 10 years of experimental N addition at a rate of 100 kg N/ha/yr at the Toolik Lake Tundra Long Term Ecological Research site (Weiss et al. 2005).

Changes in ground vegetation have been shown to occur in Swedish boreal forests at relatively low levels of atmospheric N deposition. For example, whortleberry cover decreased at N deposition ≥6 kg N/ha/yr (Strengbom et al. 2003, Nordin et al. 2005); the growth of wavy hairgrass increased at ≥5 kg N/ha/yr (Kellner and Redbo-Torstensson 1995, Nordin et al. 2005).

Samples of feather moss (*Hylocomium splendens*) are periodically collected and analyzed by the National Park Service for S, N, and heavy metals as part of the Arctic Network terrestrial monitoring program. Moss samples can be used as passive monitors to evaluate pollutant deposition. Ultimately, it is hoped that such monitoring data can be combined with dose-response data in a critical load context to guide future resource management in arctic Alaska (Linder et al. 2013).

Increased N deposition to taiga plant communities, provided the increase is sufficiently large, is expected to increase plant growth in the short term, followed by the likelihood of a longer-term change in species composition and richness (Gough et al. 2000, Geiser and Nadelhoffer 2011). In peatland bogs, species richness commonly decreases in response to N addition (Allen 2004), perhaps in part because the low pH and high water table inhibit nitrophyte invasion (Geiser and Nadelhoffer 2011).

Gordon et al. (2001) investigated the effects of N and P fertilization on arctic vegetation health, especially bryophytes that can account for up to 70% of the vegetative cover in some arctic communities (Richardson 1981). Added N reduced lichen cover but did not affect the cover of other functional groups. The combination of added N and P changed species composition. Because N and P are often colimiting, the critical load of N would be expected to be lower where P availability is greater (Gordon et al. 2001).

21.4.2 Interactions with Climate Change

The effects of increased N supply interact with effects from other perturbations, especially climatic drivers. Such interactions are pronounced in arctic environments. Loss of permafrost in response to climate warming may have dramatic effects on the distribution of wetland vegetation in Alaska, especially along the southern edge of the zone of continuous permafrost. Any changes that occur in N deposition would be expected to interact with these climate-related changes to modify vegetation structure and function.

Future climate warming could have important effects on N cycling in arctic tundra ecosystems. In the past, organic materials have accumulated in tundra soils, largely because decomposition has been slower than plant growth in this harsh environment. Climate warming may increase the rate of decomposition of soil organic matter, thereby increasing the availability of stored N (Weintraub and Schimel 2005). The distributions of woody plant species are also increasing in response to warming, with likely feedbacks on C and N cycling. For example, the dominant shrub species in the arctic tundra in Alaska is dwarf birch; it is expanding its distribution in tussock vegetation communities (Weintraub and Schimel 2005). Additionally, if the arctic climate continues to warm, widespread melting of permafrost may contribute N to surface waters. This conversion of stored N to a more easily accessible form may augment atmospherically deposited N, leading to greater eutrophication effects in the future under a warming climate.

Changes in competitive interactions among species can be just as important as growth stimulation effects caused by N inputs. For example, competition by erect vascular plants can decrease moss cover despite the N limitation of the mosses (Klanderud 2008). Such competitive effects are more likely to occur in low and mid-Arctic regions than in high Arctic regions where growth is more strongly limited by harsh climatic conditions (Cornelissen et al. 2001, van Wijk et al. 2004, Madan et al. 2007).

21.4.3 Critical Nutrient Load

Linder et al. (2013) reported the results of preliminary analysis of existing data for characterizing critical load values for N deposition for interior Alaska. They estimated the critical load for N deposition to be in the range of 1–3 kg N/ha/yr to protect lichen and bryophyte cover in tundra and to protect community composition in taiga ecosystems.

Pardo et al. (2011) compiled data on empirical critical load for protecting sensitive resources in Level I ecoregions across the conterminous United States against nutrient enrichment effects caused by atmospheric N deposition. As part of the Pardo et al. (2011) assessment, Nadelhoffer and Geiser (2011) recommended critical load in the range of 1–3 kg N/ha/yr for both taiga and arctic tundra in Alaska to protect sensitive plant communities against the effects of N enrichment. To protect shrub and grass community composition, the critical load would likely be higher, in the range of about 6 kg N/ha/yr (Strengbom et al. 2003, Nordin et al. 2005, Geiser and Nadelhoffer 2011). Data compiled by Nadelhoffer and Geiser (2011) suggested that ambient N deposition may exceed the lower limit of the expected critical load to protect against nutrient enrichment effects in each of the parks in Arctic Network. These potential exceedances were reported for the protection of lichens and herbaceous vegetation.

21.4.4 Aquatic Ecosystem Response

Relatively little information is available regarding the effects of atmospheric N inputs on fresh waters in arctic settings. Nevertheless, lake diatom flora and nutrient balances have been shown to have been altered by N deposition in the Canadian Arctic (Wolfe et al. 2006). Similar responses have been observed in lakes at high elevation in Colorado and Wyoming. Levine and Whalen (2001) reported the results of nutrient enrichment bioassays in 39 lakes and ponds in the Arctic Foothills region of Alaska. Significant ^{14}C uptake responses were observed subsequent to experimental addition of N plus P, N alone, and P alone in 83%, 35%, and 22% of the bioassays, respectively. Overall, the data suggested that N was somewhat more important than P in regulating phytoplankton production in lakes and ponds in this Arctic region. As the summer progressed, the strength of the response to nutrient addition decreased, but this decrease was not related to irradiance or water temperature. This suggests secondary limitation by a micronutrient such as iron during the latter parts of the growing season.

21.5 Ozone Injury to Vegetation

Ozone data were not available for any of the parks in the Arctic Network at the time this book was written. The risk of ozone injury to plant foliage in the Arctic Network cannot be evaluated at the time of this writing due to lack of data on ozone exposure in the parks (Kohut 2007). However, parks in in this network contain quaking aspen, an important ozone-sensitive species.

Biomass burning in Eurasia is known to affect ozone levels in western North America during spring and summer (Bertschi et al. 2004, Bertschi and Jaffe 2005). For example, during April 2008, a severe ozone episode occurred at western park locations in response to particularly active and early wildfires burning in Russia, coupled with favorable transport meteorology (Fuelberg et al. 2010, Oltmans et al. 2010). Unprecedented high ozone concentrations were recorded from northern Alaska to northern California. The highest ozone concentrations recorded in 37 years of monitoring were recorded for Barrow, Alaska, in the northernmost portion of the United States (hourly average ozone concentration >55 ppbv). At Denali National Park (DENA), an hourly average of 79 ppbv was recorded during the 8-hour period in which the average exceeded the 75 ppbv ambient air quality standard. Based on extensive trajectory calculations, biomass burning in Russia was identified as the likely source of the observed high ozone concentration (Oltmans et al. 2010). Such events might become more frequent under a warming climate.

21.6 Visibility Degradation

21.6.1 Natural Background and Ambient Visibility Conditions

Based on limited data from Bettles, AK, average haze levels are moderate (6.8 deciviews) at GAAR, increasing by a factor of 2 on the 20% haziest days. Boundary layer transport of air pollutants from Eurasia to the Arctic region during winter and spring contribute to what is called "arctic haze" (Barrie 1986).

Data indicating the composition of haze levels are available for only one year (2010) for one park (GAAR) in the Arctic Network. Based on these available data, sulfate is by far the largest contributor to haze (56%) on the 20% clearest days. For the 20% haziest days and the annual average, however, organic material is the largest contributor to haze, followed by sulfate (Figure 21.3).

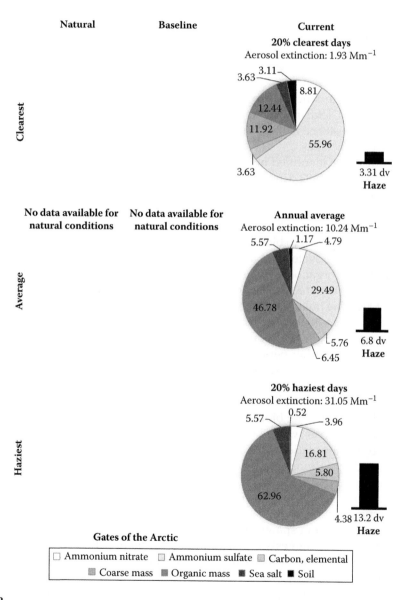

FIGURE 21.3
Estimated current levels of haze (columns) and its composition (pie charts) on the 20% clearest, annual average, and 20% haziest visibility days for GAAR. Note that data are only available for the year 2010. (From http:// views.cira.colostate.edu/fed/Tools/RegionalHazeSummary.aspx, accessed October, 2012.)

21.7 Toxic Airborne Contaminants

Biogeochemical cycling of Hg in the Arctic has been investigated, in part because observed Hg concentrations in marine animals may pose health risks for local human populations. Because of the increased solar flux to the Arctic during spring and the seasonal melting of sea ice, there may be an increased efficiency of Hg bioaccumulation in arctic food webs than would be expected based on data collected at mid-latitudes (U.S. EPA 2009). Mercury concentrations measured in WACAP exceeded thresholds for fish consumption by both humans and wildlife.

Many persistent organic pollutants have been shown to concentrate to high levels in the Arctic, at great distance from source areas. Both long-range atmospheric transport and animal migration can be involved in this transport process, in addition to biomagnification at the receptor location.

Historic-use pesticide concentrations in fish in NOAT and GAAR were midrange among WACAP study sites. Dieldrin concentrations in fish exceeded contaminant health thresholds for human subsistence fishers. Current-use pesticide concentrations in fish were low among WACAP study lakes.

Various studies have identified organochlorine compounds in remote polar ecosystems (cf., Wania and Mackay 1993, Simonich and Hites 1995, Blais 2005, Landers et al. 2008). The accumulation of these compounds at these locations has been at least partly attributed to long-range atmospheric transport. Compounds detected commonly include DDT, hexachlorocyclohexane, hexachlorobenzene, and PCBs. Such compounds tend to concentrate in aquatic biota in response to biomagnification processes and have been associated with disruption of endocrine systems in aquatic species.

Bioconcentration of organochlorines in the arctic food web has been shown in fish, seals, and polar bears (Oehme et al. 1995). Levels measured in polar bears have been especially high (AMAP 2004). The polar bear is the top predator in the Arctic and feeds preferentially on ringed seals and, to a lesser extent, on other seal species. The concentrations of DDT and hexachlorocyclohexane in polar bear tissue declined slowly between the 1960s and 2000s (Norstrom 2001). Other persistent organic pollutants, such as PCBs and chlordanes, did not show declining trends during that time period.

The long-range transport of atmospherically deposited contaminants can be augmented by biotransport. A good example of this phenomenon was documented by Ewald et al. (1998), who showed that biotransport by migrating sockeye salmon (*Oncorhynchus nerka*) in the Copper River watershed, Alaska, had a greater influence than atmospheric transport on bioaccumulation of PCBs and DDT in lake food webs. Organic pollutants accumulated by salmon during their ocean residence were effectively transferred 410 km inland to their spawning lake. Resident Arctic grayling (*Thymallus arcticus*) in a salmon spawning lake were found to contain organic pollutants more than twice as high as arctic grayling in a nearby salmon-free lake. The pollutant composition of the grayling in the salmon spawning lake was similar to that of the migrating salmon (Ewald et al. 1998). This suggests that salmon migration may have contributed to bioaccumulation of organic contaminants in other fish species in the lake used for spawning by the salmon.

Pacific sockeye salmon have been shown to transport PCBs to salmon nursery lakes in Alaska (Krümmel et al. 2003). Salmon were shown to be a major source of PCBs to lakes that receive high densities of salmon spawners. Once transported to the lake location by the migrating fish, PCBs can move to various levels of the food web.

Hageman et al. (2010) reported the results of pesticide analyses of snowpack at remote alpine, arctic, and subarctic sites in eight national parks, including NOAT/GAAR. Various current-use pesticides (dacthal, chlorpyrifos, endosulfans, and γ-hexachlorocyclohexane) and historic-use pesticides (dieldrin, α-hexachlorocyclohexane, chlordanes, and hexachlorobenzene) were commonly measured at all sites and years (2003–2005). The pesticide concentration profiles were unique for individual parks, suggesting the importance of regional sources.

Pesticide concentrations in snowpack in NOAT and GAAR, where regional cropland does not occur, were in some cases as high as those found for parks located in the lower 48 states. In most cases (dacthal and chlorpyrifos being exceptions), pesticide concentrations at NOAT/GAAR were more than six times higher than at Denali National Park (Hageman et al. 2010). The reason(s) for the observed high concentrations of pesticides at NOAT and GAAR, distant from regional cropland, is/are not known.

Flanagan Pritz et al. (2014) analyzed semivolatile organic compound data for fish collected from 14 national parks in Alaska and the western United States. The study included data for fish sampled in GAAR and NOAT in the Arctic Network. Contaminant loading was highest in fish from Alaskan and Sierra Nevadan parks. The historic-use pesticides were highest in Alaskan parks, whereas the current-use pesticides were higher in Rocky Mountain and Sierra Nevadan parks. Concentrations of the historic-use pesticides dieldrin, p,p′-DDE, and/or chlordanes in fish exceeded the EPA guidelines for wildlife (kingfisher) and for human subsistence fish consumer health thresholds at 13 of 14 parks. Additional research is needed to determine how best to manage the risk of toxic substances in these parks and to judge the roles played by anadromous fish and atmospheric contributions in defining that risk (Flanagan Pritz et al. 2014).

21.8 Summary

The Arctic Network is remote from the influence of most human activities. There are no population centers of any significant magnitude (>25,000 people) within the Arctic Network region. The sparse available atmospheric deposition data for Alaska are consistent with the general understanding that atmospheric deposition of both N and S tends to be very low at most national parklands. It can be assumed that S and N deposition across the Arctic Network would each be lower than about 1 kg/ha/yr, on average.

Data regarding the potential effects of acidic deposition on aquatic and terrestrial resources in the Arctic Network are generally lacking. However, in view of the very low levels of emissions and deposition at most locations and limited deposition data, effects are expected to be minimal. Acidification of soils and drainage waters does not appear to be an important threat. Resources might respond, however, to nutrient inputs in the form of atmospheric N deposition. In the short term, most arctic and alpine plant communities would be expected to increase productivity in response to increased N supply. In the long term, however, added N might change the species composition, perhaps resulting in a community that is less resistant to stress.

Shrub and forest vegetation communities in high-latitude locations may be highly sensitive to relatively low levels of N addition. Unfortunately, experimental data are generally lacking. Much of the land coverage in some of these Arctic Network parks is sparse vegetation or barren land, generally with few vascular plants. Lichens and mosses in the Arctic

Network are probably highly sensitive to N deposition. Lichens in the high arctic areas of Greenland and elsewhere have been found to respond to very low additions of N with changes in abundance and cover.

Various studies have identified organochlorine compounds in polar ecosystems. The accumulation of these compounds in remote northern locations has been attributed to long-range atmospheric transport. Compounds detected commonly include DDT, hexachlorocyclohexane, hexachlorobenzene, and PCBs. Such compounds tend to concentrate in aquatic biota and their mammalian predators in response to biomagnification processes and have been associated with disruption of endocrine systems in some aquatic species. NOAT and GAAR were included in the WACAP study of airborne toxics. Fish in these parks were found to have elevated Hg as well as dieldrin and other semivolatile organic compounds.

References

Allen, E. 2004. Effects of nitrogen deposition on forests and peatlands: A literature review and discussion of the potential impacts of nitrogen deposition in the Alberta Oil Sands region. Report to the Wood Buffalo Association, Fort McMurray, Alberta, Canada.

AMAP. 2004. AMAP assessment 2002: Persistent organic pollutants in the Arctic. Arctic Monitoring and Assessment Programme, Oslo, Norway.

Barrie, L.A. 1986. Arctic air pollution—An overview of current knowledge. *Atmos. Environ.* 20:643–663.

Berendse, F., N. van Breemen, H. Rydin, A. Buttler, M. Heijmans, M.R. Hoosbeek, J.A. Lee et al. 2001. Raised atmospheric CO_2 levels and increased N deposition cause shifts in plant species composition and production in Sphagnum bogs. *Glob. Change Biol.* 7(5):591–598.

Bertschi, I.T. and D.A. Jaffe. 2005. Long-range transport of ozone, carbon monoxide, and aerosols to the NE Pacific troposphere during the summer of 2003: Observations of smoke plumes from Asian boreal fires. *J. Geophys. Res.* 110:D05303.

Bertschi, I.T., D.A. Jaffe, L. Jaegle, H.U. Price, and J.B. Dennison. 2004. PHOBEA/ITCT 2002 airborne observations of transpacific transport of ozone, CO, volatile organic compounds, and aerosols to the northeast Pacific: Impacts of Asian anthropogenic and Siberian boreal fire emissions. *J. Geophys. Res.* 109:D23S12.

Blais, J.M. 2005. Biogeochemistry of persistent bioaccumulative toxicants—Processes affecting the transport of contaminants to remote areas. *Can. J. Fish. Aquat. Sci.* 62(1):236–243.

Bliss, L.C., G.M. Courtin, D.L. Pattie, R.R. Riewe, D.W.A. Whitfield, and P. Widden. 1973. Arctic tundra ecosystems. *Ann. Rev. Ecol. Syst.* 4:359–399.

Bubier, J.L., T.R. Moore, and L.A. Bldezki. 2007. Effects of nutrient addition on vegetation and carbon cycling in an ombrotrophic bog. *Glob. Change Biol.* 13:1168–1186.

Chapin, F.S., III, D.A. Johnson, and J.D. McKendrick. 1980. Seasonal movement of nutrients in plants of differing growth form in an Alaskan tundra ecosystem: Implications for herbivory. *J. Ecol.* 68:189–209.

Chapin, F.S., III and C. Körner. 1996. Arctic and alpine biodiversity: Its patterns, causes and ecosystem consequences. *In* H.A. Mooney, J.H. Cushman, E. Medina, O.E. Sala, and E.-D. Schulze (Eds.). *Functional Roles of Biodiversity: A Global Perspective.* John Wiley & Sons, Chichester, U.K., pp. 7–32.

Chapin, F.S.I., L. Moilanen, and K. Kielland. 1993. Preferential use of organic nitrogen for growth by a non-mycorrhizal arctic sedge. *Nature* 361:150–153.

Clair, T.A., D. Burns, I.R. Pérez, J. Blais, and K. Percy. 2011. Ecosystems. *In* G.M. Hidy, J.R. Brook, K.L. Demerjian, L.T. Molina, W.T. Pennell, and R.D. Scheffe (Eds.). *Technical Challenges of Multipollutant Air Quality Management*. Springer, Dordrecht, the Netherlands, pp. 139–229.

Cornelissen, J.H.C., T.V. Callaghan, J.M. Alatalo, A. Michelsen, E. Graglia, A.E. Hartley, D.S. Hik et al. 2001. Global change and arctic ecosystems: Is lichen decline a function of increases in vascular plant biomass? *J. Ecol.* 89:984–994.

Cotham, W.E. and T.F. Bidleman. 1991. Estimating the atmospheric deposition of organochlorine contaminants to the Arctic. *Chemosphere* 22:165–188.

Dietz, R., R. Bossi, F.F. Riget, C. Sonne, and E.W. Born. 2008. Increasing perfluoroalkyl contaminants in east Greenland polar bears (*Ursus maritimus*): A new toxic threat to the Arctic bears. *Environ. Sci. Technol.* 42:2701–2707.

Ellis, D.A., J.W. Martin, A.O. De Silva, S.A. Mabury, M.D. Hurley, M.P.S. Anderson, and T.J. Wallington. 2004. Degradation of fluorotelomer alcohols: A likely atmospheric source of perfluorinated carboxylic acids. *Environ. Sci. Technol.* 38:3316–3321.

Ewald, G., P. Larsson, H. Linge, L. Okla, and M. Szarzi. 1998. Biotransport of organic pollutants to an inland Alaska lake by migrating sockeye salmon (*Oncorhynchus nerka*). *Arctic* 51:40–47.

Fellin, P., L.A. Barrie, D. Dougherty, D. Toom, D. Muir, N. Grift, L. Lockhart, and B. Bileck. 1996. Air monitoring in the Arctic: Results for selected persistent organic pollutants for 1992. *Environ. Toxicol. Chem.* 15:253–261.

Flanagan Pritz, C.M., J.E. Schrlau, S.L. Massey Simonich, and T.F. Blett. 2014. Contaminants of emerging concern in fish from western U.S. and Alaskan National Parks—Spatial distribution and health thresholds. *J. Am. Water Resour. Assoc.* 50(2):309–323.

Forsius, M., M. Posch, J. Aherne, G.J. Reind, J. Christensen, and L. Hole. 2010. Assessing the impacts of long-range sulfur and nitrogen deposition on arctic and sub-arctic ecosystems. *Ambio* 39(2):136–147.

Fuelberg, H.E., D.L. Harrigan, and W. Sessions. 2010. A meteorological overview of the ARCTAS 2008 mission. *Atmos. Chem. Phys.* 10:817–842.

Geiser, L.H. and K. Nadelhoffer. 2011. Taiga. *In* L.H. Pardo, M.J. Robin-Abbott, and C.T. Driscoll (Eds.). *Assessment of Nitrogen Deposition Effects and Empirical Critical Loads of Nitrogen for Ecoregions of the United States*. General Technical Report NRS-80. U.S. Forest Service, Newtown Square, PA, pp. 49–60.

Gordon, C., J.M. Wynn, and S.J. Woodin. 2001. Impacts of increased nitrogen supply on high Arctic heath: The importance of bryophytes and phosphorus availability. *New Phytol.* 149:461–471.

Gough, L. and S.E. Hobbie. 2003. Responses of moist non-acidic arctic tundra to altered environment: Productivity, biomass, and species richness. *Oikos* 103:204–216.

Gough, L., C.W. Osenberg, K.L. Gross, and S.L. Collins. 2000. Fertilization effects on species density and primary productivity in herbaceous plant communities. *Oikos* 89:428–439.

Gough, L., P.A. Wookey, and G.R. Shaver. 2002. Dry heath arctic tundra responses to long-term nutrient and light manipulation. *Arct. Anarct. Alp. Res.* 34:211–218.

Gregor, D.I. and W. Gummar, D. 1989. Evidence of atmospheric transport and deposition of organochlorine pesticides and polychlorinated biphenyls in Canadian Arctic snow. *Environ. Sci. Technol.* 23:561–565.

Hageman, K.J., W.D. Hafner, D.H. Campbell, D.A. Jaffe, D.H. Landers, and S.L.M. Simonich. 2010. Variability in pesticide deposition and source contributions to snowpack in western U.S. national parks. *Environ. Sci. Technol.* 44(12):4452–4458.

Hageman, K.J., S.L. Simonich, D.H. Campbell, G.R. Wilson, and D.H. Landers. 2006. Atmospheric deposition of current-use and historic-use pesticides in snow at national parks in the western United States. *Environ. Sci. Technol.* 40(10):3174–3180.

Holmqvist, M., P. Stenroth, O. Berglund, P. Nystrom, K. Olsson, D. Jellyman, A.R. McIntosh, and P. Larsson. 2006. Low levels of persistent organic pollutants (POPs) in New Zealand eels reflect isolation from atmospheric sources. *Environ. Pollut.* 141:532–538.

Holt, E.A. and P.N. Neitlich. 2010. Lichen inventory synthesis. Western Arctic national parklands and Arctic Network, Alaska. Natural Resource Technical Report NPS/AKR/ARCN/NRTR-2010/385. National Park Service, Fort Collins, CO.

Kellner, P.S. and P. Redbo-Torstensson. 1995. Effects of elevated nitrogen deposition on the field-layer vegetation in coniferous forests. *Ecol. Bull.* 44:227–237.

Kielland, K. 1994. Amino acid absorption by arctic plants: Implications for plant nutrition and nitrogen cycling. *Ecology* 75:2373–2383.

Kielland, K. 1995. Landscape patterns of free amino acids in arctic tundra soils. *Biogeochemistry* 31:85–98.

Klanderud, K. 2008. Species-specific responses of an alpine plant community under simulated environmental change. *J. Veg. Sci.* 19:363–372.

Kohut, R.J. 2007. Ozone risk assessment for vital signs monitoring networks, Appalachian National Scenic Trail, and Natchez Trace National Scenic Trail. Natural Resource Technical Report NPS/NRPC/ARD/NRTR—2007/001. National Park Service, Natural Resource Program Center, Fort Collins, CO.

Körner, C. 1989. The nutritional status of plants from high altitudes. A worldwide comparison. *Oecologia* 81:379–391.

Krümmel, E., R. Macdonald, L.E. Kimpe, I. Gregory-Eaves, M.J. Demers, J.P. Smol, B. Finney, and J.M. Blais. 2003. Delivery of pollutants by spawning salmon. *Nature* 425:255–256.

Landers, D.H., S.L. Simonich, D.A. Jaffe, L.H. Geiser, D.H. Campbell, A.R. Schwindt, C.B. Schreck et al. 2008. The fate, transport, and ecological impacts of airborne contaminants in western national parks (USA). EPA/600/R-07/138. U.S. Environmental Protection Agency, Office of Research and Development, NHEERL, Western Ecology Division, Corvallis, OR.

Levine, M.A. and S.C. Whalen. 2001. Nutrient limitation of phytoplankton production in Alaskan Arctic foothill lakes. *Hydrobiologia* 455:189–201.

Lin, C.-J. and S.O. Pehkonen. 1999. The chemistry of atmospheric mercury: A review. *Atmos. Environ.* 33:2067–2079.

Lindberg, S.E., H. Zhang, A.F. Vette, M.S. Gustin, M.O. Barnette, and T. Kuiken. 2002. Dynamic flux chamber measurement of gaseous mercury emission fluxes over soils. Part 2: Effect of flushing flow rate and verification of a two-resistance exchange interface simulation mode. *Atmos. Environ.* 36:847–859.

Linder, G., W. Brumbaugh, P. Neitlich, and E. Little. 2013. Atmospheric deposition and critical loads for nitrogen and metals in Arctic Alaska: Review and current status. *J. Air Pollut.* 2:76.

Madan, N.J., L.J. Deacon, and C.H. Robinson. 2007. Greater nitrogen and/or phosphorus availability increase plant species' cover and diversity at a high Arctic polar semidesert. *Polar Biol.* 30:559–570.

Martin, J.W., D.A. Ellis, S.A. Mabury, M.D. Hurley, and T.J. Wallington. 2006. Atmospheric chemistry of perfluoroalkanesulfonamides: Kinetic and product studies of the OH radical and Cl atom initiated oxidation on N-ethyl perfluorobutanesulfonamide. *Environ. Sci. Technol.* 40:864–872.

Martin, J.W., M.M. Smithwick, B.M. Braune, P.F. Hoekstra, D.C.G. Muir, and S.A. Mabury. 2004. Identification of long chain perfluorinated acids in biota from the Canadian Arctic. *Environ. Sci. Technol.* 38:373–380.

McKane, R.B., L.C. Johnson, G.R. Shaver, K.J. Nadelhoffer, E.B. Rastetter, B. Fry, A.E. Giblin et al. 2002. Resource-based niches provide a basis for plant species diversity and dominance in arctic tundra. *Nature* 415:68–72.

Nadelhoffer, K. and L.H. Geiser. 2011. Tundra. *In* L.H. Pardo, M.J. Robin-Abbott, and C.T. Driscoll (Eds.). *Assessment of Nitrogen Deposition Effects and Empirical Critical Loads of Nitrogen for Ecoregions of the United States.* USDA Forest Service, Northern Research Station, Newtown Square, PA, pp. 37–47.

Nordin, A., J. Strengbom, J. Witzell, T. Nasholm, and L. Ericson. 2005. Nitrogen deposition and the biodiversity of boreal forests: Implication for nitrogen critical loads. *Ambio* 34:20–24.

Norstrom, R.J. 2001. Effects and trends of POPs on polar bears. *In* S. Kalhok (Ed.). *Synopsis of Research Conducted under the 2000/01 Northern Contaminants Program*. Indian and Northern Affairs Canada, Ottawa, Ontario, Canada, pp. 215–216.

Oehme, M., A. Biseth, M. Schlabach, and Ø. Wiig. 1995. Concentrations of polychlorinated dibenzo-p-dioxins, dibenzofurans and non-ortho substituted biphenyls in polar bear milk from Svalbard (Norway). *Environ. Pollut.* 90:401–407.

Oltmans, S.J., A.S. Lefohn, J.M. Harris, D.W. Tarasick, A.M. Thompson, H. Wernli, B.J. Johnson et al. 2010. Enhanced ozone over western North America from biomass burning in Eurasia during April 2008 as seen in surface and profile observations. *Atmos. Environ.* 44:4497–4509.

Pardo, L.H., M.J. Robin-Abbott, and C.T. Driscoll. 2011. Assessment of nitrogen deposition effects and empirical critical loads of nitrogen for ecoregions of the United States. General Technical Report NRS-80. U.S. Forest Service, Newtown Square, PA.

Rice, C.P. and S.M. Chernyak. 1997. Marine arctic fog: An accumulator of currently used pesticide. *Chemosphere* 35:867–878.

Richardson, D.H.S. 1981. *The Biology of Mosses*. Blackwell Scientific Publications, Oxford, U.K.

Schlesinger, W.H. and B.F. Chabot. 1977. The use of water and minerals by evergreen and deciduous shrubs in Okefenokee Swamp. *Bot. Gaz.* 138:490–497.

Shaver, G.R. and W.D. Billings. 1975. Root production and root turnover in a wet tundra ecosystem, Barrow, Alaska. *Ecology* 56:401–410.

Shaver, G.R. and F.S. Chapin. 1980. Response to fertilization by various plant growth forms in an Alaskan tundra: Nutrient accumulation and growth. *Ecology* 61:662–675.

Shaver, G.R. and F.S. Chapin, III. 1986. Effect of fertilizer on production and biomass of tussock tundra, Alaska, U.S.A. *Arct. Alp. Res.* 18:261–268.

Shaver, G.R., L.C. Johnson, D.H. Cades, G. Murray, J.A. Laundre, E.B. Rastetter, K.J. Nadelhoffer, and A.E. Giblin. 1998. Biomass and CO_2 flux in wet sedge tundras: Responses to nutrients, temperature, and light. *Ecol. Monogr.* 68:75–97.

Simonich, S.L. and R.A. Hites. 1995. Global distribution of persistent organochlorine compounds. *Science* 269:1851–1854.

Skov, H., J.H. Christensen, M.E. Goodsite, N.Z. Heidam, B. Jensen, P. Wahlin, and G. Geernaert. 2004. Fate of elemental mercury in the Arctic during atmospheric mercury depletion episodes and the load of atmospheric mercury to the Arctic. *Environ. Sci. Technol.* 38(8):2373–2382.

Strengbom, J., M. Walheim, T. Näsholm, and L. Ericson. 2003. Regional differences in the occurrence of understory species reflect nitrogen deposition in Swedish forests. *Ambio* 32:91–97.

Sullivan, T.J., G.T. McPherson, T.C. McDonnell, S.D. Mackey, and D. Moore. 2011. Evaluation of the sensitivity of inventory and monitoring national parks to acidification effects from atmospheric sulfur and nitrogen deposition. Natural Resource Report. NPS/NRPC/ARD/NRR—2011/349. U.S. Department of the Interior, National Park Service, Denver, CO.

Thomas, B. and D.F. Grigal. 1976. Phosphorus conservation by evergreenness of mountain laurel. *Oikos* 27:19–26.

U.S. Environmental Protection Agency. 2009. Integrated science assessment for particulate matter. Final report. EPA/600/R-08/139F. U.S. Environmental Protection Agency, Washington, DC.

van der Wal, R., I.S.K. Pearce, and R.W. Brooker. 2005. Mosses and the struggle for light in a nitrogen-polluted world. *Ecophysiology* 142:159–168.

van Wijk, M.T., K.E. Clemmensen, G.R. Shaver, M. Williams, T.V. Callaghan, F.S. Chapin, J.H.C. Cornelissen et al. 2004. Long-term ecosystem level experiments at Toolik Lake, Alaska, and at Abisko, Northern Sweden: Generalizations and differences in ecosystem and plant type responses to global change. *Glob. Change Biol.* 10:105–123.

Wania, F. and D. Mackay. 1993. Global fractionation and cold condensation of low volatility organochlorine compounds in polar regions. *Ambio* 22:10–18.

Weintraub, M.N. and J.P. Schimel. 2005. Nitrogen cycling and the spread of shrubs control changes in the carbon balance of the Arctic tundra ecosystems. *BioScience* 55(5):408–415.

Weiss, M., S.E. Hobbie, and G.M. Gettel. 2005. Contrasting responses of nitrogen-fixation in arctic lichens to experimental and ambient nitrogen and phosphorus availability. *Arct. Anarct. Alp. Res.* 37:396–401.

Wolfe, A.P., C.A. Cooke, and W.O. Hobbs. 2006. Are current rates of atmospheric nitrogen deposition influencing lakes in the eastern Canadian Arctic? *Arct. Anarct. Alp. Res.* 38(3):465–476.

22

Southeastern Alaska Network

22.1 Background

There are three parks in the Southeast Alaska Network: Glacier Bay National Park and Preserve (GLBA), Klondike Gold Rush National Historical Park (KLGO), and Sitka National Historical Park (SITK). Only GLBA is larger than 100 mi² (259 km²). There are no urban centers larger than 50,000 people near the Southeast Alaska Network and only one larger than 25,000 people. Figure 22.1 shows the network boundary along with locations of each park and population centers around the network that have more than 10,000 people. The predominant cover types within this network are generally forest and unvegetated areas covered by perennial ice and snow.

22.2 Atmospheric Emissions and Deposition

Air pollutant emissions are generally low throughout the Southeast Alaska Network region. However, during the height of the summer tourist season, cruise ships arrive daily in Skagway harbor (Figure 22.2) and, while docked, continue to operate their engines and boilers to provide electrical power, heat, and steam to passengers (Graw and Faure 2010). Tourist buses and trains provide historical and scenic rides around Skagway and KLGO. Waste is incinerated at the municipal incinerator. Emissions from the ships have contributed approximately 1100 pounds per hour (lb/h) of N oxides and 800 lb/h of sulfur dioxide in Skagway during busy periods (Graw et al. 2010). Monitoring in 2008–2009 found that atmospheric N oxides were 5 to 10 times higher in KLGO and Skagway near the ship docks compared to SITK and GLBA. Sulfur dioxide was also elevated in Skagway (Geiser et al. 2010, Schirokauer et al. 2014).

In 2010, many U.S. coastal areas, including the coast of southeast Alaska, were designated as Emission Control Areas under the International Convention for the Prevention of Pollution from Ships. As a result, the EPA set new emissions standards for marine ships (U.S. EPA 2010). Implementation of these new standards is expected to decrease emissions from cruise ships in the Southeast Alaska Network region (Graw et al. 2010). Model simulations predict that the new standards will reduce seasonal emissions of sulfur dioxide, oxidized N, and both fine and coarse particulate matter by 87%, 78%, 56%, and 74%, respectively. Improvements in visibility are anticipated across southeast Alaska coastal areas and in GLBA (Mölders et al. 2013). Although the standards require the use of low-S fuel in ships, in 2013 several cruise lines negotiated with the EPA to reduce S through the use of

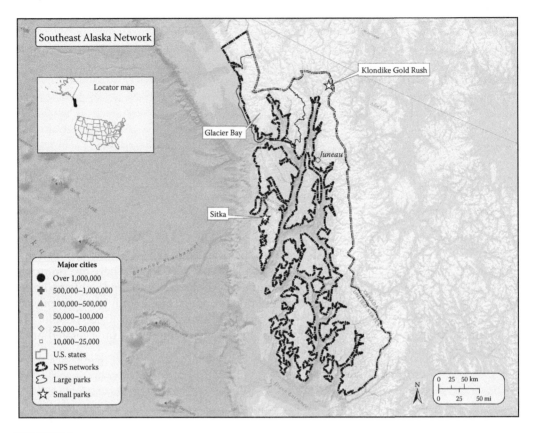

FIGURE 22.1
Network boundary and locations of parks and population centers greater than 10,000 people near the Southeast Alaska Network.

scrubbers instead of low-S fuel (http://yosemite.epa.gov/opa/admpress.nsf/e51aa292bac2 5b0b85257359003d925f/4c89a38454e4c17685257bdd004a44c6!OpenDocument&Highlight= 0,cruise,ships).

Cruise ship tourism is an important industry in coastal Alaska. Ship passengers account for more than 45% of all visitors to GLBA, and there is concern about the impacts of air pollutants on in-park biota, air quality, and visitor experience (Mölders et al. 2013). Although cruise ships contribute only marginally to overall ship traffic emissions near the Southeast Alaska Network, they often affect air quality in specific remote areas (Eijgelaar et al. 2010), where they may constitute most of the local emissions of S and N (Mölders et al. 2013). In a pristine wilderness setting, one or a few cruise ships can disproportionately impact air quality and visibility (Schembari et al. 2012).

Increased tourism in southeast Alaska has contributed to concerns regarding air pollutants emitted from cruise ships in dock and in transit to and from Skagway harbor (Geiser et al. 2010). Preliminary monitoring results suggest elevated atmospheric concentrations of N oxides and S oxides and deposition of S, Pb, Zn, and V in KLGO and the adjacent Skagway municipality. Atmospheric deposition of S and N has been higher than clean-site comparison levels (Geiser et al. 2010).

An increase in ship traffic over the previous decade has been connected with minor increases in N accumulation in epiphytic lichens at locations close to Skagway harbor

FIGURE 22.2
Cruise ships at Skagway. (National Park Service photo by Richard Graw.)

(Geiser et al. 2010). Levels of Pb, Ni, Cd, and Zn in lichens have been elevated at monitoring sites in KLGO. Geiser et al. (2010) attributed this pattern to the historical use of the harbor to transfer metal or ores from open railcars to barges.

The National Park Service and U.S. Forest Service constructed an air pollution emissions inventory for Skagway, focused on oxidized N and oxidized S (Graw and Faure 2010). The inventory included cruise ships, buses, trains, and the municipal incinerator. All are associated with the tourism industry. Cruise ships constituted the largest emissions source, accounting for an estimated 73% of local oxidized N emissions and 99% of oxidized S emissions.

Emissions near the Southeast Alaska Network were compiled by the EPA's National Emissions Inventory during a recent time period (2011). Emissions of S and N near Southeast Alaska Network parks were mostly <1 ton/mi²/yr.

Atmospheric depositions of S and N have been measured as part of the National Atmospheric Deposition Program from 2004 to the present in Juneau (site ID AK02); the site received deposition of S near 1 kg/ha/yr and N deposition near 0.5 kg/ha/yr. These rates of deposition are very low compared to most other national parks.

In 2008 and 2009, National Park Service staff deployed resin tube deposition samplers in and around Southeast Alaska Network parks during the height of the tourist season to examine fine-scale differences in deposition between areas close to cruise ships and other emissions sources and areas more remote. They found wet N deposition rates to be among the lowest in North America, generally less than 0.5 kg/ha/yr, rates much lower than the wet deposition of 1.6 kg N/ha/yr reported from relatively unimpacted areas in the Pacific Northwest (Schirokauer et al. 2014). The S deposition was also low at most sites in Southeast Alaska Network parks, from 1 to 2 kg/ha/yr, although S deposition exceeded 10 kg/ha/yr at local sites in Skagway and KLGO near the cruise ship docks. Some portion of this S deposition may be attributable to sea salt aerosols, but the higher levels in Skagway and KLGO are likely from cruise ship and port activity (Schirokauer et al. 2014).

Lichens act as passive air quality monitors, accumulating pollutants over time. They were sampled as part of the Southeast Alaska Network deposition monitoring in 2009. Lichen tissue concentrations of S, Pb, Zn, and V were higher near the Skagway docks compared to more remote areas in the network. Sulfur and V are by-products of diesel combustion, whereas Pb and Zn are indicative of ore processing and transport. Although ore transport methods were modified in 2006 to greatly reduce Pb and Zn dust, these metals will persist in the environment for a considerable period of time (Schirokauer et al. 2014). At other sites in the Southeast Alaska Network, concentrations of N, S, and heavy metals in lichens were very low, but somewhat higher than concentrations found at the more remote Tongass National Forest (Dillman et al. 2007, Schirokauer et al. 2014). Landers et al. (2008) found elevated concentrations of N and certain polycyclic aromatic hydrocarbons, all products of combustion, in lichens and conifer needles in GLBA. The concentration of N in lichens sampled at another location in GLBA in 2008 was not elevated (Geiser et al. 2010). Over the last 10 years, N content in lichen tissue in Skagway and KLGO has increased slightly, and that change in concentration was correlated with an increase in ship traffic. There may, however, be causes in addition to local emissions for the increase (e.g., increased trans-Pacific emissions, increased wildfire emissions from northern Alaska and Canada; Schirokauer et al. 2014).

The atmospheric deposition of Hg was measured as part of the National Atmospheric Deposition Program-Mercury Deposition Network from 2010 to 2013 at Bartlett Cove in GLBA; funding for the site was subsequently discontinued. Lichens can also be used as passive monitors of Hg deposition, although no effects thresholds or background levels of Hg are available. Lichens were sampled in 2008 and 2009 in the Southeast Alaska Network, and levels of Hg were found to be above detection limits at all sites. They were approximately an order of magnitude higher than levels detected in arctic national parks and were similar to those found in Oregon and Washington (Schirokauer et al. 2014).

Engstrom and Swain (1997) inferred trends in the atmospheric deposition of Hg in 210 Pb-dated sediment cores collected from three remote lakes in GLBA. The lakes are located about 2 km from the Gulf of Alaska, with no anthropogenic Hg sources in the region, little human development within several hundred km of the lakes, and the closest volcanic source more than 800 km to the northwest. Therefore, the sediment concentrations of Hg were deemed to represent global Hg pollution in the northern hemisphere, with virtually no local or regional contribution. The Alaskan sediment records indicated about a factor of 2 increase since preindustrial times in sediment Hg concentration, in agreement with studies in remote areas of northern Sweden, Finland, and Canada (Hermanson 1993, Lucotte et al. 1995).

22.3 Acidification

In addition to serving as passive indicators of atmospheric deposition, lichens can be adversely impacted by acidic or nutrient deposition. Furbish et al. (2000) reported elevated levels of S in lichens from the KLGO-Skagway area in 1999, with 67% of lichen samples exceeding levels expected at clean sites. In 2008–2009, the same sites were resampled, as well as additional sites in KLGO, GLBA, SITK, and other locations in southeast Alaska.

Based on an analysis of lichen community composition, Schirokauer et al. (2014) found that many of the study sites in the KLGO and Skagway area, and one in SITK, were potentially adversely affected by exposure to sulfur dioxide, S deposition, or acidity. The lichen community is considered to be one of the most air pollution–sensitive components of vegetation within a given ecosystem (Dillman et al. 2007).

22.4 Nutrient Nitrogen Enrichment

Lichens are also among the most sensitive bioindicators of N enrichment in terrestrial ecosystems (Bobbink et al. 2003, Glavich and Geiser 2008). Epiphytic macrolichens in southeastern Alaska have been shown to respond to even relatively low levels of N deposition. Hot spots of N deposition were first identified in 1989 in the vicinity of KLGO near downtown Skagway based on the concentrations of N in *Hypogymnia enteromorpha*, *H. inactiva*, and *Platismatia glauca* (Furbish et al. 2000, Dillman et al. 2007). In 2008–2009, the KLGO sites were resampled, and lichens were also sampled from additional sites in GLBA and SITK and other locations in southeast Alaska. Nitrogen in lichens at two sites (Dyea in KLGO and Sawyer Island in the Tracy Arm-Fords Terror Wilderness of the Tongass National Forest) exceeded clean-site reference ranges expected for western Oregon and Washington and southeastern Alaska (Schirokauer et al. 2014). Study sites with fewer N-sensitive species than 97.5% of reference sites were considered likely to be adversely affected by enhanced nutrient N; those with fewer N-sensitive species than 90% of reference plots were considered possibly impacted. Nevertheless, N-sensitive species are still present in many areas, and there is no evidence of widespread ecological effects on lichens from nutrient N enrichment to date in Southeast Alaska Network parks (Schirokauer et al. 2014).

Schirokauer et al. (2014) reported that lichen assemblages near the Skagway cruise ship terminal reflected shifts toward species that favor enhanced N and S deposition. Other sites in the Southeast Alaska Network area contained lichen species indicative of clean air. Nitrogen concentrations in epiphytic lichens did not show evidence of increase since 1999, except near local, seasonal pollution sources. The study of Schirokauer et al. (2014) suggested that local emissions sources had a greater impact on air quality in these parks than the more distant sources.

Evidence for widespread effects of N deposition in and around the Southeast Alaska Network is not conclusive. Effects on plant community composition are possible in KLGO/Skagway and SITK (Geiser et al. 2010). However, levels of N in lichens in 2008 at Bartlett Cove in GLBA were within expected clean-site ranges. In addition, Geiser et al. (2010) noted continued widespread distribution of sensitive epiphytes across the nearby Tongass National Forest.

22.5 Ozone Injury to Vegetation

Ground-level ozone is not monitored in southeast Alaska. Ozone concentrations are likely to be quite low because of the low levels of ozone precursor pollutants, N oxides and volatile organic compounds. Therefore, injury to vegetation is unlikely.

22.6 Visibility Degradation

Haze is produced in southeastern Alaska parks by a variety of fine particulate air pollutants that absorb or scatter light. Some haze is natural, resulting from sea salt, marine sulfate, and natural processes. Haze from human-caused air pollution can be a significant problem and occurs at times in Southeast Alaska Network parks. Cameras recorded episodes of haze from cruise ship emissions in GLBA during the summer of 2004. Ships were observed in the camera view on 97 of the 154 days during the monitoring season. Of the total number of observations during the study, 18% (2,185 of 12,140 daylight images) included visible vessel exhaust. Ground-based and elevated layered haze were also documented by the cameras (ARS 2005).

Cruise ship and related activities also commonly contribute to haze in KLGO. Visible haze accumulates in the morning on most days from May to September, beginning at the Skagway harbor and spreading up and down the Skagway River valley adjacent to KLGO (Geiser et al. 2010). New pollutant emissions standards required of cruise ships are expected to significantly reduce seasonal emissions of sulfur dioxide, N oxides, and both course and fine particulate matter, hopefully yielding improvements in visibility across southeast Alaska and the Southeast Alaska Network parks (Mölders et al. 2013).

22.7 Toxic Airborne Contaminants

Airborne toxic pollutants are deposited into Southeast Alaska Network ecosystems from local, regional, and global sources, including industry, power production, mining activities, and agriculture. Locally, Pb–Zn ore was transported through, and shipped out of, the port of Skagway beginning in the 1960s and continuing through 1993. A considerable amount of dust was generated by transport activities, leading to high concentrations of heavy metals in the environment around Skagway, including the KLGO Skagway Historic District (Furbish et al. 2000). Furbish et al. (2000) found that lichens collected near the port of Skagway in 1999 had significantly elevated levels of Cd, Cr, Cu, Pb, Ni, and Zn compared to lichens collected from the Tongass National Forest at sites considered to represent pristine, baseline conditions. More recently, in 2008–2009, Schirokauer et al. (2014) found levels of Pb, Zn, and V (a by-product of diesel combustion) in lichens near the Skagway ship docks elevated above background levels found in Tongass National Forest. From 1999 to 2008–2009, concentrations of some metals had decreased around Skagway and the KLGO Skagway Historic District. Aluminum, which had exceeded background thresholds in 1999, had declined to levels well below thresholds by 2008–2009; Cd and Cr also dropped below the thresholds; Ni declined but remained above thresholds at several sites. Lead levels remained high near Skagway and at other sites in KLGO, up to 150 times higher than background levels. Zinc also declined but remained above thresholds near the Skagway docks and other sites around Skagway, including Glacier Gorge in KLGO (Schirokauer et al. 2014).

Dolly Varden (*Salmo malma malma*) were sampled by Eagles-Smith et al. (2014) from two lakes and one stream in GLBA to evaluate Hg bioaccumulation in fish. Concentrations of Hg (90 ng/g wet weight) were somewhat higher than the mean across all fish and lakes sampled in the study (78 ng/g wet weight). There was considerable variation among the studied lakes, suggesting the importance of local factors in regulating fish Hg concentrations. These results for Dolly Varden in GLBA were similar to results obtained for this species elsewhere in Alaska (Jewett and Duffy 2007).

During the summer of 2007, Nagorski et al. (2011) sampled stream water, sediments, and aquatic biota in 19 streams in and near national park units in the Southeast Alaska Network. Sampled watersheds included the Indian River in SITK, the Taiya and Skagway rivers in KLGO, and 16 watersheds in and near GLBA. The highest concentrations of total Hg in stream water were found in streams that were old (not recently glaciated) and draining landscapes having peat lands. The concentration of Hg in water was associated with the stream-dissolved organic carbon concentration. The Hg concentrations increased with the time since glaciation. The levels of Hg in sampled fish were within health criteria values for the protection of piscivorous birds and mammals.

Of the 77 persistent organic pollutants analyzed in juvenile coho salmon samples collected by Nagorski et al. (2011) in and near the Southeast Alaska Network, most were below detection limits. However, some pollutants that had been banned in the United States for over three decades were present, albeit at levels below environmental and human health criteria.

22.8 Summary

Air quality in southeast Alaska is generally very good. The relatively small amount of industrial development in the vicinity of Southeast Alaska Network parks, coupled with a general absence of population centers and prevailing winds off the Pacific Ocean, all contribute to maintaining clean air at most locations in the region. However, localized air pollution from sources including marine vessels and cruise ships, wood-burning stoves, vehicle exhaust, diesel generators, mining operations, trains, incinerators, and unpaved roads, all contribute to the deterioration of air quality that can impact resources in Southeast Alaska Network parks (Dillman et al. 2007). In particular, increased cruise ship traffic in southeast Alaska has raised concerns about associated air pollution, particularly near the ports of Skagway (adjacent to KLGO), and Sitka (near SITK; Geiser et al. 2010, Schirokauer et al. 2014). Cruise ship emissions cause or contribute to noticeable haze and odors throughout downtown Skagway and the KLGO-Skagway Historic District (Geiser et al. 2010). Haze from ship emissions is also common in GLBA (Air Resource Specialists [ARS] 2005). Transport of air pollutants from Asia is also a growing concern in the region.

Lichens have been used extensively in southeast Alaska to monitor air quality. They serve as passive air pollution samplers. In addition, many lichen species are highly sensitive to air pollution. Lichens in the Klondike-Skagway area were found to have higher levels of heavy metals and S in lichen tissues than baseline values for unpolluted areas of southeast Alaska (Furbish et al. 2000). In 2008–2009, National Park Service staff resampled lichens from the KLGO sites, and from additional sites in GLBA and SITK, as well as other southeastern Alaska locations. They also sampled atmospheric concentrations of pollutants and pollutant deposition (Schirokauer et al. 2014). They found elevated levels of N and S oxides in the air, and elevated deposition of S, N, and V (a by-product of diesel combustion) in KLGO, compared to more remote sites in the Tongass National Forest (Geiser et al. 2010, Schirokauer et al. 2014). Levels of N and S in lichen tissues were elevated in KLGO and SITK compared to remote sites. Although S and N levels previously associated with adverse effects to sensitive lichens were observed in KLGO and SITK, monitoring through 2008 found no evidence of widespread adverse ecological effects from N or S deposition in Southeast Alaska Network parks (Geiser et al. 2010).

There are no significant emissions sources of Hg near Southeast Alaska Network parks. Nevertheless, models of global atmospheric Hg transport suggest that emissions from Asia may deposit in Alaska. Levels of Hg in lichens sampled in 2008 and 2009 in network parks were an order of magnitude higher than levels detected in arctic national parks, but similar to those found in Oregon and Washington (Schirokauer et al. 2014). An analysis of sediment cores from three remote lakes in GLBA indicated about a factor of 2 increase since preindustrial times in sediment Hg concentration, likely reflecting an increase in atmospheric Hg deposition. The WACAP study also reported elevated concentrations of polycyclic aromatic hydrocarbons, products of combustion, in lichens and conifer needles at Beartrack Cove, GLBA (Landers et al. 2008). Levels of Ni, Zn, Cd, and Pb, associated with past uncontained mining ore transfers in Skagway Harbor, and V from cruise ships and other sources, were found to be elevated at KLGO (Geiser et al. 2010). There are no data with which to determine the ecological effects, if any, associated with atmospheric inputs of toxic substances in the parks in this network.

References

Air Resource Specialists (ARS). 2005. Glacier Bay National Park and Preserve, Alaska, Visibility Monitoring Program: Integrated report for scene monitoring. Fort Collins, CO.

Bobbink, R., M. Ashmore, S. Braun, W. Flückiger, and I.J.J. van den Wyngaert. 2003. Empirical nitrogen critical loads for natural and semi-natural ecosystems: 2002 update. *In* B. Achermann and R. Bobbink (Eds.). *Empirical Critical Loads for Nitrogen*. Swiss Agency for Environment, Forest and Landscape SAEFL, Berne, Switzerland, pp. 43–170.

Dillman, K., L. Geiser, and G. Brenner. 2007. Air quality biomonitoring with lichens. The Tongass National Forest. Tongass National Forest, Air Resource Management Program, Petersburg, AK. Available at: http://gis.nacse.org/lichenair/?page=reports#R10 (accessed October, 2016).

Eagles-Smith, C.A., J.J. Willacker, Jr., and C.M. Flanagan Pritz. 2014. Mercury in fishes from 21 national parks in the western United States—Inter- and intra-park variation in concentrations and ecological risk. Open-File Report 2014-1051, U.S. Geological Survey, Reston, VA.

Eijgelaar, E., C. Thaper, and P. Peeters. 2010. Antarctic cruise tourism: The paradoxes of ambassadorship, "last chance tourism" and greenhouse gas emissions. *J. Sustain. Tourism* 18(3):337–354.

Engstrom, D.R. and E.B. Swain. 1997. Recent declines in atmospheric mercury deposition in the Upper Midwest. *Environ. Sci. Technol.* 31:960–967.

Furbish, C.E., L. Geiser, and C. Rector. 2000. Lichen air quality pilot study for Klondike Gold Rush National Historical Park and the city of Skagway, Alaska. Final report. U.S. Department of Interior: National Park Service, Klondike Goldrush National Historic Park, Skagway, AK.

Geiser, L., D. Schirokauer, A. Bytnerowicz, K. Dillman, and M. Fenn. 2010. Effects of cruise ship emissions on air quality and terrestrial vegetation in southeast Alaska. *Alaska Park Sci.* 9:27–31.

Glavich, D.A. and L.H. Geiser. 2008. Potential approaches to developing lichen-based critical loads and levels for nitrogen, sulfur, and metal-containing atmospheric pollutants in North America. *Bryologist* 111(4):638–649.

Graw, R. and A. Faure. 2010. Air pollution emission inventory for 2008 tourism season, Klondike Gold Rush National Historical Park, Skagway, Alaska. Report prepared for David Schirokauer, Klondike Gold Rush National Historical Park, Skagway, AK.

Graw, R., A. Faure, and D. Schirokauer. 2010. Air pollution emissions from tourist activities in Klondike Gold Rush National Historical Park. *Alaska Park Sci.* 9(2):33–35.

Hermanson, M.H. 1993. Historical accumulation of atmospherically derived pollutant trace metals in the Arctic as measured in dated sediment cores. *Water Sci. Technol.* 28(8–9):33–41.

Jewett, S.C. and L.K. Duffy. 2007. Mercury in fishes of Alaska, with emphasis on subsistence species. *Sci. Total Environ.* 387(1–3):3–27.

Landers, D.H., S.L. Simonich, D.A. Jaffe, L.H. Geiser, D.H. Campbell, A.R. Schwindt, C.B. Schreck et al. 2008. The fate, transport, and ecological impacts of airborne contaminants in western national parks (USA). EPA/600/R-07/138. U.S. Environmental Protection Agency, Office of Research and Development, NHEERL, Western Ecology Division, Corvallis, OR.

Lucotte, M., A. Mucci, C. Hillaire-Marcel, P. Pichet, and A. Grondin. 1995. Anthropogenic mercury enrichment in remote lakes of northern Québec (Canada). *Water Air Soil Pollut.* 80:467–476.

Mölders, N., S. Gende, and M. Pirhalla. 2013. Assessment of cruise–ship activity influences on emissions, air quality, and visibility in Glacier Bay National Park. *Atmos. Pollut. Res.* 4:435–445.

Nagorski, S., D. Engstrom, J. Hudson, D. Krabbenhoft, J. Dewild, E. Hood, and G. Aiken. 2011. Scale and distribution of global pollutants in Southeast Alaska Network park watersheds. NPS/SEAN/NRTR—2011/496. National Park Service, Fort Collins, CO.

Schembari, C., F. Cavalli, E. Cuccia, J. Hjorth, G. Calzolai, N. Pérez, J. Pey, P. Prati, and F. Raes. 2012. Impact of a European directive on ship emissions on air quality in Mediterranean harbours. *Atmos. Environ.* 61(0):661–669.

Schirokauer, D., L. Geiser, A. Bytnerowicz, M. Fenn, and K. Dillman. 2014. Monitoring air quality in southeast Alaska national parks and forest: Linking ambient and depositional pollutants with ecological effects. Natural Resource Technical Report NPS/SEAN/NRTR—2014/839. National Park Service, Fort Collins, CO.

U.S. Environmental Protection Agency. 2010. Control of emissions from new marine compression ignition engines at or above 30 liters per cylinder. *Federal Register*, 40 CFR 83:22895–23065. Published April 30, 2010. U.S. Government Printing Office, Washington, DC.

Index

Printed and bound by CPI Group (UK) Ltd, Croydon, CR0 4YY

24/10/2024

01778290-0016